Advances in Hyperspectral Data Exploitation

Advances in Hyperspectral Data Exploitation

Editors

Chein-I Chang
Meiping Song
Chunyan Yu
Yulei Wang
Haoyang Yu
Jiaojiao Li
Lin Wang
Hsiao-Chi Li
Xiaorun Li

MDPI • Basel • Beijing • Wuhan • Barcelona • Belgrade • Manchester • Tokyo • Cluj • Tianjin

MDPI

Editors

Chein-I Chang
Dalian Maritime University
China

Meiping Song
Dalian Maritime University
China

Chunyan Yu
Dalian Maritime University
China

Yulei Wang
Dalian Maritime University
China

Haoyang Yu
Dalian Maritime University
China

Jiaojiao Li
Xidian University
China

Lin Wang
Xidian University
China

Hsiao-Chi Li
National Taipei University of Technology
Taiwan

Xiaorun Li
Zhejiang University
China

Editorial Office
MDPI
St. Alban-Anlage 66
4052 Basel, Switzerland

This is a reprint of articles from the Special Issue published online in the open access journal *Remote Sensing* (ISSN 2072-4292) (available at: https://www.mdpi.com/journal/remotesensing/special_issues/advances_hyperspectral_data_exploitation).

For citation purposes, cite each article independently as indicated on the article page online and as indicated below:

LastName, A.A.; LastName, B.B.; LastName, C.C. Article Title. *Journal Name* **Year**, *Volume Number*, Page Range.

ISBN 978-3-0365-5795-3 (Hbk)
ISBN 978-3-0365-5796-0 (PDF)

Contents

Preface to "Advances in Hyperspectral Data Exploitation"

Hyperspectral data exploitation (HDE) has been extensively investigated in a wide range of applications. This reprint book presents a total number of 19 papers for HDE in hyperspectral image classification, hyperspectral target detection, hyperspectral unmixing, mid-wave infrared hyperspectral imaging, hyperspectral reconstruction, hyperspectral visualization, multispectral/ hyperspectral fusion. It offers a small portion of HDE that may guide those working in hyperspectral imaging to future research directions.

Chein-I Chang, Meiping Song, Chunyan Yu, Yulei Wang, Haoyang Yu , Jiaojiao Li, Lin Wang, Hsiao-Chi Li, and Xiaorun Li
Editors

remote sensing

MDPI

Editorial

Editorial for Special Issue "Advances in Hyperspectral Data Exploitation"

Chein-I Chang [1,2,*], Meiping Song [1], Chunyan Yu [1], Yulei Wang [1], Haoyang Yu [1], Jiaojiao Li [3], Lin Wang [4], Hsiao-Chi Li [5] and Xiaorun Li [6]

1 Center for Hyperspectral Imaging in Remote Sensing (CHIRS), Information and Technology College, Dalian Maritime University, Dalian 116026, China
2 Remote Sensing Signal and Image Processing Laboratory, Department of Computer Science and Electrical Engineering, University of Maryland, Baltimore, MD 21250, USA
3 The State Key Laboratory of Integrated Service Networks, Xidian University, Xi'an 710000, China
4 School of Physics and Optoelectronic Engineering, Xidian University, Xi'an 710000, China
5 Department of Electrical Engineering, National Taipei University of Technology (Taipei Tech), Taipei 10608, Taiwan
6 Department of Electrical Engineering, Zhejiang University, Hangzhou 310027, China
* Correspondence: cchang@umbc.edu

Abstract: Hyperspectral imaging (HSI) has emerged as a promising, advanced technology in remote sensing and has demonstrated great potential in the exploitation of a wide variety of application. In particular, its capability has expanded from unmixing data samples and detecting targets at the subpixel scale to finding endmembers, which generally cannot be resolved by multispectral imaging. Accordingly, a wealth of new HSI research has been conducted and reported in the literature in recent years. The aim of this Special Issue "Advances in Hyperspectral Data Exploitation" is to provide a forum for scholars and researchers to publish and share their research ideas and findings to facilitate the utility of hyperspectral imaging in data exploitation and other applications. With this in mind, this Special Issue accepted and published 19 papers in various areas, which can be organized into 9 categories, including I: Hyperspectral Image Classification, II: Hyperspectral Target Detection, III: Hyperspectral and Multispectral Fusion, IV: Mid-wave Infrared Hyperspectral Imaging, V: Hyperspectral Unmixing, VI: Hyperspectral Sensor Hardware Design, VII: Hyperspectral Reconstruction, VIII: Hyperspectral Visualization, and IX: Applications.

Keywords: hyperspectral image classification; hyperspectral imaging (HSI); hyperspectral target detection; hyperspectral reconstruction; hyperspectral unmixing

Citation: Chang, C.-I.; Song, M.; Yu, C.; Wang, Y.; Yu, H.; Li, J.; Wang, L.; Li, H.-C.; Li, X. Editorial for Special Issue "Advances in Hyperspectral Data Exploitation". *Remote Sens.* **2022**, *14*, 5111. https://doi.org/10.3390/rs14205111

Received: 15 September 2022
Accepted: 26 September 2022
Published: 13 October 2022

1. Introduction

Over the past years, hyperspectral imaging (HSI) has found in a diverse range of applications from defense, law enforcement, environmental monitoring, forestry, and agriculture to food inspection and safety and medical imaging. Through its very fine spectral resolution provided by hundreds of contiguous spectral channels, HSI is capable of uncovering and revealing many subtle material substances. This allows HSI to solve many issues that cannot be resolved by multispectral imaging (MSI), which only uses tens of discrete spectral channels such as mixed pixel classification, subpixel target detection, anomaly detection, endmember finding, and data unmixing, etc. [1–8]. The maiden flight of the Airborne Visible/InfraRed Imaging Spectrometer (AVIRIS) (https://aviris.jpl.nasa.gov/, accessed on 8 October 2022) was conducted in early 1987. Since then, AVIRIS data have been made available for extensive studies in the remote sensing community, specifically, the early development of spectral unmixing [9–14], which laid out the foundation of future developments for linear and nonlinear spectral unmixing; more detailed studies can be found in refs. [1,7]. The HYperspectral Digital Imagery Collection Experiment (HYDICE) was later developed and used to resolve the problem of targets embedded in a single pixel in August 1994, referred to as "Forest Radiance I" [15,16]. These data sets stimulated and further advanced several studies in data unmixing [1,2], subpixel target detection [1,3,8], and anomaly detection [3], with more comprehensive treatments in refs. [1,3,8]. Now, HSI has deviated from its original goals of military applications to civilian applications, specifically from unmixing and subpixel analyses to endmember finding [2,3], classification [17], compression [2], progressive processing [3], real-time processing [3,4], parallel computing [18], and fusion, etc. This Special Issue "Advances in Hyperspectral Data Exploitation" (https://www.mdpi.com/journal/remotesensing/special_issues/advances_hyperspectral_data_exploitation, accessed on 8 October 2022) intends to provide a forum for this fast-growing area to publish new ideas and technologies to facilitate hyperspectral imaging in data exploitation and to further explore its potential in different applications.

2. Overview of Published Papers

This Special Issue consists of 19 papers in various areas, which can be organized into nine categories; the number of papers published in each category is shown in its respective parentheses.

I. Hyperspectral Classification (three papers)
II. Hyperspectral Target Detection (three papers)
III. Hyperspectral and Multispectral Fusion (three papers)
IV. Mid-wave Infrared Hyperspectral Imaging (two papers)
V. Hyperspectral Unmixing (one paper)
VI. Hyperspectral Sensor Hardware Design (one paper)
VII. Hyperspectral Reconstruction (one paper)
VIII. Hyperspectral Image Visualization (one paper)
IX. Applications (four papers)

A short descriptive summary is provided for each paper so that readers can quickly discern their respective contents and more quickly find what they are interested in.

I. Hyperspectral Image Classification (three papers)

remotesensing-12-00923

Deep Relation Network for Hyperspectral Image Few-Shot Classification

Kuiliang Gao [1,*], Bing Liu [1], Xuchu Yu [1], Jinchun Qin [2], Pengqiang Zhang [1] and Xiong Tan [1]

[1] Information Engineering University, Zhengzhou 450001, China
[2] Xi'an Research Institute of Surveying and Mapping, Xi'an 710054, China
* Correspondence: 311405000803@home.hpu.edu.cn

This paper developed a few-shot hyperspectral images classification approach using only a few labeled samples. It consists of two modules, i.e., a feature learning module and a relation learning module to capture the spatial–spectral information in hyperspectral images and then carry out relation learning by comparing the similarity between samples. It is followed by a task-based learning strategy to enhance its ability in terms of learning with a large number of tasks randomly generated from different data sets. Accordingly, the proposed method has excellent generalization ability and can achieve satisfactory classification with only a few labeled samples. The experimental results indicated that the proposed method can perform better than the traditional, semisupervised support vector machine and semisupervised deep learning models.

remotesensing-12-01780-v3

Hyperspectral Image Classification Based on a Shuffled Group Convolutional Neural Network with Transfer Learning

Yao Liu [1], Lianru Gao [2,*], Chenchao Xiao [1], Ying Qu [3], Ke Zheng [2] and Andrea Marinoni [4]

[1] Land Satellite Remote Sensing Application Center, Ministry of Natural Resources of China; Beijing 100048, China; liuyao@lasac.cn (Y.L.); xiaochenchao@lasac.cn (C.X.)
[2] Key Laboratory of Digital Earth Science, Aerospace Information Research Institute, Chinese Academy of Sciences, Beijing 100094, China; zhengkevic@aircas.ac.cn
[3] Department of Electrical Engineering and Computer Science, The University of Tennessee, Knoxville, TN 37996, USA; yqu3@vols.utk.edu
[4] Department of Physics and Technology, UiT The Arctic University of Norway, NO-9037 Tromsø, Norway; andrea.marinoni@uit.no
* Correspondence: gaolr@aircas.ac.cn

This paper proposed a novel, lightweight, shuffled group convolutional neural network (abbreviated as SG-CNN) to achieve efficient training with a limited training dataset in HSI classification. It consists of SG conv units that employ conventional and atrous convolution in different groups, followed by channel shuffle operation and shortcut connection. As a result, SG-CNNs have less trainable parameters, whilst they can still be accurately and efficiently trained with fewer labeled samples. In addition, transfer learning between different HIS datasets was also applied to the SG-CNN to further improve the classification accuracy. The experimental results demonstrated that SG-CNNs can achieve a competitive classification performance when the amount of labeled data for training is poor, as well as efficiently provide satisfying classification results.

remotesensing-12-03342-v2

Hyperspectral Imagery Classification Based on Multiscale Superpixel-Level Constraint Representation

Haoyang Yu [1], Xiao Zhang [1], Meiping Song [1,*], Jiaochan Hu [2], Qiandong Guo [3] and Lianru Gao [4]

[1] Center of Hyperspectral Imaging in Remote Sensing, Information Science and Technology College, Dalian Maritime University, Dalian 116026, China; yuhy@dlmu.edu.cn (H.Y.), xiaozhang@dlmu.edu.cn (X.Z.)
[2] College of Environmental Sciences and Engineering, Dalian Maritime University, Dalian 116026, China; hujc@dlmu.edu.cn
[3] School of Geosciences, University of South Florida, Tampa, FL 33620, USA, guo1@mail.usf.edu
[4] The Key Laboratory of Digital Earth Science, Aerospace Information Research Institute, Chinese Academy of Sciences, Beijing 100094, China; gaolr@aircas.ac.cn
* Correspondence: smping@dlmu.edu.cn

This paper developed a spectral–spatial classification method called superpixel-level constraint representation (SPCR) that uses the participation degree (PD) with respect to the sparse coefficient from the constraint representation (CR) model and then transforms the individual PD to a united activity degree (UAD)-driven mechanism via a spatial constraint generated by the superpixel segmentation algorithm. The final classification is determined based on the UAD-driven mechanism. Considering that the SPCR is susceptible to the segmentation scale, an improved multiscale superpixel-level constraint representation (MSPCR) was further proposed through the decision fusion process of SPCR at different scales with the final decision of each test pixel determined by the maximum number of the predicated labels among the classification results at each scale. Experimental results on four real hyperspectral datasets including a GF-5 satellite data verified the efficiency and practicability of the proposed methods.

II. **Hyperspectral Target Detection (three papers)**

remotesensing-12-01056-v2

Underwater Hyperspectral Target Detection with Band Selection

Xianping Fu [1,2], Xiaodi Shang [1], Xudong Sun [1,2], Haoyang Yu [1], Meiping Song [1,*] and Chein-I Chang [1,3,4]

[1] Information Science and Technology College, Dalian Maritime University, Dalian 116026, China; fxp@dlmu.edu.cn (X.F.); shangxd329@dlmu.edu.cn (X.S.); sxd@dlmu.edu.cn (X.S.); yuhy@dlmu.edu.cn (H.Y.); cchang@umbc.edu (C.-I C.)
[2] Peng Cheng Laboratory, Shengzhen 518000, China
[3] Remote Sensing Signal and Image Processing Laboratory, Department of Computer Science and Electrical Engineering, University of Maryland, Baltimore, MD 21250, USA
[4] Department of Computer Science and Information Management, Providence University, Taichung 02912, Taiwan
* Correspondence: smping@dlmu.edu.cn

This paper presented a fast, hyperspectral, underwater target-detection approach using band selection (BS). Due to the high data redundancy, slow imaging speed, and long processing of hyperspectral imagery, the direct use of hyperspectral images in detecting targets cannot meet the needs of the rapid detection of underwater targets. To resolve this issue, the proposed method first developed constrained-target optimal index factor (OIF) band selection (CTOIFBS) to select a band subset with spectral wavelengths specifically responding to the targets of interest. Then, an underwater spectral imaging system integrated with the best-selected band subset was constructed for underwater target image acquisition. Finally, a constrained energy minimization (CEM) target detection algorithm was used to detect the desired underwater targets. The experimental results demonstrated that the acquisition and detection speed of the designed underwater spectral acquisition system using CTOIFBS could be significantly improved over the original underwater hyperspectral image system without BS.

remotesensing-13-04927-v2

Attention-Based Spatial and Spectral Network with PCA-Guided Self-Supervised Feature Extraction for Change Detection in Hyperspectral Images

Zhao Wang [1], Fenlong Jiang [1], Tongfei Liu [1], Fei Xie [2,*] and Peng Li [1]

[1] Key Laboratory of Electronic Information Countermeasure and Simulation Technology of Ministry of Education, School of Electronic Engineering, Xidian University, No. 2 South TaiBai Road, Xi'an 710075, China; wangzhao@xidian.edu.cn (Z.W.); fljiang@stu.xidian.edu.cn (F.J.); ltfei@stu.xidian.edu.cn (T.L.); penglixd@xidian.edu.cn (P.L.)
[2] Academy of Advanced Interdisciplinary Research, Xidian University, No. 2 South TaiBai Road, Xi'an 710068, China
* Correspondence: fxie@xidian.edu.cn

This paper proposed an attention-based spatial and spectral network with a PCA-guided, self-supervised feature extraction mechanism to detect changes in hyperspectral images. It consists of two steps: a self-supervised mapping from each patch of the difference map to the principal components of the central pixel of each patch with spatial features of differences extracted by a multilayer convolutional neural network in the first step, followed by an attention mechanism which calculates adaptive weights between spatial and spectral features of each pixel from concatenated spatial and spectral features in the second step. Finally, a joint analysis of the weighted spatial and spectral features was used to detect the changes of pixels in different positions. Experimental results on several real hyperspectral change detection data sets showed the effectiveness and advancement of the proposed method.

remotesensing-12-02783-v2

A Constrained Sparse-Representation-Based Spatio-Temporal Anomaly Detector for Moving Targets in Hyperspectral Imagery Sequences

Zhaoxu Li †, Qiang Ling †, Jing Wu, Zhengyan Wang and Zaiping Lin *

College of Electronic Science and Technology, National University of Defense Technology, Changsha 410073, China; lizhaoxu@nudt.edu.cn (Z.L.); lingqiang16@nudt.edu.cn (Q.L.); jingwu@nudt.edu.cn (J.W.); wangzhengyan@nudt.edu.cn (Z.W.)
* Correspondence: linzaiping@nudt.edu.cn
† These authors contributed equally to this work.

This paper proposed a constrained sparse representation-based spatio-temporal anomaly detection approach which extends AD from the spatial domain to the spatio-temporal domain. It includes a spatial detector to suppress moving background regions and a temporal detector to suppress non-homogeneous background and stationary objects, both of which maintain the effectiveness of the temporal detector for multiple targets in complex motion situations. Moreover, the smoothing and fusion of the spatial and temporal detection maps could adequately suppress background clutter and false alarms on the maps. Experiments conducted on a real dataset and a synthetic dataset showed that the proposed algorithm could accurately detect multiple targets with different velocities and dense targets with the same trajectory and that it also outperforms other state-of-the-art algorithms in high-noise scenarios.

III. **Hyperspectral and Multispectral Fusion (three papers)**

remotesensing-12-02535-v2

Toward Super-Resolution Image Construction Based on Joint Tensor Decomposition

Xiaoxu Ren [1], Liangfu Lu [2,*] and Jocelyn Chanussot [3]

1 College of Intelligence and Computing, Tianjin University, Tianjin 300350, China; xiaoxuren@tju.edu.cn
2 School of Mathematics, Tianjin University, Tianjin 300350, China
3 LJK, CNRS, Inria, Grenoble INP, Université Grenoble Alpes, 38000 Grenoble, France; jocelyn.chanussot@grenoble-inp.fr
* Correspondence: liangfulv@tju.edu.cn

This paper proposed an image-fusion method based on joint-tensor decomposition (JTF), which is more effective and applicable when degenerate operators are unknown or tough to gauge. Specifically, the proposed JTF method considers a super-resolution image (SRI) as a three-dimensional tensor and redefines a fusion problem as the joint estimation of the coupling factor matrix, which can also be expressed as a joint-tensor decomposition problem for the hyperspectral image tensor, multispectral image tensor, and noise regularization term. The JTF algorithm was then utilized to fuse HSI and MSI so as to explore the problem of SRI reconstruction. The experimental results showed the superior performance of the proposed method in comparison with the current popular schemes.

remotesensing-13-01260

Hyperspectral Image Super-Resolution with Self-Supervised Spectral-Spatial Residual Network

Wenjing Chen [1,2], Xiangtao Zheng [1,*] and Xiaoqiang Lu [1]

[1] Key Laboratory of Spectral Imaging Technology CAS, Xi'an Institute of Optics and Precision Mechanics, Chinese Academy of Sciences, Xi'an 710119, China; chenwenjing2017@opt.cn (W.C.); luxiaoqiang@opt.ac.cn (X.L.)
[2] University of Chinese Academy of Sciences, Beijing 100049, China
* Correspondence: zhengxiangtao@opt.cn

This paper proposed a self-supervised, spectral–spatial residual network (SSRN) to fuse a low-spatial-resolution (LR) hyperspectral image (HSI) with a high-spatial-resolution (HR) multispectral image (MSI) to obtain HR HSIs. In particular, SSRN does not require HR HSIs as supervised information in training. SSRN considers the fusion of HR MSIs and LR HSIs as a pixel-wise spectral mapping problem wherein the spectral mapping between HR MSIs and HR HSIs can be approximated by the spectral mapping between LR MSIs (derived from HR MSIs) and LR HSIs. Then the spectral mapping between LR MSIs and LR HSIs was further explored by SSRN. Finally, a self-supervised fine-tuning strategy was proposed to transfer the learned spectral mapping to generate HR HSIs. Simulated and real hyperspectral databases were utilized to verify the performance of SSRN.

remotesensing-13-04967-v2

A Spatial-Enhanced LSE-SFIM Algorithm for Hyperspectral and Multispectral Images Fusion

Yulei Wang [1,2], Qingyu Zhu [1], Yao Shi [1], Meiping Song [1,*] and Chunyan Yu [1]

[1] Center of Hyperspectral Imaging in Remote Sensing, Information Science and Technology College, Dalian Maritime University, Dalian 116026, China; wangyulei@dlmu.edu.cn (Y.W.); zhuqingyu@dlmu.edu.cn (Q.Z.); 1120180233@dlmu.edu.cn (Y.S.); yucy@dlmu.edu.cn (C.Y.)
[2] State Key Laboratory of Integrated Services Networks, Xidian University, Xi'an 710000, China
* Correspondence: smping@dlmu.edu.cn

This paper developed a spatial, filter-based, least squares estimation (LSE)-smoothing filter-based intensity modulation (SFIM) algorithm to fuse a hyperspectral image (HSI) and a multispectral image (MSI). It first combines the LSE algorithm with the SFIM method to effectively improve the spatial information quality of the fused image. At the same time, in order to better maintain the spatial information, four spatial filters (mean, median, nearest, and bilinear) were used for the simulated MSI image to extract fine spatial information. Experimental results of three HSI-MSI data sets using six image quality indexes showed the effectiveness of the proposed algorithm compared with three state-of-the-art HSI-MSI fusion algorithms (CNMF, HySure, and FUSE), while the computing time was much shorter.

IV. **Mid-wave infrared hyperspectral imaging (two papers)**

remotesensing-12-01860

Novel Air Temperature Measurement Using Midwave Hyperspectral Fourier Transform Infrared Imaging in the Carbon Dioxide Absorption Band

Sungho Kim

Department of Electronic Engineering, Yeungnam University, 280 Daehak-Ro, Gyeongsan, Gyeongbuk 38541, Korea; sunghokim@ynu.ac.kr; Tel.: +82-53-810-3530

This paper presented an approach to measuring air temperature from mid-wave hyperspectral Fourier transform infrared (FTIR) imaging in the carbon dioxide absorption band (between 4.25 and 4.35 mm) where the accurate visualization of air temperature distribution can be useful for various thermal analyses in many fields such as human health and the heat transfer of local areas. The proposed visual-air temperature (VisualAT) measurement is based on the observation that the carbon dioxide band shows zero transmissivity at short distances. Based on the analysis of the radiative transfer equation in this band, only the path radiance by air temperature survives. The brightness temperature of the received radiance can provide the raw air temperature and spectral average followed by a spatial median–mean filter that can produce final air temperature images.

remotesensing-13-01249-v2

AT²ES: Simultaneous Atmospheric Transmittance-Temperature-Emissivity Separation Using Online Upper Midwave Infrared Hyperspectral Images

Sungho Kim [1,*], Jungsub Shin [2] and Sunho Kim [2]

[1] Department of Electronic Engineering, Yeungnam University, 280 Daehak-Ro, Gyeongsan, Gyeongbuk 38541, Korea
[2] Agency for Defense Development, 488-160 Bukyuseong-Daero, Yuseong, Daejeon 34186, Korea; jss@add.re.kr (J.S.); edl423@add.re.kr (S.K.)
* Correspondence: sunghokim@yu.ac.kr; Tel.: +82-53-810-3530

This paper presented a method for atmospheric transmittance–temperature–emissivity separation (AT^2ES) using online mid-wave infrared hyperspectral images. Conventionally, temperature and emissivity separation (TES) is a well-known problem in the remote sensing domain. However, previous approaches have used the atmospheric correction process before TES using MODTRAN in the long-wave infrared band. Simultaneous online atmospheric transmittance–temperature–emissivity separation starts with approximation of the radiative transfer equation in the upper mid-wave infrared band. The highest atmospheric band was used to estimate surface temperature, assuming high emissive materials. The lowest atmospheric band (CO2 absorption band) was used to estimate air temperature. Through onsite hyperspectral data regression, atmospheric transmittance was obtained from the y-intercept and emissivity was separated using the observed radiance, the separated object temperature, the air temperature, and atmospheric transmittance. The novelty of the proposed method is in that it is the first attempt at simultaneous AT^2ES and online separation without any prior knowledge and pre-processing. Mid-wave Fourier transform infrared (FTIR)-based outdoor experimental results validated the feasibility of the proposed AT^2ES method.

V. **Hyperspectral Unmixing (one paper)**

remotesensing-12-04117-v2

Hyperspectral Nonlinear Unmixing by Using Plug-and-Play Prior for Abundance Maps

Zhicheng Wang [1,2], Lina Zhuang [3], Lianru Gao [1,*], Andrea Marinoni [4], Bing Zhang [1,2] and Michael K. Ng [5]

[1] Key Laboratory of Digital Earth Science, Aerospace Information Research Institute, Chinese Academy of Sciences, Beijing 100094, China; wangzc@radi.ac.cn (Z.W.); zb@radi.ac.cn (B.Z.)
[2] School of Electronic, Electrical and Communication Engineering, University of Chinese Academy of Sciences, Beijing 100049, China
[3] Department of Mathematics, Hong Kong Baptist University, Hong Kong, China; linazhuang@hkbu.edu.hk
[4] Department of Physics and Technology, UiT The Arctic University of Norway, NO-9037 Tromsø, Norway; andrea.marinoni@uit.no
[5] Department of Mathematics, The University of Hong Kong, Hong Kong, China; mng@maths.hku.hk
* Correspondence: gaolr@aircas.ac.cn

This paper proposed a new nonlinear unmixing method based on a general bilinear model. Instead of investing effort in designing more regularizing abundance fractions, a plug-and-play prior technique was developed to exploit the spatial correlation of abundance maps and nonlinear interaction maps. The numerical results in simulated data and a real hyperspectral dataset showed that the proposed method could improve the estimation of abundances dramatically compared with state-of-the-art nonlinear unmixing methods.

VI. **Hyperspectral Sensor Hardware Design (one paper)**

remotesensing-13-03295-v2

A Particle Swarm Optimization Based Approach to Pre-tune Programmable Hyperspectral Sensors

Bikram Pratap Banerjee [1,2] and Simit Raval [2,*]

[1] Agriculture Victoria, Grains Innovation Park, 110 Natimuk Road, Horsham, VIC 3400, Australia; bikram.banerjee@agriculture.vic.gov.au
[2] School of Minerals and Energy Resources Engineering, University of New South Wales, Sydney, NSW 2052, Australia
* Correspondence: simit@unsw.edu.au; Tel.: +61-(2)-9385-5005.

This paper designed an innovative workflow that can be implemented to simplify the process of in-field spectral sampling and its real-time analysis for the identification of optimal spectral wavelengths, specifically for programmable hyperspectral sensors mounted on unmanned aerial vehicles (UAV-hyperspectral systems), which requires a pre-selection of optimal bands when mapping new environments with new target classes with unknown spectra. The proposed band-selection optimization workflow involves particle-swarm optimization with minimum estimated abundance covariance (PSO-MEAC) for the identification of a set of bands most appropriate for UAV-hyperspectral imaging in a given environment, where the criterion function, MEAC, greatly simplifies the in-field spectral data acquisition process by requiring a few target class signatures and not requiring extensive training samples for each class. The metaheuristic method was tested on an experimental site with a diversity of vegetation species and communities. The optimal set of bands was found to suitably capture the spectral variations between target vegetation species and communities. The approach streamlines the pre-tuning of wavelengths in programmable hyperspectral sensors in mapping applications. This further reduces the total flight time in UAV-hyperspectral imaging, as obtaining information for an optimal subset of wavelengths is more efficient and requires less data storage and computational resources for post-processing the data.

VII. **Hyperspectral Reconstruction (one paper)**

remotesensing-13-00115

Residual Augmented Attentional U-Shaped Network for Spectral Reconstruction from RGB Images

Jiaojiao Li †, Chaoxiong Wu *, Rui Song †, Yunsong Li and Weiying Xie

The State Key Laboratory of Integrated Service Networks, Xidian University, Xi'an 710000, China; jjli@xidian.edu.cn (J.L.); rsong@xidian.edu.cn (R.S.); ysli@mial.xidian.edu.cn (Y.L.); wyxie@xidian.edu.cn (W.X.)
* Correspondence: cxwu@stu.xidian.edu.cn; Tel.: +86-155-2960-9856
† These authors contributed equally to this work.

This paper proposed a deep residual-augmented attentional u-shape network (RA^2UN) for spectral reconstruction (SR) using several double-improved residual blocks (DIRB) instead of paired plain convolutional units. Specifically, a trainable spatial augmented attention (SAA) module was developed to bridge the encoder and decoder to emphasize the features in the informative regions. Furthermore, a channel-augmented attention (CAA) module embedded in the DIRB was also introduced to adaptively rescale and enhance residual learning by using first-order and second-order statistics for stronger feature representations. Finally, a boundary-aware constraint was employed to focus on the salient edge information and recover more accurate high-frequency details. Experimental results on four benchmark datasets demonstrated that the proposed RA^2UN network outperformed the state-of-the-art SR) methods in terms of quantitative measurements and perceptual comparison.

VIII. **Hyperspectral Image Visualization (one paper)**

remotesensing-12-02479-v2

Linear and Non-Linear Models for Remotely-Sensed Hyperspectral Image Visualization

Radu-Mihai Coliban *, Maria Marincas, Cosmin Hatfaludi and Mihai Ivanovici

Electronics and Computers Department, Transilvania University of Bra¸sov, 500036 Bra¸sov, Romania; maria.marincas@student.unitbv.ro (M.M.); cosmin.hatfaludi@student.unitbv.ro (C.H.); mihai.ivanovici@unitbv.ro (M.I.)
* Correspondence: coliban.radu@unitbv.ro

This paper proposed the use of a linear model for color formation to emulate the image acquisition process by a digital color camera and investigated the impact of the choice of spectral sensitivity curves on the visualization of hyperspectral images as RGB color images. In addition, a non-linear model based on an artificial neural network was also proposed. With the proposed linear and nonlinear models, the impact and the intrinsic quality of the hyperspectral image visualization could be assessed based on the amount of information present in the image quantified by color entropy and scene complexity measured by color fractal dimension, both of which provide an indication of detail and texture characteristics of the image. The experiments compared four other methods and the superiority of the proposed method was demonstrated.

IX. **Applications (four papers)**

remotesensing-13-02243-v3

Generative Adversarial Network Synthesis of Hyperspectral Vegetation Data

Andrew Hennessy *, Kenneth Clarke and Megan Lewis

School of Biological Sciences, The University of Adelaide, Adelaide 5000, Australia; kenneth.clarke@adelaide.edu.au (K.C.); megan.lewis@adelaide.edu.au (M.L.)
* Correspondence: andrew.hennessy@adelaide.edu.au

This paper applied advances in generative deep learning models to produce realistic synthetic hyperspectral vegetation data whilst maintaining class relationships. Specifically, a Generative Adversarial Network (GAN) was trained using the Cramér distance on two vegetation hyperspectral datasets, demonstrating the ability to approximate the distribution of the training samples. The creation of an augmented dataset consisting of synthetic and original samples was used to train multiple classifiers, with increases in classification accuracy observed in almost all circumstances. Both datasets showed improvements in classification accuracy ranging from a modest 0.16% for the Indian Pines set to a substantial increase of 7.0% for the New Zealand vegetation.

remotesensing-13-03207

Rice Leaf Blast Classification Method Based on Fused Features and One-Dimensional Deep Convolutional Neural Network

Shuai Feng [1], Yingli Cao [1,2], Tongyu Xu [1,2,*], Fenghua Yu [1,2], Dongxue Zhao [1] and Guosheng Zhang [1]

[1] College of Information and Electrical Engineering, Shenyang Agricultural University, Shenyang 110866, China; fengshuai@stu.syau.edu.cn (S.F.); caoyingli@syau.edu.cn (Y.C.); adan@syau.edu.cn (F.Y.); zhaoDX@stu.syau.edu.cn (D.Z.); gszhang@stu.syau.edu.cn (G.Z.)
[2] Liaoning Engineering Research Center for Information Technology in Agriculture, Shenyang 110866, China
* Correspondence: xutongyu@syau.edu.cn; Tel.: +86-024-8848-7121

This paper developed seven one-dimensional deep convolutional neural network (DCNN) models to determine the best classification features and classification models for the five disease classes of leaf blast in order to improve the accuracy of grading the disease. It first pre-processed the hyperspectral imaging data to extract rice leaf samples of five disease classes, and the number of samples was increased by data-augmentation methods; then, spectral feature wavelengths, vegetation indices, and texture features were obtained based on the amplified sample data, which were used to construct CNN-based models. Finally, the proposed models were compared and analyzed with the Inception V3, ZF-Net, TextCNN, and bidirectional gated recurrent unit (BiGRU); support vector machine (SVM); and extreme learning machine (ELM) models in order to determine the best classification features and classification models for different disease classes of leaf blast. The experimental results also showed that the DCNN models provided better classification capability for disease classification than the Inception V3, ZF-Net, TextCNN, BiGRU, SVM, and ELM classification models. The SPA + TFs-DCNN achieved the best classification accuracy with an overall accuracy (OA) and Kappa of 98.58% and 98.22%, respectively. In terms of the classification of the specific different disease classes, the F1-scores for diseases of classes 0, 1, and 2 were all 100%, while the F1-scores for diseases of classes 4 and 5 were 96.48% and 96.68%, respectively. This study provides a new method for the identification and classification of rice leaf blast and a research basis for assessing the extent of the disease in the field.

remotesensing-13-04587-v2

Detection and Classification of Rice Infestation with Rice Leaf Folder (Cnaphalocrocis medinalis) Using Hyperspectral Imaging Techniques

Gui-Chou Liang [1], Yen-Chieh Ouyang [2] and Shu-Mei Dai [1,*]

[1] Department of Entomology, National Chung Hsing University, Taichung 402, Taiwan; g107036008@mail.nchu.edu.tw
[2] Department of Electrical Engineering, National Chung Hsing University, Taichung 402, Taiwan; ycouyang@nchu.edu.tw
* Correspondence: sdai5497@dragon.nchu.edu.tw; Tel.: +886-0963-234-136

This paper developed a hyperspectral image technique that combines constrained energy minimization (CEM) and deep neural networks to detect defects in the spectral images of infected rice leaves and compare the performance of each in the full spectral band, selected bands, and band expansion process (BEP) to compressed spectral information for the selected bands. A total of 339 hyperspectral images were collected in this study; the results showed that six bands were sufficient for detecting early infestations of rice leave folder (RLF), with a detection accuracy of 98% and a Dice similarity coefficient of 0.8, which provides advantages in the commercialization of this field.

remotesensing-12-02348-v2

Detection of Insect Damage in Green Coffee Beans Using VIS-NIR Hyperspectral Imaging

Shih-Yu Chen [1,2,*], Chuan-Yu Chang [1,2], Cheng-Syue Ou [1] and Chou-Tien Lien [1]

[1] Department of Computer Science and Information Engineering, National Yunlin University of Science and Technology, Yunlin 64002, Taiwan; chuanyu@gemail.yuntech.edu.tw (C.-Y.C.); m10717033@gemail.yuntech.edu.tw (C.-S.O.); m10617013@gemail.yuntech.edu.tw (C.-T.L.)
[2] Artificial Intelligence Recognition Industry Service Research Center, National Yunlin University of Science and Technology, Yunlin 64002, Taiwan
* Correspondence: sychen@yuntech.edu.tw

This paper developed a hyperspectral insect damage-detection algorithm (HIDDA) that can automatically detect insect-damaged beans using only a few bands and one spectral signature. It used a push-broom visible-near infrared (VIS-NIR) hyperspectral sensor to obtain images of coffee beans. It takes advantage of recently developed constrained energy minimization (CEM)-based band selection methods coupled with two classifiers, support vector machine (SVM) and convolutional neural networks (CNN), to select bands. The experiments showed that 850–950 nm is an important wavelength range for accurately identifying insect damaged beans, and HIDDA can indeed detect insect damaged beans with only one spectral signature, which will provide an advantage in terms of practical applications and commercialization in the future.

3. Conclusions

The success of this Special Issue is owed to many researchers who are willing to share their original ideas and findings. All guest editors would like to express their sincere gratitude to the researchers for their time and efforts devoted to make this Special Issue a reality. Lastly, we extend a special thanks to the anonymous reviewers for their hard work in valuable and insightful comments to help the authors improve their presentation and quality of their papers. Without their contributions, this Special Issue could not have been completed.

Acknowledgments: The work of M. Song was supported by National Nature Science Foundation of China (61971082, 61890964). The work of H. Yu was supported by the National Natural Science Foundation of China under Grant 42101350, and the China Postdoctoral Science Foundation under Grant 2022T150080 and Grant 2020M680925. The work of H. Li was supported by the National Science and Technology Council, Taiwan under Grant No. MOST 109-2221-E-027-124-MY3.

Conflicts of Interest: The Guest editors declare no conflict of interest.

References

1. Chang, C.-I. *Hyperspectral Imaging: Techniques for Spectral Detection and Classification*; Kluwer Academic/Plenum Publishers: New York, NY, USA, 2003.
2. Chang, C.-I. *Hyperspectral Data Processing: Algorithm Design and Analysis*; Wiley: Hoboken, NJ, USA, 2013.
3. Chang, C.-I. *Real-Time Progressive Hyperspectral Image Processing: Endmember Finding and Anomaly Detection*; Springer: Berlin/Heidelberg, Germany, 2016.
4. Chang, C.-I. *Real-Time Recursive Hyperspectral Sample and Band Processing: Algorithm Architecture and Implementation*; Springer: Berlin/Heidelberg, Germany, 2017.

5. Chang, C.-I. (Ed.) *Hyperspectral Data Exploitation: Theory and Applications*; John Wiley & Sons: Hoboken, NJ, USA, 2007.
6. Chang, C.-I. (Ed.) *Advances in Hyperspectral Image Processing Techniques*; Wiley: Hoboken, NJ, USA, 2022; *to be publihed*; ISBN-13 978-1119687764, ISBN-10 1119687764.
7. Song, M.; Chang, C.-I. *Hypersspectral Data Unmixing: Algorithm Design and Analysis*; Hubei Science and Technology Press: Wuhan, China, 2021.
8. Chang, C.-I.; Wang, Y.; Xue, B.; Wang, L.; Yu, C.; Song, M. *Hyperspectral Target Detection: Algorithm Design and Analysis*; Hubei Science and Technology Press: Wuhan, China, 2021.
9. Adams, J.B.; Smith, M.O.; Gillepie, A.R. Simple models for complex natural surfaces: A strategy for hyperspectral era of remote sensing. *Proc. IEEE Int. Geosci. Remote Sens. Symp.* **1989**, *1*, 16–21.
10. Adams, J.B.; Smith, M.O.; Gillespie, A.R. Image spectroscopy: Interpretation based on spectral mixture analysis. In *Remote Geochemical Analysis: Elemental and Mineralogical Composition*; Pieters, C.M., Englert, P.A., Eds.; Cambridge University Press: Cambridge, UK, 1993; pp. 145–166.
11. Smith, M.O.; Adams, J.B.; Sabol, D.E. Spectral mixture analysis-new strategies for the analysis of multispectral data. In *Image Spectroscopy—A Tool for Environmental Observations*; Hill, J., Mergier, J., Eds.; Springer: Berlin/Heidelberg, Germany, 1994; pp. 125–143.
12. Smith, M.O.; Roberts, D.A.; Hill, J.; Mehl, W.; Hosgood, B.; Verdebout, J.; Schmuck, G.; Koechler, C.; Adams, J.B. A new approach to quantifying abundances of materials in multispectral images. *Proc. IEEE Int. Geosci. Remote Sens. Symp.* **1944**, *4*, 2372–2374.
13. Gillespie, A.R.; Smith, M.O.; Adams, J.B.; Willis, S.C.; Fischer, A.F.; Sabol, D.E., III. Interpretation of residual images: Spectral mixture analysis of AVIRIS images, Owens valley, California. In Proceedings of the Second Airborne Visible/Infrared Imaging Spectrometer (AVIRIS) Workshop, Pasadena, CA, USA, 6–8 May 1990; pp. 243–270.
14. Goetz, A.F.H.; Boardman, J.W. Quantitative determination of imaging spectrometer specifications based on spectral mixing models. *Proc. IEEE Int. Geosci. Remote Sens. Symp.* **1989**, *1*, 1036–1039.
15. Basedow, R.; Silverglate, P.; Rappoport, W.; Rockwell, R.; Rosenberg, D.; Shu, K.; Whittlesey, R.; Zalewski, E. The HYDICE instrument design. *Proc. Int. Symp. Spectr. Sens. Res.* **1992**, *1*, 430–445.
16. Manolakis, D.; Shaw, G. Detection algorithms for hyperspectral imaging applications. *IEEE Signal Process. Mag.* **2002**, *19*, 29–43. [CrossRef]
17. Benediktsson, J.A.; Ghamisi, P. *Spectral-Spatial Classification of Hyperspectral Remote Sensing Images*; Artech House: Norwood, MA, USA, 2015.
18. Plaza, A.; Chang, I.-C. (Eds.) *High Performance Computing in Remote Sensing*; Chapman & Hall/CRC Press: Boca Raton, FL, USA, 2007.

remote sensing

MDPI

Article

Deep Relation Network for Hyperspectral Image Few-Shot Classification

Kuiliang Gao [1,*], Bing Liu [1], Xuchu Yu [1], Jinchun Qin [2], Pengqiang Zhang [1] and Xiong Tan [1]

1 Information Engineering University, Zhengzhou 450001, China
2 Xi'an Research Institute of Surveying and Mapping, Xi'an 710054, China
* Correspondence: 311405000803@home.hpu.edu.cn

Received: 20 February 2020; Accepted: 10 March 2020; Published: 13 March 2020

Abstract: Deep learning has achieved great success in hyperspectral image classification. However, when processing new hyperspectral images, the existing deep learning models must be retrained from scratch with sufficient samples, which is inefficient and undesirable in practical tasks. This paper aims to explore how to accurately classify new hyperspectral images with only a few labeled samples, i.e., the hyperspectral images few-shot classification. Specifically, we design a new deep classification model based on relational network and train it with the idea of meta-learning. Firstly, the feature learning module and the relation learning module of the model can make full use of the spatial–spectral information in hyperspectral images and carry out relation learning by comparing the similarity between samples. Secondly, the task-based learning strategy can enable the model to continuously enhance its ability to learn how to learn with a large number of tasks randomly generated from different data sets. Benefitting from the above two points, the proposed method has excellent generalization ability and can obtain satisfactory classification results with only a few labeled samples. In order to verify the performance of the proposed method, experiments were carried out on three public data sets. The results indicate that the proposed method can achieve better classification results than the traditional semisupervised support vector machine and semisupervised deep learning models.

Keywords: hyperspectral image few-shot classification; deep learning; meta-learning; relation network; convolutional neural network

1. Introduction

Hyperspectral remote sensing, as an important means of earth observation, is one of the most important technological advances in the field of remote sensing. Utilizing the imaging spectrometer with very high spectral resolution, hyperspectral remote sensing can obtain abundant spectral information on the observation area so as to produce hyperspectral images (HSI) with a three-dimensional data structure. As HSI have the unique advantage of "spatial–spectral unity" (HSI contain both abundant spectral and spatial information), hyperspectral remote sensing has been widely used in fine agriculture, land-use planning, target detection, and many other fields.

HSI classification is one of the most important steps in HSI analysis and application, the basic task of which is to determine a unique category for each pixel. In early research, the working mode of feature extraction combined with classifiers such as support vector machines (SVM) [1] and random forest (RF) [2] was dominant at the time. Initially, in order to alleviate the Hughes phenomenon caused by band redundancy, researchers introduced a series of feature extraction methods to extract spectral features conducive to classification from abundant spectral information. Common spectral feature extraction methods include principal component analysis (PCA) [3], independent component analysis (ICA) [4], linear discriminant analysis (LDA) [5], and other linear methods, as well as kernel principal component analysis (KPCA) [6], locally linear embedding (LLE) [7], t-distributed stochastic neighbor embedding (t-SNE) [8], and other nonlinear methods. Admittedly, the above feature extraction method can achieve some results, but ignoring spatial structure information in HSI still seriously hinders the increase of classification accuracy. To this end, a series of spatial information utilization methods are introduced, such as extended morphological profile (EMP) [9], local binary patterns (LBP) [10], 3D Gabor features [11], Markov random field (MRF) [12], spatial filtering [13], variants of non-negative matrix underapproximation (NMU) [14], and so on. The extraction and utilization of spatial features can effectively improve classification accuracy. However, due to the separation of feature extraction process and classification in traditional classification mode, the adaptability between them cannot be fully considered [15]. In addition, the classification results of traditional methods largely depend on the accumulated experience and parameter setting, which lacks stability and robustness.

In recent years, with the development of artificial intelligence, deep learning has been applied to the field of remote sensing [16]. Compared to traditional methods, deep learning can automatically learn the required features from the data by establishing a hierarchical framework. Moreover, these features are often more discriminative and more conducive to the classification. Stacked AutoEncoder (SAE) [17], recurrent neural network (RNN) [18,19], and deep belief networks (DBN) [20,21] are first applied to HSI classification. These methods can achieve higher classification accuracy than traditional methods under certain conditions. Nevertheless, some necessary preprocessing must be performed to transform HSI into a one-dimensional vector for feature extraction, which destroys the spatial structure information in HSI. Compared with the above deep learning models, convolutional neural networks (CNNs) are more suitable for HSI processing and feature extraction. At present, 2D-CNN and 3D-CNN are two basic models widely used in HSI classification [22]. By means of two-dimensional and three-dimensional convolution operation, 2D-CNN and 3D-CNN can both fully extract and utilize the spatial–spectral features in HSI. Yue et al. take the lead in exploring the effect of 2D-CNN in HSI classification. Subsequently, many improved models based on 2D-CNN have been proposed and refresh classification accuracy constantly, such as DR-CNN [23], contextual deep CNN [24], DCNN [25], DC-CNN [26], and so on. Most 2D-CNN-based methods use PCA to reduce the dimension of HSI in order to reduce the number of channels in the convolution operation. However, this practice inevitably loses important detail information in HSI. The advantage of 3D-CNN is that it can directly perform three-dimensional convolution operation on HSI without any preprocessing and can make full use of spatial–spectral information to further improve classification accuracy. Chen et al. take the lead in utilizing 3D-CNN for HSI classification and have conducted detailed studies on the number of network layers, number of convolution kernels, size of the neighborhood, and other hyperparameters [27]. On this basis, methods such as residual learning [28], attention mechanism [29], dense network [30], and multiscale convolution [31] are combined with 3D-CNN, resulting in higher classification accuracy. In addition, CNN is combined with other methods such as active learning [32], capsule network [33], superpixel segmentation [34], and so on, which can achieve promising classification results when the training samples are sufficient.

Indeed, deep learning has seen great success in HSI classification. However, there is still a serious contradiction between the huge parameter space of the deep learning model and the limited labeled samples of HSI. In other words, the deep learning model must have enough labeled samples as a guarantee, so as to give full play to the its classification performance. Nevertheless, it is difficult to obtain enough labeled samples in practice, because the acquisition of labeled samples is time-consuming and laborious. In order to improve classification accuracy under the condition of limited labeled samples, semisupervised learning and data augmentation are widely applied. In [35,36], CNN was combined with semisupervised classification. In [37], Kang et al. first extracted PCA, EMP, and edge-preserving features (EPF), then carried out classification by combining semisupervised method and decision confusion strategy. In [27], Chen et al. generated virtual training samples by adding noise to the original labeled samples, while in [38,39], the number of training samples were increased by constructing training sample pairs. In recent years, with the emergence of generative adversarial networks (GANs), many researchers have utilized the synthetic sample generated by GAN to assist in training networks [40–42]. It is true that the above methods can improve classification accuracy under the condition of limited labeled samples, but they either further explore the feature of the insufficient labeled samples or utilize the information of unlabeled samples in the HSI being classified to further train the model. In other words, the HSI used to train model are exactly identical to the target HSI used to test the model. This means that when processing a new HSI, the model must be retrained from scratch. However, it is impossible to train a classifier for each HSI, which will incur significant overhead in practice.

Few-shot learning is when a model can effectively distinguish the categories in the new data set with only a very few labeled samples processing a new data set [43]. The availability of very few samples challenges the standard training practice in deep learning [44]. Different from the existing deep learning model, however, humans are very good at few-shot learning, because they can effectively utilize the previous learning experience and have the ability to learn how to learn, which is the concept of meta-learning [45,46]. Therefore, we should effectively utilize transferable knowledge in the collected HSI to further classify other new HSI, so as to reduce cost as much as possible. Different HSI contain different types and quantities of ground objects, so it is difficult for the general transfer learning [47,48] to obtain satisfactory classification accuracy with a few labeled sample. According to the idea of meta-learning, the model not only needs to learn transferable knowledge that is conducive to classification but also needs to learn the ability to learn.

The purpose of this paper is to explore how to accurately classify new HSI which are absolutely different from the HSI used for training with only a few labeled samples (e.g., five labeled samples per class). More specifically, this paper designs a new model based on a relation network [49] for HSI few-shot classification (RN-FSC) and trains it with the idea of meta-learning. The designed model is an end-to-end framework, including two modules: feature learning module and relation learning module, which can effectively simplify the classification process. The feature learning module is responsible for extracting deep features from samples in HSI, while the relation learning module carries on relation learning by comparing the similarity between different samples, that is, the relation score between samples belonging to the same class is high, and the relation score between samples belonging to different class is low. From the perspective of workflow, the proposed RN-FSC method consists of three steps. In the first step, we use the designed network model to carry out meta-learning on the source HSI data set, so that the model can fully learn the transferable feature knowledge and relation comparison ability, i.e., the ability to learn how to learn. In the second step, the network model is fine-tuned with only a few labeled samples in the target HSI data set so that the model can quickly adapt to new classification scenarios. In the third step, the target HSI data sets are used to test the classification performance of the proposed method. It is important to note that the target HSI data set for classification and the source HSI data set for meta-learning are completely different.

The main contributions of this paper are as follows:

1. The RN-FSC method is proposed to carry out classification on the new HSI with only a few labeled samples. The RN-FSC method has the ability to learn how to learn through meta-learning on the source HSI data set, so it can accurately classify the new HSI;
2. The network model containing the feature learning module and relation learning module is designed for HSI classification. Specifically, 3D convolution is utilized for feature extraction to make full use of spatial–spectral information in HSI, and the 2D convolution layer and fully connected layer are utilized to approximate the relationship between sample features in an abstract nonlinear approach;
3. Experiments are conducted on three well-known HSI data sets, which demonstrate that the proposed method can outperform conventional semisupervised methods and the semisupervised deep learning model with a few labeled samples.

The remainder of this paper is structured as follows. In Section 2, HSI few-shot classification is introduced. In Section 3, the design relation network model is described in detail. In Section 4, experimental results and analysis on three public available HSI data sets are presented. Finally, conclusions are provided in Section 5.

2. HSI Few-Shot Classification

In this section, we first explain the definition of few-shot classification, then describe the task-based learning strategy in detail, and finally give the complete process of HSI few-shot classification.

2.1. *Definition of Few-Shot Classification*

In order to explain the definition of few-shot classification, we must first distinguish several concepts: source data set, target data set, fine-tuning data set, and testing data set. Both the fine-tuning data set and the testing data set are subsets of the target data set, sharing the same label space, while the source data set and the target data set are totally different. With reference to most of the existing deep learning models, we can only utilize the fine-tuning data set to train a classifier. However, the classification performance of this classifier is very poor due to the very small fine-tuning data set. Therefore, we need to use the idea of meta-learning to carry out the classification task (as shown in Figure 1). The model first performs meta-learning on the source data set to extract the transferable feature knowledge and cultivate the ability of learning to learn. After meta-learning, the model can acquire enough generalization knowledge. Then, the model is fine-tuned on the fine-tuning data set to extract individual knowledge, so as to adapt to the new classification scenario quickly. The fine-tuning data set is very small compared to the testing data set, so the process of fine-tuning can be called few-shot learning. If the fine-tuning data set contains C unique classes and each class includes K labeled samples, the classification problem can be called C-way K-shot. Finally, the model is utilized to classify the testing data set.

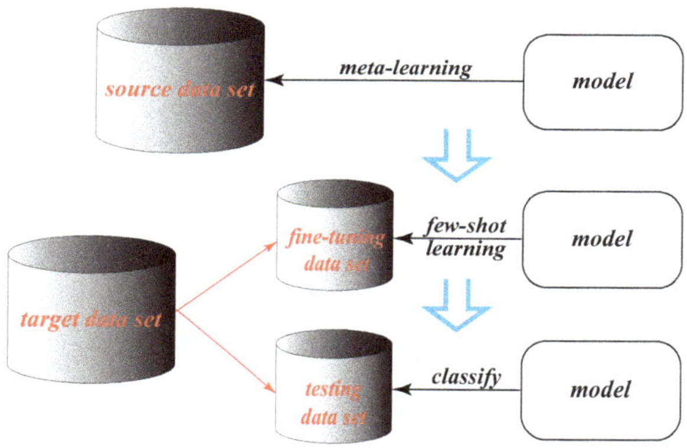

Figure 1. Definition of few-shot classification.

2.2. *Task-Based Learning Strategy*

At present, batch-based training strategy is widely used in deep learning models, as shown in Figure 2a. In the training process, each batch contains a certain amount of samples with specific labels. The training process of the model is actually based on samples to calculate the loss and update the network parameters. General transfer learning also uses this strategy for model training.

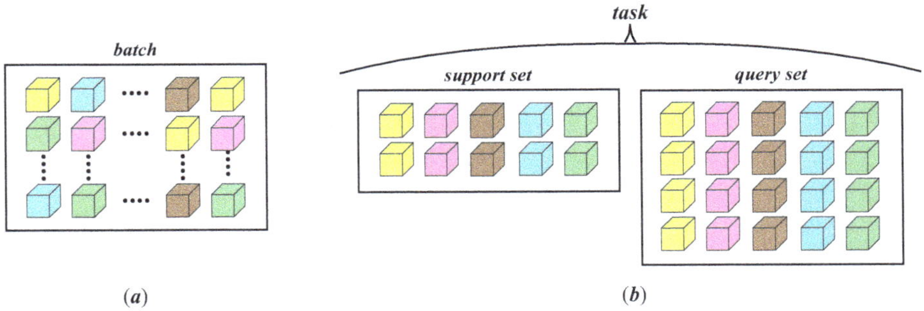

Figure 2. Different training and learning strategy where color represents class. (**a**) Batch-based training strategy used widely in deep learning. (**b**) Task-based learning strategy used in meta-learning.

Meta-learning can also be regarded as a learning process of transferring feature knowledge. The key of meta-learning allowing the model to acquire more outstanding learning ability than general transfer learning is the task-based learning strategy. In meta-learning, tasks are treated as the basic unit for training [45,49]. As shown in Figure 2b, a task contains a support set and a query set. The support set and the query set are sampled randomly from the same data set and share the same label space. The sample x in the support set are clearly labeled by y, while the labels of samples in the query set are regarded as unknown. The model predicts the labels of samples in the query set under the supervision of the support set and calculates the loss by comparing the predictive labels with the real labels, thus realizing the update of parameters.

The model runs on the basis of the task-based learning strategy, whether in the meta-learning phase, the few-shot learning phase, or the classification phase. One task is actually a training iteration. Take meta-learning on a source data set containing C_{src} classes as an example. During each iteration, a task is generated by randomly selecting C classes and K samples per class from the source data set. Thus, the support set can be denoted as $\mathcal{S} = \{(x_i, y_i)\}_{i=1}^{C\times K}$. Similarly, $C \times N$ samples are randomly sampled from the same C classes to form a query set $\mathcal{Q} = \{(x_j, y_j)\}_{j=1}^{C\times N}$. It is important to note that there is no intersection between $\mathcal{S}$ and $\mathcal{Q}$. In practice, we usually set $C < C_{src}$, which can guarantee the richness of tasks and thus improve the robustness of the model. In theory, N tends to be much larger than K, so as to mimic the actual few-shot classification scenario. In summary, through the above description, a C-way K-shot N-query learning task has been builded on the source data set.

2.3. HSI Few-Shot Classification

In the previous sections, we explained in detail the few-shot classification and its learning strategy. It is not difficult to apply it to HSI classification. We only need to utilize the collected HSI as the source data set, e.g., the Botswana and Houston data sets, and utilize other HSI as the target data set, e.g., the Pavia Center data set. The complete HSI few-shot classification process based on the task-based learning strategy can be summarized as follows.

(1) In the first phase, learning tasks are built on the source data set, and the model performs meta-learning;
(2) In the second phase, learning tasks are built on the fine-tuning data set, and the model performs few-shot learning;
(3) In the third phase, the entire fine-tuning data set is regarded as the support set, and the testing data set is regarded as the query set, so as to build tasks for HSI classification.

3. The Designed Relation Network Model

This section introduces the designed relation network model for the HSI few-shot classification. The designed model consists of two core modules, feature learning module and relation learning module, which are introduced in detail. In addition, we explain how the model acquires the ability to learn how to learn from three different perspectives.

3.1. Model Overview

The designed relation network model for HSI few-shot classification consists of three parts: feature learning, feature concatenation, and relation learning, as illustrated in Figure 3. The model is an end-to-end framework, with tasks as inputs and predictive labels as outputs.

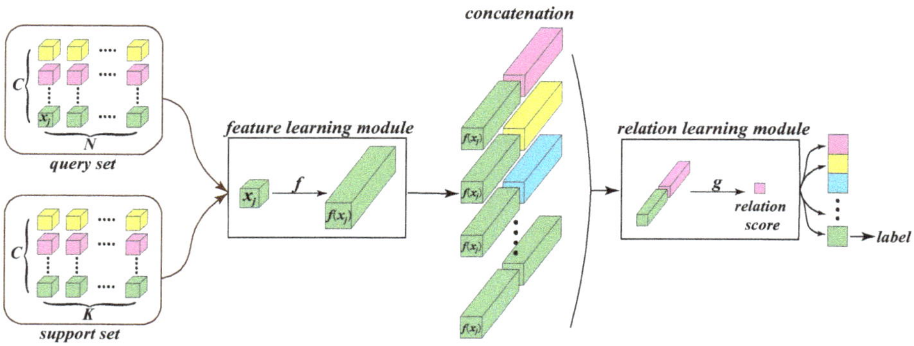

Figure 3. Visual representation of the designed relation network model for HSI few-shot classification.

Specifically, we select the data cubes belonging to each pixel in HSI as the samples in the task. As defined in Section 2.2, the sample in the support set is denoted as x_i, and the sample in the query set is denoted as x_j. The feature learning module is equivalent to a nonlinear embedding function f, which maps samples x_i and x_j in the data space to abstract features $f(x_i)$ and $f(x_j)$ in the feature space. Then, features $f(x_i)$ and $f(x_j)$ are concatenated in the depth dimension, which can be denoted as $\mathcal{C}(f(x_i), f(x_j))$. Of course, there is more than one way to perform concatenation. It should be noted, however, that each sample feature in the query set should be concatenated to each feature generated by the support set. In addition, in order to simplify the following calculation and improve the robustness of the model, the sample features belonging to the same class in the support set are averaged. Consequently, the number of features generated from the support set is always equal to C. This means that, for the support set $\mathcal{S} = \{(x_i, y_i)\}_{i=1}^{C\times K}$ and the query set $\mathcal{Q} = \{(x_j, y_j)\}_{j=1}^{C\times N}$, $C \times C \times K$ concatenations would be generated. The relation learning module can also be regarded as a nonlinear function g, which maps each concatenation to a relation score $r_{i,j} = g[\mathcal{C}(f(x_i), f(x_j))]$ representing the similarity between x_i and x_j. If samples x_i and x_j belong to the same class, the relation score will be close to 1, otherwise the relation score will be close to 0. Finally, the maximum score is obtained from the relation score set $\mathcal{R} = \{r_{l,j}\}(l = 1, \ldots, C)$ of sample x_j, so as to decide the predictive label.

The model is trained with mean square error (MSE) as loss function (Equation (1)). MSE is easy to calculation and sufficient for training. If y_i and y_j belong to the same class, $(y_i == y_j)$ is 1, otherwise 0, which can effectively achieve relation learning.

$$L_{MSE} = \sum_{i=1}^{C\times K} \sum_{j=1}^{C\times N} (r_{i,j} - 1 \cdot (y_i == y_j))^2. \tag{1}$$

3.2. The Feature Learning Module

The goal of the feature learning module is to extract more discriminative features from the input data cubes. Theoretically, any network structure can be built in this module for feature learning. A large number of studies have shown that 3D convolution is more suitable for the spatial–spectral features extraction because of the close correlation between the spatial domain and spectral domain in HSI, . Therefore, we take the 3D convolutional layer as the core and construct the feature learning network as shown in Figure 4.

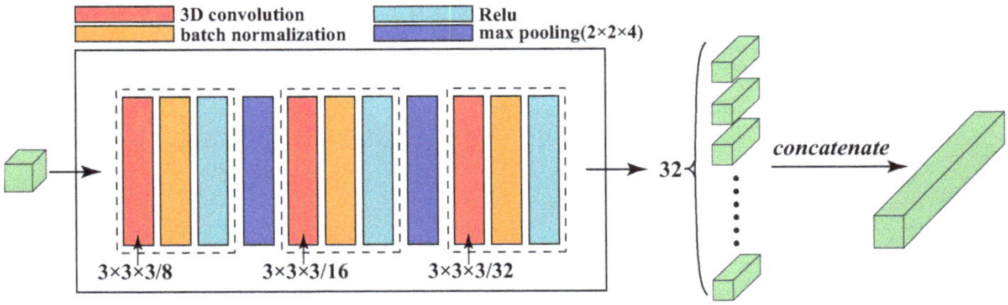

Figure 4. Visual representation of the feature learning module.

The feature learning module consists of a 3D convolutional layer, batch normalization layer, Relu activation function, maximum pooling layer, and concatenation operation. 3D convolution can process the input data cubes directly without any preprocessing. Compared with the general 2D convolution, 3D convolution can extract more discriminative spatial–spectral features by cascading spectral information of adjacent bands. Specifically, the 3D convolution kernel is set as $3 \times 3 \times 3$, and the number of convolution kernel increases from 8 to 32 by multiples, which is consistent with the experience in the field of computer vision. Batch normalization layers are added after each 3D convolutional layer, which can effectively alleviate the problem of vanishing gradient and enhance the generalization ability of the model. Relu activation function, one of the most widely used activation functions in deep learning, can increase the nonlinearity of the model and speed up the convergence. The 3D convolutional layer, batch normalization layer, and Relu layer can be considered as a basic unit. Each unit is connected via maximum pooling layer. Considering the characteristics of HSI, the maximum pooling layer is set to $2 \times 2 \times 4$ to deal with spectral redundancy.

After three convolution operations, the input samples become data cubes with 32 channels. To facilitate the subsequent operation in the feature concatenation phase, we first concatenate the 32 data cubes in the channel dimension. Given that the dimension of the data cubes is $(32, H, W, D)$, it becomes $(H, W, D \times 32)$ after channel concatenation.

3.3. The Relation Learning Module

Under the combined action of the first two phases, the data cubes are transformed into different concatenations which are the input of the relation learning module (Figure 5). The purpose of the relation learning module is to map each concatenation to a relation score measuring the similarity between the two samples, i.e., the relationship.

In order to speed up computation, 2D convolution is regarded as the core to build the relation learning module. Therefore, the dimension of the concatenations can be regarded as (H, W, C), where C stands for the channel dimension. Considering that the channel dimension is much larger than the spatial dimension, the 1×1 2D convolution [50] is first adopted, which can extract the cascaded features across the channel while reducing the dimension effectively. After 1×1 convolution, 128 convolution kernels of 3×3 are utilized to ensure the diversity of features. In order to fully train the network, the batch normalization layer and Relu activation function are also applied after each convolution. Finally, two fully connected layers of 128 and 1 are added, so as to transform the feature maps into relation scores. Dropout is introduced between the fully connected layer to further enhance the generalization capability. In addition, sigmoid activation function is used to limit the output to the interval [0, 1].

Relation score is not only the final result of relation learning, but also a kind of similarity measure. If the two samples belong to the same class, the relation score is close to 1, otherwise 0. Therefore, the classes of samples in the query set will be determined according to the relation score.

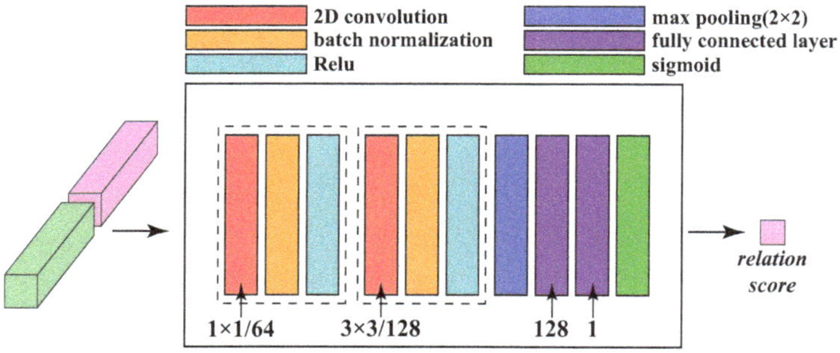

Figure 5. Visual representation of the relation learning module.

3.4. The Ability of Learning to Learn

Our proposed method, RN-FSC, is essentially a meta-learning-based method for HSI few-shot classification. The core idea of meta-learning is to cultivate the ability of learning to learn. In this section, we expound this ability of RN-FSC from the following three aspects:

(1) Learning process
General deep learning models are trained based on the unique correspondence between data and labels and can only be trained in one specific task space at a time. However, the proposed method is task-based learning at any phase. The model focuses not on the specific classification task but on the learning ability with many different tasks;

(2) Scalability
The proposed method performs meta-learning on the source data set to extract the transferable feature knowledge and cultivate the ablitity of learning to learn. From the perspective of knowledge transfer, the richer the categories in the source data set, the stronger the acquired learning ability, which is consistent with the human learning experience. Therefore, we can appropriately extend the source data set to enhance the generalization ability of the model;

(3) Core mechanism
The proposed method is not to learn how to classify a specific data set, but to learn a deep metric space with the help of many tasks from different data sets, in which relation learning is performed by comparison. In a data-driven way, this metric space is nonlinear and transferrable. By comparing the similarity between the support samples and the query samples in the deep metric space, the classification is realized indirectly.

4. Experiments and Discussion

All experiments were carried out on a laptop with an Intel Core i7-9750H, 2.60 GHz and an Nvidia GeForce RTX 2070. The laptop's memory is 16 GB. All programs are developed and implemented based on Pytorch library.

4.1. Experimental Data Sets

4.1.1. Source Data Sets

Four publicly available HSI data sets were collected to build the source data sets, which are Houston, Botswana, Kennedy Space Center (KSC), and Chikusei. The four data sets were photographed by different imaging spectrometers on different regions, with different ground sample distance and spectral range (as shown in Table 1). This can ensure the diversity and richness of samples, which is conducive to meta-learning. There are 76 different ground objects contained in the four data sets, and the distribution of their respective labeled samples can be seen in Figures 6–9. We exclude the classes with less samples and only select the 54 classes with more than 200 samples to build the source data set. In addition, 100 bands are selected on each data set via the graph-representation-based band selection (GRBS) [51] instead of all bands, so as to reduce spectral redundancy and guarantee the uniformity of the number of bands (Table 2). GRBS, an unsupervised band selection method based on graph representation, can perform better in both accuracy and efficiency. The spatial neighborhood of each pixel is set to 9 × 9 with reference to [25,39,48]. After the above processing, each HSI is transformed into a number of 9 × 9 × 100 data cubes, so as to standardize the data dimensions and optimize the learning process.

Table 1. Detailsof the source data sets. Kennedy Space Center (KSC), ground sample distance (GSD)(m), spatial size (pixel), spectral range (nm), airborne visible infrared imaging spectrometer (AVIRIS).

	Houston	**Botswana**	**KSC**	**Chikusei**
Spatial size	349 × 1905	1476 × 256	512 × 614	2517 × 2335
Spectral range	380–1050	400–2500	400–2500	363–1018
No. of bands	144	145	176	128
GSD	2.5	30	18	2.5
Sensor type	ITRES-CASI 1500	EO-1	AVIRIS	Hyperspec-VNIR-C
Areas	Houston	Botswana	Florida	Chikusei
No. of classes	30	14	13	19
Labeled samples	15029	3248	5211	77592

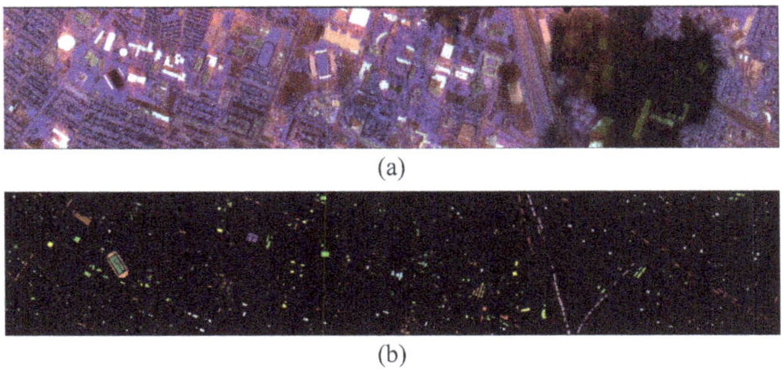

Figure 6. Houston data set. (**a**) Pseudocolor image. (**b**) Ground-truth map.

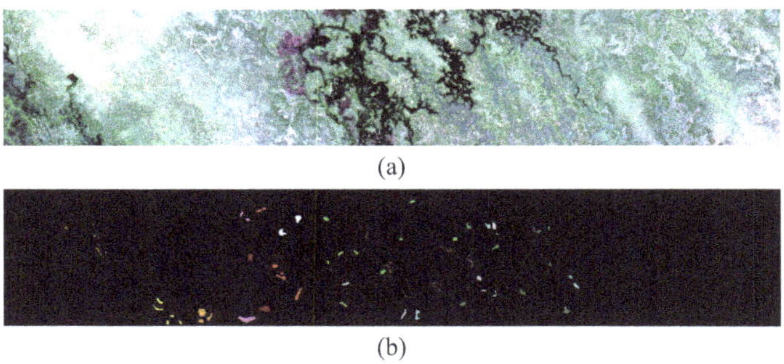

Figure 7. Botswana data set. (**a**) Pseudocolor image. (**b**) Ground-truth map.

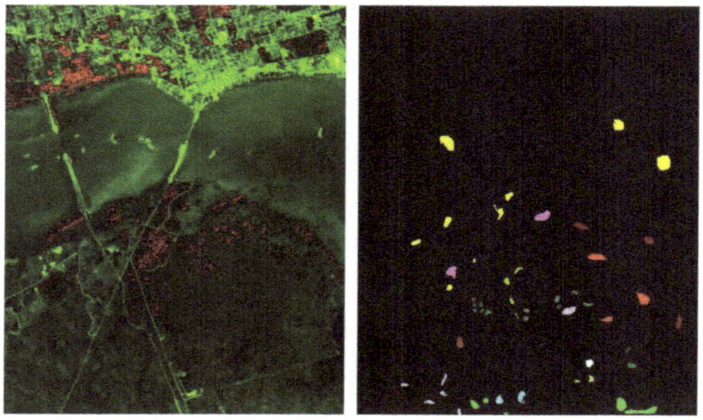

(a) (b)

Figure 8. Kennedy Space Center (KSC) data set. (**a**) Pseudocolor image. (**b**) Ground-truth map.

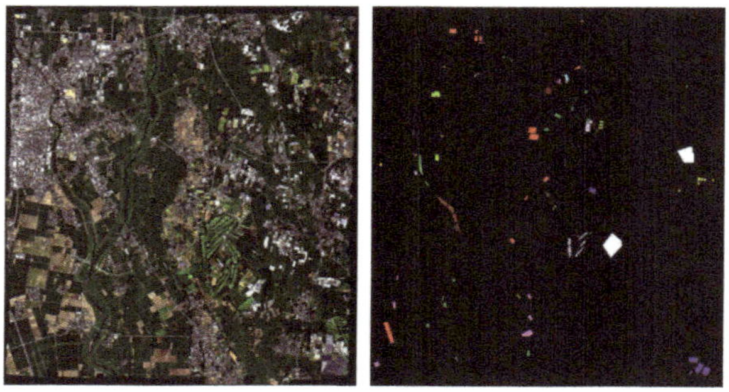

(a) (b)

Figure 9. Chikusei data set. (**a**) Pseudocolor image. (**b**) Ground-truth map.

Table 2. The selected bands on the source data sets via graph-representation-based band selection (GRBS). Kennedy Space Center (KSC).

	The Selected Bands
Houston	2 3 4 5 6 7 8 9 10 11 12 13 14 15 16 17 18 19 20 21 22 23 24 25 26 27 28 29 30 31 32 33 34 35 36 37 38 39 40 41 42 43 44 45 46 47 48 49 51 52 53 54 55 56 57 58 59 60 61 62 63 64 65 66 67 68 69 70 71 72 77 107 109 110 111 112 113 114 115 116 117 118 119 120 121 122 123 124 125 126 127 128 129 130 132 133 134 135 143 144
Botswana	2 3 4 5 6 7 8 9 10 11 12 13 14 15 16 17 18 19 20 21 22 23 24 25 26 27 28 29 30 31 32 33 34 35 36 37 38 39 40 41 42 43 44 45 46 47 48 49 50 51 52 53 55 56 57 58 59 60 61 62 63 64 65 66 67 68 69 70 71 72 73 88 110 111 112 113 114 115 116 117 118 119 120 121 122 123 124 125 126 127 128 137 138 139 140 141 142 143 144 145
KSC	2 3 4 5 6 7 8 9 10 11 12 13 14 15 16 17 18 19 20 21 22 23 24 25 26 28 29 31 32 33 35 36 37 39 40 41 42 43 44 45 46 47 48 49 50 51 52 53 54 55 56 57 58 59 60 61 62 63 64 65 66 67 68 71 72 73 74 75 76 77 78 79 80 81 82 83 84 85 86 87 88 89 90 95 101 120 132 143 144 145 146 147 148 149 150 151 155 167 175 176
Chikusei	1 2 3 4 5 6 7 8 9 10 11 12 13 14 15 16 17 18 19 20 21 22 23 24 25 26 27 28 29 30 31 32 33 34 35 36 37 38 39 40 41 42 43 44 45 46 47 48 49 50 51 52 53 54 55 56 57 58 59 60 61 62 63 65 66 67 94 95 96 97 98 99 100 101 102 103 104 105 106 107 108 109 110 111 112 113 114 116 117 118 119 120 121 122 123 124 125 126 127 128

4.1.2. Target Data Sets

Three well-known HSI data sets, i.e., the University of Pavia (UP), the Pavia Center (PC), and Salinas, were selected to build the target data sets. Table 3 shows the detailed information. In order to standardize data dimensions, we still used the GRBS method to select 100 bands for each HSI (Table 4) and set the spatial neighborhood as 9×9. Furthermore, five labeled samples per class were selected to build the fine-tuning data set, and the remaining samples were used as the testing data set. Consequently, we used three different HSI to build three different target data sets. The proposed method performs few-shot classification on the three target data sets respectively, so as to verify its effectiveness.

In summary, Houston, Botswana, KSC, and Chikusei were used to build the source data sets, and UP, PC, and Salinas were used to build the target data sets. Therefore, the source data set and the target data set are completely different. In the target data sets, only a few labeled samples (five samples per class) were used to build the fine-tuning data sets to fine-tune the designed model. In order to make a fair comparison with other classification methods, fine-tuning data sets were also used for supervised training in comparison experiments (Section 4.3).

Table 3. Details of three target data sets. University of Pavia (UP), Pavia Center (PC), ground sample distance (GSD) (m), spatial size (pixel), spectral range (nm), reflective optics system imaging spectrometer (ROSIS), airborne visible infrared imaging spectrometer (AVIRIS).

	UP	PC	Salinas
Spatial size	610×340	1096×715	512×217
Spectral range	430–860	430–860	400–2500
No. of bands	103	102	204
GSD	1.3	1.3	3.7
Sensor type	ROSIS	ROSIS	AVIRIS
Areas	Pavia	Pavia	California
No. of classes	9	9	16
Labeled samples	42776	148152	54129

Table 4. The selected bands on the target data sets via GRBS. University of Pavia (UP), Pavia Center (PC).

	The Selected Bands
UP	2 3 4 5 6 7 8 9 10 11 12 13 14 15 16 17 18 19 20 21 23 24 25 26 27 28 29 30 31 32 33 34 35 36 37 38 39 40 41 42 43 44 45 46 47 48 49 50 51 53 54 55 56 57 58 59 60 61 62 63 64 65 66 67 68 69 70 71 72 73 74 75 76 77 78 79 80 81 82 83 84 85 86 87 88 89 90 91 92 93 94 95 96 97 98 99 100 101 102 103
PC	1 2 3 4 5 6 7 8 9 10 11 12 13 14 15 16 17 18 19 21 22 23 24 25 26 27 28 29 30 31 32 33 34 35 36 37 38 39 40 41 42 43 44 45 46 47 48 49 50 51 52 53 54 55 56 57 58 60 61 62 63 64 65 66 67 68 69 70 71 72 73 74 75 76 77 78 79 80 81 82 83 84 85 86 87 88 89 90 91 92 93 94 95 96 97 98 99 100 101 102
Salinas	2 3 4 5 6 7 8 9 10 11 12 13 14 15 16 17 18 19 20 21 22 23 24 25 26 27 28 29 31 33 34 35 36 37 38 39 40 41 42 43 44 45 46 47 48 49 50 51 52 53 54 55 56 57 58 59 60 61 62 63 64 65 66 67 68 69 70 71 72 73 74 75 76 77 78 79 80 81 82 83 84 85 86 87 88 89 90 91 92 93 94 95 96 97 98 99 100 126 139 204

4.2. Experimental Setup

Meta-learning is a very important phase for the proposed method. The main hyperparameters in meta-learning include the number of class in each task C, the number of support samples per class K, and the number of query samples per class N, which are directly related to building the learning task. Therefore, we first carried out experiments to explore the influence of C, K, N on classification results.

The hyperparameters C determine the number of classes in each learning task, i.e., the complexity of the task. As described in Section 4.1, the source data set consists of 54 classes, so we explored the influence of C at 10, 20, 30, and 40. Figure 10 shows the experimental results. It can be seen that on three different target data sets, the model can always obtain the highest classification accuracy when C is 20. This indicates that when the number of classes in task is too small, the model cannot carry on sufficient learning. Given a class contained in the source data sets, if C is too small, this class will appear less often in the task, which reduces the chances of model learning from this class. Otherwise, when C is equal to 30 or 40, the complexity of the task exceeds the representation ability of the model, resulting in a significant decrease in classification accuracy.

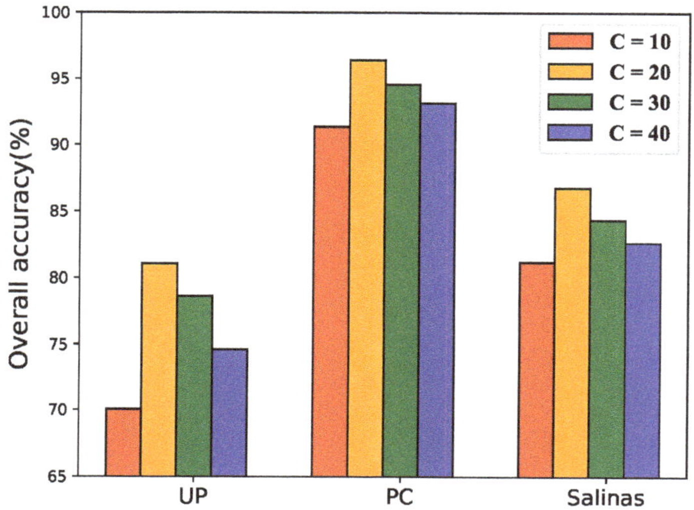

Figure 10. Overall accuracy under different C.

K and N together determine the diversity and richness of samples in the task and directly affect the size of the task. With reference to [49], we fixed the size of task as 20 samples per class and explored the influence of K and N on the classification results by trying different combinations. Table 5 shows the experimental results. It can be found that with the increase of K, the classification accuracy decreases gradually. When K is 1, the highest classification accuracy is obtained for all three different data sets. This experimental result verifies the theory described in Section 2.2, i.e., setting $K < N$ in the meta-learning phase can imitate the subsequent few-shot classification process, so as to obtain better classification results.

Table 5. Overall accuracy (%) with different combinations of K and N.

	$K = 1, N = 19$	$K = 5, N = 15$	$K = 10, N = 10$	$K = 15, N = 5$
UP	81.94	78.84	76.26	74.55
PC	96.36	96.03	95.53	94.89
Salinas	86.99	85.97	84.61	82.95

Through the above experimental exploration, the optimal task setting in the meta-learning phase has been found, i.e, the 20-way 1-shot 19-query learning task. In order to further optimize the meta-learning process, the appropriate value of learning rate is analyzed. With reference to relevant experience, we analyzed the influence of learning rate at 0.01 and 0.001 on the loss function value, as shown in Figure 11. It can be seen that the loss value obviously fluctuates, due to the diversity of source data set and the randomness of task. Nevertheless, after approximately 2000 episodes, the 0.001 learning rate is able to acquire a lower loss value, indicating that the 0.001 learning rate can enable the model to learn fully.

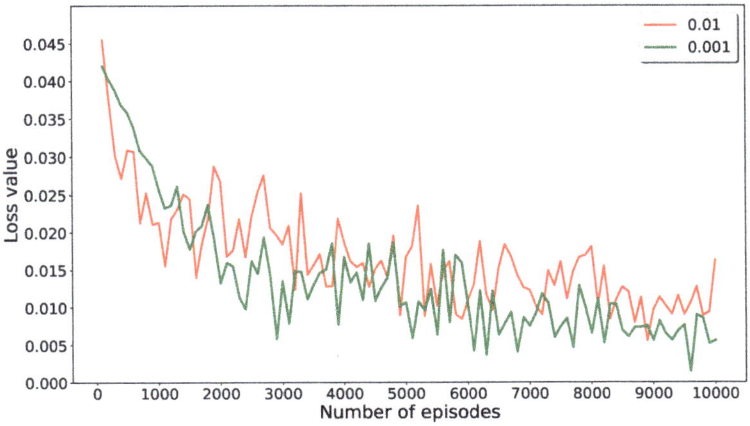

Figure 11. Loss value under different learning rates.

In addition, we utilized UP as the target data set to explore the influence of different network structure settings on classification results. Table 6 lists the specific structures of the feature learning module and the relation learning module and their corresponding classification accuracy. It should be noted that only the changed structure settings are listed in Table 6, while other basic settings, such as the batch normalization layer and Relu activation function, are set in accordance with Section 3. The exploration for network settings can be divided into two parts: *NO*.1 to *NO*.4 settings change the feature learning module, and *NO*.5 to *NO*.7 settings change the relation learning module. It can be found that *NO*.2 network settings can achieve the best classification effect, the specific structure of which is consistent with the description in Sections 3.2 and 3.3. According to the experimental results in the table, it is not difficult to obtain the following three observations:

(1) The number of convolutional layers has an important influence on the classification results. From *NO*.4 to *NO*.1, the number of convolutional layers in the feature learning module increases gradually, and the corresponding classification accuracy increases first and then decreases gradually. This indicates that the appropriate number of convolutional layers can obtain the best classification results, while too much or too little will reduce the effect of feature learning. In addition, a comparison between *NO*.2 and *NO*.5 can also verify a similar conclusion;
(2) By comparing *NO*.2 and *NO*.6 network settings, it can be found that the 1×1 convolution in the relation learning module can effectively improve the classification accuracy by 3.57%. The 1×1 convolution is mainly used to extract cross-channel cascaded features and reduce the dimension of concatenations, which is conducive to relation learning;
(3) The experimental results of *NO*.7 setting show that the classification effect of only applying the fully connected layer in the relational learning module is very poor, which directly proves the importance of the convolutional layer in relation learning.

In addition to the hyperparameters explored above, other basic experimental settings are given directly by referring to the existing deep learning model. We used *Adam* as the optimization algorithm and set the number of episodes in the meta-learning phase to 10,000, and the number of episodes in the few-shot learning phase to 1000. In the Dropout layer, the probability of random discard is 50%. All convolution kernels are initialized by Xavier [52].

Table 6. Overall classification accuracy (OA, %) on the UP data set under different network structure settings. The feature learning module (FLM), the relation learning module (RLM), max pooling (MP), fully connected layer (FC).

No.	1	2	3	4	5	6	7
FLM	3 × 3 × 3 (8) 3 × 3 × 3 (16) 2 × 2 × 4 MP 3 × 3 × 3 (32) 3 × 3 × 3 (64) 2 × 2 × 4 MP	3 × 3 × 3 (8) 2 × 2 × 4 MP 3 × 3 × 3 (16) 2 × 2 × 4 MP 3 × 3 × 3 (32)	3 × 3 × 3 (8) 2 × 2 × 4 MP 3 × 3 × 3 (16) 2 × 2 × 4 MP	3 × 3 × 3 (8) 4 × 4 × 8 MP	3 × 3 × 3 (8) 2 × 2 × 4 MP 3 × 3 × 3 (16) 2 × 2 × 4 MP 3 × 3 × 3 (32)	3 × 3 × 3 (8) 2 × 2 × 4 MP 3 × 3 × 3 (16) 2 × 2 × 4 MP 3 × 3 × 3 (32)	3 × 3 × 3 (8) 2 × 2 × 4 MP 3 × 3 × 3 (16) 2 × 2 × 4 MP 3 × 3 × 3 (32)
RLM	1 × 1 (64) 3 × 3 (128) 2 × 2 MP 128 FC 1 FC	1 × 1 (64) 3 × 3 (128) 2 × 2 MP 128 FC 1 FC	1 × 1 (64) 3 × 3 (128) 2 × 2 MP 128 FC 1 FC	1 × 1 (64) 3 × 3 (128) 2 × 2 MP 128 FC 1 FC	1 × 1 (64) 3 × 3 (128) 3 × 3 (256) 2 × 2 MP 128 FC 1 FC	3 × 3 (128) 2 × 2 MP 128 FC 1 FC	1024 FC 512 FC 1 FC
OA	80.37	81.94	79.50	75.43	77.83	78.37	26.55

4.3. Comparison and Analysis

In order to verify the effectiveness of the proposed method in HSI few-shot classification, we compared the experimental results of RN-FSC with the widely used SVM, two classical semisupervised methods LapSVM and TSVM provided in [53], the deep learning model Res-3D-CNN [54], two semisupervised deep models SS-CNN [35] and DCGAN+SEMI [55], and the graph convolutional network (GCN) [56] model. SVM can map nonlinear data to linearly separable high-dimensional feature spaces utilizing the kernel method, so it can obtain a better classification effect than other traditional classifiers when processing high-dimensional HSI. LapSVM and TSVM are both classical semisupervised support vector machines. Res-3D-CNN constructs a deep classification model with the 3D convolutional layer and residual structure, which can make full use of the spatial–spectral information in HSI. By combining CNN and DCGAN with semisupervised learning, respectively, SS-CNN and DCGAN+SEMI can use the information of unlabeled samples for classification. GCN is also an advanced semisupervised classification model.

In order to quantitatively compare the classification performance of the above different methods, the overall accuracy (OA), classification accuracy per class, average accuracy (AA), and *Kappa* coefficient are used as evaluation indicators. The overall accuracy is the percentage of samples classified correctly in all samples, and the average accuracy is the average of classification accuracy per class. It should be noted that for RN-FSC, five labeled samples per class in the target data set were used for fine-tuning, and for other methods, five labeled samples per class were used for training. Tables 7–9 summarize the experimental results on the three different target data sets, from which the following five observations can be obtained:

(1) In general, the performance of the traditional SVM classifier is better than that of the supervised deep learning model. Deep learning models need sufficient training samples for parameter optimization. However, in the HSI few-shot classification problem, limited labeled samples cannot provide guarantee for enough training, so the performance of supervised deep learning models is worse than that of SVM. For example, the OA of SVM is 6.04% higher than that of Res-3D-CNN on the Salinas data set;

(2) By comparing SVM and semisupervised SVM, Res-3D-CNN, and other semisupervised deep models, it can be found that the classification performance of the methods trained with only the labeled samples is poor. In this case, the semisupervised method can further improve the classification accuracy by utilizing the information of unlabeled samples;

(3) The classification performance of the semisupervised deep model is always better than that of the traditional semisupervised SVM. Deep learning models can extract more discriminative

features from labeled and unlabeled samples by building an end-to-end hierarchical framework, so they can obtain better classification results;

(4) Compared with other methods, RN-FSC has the best classification performance, with the highest OA, AA, and *Kappa* in all target data sets. The OA of RN-FSC is about 8.5%, 5%, and 6% higher than DCGAN+SEMI and GCN, which have similar performances on the three data sets. The most significant difference between RN-FSC and other methods is that other methods only perform training and classification on specific target data sets, while RN-FSC performs meta-learning on the collected source data sets through a large number of different tasks. Therefore, when processing new target data sets, RN-FSC has stronger generalization ability and can obtain better classification results with only a few labeled samples;

(5) For the classes that other methods do not recognize accurately, RN-FSC can obtain better results, such as Bricks, Bare Soil and Gravel in UP, and Corn_senesced_green_weeds, Fallow in Salinas. Benefitting from meta-learning and network design, RN-FSC can acquire the ability to learn how to learn in the form of comparison. By comparing similarities between samples in the deep metric space, RN-FSC can take advantage of more abstract features. Therefore, RN-FSC can accurately recognize the uneasily distinguished classes.

Table 7. Classification results of the different methods on the UP data set (5 samples per class in the fine-tuning data set for RN-FSC; 5 samples per class are used for training for other methods; bold values represent the best results among these methods).

Class	SVM	LapSVM	TSVM	Res-3D-CNN	SS-CNN	DCGAN+SEMI	GCN	RN-FSC
Asphalt	94.08	98.12	96.55	71.67	89.89	92.18	96.00	87.28
Meadows	79.03	81.57	80.47	88.96	84.40	90.32	93.39	84.33
Gravel	27.67	30.97	11.11	23.30	59.94	41.80	50.71	90.42
Trees	57.71	62.47	48.71	88.86	57.94	86.39	95.85	78.09
Metal Sheets	91.67	91.39	94.92	89.39	97.11	83.30	99.19	99.56
Bare Soil	21.10	37.78	37.91	37.88	53.01	43.63	37.54	63.25
Bitumen	35.33	37.67	20.50	38.62	36.15	44.54	57.26	52.09
Bricks	57.31	60.47	55.36	42.59	72.70	62.11	73.31	84.81
Shadow	99.79	99.89	99.89	63.13	48.65	66.33	98.13	95.94
OA	55.79	67.06	61.92	65.44	71.73	73.52	73.40	**81.94**
AA	62.63	66.70	60.60	60.49	66.64	67.84	77.93	**81.75**
Kappa	46.60	57.90	51.44	55.63	63.37	66.07	66.96	**75.84**

Table 8. Classification results of the different methods on the PC data set (5 samples per class in the fine-tuning data set for RN-FSC; 5 samples per class are used for training for other methods; bold values represent the best results among these methods).

Class	SVM	LapSVM	TSVM	Res-3D-CNN	SS-CNN	DCGAN+SEMI	GCN	RN-FSC
Water	99.95	99.99	95.12	99.99	99.17	98.13	99.74	100.00
Trees	94.68	94.75	92.22	74.17	93.34	98.15	99.36	99.53
Meadows	40.86	60.84	40.12	80.24	75.17	65.81	61.53	67.60
Bricks	56.47	14.57	8.12	27.11	68.85	55.64	68.22	72.43
Bare Soil	19.51	65.47	27.15	23.08	38.25	53.42	42.77	96.91
Asphalt	63.66	61.85	46.87	67.69	81.42	84.21	81.62	85.86
Bitumen	78.21	92.83	1.38	77.38	75.82	99.37	91.61	85.55
Tile	88.66	94.55	97.14	98.88	99.57	99.02	99.06	99.94
Shadow	99.76	99.86	93.17	87.61	95.60	77.46	98.00	91.87
OA	83.11	86.43	67.60	80.03	89.27	91.85	90.65	**96.36**
AA	71.31	76.08	55.70	70.69	80.80	81.24	82.43	**88.86**
Kappa	76.62	81.22	56.60	73.16	88.30	91.02	89.79	**95.98**

Table 9. Classification results of the different methods on the Salinas data set (5 samples per class in the fine-tuning data set for RN-FSC; 5 samples per class are used for training for other methods; bold values represent the best results among these methods).

Class	SVM	LapSVM	TSVM	Res-3D-CNN	SS-CNN	DCGAN+SEMI	GCN	RN-FSC
Brocoli_green_weeds_1	85.60	78.59	80.50	39.47	93.02	56.94	100.00	99.26
Brocoli_green_weeds_2	98.54	98.99	98.08	74.02	92.51	71.53	81.95	100.00
Fallow	65.38	82.96	65.19	49.33	84.31	87.44	83.50	97.87
Fallow_rough_plow	95.82	96.64	95.46	88.71	86.43	76.45	96.99	99.50
Fallow_smooth	95.83	88.09	64.25	77.50	90.91	94.95	96.96	97.81
Stubble	99.92	100.00	99.95	97.52	99.55	99.47	99.82	99.35
Celery	95.29	89.61	85.10	61.53	97.54	89.63	94.66	100.00
Grapes_untrained	57.00	63.87	44.29	68.93	73.52	70.93	86.00	66.24
Soil_vinyard_develop	90.64	79.49	74.06	92.83	93.81	92.89	95.65	97.34
Corn_senesced_green_weeds	85.87	56.55	64.71	69.33	77.21	63.58	81.31	93.66
Lettuce_romaine_4wk	38.32	38.02	47.56	59.07	42.37	83.81	60.05	73.96
Lettuce_romaine_5wk	87.56	92.71	92.56	70.59	95.85	97.33	95.65	99.84
Lettuce_romaine_6wk	88.66	46.88	47.87	75.38	99.23	97.53	89.39	100.00
Lettuce_romaine_7wk	87.87	93.26	86.81	89.12	92.98	87.09	86.41	96.39
Vinyard_untrained	33.18	49.84	32.31	47.62	50.37	74.78	51.00	68.85
Vinyard_vertical_trellis	81.64	91.00	54.24	88.90	80.54	77.17	95.07	99.89
OA	73.64	74.99	64.63	67.60	79.23	80.11	80.90	**86.99**
AA	80.45	77.91	70.81	71.87	84.38	82.60	97.15	**93.12**
Kappa	70.70	72.05	60.95	64.28	77.04	77.86	78.95	**85.44**

In order to better compare and analyze the classification results of the above methods, Figures 12–14 respectively show their classification maps on the three target data sets. With the continuous improvement of the classification accuracy, the noise and misclassification phenomena gradually decrease, and the classification map gradually approaches the ground-truth map. In fact, the results of Figures 12–14 and Tables 7–9 are the same, both of which can prove the effectiveness of the proposed method.

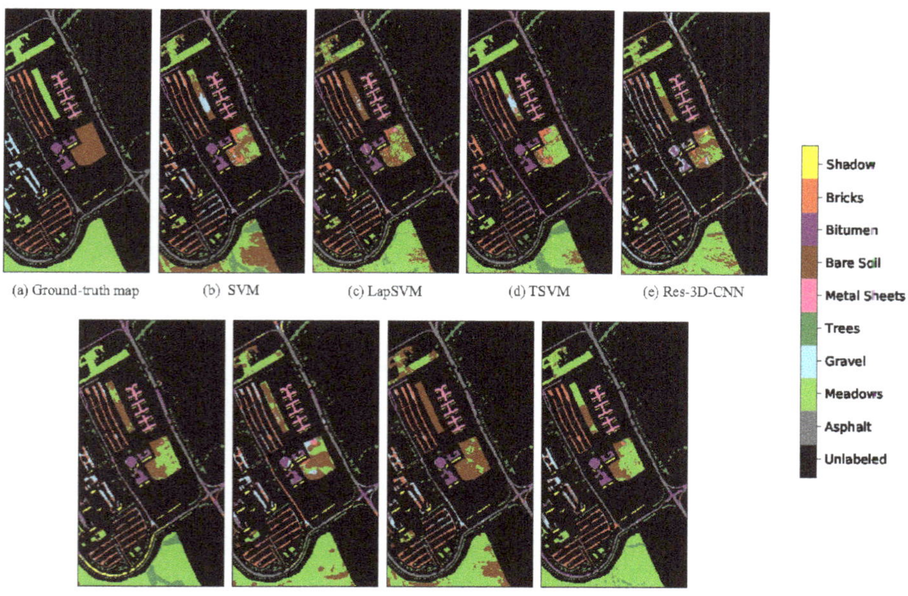

Figure 12. Classification maps resulting from different methods on the UP data set.

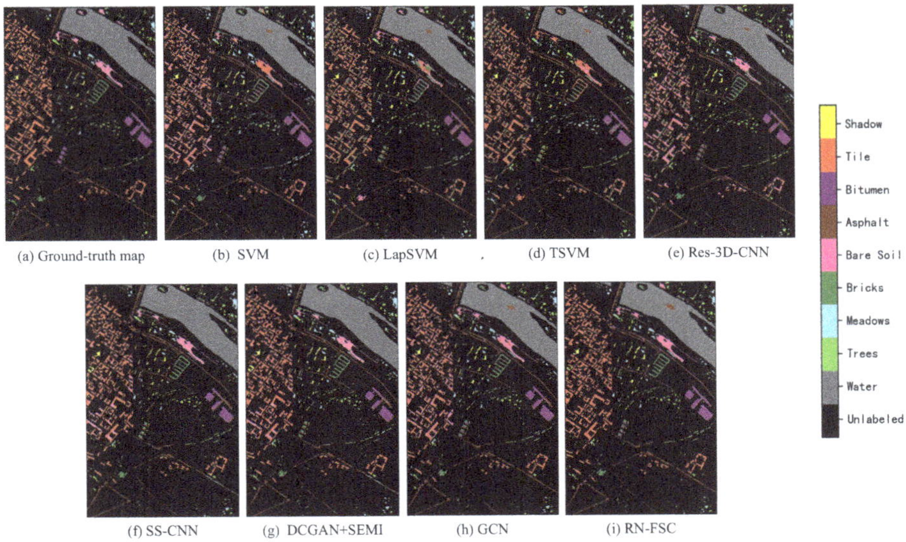

Figure 13. Classification maps resulting from different methods on the PC data set.

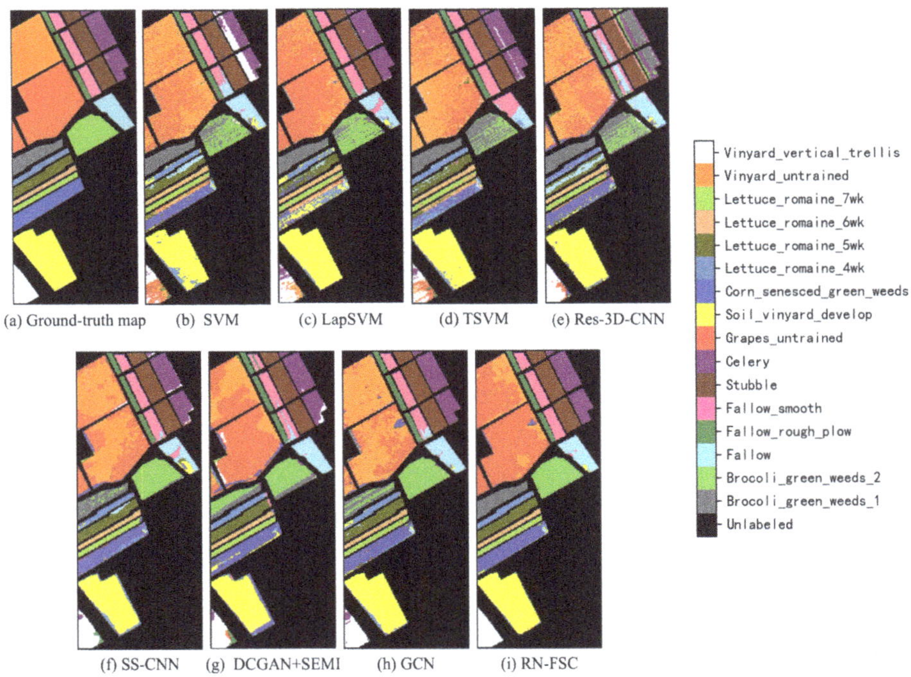

Figure 14. Classification maps resulting from different methods on the SA data set.

In order to further verify that the observed increase in classification accuracy is statistically significant, we repeated the experiment 20 times for different methods and carried out the paired t-test on OA. The paired t-test is a widely used statistical method to verify whether there is a significant difference between the two groups of related samples [17,39]. In our test, if the result t is greater than 3.57, it indicates that there is a significant difference between the two groups of samples at the 99.9% confidence level. As seen from Table 10, all the results are greater than 3.57, indicating that the increase in classification accuracy is statistically significant.

Table 10. Results of the paired t-test on three target data sets.

University of Pavia	Pavia Center	Salinas
t/significant?	t/significant?	t/significant?
	RN-FSC vs. SVM	
75.53/yes	26.34/yes	34.40/yes
	RN-FSC vs. LapSVM	
24.36/yes	34.79/yes	33.75/yes
	RN-FSC vs. TSVM	
47.44/yes	71.68/yes	37.29/yes
	RN-FSC vs. Res-3D-CNN	
35.62/yes	30.08/yes	29.19/yes
	RN-FSC vs. SS-CNN	
27.17/yes	22.60/yes	19.33/yes
	RN-FSC vs. DCGAN+SEMI	
23.56/yes	19.88/yes	16.86/yes
	RN-FSC vs. GCN	
21.05/yes	20.05/yes	15.98/yes

4.4. *Influence of the Number of Labeled Samples*

The objective of the experiments is to verify the classification effect of the proposed method on new HSI with only a few labeled samples. Therefore, it is necessary to explore the classification effect of the proposed method under different numbers of labeled samples. To this end, we randomly selected 5, 10, 15, 20, and 25 labeled samples per class to build the fine-tuning data set. Accordingly, we explored the classification results of other methods with 5, 10, 15, 20, and 25 labeled samples per class for training. Figure 15 shows the experimental results. It can be seen that the OA of all methods presents an increasing trend with the increase in the number of labeled samples. RN-FSC always has the highest classification accuracy, which indicates that it has the best adaptability to the number of labeled samples.

Experimental results from Tables 7–9 and Figure 15 have shown that the proposed method can achieve better classification results when classifying new HSI with only a few labeled samples. In order to further explore the influence of the number of labeled samples on the classification effect of RN-FSC, we conducted comparative experiments on Salinas and Indian Pines data sets with reference to [57–59]. The Indian Pines data set, containing 16 classes of the Indian Pine test site in Northwestern Indiana, was collected by AVIRIS. Salinas and Indian Pines both contain 16 classes, and Indian Pines contains 4 small classes with less than 100 labeled samples, which can further verify the effectiveness of the classification method. In the experiments, 10% and 2% labeled samples were randomly selected to build the fine-tuning data set (1083 labeled samples for Salinas and 1025 labeled samples for Indian Pines), which is far more than that of the previous experiments. It should be noted that the selection of labeled samples per class is exactly the same as in [57–59]. EPF-B-g, EPF-B-c, EPF-G-g, EPF-G-c, and IEPF-G-g provided in [57–59] were selected to make a comparison with the proposed method. Table 11 shows the experimental results. In the Salinas data set, the OA and AA of RN-FSC are higher than those of other methods. In the Indian Pines data set, the classification results of IEPF-G-g are the best, followed by those of RN-FSC. Overall, when the labeled samples are further increased (approximately 1000–1100 labeled samples for each data set), the proposed method can still obtain satisfactory results.

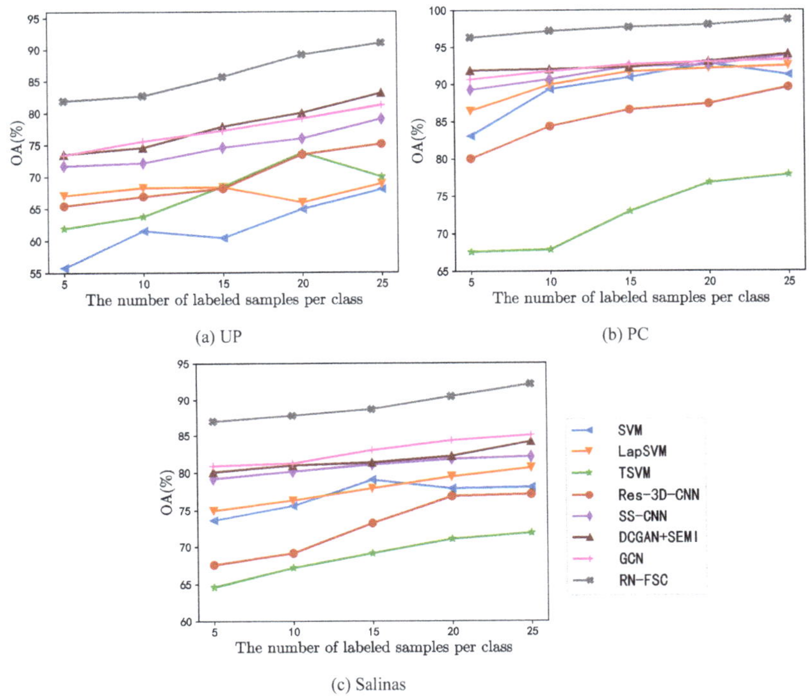

Figure 15. Classification accuracy under different number of labeled samples on three target data sets.

Table 11. Classification results of the different methods on Salinas and Indian Pines.

The Target Data Set		EPF-B-g	EPF-B-c	EPF-G-g	EPF-G-c	IEPF-G-g	RN-FSC
Salinas	*OA*	96.46	96.23	96.61	97.11	98.36	98.86
	AA	98.17	98.08	98.24	98.56	99.19	99.26
Indian Pines	*OA*	94.82	95.30	93.94	94.87	98.19	97.46
	AA	95.96	95.44	96.64	96.52	98.67	96.68

4.5. Exploration on the Effectiveness of Meta-Learning

The learning process of the proposed method, RN-FSC, can be divided into two phases: meta-learning on the source data set and few-shot learning on the fine-tuning data set. As mentioned in the previous sections, the reason RN-FSC has a better classification effect in the HSI few-shot classification is that it has acquired a large amount of feature knowledge and mastered the ability to learn how to learn through meta-learning. To verify this point, we carried out experiments to explore the influence of the meta-learning phase on the final classification results. Table 12 lists the overall accuracy with and without meta-learning with a different number of labeled samples. The model without meta-learning can only perform supervised training with a few labeled samples in the fine-tuning data set, so its classification results are poor. On the UP, PC, and Salinas data sets, the meta-learning phase can increase the classification accuracy by 20.20%, 10.73%, and 15.91%, respectively, when $L = 5$, which fully proves the effectiveness of meta-learning in HSI few-shot classification. In addition, it can be found that with the increase in the number of labeled samples, the difference between the results with and without meta-learning shows a decreasing trend. For example, on the UP data set, the difference is 20.20% when $L = 5$, and 10.83% when $L = 25$.

Table 12. Influence of the meta-learning phase on classification accuracy (OA, %). L is the number of labeled samples per class.

Target Data Set	Meta-Learning	$L = 5$	$L = 10$	$L = 15$	$L = 20$	$L = 25$
UP	yes	81.94	82.71	85.70	89.20	91.05
	not	61.74	69.56	74.12	78.93	80.22
PC	yes	96.36	97.21	97.70	98.03	98.72
	not	85.63	86.46	87.41	90.53	92.10
Salinas	yes	86.99	87.79	88.68	90.49	92.15
	not	71.08	74.08	75.96	77.34	80.45

4.6. Execution Time Analysis

The execution time of general deep learning models usually consists of training time and testing time. As described in Section 2.3, the proposed method consists of three phases: meta-learning, few-shot learning, and classification. The biggest difference between RN-FSC and other general deep models for HSI classification is that it first performs meta-learning on the previously collected source data sets and then classifies the new HSI data sets, which are absolutely different from the source data sets. In other words, only performing meta-learning in advance one time, RN-FSC can quickly classify all other new data sets, which is of great significance in practical applications. In our experiment, it takes approximately 12.83 h for the model to perform meta-learning. In practice, the model used to perform the classification task should have completed meta-learning. Therefore, the model needs only to perform few-shot learning and classification when processing the target HSI. Table 13 lists the execution times of DCGAN+SEMI, GCN, and RN-FSC on three different target data sets, because they present better classification results than other methods. DCGAN+SEMI and GCN include training and testing time, while RN-FSC includes few-shot learning time and classification time. DCGAN+SEMI needs to train the generator and the discriminator, respectively, while GCN utilizes all the labeled samples for graph construction, so their training time is longer. RN-FSC only utilizes a few labeled samples for fine-tuning, so the few-shot learning time is shorter. However, since RN-FSC needs to calculate the relation score through comparison, its classification time is longer. Generally speaking, the execution time of RN-FSC is shorter than that of DCGAN+SEMI and GCN, which indicates RN-FSC has better work efficiency.

Table 13. Execution times on three target data sets (5 samples per class are used as labeled samples).

Target Data Set	DCGAN+SEMI	GCN	RN-FSC
UP	$1355.86(s) + 2.57(s)$	$1915.29(s) + 0.98(s)$	$217.27(s) + 72.57(s)$
PC	$1401.31(s) + 8.13(s)$	$3042.18(s) + 1.44(s)$	$214.98(s) + 198.09(s)$
Salinas	$2386.74(s) + 3.03(s)$	$1224.03(s) + 1.10(s)$	$632.98(s) + 81.23(s)$

4.7. Discussion

It is difficult for deep learning models to be fully trained and achieve promising classification results with a few labeled samples. At the same time, for complex and diverse HSI, the working mode that general deep learning models need to be trained from scratch every time is very inefficient and not desirable in practice. However, our method can obtain better classification results with only a few labeled samples (five samples per class) when processing new HSI. The root cause is the implementation of meta-learning, the core of which is the ability to learn how to learn. In our method, this ability is demonstrated in the form of comparison. Firstly, the model maps the data space to a deep metric space, where it performs relation learning by comparing the similarity of sample features, i.e., the similarity between samples belonging to the same class is high and the relation score is high, whereas the similarity between samples belonging to the different class is low and the relation score is low. In fact, the form of the ability to learn how to learn is not unique in the field of meta-learning, which largely depends on the specific network structure and loss function.

The task-based learning strategy is key to performing meta-learning. Lots of randomly generated tasks from different HSI can effectively enhance the generalization ability of the model, because the model learns how to compare with different tasks instead of how to classify a specific data set. To acquire the best learning effect, we explored the optimal task setting, including the number of classes, the number of support samples, and the number of query samples in the task. Experiments showed that the support samples should be much fewer than the query samples, so as to fully simulate the situation of HSI few-shot classification. In addition, experiments were conducted to explore the influence of learning rate to further optimize the meta-learning process. At the same time, the network structure can directly affect the classification results. A new deep model based on relation network was designed for HSI few-shot classification. In the feature learning module, the 3D convolutional layer can effectively utilize the spatial–spectral information to extract the highly discriminant features. In addition, we found that the convolutional layer is necessary in the relation learning module, which can guarantee the comparison ability of the model to some extent.

Through detailed comparison and analysis, it can be demonstrated that the proposed method outperforms SVM, semisupervised SVM, and several supervised and semisupervised deep learning models with a few labeled samples. Moreover, the proposed method has better adaptability to the number of samples. The paired t-test shows that the increase in classification accuracy is statistically significant and not accidental. In addition, by comparing the results of the model with and without meta-learning, the importance of the meta-learning phase is directly proved again. Finally, the efficiency of different methods was compared, indicating the potential value of the proposed method in practical application.

5. Conclusions

Although the deep learning model has achieved great success in HSI classification, it still faces great difficulties in classifying new HSI with a few labeled samples. To this end, this paper proposes a new classification method based on a relation network for HSI few-shot classification. Meta-learning is the core of this method, and the network settings realize the ability to learn how to learn in the form of comparison in deep metric space, that is, the relation score between samples belonging to the same class is high, while the relation score between samples belonging to different classes is low. Benefitting from a large number of tasks generated from different data sets, the generalization ability of the model is constantly enhanced. Experiments on three different target data sets show that the proposed method outperforms traditional semisupervised SVM and semisupervised deep learning methods when only a few labeled samples are available.

Author Contributions: Methodology, K.G. and B.L.; investigation, X.Y., J.Q., and P.Z.; resources, X.Y., P.Z., and X.T.; writing—original draft preparation, K.G.; writing—review and editing, K.G. and B.L.; visualization, J.Q. and X.T.; supervision, X.Y., P.Z., and X.T. All authors have read and agreed to the published version of the manuscript.

Funding: This research was funded by National Natural Science Foundation of China under Grant 41801388 .

Acknowledgments: The authors would like to thank Yokoya for providing the data used in this study. The authors would also like to thank all the professionals for kindly providing the codes associated with the experiments.

Conflicts of Interest: The authors declare no conflict of interest.

References

1. Sun, S.; Zhong, P.; Xiao, H.; Liu, F.; Wang, R. An active learning method based on SVM classifier for hyperspectral images classification. In Proceedings of the 7th Workshop on Hyperspectral Image and Signal Processing: Evolution in Remote Sensing (WHISPERS), Tokyo, Japan, 2–5 June 2015; pp. 1–4, doi:10.1109/WHISPERS.2015.8075484. [CrossRef]
2. Yuemei, R.; Yanning, Z.; Wei, W.; Lei, L. A spectral-spatial hyperspectral data classification approach using random forest with label constraints. In Proceedings of the IEEE Workshop on Electronics, Computer and Applications, Ottawa, ON, Canada, 8–9 May 2014; pp. 344–347, doi:10.1109/IWECA.2014.6845627. [CrossRef]
3. Agarwal, A.; El-Ghazawi, T.; El-Askary, H.; Le-Moigne, J. Efficient Hierarchical-PCA Dimension Reduction for Hyperspectral Imagery. In Proceedings of the IEEE International Symposium on Signal Processing and Information Technology, Cairo, Egypt, 15–18 December 2007; pp. 353–356, doi:10.1109/ISSPIT.2007.4458191. [CrossRef]
4. Falco, N.; Bruzzone, L.; Benediktsson, J.A. An ICA based approach to hyperspectral image feature reduction. In Proceedings of the IEEE Geoscience and Remote Sensing Symposium, Quebec City, QC, Canada, 13–18 July 2014; pp. 3470–3473, doi:10.1109/IGARSS.2014.6947229. [CrossRef]
5. Li, C.; Chu, H.; Kuo, B.; Lin, C. Hyperspectral image classification using spectral and spatial information based linear discriminant analysis. In Proceedings of the IEEE International Geoscience and Remote Sensing Symposium, Vancouver, BC, Canada, 4–29 July 2011; pp. 1716–1719, doi:10.1109/IGARSS.2011.6049566. [CrossRef]
6. Liao, W.; Pizurica, A.; Philips, W.; Pi, Y. A fast iterative kernel PCA feature extraction for hyperspectral images. In Proceedings of the IEEE International Conference on Image Processing, Hong Kong, China, 26–29 September 2010; pp. 1317–1320, doi:10.1109/ICIP.2010.5651670. [CrossRef]
7. Chen, Y.; Qu, C.; Lin, Z. Supervised Locally Linear Embedding based dimension reduction for hyperspectral image classification. In Proceedings of the IEEE International Geoscience and Remote Sensing Symposium—IGARSS, Melbourne, Australia, 21–26 July 2013; pp. 3578–3581, doi:10.1109/IGARSS.2013.6723603. [CrossRef]
8. Gao, L.; Gu, D.; Zhuang, L.; Ren, J.; Yang, D.; Zhang, B. Combining t-Distributed Stochastic Neighbor Embedding With Convolutional Neural Networks for Hyperspectral Image Classification. *IEEE Geosci. Remote Sens. Lett.* **2019**, 1–5, doi:10.1109/LGRS.2019.2945122. [CrossRef]
9. Quesada-Barriuso, P.; Argüello, F.; Heras, D.B. Spectral–Spatial Classification of Hyperspectral Images Using Wavelets and Extended Morphological Profiles. *IEEE J. Sel. Topics Appl. Earth Obs. Remote Sens.* **2014**, *7*, 1177–1185, doi:10.1109/JSTARS.2014.2308425. [CrossRef]
10. Jia, S.; Hu, J.; Zhu, J.; Jia, X.; Li, Q. Three-Dimensional Local Binary Patterns for Hyperspectral Imagery Classification. *IEEE Trans. Geosci. Remote Sens.* **2017**, *55*, 2399–2413, doi:10.1109/TGRS.2016.2642951. [CrossRef]
11. Bau, T.C.; Sarkar, S.; Healey, G. Hyperspectral Region Classification Using a Three-Dimensional Gabor Filterbank. *IEEE Trans. Geosci. Remote Sens.* **2010**, *48*, 3457–3464, doi:10.1109/TGRS.2010.2046494. [CrossRef]
12. Xu, Y.; Wu, Z.; Wei, Z. Markov random field with homogeneous areas priors for hyperspectral image classification. In Proceedings of the IEEE Geoscience and Remote Sensing Symposium, Quebec City, QC, Canada, 13–18 July 2014; pp. 3426–3429, doi:10.1109/IGARSS.2014.6947218. [CrossRef]
13. He, L.; Chen, X. A three-dimensional filtering method for spectral-spatial hyperspectral image classification. In Proceedings of the IEEE International Geoscience and Remote Sensing Symposium (IGARSS), Beijing, China, 10–15 July 2016; pp. 2746–2748, doi:10.1109/IGARSS.2016.7729709. [CrossRef]
14. Casalino, G.; Gillis, N. Sequential dimensionality reduction for extracting localized features. *Pattern Recognit.* **2017**, *63*, 15–29, doi:10.1016/j.patcog.2016.09.006. [CrossRef]
15. Yin, J.; Li, S.; Zhu, H.; Luo, X. Hyperspectral Image Classification Using CapsNet With Well-Initialized Shallow Layers. *IEEE Geosci. Remote Sens. Lett.* **2019**, *16*, 1095–1099, doi:10.1109/LGRS.2019.2891076. [CrossRef]
16. Zhang, L.; Zhang, L.; Du, B. Deep Learning for Remote Sensing Data: A Technical Tutorial on the State of the Art. *IEEE Geosci. Remote Sens. Mag.* **2016**, *4*, 22–40, doi:10.1109/MGRS.2016.2540798. [CrossRef]
17. Chen, Y.; Lin, Z.; Zhao, X.; Wang, G.; Gu, Y. Deep Learning-Based Classification of Hyperspectral Data. *IEEE J. Sel.Topics Appl. Earth Obs. Remote Sens.* **2014**, *7*, 2094–2107, doi:10.1109/JSTARS.2014.2329330. [CrossRef]
18. Mou, L.; Ghamisi, P.; Zhu, X.X. Deep Recurrent Neural Networks for Hyperspectral Image Classification. *IEEE Trans. Geosci. Remote Sens.* **2017**, *55*, 3639–3655, doi:10.1109/TGRS.2016.2636241. [CrossRef]

19. Hang, R.; Liu, Q.; Hong, D.; Ghamisi, P. Cascaded Recurrent Neural Networks for Hyperspectral Image Classification. *IEEE Trans. Geosci. Remote Sens.* **2019**, *57*, 5384–5394, doi:10.1109/TGRS.2019.2899129. [CrossRef]
20. Chen, Y.; Zhao, X.; Jia, X. Spectral–Spatial Classification of Hyperspectral Data Based on Deep Belief Network. *IEEE J. Sel. Topics Appl. Earth Obs. Remote Sens.* **2015**, *8*, 2381–2392, doi:10.1109/JSTARS.2015.2388577. [CrossRef]
21. Mughees, A.; Tao, L. Multiple deep-belief-network-based spectral-spatial classification of hyperspectral images. *Tsinghua Sci. Technol.* **2019**, *24*, 183–194, doi:10.26599/TST.2018.9010043. [CrossRef]
22. He, L.; Li, J.; Liu, C.; Li, S. Recent Advances on Spectral–Spatial Hyperspectral Image Classification: An Overview and New Guidelines. *IEEE Trans. Geosci. Remote Sens.* **2018**, *56*, 1579–1597, doi:10.1109/TGRS.2017.2765364. [CrossRef]
23. Zhang, M.; Li, W.; Du, Q. Diverse Region-Based CNN for Hyperspectral Image Classification. *IEEE Trans. Image Process.* **2018**, *27*, 2623–2634, doi:10.1109/TIP.2018.2809606. [CrossRef]
24. Lee, H.; Kwon, H. Going Deeper With Contextual CNN for Hyperspectral Image Classification. *IEEE Trans. Image Process.* **2017**, *26*, 4843–4855, doi:10.1109/TIP.2017.2725580. [CrossRef]
25. Zhi, L.; Yu, X.; Liu, B.; Wei, X. A dense convolutional neural network for hyperspectral image classification. *Remote Sens. Lett.* **2019**, *10*, 59–66, doi:10.1080/2150704X.2018.1526424. [CrossRef]
26. Zhang, H.; Li, Y.; Zhang, Y.; Shen, Q. Spectral-spatial classification of hyperspectral imagery using a dual-channel convolutional neural network. *Remote Sens. Lett.* **2017**, *8*, 438–447, doi:10.1080/2150704X.2017.1280200. [CrossRef]
27. Chen, Y.; Jiang, H.; Li, C.; Jia, X.; Ghamisi, P. Deep Feature Extraction and Classification of Hyperspectral Images Based on Convolutional Neural Networks. *IEEE Trans. Geosci. Remote Sens.* **2016**, *54*, 6232–6251, doi:10.1109/TGRS.2016.2584107. [CrossRef]
28. Zhong, Z.; Li, J.; Luo, Z.; Chapman, M. Spectral–Spatial Residual Network for Hyperspectral Image Classification: A 3-D Deep Learning Framework. *IEEE Trans. Geosci. Remote Sens.* **2018**, *56*, 847–858, doi:10.1109/TGRS.2017.2755542. [CrossRef]
29. Fang, B.; Li, Y.; Zhang, H.; Chan, J. Hyperspectral Images Classification Based on Dense Convolutional Networks with Spectral-Wise Attention Mechanism. *Remote Sens.* **2019**, *11*, 159, doi:10.3390/rs11020159. [CrossRef]
30. Li, A.; Shang, Z. A new Spectral-Spatial Pseudo-3D Dense Network for Hyperspectral Image Classification. In Proceedings of the International Joint Conference on Neural Networks (IJCNN), Budapest, Hungary, 14–19 July 2019; pp. 1–7, doi:10.1109/IJCNN.2019.8851917. [CrossRef]
31. Xu, Q.; Xiao, Y.; Wang, D.; Luo, B. CSA-MSO3DCNN: Multiscale Octave 3D CNN with Channel and Spatial Attention for Hyperspectral Image Classification. *Remote Sens.* **2020**, *12*, 188, doi:10.3390/rs12010188. [CrossRef]
32. Jamshidpour, N.; Aria, E.H.; Safari, A.; Homayouni, S. Adaptive Self-Learned Active Learning Framework for Hyperspectral Classification. In Proceedings of the 10th Workshop on Hyperspectral Imaging and Signal Processing: Evolution in Remote Sensing (WHISPERS), Amsterdam, The Netherlands, 24–26 September 2019; pp. 1–5, doi:10.1109/WHISPERS.2019.8921298. [CrossRef]
33. Paoletti, M.E.; Haut, J.M.; Fernandez-Beltran, R.; Plaza, J.; Plaza, A.; Li, J.; Pla, F. Capsule Networks for Hyperspectral Image Classification. *IEEE Trans. Geosci. Remote Sens.* **2019**, *57*, 2145–2160, doi:10.1109/TGRS.2018.2871782. [CrossRef]
34. Fan, J.; Tan, H.L.; Toomik, M.; Lu, S. Spectral-spatial hyperspectral image classification using super-pixel-based spatial pyramid representation. In *Image and Signal Processing for Remote Sensing XXII*; Bruzzone, L., Bovolo, F., Eds.; International Society for Optics and Photonics, SPIE: San Francisco, CA, USA, 2016; Volume 10004, pp. 315–321, doi:10.1117/12.2241033. [CrossRef]
35. Liu, B.; Yu, X.; Zhang, P.; Tan, X.; Yu, A.; Xue, Z. A semi-supervised convolutional neural network for hyperspectral image classification. *Remote Sens. Lett.* **2017**, *8*, 839–848, doi:10.1080/2150704X.2017.1331053. [CrossRef]
36. Wu, Y.; Mu, G.; Qin, C.; Miao, Q.; Ma, W.; Zhang, X. Semi-Supervised Hyperspectral Image Classification via Spatial-Regulated Self-Training. *Remote Sens.* **2020**, *12*, 159, doi:10.3390/rs12010159. [CrossRef]
37. Kang, X.; Zhuo, B.; Duan, P. Semi-supervised deep learning for hyperspectral image classification. *Remote Sens. Lett.* **2019**, *10*, 353–362, doi:10.1080/2150704X.2018.1557787. [CrossRef]
38. Li, W.; Wu, G.; Zhang, F.; Du, Q. Hyperspectral Image Classification Using Deep Pixel-Pair Features. *IEEE Trans. Geosci. Remote Sens.* **2017**, *55*, 844–853, doi:10.1109/TGRS.2016.2616355. [CrossRef]
39. Liu, B.; Yu, X.; Zhang, P.; Yu, A.; Fu, Q.; Wei, X. Supervised Deep Feature Extraction for Hyperspectral Image Classification. *IEEE Trans. Geosci. Remote Sens.* **2018**, *56*, 1909–1921, doi:10.1109/TGRS.2017.2769673. [CrossRef]

40. Xu, S.; Mu, X.; Chai, D.; Zhang, X. Remote sensing image scene classification based on generative adversarial networks. *Remote Sens. Lett.* **2018**, *9*, 617–626, doi:10.1080/2150704X.2018.1453173. [CrossRef]
41. Qin, J.; Zhan, Y.; Wu, K.; Liu, W.; Yang, Z.; Yao, W.; Medjadba, Y.; Zhang, Y.; Yu, X. Semi-Supervised Classification of Hyperspectral Data for Geologic Body Based on Generative Adversarial Networks at Tianshan Area. In Proceedings of the IGARSS 2018—2018 IEEE International Geoscience and Remote Sensing Symposium, Valencia, Spain, 22–27 July 2018; pp. 4776–4779, doi:10.1109/IGARSS.2018.8518946. [CrossRef]
42. Wang, H.; Tao, C.; Qi, J.; Li, H.; Tang, Y. Semi-Supervised Variational Generative Adversarial Networks for Hyperspectral Image Classification. In Proceedings of the IGARSS 2019—2019 IEEE International Geoscience and Remote Sensing Symposium, Yokohama, Japan, 28 July–2 August 2019; pp. 9792–9794, doi:10.1109/IGARSS.2019.8900073. [CrossRef]
43. Ren, M.; Triantafillou, E.; Ravi, S.; Snell, J.; Swersky, K.; Tenenbaum, J.; Larochelle, H.; Zemel, R. Meta-Learning for Semi-Supervised Few-Shot Classification. *arXiv* **2018**, arXiv:1803.00676.
44. Finn, C.; Abbeel, P.; Levine, S. Model-Agnostic Meta-Learning for Fast Adaptation of Deep Networks. *arXiv* **2017**, arXiv:1703.03400.
45. Andrychowicz, M.; Denil, M.; Gómez, S.; Hoffman, M.; Pfau, D.; Schaul, T.; Freitas, N. Learning to learn by gradient descent by gradient descent. *arXiv* **2016**, arXiv:1606.04474.
46. Li, Z.; Zhou, F.; Chen, F.; Li, H. Meta-SGD: Learning to Learn Quickly for Few Shot Learning. *arXiv* **2017**, arXiv:1707.09835.
47. Liang, H.; Fu, W.; Yi, F. A Survey of Recent Advances in Transfer Learning. In Proceedings of the IEEE 19th International Conference on Communication Technology (ICCT), Xi'an, China, 16–19 October 2019; pp. 1516–1523, doi:10.1109/ICCT46805.2019.8947072. [CrossRef]
48. Liu, B.; Yu, X.; Yu, A.; Wan, G. Deep convolutional recurrent neural network with transfer learning for hyperspectral image classification. *J. Appl. Remote Sens.* **2018**, *12*, 1, doi:10.1117/1.JRS.12.026028. [CrossRef]
49. Sung, F.; Yang, Y.; Zhang, L.; Xiang, T.; Torr, P.; Hospedales, T. Learning to Compare: Relation Network for Few-Shot Learning. In Proceedings of the IEEE/CVF Conference on Computer Vision and Pattern Recognition, Salt Lake City, UT, USA, 18–23 June 2018; pp. 1199–1208, doi:10.1109/CVPR.2018.00131. [CrossRef]
50. Lin, M.; Chen, Q.; Yan, S. Network In Network. *arXiv* **2013**, arXiv:1312.4400.
51. Sun, V.; Geng, X.; Chen, J.; Ji, L.; Tang, H.; Zhao, Y.; Xu, M. A robust and efficient band selection method using graph representation for hyperspectral imagery. *Int. J. Remote Sens.* **2016**, *37*, 4874–4889, doi:10.1080/01431161.2016.1225173. [CrossRef]
52. Glorot, X.; Bengio, Y. Understanding the difficulty of training deep feedforward neural networks. *J. Mach. Learn. Res. Proc. Track* **2010**, *9*, 249–256.
53. Wang, L.; Hao, S.; Wang, Q.; Wang, Y. Semi-supervised classification for hyperspectral imagery based on spatial-spectral Label Propagation. *ISPRS J. Photogramm. Remote Sens.* **2014**, *97*, 123–137, doi:10.2495/ISME20141481. [CrossRef]
54. Liu, B.; Yu, X.; Zhang, P.; Tan, X. Deep 3D convolutional network combined with spatial-spectral features for hyperspectral image classification. *Cehui Xuebao/Acta Geodaetica et Cartographica Sinica* **2019**, *48*, 53–63, doi:10.11947/j.AGCS.2019.20170578. [CrossRef]
55. Salimans, T.; Goodfellow, I.; Zaremba, W.; Cheung, V.; Radford, A.; Chen, X. Improved Techniques for Training GANs. *arXiv* **2016**, arXiv:1606.03498.
56. Kipf, T.; Welling, M. Semi-Supervised Classification with Graph Convolutional Networks. *arXiv* **2016**, arXiv:1609.02907.
57. Kang, X.; Li, S.; Benediktsson, J.A. Spectral-Spatial Hyperspectral Image Classification With Edge-Preserving Filtering. *IEEE Transm Geoscim Remote Sens.* **2014**, *52*, 2666–2677. [CrossRef]
58. Zhong, S.; Chang, C.I.; Zhang, Y. Iterative Edge Preserving Filtering Approach to Hyperspectral Image Classification. *IEEE Geosci. Remote Sens. Lett.* **2018**, doi:10.1109/LGRS.2018.2868841. [CrossRef]
59. Zhong, S.; Chang, C.; Li, J.; Shang, X.; Chen, S.; Song, M.; Zhang, Y. Class Feature Weighted Hyperspectral Image Classification. *IEEE J. Sel. Topics Appl. Earth Obs. Remote Sens.* **2019**, *12*, 4728–4745, doi:10.1109/JSTARS.2019.2950876. [CrossRef]

remote sensing

MDPI

Article

Hyperspectral Image Classification Based on a Shuffled Group Convolutional Neural Network with Transfer Learning

Yao Liu [1], Lianru Gao [2,*], Chenchao Xiao [1], Ying Qu [3], Ke Zheng [2] and Andrea Marinoni [4]

1 Land Satellite Remote Sensing Application Center, Ministry of Natural Resources of China, Beijing 100048, China; liuyao@lasac.cn (Y.L.); xiaochenchao@lasac.cn (C.X.)
2 Key Laboratory of Digital Earth Science, Aerospace Information Research Institute, Chinese Academy of Sciences, Beijing 100094, China; zhengkevic@aircas.ac.cn
3 Department of Electrical Engineering and Computer Science, The University of Tennessee, Knoxville, TN 37996, USA; yqu3@vols.utk.edu
4 Department of Physics and Technology, UiT The Arctic University of Norway, NO-9037 Tromsø, Norway; andrea.marinoni@uit.no
* Correspondence: gaolr@aircas.ac.cn

Received: 4 May 2020; Accepted: 27 May 2020; Published: 1 June 2020

Abstract: Convolutional neural networks (CNNs) have been widely applied in hyperspectral imagery (HSI) classification. However, their classification performance might be limited by the scarcity of labeled data to be used for training and validation. In this paper, we propose a novel lightweight shuffled group convolutional neural network (abbreviated as SG-CNN) to achieve efficient training with a limited training dataset in HSI classification. SG-CNN consists of SG conv units that employ conventional and atrous convolution in different groups, followed by channel shuffle operation and shortcut connection. In this way, SG-CNNs have less trainable parameters, whilst they can still be accurately and efficiently trained with fewer labeled samples. Transfer learning between different HSI datasets is also applied on the SG-CNN to further improve the classification accuracy. To evaluate the effectiveness of SG-CNNs for HSI classification, experiments have been conducted on three public HSI datasets pretrained on HSIs from different sensors. SG-CNNs with different levels of complexity were tested, and their classification results were compared with fine-tuned ShuffleNet2, ResNeXt, and their original counterparts. The experimental results demonstrate that SG-CNNs can achieve competitive classification performance when the amount of labeled data for training is poor, as well as efficiently providing satisfying classification results.

Keywords: lightweight convolutional neural networks; deep learning; hyperspectral imagery classification; transfer learning

1. Introduction

Hyperspectral sensors are able to grasp detailed information of objects and phenomena on Earth's surface by severing their spectral characteristics in a large number of channels (bands) over a wide portion of the electromagnetic spectrum. Such rich spectral information allows hyperspectral imagery (HSI) to be used for interpretation and analysis of surface materials in a more thorough way. Accordingly, hyperspectral remote sensing has been widely used in several research fields, such as environmental monitoring [1–3], land management [4–6], and agriculture [7–9].

Land cover classification is an important HSI analysis task that aims to label every pixel in the HSI image with its unique type [10]. In the past several decades, various classification methods have been developed based on spectral features [11,12] or spatial-spectral features [13–15]. Recently, deep-learning (DL)-based methods have attracted increasing attention for HSI classification [16]. Compared to traditional methods that require sophisticated feature extraction methods [17], DL methods allow models to automatically extract hidden features and learn parameters from labeled samples. Existing DL methods include fully connected feedforward neural networks [18–20], convolutional neural networks (CNNs) [21–23], recurrent neural networks (RNNs) [24,25], and so on. Among these networks, CNN has become the major deep learning framework applied for hyperspectral image classification, as it can maintain the local invariance of the image and has a relatively small number of coefficients to be tuned [26].

For HSI classification, the scarcity of labeled data to be used for training is a common problem [27]. Nonetheless, supervised DL methods require large training datasets to achieve accurate classification results [28]. Since data labeling is time-consuming and costly, many techniques have been developed to deal with HSI classification of small datasets, such as data augmentation [29–31] and transfer learning [32–38]. Data augmentation is an effective technique that artificially enlarges the size of a training dataset by creating its modified versions, e.g., by flipping and rotating the original sample image [30]. On the other hand, transfer learning reuses a trained model and adapts it to a related new task, alleviating the requirement on large-scale labeled samples for effective training. In [32,33], transfer learning has been employed between HSI records acquired by the same sensor. Recently, HSI classification based on cross-sensor transfer learning has become a hot topic within the scientific community, since it allows to achieve high accuracy by combining the information retrieved from multiple hyperspectral images [34–38]. In these studies, efficient network architecture was proposed with units that have only a few parameters to be tuned (e.g., separable convolutions [34], bottleneck unit[36]) and deeper layers that can accurately extract complex features (e.g., VGGNet in [35], ResNet in [36]). However, with tens of layers in these CNNs, the number of parameters can easily reach several hundred thousands, or even millions, and hyperparameters need to be carefully tuned for these networks to avoid overfitting. When labeled samples are scarce (either in terms of quality, reliability, or size), a simpler structure is suitable to avoid the risk of overfitting. Accordingly, we propose a new CNN called shuffled group convolutional neural network (SG-CNN). SG-CNN has efficient building blocks called SG conv units and does not contain a large number of parameters. In addition, we applied SG-CNN with transfer learning between HSI of different sensors to improve the classification performance with limited samples.

The main contributions of this study are summarized as follows.

(1) We propose a DL-based method that brings improvement to HSI classification with limited samples through transfer learning on the new proposed SG-CNN. The SG-CNN reduces the number of parameters and computation time whilst guaranteeing high classification accuracy.

(2) To conduct transfer learning, a simple dimensionality reduction strategy is put forward to keep the dimensions of input data consistent. This strategy is very easily and quickly performed and requires no labeled samples from the HSIs. The bands of original HSI datasets are selected according to this strategy to ensure both the source data and target data have the same number of bands to be the SG-CNN inputs.

The remainder of this paper is organized as follows. Section 2 gives a detailed illustration of the proposed framework for classification, including the structure of the network and the new proposed SG conv unit. Datasets, experimental setup, as well as classification results and analysis are given in Section 3. Finally, conclusions are presented in Section 4.

2. Proposed Method

As previously mentioned, DL models have been applied in HSI classification with satisfying performance. However, as a lack of sufficient samples is typical for HSI, there is still room for improvement of DL-based classification methods. Inspired by the lightweight networks [39,40] and the effects of atrous convolution in semantic segmentation tasks [41–43], we developed a new lightweight CNN for HSI classification. In this section, the structure of this new proposed network as well as how it is applied to transfer learning is given next.

2.1. A SG-CNN-Based Classification Framework

The framework of the proposed classification is shown in Figure 1. It consists of three parts: (1) dimensionality reduction (DR), (2) sample generation, and (3) SG-CNN for feature extraction and classification.

First, DR is conducted to ensure that the SG-CNN input data from both the source and target HSIs have the same dimensions. Considering that typical HSIs have 100–200 bands and generally require less than 20 bands to summarize the most informative spectral features [44], a simple band reduction strategy is implemented, and the number of bands is fixed to 64 for the CNN input data. These 64 bands are selected at equal intervals from the original HSI. Specifically, given HSI data with N_b bands, the number of bands and intervals are determined as follows.

(1) Two intervals are used and respectively set to $\lfloor N_b/64 \rfloor$ and $\lfloor N_b/64 \rfloor + 1$, where $\lfloor\rfloor$ represents the floor operation of its input.

(2) Assume x and y are the number of bands selected respectively at these two intervals. Then we can have equations as follows:

$$\begin{cases} x + y = 64 \\ \lfloor N_b/64 \rfloor * x + (\lfloor N_b/64 \rfloor + 1) * y = N_b \end{cases} \quad (1)$$

where x and y are solved using these linear equations. The 64 selected bands of both source and target data are thus determined. Compared with band selection methods, this DR strategy retains more bands but is very easy and fast to implement.

Second, a $S \times S \times 64$-sized cube is extracted as a sample from a window centered around a labeled pixel. S is the window size, and 64 is the number of bands. The label of the center pixel in the cube is used as the sample's label. In addition, we used the mirroring preprocessing in [23] to ensure sample generation for pixels belonging to image borders.

Finally, samples are fed to the SG-CNN that mainly consists of two parts to achieve classification: (1) the input data are put through SG conv units for feature extraction; (2) the output of the last SG conv unit is subject to global average pooling and then fed to a fully connected (FC) layer, further predicting the sample class using the softmax activation function.

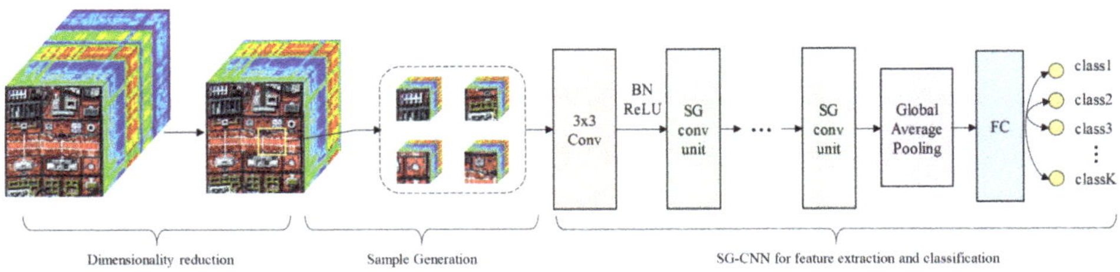

Figure 1. Shuffled group convolutional neural network (SG-CNN)-based hyperspectral imagery (HSI) classification framework.

2.2. SG Conv Unit

Networks with a large number of training parameters can be prone to overfitting. To tackle this issue, we designed a lightweight SG conv unit inspired by the structure in ResNeXt [45]. In the SG conv units, group convolution is used to decrease the number of parameters. We used not only conventional convolution, but we also introduced atrous convolution into the group convolution, which was followed by a channel shuffle operation; this is a major difference with respect to the ResNeXt structure. To further boost the training efficiency, batch normalization [46] and short connection [47] were also included in this unit.

The details of this unit are displayed in Figure 2. From top to bottom, this unit mainly contains a 1 × 1 convolution, group convolution layers followed by channel shuffle, and another 1 × 1 convolution, which is added to the input of this unit and then fed to the next SG conv unit or global average pooling layer. Specifically, in the group convolution, half the groups perform conventional convolutions, while the other half employ subsequent convolutional layers that have different dilation rates. The inclusion of atrous convolution is motivated by its ability to enlarge the respective field without increasing the number of parameters. Moreover, atrous convolution has shown outstanding performance in semantic segmentation [41–43], whose task is similar to HSI classification, i.e., to label every pixel with a category. In addition, since stacked group convolutions only connect to a small fraction of input channels, channel shuffle (Figure 2b) is performed to make the group convolution layers more powerful through connections with different groups [39,40].

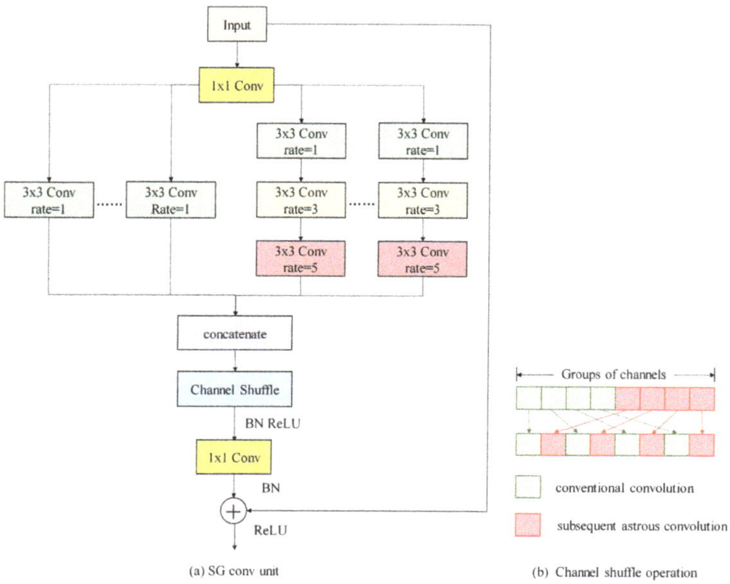

Figure 2. SG conv unit: (**a**) A SG conv unit has a 1x1 convolution, group convolution layers followed by channel shuffle, another 1x1 convolution, and a shortcut connection. (**b**) Channel shuffle operation in the SG conv unit mixes groups that have conventional convolution and atrous convolution.

2.3. *Transfer Learning between HSIs of Different Sensors*

In order to improve the classification results for HSI data with limited samples, transfer learning was applied to the SG-CNN. As shown in Figure 3, this process consisted of two stages: pretraining and fine-tuning. Specifically, the SG-CNN was first trained on the source data that had a large number of samples, and then it was fine-tuned on the target data with fewer samples. In the fine-tuning stage, apart from parameters in the FC layer, all other parameters from the pretrained network were used in the initialization to train the SG-CNN; parameters in the FC layer were randomly initialized.

3. Experimental Results

Extensive experiments were conducted on public hyperspectral data to evaluate the classification performance of our proposed transfer learning method.

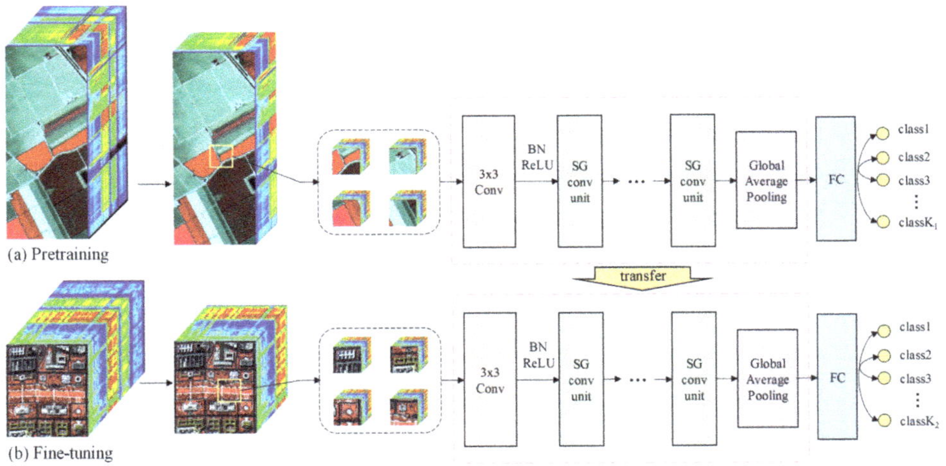

Figure 3. Transfer learning process: (**a**) pretrain the SG-CNN with samples from source HSI data, (**b**) fine-tune the SG-CNN for target HSI data classification.

3.1. Datasets

Six widely known hyperspectral datasets were used in this experiment. These hyperspectral scenes included Indian Pines, Botswana, Salinas, DC Mall, Pavia University (i.e., PaviaU), and Houston from the 2013 IEEE Data Fusion Contest (referred as Houston 2013 hereafter). The Indian Pines and Salinas were collected by the 224-band Airborne Visible/Infrared Imaging Spectrometer (AVIRIS). Botswana was acquired by the Hyperion sensor onboard the EO-1 satellite, with the data acquisition ability of 242 bands covering the 0.4–2.5 µm. DC Mall was gathered by the Hyperspectral digital imagery collection experiment (HYDICE). PaviaU and Houston 2013 were acquired by the ROSIS and CASI sensor, respectively. Detailed information about these data are listed in Table 1: uncalibrated or noisy bands covering the region of water absorption have been removed from these datasets.

Three pairs of transfer learning experiments were designed using these six datasets: (1) pretrain on the Indian Pines, and fine-tune on the Botswana scene; (2) pretrain on the PaviaU scene, and fine-tune on the Houston 2013 scene; (3) pretrain on the Salinas scene, and fine-tune on the DC Mall scene. The experiments were designed as above for two reasons: (1) the source data and target data were collected by different sensors, but they were similar in terms of spatial resolution and the spectral range; (2) the source data have more labeled samples in each class than those of the target data. Despite that slight differences of band wavelengths may exist between the source and target data, SG-CNNs will automatically adapt its parameters to extract spectral features for the target data in the fine-tuning process.

Table 1. Hyperspectral datasets used in the experiment.

No.	Data Usage	Scene	Sensor	Image Size	Spectral Range (µm)	Number of Bands	Spatial Resolution (m)	Number of Classes
1	Source	Indian Pines	AVIRIS	145 × 145	0.4–2.5	200	20	9 *
	Target	Botswana	Hyperion	1476 × 256	0.4–2.5	145	30	14
2	Source	PaviaU	ROSIS	610 × 340	0.43–0.86	103	1.3	9
	Target	Houston 2013	CASI	1905 × 349	0.38–1.05	144	2.5	15
3	Source	Salinas	AVIRIS	512 × 217	0.4–2.5	204	3.7	16
	Target	DC Mall	HYDICE	280 × 307	0.4–2.5	191	3	6

* Only nine classes having most labeled samples were used from the Indian Pines data. Other classes with fewer training samples were excluded from the experiment.

3.2. Experimental Setup

To evaluate the performance of the proposed classification framework, classification results of three target datasets were compared with those predicted from two baseline models, i.e., ShuffleNet V2 (abbreviated as ShuffleNet2) [40] and ResNeXt [45]. ShuffleNet2 is well-known for its speed and accuracy tradeoff. ResNeXt consists of building blocks with group convolution and shortcut connections, which are also used in the SG-CNN. It is worth noting that we used ShuffleNet2 and ResNeXt with fewer building blocks rather than their original models, considering the limited samples of HSIs. Specifically, convolution layers in Stages 3 and 4 of ShuffleNet2 were removed, and output channels was set to 48 for Stage 2 layers; for the ResNeXt model, only one building block was retained. For further details on ShuffleNet2 and ResNeXt architectures, the reader is referred to [40,45]. In addition, simplified ShuffleNet2 and ResNeXt were both trained on the original target HSI data as well as fine-tuned on the 64-band target data using a corresponding pretrained network from the 64-band source data. Classification results obtained from the transfer learning of baseline models were referred to ShuffleNet2_T and ResNeXt_T, respectively. In addition, we performed transfer learning with SG-CNNs throughout the experiment.

Three SG-CNNs with three levels of complexity were tested for evaluation (see Table 2). SG-CNN-X represents the SG-CNN with X layers of convolution. It is worth noting that ResNeXt and SG-CNN-8 have the same number of layers, and the only difference between their structure is the introduction of atrous convolution for half the groups and shuffle operation in the SG-CNN-8 model. The number of groups was fixed to eight for both the SG-CNNs and ResNeXt, and the sample size was set to 19 × 19. In the SG conv unit, the dilation rates of three atrous convolutions were set to 1, 3, and 5 to get a receptive field of 19 (i.e., the full size of a sample).

Table 2. Overall SG-CNN architecture with different levels of complexity.

Basic Block	Channel Number	SG-CNN-7	SG-CNN-8	SG-CNN-12
Image	64	64	64	64
Conv	64	-	3 × 3, 64	-
SG conv unit 1	128	1 × 1, 64 3 × 3, 64, r = 1 3 × 3, 64, r = 3 3 × 3, 64, r = 5 1 × 1, 128	1 × 1,64 3 × 3, 64, r = 1 3 × 3, 64, r = 3 3 × 3, 64, r = 5 1 × 1, 128	1 × 1, 64 3 × 3, 64, r = 1 3 × 3, 64, r = 3 3 × 3, 64, r = 5 1 × 1, 128
SG conv unit 2	256			1 × 1,128 3 × 3, 128, r = 1 3 × 3, 128, r = 3 3 × 3, 128, r = 5 1 × 1, 256
FC	14/15/6			
No. of trainable parameters		~70,000	~100,000	~140,000

Groups that have conventional convolution in SG conv units are omitted in the table, as this operation is the same as the first layer of subsequent atrous convolution layers with a dilation rate of 1 (i.e., r = 1).

Before network training, original data were normalized to guarantee input values within 0 to 1. Data augmentation techniques (including horizontal and vertical flip) were used to increase the training samples. All classification methods were implemented using python code with high-level APIs Tensorflow [48] and Keras. To further alleviate possible overfitting, the sum of multi-class cross entropy and L2 regularization term was taken as the loss function, and we set the weight decay to 5×10^{-4} in the L2 regularizer. The Adam optimizer [49] was adopted with an initial learning rate of 0.001, and the learning rate would be reduced to one-fifth of its value if the validation loss function did not decrease for 10 epochs. We used the Adam optimizer with a mini-batch size of 32 on a NVIDIA GEFORCE RTX 2080Ti GPU. The number of epochs was set to 150–250 for different datasets, and it is determined based on the number of training samples.

3.3. Experiments on Indian Pines and Botswana Scenes

The false-color composites of the Indian Pines and Botswana scenes are displayed in Figures 4 and 5, with their corresponding ground truth. In the pre-training and fine-tuning stage, Table 3 gives the number of labeled pixels that were randomly selected for training, and the remaining labeled samples were used for the test.

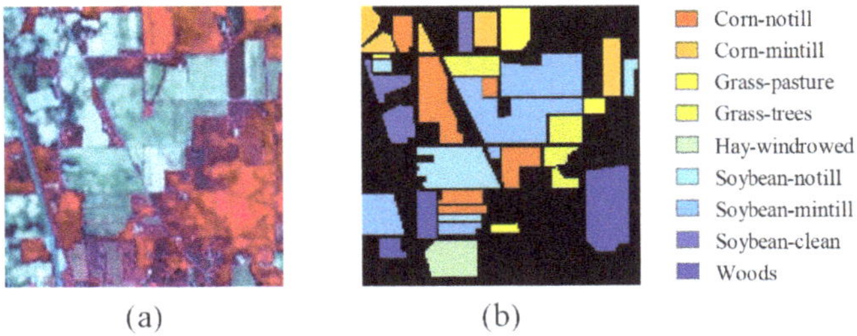

Figure 4. The Indian Pines scene: (**a**) false-color composite image; (b) ground truth.

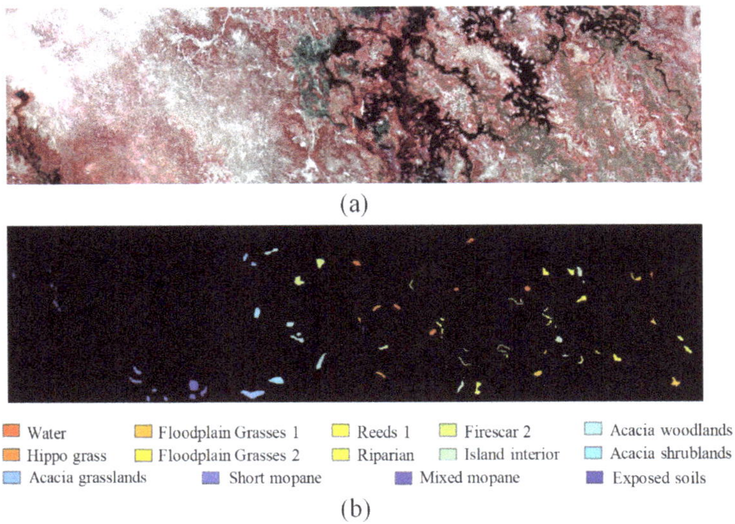

Figure 5. The Botswana scene: (**a**) false-color composite image; (**b**) ground truth.

Table 3. The number of training and test samples used in Indiana Pines and Botswana datasets.

No.	Indian Pines			Botswana		
	Class Name	Train	Test	Class Name	Train	Test
1	Corn-notill	200	1228	Water	30	240
2	Corn-mintill	200	630	Hippo grass	30	71
3	Grass-pasture	200	283	Floodplain Grasses 1	30	221
4	Grass-trees	200	530	Floodplain Grasses 2	30	185
5	Hay-windrowed	200	278	Reeds 1	30	239
6	Soybean-notill	200	772	Riparian	30	239
7	Soybean-mintill	200	2255	Firescar 2	30	229
8	Soybean-clean	200	393	Island interior	30	173
9	Woods	200	1065	Acacia woodlands	30	284
10				Acacia shrublands	30	218
11				Acacia grasslands	30	275
12				Short mopane	30	151
13				Mixed mopane	30	238
14				Exposed soils	30	65

The loss function of SG-CNNs converged in the 150 epochs of training, indicating no overfitting during the fine-tuning process (see Figure 6). Classification results obtained by SG-CNNs were then compared with other methods in Table 4 for the Botswana scene. A range of criteria, including overall accuracy (OA), average accuracy (AA), and Kappa coefficient (K), were all reported as well as the classification accuracy of each class and training time. OA and AA are defined as below:

$$OA = \frac{\sum_{i=1}^{n} C_i}{\sum_{i=1}^{n} S_i} \tag{2}$$

$$AA = \frac{1}{n} \sum_{i=1}^{n} \frac{C_i}{S_i} \tag{3}$$

where C_i is the number of correctly predicted samples out of S_i samples in class i, and n is the number of classes.

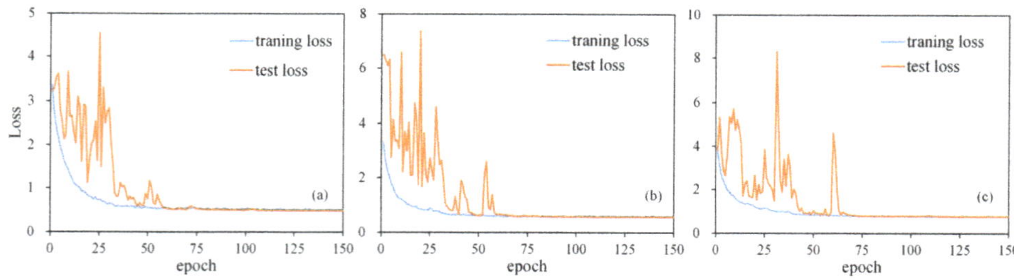

Figure 6. Convergence curves during the fine-tuning process of the Botswana scene: (**a**) SG-CNN-7, (**b**) SG-CNN-8, (**c**) SG-CNN-12.

Based on the results in Table 4, several preliminary conclusions can be drawn as follows.

(1) Compared with baseline models, SG-CNNs typically achieve better classification performance, providing higher accuracy and spending relatively less training time. Specifically, the overall accuracy of SG-CNNs was 98.97–99.65%, which was approximately ~1% and ~3.5% higher, on average, than ResNeXt and ShuffleNet2 models, respectively. In addition, SG-CNN-7 and SG-CNN-8 were shown to be quite efficient, as the execution time of their fine-tuning process was comparable to that of ShuffleNet2_T and ResNeXt_T. As an effect of its complicated structure with more trainable parameters, SG-CNN-12 required a longer period of time to fine-tune.

(2) As mentioned in Section 3.2, SG-CNN-8 can be seen as the baseline ResNeXt model that introduces atrous convolution and channel shuffle into its group convolution. Comparing the classification results of these two models, we can appreciate that the inclusion of atrous convolution and channel shuffle improved the classification.

(3) For the baseline models, both ShuffleNet2_T and ResNeXt_T, which were fine-tuned on the 64-band target data, obtained similar accuracy with much lower execution time, compared with their counterparts that were directly trained from original HSIs. This indicates that the simple band selection strategy applied in transfer learning can generally help to enhance the training efficiency.

Table 4. Classification accuracy (%) and computation time of the Botswana scene. A total of 420 labeled samples (30 per class) were used for fine-tuning. The No. column refers to the corresponding class in Table 3. The best results are in **bold**.

No.	ShuffleNet2	ShuffleNet2_T	ResNeXt	ResNeXt_T	SG-CNN-7	SG-CNN-8	SG-CNN-12
1	94.12	95.65	91.53	93.28	98.36	97.17	**99.17**
2	75.53	81.61	95.95	92.21	**100.00**	**100.00**	**100.00**
3	**100.00**	**100.00**	**100.00**	**100.00**	**100.00**	**100.00**	**100.00**
4	87.68	87.68	93.43	93.91	93.91	**98.40**	97.88
5	89.27	88.73	93.55	91.70	**99.11**	98.31	99.57
6	97.42	98.33	**100.00**	**100.00**	**100.00**	**100.00**	**100.00**
7	97.86	94.24	99.13	**100.00**	97.45	**100.00**	**100.00**
8	94.02	97.19	**100.00**	97.19	**100.00**	**100.00**	99.43
9	**100.00**	**100.00**	**100.00**	**100.00**	**100.00**	**100.00**	**100.00**
10	**100.00**	88.26	**100.00**	**100.00**	99.54	**100.00**	**100.00**
11	**100.00**	**100.00**	**100.00**	99.64	98.56	98.57	99.28
12	85.80	**100.00**	99.34	**100.00**	**100.00**	**100.00**	**100.00**
13	**100.00**	99.58	99.58	**100.00**	**100.00**	**100.00**	**100.00**
14	**100.00**	**100.00**	**100.00**	**100.00**	**100.00**	**100.00**	**100.00**
OA	95.33	95.44	98.06	97.91	98.97	99.36	**99.65**
AA	94.41	95.09	98.04	97.71	99.07	99.46	**99.67**
K	94.94	95.05	97.89	97.74	98.89	99.31	**99.62**
Time(s)	626.61	460.77	1591.27	**375.60**	524.25	389.06	1459.72

For the SG-CNNs, all classification results are obtained with fine-tuning on the target data based on a pretrained model using the source data.

Our second test with the Botswana scene evaluated the classification performance of transfer learning with SG-CNNs using varying sizes of samples. Specifically, 15, 30, 45, 60, and 75 samples per class from the Botswana scene were used, respectively, to fine-tune the pretrained SG-CNNs, and their classification performances were evaluated from OAs of the corresponding remaining samples (i.e., the test samples). Meanwhile, the same samples used for fine-tuning SG-CNNs were utilized to train ShuffleNet2 and ResNext and fine-tune ShuffleNet2_T and ResNext_T. These models were also assessed with OA of test samples. Figure 7 displays OAs in the test dataset from different classification methods with different numbers of training samples. Several conclusions can be drawn:

(1) Compared with ShuffleNet2, ShuffleNet2_T, and ResNeXt, SG-CNNs showed a remarkable improvement for classification by providing a higher classification accuracy, especially when labeled samples were relatively small (i.e., 15–60 samples per class).

(2) Compared with ResNeXt_T, SG-CNNs generally yielded better classification results when the training samples were limited (i.e., 15–45 per class). As the number of samples increased to 60–75 for each class, ResNeXt_T provided comparable accuracy.

(3) Although SG-CNN-12 generally achieved the best performance, its classification accuracy was merely 0.1–0.7% higher than that of SG-CNN-7 and SG-CNN-8. However, the latter two showed smaller values of execution time for the fine-tuning than the former. In other words, SG-CNN-7 and SG-CNN-8 had better tradeoffs between classification accuracy and efficiency.

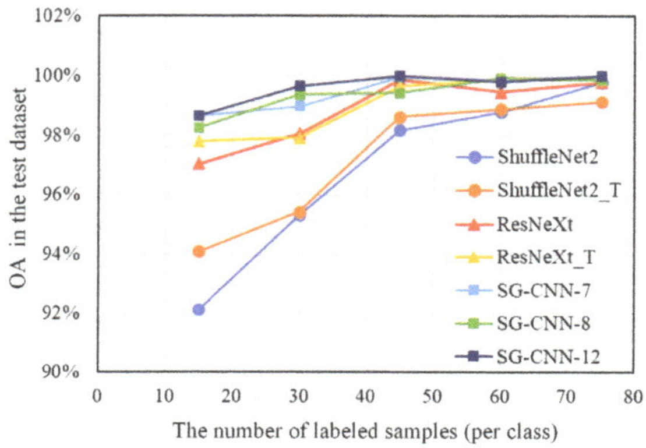

Figure 7. Overall classification accuracies of the test samples based on various methods trained/fine-tuned with 15–75 labeled samples for the Botswana scene.

3.4. Experiments on PaviaU and Houston 2013 Scenes

PaviaU and Houston 2013 datasets are displayed with their labeled sample distributions in Figures 8 and 9. Figure 8 shows that the PaviaU scene contained five manmade types, two types of vegetation, and one type for soil and shadow. As shown in Figure 9, the Houston 2013 scene had nine manmade types, four types of vegetation, and one type for soil and water. Surface types distributions were similar in these two scenes. ShuffleNet2, ResNeXt, and SG-CNNs were fine-tuned on the Houston 2013 scene, with pretrained models acquired from training with the PaviaU dataset. Table 5 displays the number of samples used in the experiment, respectively. Six hundred labeled samples per class in the PaviaU scene were utilized to pretrain the models, whereas 100 randomly selected samples per class in the Houston scene were used for fine-tuning.

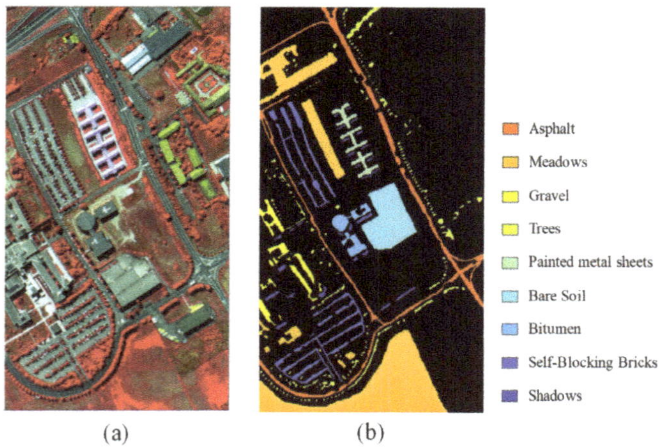

Figure 8. The PaviaU scene: (**a**) false-color composite image; (**b**) ground truth.

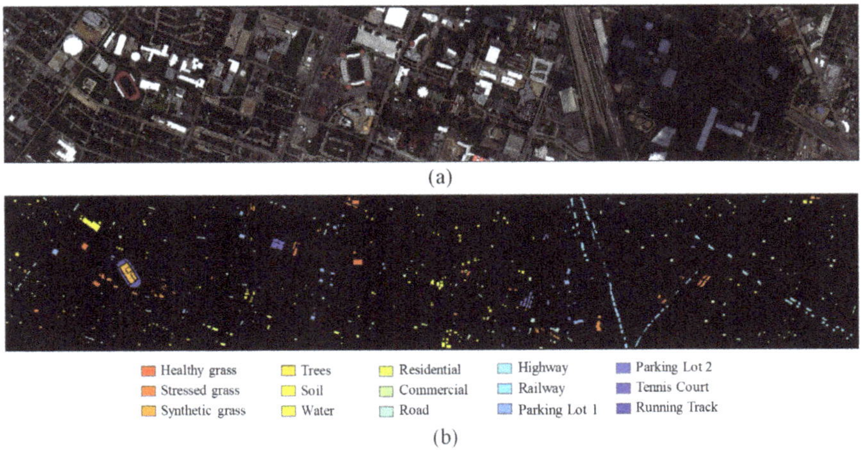

Figure 9. Houston 2013 scene: (**a**) true-color composite image; (**b**) ground truth.

Table 5. The number of training and test samples for PaviaU and Houston 2013 datasets.

No.	PaviaU			Houston 2013		
	Class Name	Train	Test	Class Name	Train	Test
1	Asphalt	600	6031	Healthy grass	100	1151
2	Meadows	600	18,049	Stressed grass	100	1154
3	Gravel	600	1499	Synthetic grass	100	597
4	Trees	600	2464	Trees	100	1144
5	Painted metal sheets	600	745	Soil	100	1142
6	Bare soil	600	4429	Water	100	225
7	Bitumen	600	730	Residential	100	1168
8	Self-Blocking Bricks	600	3082	Commercial	100	1144
9	Shadows	600	347	Road	100	1152
10				Highway	100	1127
11				Railway	100	1135
12				Parking Lot 1	100	1133
13				Parking Lot 2	100	369
14				Tennis Court	100	328
15				Running Track	100	560

Convergence curves of the loss function are shown in Figure 10 for the fine-tuning of SG-CNNs applied to the Houston 2013 scene. Classification results acquired from SG-CNNs and baseline models are detailed in Table 6. As shown in Table 6, SG-CNNs with different levels of complexity achieved higher classification accuracies than those of ShuffleNet2, ShuffleNet2_T, ResNeXt, and ResNeXt_T. Specifically, SG-CNN-12 provided the best classification results with the highest OA (99.45%), AA (99.40%), and Kappa coefficient (99.35%), and it also achieved the highest classification accuracy for eight classes in the test samples. Comparing the results from SG-CNN-8 and ResNeXt_T, the former obtained a slightly higher OA than the latter but spent less than half the training time, indicating the SG conv unit's effectiveness for classification improvement. In addition, fine-tuned ResNeXt_T and ShuffleNet2_T yielded better results than the original ResNeXt and ShuffleNet2. Hence, this confirms the previous conclusion that our band selection strategy applied in transfer learning boosts the classification performance.

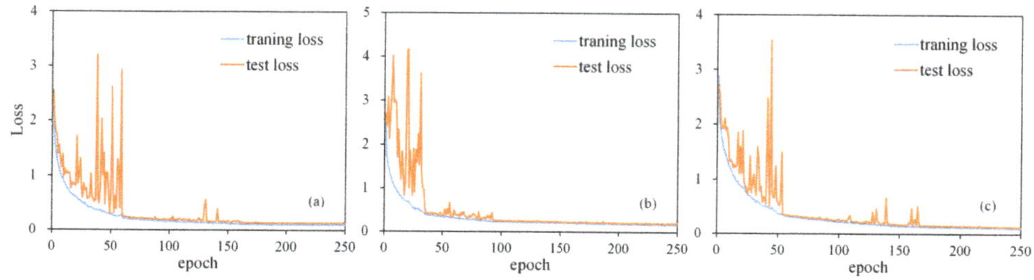

Figure 10. Convergence curves during the fine-tuning process of the Houston 2013 scene: (**a**) SG-CNN-7, (**b**) SG-CNN-8, and (**c**) SG-CNN-12.

Table 6. Classification accuracy (%) and computation time of the Houston 2013 scene. A total of 1500 labeled samples (100 per class) were used for fine-tuning. The No. column refers to the corresponding class in Table 5. The best results are in **bold**.

No.	ShuffleNet2	ShuffleNet2_T	ResNeXt	ResNeXt_T	SG-CNN-7	SG-CNN-8	SG-CNN-12
1	90.09	91.54	84.71	92.65	**99.83**	97.62	99.74
2	92.33	99.28	97.77	96.72	**99.65**	**99.65**	99.40
3	90.73	99.66	99.66	99.83	**100.00**	99.83	**100.00**
4	97.28	99.22	96.87	99.08	99.91	99.82	**100.00**
5	**100.00**	98.87	99.22	99.22	**100.00**	99.65	99.22
6	89.36	97.38	83.08	93.75	95.34	95.34	97.40
7	87.18	94.65	92.60	94.84	98.29	**100.00**	**100.00**
8	99.30	97.99	98.84	99.46	**100.00**	89.22	99.82
9	86.49	93.46	88.50	96.99	97.62	96.69	**97.86**
10	92.15	96.24	94.15	94.47	99.20	98.41	**99.64**
11	95.37	94.00	97.07	97.88	**100.00**	**100.00**	**100.00**
12	92.50	96.65	**100.00**	97.00	95.94	89.26	99.47
13	97.43	93.26	95.65	**100.00**	**100.00**	**100.00**	**100.00**
14	**100.00**	84.75	**100.00**	**100.00**	**100.00**	**100.00**	**100.00**
15	95.87	97.38	96.54	96.88	97.22	**97.90**	97.73
OA	93.27	95.92	94.95	97.02	98.98	97.18	**99.45**
AA	93.74	95.62	94.98	97.25	98.87	97.56	**99.40**
K	92.71	95.58	94.53	96.77	98.90	96.94	**99.35**
Time(s)	2068.42	1614.16	5120.20	2309.30	2088.32	**1035.15**	2957.94

Classification experiments with varying numbers of training samples were also conducted. Specifically, 50–250 samples per class in the Houston scene were used for fine-tuning the SG-CNNs, as well as for training or fine-tuning the baseline networks. OAs of the remaining test samples are shown in Figure 11 for all the methods. Some conclusions can be reached from making comparisons between these results:

(1) As training samples varied from 50 to 250 per class, SG-CNNs outperformed ShuffleNet2, ShuffleNet2_T, and ResNeXt for the Houston 2013 scene classification. The accuracies of the fine-tuned SG-CNNs are ∼1.3–7.4% higher than that of the other three baseline networks, indicating that SG-CNNs greatly improved the classification performance with both limited and sufficient samples.

(2) Comparing with ResNeXt_T, SG-CNNs obtained better results when few samples were provided (i.e., 50–100 per class). As the number of samples increased to 150–250 per class, the ResNeXt_T and SG-CNNs achieved comparable accuracy. This suggests that SG-CNNs have better performance with limited samples.

(3) In general, SG-CNN-12 provided the highest classification accuracy among the three SG-CNNs. However, as the number of training samples increased, the performance of SG-CNN-12 showed no obvious improvement compared to SG-CNN-7 and SG-CNN-8, which are more efficient and require less computing time.

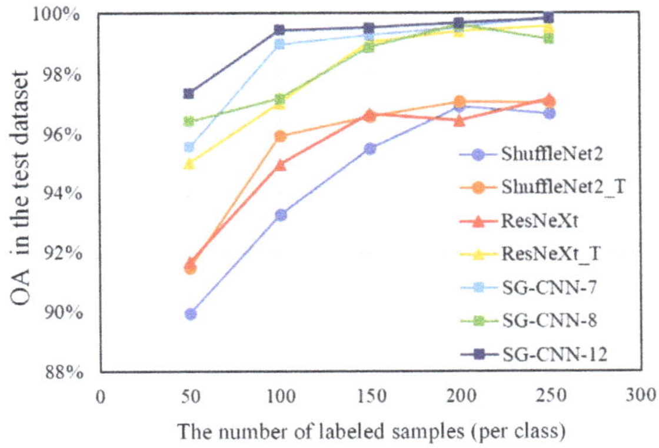

Figure 11. Overall classification accuracies of the test samples based on various methods trained/fine-tuned with 50–250 labeled samples for the Houston 2013 scene.

3.5. Experiments on Salinas and DC Mall Scenes

Salinas and DC Mall images and their labeled samples are shown in Figures 12 and 13, respectively. It is important to note that surface types were quite different between these two scenes. The Salinas scene mainly consisted of natural materials (i.e., vegetation and three types of fallow), whereas the DC Mall scene included grass, trees, shadows, and three manmade materials. Table 7 provides the number of samples used as training and test datasets. Five hundred samples of each class in the Salinas scene were randomly selected for base network training, whereas 100 samples of each class in the DC Mall scene were used for fine-tuning.

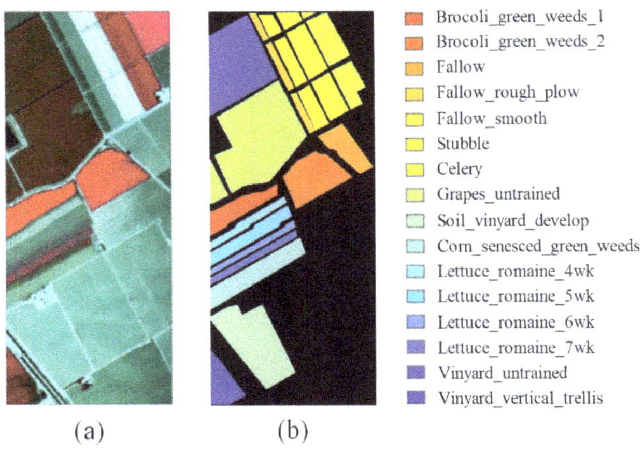

Figure 12. The Salinas scene: (**a**) false-color composite image; (**b**) ground truth.

The loss function of SG-CNNs converged during the fine-tuning for the DC Mall scene (see Figure 14). The classification results of both baseline models and SG-CNNs are listed in Table 8 with their corresponding training time. As shown in Table 8, similar conclusions can be reached from the DC Mall experiment. First, SG-CNNs outperformed the baseline models in terms of classification results. Moreover, SG-CNN-8 had an OA nearly 10% higher than that of ResNeXt_T, indicating the improvement brought by the proposed SG conv unit. Furthermore, although the target data and source data had different surface types, transfer learning on the SG-CNNs led to major improvement in the classification accuracy.

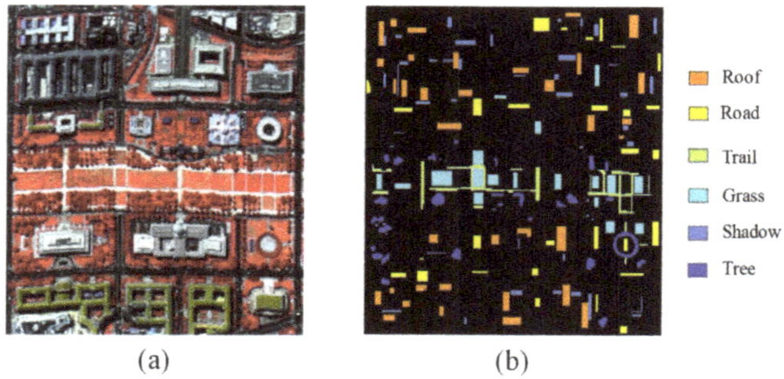

Figure 13. The DC Mall scene: (**a**) false-color composite image; (**b**) ground truth.

Table 7. The number of training and test samples for Salinas and DC Mall datasets.

No.	Salinas			DC Mall		
	Class Name	Train	Test	Class Name	Train	Test
1	Brocoli_green_weeds_1	500	1309	Roof	100	2816
2	Brocoli_green_weeds_2	500	3226	Grass	100	1719
3	Fallow	500	1476	Road	100	1164
4	Fallow_rough_plow	500	1194	Trail	100	1690
5	Fallow_smooth	500	2178	Tree	100	1020
6	Stubble	500	3459	Shadow	100	1181
7	Celery	500	3079			
8	Grapes_untrained	500	10,771			
9	Soil_vinyard_develop	500	5703			
10	Corn_senesced_green_weeds	200	2778			
11	Lettuce_romaine_4wk	500	568			
12	Lettuce_romaine_5wk	500	1327			
13	Lettuce_romaine_6wk	500	416			
14	Lettuce_romaine_7wk	500	570			
15	Vinyard_untrained	500	6768			
16	Vinyard_vertical_trellis	500	1307			

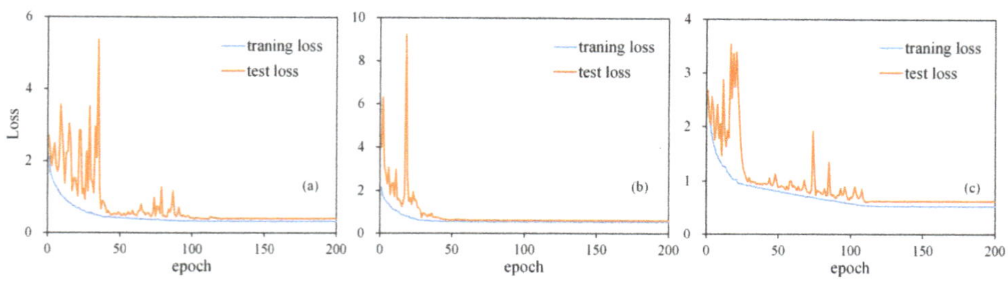

Figure 14. Convergence curves during the fine-tuning process for the DC Mall scene: (**a**) SG-CNN-7, (**b**) SG-CNN-8, and (**c**) SG-CNN-12.

Analogously, our second test on the DC Mall scene evaluated the classification performance of the proposed method with varying sizes of labeled samples. We used 50–250 samples per class at an interval of 50 to train ShuffleNet2 and ResNeXt and to fine-tune SG-CNNs, ShuffleNet2_T, and ResNeXt_T. Figure 15 shows the OAs for the test samples from all methods. In the DC Mall experiment, SG-CNNs outperformed all baseline models, including the ResNeXt_T, when a large number of training samples (e.g., 250 samples per class) was provided. Specifically, the OA of SG-CNNs was higher than that of other methods by 5.3–18.2%, which confirmed the superiority of our proposed method. For the DC Mall dataset, SG-CNN-12 achieved better results when samples were relatively limited (i.e., 50–150 samples per class). With 200–250 training samples in each category, SG-CNN-7 and SG-CNN-8 required less time to obtain a comparable accuracy to that of SG-CNN-12.

Table 8. Classification accuracy (%) and computation time of the DC Mall scene. A total of 600 labeled samples (100 per class) were used for fine-tuning. The No. column refers to the corresponding class in Table 7. The best results are in **bold**.

No.	ShuffleNet2	ShuffleNet2_T	ResNeXt	ResNeXt_T	SG-CNN-7	SG-CNN-8	SG-CNN-12
1	90.90	91.50	89.65	96.65	98.46	**99.77**	99.47
2	92.03	91.02	90.96	92.14	93.47	92.77	**94.85**
3	77.57	76.34	66.87	78.18	92.49	**95.53**	93.37
4	94.21	92.16	89.44	92.20	99.19	**99.51**	99.45
5	50.53	52.23	51.79	65.93	80.67	90.19	**92.63**
6	92.17	91.69	89.85	95.34	97.42	99.24	**99.58**
OA	83.89	83.22	80.67	88.18	94.60	96.68	**97.06**
AA	82.90	82.49	79.76	86.74	93.62	96.17	**96.56**
K	80.31	79.53	76.39	85.53	93.36	95.92	**96.38**
Time(s)	2535.16	1660.96	4310.51	2670.86	1133.61	**885.03**	2324.81

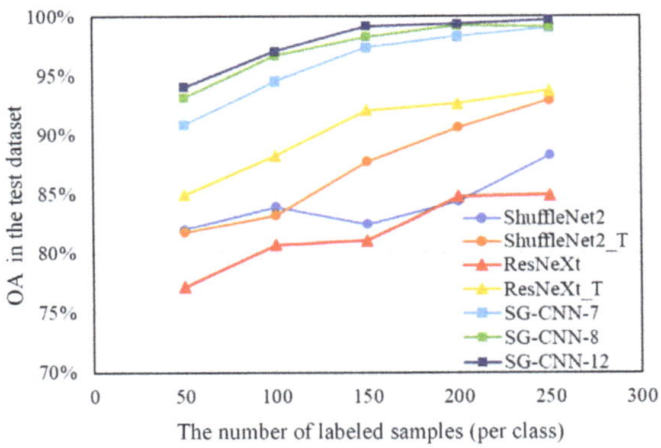

Figure 15. Overall classification accuracies of the test samples based on various methods trained/fine-tuned with 50–250 labeled samples for the DC Mall scene.

4. Conclusions

Typically, only limited labeled samples are available for HSI classification. To improve the HSI classification for such conditions, we proposed a new CNN-based classification method that performed transfer learning between different HSI datasets on a proposed lightweight CNN. This scheme, named SG-CNN, consisted of SG conv units, which combined group convolution, atrous convolution, and channel shuffle operation. In the SG conv unit, group convolution was utilized to reduce the number of parameters, while channel shuffle was employed to connect information in different groups. Also, atrous convolution was introduced in addition to conventional convolution in the groups so that the receptive field was enlarged. To further improve the classification performance with limited samples, transfer learning was applied on SG-CNNs, with a simple dimensionality reduction implemented to keep the dimensions of input data consistent for both the source and target data.

To evaluate the classification performance of the proposed method, transfer learning experiments were performed on SG-CNNs between three pairs of public HSI scenes. Specifically, three SG-CNNs with different levels of complexity were tested. Compared with ShuffleNet-V2, ResNeXt, and their fine-tuned models, the proposed method considerably improved classification results when the training samples were limited, and it also enhanced model efficiency by reducing the computing cost for the training process. It suggests that the combination of atrous convolution with group convolution is effective for training with limited samples, and the band selection method can be helpful for transfer learning.

Author Contributions: Conceptualization, Y.L.; Funding acquisition, Y.L. and A.M.; Resources, C.X.; Supervision, L.G.; Writing—original draft, Y.L.; Writing—review & editing, Y.Q., K.Z. and A.M. All authors have read and agreed to the published version of the manuscript.

Funding: This work was supported by the National Natural Science Foundation of China under Grant No. 41901304, No. 41722108, and also funded in part by the Centre for Integrated Remote Sensing and Forecasting for Arctic Operations (CIRFA) and the Research Council of Norway (RCN Grant no. 237906), and by the Fram Center under the Automised Large-scale Sea Ice Mapping (ALSIM) "Polhavet" flagship project.

Acknowledgments: The authors would like to thank http://www.ehu.eus/ for providing the original remote sensing images.

Conflicts of Interest: The authors declare no conflicts of interest.

Abbreviations

The following abbreviations are used in this manuscript:

AA	Average Accuracy
AVIRIS	Airborne Visible/Infrared Imaging Spectrometer
CNN	Convolutional Neural Network
DR	Dimensionality Reduction
HSI	Hyperspectral Image
HYDICE	Hyperspectral Digital Imagery Collection Experiment
K	Kappa coefficient
OA	Overall Accuracy

References

1. Zhang, B.; Wu, D.; Zhang, L.; Jiao, Q.; Li, Q. Application of hyperspectral remote sensing for environment monitoring in mining areas. *Environ. Earth Sci.* **2012**, *65*, 649–658. [CrossRef]
2. Kudela, R.M.; Palacios, S.L.; Austerberry, D.C.; Accorsi, E.K.; Guild, L.S.; Torres-Perez, J. Application of hyperspectral remote sensing to cyanobacterial blooms in inland waters. *Remote Sens. Environ.* **2015**, *167*, 196–205. [CrossRef]
3. Sankey, T.; Donager, J.; McVay, J.; Sankey, J.B. UAV lidar and hyperspectral fusion for forest monitoring in the southwestern USA. *Remote Sens. Environ.* **2017**, *195*, 30–43. [CrossRef]
4. Olmanson, L.G.; Brezonik, P.L.; Bauer, M.E. Airborne hyperspectral remote sensing to assess spatial distribution of water quality characteristics in large rivers: The Mississippi River and its tributaries in Minnesota. *Remote Sens. Environ.* **2013**, *130*, 254–265. [CrossRef]
5. Yokoya, N.; Chan, J.C.W.; Segl, K. Potential of resolution-enhanced hyperspectral data for mineral mapping using simulated EnMAP and Sentinel-2 images. *Remote Sens.* **2016**, *8*, 172. [CrossRef]
6. Makki, I.; Younes, R.; Francis, C.; Bianchi, T.; Zucchetti, M. A survey of landmine detection using hyperspectral imaging. *ISPRS J. Photogramm. Remote Sens.* **2017**, *124*, 40–53. [CrossRef]
7. Datt, B.; McVicar, T.R.; Van Niel, T.G.; Jupp, D.L.; Pearlman, J.S. Preprocessing EO-1 Hyperion hyperspectral data to support the application of agricultural indexes. *IEEE Trans. Geosci. Remote Sens.* **2003**, *41*, 1246–1259. [CrossRef]
8. Gevaert, C. M.; Suomalainen, J.; Tang, J.; Kooistra, L. Generation of spectral–temporal response surfaces by combining multispectral satellite and hyperspectral UAV imagery for precision agriculture applications. *IEEE J. Sel. Top. Appl. Earth Obs. Remote Sens.* **2015**, *8*, 3140–3146. [CrossRef]
9. Adão, T.; Hruška, J.; Pádua, L.; Bessa, J.; Peres, E.; Morais, R.; Sousa, J.J. Hyperspectral imaging: A review on UAV-based sensors, data processing and applications for agriculture and forestry. *Remote Sens.* **2017**, *9*, 1110. [CrossRef]

10. Gewali, U.B.; Monteiro, S.T.; Saber, E. Machine learning based hyperspectral image analysis: A survey. *arXiv* **2018**, arXiv:1802.08701.
11. Melgani, F.; Bruzzone, L. Classification of hyperspectral remote sensing images with support vector machines. *IEEE Trans. Geosci. Remote Sens.* **2004**, *42*, 1778–1790. [CrossRef]
12. Kuching, S. The performance of maximum likelihood, spectral angle mapper, neural network and decision tree classifiers in hyperspectral image analysis. *J. Comput. Sci.* **2007**, *3*, 419–423.
13. Fauvel, M.; Tarabalka, Y.; Benediktsson, J.A.; Chanussot, J.; Tilton, J.C. Advances in spectral-spatial classification of hyperspectral images. *Proc. IEEE* **2012**, *101*, 652–675. [CrossRef]
14. Yu, H.; Gao, L.; Li, J.; Li, S.S.; Zhang, B.; Benediktsson, J.A. Spectral-spatial hyperspectral image classification using subspace-based support vector machines and adaptive markov random fields. *Remote Sens.* **2016**, *8*, 355. [CrossRef]
15. Yu, H.; Gao, L.; Liao, W.; Zhang, B.; Zhuang, L.; Song, M.; Chanussot, J. Global spatial and local spectral similarity-based manifold learning group sparse representation for hyperspectral imagery classification. *IEEE Trans. Geosci. Remote Sens.* **2020**, *58*, 3043–3056. [CrossRef]
16. Li, S.; Song, W.; Fang, L.; Chen, Y.; Ghamisi, P.; Benediktsson, J.A. Deep learning for hyperspectral image classification: An overview. *IEEE Trans. Geosci. Remote Sens.* **2019**, *57*, 6690–6709. [CrossRef]
17. Zhang, L.; Zhang, L.; Tao, D.; Huang, X. Tensor discriminative locality alignment for hyperspectral image spectral–spatial feature extraction. *IEEE Trans. Geosci. Remote Sens.* **2012**, *51*, 242–256. [CrossRef]
18. Chen, Y.; Lin, Z.; Zhao, X.; Wang, G.; Gu, Y. Deep learning-based classification of hyperspectral data. *IEEE J. Sel. Top. Appl. Earth Obs. Remote Sens.* **2014**, *7*, 2094–2107. [CrossRef]
19. Liu, Y.; Cao, G.; Sun, Q.; Siegel, M. Hyperspectral classification via deep networks and superpixel segmentation. *Int. J. Remote Sens.* **2015**, *36*, 3459–3482. [CrossRef]
20. Ma, X.; Wang, H.; Geng, J. Spectral–spatial classification of hyperspectral image based on deep auto-encoder. *IEEE J. Sel. Top. Appl. Earth Obs. Remote Sens.* **2016**, *9*, 4073–4085. [CrossRef]
21. Hu, W.; Huang, Y.; Wei, L.; Zhang, F.; Li, H. Deep convolutional neural networks for hyperspectral image classification. *J. Sens.* **2015**, *2015*, 1–12. [CrossRef]
22. Chen, Y.; Jiang, H.; Li, C.; Jia, X.; Ghamisi, P. Deep feature extraction and classification of hyperspectral images based on convolutional neural networks. *IEEE Trans. Geosci. Remote Sens.* **2016**, *54*, 6232–6251. [CrossRef]
23. Paoletti, M.E.; Haut, J.M.; Plaza, J.; Plaza, A. A new deep convolutional neural network for fast hyperspectral image classification. *ISPRS J. Photogramm. Remote Sens.* **2018**, *145*, 120–147. [CrossRef]
24. Mou, L.; Ghamisi, P.; Zhu, X.X. Deep recurrent neural networks for hyperspectral image classification. *IEEE Trans. Geosci. Remote Sens.* **2017**, *55*, 3639–3655. [CrossRef]
25. Liu, Q.; Zhou, F.; Hang, R.; Yuan, X. Bidirectional-convolutional LSTM based spectral-spatial feature learning for hyperspectral image classification. *Remote Sens.* **2017**, *9*, 1330. [CrossRef]
26. Krizhevsky, A.; Sutskever, I.; Hinton, G. E. ImageNet classification with deep convolutional neural networks. In Proceedings of the 26th Annual Conference on Neural Information Processing Systems (NIPS), Lake Tahoe, NV, USA, 3–6 December 2012; pp. 1097–1105.
27. Makantasis, K.; Karantzalos, K.; Doulamis, A.; Doulamis, N. Deep supervised learning for hyperspectral data classification through convolutional neural networks. In Proceedings of the 2015 IEEE International Geoscience and Remote Sensing Symposium (IGARSS), Milan, Italy, 26–31 July 2015; pp. 4959–4962.
28. Liu, B.; Wei, Y.; Zhang, Y.; Yang, Q. Deep neural networks for high dimension, low sample size data. In Proceedings of the 21 International Joint Conference on Artificial Intelligence (IJCAI), Melbourne, Australia, 19–25 August 2017; pp. 2287–2293.
29. Li, W.; Wu, G.; Zhang, F.; Du, Q. Hyperspectral image classification using deep pixel-pair features. *IEEE Trans. Geosci. Remote Sens.* **2016**, *55*, 844–853. [CrossRef]
30. Zhang, H.; Li, Y.; Zhang, Y.; Shen, Q. Spectral-spatial classification of hyperspectral imagery using a dual-channel convolutional neural network. *Remote Sens. Lett.* **2017**, *8*, 438–447. [CrossRef]
31. Li, W.; Chen, C.; Zhang, M.; Li, H.; Du, Q. Data augmentation for hyperspectral image classification with deep cnn. *IEEE Geosci. Remote Sens. Lett.* **2018**, *16*, 593–597. [CrossRef]
32. Yang, J.; Zhao, Y. Q.; Chan, J. C. W. Learning and transferring deep joint spectral–spatial features for hyperspectral classification. *IEEE Trans. Geosci. Remote Sens.* **2017**, *55*, 4729–4742. [CrossRef]

33. Liu, X.; Sun, Q.; Meng, Y.; Fu, M.; Bourennane, S. Hyperspectral image classification based on parameter-optimized 3D-CNNs combined with transfer learning and virtual samples. *Remote Sens.* **2018**, *10*, 1425. [CrossRef]
34. Jiang, Y.; Li, Y.; Zhang, H. Hyperspectral image classification based on 3-D separable ResNet and transfer learning. *IEEE Geosci. Remote Sens. Lett.* **2019**, *16*, 1949–1953. [CrossRef]
35. He, X.; Chen, Y.; Ghamisi, P. Heterogeneous transfer learning for hyperspectral image classification based on convolutional neural network. *IEEE Trans. Geosci. Remote Sens.* **2019**, *58*, 3246–3263. [CrossRef]
36. Zhang, H.; Li, Y.; Jiang, Y.; Wang, P.; Shen, Q.; Shen, C. Hyperspectral classification based on lightweight 3-D-CNN with transfer learning. *IEEE Trans. Geosci. Remote Sens.* **2019**, *57*, 5813–5828. [CrossRef]
37. Nalepa, J.; Myller, M.; Kawulok, M. Transfer learning for segmenting dimensionally reduced hyperspectral images. *IEEE Geosci. Remote Sens. Lett.* **2019**. [CrossRef]
38. Zhao, X.; Liang, Y.; Guo, A. J.; Zhu, F. Classification of small-scale hyperspectral images with multi-source deep transfer learning. *Remote Sens. Lett.* **2020**, *11*, 303–312. [CrossRef]
39. Zhang, X.; Zhou, X.; Lin, M.; Sun, J. ShuffleNet: An extremely efficient convolutional neural network for mobile devices. In Proceedings of the IEEE Conference on Computer Vision and Pattern Recognition (CVPR), Salt Lake City, UT, USA, 18–22 June 2018; pp. 6848–6856.
40. Ma, N.; Zhang, X.; Zheng, H. T.; Sun, J. ShuffleNet v2: Practical guidelines for efficient CNN architecture design. In *Proceedings of the European Conference on Computer Vision (ECCV)*, Munich, Germany, 8–14 September 2018; pp. 116–131.
41. Chen, L. C.; Papandreou, G.; Kokkinos, I.; Murphy, K.; Yuille, A. L. Semantic image segmentation with deep convolutional nets and fully connected CRFs. *arXiv* **2014**, arXiv:1412.7062.
42. Chen, L.C.; Papandreou, G.; Kokkinos, I.; Murphy, K.; Yuille, A.L. Deeplab: Semantic image segmentation with deep convolutional nets, atrous convolution, and fully connected CRFs. *IEEE Trans. Pattern Anal. Mach. Intell.* **2017**, *40*, 834–848. [CrossRef]
43. Chen, L.C.; Papandreou, G.; Schroff, F.; Adam, H. Rethinking atrous convolution for semantic image segmentation. *arXiv* **2017**, arXiv:1706.05587.
44. Gao, J.; Du, Q.; Gao, L.; Sun, X.; Zhang, B. Ant colony optimization-based supervised and unsupervised band selections for hyperspectral urban data classification. *J. Appl. Remote Sens.* **2014**, *8*, 085094. [CrossRef]
45. Xie, S.; Girshick, R.; Dollár, P.; Tu, Z.; He, K. Aggregated residual transformations for deep neural networks. In *Proceedings of the IEEE Conference on Computer Vision and Pattern Recognition (CVPR)*, Honolulu, HI, USA, 21–26 July 2017; pp. 1492–1500.
46. Ioffe, S.; Szegedy, C. Batch Normalization: Accelerating deep network training by reducing internal covariate shift. *arXiv* **2015**, arXiv:1502.03167.
47. He, K.; Zhang, X.; Ren, S.; Sun, J. Deep residual learning for image recognition. In Proceedings of the IEEE Conference on Computer Vision and Pattern Recognition (CVPR), Las Vegas, NV, USA, 26 June–1 July 2016; pp. 770–778.
48. Abadi, M.; Agarwal, A.; Barham, P.; Brevdo, E.; Chen, Z.; Citro, C.; Corrado, G.S.; Davis, A.; Dean, J.; Devin, M.; et al. Tensorflow: Large-scale machine learning on heterogeneous distributed systems. *arXiv* **2016**, arXiv:1603.04467.
49. Kingma, D.P.; Ba, J. Adam: A method for stochastic optimization. *arXiv* **2014**, arXiv:1412.6980.

remote sensing

MDPI

Article

Hyperspectral Imagery Classification Based on Multiscale Superpixel-Level Constraint Representation

Haoyang Yu [1], Xiao Zhang [1], Meiping Song [1,*], Jiaochan Hu [2], Qiandong Guo [3] and Lianru Gao [4]

1 Center of Hyperspectral Imaging in Remote Sensing, Information Science and Technology College, Dalian Maritime University, Dalian 116026, China; yuhy@dlmu.edu.cn (H.Y.); xiaozhang@dlmu.edu.cn (X.Z.)

2 College of Environmental Sciences and Engineering, Dalian Maritime University, Dalian 116026, China; hujc@dlmu.edu.cn

3 School of Geosciences, University of South Florida, Tampa, FL 33620, USA; guo1@mail.usf.edu

4 The Key Laboratory of Digital Earth Science, Aerospace Information Research Institute, Chinese Academy of Sciences, Beijing 100094, China; gaolr@aircas.ac.cn

* Correspondence: smping@dlmu.edu.cn

Received: 2 September 2020; Accepted: 10 October 2020; Published: 13 October 2020

Abstract: Sparse representation (SR)-based models have been widely applied for hyperspectral image classification. In our previously established constraint representation (CR) model, we exploited the underlying significance of the sparse coefficient and proposed the participation degree (PD) to represent the contribution of the training sample in representing the testing pixel. However, the spatial variants of the original residual error-driven frameworks often suffer the obstacles to optimization due to the strong constraints. In this paper, based on the object-based image classification (OBIC) framework, we firstly propose a spectral–spatial classification method, called superpixel-level constraint representation (SPCR). Firstly, it uses the PD in respect to the sparse coefficient from CR model. Then, transforming the individual PD to a united activity degree (UAD)-driven mechanism via a spatial constraint generated by the superpixel segmentation algorithm. The final classification is determined based on the UAD-driven mechanism. Considering that the SPCR is susceptible to the segmentation scale, an improved multiscale superpixel-level constraint representation (MSPCR) is further proposed through the decision fusion process of SPCR at different scales. The SPCR method is firstly performed at each scale, and the final category of the testing pixel is determined by the maximum number of the predicated labels among the classification results at each scale. Experimental results on four real hyperspectral datasets including a GF-5 satellite data verified the efficiency and practicability of the two proposed methods.

Keywords: hyperspectral remote sensing; image classification; constraint representation; superpixel segmentation; multiscale decision fusion

1. Introduction

Hyperspectral remote sensing is a leading technology developed from remote sensing (RS) in the field of Earth observation, which accesses multidimensional information by combining imaging technology and spectral technology [1,2]. Hyperspectral image (HSI) can be viewed as a data cube with a diagnostic continuous spectrum, providing abundant spectral–spatial information, and different substances usually exhibit diverse spectral curves [3,4]. Because of the ability of characterization and discrimination of ground objects, HSI has become an indispensable technology in a wide range of applications such as civil construction and military fields [5,6]. As one of the popular applications in remote sensing, HSI classification (HSIC) is to use a mapping function to assign each pixel with a class label via its spectral characteristic and spatial information [7–9]. At present, a large number of HSIC methods have been proposed successively, mainly including the following two aspects: one is the classification based on the spectral information, which mainly focuses on the study of spectral features and spectral classifiers, such as support vector machines (SVM) and the maximum likelihood classifier (MLC). The other is realized by extracting spatial features to assist the discrimination, for example, SVM-based Markov Random Field (SVM-MRF) and some segmentation-based classification frameworks [10–13]. However, due to the high dimensionality of HSI, the high correlation and redundancy have been discovered in both the spectral and spatial domains, it can be inferred that HSI is mainly low-rank and can be represented sparsely, though the original HSI is not sparse [14,15].

In this context, sparse representation (SR)-based methods have been widely applied for HSIC and accompanied a state-of-the-art performance [16]. The classic SR-based classifier (SRC) is to use as few samples as possible to better represent the testing pixel [17]. Concretely, SRC firstly constructs a dictionary by labeling samples in different classes, and then represents the testing pixel by a mean of a linear combination of the dictionary and a weight coefficient under a sparse constraint. After obtaining the approximation of the testing pixel, the classification can be realized by analyzing which class yields the least reconstruction error [18]. However, this residual error-driven mechanism ignores the underlying significance and property of the sparse coefficient to a certain extent. The sparse coefficient plays a decisive role in the constraint representation (CR) model, and the category of the testing pixel is determined by the maximum participant degree (PD) in CR, of which PD is the contribution of labeled samples from different classes in representing the testing pixel. The CR model makes full use of a sparse principle to deal with the sparse coefficient, and achieves an equivalent and simplified effect to the classic SRC. As a powerful pattern recognition, both the SRC model and the CR model are the effective representational-based model, and generate a rather accurate result compared with SVM and some other spectral classifiers [19].

However, due to the sparse coefficients are susceptible to suffer spectral variability, some joint representation (JR)-based frameworks have successively appeared with consideration of the local spatial consistency, such as the joint sparse representational-based classifier (JSRC) and the joint collaborate representative-based classifier (JCRC) [20,21]. Similarly, based on the concept of PD and the PD-driven decision mechanism, adjacent CR (ACR) utilizes the PD of adjacent pixels as class-dependent constraints to classify the testing pixel. The adjacent pixels are defined in a fixed window in ACR, lacking consideration of the correlation of ground object, although there is no strong constraint in comparison with JSRC model. Therefore, in order to better characterize the image for classification, it is reasonable to utilize various features from spectral and spatial domains in the image [22,23].

Object-based image classification (OBIC) is a widely adopted classification framework with spatial discriminant characteristics. OBIC usually performs classification after segmentation [24]. Segmentation technology divides an image into several non-overlapping homogeneous regions according to the agreed similarity criteria. Some segmentation algorithms have shown an effective result in HSI, such as partitioned clustering and watershed segmentation [25–28]. In particular, the combination of the vector quantization clustering methods and the representation based has shown a well classification performance in some related literatures [29]. Therefore, the OBIC is a well-established framework, which can be widely applied for the HSIC tasks.

In this paper, a superpixel-level constraint representation (SPCR) model is proposed, combining a spatial constraint, simple linear iterative clustering (SLIC) superpixel segmentation, to the CR model [30]. Differing from the ACR model, the proposed SPCR method extracts the spectral–spatial information of pixels inside the superpixel block, preserves most of the edge information in the image, and estimates the real distribution of ground objects [31]. In general, the SPCR model utilizes the spectral feature of adjacent pixels, and transforms the individual PD to united activity degree (UAD) through a relaxed and adaptive constraint. As shown on the right side of Figure 1, the decision mechanism of the SPCR model is to classify the testing pixel into the category with the maximum UAD. However, like most OBIC-based methods, the constrained representation classification with a single fixed scale needs to find the optimal scale. To address this issue, it is necessary to propose a multiscale OBIC framework to comprehensively utilize image information [32]. As illustrated in Figure 1, we proposed an improved version based on the above SPCR model, called the multiscale superpixel-level constraint representation (MSPCR) method.

The MSPCR merges the classification maps generated by SPCR at different superpixel segmentation scales, which is implemented in three steps: (1) a segmentation step, in which the processed hyperspectral image is segmented into superpixel images with different scales by the SLIC algorithm; (2) a classification step, in which the PD of pixels inside the superpixel is utilized to shape the class-dependent constraint of the testing pixel; and (3) a decision fusion step, in which the final classification map of MSPCR is obtained through the decision fusion processed, based on the classification result of SPCR at each scale.

As mentioned above, the CR model classifies the testing pixel based on the PD-driven decision mechanism, and obtains a reliable performance with relatively low computational time. Considering the influence of the spectral variability, the ACR model adopts the PD of adjacent pixels to obtain the category of the testing pixel. However, the ACR only regards the pixels within a fixed window as adjacent pixels, lacking consideration to the correlation of ground objects. To address this issue, the SPCR model is firstly established by joining the CR model with the SLIC superpixel segmentation algorithm. Then, the MSPCR approach is successively proposed to alleviate the impact of the segmentation scale on the classification result of the SPCR method, and obtains high accuracies. Experimental results on four real hyperspectral datasets including a GF-5 satellite data are used to evaluate the classification performance of the proposed SPCR and MSPCR methods.

The rest of this paper is organized as follows. Section 2 reviews the related models, including classic representation-based classification methods and superpixel segmentation algorithm, i.e., SLIC that we used in this paper. Section 3 presents our proposed methods, firstly introduces the CR method and the ACR classifier, then presents the SPCR model and the MSPCR method proposed in this paper. Section 4 evaluates the classification performance of our proposed methods and other related methods via the experimental results on three real hyperspectral datasets. Section 5 takes a practical application and analysis to our proposed methods via the experiment on a GF-5 satellite data. Section 6 concludes this paper with some remarks.

2. Related Methods

In this section, we introduce several related methods of our framework. The classic sparse representation (SR)-based model and the joint representation (JR)-based framework are firstly reviewed in Section 2.1. Then the simple linear iterative clustering (SLIC) is presented in Section 2.2.

2.1. Representation-Based Classification Methods

Defining a testing pixel $\mathbf{x}_{i,j} \in \mathbf{X}$ in the location (i, j) of HSI $\mathbf{X}$ which contains B spectral bands and $N = r \times c$ pixels (r and c index the row and column of scene). The dictionary can be denoted as $\mathbf{D} = (\mathbf{D}_1, \ldots, \mathbf{D}_K) \in \mathbf{X}$, in which each column of $\mathbf{D}_k$ is the samples selected from class $k \in [1, K]$ (K is the number of classes).

2.1.1. Sparse Representation-Based Model

Since pixels in HSI can be represented sparsely, representation-based methods have been widely applied to process HSI due to their no assumption of data density distribution [33]. The SRC is a classic SR-based model, implementing classification based on several steps as follows. Firstly, it constructs a dictionary by training the available labeled samples, then represents the testing pixel by a sparse linear combination of the dictionary. Moreover, in order to use as few labeled samples as possible to represent the testing pixel, the weighted coefficients used in representation are sparsely constrained. Finally, the classification is conducted by a residual error-driven decision mechanism, which classifies the testing pixel as the class with minimum class-dependent residual error using the following formula:

$$\begin{cases} \hat{\boldsymbol{\alpha}}_{i,j} = \underset{\boldsymbol{\alpha}_{i,j}}{\operatorname{argmin}}\left\{\|\mathbf{x}_{i,j} - \mathbf{D}\boldsymbol{\alpha}_{i,j}\|_2^2 + \lambda\|\boldsymbol{\alpha}_{i,j}\|_1\right\} \\ class(\mathbf{x}_{i,j}) = \underset{k}{\operatorname{argmin}}\left\{\|\mathbf{x}_{i,j} - \mathbf{D}\delta_k(\hat{\boldsymbol{\alpha}}_{i,j})\|_2^2\right\} \end{cases}, \tag{1}$$

where $\|\boldsymbol{\alpha}_{i,j}\|_1 = \sum_{m=1}^{n}|\alpha_m|$ denotes the l_1-norm and $\|\cdot\|_2$ is the l_2-norm, due to the optimization of l_0-norm is a combinatorial NP-hard problem, the sparse constraint of weight coefficients $\boldsymbol{\alpha}_{i,j}$ adopts l_1-norm to substitute l_0-norm, where l_1-norm is the closet convex function to the l_0-norm. Moreover, λ is a scalar regularization parameter. As an indicator function, $\delta_k(\hat{\boldsymbol{\alpha}}_{i,j})$ can assign zero to the element that does not belong to the class k. The weight vector, $\hat{\boldsymbol{\alpha}}_{i,j}$, can be optimized by the basis pursuit (BP) or basis pursuit denoising (BPDN) algorithm.

2.1.2. Joint Representation-Based Framework

HSIC initially focused on the spectral information because of its data characteristic, while the spatial information can be further exploited to reduce classification errors, according to the similar spectral characteristic among neighborhood pixels. As the second generation of SRC, the joint SRC (JSRC) is introduced under the JR-based framework, which has a solid classification performance after integrating spectral information with the local spatial coherence.

Based on the local spatial consistency, the fundamental assumption of JSRC is that the sparse vectors related with the adjacent pixels could share a common sparsity support [34]. In the JSRC, both the testing pixel and its neighboring pixels are stacked into the joint signal matrix, and sparsely represented using the dictionary and a row-sparse coefficient matrix [35]. The final classification result of JSRC is obtained by calculating the minimum total residual error as follows:

$$\begin{cases} \hat{\mathbf{A}}_{i,j} = \underset{\mathbf{A}_{i,j}}{\operatorname{argmin}}\left\{\|\mathbf{X}_{i,j} - \mathbf{D}\mathbf{A}_{i,j}\|_F^2 + \lambda\|\mathbf{A}_{i,j}\|_{1,2}\right\} \\ class(\mathbf{x}_{i,j}) = \underset{k}{\operatorname{argmin}}\left\{\|\mathbf{X}_{i,j} - \mathbf{D}\delta_k(\hat{\mathbf{A}}_{i,j})\|_F^2\right\} \end{cases}, \tag{2}$$

where $\mathbf{X}_{i,j} = (\mathbf{x}_{i-w,j-w}, \cdot, \mathbf{x}_{i,j}, \cdot, \mathbf{x}_{i+w,j+w})$ is a $\widehat{w} \times \widehat{w}$ pixel-sized square neighborhood centered on $\mathbf{x}_{i,j}$, and $\mathbf{A}_{i,j}$ is the corresponding coefficient matrix. $\|\cdot\|_F$ is the Frobenius norm and $\|\mathbf{A}_{i,j}\|_{1,2} = \sum_{s=1}^{n}\|\mathbf{a}^s\|_2$ is the $l_{1,2}$-norm, in which $\mathbf{a}^s$ is the s-th row of $\mathbf{A}_{i,j}$.

2.2. *Simple Linear Iterative Clustering*

The OBIC is a widely used spectral–spatial classification framework, and it utilizes the spatial information after the procedure of segmentation [36]. As one of the widely used segmentation methods, the SLIC algorithm identifies superpixels by the over-segmentation approach. The idea of SLIC is to locally apply the K-means algorithm to obtain an effectively cluster segmentation result. Specifically, it measures the distance from each cluster center to pixels within a $2S \times 2S$ block, where $S = \sqrt{N/P}$. Here, N is the number of pixels, and P is the number of clustering centers which equals to the total number of superpixels [37].

In general, the SLIC algorithm can be implemented in several steps as follows: the first step is to select P initial clustering centers from the original image. Then it classifies each pixel to the nearest clustering center, and constructs various clusters respectively. The iterative clustering process is performed until the position of the cluster center became stable. As stated above, the original K-means algorithm calculates the distance from the whole map, while the searching area of SLIC is in the local area of each superpixel, thereby the SLIC algorithm alleviates the computation complexity to a great extent. The distance in SLIC is defined as follows:

$$\mathrm{D}_{SLIC} = D_{spectral} + \frac{m}{\rho}D_{spatial}, \tag{3}$$

where $D_{spectral}$ is a spectral distance, which is used to ensure the homogeneity inside the superpixel, and the spectral distance between pixel i and pixel j is described as follows:

$$D_{spectral} = \sqrt{\sum_{d=1}^{D} (x_{i,d} - x_{j,d})^2}, \tag{4}$$

where $x_{i,d}$ is the value of pixel i in band d, and $D_{spatial}$ represents the spatial distance, which is used to control the compact and regularity of the superpixels, the spatial distance between pixel i and pixel j is defined as follows:

$$D_{spatial} = \sqrt{(a_i - a_j)^2 + (b_i - b_j)^2}, \tag{5}$$

where (a_i, b_i) is the location of pixel i, m, and ρ in Equation (3) are the scale parameter of superpixels.

3. Proposed Approach

As introduced in Section 2.1, both the classic SR-based model and the variant JR-based method conduct the classification using the class-dependent minimum residual error between the original observation and the approximate representation value. However, the residual error-based decision mechanism in the SR-based and JR-based frameworks ignore the importance of sparse coefficients. Section 3.1 introduces that the CR method and the ACR classifier can exploit the characteristic of the sparse coefficient. After that, we present the details of SPCR and the MSPCR in Section 3.2. Both methods are generally based on the spatial correlation. Specifically, the SPCR utilizes the spectral consistency feature among adjacent pixels in ACR, and then MSPCR achieves comprehensive utilization of various regional distribution.

3.1. Constraint Representation (CR) and Adjacent CR (ACR)

3.1.1. CR Model

According to the principle of representation-based model, it can be regarded as representing the testing pixel via a sparse linear combination of the labeled samples. For the sake of understanding, a simple case can be assumed as Equation (6). The testing pixel is represented by a single element with nonzero coefficient $(\alpha_p, \alpha_q, \alpha_m, \ldots, \alpha_n)$ from some certain classes $(k, k+1, \ldots, k^* \in [1, K])$ as follows [38]:

$$\mathbf{x}_{i,j} \approx \alpha_p \mathbf{x}_t^k + \alpha_q \mathbf{x}_{t_1}^{k+1} + \alpha_m \mathbf{x}_{t_2}^{k+2} + \ldots + \alpha_n \mathbf{x}_{t_n}^{k^*} \tag{6}$$

Since $\hat{\boldsymbol{\alpha}}_{i,j}$ is sparsely constrained, the labeled samples which contributes to representing the testing pixel are the ones whose coefficients are not zero. In the process of representation, the larger measurement of the coefficient value, the higher contribution in representing the testing pixel, such that the testing pixel more likely belongs to the corresponding category. Therefore, CR directly exploits the sparse coefficient to conduct the classification, which is concise and equivalent to the residual error-driven determination mechanism. Specifically, it defines the participant degree (PD) from the perspective of the sparse coefficient, which estimates the contribution of labeled samples from different classes in representing the testing pixel $\mathbf{x}_{i,j}$. The PD of each class is calculated by the corresponding weight vector with l_d-normed measurement (d = 1 or d = 2) as follows:

$$\mathrm{PD}_k = \|\boldsymbol{\alpha}_{i,j}^k\|_d \tag{7}$$

The PD-driven decision mechanism of CR is to determine the category with the maximum PD, which can be expressed in Equation (8):

$$class(\mathbf{x}_{i,j}) = \max_k \left\{ \mathrm{PD}_{i,j}^k \middle| k \in [1, K] \right\}. \tag{8}$$

3.1.2. ACR Model

Based on the PD-driven mechanism, an improved version, ACR has been proposed to correct spectral variation by imposing spatial constraints during the classification. According to the spectral similarity characteristic among the adjacent pixels, the adjacent pixels more likely belong to the same class [39]. In this context, the ACR brings better classification performance than that of the CR model through innovating the PD-driven mechanism with the spatial consistency of the adjacent pixels. The main principle of ACR is to use the PD of adjacent pixels as a constraint to determine the category of the testing pixel. Specifically, the ACR firstly defines adjacent pixels within a $\widehat{w} \times \widehat{w}$ pixel-sized window centered on the testing pixel, then constructs a k-dimensional PD image, and each dimensionality of the PD image shows the PD values of pixels in one class. The class-dependent activity degree (CAD) of each element is obtained after successively normalizing the PD image at each dimensionality, which could be expressed as follows:

$$\mathrm{CAD}_{i,j}^{k} = \mathrm{PD}_{i,j}^{k} / \sum_{k*=1}^{K} \mathrm{PD}_{i,j}^{k*}, \tag{9}$$

where $k \in [1, K]$ denotes the class index, and (i, j) are the location of the testing pixel. With consideration of the spatial constraint of the adjacent pixels, the relative activity degree (RAD) is generated by combining the CAD of the testing pixel with the inactivity degree of its adjacent pixels through a scale compensation parameter τ, where the index of the adjacent pixels is $v \in [1, \widehat{w}^2]$. The ACR uses the RAD as the final contribution degree in representing the testing pixel $\mathbf{x}_{i,j}$, and the class of $\mathbf{x}_{i,j}$ can be determined by the maximum RAD as follows:

$$\begin{cases} \mathrm{RAD}_{i,j}^{k} = \mathrm{CAD}_{i,j}^{k} - \tau \sum_{v=1}^{\widehat{w}^2} (1 - \mathrm{CAD}_{v}^{k}) \\ class(\mathbf{x}_{i,j}) = \max\limits_{k} \left\{ \mathrm{RAD}_{i,j}^{k} \middle| k \in [1, K] \right\} \end{cases} \tag{10}$$

3.2. Superpixel-Level CR (SPCR) and Multiscale SPCR (MSPCR)

3.2.1. SPCR Model

As mentioned above, the ACR model defines the adjacent pixels as pixels within a fixed pixel-sized window centered on the testing pixel. However, it does not consider the real distribution of ground objects. The superpixel block obtained by the superpixel segmentation algorithm is made up of some neighborhood pixels with similar spatial characteristics. Through combing the superpixel segmentation algorithm, we establish the SPCR model to further utilize the spectral consistency feature from the subset of adjacent pixels. In this way, the SPCR model conducts class-dependent constrained represent according to the PD of pixels inside the superpixel block centered on the testing pixel, which preserves most edge information of image in comparison to the sample selection in fixed window in ACR, and has a more objective consideration to the spatial distribution of the testing pixel. As illustrated in Figure 1, the schematic diagram of SPCR model is equal to MSPCR at a single segment scale, which can be implemented in several steps as follows.

Firstly, we obtain superpixel blocks by the SLIC algorithm. Since the SLIC can only process an image in the CIELAB color space, it is necessary to convert an HSI to a three bands image before processed by the SLIC algorithm. Therefore, the principal component analysis (PCA) method is adopted to reduce the spectral dimensionality in the SPCR method, which selects the first three components as the input of SLIC to generate a stable superpixel segmentation result [40]. Then, the category of the testing pixel can be measured by calculating the PD values of pixels inside the superpixel where the testing pixel is located. Specifically, using the PD values of pixels at the corresponding position of the superpixel, we built a SPD image surrounding $\mathbf{x}_{i,j}$ with K dimension, and each dimension of SPD image shows the PD values of pixels in one class. Similar to ACR, the normalized value of each pixel in the kth SPD image is defined as the class-dependent activity degree (CAD) with regard to the class k.

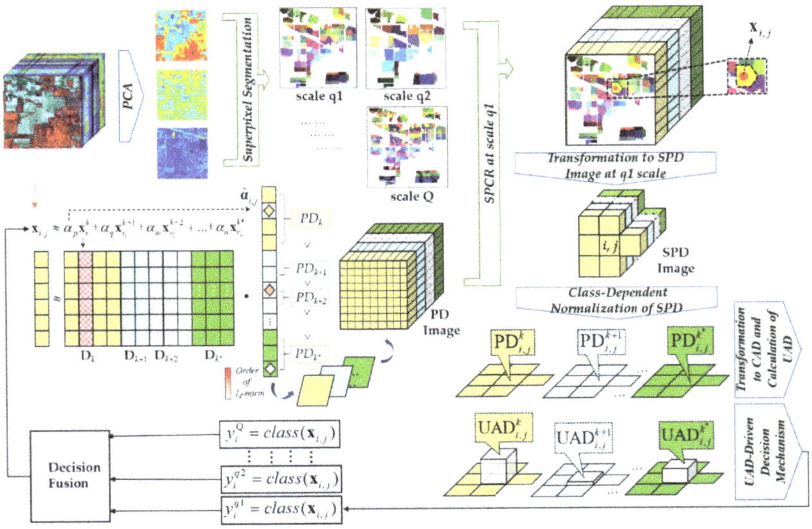

Figure 1. The workflow of multiscale superpixel-level constraint representation (MSPCR).

In order to further utilize the correlation of ground objects, SPCR combines the CAD of $\mathbf{x}_{i,j}$ with CAD of other pixels insides superpixel through the scale compensation parameter, such that other pixels can give a properly constraint in classifying the testing pixel $\mathbf{x}_{i,j}$. Compared to the constraint with the local spatial information in RAD shown in formula (10), the united activity degree (UAD) utilizes the correlation of ground object via a similar combination, represented as follows:

$$\mathrm{UAD}_{i,j}^{k} = \mathrm{CAD}_{i,j}^{k} + \gamma \sum\nolimits_{e=1}^{l} \mathrm{CAD}_{e}^{k}, \tag{11}$$

where $e \in [1, l]$ indicates the element index in superpixel block, γ represents a scale compensation parameter. Moreover, the SPCR model classifies $\mathbf{x}_{i,j}$ by analyzing which class leads to the maximum $\mathrm{UAD}_{i,j}$ as follows:

$$class(\mathbf{x}_{i,j}) = \max_{k}\left\{UAD_{i,j}^{k} \middle| k \in [1, K]\right\} \tag{12}$$

3.2.2. MSPCR Model

As shown in the aforementioned algorithm, the proposed SPCR method based on the OBIC framework generates solid performance through exploiting the spatial contextual information. However, as the classification results of SPCR with different segmentation scales are not the same, the superpixel segmentation-based HSI classification may not generate a comprehensive and stable result under a fixed segmentation scale. Thus, in particular, the performance of SPCR is highly affected by the scale level [41]. In order to solve these problems, it is reasonable to propose multiscale OBIC framework to comprehensively utilize image information. In this paper, MSPCR is firstly proposed by means of decision fusion with the classification result maps obtained by SPCR method at different segmentation scales. Compared with SPCR, the improved MSPCR not only uses multiple scales to balance the different size and distribution of ground objects, but also solves the problem of selecting the optimal segmentation scale.

Specifically, Figure 1 and Algorithm 1 illustrate the general schematic diagram and pseudo procedures of the MSPCR method, respectively. Firstly, similar to the workflow of the SPCR method, we simultaneously obtain the classification results of the testing pixel at different superpixel segmentation scales. In this process, the superpixel block is generated by inputting the result of PCA into the SLIC algorithm, then classify the testing pixel by a relaxed and adaptive constraint inside the superpixel. After performing the SPCR method at each scale, a decision fusion process is applied to obtain the classification result of MSPCR, in which the category of the testing pixel $\mathbf{y}_i$ is determined by the maximum number of labels of the testing pixel $\mathbf{x}_{i,j}$ among the classification results at each scale, and the decision fusion process is expressed as follows:

$$class(\mathbf{y}_i) = \arg \underset{q=q_1,..Q}{\mathrm{mod}}\, class(\mathbf{y}_i^q), \tag{13}$$

where $\mathbf{y}_i$ is denoted as the final class label of $\mathbf{x}_{i,j}$, $\mathbf{y}_i^q$ represents the classification result of $\mathbf{x}_{i,j}$ when the segmentation scale parameter is described as q, and mod is a modular function which defines $\mathbf{y}_i$ with the most frequency category in $[\mathbf{y}_i^{q_1}, \ldots, \mathbf{y}_i^{Q}]$.

Algorithm 1. The proposed MSPCR method

Input: A hyperspectral image (HSI) image $\mathbf{X}$, dictionary $\mathbf{D}$, the testing pixel $\mathbf{x}_{i,j}$, regularization parameter λ, scale compensation parameter γ.
Step 1: Reshape $\mathbf{X}$ into a color image by compositing the first three principal component analysis (PCA) bands.
Step 2: Obtain multiscale superpixel segmentation images $S^Q = \{S^q\}_{q=1}^{Q}$ of $\mathbf{X}$ according to SLIC in Equations (3) to (5).
Step 3: Obtain the participation degree (PD) image of $\mathbf{X}$ according to Equation (7).
Step 4: Extract superpixel centered on the testing pixel $\mathbf{x}_{i,j}$ from the PD image of $\mathbf{X}$ to get multiscale SPD image.
Step 5: Class-dependent normalization at each scale according to Equation (9).
Step 6: Calculate the united activity degree (UAD) according to Equation (11).
Step 7: Assign the class of $\mathbf{x}_{i,j}$ at each scale according to Equation (12).
Step 8: Determine the final class label by the decision fusion according to Equation (13).
Output: The class labels $\mathbf{y}$.

4. Experimental Results and Analysis

In this section, we investigated the effectiveness of the proposed SPCR and MSPCR models with three hyperspectral datasets. The detailed description of the applied datasets is given in Section 4.1. The parameter tuning related to the proposed models and other compared methods is presented in Section 4.2. We evaluate the performance of two proposed methods in comparison with the methods in the spectral domain and the spectral–spatial domain. The classic SR-based method, including SRC as well as its simplified model CR, and the classic SVM are firstly selected in the comparative experiments in the spectral domain. Then, the classic JR-based model JSRC, the typical model with post-processing of spatial information SVM-MRF, and the previously proposed ACR are further tested in the spectral–spatial domain. We randomly selected training samples 20 times in each experiment and calculated the overall accuracy (OA) and class-dependent accuracy (CA). We analyzed the experimental results of the two proposed methods and other related methods in Sections 4.3–4.5.

4.1. Experimental Data Description

4.1.1. Airborne Visible/Infrared Imaging Spectrometer (AVIRIS) Indian Pines Scene

The first data are of the Indian Pines scene acquired by the Airborne Visible/Infrared Imaging Spectrometer (AVIRIS) sensors in the Northwestern Indiana, with a spatial resolution of 20 m. The scene covers 220 spectral bands ranging from 0.4 to 2.5 μm, and the size of the image is 145 × 145. In order to satisfy the sparse thought, eight ground-truth classes with a total of 8624 labeled samples are extracted from the original sixteen categories reference data. Figure 2a,b shows the false-color composite image and the reference map of this scene, respectively.

4.1.2. Reflective Optics Spectrographic Imaging System (ROSIS) University of Pavia Scene

The second data are of the University of Pavia scene collected by the Reflective Optics Spectrographic Imaging System (ROSIS) over a downtown area near the University of Pavia in Italy, with a spatial resolution of 1.3 m. After removing 12 bands with high noise and water absorption, the scene has 103 spectral bands ranging from 0.43 to 0.86 μm, with 610 × 340 pixels. Nine ground-truth classes with a total of 42,776 labeled samples are contained in the reference data. Figure 3a,b shows the false-color composite image and the reference map of this scene, respectively.

4.1.3. Hyperspectral Digital Image Collection Experiment (HYDICE) Washington, DC, National Mall Scene

The third data are of the Washington, DC, National Mall scene captured by the Hyperspectral Digital Image Collection Experiment (HYDICE) sensor over the Washington, DC, in USA, with a spatial resolution of 3 m. The original scene contains 210 spectral bands ranging from 0.4 to 2.5 μm, with 280 × 307 pixels. After removing the atmospheric absorption bands from 0.9 to 1.4 μm, 191 bands were remaining. Six ground-truth classes with a total of 10190 labeled samples were included in the reference data. Figure 4a,b shows the false-color composite image and the reference map of this scene, respectively.

(a) (b)

Figure 2. The Airborne Visible/Infrared Imaging Spectrometer (AVIRIS) Indian Pines scene: (**a**) false-color composite image and (**b**) reference map.

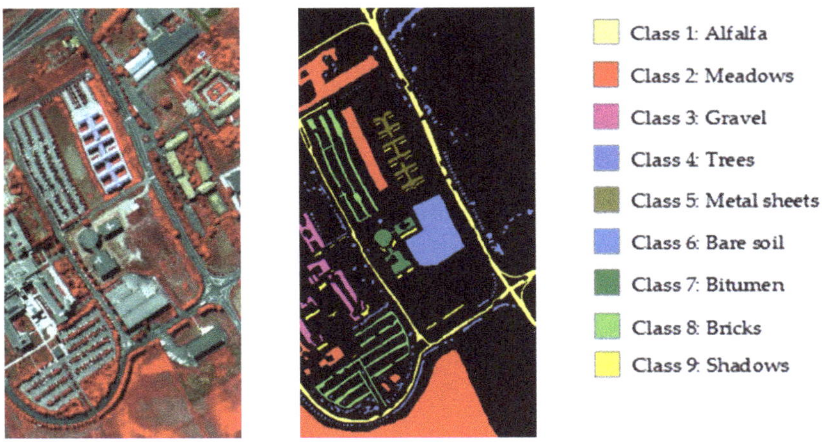

(a) (b)

Figure 3. The Reflective Optics Spectrographic Imaging System (ROSIS) University of Pavia scene: (**a**) false-color composite image and (**b**) reference map.

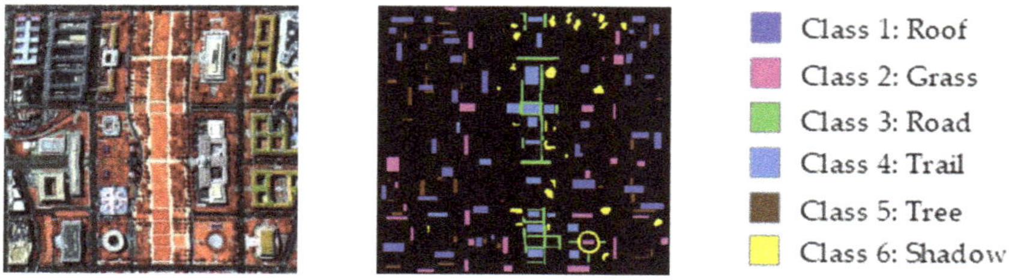

(**a**) (**b**)

Figure 4. The Hyperspectral Digital Image Collection Experiment (HYDICE) Washington, DC, National Mall scene: (**a**) false-color composite image; (**b**) reference map.

4.2. Parameter Tuning

In the experiment of this paper, the regularization parameter λ for all SR-based models was selected from 10^{-3} to 10^{-1}. For the scale compensation parameter τ and γ in ACR and SPCR-based methods, we set them in a properly range according to the value of $\widehat{w}$ and the number of superpixels P. Due to the different value of $\widehat{w}$, the distributions of the ground objects in the $\widehat{w} \times \widehat{w}$ pixel-sized window centered on the testing pixel are different. This fact produces a critical constraint based on the assumption that the adjacent pixels inside the window belong to the same class. Referring to the definition of $\widehat{w}$ in [22], each scene usually has a proper $\widehat{w}$ with a consideration of the spatial consistency, and the exceeded size could influence the result. Therefore, in order to obtain a high classification accuracy, we optimized the size of the window $\widehat{w}$ in each experimental scene.

In addition, the number of superpixels P in SPCR and MSPCR classifier is decided by the segmentation scale S and the number of the pixels N via $P = \sqrt{N/S}$. The corresponding experimental analysis about P and the classification accuracy is illustrated in Figures 5 and 6. We can infer the relationship between the segmentation scale S and the classification accuracy, which is equal to the relationship of P and the classification results. Firstly, Figure 5 shows the impact of the number of superpixels on the classification accuracy (50 samples per class). We mainly select five and four classes to display from the AVIRIS Indian Pines dataset and HYDICE Washington, DC, National Mall dataset, respectively. As illustrated in Figure 5a, the result indicates that the optimal segmentation scale is various for different classes. For example, the optimal segmentation scale of the class 2 is distinct from the other three classes in Figure 5b. In addition, the relationship of the number of superpixels, overall accuracy and the number of the labeled samples is shown in Figure 6. Generally, the overall accuracy increased with the number of labeled samples at each segmentation scale. It is notable that under different number of labeled samples, the segmentation scale is various in order to achieve the highest classification accuracy. Like the most OBIC frameworks, the proposed SPCR method also needs to set the optimal segmentation scale, while the improved MSPCR method can overcome this drawback through taking fusion the spatial–spectral characteristics of HSI at different segmentation scales.

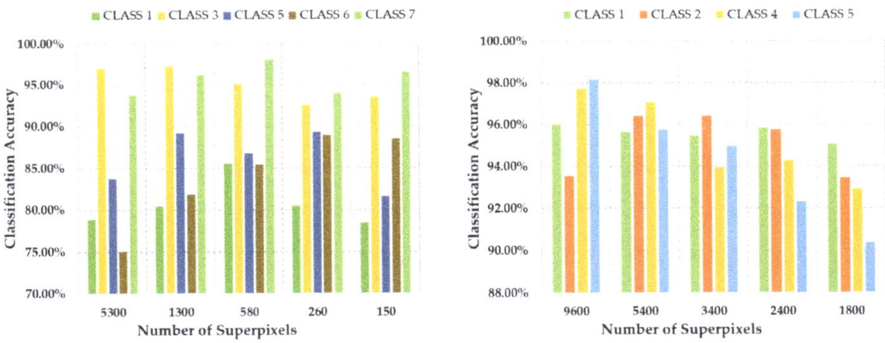

(a) (b)

Figure 5. The sensitivity analysis of the number of superpixels on classification accuracy (50 samples per class). (**a**) the AVIRIS Indian Pines dataset. (**b**) the HYDICE Washington, DC, National Mall dataset.

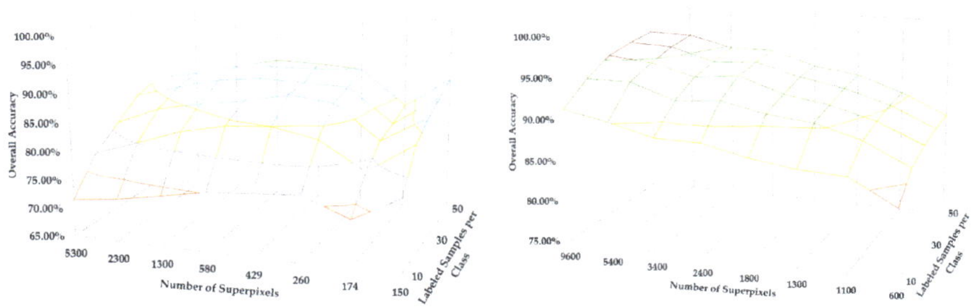

(a) (b)

Figure 6. The sensitivity analysis of the number of superpixels versus training size. (**a**) the AVIRIS Indian Pines dataset. (**b**) the HYDICE Washington, DC, National Mall dataset.

4.3. Experiments with the AVIRIS Indian Pines Scene

In the first experiment with the AVIRIS Indian Pines hyperspectral scene, we randomly selected 90 labeled samples per class with a total of 720 samples to construct a dictionary and the training model. The selected training samples constitutes the approximately 8.35% of the labeled samples in the reference map, and the other remained samples are used in validation. As illustrated in Table 1, the OAs and the CAs of different methods are calculated, and the corresponding classification maps are presented in Figure 7. We analyzed the classification results as follows:

1 As a widely applied supervised classification framework, the SVM classifier has a feasible performance in the classification of HSI. However, there are some isolated pixels appeared in the result due to the noise and spectral variability, as shown in Figure 7. Compared with the SVM, the classic SRC method gains a better classification result, which proves that the SR-based classifier is suitable for the hyperspectral image classification tasks. Compared with the SRC, the CR model obtains an approximate equivalent classification result with a lower computational cost than that of SRC. The result not only underlines the CR model simplified the SRC model via an improved procedure without the calculation of residual error, but also verifies the effectiveness of the PD-driven decision mechanism in the process of HSIC.

2 In the spectral–spatial domain, as shown in Figure 7, SVM-MRF model outperforms the SVM classifier, which demonstrates the exploration of the spatial information can bring a further

improvement on the spectral classifiers. Similarly, since the JSRC conducts the classification by sharing a common sparsity support among all neighborhood pixels, the improvement of overall accuracy also appeared in JSRC compared to SRC. Compared with the CR model, the ACR classifier obtains a significant improvement. It solves the spectral variability problem in CR by setting a spatial constraint, and proves that the innovation of decision mechanism from PD-driven to RAD-driven is effective for the HSIC tasks. As mentioned above, the improvements of SVM-MRF, JSRC, and ACR models relative to their original counterparts SVM, SRC, and CR confirm the effectiveness of introducing spatial information into the spectral domain classifiers.

3 From Figure 7, the JSRC has a better classification performance than the SVM-MRF in the AVIRIS Indian Pines scene. As illustrated in Table 1, the ACR classifier achieves a better classification result in comparison to JSRC and SVM-MRF, of which the overall accuracy is 2.38% higher than that of JSRC and 6.11% higher than that of SVM-MRF. On one hand, the RAD-driven mechanism in ACR is more effective than the hybrid norm constraint in JSRC. On the other hand, the post-processing of spatial information in SVM-MRF takes more emphasis on adjusting the initial classification result generated from spectral features, lacking an effective strategy integrating spatial information with spectral information.

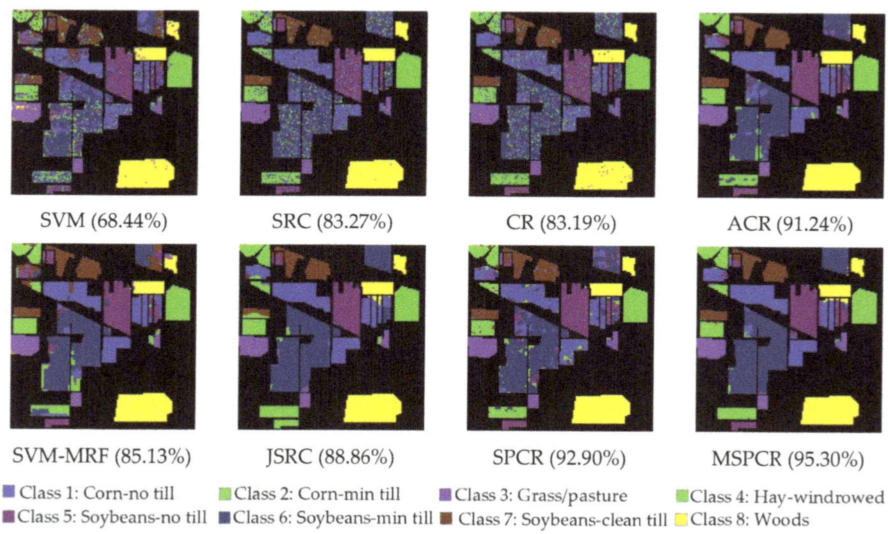

Figure 7. Classification maps obtained by the different tested methods with 90 samples per class for the AVIRIS Indian Pines dataset (overall accuracy (OA) is in parentheses). SVM = support vector machine; MRF = Markov Random Field; SRC = sparse-representation-based classifier; CR = constraint representation; ACR = adjacent constraint representation; JSRC = joint sparse representational-based classifier; SPCR = superpixel-level constraint representation.

4 Compared with the ACR, the proposed SPCR has a slightly higher OA. Table 1 demonstrates the effectiveness of introducing the superpixel segmentation, which preserves the edge information and fully considers the distribution of ground object. In addition, the practicability and reliability of the sparse coefficient, which plays an important role in the PD-driven decision mechanism and the UAD-driven decision mechanism. Thus, the combination of superpixel segmentation and sparse coefficients is effective, the overall accuracy of SPCR reaches to 92.90%, which is 1.66%, 4.04%, and 7.77% higher than ACR, JSRC, and SVM-MRF, respectively.

5 Compared with the SPCR, the proposed MSPCR model brings an improvement. Firstly, it verifies that the MSPCR performs better than the SPCR via alleviating the impact of superpixel

segmentation scale on the classification results. Then, it also indicates that the decision fusion takes a comprehensive consideration to the different spatial features and distributions of various categories of objects, which elevates the final classification accuracy.

Table 1. Overall and classification accuracies (in percent) obtained by the different tested methods for the AVIRIS Indian Pines scene. In all cases, 720 labeled samples in total (90 samples per class) were used for training.

Class	Samples	SVM	SRC	CR	SVM-MRF	JSRC	ACR	SPCR	MSPCR
1	1460	55.60%	77.29%	77.00%	73.51%	82.29%	83.97%	87.53%	89.74%
2	834	57.82%	83.62%	84.36%	82.77%	91.92%	93.53%	94.24%	97.84%
3	497	88.99%	97.53%	97.38%	95.52%	98.79%	98.98%	96.38%	97.44%
4	489	98.90%	99.84%	99.88%	99.34%	100.00%	100.00%	99.39%	99.82%
5	968	71.45%	81.94%	81.60%	89.00%	92.13%	94.41%	87.93%	94.12%
6	2468	56.22%	70.47%	70.19%	77.95%	78.76%	81.23%	91.75%	93.41%
7	614	68.72%	91.35%	91.68%	95.73%	96.06%	96.81%	93.55%	99.49%
8	1294	94.41%	99.61%	99.62%	98.36%	99.66%	99.85%	99.91%	99.92%
OA		68.44%	83.27%	83.19%	85.13%	88.86%	91.24%	92.90%	95.30%

In general, the proposed MSPCR obtains an overall accuracy of 95.30%, which is 2.40% and 4.06% higher than SPCR and ACR, and also 12.11% higher than CR, respectively. For individual class accuracy, it provides great results, especially for the classes 2, 6, and 7. The classification maps in Figure 7 verify the improvement achieved by the MSPCR.

In the second test with the AVIRIS Indian Pines scene, we randomly selected 10 to 90 samples per class as the training samples to measure the proposed SPCR and MSPCR. Figure 8 shows the overall classification accuracies acquired by different methods with different number of labeled samples. The results can be summarized as follows:

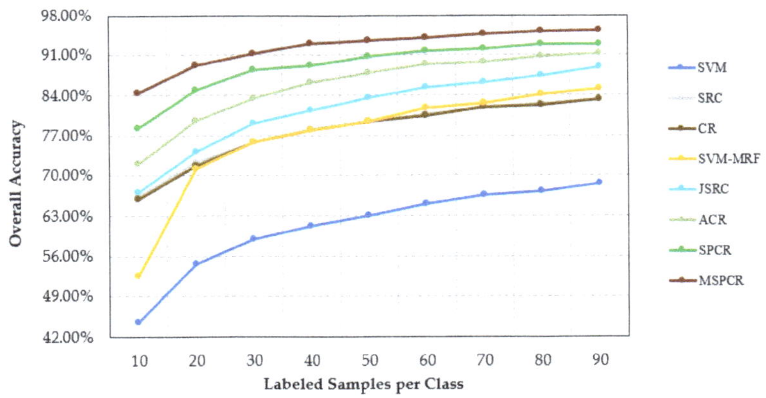

Figure 8. Overall classification accuracy obtained by different tested methods with different numbers of labeled samples for the AVIRIS Indian Pines scene.

1. The classification results demonstrate that the overall accuracy has a positive relationship with the number of the labeled samples, the overall accuracy is increased by the number of labeled samples. Besides, this phenomenon only be satisfied under a certain number of the labeled samples, the growth trend would be stopped when the labeled samples reach a certain number.
2. The integration of the spatial and spectral information benefits precision classification than the pixel-based classification method, which can be verified by the improvement of SVM-MRF, JSRC, ACR, SPCR, and MSPCR relative to their original counterparts, i.e., SVM, SRC, and CR.
3. Compared to the traditional classifiers, the PD-driven classifiers provide a better classification performance. This can be confirmed by the overall accuracies of ACR and SPCR toward JSRC

and SVM-MRF, as well as CR toward SVM. Moreover, the proposed MSPCR achieved the best performance among these classifiers.

4.4. Experiments with the ROSIS University of Pavia Scene

In the first test of the experiment with the ROSIS University of Pavia scene, we select 90 labeled samples per class with a total of 810 samples (which constitutes approximately 1.89% of the available labeled samples in the reference map), and the remaining labeled samples are used for validation. Table 2 and Figure 9 show the OAs and CAs for the classifiers, and the corresponding classification maps. From the experimental results, we have similar results with those obtained under the AVIRIS Indian Pines dataset: First, SRC and CR achieved similar classification results, with comparative result in comparison with the SVM in the spectral domain. In the spatial domain, SVM-MRF, JSRC, and ACR bring significant improvement to the SVM, SRC, and CR model by integrating the spatial information. Moreover, SVM-MRF owns a better classification accuracy than JSRC, different from the performance of these two methods in AVIRIS Indian Pines dataset. In comparison with the ACR, the introduction of the superpixel segmentation algorithm contributes to a higher accuracy in SPCR. Last but not the least, the proposed MSPCR achieves the best classification result with the overall accuracy of 96.90%, which is 3.64% and 4.71% higher than SPCR and ACR, and also 16.7% higher than CR, respectively. Additionally, it brings considerable improvements for individual class accuracy, especially for class 2 and class 4, which can be proved by the classification maps shown in Figure 9.

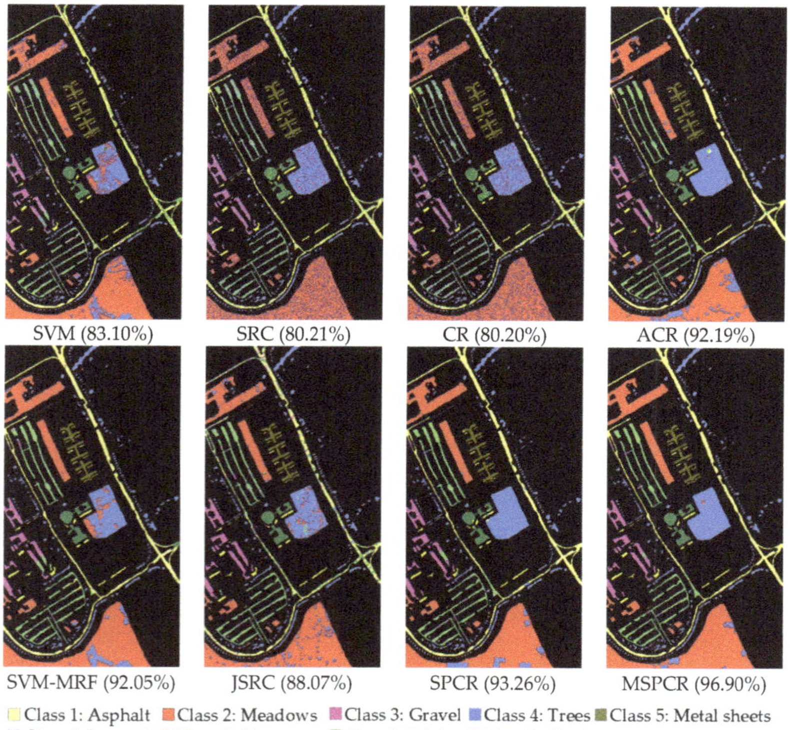

Figure 9. Classification maps obtained by the different tested methods with 90 samples per class for the ROSIS University of Pavia dataset (OAs are in parentheses).

Table 2. Overall and classification accuracies (in percent) obtained by the different tested methods for the ROSIS University of Pavia scene. In all cases, 810 labeled samples in total (90 samples per class) were used for training.

Class	Samples	SVM	SRC	CR	SVM-MRF	JSRC	ACR	SPCR	MSPCR
1	6631	75.04%	76.12%	75.76%	91.06%	65.12%	93.94%	90.62%	94.79%
2	18649	80.69%	78.43%	78.83%	88.76%	92.35%	87.98%	93.37%	96.83%
3	2099	80.50%	78.04%	78.91%	89.26%	95.90%	92.09%	89.48%	93.67%
4	3064	94.31%	94.96%	95.47%	97.06%	92.20%	97.03%	89.15%	98.18%
5	1345	99.21%	99.80%	99.82%	99.55%	100.00%	100.00%	97.62%	99.93%
6	5029	87.32%	80.06%	79.58%	96.08%	84.91%	98.19%	98.71%	98.86%
7	1330	92.82%	89.28%	89.83%	96.37%	99.85%	97.59%	96.42%	99.83%
8	3682	83.07%	70.65%	68.66%	94.18%	93.29%	91.47%	95.43%	96.96%
9	947	99.86%	98.27%	98.34%	99.90%	96.62%	99.68%	83.44%	96.96%
OA		83.10%	80.21%	80.20%	92.05%	88.07%	92.19%	93.26%	96.90%

Our second test of the ROSIS University of Pavia scene measured the proposed SPCR and MSPCR with various sizes of labeled samples (from 10 to 90 samples per class). Figure 10 shows the overall classification accuracies obtained by different testing methods, under different number of training samples. With the number of the labeled sample increased, most of measured methods have an increase trend in accuracy. In comparison to the overall classification accuracy of SVM, the SRC and CR firstly have better performances, then perform worse as the number of the labeled samples increased. Considering the correlation of ground object, the classification performance of ACR and SVM-MRF, achieved significant improvements with the increase of the number of samples, with a higher classification accuracy than the JSRC in most cases. In addition, the combination of the PD-decision mechanism and the superpixel segmentation algorithm brings reliable and stable improvement, which can be confirmed by the overall classification accuracies obtained by SPCR method in all cases. From Figure 10, MSPCR method achieves the best classification result among these compared methods, as a result of applying the decision fusion which alleviates the challenge of adapting the fixed single segmentation scale to the spatial characteristic of all categories in the image.

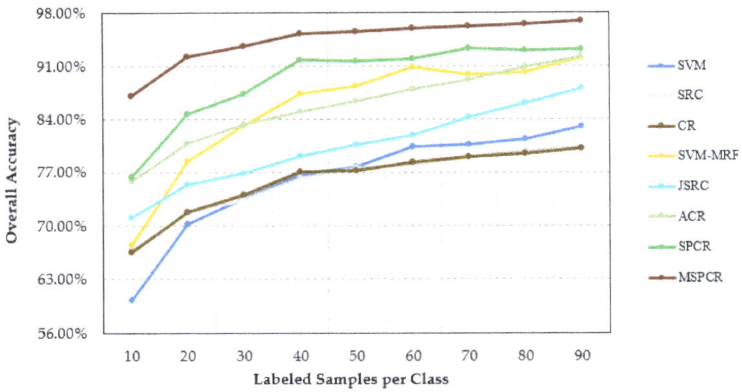

Figure 10. Overall classification accuracy obtained by the different tested methods with different numbers of labeled samples for the ROSIS University of Pavia scene.

4.5. Experiments with the HYDIC Washington, DC, National Mall Scene

In our first test with the HYDICE Washington, DC, National Mall scene, we first randomly select 50 labeled samples per class with a total of 300 samples for training and dictionary construction (which constitutes approximately 2.94% of the available labeled samples), the remaining samples are applied for validation. Table 3 shows the OAs and CAs obtained in different tested methods, and Figure 11 shows the corresponding classification maps. In the spectral domain, the traditional SRC provides an approximately equivalent result to CR, and both of them outperform the traditional SVM method, once again proving that the sparse coefficient is powerful to represent the spectral characteristics. In the spectral–spatial domain, the SVM-MRF, ACR, and SPCR perform well toward their original counterparts, i.e., SVM and CR. In addition, it also can be seen from the overall accuracies of the SRC method and the JSRC model that an improperly spatial constraint may have a negative impact on the classification performance. Distinct from the classification results in the above two datasets, the ACR gains a better classification performance than the proposed SPCR method in the HYDICE Washington, DC, National Mall scene, indicating that the SPCR model is susceptible to the superpixel segmentation scale. That is the original intention for us to propose MSPCR method, which eliminates the impact of the number of superpixels on classification by fusing the classification results at different segmentation scales. Furthermore, it can be found that the proposed MSPCR method achieves the highest accuracy 98.32%, which is similar with the results in the AVIRIS Indian Pines hyperspectral scene and the ROSIS University of Pavia scene. In addition, the proposed MSPCR provides reliable individual classification accuracy for each class, especially for class 1 and 2, which can be seen from the classification maps in Figure 11.

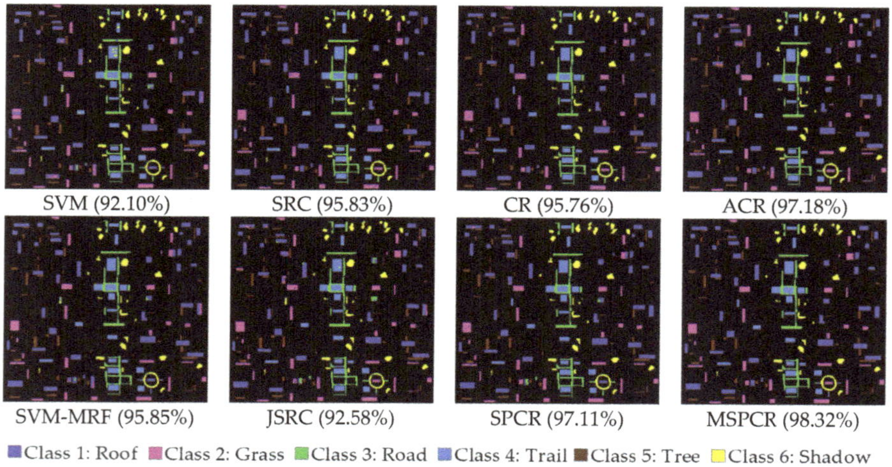

Figure 11. Classification maps obtained by the different tested methods with 50 samples per class for the HYDICE Washington, DC, National Mall dataset (OAs are in parentheses).

Table 3. Overall and classification accuracies (in percent) obtained by the different tested methods for the HYDICE Washington, DC, National Mall scene. In all cases, 300 labeled samples in total (50 samples per class) were used for training.

Class	Samples	SVM	SRC	CR	SVM-MRF	JSRC	ACR	SPCR	MSPCR
1	2916	85.56%	94.47%	93.54%	93.08%	85.08%	92.77%	95.79%	98.36%
2	1819	88.47%	90.22%	91.32%	94.33%	94.48%	95.61%	94.22%	97.81%
3	1264	96.12%	98.72%	98.54%	97.00%	95.81%	97.24%	98.57%	97.50%
4	1790	96.96%	98.73%	98.89%	98.32%	96.91%	99.22%	98.77%	98.32%
5	1120	98.38%	99.51%	99.49%	98.87%	94.30%	92.35%	99.42%	98.13%
6	1281	96.22%	96.81%	96.74%	97.25%	96.24%	97.58%	98.43%	99.89%
OA		92.10%	95.83%	95.76%	95.85%	92.58%	97.18%	97.11%	98.32%

In our second test with the HYDICE Washington, DC, National Mall scene, we evaluated the classification performance of our proposed methods from the spectral–spatial domain with different numbers of training samples. As shown in Figure 12, the classification result shows a rising tendency with the increase of the number of training samples, and curve tends to be flat when the number of training samples reaches to a certain amount. Firstly, the SRC and CR gain a better classification results toward SVM with the increase of the number of the labeled samples in the spectral domain. Though JSRC obtains relatively poor results than SRC, the SVM-MRF, ACR, and SPCR still provide competitive classification performances toward the SVM and CR with the increase of the number of training samples, which proves the integration of the spectral feature discrimination and spatial coherence is a reliable processing framework for the HSIC in most cases. On the other hand, improvement also appeared by the combination of the PD-driven and spatial constraint, which is indicated by the performance of ACR and SPCR-based method versus SVM-MRF and JSRC. In the spectral–spatial domain for all cases, the proposed MSPCR yields the best overall accuracy in comparison with the other related methods, and makes a significant improvement in comparison to the proposed SPCR.

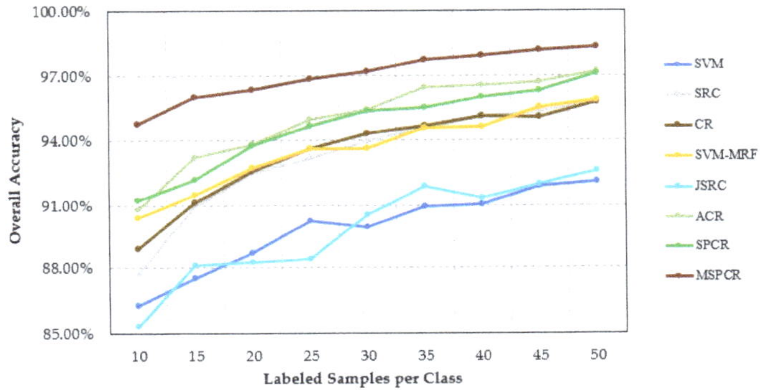

Figure 12. Overall classification accuracy obtained by the different tested methods with different numbers of labeled samples for the HYDICE Washington, DC, National Mall scene.

In addition, we compared the calculation cost of some spectral–spatial-based methods in the above three hyperspectral datasets, and the setting of the labeled samples corresponds to the cases in Tables 1–3. As shown in Figure 13, for the experiments on the above three datasets, the JSRC has the fastest speed but with the lowest classification accuracy. The proposed MSPCR not only achieves the best classification accuracy, which also has an increase in the time-consuming (about five times), as compared to the SPCR, due to the decision fusion process. On the ROSIS University of Pavia dataset and the AVIRIS Indian Pines dataset, the SPCR is the second best with an approximately equivalent time-consuming to ACR. On the HYDIC Washington, DC, National Mall dataset, the ACR achieves the second highest classification accuracy with a similar speed to SPCR.

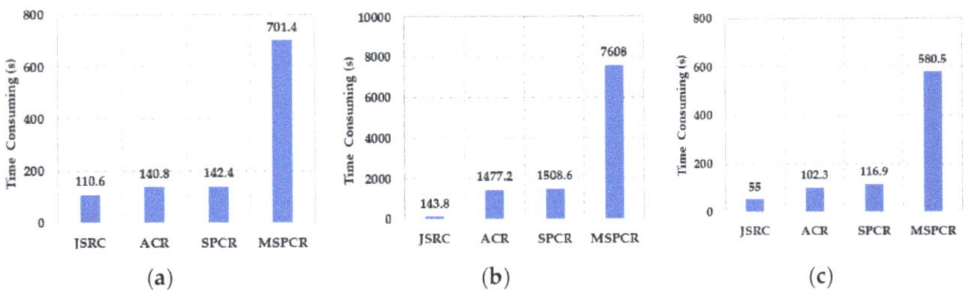

Figure 13. Calculation time-consuming comparison schematic diagram of different tested methods for (**a**) the AVIRIS Indian Pines dataset, (**b**) the ROSIS University of Pavia dataset, and (**c**) the HYDICE Washington, DC, National Mall dataset. The experiments are carried out using MATLAB on Intel(R) Core (TM) i7-6700K CPU machine with 16 GB of RAM.

Synthesizing the above experimental results and analysis, the firstly proposed SPCR method obtains a considerable overall and individual classification accuracy. The improved MSPCR gets better classification performance than the SPCR method. Moreover, the experimental results in different datasets also show that MSPCR outperforms several other related methods. Furthermore, the classification experimental results under different number of training samples also indicate the superiority and practicability of the proposed SPCR and MSPCR methods.

It should also be noted that the computational cost of the proposed MSPCR is relatively high, which is also the part of optimization in the future. Moreover, there are some potential points, for instance, the sample selection mechanism with related to the adaptive capability of method could be the follow-up research line.

5. Practical Application and Analysis

Different from the above three experimental datasets, we adopt the hyperspectral image data collected by the GF-5 satellite, to measure the practicability of the proposed SPCR and MSPCR method. GF-5 is the first hyperspectral comprehensive observation satellite of China, with a spatial resolution of 30 m. There are six payloads on GF-5, including two land imagers and four atmospheric sounders. In this paper, we select a scene from the hyperspectral image data obtained by visible short wave infrared hyperspectral camera.

First, we select the range of visible light to near infrared spectrum in the original data. After the atmospheric correction and radiation correction processing, the scene covers 150 spectral bands ranging from 0.4 to 2.5 μm, and the size of the image is 200 × 200. Six ground-truth classes with a total of 2216 labeled samples are contained in the reference data. Figure 14 shows the false-color composite image and the reference map of this scene.

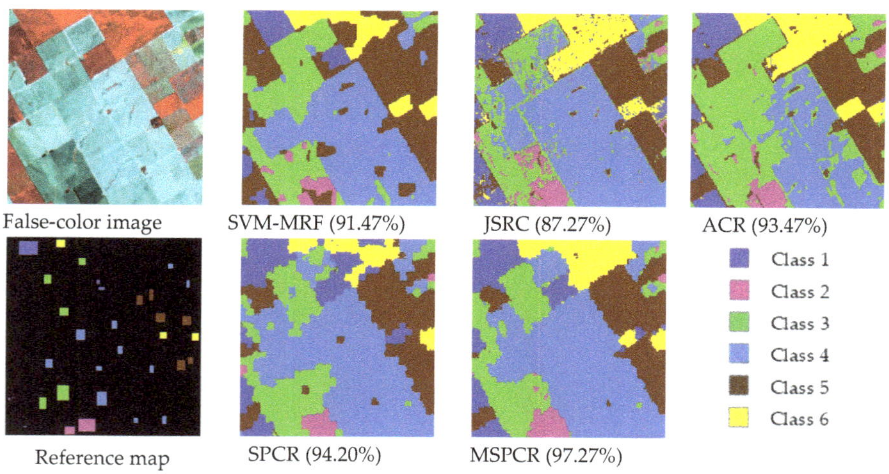

Figure 14. Classification maps obtained by the different tested methods with 5 samples per class for the GF-5 satellite dataset (OAs are in parentheses).

In the experiment with the GF-5 satellite dataset, we randomly selected five labeled samples per class with a total of 30 samples to construct a dictionary and the training model. The selected training samples constitute the approximately 1.35% of the labeled samples in the reference map, and the other remaining samples are used in validation. Figure 14 displays the classification maps of different methods. We analyzed the classification results as follows:

Compared with the SVM-MRF and JSRC, the ACR has a better classification performance, of which the overall accuracy is 6.20% higher than that of JSRC and 2.00% higher than that of SVM-MRF.

It confirms that the PD-driven-based decision mechanism plays an important role in classification. Compared with the ACR, the SPCR method obtains a better classification result, which verifies the effectiveness of integrating the PD-driven mechanism with the superpixel segmentation algorithm. The MSPCR outperforms the SPCR and yields the best accuracy in comparison to other related methods, which not only proves the MSPCR alleviates the impact of superpixel segmentation scale on the classification effect, but also indicates the decision fusion processing plays a decisive role in adapting different spatial characteristics of various categories of objects.

6. Conclusions

In this paper, a novel classification framework based on sparse representation, called the superpixel-level constraint representation (SPCR), was firstly proposed for hyperspectral imagery classification. SPCR uses the characteristics of spectral consistency of pixels inside the superpixel to determine the category of the testing pixel. Besides this, we proposed an improved multiscale superpixel-level constraint representation (MSPCR) method, obtaining the final classification result through fusing the classification maps of SPCR at different segmentation scales. The proposed SPCR method exploits the latent property of sparse coefficient and improves the contextual constraint, with consideration of spatial characterization. Moreover, the proposed MSPCR achieves comprehensive utilization of various regional distribution, resulting in strong classification performance. The experimental results with four real hyperspectral datasets including a GF-5 satellite data demonstrated that the SPCR outperforms several other classification methods, and the MSPCR yields a better classification accuracy than SPCR.

Author Contributions: Conceptualization, H.Y. and J.H.; formal analysis, M.S.; methodology, H.Y. and X.Z.; writing—original draft preparation, H.Y. and X.Z.; writing—review and editing, Q.G. and L.G. All authors have read and agreed to the published version of the manuscript.

Funding: This research was funded by the National Nature Science Foundation of China, grant numbers 61971082 and 61890964; Fundamental Research Funds for the Central Universities, grant numbers 3132020218 and 3132019341.

Acknowledgments: The authors would like to thank the Key Laboratory of Digital Earth Science, Aerospace Information Research Institute, Chinese Academy of Sciences for generously providing the GF-5 satellite data.

Conflicts of Interest: The authors declare no conflict of interest.

References

1. Hong, D.; Gao, L.; Yokoya, N.; Yao, J.; Chanussot, J.; Du, Q.; Zhang, B. More Diverse Means Better: Multimodal Deep Learning Meets Remote-Sensing Imagery Classification. *IEEE Trans. Geosci. Remote Sens.* **2020**. [CrossRef]
2. Hong, D.; Gao, L.; Yao, J.; Zhang, B.; Plaza, A.; Chanussot, J. Graph Convolution Networks for Hyperspectral Image Classification. *IEEE Trans. Geosci. Remote Sens.* **2020**. [CrossRef]
3. He, L.; Li, J.; Liu, C.; Li, S. Recent Advances on Spectral-Spatial Hyperspectral Image Classification: An Overview and New guidelines. *IEEE Trans. Geosci. Remote Sens.* **2018**, *56*, 1579–1597. [CrossRef]
4. Tong, F.; Tong, H.; Jiang, J.; Zhang, Y. Multiscale Union Regions Adaptive Sparse Representation for Hyperspectral Image Classification. *Remote Sens.* **2017**, *9*, 872. [CrossRef]
5. Cui, B.; Cui, J.; Lu, Y.; Guo, N.; Gong, M. A Sparse Representation-Based Sample Pseudo-Labeling Method for Hyperspectral Image Classification. *Remote Sens.* **2020**, *12*, 664. [CrossRef]
6. Gao, L.; Yao, D.; Li, Q.; Zhuang, L.; Zhang, B.; Bioucas-Dias, J.M. A New Low-Rank Representation Based Hyperspectral Image Denoising Method for Mineral Mapping. *Remote Sens.* **2017**, *9*, 1145. [CrossRef]
7. Ghamisi, P.; Maggiori, E.; Li, S.; Souza, R.; Tarablaka, Y.; Moser, G.; De Giorgi, A.; Fang, L.; Chen, Y.; Chi, M.; et al. New Frontiers in Spectral-Spatial Hyperspectral Image Classification: The Latest Advances Based on Mathematical Morphology, Markov Random Fields, Segmentation, Sparse Representation, and Deep Learning. *IEEE Geosci. Remote Sens. Mag.* **2018**, *6*, 10–43. [CrossRef]

8. Rasti, B.; Hong, D.; Hang, R.; Ghamisi, P.; Kang, X.; Chanussot, J.; Benediktsson, J.A. Feature Extraction for Hyperspectral Imagery: The Evolution from Shallow to Deep (Overall and Toolbox). *IEEE Geosci. Remote Sens. Mag.* **2020**. [CrossRef]
9. Gao, L.; Li, J.; Khodadadzadeh, M.; Plaza, A.; Zhang, B.; He, Z.; Yan, H. Subspace-Based Support Vector Machines for Hyperspectral Image Classification. *IEEE Geosci. Remote Sens. Lett.* **2015**, *12*, 349–353.
10. Wang, K.; Cheng, L.; Yong, B. Spectral-Similarity-Based Kernel of SVM for Hyperspectral Image Classification. *Remote Sens.* **2020**, *12*, 2154. [CrossRef]
11. Paoletti, M.E.; Haut, J.M.; Tao, X.; Miguel, J.P.; Plaza, A. A New GPU Implementation of Support Vector Machines for Fast Hyperspectral Image Classification. *Remote Sens.* **2020**, *12*, 1257. [CrossRef]
12. Hu, S.; Peng, J.; Fu, Y.; Li, L. Kernel Joint Sparse Representation Based on Self-Paced Learning for Hyperspectral Image Classification. *Remote Sens.* **2019**, *11*, 1114. [CrossRef]
13. Yu, H.; Gao, L.; Li, J.; Li, S.S.; Zhang, B.; Benediktsson, J.A. Spectral-Spatial Hyperspectral Image Classification Using Subspace-Based Support Vector Machines and Adaptive Markov Random Fields. *Remote Sens.* **2016**, *8*, 355. [CrossRef]
14. Jia, S.; Deng, B.; Zhu, J.; Jia, X.; Li, Q. Superpixel-Based Multitask Learning Framework for Hyperspectral Image Classification. *IEEE Trans. Geosci. Remote Sens.* **2017**, *55*, 2575–2588. [CrossRef]
15. Gao, L.; Hong, D.; Yao, J.; Zhang, B.; Gamba, P.; Chanussot, J. Spectral Superresolution of Multispectral Imagery with Joint Sparse and Low-Rank Learning. *IEEE Trans. Geosci. Remote Sens.* **2020**. [CrossRef]
16. Yang, J.; Li, Y.; Chan, J.C.-W.; Shen, Q. Image Fusion for Spatial Enhancement of Hyperspectral Image via Pixel Group Based Non-Local Sparse Representation. *Remote Sens.* **2017**, *9*, 53. [CrossRef]
17. Gao, Q.; Lim, S.; Jia, X. Improved Joint Sparse Models for Hyperspectral Image Classification Based on a Novel Neighbour Selection Strategy. *Remote Sens.* **2018**, *10*, 905. [CrossRef]
18. Zhang, S.; Li, S.; Fu, W.; Fang, L. Multiscale Superpixel-Based Sparse Representation for Hyperspectral Image Classification. *Remote Sens.* **2017**, *9*, 139. [CrossRef]
19. Li, W.; Du, Q. Collaborative Representation for Hyperspectral Anomaly Detection. *IEEE Trans. Geosci. Remote Sens.* **2015**, *53*, 1463–1474. [CrossRef]
20. Wang, J.; Jiao, L.; Wang, S.; Hou, B.; Liu, F. Adaptive Nonlocal Spatial–Spectral Kernel for Hyperspectral Imagery Classification. *IEEE J. Sel. Top. Appl. Earth Obs. Remote Sens.* **2016**, *9*, 4086–4101. [CrossRef]
21. Geng, J.; Wang, H.; Fan, J.; Ma, X.; Wang, B. Wishart Distance-Based Joint Collaborative Representation for Polarimetric SAR Image Classification. *IET Radar Sonar Navig.* **2017**, *11*, 1620–1628. [CrossRef]
22. Yu, H.; Shang, X.; Zhang, X.; Gao, L.; Song, M.; Hu, J. Hyperspectral Image Classification Based on Adjacent Constraint Representation. *IEEE Geosci. Remote Sens. Lett.* **2020**. [CrossRef]
23. Liang, J.; Zhou, J.; Qian, Y.; Wen, L.; Bai, X.; Gao, Y. On the Sampling Strategy for Evaluation of Spectral-Spatial Methods in Hyperspectral Image Classification. *IEEE Trans. Geosci. Remote Sens.* **2017**, *55*, 862–880. [CrossRef]
24. Yu, H.; Gao, L.; Liao, W.; Zhang, B.; Pižurica, A.; Philips, W. Multiscale Superpixel-Level Subspace-Based Support Vector Machines for Hyperspectral Image Classification. *IEEE Geosci. Remote Sens. Lett.* **2017**, *14*, 2142–2146. [CrossRef]
25. Wang, J.; Zhu, C.; Zhou, Y.; Zhu, X.; Wang, Y.; Zhang, W. From Partition-Based Clustering to Density-Based Clustering: Fast Find Clusters with Diverse Shapes and Densities in Spatial Databases. *IEEE Access.* **2018**, *6*, 1718–1729. [CrossRef]
26. Garg, I.; Kaur, B. Color Based Segmentation Using K-Mean Clustering and Watershed Segmentation. In Proceedings of the 2016 3rd International Conference on Computing for Sustainable Global Development (INDIACom), New Delhi, India, 16–18 March 2016; pp. 3165–3169.
27. Sun, H.; Ren, J.; Zhao, H.; Yan, Y.; Zabalza, J.; Marshall, S. Superpixel based Feature Specific Sparse Representation for Spectral-Spatial Classification of Hyperspectral Images. *Remote Sens.* **2019**, *11*, 536. [CrossRef]
28. Sharma, J.; Rai, J.K.; Tewari, R.P. A Combined Watershed Segmentation Approach Using K-Means Clustering for Mammograms. In Proceedings of the 2015 2nd International Conference on Signal Processing and Integrated Networks (SPIN), Noida, India, 19–20 February 2015; pp. 109–113.
29. Jia, S.; Deng, B.; Jia, X. Superpixel-Level Sparse Representation-Based Classification for Hyperspectral Imagery. *IGARSS* **2016**, 3302–3305. [CrossRef]
30. Csillik, O. Fast Segmentation and Classification of Very High Resolution Remote Sensing Data Using SLIC Superpixels. *Remote Sens.* **2017**, *9*, 243. [CrossRef]

31. Jia, S.; Deng, B.; Zhu, J.; Jia, X.; Li, Q. Local Binary Pattern-Based Hyperspectral Image Classification with Superpixel Guidance. *IEEE Geosci. Remote Sens.* **2018**, *56*, 749–759. [CrossRef]
32. Li, G.; Li, L.; Zhu, H.; Liu, X.; Jiao, L. Adaptive Multiscale Deep Fusion Residual Network for Remote Sensing Image Classification. *IEEE Trans. Geosci. Remote Sens.* **2019**, *57*, 8506–8521. [CrossRef]
33. Shao, Z.; Wang, L.; Wang, Z.; Deng, J. Remote Sensing Image Super-Resolution Using Sparse Representation and Coupled Sparse Autoencoder. *IEEE J. Sel. Top. Appl. Earth Obs. Remote Sens.* **2019**, *12*, 2663–2674. [CrossRef]
34. Li, W.; Du, Q.; Xiong, M. Kernel Collaborative Representation with Tikhonov Regularization for Hyperspectral Image Classification. *IEEE Geosci. Remote Sens. Lett.* **2015**, *12*, 48–52.
35. Zhang, H.; Li, J.; Huang, Y.; Zhang, L. A Nonlocal Weighted Joint Sparse Representation Classification Method for Hyperspectral Imagery. *IEEE J. Sel. Top. Appl. Earth Obs. Remote Sens.* **2014**, *7*, 2056–2065. [CrossRef]
36. Zou, B.; Xu, X.; Zhang, L. Object-Based Classification of PolSAR Images Based on Spatial and Semantic Features. *IEEE J. Sel. Top. Appl. Earth Obs. Remote Sens.* **2020**, *13*, 609–619. [CrossRef]
37. Xie, F.; Lei, C.; Yang, J.; Jin, C. An Effective Classification Scheme for Hyperspectral Image Based on Superpixel and Discontinuity Preserving Relaxation. *Remote Sens.* **2019**, *11*, 1149. [CrossRef]
38. Zhu, L.; Wen, G. Hyperspectral Anomaly Detection via Background Estimation and Adaptive Weighted Sparse Representation. *Remote Sens.* **2018**, *10*, 272.
39. Yu, H.; Gao, L.; Liao, W.; Zhang, B.; Zhuang, L.; Song, M.; Chanussot, J. Global Spatial and Local Spectral Similarity-Based Manifold Learning Group Sparse Representation for Hyperspectral Imagery Classification. *IEEE Trans. Geosci. Remote Sens.* **2020**, *58*, 3043–3056. [CrossRef]
40. Li, S.; Ni, L.; Jia, X.; Gao, L.; Zhang, B.; Peng, M. Multi-Scale Superpixel Spectral–Spatial Classification of Hyperspectral Images. *Int. J. Remote Sens.* **2016**, *37*, 4905–4922. [CrossRef]
41. Jia, S.; Deng, X.; Zhu, J.; Xu, M.; Zhou, J.; Jia, X. Collaborative Representation-Based Multiscale Superpixel Fusion for Hyperspectral Image Classification. *IEEE Trans. Geosci. Remote Sens.* **2019**, *57*, 7770–7784. [CrossRef]

remote sensing

MDPI

Article

Underwater Hyperspectral Target Detection with Band Selection

Xianping Fu [1,2], Xiaodi Shang [1], Xudong Sun [1,2], Haoyang Yu [1], Meiping Song [1,*] and Chein-I Chang [1,3,4]

1 Information Science and Technology College, Dalian Maritime University, Dalian 116026, China; fxp@dlmu.edu.cn (X.F.); shangxd329@dlmu.edu.cn (X.S.); sxd@dlmu.edu.cn (X.S.); yuhy@dlmu.edu.cn (H.Y.); cchang@umbc.edu (C.-I.C.)
2 Peng Cheng Laboratory, Shengzhen 518000, China
3 Remote Sensing Signal and Image Processing Laboratory, Department of Computer Science and Electrical Engineering, University of Maryland, Baltimore, MD 21250, USA
4 Department of Computer Science and Information Management, Providence University, Taichung 02912, Taiwan
* Correspondence: smping@dlmu.edu.cn

Received: 24 January 2020; Accepted: 20 March 2020; Published: 25 March 2020

Abstract: Compared to multi-spectral imagery, hyperspectral imagery has very high spectral resolution with abundant spectral information. In underwater target detection, hyperspectral technology can be advantageous in the sense of a poor underwater imaging environment, complex background, or protective mechanism of aquatic organisms. Due to high data redundancy, slow imaging speed, and long processing of hyperspectral imagery, a direct use of hyperspectral images in detecting targets cannot meet the needs of rapid detection of underwater targets. To resolve this issue, a fast, hyperspectral underwater target detection approach using band selection (BS) is proposed. It first develops a constrained-target optimal index factor (OIF) band selection (CTOIFBS) to select a band subset with spectral wavelengths specifically responding to the targets of interest. Then, an underwater spectral imaging system integrated with the best-selected band subset is constructed for underwater target image acquisition. Finally, a constrained energy minimization (CEM) target detection algorithm is used to detect the desired underwater targets. Experimental results demonstrate that the band subset selected by CTOIFBS is more effective in detecting underwater targets compared to the other three existing BS methods, uniform band selection (UBS), minimum variance band priority (MinV-BP), and minimum variance band priority with OIF (MinV-BP-OIF). In addition, the results also show that the acquisition and detection speed of the designed underwater spectral acquisition system using CTOIFBS can be significantly improved over the original underwater hyperspectral image system without BS.

Keywords: constrained-target optimal index factor band selection (CTOIFBS); hyperspectral image; underwater spectral imaging system; underwater hyperspectral target detection; band selection (BS); constrained energy minimization (CEM)

1. Introduction

Underwater target detection using the images acquired by traditional red-green-blue (RGB) cameras has become more and more mature where traditional image processing methods [1,2] and target detection algorithms based on deep learning, such as Faster Region-based Convolutional Neural Networks (Faster R-CNN) [3] and You Only Look Once (YOLO) [4], have been widely applied to underwater target detection. In an ideal underwater imaging environment, the detection speed and accuracy of various algorithms can reach a high level of performance. However, the traditional

RGB image detection technology suffers from a series of problems. When the underwater imaging environment is poor and marine animals have their protective color mechanism, it is difficult to detect and identify targets of interest effectively from the complex background [5,6].

Hyperspectral imaging technology can provide a higher spectral resolution than RGB images, and its band coverage can range from ultraviolet, visible, near-infrared to mid-infrared bands and provides wealthy spectral information. Hyperspectral data is generally acquired by hundreds of contiguous narrow spectral bands, which can resolve the problems encountered in traditional RGB image detection technology and also make it have a good ability to identify targets and distinguishing similar targets. Classical hyperspectral target detection algorithms include an anomaly detector developed by Reed and Xiaoli, called the RXD algorithm [7], kernel RXD (KRXD) algorithm [8], orthogonal subspace projection (OSP) algorithm [9], and constrained energy minimization (CEM) algorithm [10]. Among them, CEM is a subpixel target detection algorithm that has been shown to be an effective and promising technique when only the target spectrum of interest is known and the background spectrum is unknown. So, it is quite suitable for target detection in a complicated underwater background and environment with insufficient prior knowledge.

At the present time, only a few studies on hyperspectral underwater target detection are available in the literature and most of them mainly focused on three aspects. First, in order to ensure that the hyperspectral imager can accurately extract key information in the complex marine environment when collecting underwater images, the designed key technologies are different. Second, since a hyperspectral image has a large number of spectral bands with very high spectral resolution, it has good target recognition ability. However, this is also traded for a slow imaging speed, enormous data volume, long transmission cycle, and slow calculation speed, all of which cannot be suitable for the remote operated vehicle (ROV) platform and real-time underwater target detection [11,12]. Third, hyperspectral underwater target detection technology tends to have strategic significance in both military and economic aspects. So, the degree of technological openness is extremely limited.

Because of the low imaging speed and long processing time of underwater hyperspectral images, the current research on underwater spectral imaging and detection is mainly focused on the detection of underwater pipelines, the distribution and species detection of underwater plants and microorganisms, etc. [13–17], but the capability of real-time detection is low. Some researchers have complied a spectral library for recognition and detection of different underwater targets. Kuniaki Uto et al. [18] classified the objects of interest by measuring their average spectral curves of cauliflower and sand to calculate their resultant correlation coefficient. Tegdan [19] et al. used a spectral library of some known objects of interest to achieve automatic recognition of other objects. An underwater hyperspectral imaging (UHI) system, jointly developed by Norwegian company Ecotone and Norwegian Underwater application robotics, is an optimized underwater hyperspectral imaging system, which can be used for underwater hyperspectral remote sensing. This system is capable of collecting information in the full color spectrum (370–800 nm).

In this paper, sea cucumbers are selected as our primary underwater targets due to its economic value on gross domestic product (GDP) growth in Dalian, China. Marine aquaculture is one of Dalian's pillar industries with an annual output value of more than 3.5 billion US dollars. Sea cucumbers are the major seafood products to account for the most revenue. At present, the main methods for fishing sea cucumbers, abalone, and other sea treasures rely on diver operation and submarine trawl operation. However, such diver operation is inefficient, and the deep-sea environment is extremely harmful to the health of divers. On the other hand, the submarine trawl operation generally causes severe damage to the underwater ecological environment. Therefore, autonomous fishing of seafood using underwater robots has become the most effective solution, and the rapid detection of underwater objects is a key issue that needs to be solved urgently.

Comparing to other targets, sea cucumber detection has more difficulty and greater challenges because sea cucumbers have a strong protective color mechanism. It is difficult to observe using color and texture characteristics when ordinary RGB cameras are used for underwater observation.

However, the sea cucumber exhibits relatively obvious reflectance characteristics in some special bands, which is the exact reason why we use hyperspectral technology to solve this problem.

The methods described above can effectively apply hyperspectral imaging technology to underwater biological classification and detection but cannot achieve real-time detection of underwater targets [20]. For the target to be detected, if its sensitive bands can be selected for detection in advance, the image processing speed can be increased to satisfy the real-time requirements. Gleason [21] found that the bands of 546, 568, and 589 nm could more easily separate corals and algae from other background objects. So, a multi-spectral camera could be constructed by six bands for fast acquisition of images for target detection. Experiments show that compared to the traditional RGB cameras, the six-band multi-spectral cameras had better performance in detecting submarine corals. However, the selected bands used for coral detection in the experiments were obtained as a by-product of other experiments, which are not applicable to other underwater targets and are not universal. Therefore, a reliable BS method needs to be designed so that it can select representative band subsets for different targets.

The researchers put forward some effective methods for BS. For example, information divergence (ID) selects bands according to the difference between the probability distributions of a measured band and its corresponding Gaussian probability distribution. The maximum-variance principal component analysis (MVPCA) developed in [22] first performed PCA transformation on the original data and then constructed the loading factor matrix from the obtained eigenvectors and eigenvalues. The priority of a band was determined by the variance of its corresponding loading factor. However, the bands selected according to such band prioritization methods were usually highly correlated. By factoring band correlation into consideration, the optimal index factor (OIF) [23] method was developed to find the largest OIF index. Yang et al. [24] proposed a BS method based on linear prediction, which used linear prediction as a similarity measure to find the next least similar band by sequential forward selection. All of the described methods select band subsets in accordance with the characteristics of the data itself and are not designed to select an optimal band subset for a specific target.

For target detection, Yuan et al. [25] proposed a multigraph determinantal point process (MDPP) model to effectively search for discriminative band sets. Wang [26] proposed the multi-band selection (MBS) method, which did not require prioritizing the bands but relied on a specific application to select desired bands. Based on the concept of CEM, Geng [27] proposed a sparse constrained band selection (SCBS), which is convenient for solving the global optimal solution and avoids the complicated subset search process. Wang et al. [28] proposed a new multi-target detection BS method, MinV-BP, which minimized the variance generated by the target of interest to measure the priority of the band.

This paper proposes a real-time detection method for hyperspectral underwater targets based on BS. First of all, in order to solve the problems suffering from a large amount of redundant data and slow acquisition and processing speed of hyperspectral image data, a BS method is designed in combination with MinV-BP [28] and OIF [23] to select an optimal band subset with strong ability in characterizing specific targets, called constrained-target OIF band selection (CTOIFBS). Then, an underwater multi-spectral sensor composed of the selected bands is particularly designed to collect images to overcome the difficulty of long transmission time of the complete hyperspectral image. Finally, CEM is used to detect underwater targets. The proposed CTOIFBS not only can extract a set of bands more suitable for specific targets to improve detection performance but can also meet the real-time requirements of underwater image acquisition.

2. Materials and Methods

2.1. MinV-BP

The idea of the Minimum Variance Band Prioritization (MinV-BP) is based on CEM, which was derived from the linearly constrained minimum variance beamformer in the field of digital signal processing. It detects signals in a specific direction and minimizes signal interference in other directions, thereby achieving target detectability from the image and suppressing the background [10].

Suppose $\{\mathbf{r}_1, \mathbf{r}_2, \ldots, \mathbf{r}_N\}$ is a hyperspectral image with N pixels. N is the total number of pixels in the image. Each pixel, $\mathbf{r}_i = (\mathbf{r}_{i1}, \mathbf{r}_{i2}, \ldots, \mathbf{r}_{iL})^T$, is an L-dimensional column vector, where L is the number of bands. Define $\mathbf{d}$ as the target spectral signal to be detected, which is known prior information. The purpose of CEM is to design a linear FIR filter $\mathbf{w} = [\mathbf{w}_1, \mathbf{w}_2, \ldots, \mathbf{w}_L]^T$ so that its output energy is minimized under the constraint term (1):

$$\mathbf{d}^T\mathbf{w} = \sum\nolimits_{l=1}^{L} \mathbf{d}_l\mathbf{w}_l = 1 \quad (1)$$

where $\mathbf{w} = [\mathbf{w}_1, \mathbf{w}_2, \ldots, \mathbf{w}_L]^T$ is an L-dimensional column vector formed by the filter coefficient. Suppose the output of the FIR filter corresponding to the input pixel $\mathbf{r}_i$ is $\mathbf{y}_i$ defined in Equation (2):

$$\mathbf{y}_i = \sum\nolimits_{i=1}^{L} \mathbf{w}_l\mathbf{r}_{il} = \mathbf{w}^T\mathbf{r}_i = \mathbf{r}_i{}^T\mathbf{w} \quad (2)$$

Then, for all input $\{\mathbf{r}_1, \mathbf{r}_2, \ldots, \mathbf{r}_N\}$, the average energy of the filter output is:

$$E = \frac{1}{N}\sum_{i=1}^{N} \mathbf{y}_i^2 = \frac{1}{N}(\mathbf{r}_i^T\mathbf{w})^T\mathbf{r}_i^T\mathbf{w} = \frac{1}{N}\sum_{i=1}^{N} \mathbf{w}^T\mathbf{r}_i\mathbf{r}_i^T\mathbf{w} = \mathbf{w}^T\left(\frac{1}{N}\sum_{i=1}^{N} \mathbf{r}_i\mathbf{r}_i^T\right)\mathbf{w} = \mathbf{w}^T\mathbf{R}\mathbf{w} \quad (3)$$

where $\mathbf{R} = \left(\frac{1}{N}\right)\sum_{i=1}^{N} \mathbf{r}_i\mathbf{r}_i^T$ represents the sample autocorrelation matrix of the $L \times L$ dimension. CEM can be expressed as the following linear constrained optimization problem:

$$\min_{\mathbf{w}}\{E\} = \min_{\mathbf{w}}\{\mathbf{w}^T\mathbf{R}\mathbf{w}\} \ s.t.\mathbf{d}^T\mathbf{w} = 1 \quad (4)$$

By using the Lagrange multiplier method, the optimal solution and CEM error of Equation (4) are obtained as follows:

$$\mathbf{w}_{CEM} = \frac{\mathbf{R}^{-1}\mathbf{d}}{\mathbf{d}^T\mathbf{R}^{-1}\mathbf{d}} \quad (5)$$

and:

$$\min_{\mathbf{w}}\mathbf{w}\mathbf{R}^{-1}\mathbf{w} = \left(\mathbf{w}^{\mathrm{CEM}}\right)^T\mathbf{R}^{-1}\mathbf{w}^{\mathrm{CEM}} = \left(\mathbf{d}^T\mathbf{R}^{-1}\mathbf{d}\right)^{-1} \quad (6)$$

The CEM filter is obtained from Equation (5):

$$\delta_{CEM}(\mathbf{r}) = (\mathbf{w}_{CEM})^{\mathrm{T}}\mathbf{r} = \left(\frac{\mathbf{R}^{-1}\mathbf{d}}{\mathbf{d}^T\mathbf{R}^{-1}\mathbf{d}}\right)^T\mathbf{r} = \frac{\mathbf{d}^T\mathbf{R}^{-1}\mathbf{r}}{\mathbf{d}^T\mathbf{R}^{-1}\mathbf{d}} \quad (7)$$

The CEM operator is applied to every pixel in the image to minimize the output energy caused by other unknown signals so that the target $\mathbf{d}$ of interest can be detected to achieve the purpose of detection.

According to the CEM algorithm, single band minimum variance band prioritization (MinV-BP) can further use the variance generated by the target of interest to measure the priority of the band to obtain the band with the best characterization ability for the specific target. Suppose $\{\mathbf{b}_l\}_{l=1}^{L}$ is the band set of hyperspectral image, where $\mathbf{b}_l$ is a column vector, $\mathbf{b}_l = (b_{l1}, b_{l2}, \cdots, b_{lN})^T$, representing the image of the l-th band. $\{b_{li}\}_{i=1}^{N}$ is the set of all N pixels on the l-th band image $\mathbf{b}_l$. According to the CEM error derived from Equation (6), MinV-BP is defined as:

$$\mathrm{V}(\mathbf{b}_l) = \left(\mathbf{d}_{\mathbf{b}_l}^T\mathbf{R}_{\mathbf{b}_l}^{-1}\mathbf{d}_{\mathbf{b}_l}\right)^{-1} \quad (8)$$

Using Equation (8), MinV-BP can obtain the band priority sequence for the target of interest. Where, the smaller the variance, the higher the priority. The larger the variance, the lower the priority.

In short, the advantage of MinV-BP is that it can give higher priority to the band with strong target characterization ability through the minimum variance criterion. However, when MinV-BP prioritizes the bands, it only considers the ability of the bands to represent the target vector but does not consider the strong correlation and redundancy between the bands. As a result, the bands with high priority in the resulting sequence are largely adjacent bands with a strong correlation. Therefore, how to de-correlate the priority bands and obtain a band set with weak correlation and stronger discrimination ability is a subsequent problem to be solved.

2.2. OIF

Chavez et al. [23] proposed the optimum index factor (OIF) defined as:

$$\mathrm{OIF} = \sum_{i=1}^{L} \mathbf{S}_i / \sum_{i=1}^{L} \sum_{j=i+1}^{L} \left|\mathbf{R}_{ij}\right| \tag{9}$$

to evaluate the amount of information in a dataset where $\mathbf{S}_i$ and $\mathbf{R}_{ij}$ represent the standard deviation of the l-th band and the correlation coefficient between band i and j, respectively, and L is the total number of bands. The standard deviation is used to represent the amount of image information. Based on the ratio of the amount of information in the band set to the correlation coefficient between the bands defined by:

$$\mathbf{R}_{ij} = \frac{\mathbf{S}_{ij}^2}{\mathbf{S}_i \times \mathbf{S}_j} \tag{10}$$

A band subset with a large amount of information and a small correlation can be selected as a band subset. In Equation (10), $\mathbf{S}_{ij}$ represents the covariance of bands i and j, and:

$$\mathbf{S}_{ij}^2 = Cov(i,j) = \frac{1}{n} \sum_{w=1}^{n} (\mathbf{x}_{iw} - \overline{\mathbf{x}}_i)(\mathbf{y}_{jw} - \overline{\mathbf{y}}_j) \tag{11}$$

where $\mathbf{x}_i$ represents the spectral grayscale value for the i-th band; $\mathbf{x}_{iw}$ represents the gray value of the w-th pixel in the i-th band; $\overline{\mathbf{y}}_i$ represents the spectral grayscale value for the j-th band; $\mathbf{y}_{jw}$ represents the gray value of the w-th pixel in the j-th band; $\mathbf{N}$ represents the number of pixels in a single band and n is the n-th pixel in the band, $1 \leq n \leq N$.

In other words, for a hyperspectral image containing L bands, the standard deviation of the single-band image and the correlation coefficient matrix of each band are calculated first, and then the OIF index corresponding to all possible band subsets are calculated subsequently, and the optimal band subset is finally selected according to the index value.

2.3. Constrained-Target OIF Band Selection

Hyperspectral data generally have very high band correlation and data redundancy. In order to mitigate this problem, a BS method with target constraints, called constrained-target optimum index factor BS (CTOIFBS), is developed in this paper. It first prioritizes all bands by MinV-BP to obtain a band priority sequence. The smaller the variance, the higher the priority of the band, and the stronger the ability of the band to represent the target. It is then followed by estimating virtual dimensionality (VD) [10,29–31] to determine the required number of bands, n_{BS}, where VD is defined as the number of spectrally distinct signal sources present in the data that can effectively characterize the hyperspectral data from a perspective view of target detection and classification. In this case, the first n bands with higher priorities in the sequence are clustered into n_{BS} clusters by a K-means method to remove the band correlation. As a result, the band correlation in the same cluster will be high, while the band correlation between different clusters will be low. Finally, a band is selected from each cluster to form a band subset. The OIF value of the band subset is then calculated. The band subset with the largest OIF value is selected as the best band subset. The CTOIFBS process is as follows.

Algorithm CTOIFBS

Input: Hyperspectral image data $\mathbf{\Omega}$
Output: The optimal band set $\mathbf{\Omega}^*_{n_{\mathrm{BS}}}$

1. According to (8), all bands $\mathbf{b}_l$ in $\mathbf{\Omega}$ are ranked to obtain the priority sequence of bands, $\mathbf{b}_{l_1} > \mathbf{b}_{l_2} > \cdots > \mathbf{b}_{l_L}$, where $\mathbf{b}_{l_j} > \mathbf{b}_{l_k} \Leftrightarrow \mathrm{V}(\mathbf{b}_{l_j}) < \mathrm{V}(\mathbf{b}_{l_k})$, the notation "$>$" is used to indicate "superior to".

2. The required number of bands n_{BS} is determined by VD.

3. The first n bands of priority sequence obtained in step 1 were divided into n_{BS} bands set $\mathbf{\Omega}_k$ by K-means, where $\mathbf{\Omega}_k = \{\mathbf{b}^k_1, \mathbf{b}^k_2, \ldots, \mathbf{b}^k_{n_k}\}$, $1 \le k \le n_{\mathrm{BS}}$, n_k denotes the number of bands included in $\mathbf{\Omega}_{n_{\mathrm{BS}}}$, $n = \sum_{i=1}^{n_{\mathrm{BS}}} n_k$.

4. Combining the bands in $\mathbf{\Omega}_k$, $\mathbf{\Omega}^* = \{\mathbf{\Omega}_1 \times \cdots \times \mathbf{\Omega}_K\}$, where "$\times$" stands for cartesian product. $\mathbf{\Omega}^*$ contains M band sets, $M = n_1 \times n_2 \times \cdots n_K$. Then calculate the OIF value of each band set in $\mathbf{\Omega}^*$.

5. The maximum OIF value is selected as the optimal band set $\mathbf{\Omega}^*_{n_{\mathrm{BS}}}$.

A flowchart implementing CTOIFBS is shown in Figure 1.

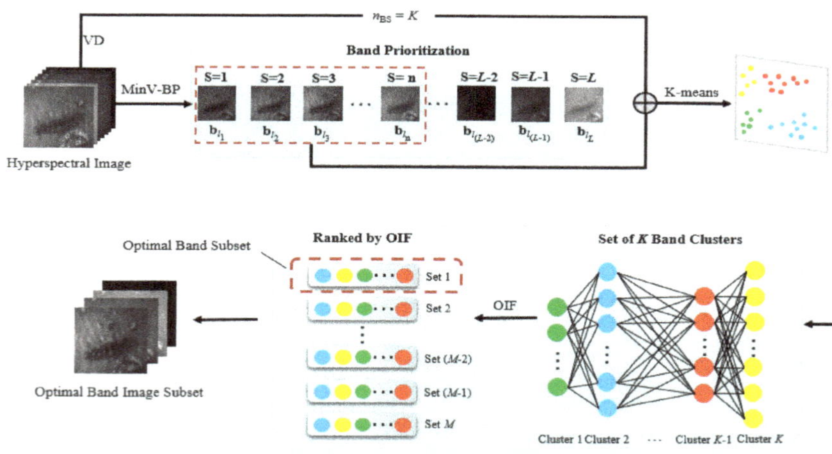

Figure 1. A flowchart of implementing CTOIFBS (constrained-target optimal index factor band selection).

Using the MinV-BP criterion, a band priority sequence for the target of interest can be obtained, and then bands with strong characterization of the target can be selected from all the band sequence. However, there is still a problem, which is high inter-band correlation in this band sequence. OIF takes two factors into account: variance and correlation coefficient. Theoretically, the optimal band subset with large information amount and small inter-band correlation can be obtained by optimizing the priority sequence of the band using OIF. However, it has been found in experiments that the use of OIF alone to process band priority sequences was not effective since a band subset with high correlation will still be selected. This is because OIF strives to make the standard deviation of the selected bands as large as possible, while the correlation coefficient between the bands is as small as possible. Unfortunately, it is difficult to achieve the best of both measures [15]. Therefore, instead of selecting the first n bands of the priority sequence directly by the OIF index as a band subset, CTOIFBS is developed to use clusters to perform band de-correlation prior to using OIF. That is, the selected candidate bands are divided into several subsets to further reduce the band correlation and band redundancy. The advantages of such cluster-based band de-correlation have two advantages. One is the pre-grouping process, which reduces the total number of band subset to be compared so that computational complexity can be greatly reduced. The other is clustering by a K-means method in advance to effectively remove band redundancy so as to improve subsequent detection performance.

2.4. Underwater Spectral Imaging System

Using an underwater spectrum camera composed of a best-selected band subset to collect the target image can greatly reduce data redundancy and solve the problem of long transmission time of a complete hyperspectral image. However, due to the complicated underwater imaging environment on the one hand and the difficulty in finding the proper loader or vehicle on the other hand, the development of underwater spectral imaging technology is still far from that of atmospheric spectral imaging. Therefore, how to design a suitable underwater spectral imaging (USI) system is the very key to success in realizing the rapid detection of hyperspectral underwater targets.

The core of the spectral imaging system is the optical splitting system. The spectroscopic techniques currently being used are based on dispersion, filtering, and interferometry, and commonly used optical splitting components include gratings, prisms, and various filters. This paper develops a filter wheel spectral camera to collect spectral images. There are several reasons. First of all, it has a wheel with multiple single band-pass filters to collect spectral information of different bands, which is suitable for the case of fewer bands needed. Second, a narrow band filter has a high transmittance, so it is suitable for the special light conditions under water. Third, it adapts to different filter combinations that can be changed according to different objects. Fourth, this type of camera is much cheaper than the commonly used liquid crystal tunable filter (LCTF) spectral camera.

Therefore, this paper builds an underwater spectral imaging system based on a filter wheel spectral camera, as shown in Figure 2. Its main components include a FLIR Blackfly S USB3 CCD camera and its corresponding lens, electric filter wheel, and single band-pass filters with the wavelengths between 400 and 830 nm at intervals of 10 nm. These filters have a bandwidth of 14nm and a cut-off depth of OD3 and a single chip microcomputer for controlling the camera and filter wheel. All the above parts are packed in a watertight enclosure. This system uses electric filter wheels to collect single-band images in different bands and synthesize the target's spectral image. It is also possible to obtain spectral images of different band subsets by replacing the filter combinations on the filter wheel. It is important to note that the spectral filter wheel designed is not limited to the USI system and can be applicable to various beam splitters, such as LCTF, acousto-optic tunable filter (AOTF), or spectral filter array (SFA) according to their application scenarios and costs.

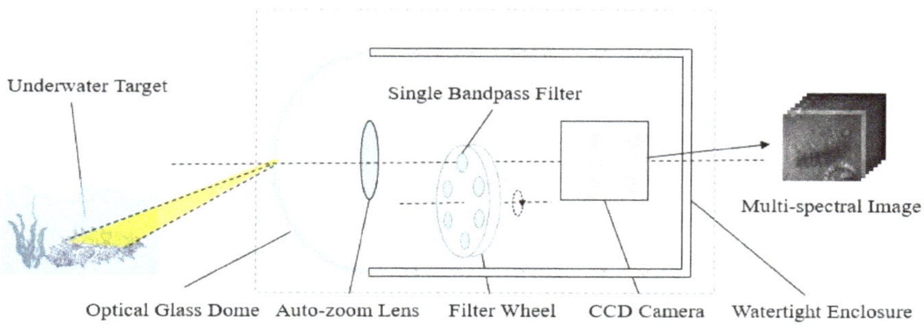

Figure 2. Diagram of the underwater spectral imaging system.

3. Results and Discussion

The experiments conducted in this section are divided into three parts. The first part is to validate the performance of the CTOIFBS on a real hyperspectral image, i.e., hyperspectral digital imagery collection experiment (HYDICE) data. A second part is to apply CTOIFBS to real underwater hyperspectral data and to use the calibrated image to select a band subset to validate the CTOIFBS used for the test image. A third part is to design an underwater spectral imaging system to be used to collect the band images of underwater targets according to bands selected by CTOIFBS for detection to verify the feasibility of the USI system for rapid detection of underwater targets and the superiority of CTOIFBS to other BS methods. To further justify the three BS methods, UBS, MinV-BP, and MinV-BP-OIF along with full bands are compared in the experiments where MinV-BP-OIF uses OIF to directly select the optimal band subset for the first n bands selected by MinV-BP. The main difference between CTOIFBS and MinV-BP-OIF is that prior to calculating the OIF value, CTOIFBS uses the K-means method to divide the first n bands selected by MinV-BP into n_{BS} spectral low-relevance clusters. Then, CTOIFBS combines each band from various clusters to form a band subset and then selects a band subset with the largest OIF value as the desired band subset. Comparing to MinV-BP-OIF, the correlation among the bands selected by CTOIFBS is lower than MinV-BP-OIF. In addition, the required number of bands for HYDICE and real underwater hyperspectral data of sea cucumbers were determined by virtual dimensionality (VD) [10,29], which are six and five, respectively. Finally, visual inspection and quantitative analysis are also used to analyze and compare the performance of various BS methods.

Specifically, a 3D receiver operating characteristic (ROC) analysis-based quantitative analysis developed in [32,33] was conducted by calculating the area under the curve (AUC) for the 2D ROC curves of (P_D, P_F), (P_D, τ), and (P_F, τ) widely used in target detection where P_D and P_F represent the detection probability and the false alarm probability defined in [34], respectively, which were produced by using a different τ range from 0 to 1 to binarize the normalized detection result. The AUC values of (P_D, P_F), (P_D, τ), and (P_F, τ) were used to measure the overall detection performance, target detection capability, and background suppression ability of a detector, respectively. It should be noted that the higher the AUC values of (P_D, P_F) and (P_D, τ) are, the better the detection performance of the detector is. Conversely, the smaller the AUC value of (P_F, τ), the better the suppression ability of the background.

3.1. Real HYDICE Image

This real HYDICE scene has been widely used in target detection. It has a spatial resolution of 1.56 m and contains 169 spectral bands with a size of 64 × 64. There are 15 panels divided into five types of targets, $\mathbf{p}_1$, $\mathbf{p}_2$, $\mathbf{p}_3$, $\mathbf{p}_4$, and $\mathbf{p}_5$, which are distributed on each row with three different sizes, 3 × 3 m, 2 × 2 m, and 1 × 1 m, respectively shown in Figure 3a. Figure 3b shows their precise spatial locations with the pixels in yellow (Y pixels), indicating panel pixels mixed with the BKG. In addition, there are a total of 19 panel pixels highlighted by red, which are the target pixels to focus on.

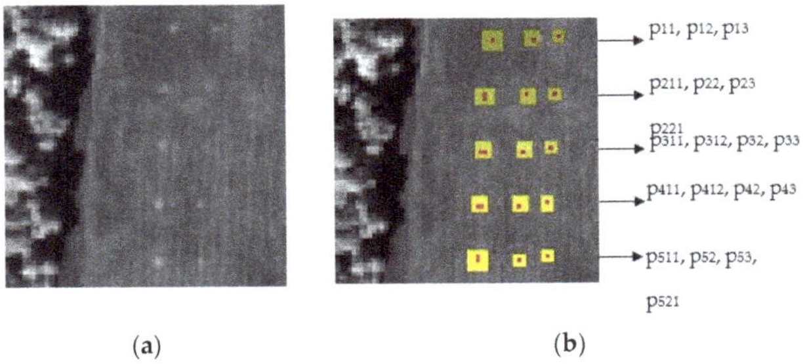

Figure 3. (**a**) Hyperspectral digital imagery collection experiment (HYDICE) scene. (**b**) Ground truth map of the 15 panels.

Table 1 shows the band subsets selected by four BS methods along with full bands for target $\mathbf{p}_1$, $\mathbf{p}_2$, $\mathbf{p}_3$, $\mathbf{p}_4$, and $\mathbf{p}_5$ in the HYDICE image. Unlike UBS, which is independent of targets, when the desired targets are different, the bands selected by three BS methods for target detection, MinV-BP, MinV-BP-OIF, and CTOIFBS, are also different. Figure 4 shows the detection results of each target under different sets of bands using CEM. From the intuitive detection results, it can be seen that the detection results are best when using the full bands with the background well suppressed. When using the set of bands selected by MinV-BP and UBS to detect targets, undesired targets respond strongly and are clearly detected. Moreover, the detection results of UBS showed that the band selected by UBS had a weak suppression ability on the background. Finally, compared with the MinV-BP-OIF and CTOIFBS methods, it can be obtained that CTOIFBS has a better ability to detect targets and has a good background suppression effect.

Table 1. Optimal band subsets selected by four BS (band selection) methods along with full bands.

Target	Method	Band Set
	Full bands	1:1:169
	UBS	1 29 57 86 114 142
$\mathbf{p}_1$	MinV-BP	169 122 123 168 167 166
	MinV-BP-OIF	133 134 98 99 135 100
	CTOIFBS	122 169 131 98 162 149
$\mathbf{p}_2$	MinV-BP	122 169 123 132 133 131
	MinV-BP-OIF	136 98 137 138 99 100
	CTOIFBS	159 100 137 128 122 98
$\mathbf{p}_3$	MinV-BP	122 123 132 133 124 131
	MinV-BP-OIF	52 53 51 54 99 100
	CTOIFBS	124 98 128 53 101 169
$\mathbf{p}_4$	MinV-BP	123 122 124 125 127 128
	MinV-BP-OIF	99 100 101 102 103 104
	CTOIFBS	99 103 124 137 127 122
$\mathbf{p}_5$	MinV-BP	122 123 124 125 126 168
	MinV-BP-OIF	127 134 135 98 138 145
	CTOIFBS	167 159 98 157 128 163

UBS: uniform band selection; MinV-BP: minimum variance band priority; MinV-BP-OIF: minimum variance band priority with OIF; CTOIFBS: constrained-target optimal index factor (OIF) band selection.

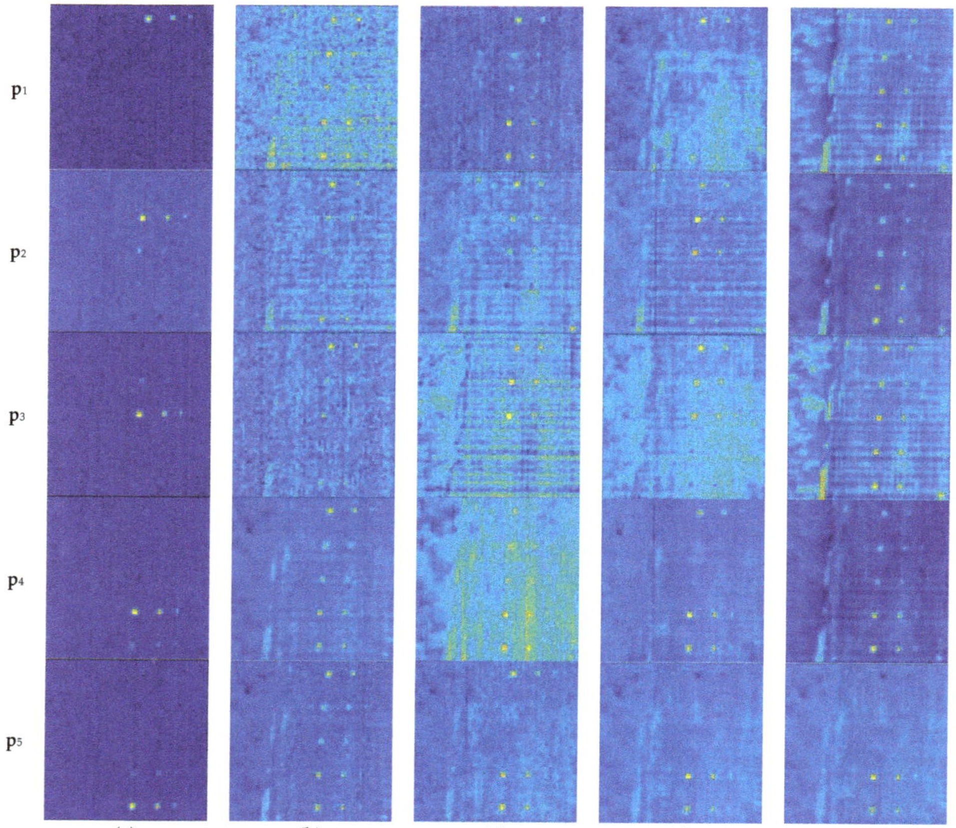

(**a**) (**b**) (**c**) (**d**) (**e**)

Figure 4. CEM (constrained energy minimization) detection map results using different band subsets selected by four BS (band selection) methods along with full bands: (**a**) Full bands; (**b**) MinV-BP: minimum variance band priority; (**c**) MinV-BP-OIF: minimum variance band priority with OIF; (**d**) CTOIFBS: constrained-target optimal index factor (OIF) band selection; (**e**) UBS: uniform band selection.

In addition to analyzing the performance of various BS methods by visual inspection, the experiment also performed quantitative analysis. Table 2 tabulates the AUC values of the five BS methods where the best and worst results are highlighted by red and green, respectively. The higher the AUC value, the better the detection, that is, the better the selected band subset to represent the target. As expected, the results using full bands were the best. However, among all the four BS methods, CTOIFBS generally outperformed the other three BS methods in terms of (P_D, P_F). In order to further demonstrate the effectiveness of CTOIFBS, Table 3 ranks the AUC value of (P_D, P_F) of various methods. The last row of Table 3 ranks the total target detection capability by the BS methods. The smaller the value, the better detection capability of the selected band subset. Among them, the value of full bands is five, ranking first, and the detection capability is the best. CTOIFBS scores 13, which is only worse than full bands. Although CTOIFBS is slightly inferior to using the full bands in detection performance, its transmission time and processing time are much lower than using the full bands due to the reduced data dimensionality. In addition, CTOIFBS performed better than MinV-BP, MinV-BP-OIF, and UBS assuming that the same number of selected bands was used.

Table 2. AUC (area under the curve) values of detection results for target $\mathbf{p}_1$, $\mathbf{p}_2$, $\mathbf{p}_3$, $\mathbf{p}_4$, and $\mathbf{p}_5$ using four BS (band selection) methods along with full bands: (a) AUC values of detection results for target $\mathbf{p}_1$; (**b**) AUC values of detection results for target $\mathbf{p}_2$; (**c**) AUC values of detection results for target $\mathbf{p}_3$; (**d**) The AUC values of detection results for target $\mathbf{p}_4$; (**e**) AUC values of detection results for target $\mathbf{p}_5$.

(a)

Method	(P_D, P_F)	(P_D, τ)	(P_F, τ)
Full bands	0.9993	0.6017	0.0328
MinV-BP	0.8179	0.7017	0.3120
MinV-BP-OIF	0.9794	0.5817	0.1603
CTOIFBS	0.9274	0.6583	0.2418
UBS	0.9784	0.5550	0.2170

(b)

Method	(P_D, P_F)	(P_D, τ)	(P_F, τ)
Full bands	0.9998	0.8375	0.1012
MinV-BP	0.9847	0.4975	0.2378
MinV-BP-OIF	0.9943	0.5875	0.2534
CTOIFBS	0.9978	0.8200	0.2266
UBS	0.9837	0.3725	0.1160

(c)

Method	(P_D, P_F)	(P_D, τ)	(P_F, τ)
Full bands	0.9997	0.7425	0.0519
MinV-BP	0.9914	0.4850	0.2113
MinV-BP-OIF	0.9937	0.7650	0.3161
CTOIFBS	0.9968	0.5775	0.2989
UBS	0.9895	0.6775	0.2412

(d)

Method	(P_D, P_F)	(P_D, τ)	(P_F, τ)
Full bands	0.9998	0.7750	0.0546
MinV-BP	0.9953	0.5700	0.1868
MinV-BP-OIF	0.9944	0.8150	0.3585
CTOIFBS	0.9985	0.7775	0.1918
UBS	0.9954	0.5925	0.1007

(e)

Method	(P_D, P_F)	(P_D, τ)	(P_F, τ)
Full bands	0.9998	0.7000	0.0495
MinV-BP	0.9960	0.6175	0.1574
MinV-BP-OIF	0.9952	0.7500	0.2148
CTOIFBS	0.9935	0.5625	0.2187
UBS	0.9954	0.7000	0.0988

MinV-BP: minimum variance band priority; MinV-BP-OIF: minimum variance band priority with OIF; CTOIFBS: constrained-target optimal index factor (OIF) band selection; UBS: uniform band selection.

Table 3. Order of the AUC (area under the curve) values of (P_D, P_F) of four BS (band selection) methods along with full bands.

	Full Bands	MinV-BP	MinV-BP-OIF	CTOIFBS	UBS
$\mathbf{p}_1$	1	5	2	4	3
$\mathbf{p}_2$	1	4	3	2	5
$\mathbf{p}_3$	1	4	3	2	5
$\mathbf{p}_4$	1	4	5	2	3
$\mathbf{p}_5$	1	2	4	5	3
SUM	5	19	17	13	19

MinV-BP: minimum variance band priority; MinV-BP-OIF: minimum variance band priority with OIF; CTOIFBS: constrained-target optimal index factor (OIF) band selection; UBS: uniform band selection.

3.2. *Underwater Hyperspectral Image*

In this section, real hyperspectral data were collected and conducted for sea cucumber detection to validate the performance of CTOIFBS. To demonstrate the effectiveness of CTOIFBS, several state-of-the-art BS methods, full bands, UBS, MinV-BP, and MinV-BP-OIF are compared by experiments where the required number of bands is five determined by VD. Finally, detection results and quantitative analysis were used to analyze and compare the performance of various BS methods. Specifically, quantitative analysis was conducted by the area under the curve (AUC) widely used in target detection.

The data used in our experiments were underwater sea cucumber images collected by a hyperspectral imager, covering 256 bands with a spectral range of 0.4 to 1.05 nm. Due to the fast attenuation of infrared bands in underwater, the sensor could not collect enough information from infrared bands. So, part of the infrared bands (171–256) were removed, and only 1–170 bands were analyzed for experiments with a spectral coverage of 0.4~0.825 nm. Shown in Figure 5a,b are the RGB images of the calibrated data and their corresponding mask image, respectively.

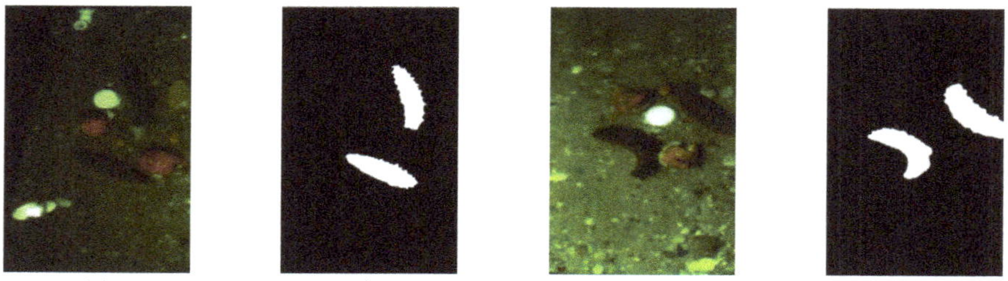

(a) (b) (c) (d)

Figure 5. Sea cucumber data for experiments: (**a**) RGB image of calibrated data; (**b**) ground truth map of calibrated data; (**c**) RGB image of validated data; (**d**) ground truth map of validated data.

We have plotted the spectra for five types of ground features, including the sea cucumber, sand, pebble, clam, and scallop from calibrated data, as shown in Figure 5a, where the sea cucumber was selected as the target of interest and the other four features as the background. The obtained spectra were used to mark the spectral bands location (points) selected by the four BS methods in Table 4, which is shown in Figure 6 using red vertical dashed lines for visual inspection and comparison among correlation of the selected band sets.

Table 4. Band subsets selected by four BS (band selection) methods along with full bands.

Method	Band Set
Full bands	1:1:170
MinV-BP	170 168 43 46 36
MinV-BP-OIF	170 168 169 167 29
CIOIFBS	34 43 29 58 170
UBS	1 35 69 103 137

MinV-BP: minimum variance band priority; MinV-BP-OIF: minimum variance band priority with OIF; CTOIFBS: constrained-target optimal index factor (OIF) band selection; UBS: uniform band selection.

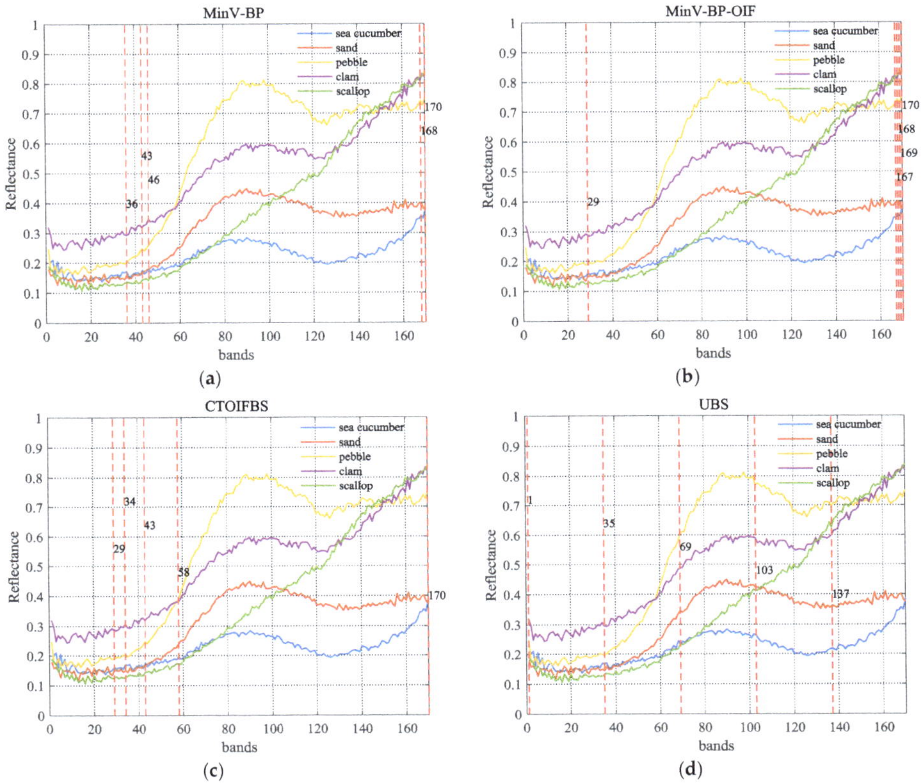

Figure 6. Bands selected by four BS (band selection) methods: (**a**) MinV-BP: minimum variance band priority; (**b**) MinV-BP-OIF: minimum variance band priority with OIF; (**c**) CTOIFBS: constrained-target optimal index factor (OIF) band selection; (**d**) UBS: uniform band selection.

On the one hand, comparing to MinV-BP and MinV-BP-OIF, CTOIFBS took the correlation among bands into consideration. As a result, the bands selected by CTOIFBS were more dispersed and contained more spectral information. On the other hand, although the distribution of band selected by UBS was more dispersed than the other three methods, the detection results were not satisfactory. This is because UBS did not consider the special relationship between the target and its selected bands. Consequently, it was unable to select bands pertaining to target information compared to the band set selected by CTOIFBS, which can effectively avoid high correlation between bands and can be further used to characterize targets of interest.

Table 5 shows the correlation coefficient among bands in each band subset selected by a different BS method where the greater the value between two bands in a band subset, the higher the correlation between these two bands. So, a better band subset should have less correlation among its bands. Furthermore, Table 6 shows the mean correlation coefficients among bands selected by different BS methods.

Table 5. Correlation coefficient matrices of the band subset selected by four BS (band selection) methods: (**a**) correlation coefficient matrix of the band subset selected by MinV-BP; (**b**) correlation coefficient matrix of the band subset selected by MinV-BP-OIF; (**c**) correlation coefficient matrix of the band subset selected by CTOIFBS; (**d**) correlation coefficient matrix of the band subset selected by UBS.

(a)

Band no.	170	168	43	46	36
170	1				
168	0.9913	1			
43	0.8840	0.8887	1		
46	0.8958	0.9018	0.9928	1	
36	0.8420	0.8440	0.9809	0.9651	1

(b)

Band no.	170	168	169	167	29
170	1				
168	0.9913	1			
169	0.9918	0.9923	1		
167	0.9904	0.9928	0.9919	1	
29	0.8016	0.8023	0.8021	0.8024	1

(c)

Band no.	34	43	29	58	170
34	1				
43	0.9711	1			
29	0.9920	0.9518	1		
58	0.8261	0.9237	0.7890	1	
170	0.8269	0.8840	0.8016	0.8958	1

(d)

Band no.	1	35	69	103	137
1	1				
35	0.9212	1			
69	0.5197	0.7005	1		
103	0.4206	0.6024	0.9694	1	
137	0.6408	0.7946	0.9405	0.9227	1

Table 6. Mean correlation coefficients of four BS (band selection) methods.

Method	MinV-BP	MinV-BP-OIF	CTOIFBS	UBS
Mean correlation coefficient	0.9186	0.9158	0.8862	0.7432

MinV-BP: minimum variance band priority; MinV-BP-OIF: minimum variance band priority with OIF; CTOIFBS: constrained-target optimal index factor (OIF) band selection; UBS: uniform band selection.

From Table 6, it can be seen that compared to the other two target-constrained BS methods, the mean correlation coefficient among the bands selected by CTOIFBS is the smallest, which validates the advantage of CTOIFBS in reducing correlation between bands during the BS. It is worth noting that although the mean correlation coefficient among the bands selected by UBS is the smallest, its detection results were poor due to its inability to select effective bands to characterize the target.

According to the band subsets selected by different BS methods in Table 4, their corresponding band images of the calibrated data shown in Figure 5a were synthesized. CEM was then used to detect sea cucumbers, and the detection results of using full bands and band subsets selected by four BS methods were shown in Figure 7. The brighter a pixel in the image is, the higher the probability that the pixel is considered to be more likely a target by the detector. It is also observed that the target pixels detected with a band set selected by UBS were not obvious and have been buried in the background. Furthermore, the AUC values calculated in Table 7 were also used to quantitatively analyze the effect of different BS methods on detection performance where the best and worst results are highlighted by red and green, respectively. Comparing to the AUC values of (P_D, P_F), the full band was the best followed by CTOIFBS, MinV-BP-OIF, and MinV-BP, and finally, UBS.

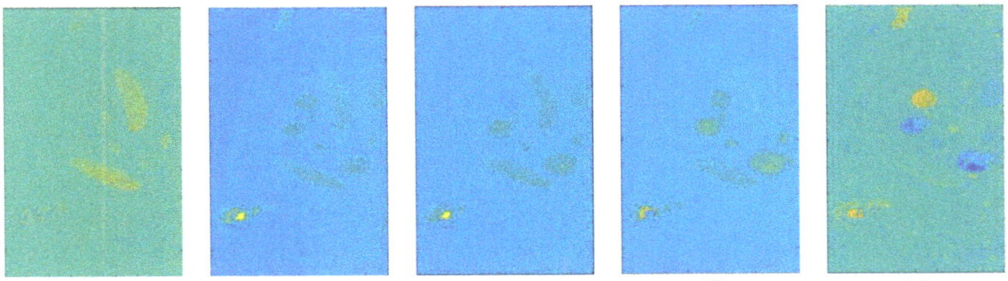

(a) (b) (c) (d) (e)

Figure 7. Detection results of the calibrated data of the RGB image and ground truth map shown in Figure 5a,b by full bands and four BS methods: (**a**) Full bands; (**b**) MinV-BP: minimum variance band priority; (**c**) MinV-BP-OIF: minimum variance band priority with OIF; (**d**) CTOIFBS: constrained-target optimal index factor (OIF) band selection; (**e**) UBS: uniform band selection.

Table 7. AUC (area under the curve) values of five BS (band selection) methods.

Method	(P_D, P_F)	(P_D, τ)	(P_F, τ)
Full bands	0.9022	0.5783	0.4917
MinV-BP	0.7315	0.3511	0.2998
MinV-BP-OIF	0.7577	0.3666	0.3145
CTOIFBS	0.7961	0.3522	0.3176
UBS	0.6148	0.4717	0.4601

MinV-BP: minimum variance band priority; MinV-BP-OIF: minimum variance band priority with OIF; CTOIFBS: constrained-target optimal index factor (OIF) band selection; UBS: uniform band selection.

In order to further validate the effectiveness of CTOIFBS in detecting underwater targets, an additional experimental image was also selected for testing the performance of various BS methods. Figure 8 shows the detection results of sea cucumbers on the test image using a set of bands selected in Table 4. Table 8 tabulates their AUC values where the best and worst results are highlighted by red and green, respectively. According to the AUC values of (P_D, P_F) in Table 8, the detection result of CTOIFBS was higher than that of other BS methods, MinV-BP, MinV-BP-OIF, and UBS using the same number of bands. As expected, the CTOIFBS result was only worse than that of using full bands. This proves that it is feasible to use the band subset selected by CTOIFBS for underwater target detection.

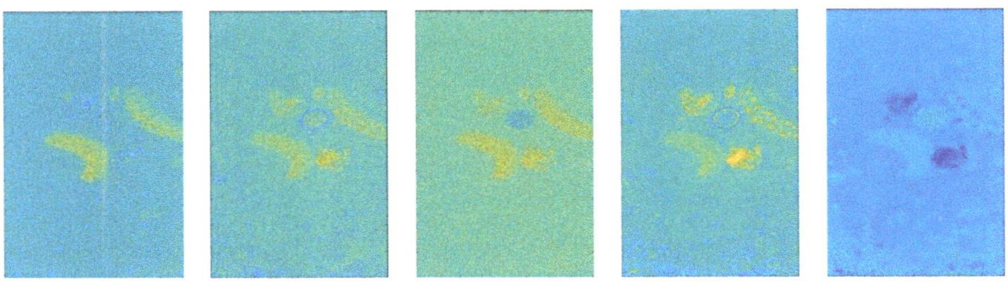

(a) (b) (c) (d) (e)

Figure 8. Detection results of the validated data of the RGB image and ground truth map shown in Figure 5c,d by full bands and four BS methods: (**a**) Full bands; (**b**) MinV-BP: minimum variance band priority; (**c**) MinV-BP-OIF: minimum variance band priority with OIF; (**d**) CTOIFBS: constrained-target optimal index factor (OIF) band selection; (**e**) UBS: uniform band selection.

Table 8. AUC (area under the curve) values of four BS (band selection) methods along with full bands.

Method	(P_D, P_F)	(P_D, τ)	(P_F, τ)
Full bands	0.9396	0.5494	0.4036
MinV-BP	0.8482	0.5566	0.4542
MinV-BP-OIF	0.8442	0.6199	0.5181
CTOIFBS	0.9318	0.5349	0.4357
UBS	0.7460	0.3099	0.2914

MinV-BP: minimum variance band priority; MinV-BP-OIF: minimum variance band priority with OIF; CTOIFBS: constrained-target optimal index factor (OIF) band selection; UBS: uniform band selection.

The above real image sea cucumber image experiments also proved that it was feasible to use the band subset selected by CTOIFBS for underwater target detection. Although the detection result of CTOIFBS is slightly worse than that of using full bands, the acquisition and transmission speeds are considerably faster than using full bands because a smaller number of bands were used, and the smaller amount of image data is being processed. Table 9 shows the detection speeds of using full bands and CTOIFBS under the same experimental environment.

Table 9. Comparison of the average speed of two methods for detecting a single image.

Method	Detection Speed (ms)
Full bands	494
CTOIFBS	98

CTOIFBS: constrained-target optimal index factor (OIF) band selection.

From Table 9, the process of using full bands consumed a great deal of time, which was reflected in imaging, transmission, and processing. Under the effect of water flow, target movement, and other factors, a USI system needs to detect the target quickly. Obviously, a USI system using full bands cannot meet the requirement for rapid detection of an underwater target. In addition, studies have found that using full bands may incur an issue of the Hughes phenomenon [35], that is, high dimensionality may decrease the detection accuracy. Furthermore, the experiments further demonstrated that the detection results of CTOIFBS could be very close to that obtained using the full bands. With all things considered above, a USI system with full bands is not suitable for underwater rapid target detection.

3.3. *Underwater Spectral Imaging System*

In order to verify that the collected target spectral data by the constructed underwater spectral imaging (USI) system can accurately detect underwater targets, two experiments were set up in this section. The first experiment was conducted by comparing the hyperspectral data using the selected band subset to the multi-spectral data collected by the USI system using the same band subset under similar scenes to prove that the multi-spectral data collected by the USI system has consistent feature expression capability with the hyperspectral images. A second experiment was also conducted under the same scenes to compare the detection performance of data collected by the USI system using different BS methods to verify the detection capability of CTOIFBS.

3.3.1. First Experiment: Compatibility of USI to HSI

In order to show that the multi-spectral data collected by the USI system have the same feature expression ability as the hyperspectral images, the experiment collected the hyperspectral data and the filter bands corresponding to the band subset selected by CTOIFBS in similar scenes. Because the bands selected by CTOIFBS are 470, 480, 500, 540, and 830 nm, the band images corresponding to the hyperspectral data were extracted to form a band subset for subsequent target detection. Figure 9 shows the images collected by two methods and their corresponding detection results of sea cucumbers in similar scenes.

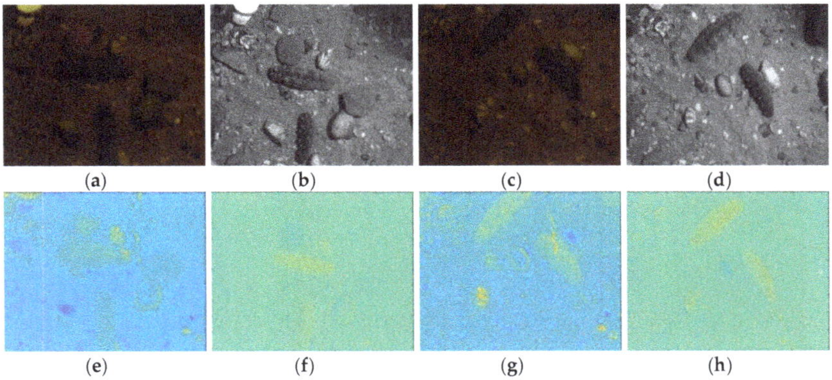

Figure 9. Images collected by two methods and corresponding detection map results in similar scenes. HSI-01 (**a**), HSI-02 (**c**) are hyperspectral images, USI-01 (**b**), USI-02 (**d**) are images collected by the USI system; (**e**), (**f**), (**g**), (**h**) are the detection results corresponding to (**a**), (**b**), (**c**), (**d**).

According to the detection results, both methods are capable of detecting sea cucumbers. From the performance of suppressing non-target pixels, although the image extracted from the HSI data can suppress the main background, which is sand, it has a high response to interference targets, such as stones and clams. By contrast, the data collected by the USI system can suppress non-target pixels more effectively. From the AUC values of (P_D, P_F) in Table 10, the AUC value detected using the data collected by the USI system is higher than that using HSI data, indicating that its ability to detect targets is higher. Of course, due to the difference in the performance of the sensors used by the two methods, this experiment may not have sufficient evidence to conclude that the detection results based on the data collected by the USI system must be better than the data using the corresponding band of HSI. Nevertheless, it can prove that the data collected using the USI system has the same feature expression ability as the hyperspectral images and can be used for underwater spectral data collection and target detection.

Table 10. AUC (area under the curve) values for CEM (constrained energy minimization) detection map results using four images shown in Figure 9.

Data	(P_D, P_F)	(P_D, τ)	(P_F, τ)
HSI-01	0.7391	0.1459	0.0831
USI-01	0.8673	0.0901	0.0318
HSI-02	0.8451	0.2173	0.0940
USI-02	0.9274	0.1329	0.0432

3.3.2. Second Experiment: USI System using CTOIFBS

This section uses the data collected by the USI system to compare the performance of the CTOIFBS with four BS methods. MinV-BP, MinV-BP-OIF, and UBS with their corresponding band subsets tabulated in Table 11. Then, the single-band images are collected by the USI system, as shown in Figure 10. Finally, the collected single-band images are integrated into multi-spectral image cubes for target detection. It should be noted that the single band image-constructed multi-spectral image data has indeed a spectral resolution of approximate 10 nm, and thus, the filters actually used are rounded to 10nm.

Table 11. Band subsets selected by four BS (band selection) methods.

Methods	Selected Bands (nm)
MinV-BP	825.6 820.4 506.0 513.5 489.1
MinV-BP-OIF	825.6 820.4 823.0 817.8 472.1
CTOIFBS	484.2 506.0 472.1 542.9 825.6
UBS	400.0 486.7 570.0 654.6 740.7

MinV-BP: minimum variance band priority; MinV-BP-OIF: minimum variance band priority with OIF; CTOIFBS: constrained-target optimal index factor (OIF) band selection; UBS: uniform band selection.

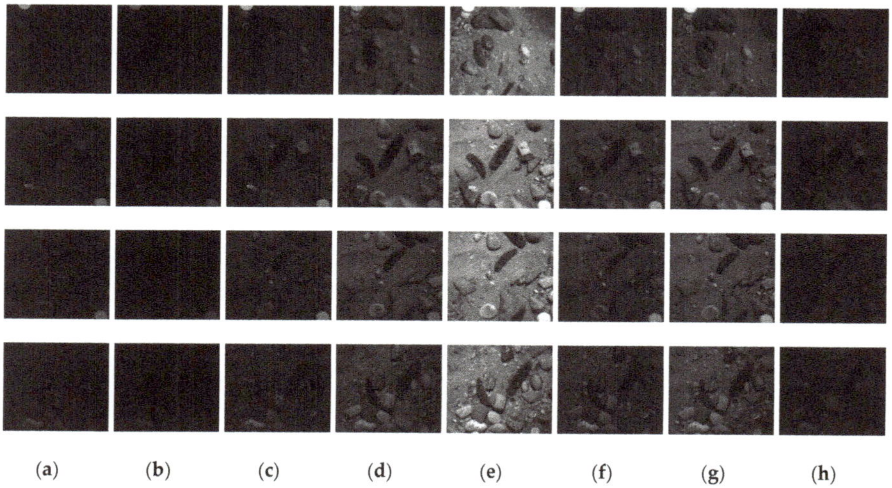

Figure 10. Image acquisition with different bands: (**a**) 470; (**b**) 490; (**c**) 510; (**d**) 540; (**e**) 570; (**f**) 650; (**g**) 740; (**h**) 820 nm.

CEM was used to detect the sea cucumbers in the composite image of each band subset. The detection results corresponding to each method are shown in Figure 11.

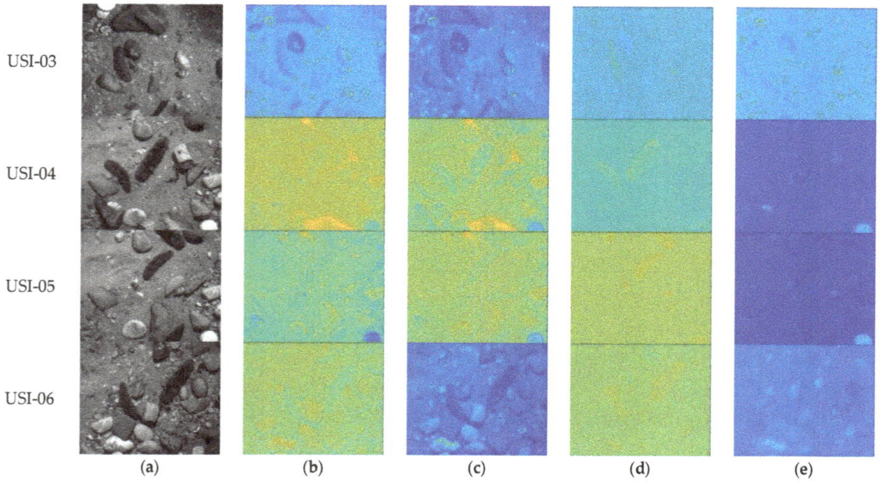

Figure 11. Detection map results using four BS (band selection) methods: (**a**) original image; (**b**) MinV-BP: minimum variance band priority; (**c**) MinV-BP-OIF: minimum variance band priority with OIF; (**d**) CTOIFBS: constrained-target optimal index factor (OIF) band selection; (**e**) UBS: uniform band selection.

The detection results shown in Figure 11 illustrated that when the set of bands selected by CTOIFBS was used to detect sea cucumbers, non-target pixels could be removed more effectively compared to other BS methods. On the contrary, MinV-BP and MinV-BP-OIF had poor ability in distinguishing the targets from the background, and the response to non-target pixels was also high when the target was detected. Table 12 shows the AUC values of the detection, and we also highlight the best and worst results by red and green. According to the AUC values of (P_D, P_F) in Table 12, UBS has the worst performance on all four test images. This shows that BS methods based on a constrained-target are more conducive to target detection. Furthermore, except for image USI-06, the AUC value of CTOIFBS is the highest. This proves that compared to other BS methods based on a constrained-target, MinV-BP, and MinV-BP-OIF, CTOIFBS has a better ability to characterize targets.

Table 12. AUC (area under the curve) values of detection using four BS (band selection) methods.

Data	Method	USI-03	USI-04	USI-05	USI-06
(P_D, P_F)	MinV-BP	0.6343	0.6008	0.5335	0.7845
	MinV-BP-OIF	0.7061	0.7007	0.6414	0.8394
	CTOIFBS	0.8603	0.8500	0.7727	0.7859
	UBS	0.5997	0.5236	0.5915	0.5460
(P_D, τ)	MinV-BP	0.8644	0.9074	0.8831	0.8657
	MinV-BP -OIF	0.8660	0.8316	0.9042	0.8893
	CTOIFBS	0.9123	0.9024	0.9177	0.9225
	UBS	0.9134	0.9611	0.9634	0.9609
(P_F, τ)	MinV-BP	0.8349	0.8868	0.8715	0.7950
	MinV-BP -OIF	0.8262	0.7641	0.8707	0.8189
	CTOIFBS	0.9367	0.9614	0.9579	0.9606
	UBS	0.9244	0.9573	0.9528	0.9532

MinV-BP: minimum variance band priority; MinV-BP-OIF: minimum variance band priority with OIF; CTOIFBS: constrained-target optimal index factor (OIF) band selection; UBS: uniform band selection.

4. Conclusions

Hyperspectral imaging technology has advantages of high spectral resolution and abundant spectral information. Its applications to underwater object detection can help overcome the problems of a poor underwater imaging environment and complex background. The fast processing of detecting underwater hyperspectral targets can be achieved by CTOIFBS, while retaining crucial spectral information. In the meantime, CTOIFBS also overcomes the imaging and processing speed problems. Experiments show that the detection performance of the band subset selected by CTOIFBS is better than that by using other BS methods.

Author Contributions: Conceptualization, X.F. and M.S.; formal analysis, H.Y.; methodology, X.S. (Xiaodi Shang) and X.S. (Xudong Sun); writing—original draft, X.S. (Xiaodi Shang) and X.S. (Xudong Sun); writing—review and editing, C.-IC. All authors have read and agreed to the published version of the manuscript.

Funding: This research was funded by the National Nature Science Foundation of China, grant number 61601077, 61971082, 61890964; Fundamental Research Funds for the Central Universities, grant number 3132019341; State Administration of Foreign Experts Affairs, grant number ZD20180073.

Conflicts of Interest: The authors declare no conflict of interest.

References

1. Christian, B.; Ronald, P. A fully automated method to detect and segment a manufactured object in an underwater color image. *EURASIP J. Adv. Signal Process.* **2010**, 1–11.
2. Li, C.; Guo, J.-C.; Cong, R.; Pang, Y.-W.; Wang, B. Underwater Image Enhancement by Dehazing With Minimum Information Loss and Histogram Distribution Prior. *IEEE Trans. Image Process.* **2016**, *25*, 5664–5677. [CrossRef] [PubMed]

3. Ren, S.; He, K.; Girshick, R.; Sun, J. Faster r-cnn: Towards real-time object detection with region proposal networks. *IEEE Transactions PAMI* **2015**, *39*, 1137–1149. [CrossRef] [PubMed]
4. Redmon, J.; Divvala, S.; Girshick, R.; Farhadi, A. You Only Look Once: Unified, Real-Time Object Detection. In Proceedings of the 2016 IEEE Conference on Computer Vision and Pattern Recognition (CVPR), Las Vegas, NV, USA, 27–30 June 2016; pp. 779–788.
5. Jaffe, J.S. Underwater Optical Imaging: The Past, the Present, and the Prospects. *IEEE J. Ocean. Eng.* **2015**, *40*, 683–700. [CrossRef]
6. Johnsen, S. Transparent animals. *Sci. Am.* **2000**, *282*, 62–71. [CrossRef] [PubMed]
7. Reed, I.; Yu, X. Adaptive multiple-band CFAR detection of an optical pattern with unknown spectral distribution. *IEEE Trans. Acoust. Speech, Signal Process.* **1990**, *38*, 1760–1770. [CrossRef]
8. Kwon, H.; Nasrabadi, N. Kernel RX-algorithm: A nonlinear anomaly detector for hyperspectral imagery. *IEEE Trans. Geosci. Remote. Sens.* **2005**, *43*, 388–397. [CrossRef]
9. Chang, C.-I. Orthogonal subspace projection (OSP) revisited: A comprehensive study and analysis. *IEEE Trans. Geosci. Remote Sens.* **2005**, *43*, 502–518. [CrossRef]
10. Chang, C.-I. *Hyperspectral Imaging: Techniques for Spectral Detection and Classification*; Plenum Publishing Co.: New York, NY, USA, 2003.
11. Johnsen, G.; Ludvigsen, M.; Sørensen, A.; Aas, L.M.S. The use of underwater hyperspectral imaging de-ployed on remotely operated vehicles—Methods and applications. *IFAC-PapersOnLine* **2016**, *49*, 476–481. [CrossRef]
12. Sture, Ø.; Ludvigsen, M.; Soreide, F.; Aas, L.M.S. Autonomous underwater vehicles as a platform for underwater hyperspectral imaging. *OCEANS Aberdeen* **2017**, 1–8.
13. Klonowski, W.M.; Fearns, P.; Lynch, M.J. Retrieving key benthic cover types and bathymetry from hyperspectral imagery. *J. Appl. Remote. Sens.* **2007**, *1*, 011505. [CrossRef]
14. Fearns, P.R.C.; Klonowski, W.; Babcock, R.C.; England, P.; Phillips, J. Shallow water sub-strate mapping using hyperspectral remote sensing. *Cont. Shelf Res.* **2011**, *31*, 1249–1259. [CrossRef]
15. Dierssen, H.M. Overview of hyperspectral remote sensing for mapping marine benthic habitats from airborne and underwater sensors. *Opt. Eng. Appl.* **2013**, 88700.
16. Dierssen, H.; Chlus, A.; Russell, B. Hyperspectral discrimination of floating mats of seagrass wrack and the macroalgae Sargassum in coastal waters of Greater Florida Bay using airborne remote sensing. *Remote. Sens. Environ.* **2015**, *167*, 247–258. [CrossRef]
17. Dumke, I.; Purser, A.; Marcon, Y.; Nornes, S.M.; Johnsen, G.; Ludvigsen, M.; Søreide, F. Underwater hyperspectral imaging as an in situ taxonomic tool for deep-sea megafauna. *Sci. Rep.* **2018**, *8*, 12860. [CrossRef]
18. Uto, K.; Seki, H.; Saito, G.; Kosugi, Y.; Komatsu, T. Development of a Low-Cost Hyperspectral Whiskbroom Imager Using an Optical Fiber Bundle, a Swing Mirror, and Compact Spectrometers. *IEEE J. Sel. Top. Appl. Earth Obs. Remote. Sens.* **2016**, *9*, 3909–3925. [CrossRef]
19. Tegdan, J.; Ekehaug, S.; Hansen, I.M.; Aas, L.M.S.; Steen, K.J.; Pettersen, R.; Beuchel, F.; Camus, L. Underwater hyperspectral imaging for environmental mapping and monitoring of seabed habitats. *OCEANS Genova* **2015**, 1–6.
20. Keshava, N. Distance metrics and band selection in hyperspectral processing with applications to material identification and spectral libraries. *IEEE Trans. Geosci. Remote. Sens.* **2004**, *42*, 1552–1565. [CrossRef]
21. Gleason, A.C.R.; Reid, R.P.; Voss, K. Automated classification of underwater multispectral imagery for coral reef monitoring. *OCEANS* **2007**, 1–8.
22. Chang, C.-I.; Du, Q.; Sun, T.-L.; Althouse, M. A joint band prioritization and band-decorrelation approach to band selection for hyperspectral image classification. *IEEE Trans. Geosci. Remote. Sens.* **1999**, *37*, 2631–2641. [CrossRef]
23. Chavez, P.S.; Berlin, G.L.; Sowers, L.B. Statistical method for selecting Landsat MSS ratios. *J. Appl. Photogr. Eng.* **1982**.
24. Yang, H.; Du, Q.; Su, H.; Sheng, Y. An Efficient Method for Supervised Hyperspectral Band Selection. *IEEE Geosci. Remote. Sens. Lett.* **2010**, *8*, 138–142. [CrossRef]
25. Yuan, Y.; Zheng, X.; Lu, X. Discovering Diverse Subset for Unsupervised Hyperspectral Band Selection. *IEEE Trans. Image Process.* **2016**, *26*, 51–64. [CrossRef]

26. Wang, L.; Chang, C.-I.; Lee, L.-C.; Wang, Y.; Xue, B.; Song, M.; Yu, C.; Li, S. Band Subset Selection for Anomaly Detection in Hyperspectral Imagery. *IEEE Trans. Geosci. Remote. Sens.* **2017**, *55*, 4887–4898. [CrossRef]
27. Geng, X.; Sun, K.; Ji, L. Band selection for target detection in hyperspectral imagery using sparse CEM. *Remote. Sens. Lett.* **2014**, *5*, 1022–1031. [CrossRef]
28. Wang, Y.; Wang, L.; Yu, C.; Zhao, E.; Song, M.; Wen, C.-H.; Chang, C.-I. Constrained-Target BS for multiple-target detection. *IEEE Trans. Geosci. Remote Sens.* **2019**, *57*, 6079–6103. [CrossRef]
29. Chang, C.-I.; Du, Q. Estimation of Number of Spectrally Distinct Signal Sources in Hyperspectral Imagery. *IEEE Trans. Geosci. Remote. Sens.* **2004**, *42*, 608–619. [CrossRef]
30. Yu, C.; Lee, L.-C.; Chang, C.-I.; Xue, B.; Song, M.; Chen, J. Band-Specified Virtual Dimensionality for Band Selection: An Orthogonal Subspace Projection Approach. *IEEE Trans. Geosci. Remote. Sens.* **2018**, *56*, 2822–2832. [CrossRef]
31. Chang, C.-I. A review of virtual dimensionality for hyperspectral imagery. *IEEE J-STARS* **2018**, *11*, 1285–1305. [CrossRef]
32. Chang, C.-I. Multiple-parameter receiver operating characteristic analysis for signal detection and classification. *IEEE Sens. J.* **2019**, *10*, 423–442. [CrossRef]
33. Chang, C.-I. *Hyperspectral Data Processing: Algorithm Design and Analysis*; Wiley: New Jersey, NJ, USA, 2013.
34. Poor, H.V. *An Introduction to Detection and Estimation Theory*, 2nd ed.; Springer: New York, NY, USA, 1994.
35. Hughes, G. On the mean accuracy of statistical pattern recognizers. *IEEE Trans. Inf. Theory* **1968**, *14*, 55–63. [CrossRef]

remote sensing

MDPI

Article

Attention-Based Spatial and Spectral Network with PCA-Guided Self-Supervised Feature Extraction for Change Detection in Hyperspectral Images

Zhao Wang [1], Fenlong Jiang [1], Tongfei Liu [1], Fei Xie [2,*] and Peng Li [1]

[1] Key Laboratory of Electronic Information Countermeasure and Simulation Technology of Ministry of Education, School of Electronic Engineering, Xidian University, No. 2 South TaiBai Road, Xi'an 710075, China; wangzhao@xidian.edu.cn (Z.W.); fljiang@stu.xidian.edu.cn (F.J.); ltfei@stu.xidian.edu.cn (T.L.); penglixd@xidian.edu.cn (P.L.)

[2] Academy of Advanced Interdisciplinary Research, Xidian University, No. 2 South TaiBai Road, Xi'an 710068, China

* Correspondence: fxie@xidian.edu.cn

Abstract: Joint analysis of spatial and spectral features has always been an important method for change detection in hyperspectral images. However, many existing methods cannot extract effective spatial features from the data itself. Moreover, when combining spatial and spectral features, a rough uniform global combination ratio is usually required. To address these problems, in this paper, we propose a novel attention-based spatial and spectral network with PCA-guided self-supervised feature extraction mechanism to detect changes in hyperspectral images. The whole framework is divided into two steps. First, a self-supervised mapping from each patch of the difference map to the principal components of the central pixel of each patch is established. By using the multi-layer convolutional neural network, the main spatial features of differences can be extracted. In the second step, the attention mechanism is introduced. Specifically, the weighting factor between the spatial and spectral features of each pixel is adaptively calculated from the concatenated spatial and spectral features. Then, the calculated factor is applied proportionally to the corresponding features. Finally, by the joint analysis of the weighted spatial and spectral features, the change status of pixels in different positions can be obtained. Experimental results on several real hyperspectral change detection data sets show the effectiveness and advancement of the proposed method.

Keywords: hyperspectral images; change detection; self-supervised learning; attention mechanism

Citation: Wang, Z.; Jiang, F.; Liu, T.; Xie, F.; Li, P. Attention-Based Spatial and Spectral Network with PCA-Guided Self-Supervised Feature Extraction for Change Detection in Hyperspectral Images. *Remote Sens.* **2021**, *13*, 4927. https://doi.org/10.3390/rs13234927

Academic Editors: Chein-I Chang, Meiping Song, Chunyan Yu, Yulei Wang, Haoyang Yu, Jiaojiao Li, Lin Wang, Hsiao-Chi Li and Xiaorun Li

Received: 13 October 2021
Accepted: 30 November 2021
Published: 4 December 2021

1. Introduction

Change detection (CD) has been a popular research and application in the field of remote sensing in recent years, which aims to acquire the change information from multi-temporal images in the same geographical area. The change information is vital in many applications, such as disaster detection and assessment [1], environmental governance [2], ecosystem monitoring [3], urban sustainable development [4,5], etc.

With the advances in sensing and imaging technology, hyperspectral images (HSIs) have attracted increasing attention and been widely utilized in earth observation applications [4,6]. Some characteristics of HSIs should be noticed: unlike multispectral images and SAR images, HSIs typically have hundreds of spectral bands, and this rich spectral information helps detect finer changes for CD. Although HSIs bring some key advantages, redundant spectral bands may introduce interference information as adjacent bands have similar spectral values, which are continuously measured by the hyperspectral sensor [4]. Moreover, the high-dimensional spectral band also leads to a significant increase in the storage and computational complexity of HSIs processing and analysis [7]. In addition, for HSIs, spatial feature extraction is more challenging than multispectral image as the serious mixed pixels phenomenon caused by low spatial resolution [8]. Furthermore, it is very difficult to obtain enough labeled training samples in HSIs analysis.

In view of the characteristics of HSIs, many approaches have been proposed for CD in HSIs. These methods can be mainly summarized into two categories:

(1) One is to directly use spectral features to obtain change information for multi-temporal HSIs. For example, Liu et al. promoted a sequential spectral change vector analysis to detect multiple changes for HSIs [9], which employs an adaptive spectral change vector representation to identify changes. Liu et al. employed spectral change information to detect change classes for achieving unsupervised HSIs change detection [10]. Different from the common method by reducing or selecting the band to reduce the band redundancy for CD in HSI, in [11], change information of each band is utilized to construct the hyperspectral change vectors for detecting multiple types of change. Recently, a general end-to-end convolutional neural network (CNN) has been proposed for HSI CD in [6], named GETNET, which introduces a unmixing-based subpixel representation to fuse multi-source information. The performance of these methods is often hindered as they usually utilize change vector analysis of spectral feature to generate directly change magnitude between multi-temporal HSIs.

(2) However, only using spectral features is bound to ignore spatial contextual information [12]. Therefore, joint spatial-spectral analysis is a common technical means in HSI-based tasks [13–17]. Therefore, the other is to obtain changes and improve detection accuracy through joint analysis of spectral and spatial features of HSI. For instance, Wu et al. stacked first multi-temporal HSIs, and then the local spatial information around the pixel is presented through joint sparse representation for hyperspectral anomalous CD [18]. Recently, a CD approach based on multiple morphological profiles has been proposed in HSIs [19]. This approach employed multiple morphological profiles to extract spatial information, and then a spectral angle weighted-based local absolute distance and an absolute distance are used to obtain changes. In addition, some deep learning-based techniques can help improve the performance of CD due to its ability to effectively capture and fuse spectral and spatial features. A recurrent 3D fully convolutional networks is designed to capture spectral-spatial features of HSIs simultaneously for CD in [12]. Zhan et al. promoted a three-directions spectral-spatial convolution neural network (TDSSC) in [20], which can capture representative spectra-spatial features by concatenating the feature of spectral direction and two spatial directions, and thus improving detection performance. Such methods are usually weighted to equalize spatial and spectral features to conduct joint analysis and classification, and have achieved good performance, but they usually have the following common problems:

- The spatial features extracted by existing methods may not target for CD. For example, some methods require transfer learning from other tasks such as classification, segmentation, etc. These tasks require large-scale labeled data sets for supervised training, which increases the cost of use. There are also some methods that use autoencoders to extract the deep expression of each image. The features extracted by these two methods may not be suitable for CD. Therefore, how to extract sufficiently good spatial differential representations for CD tasks is a very critical issue.
- Most methods adopt a uniform global weight factor when combining spatial and spectral features, that is, spatial and spectral features are analyzed according to the same ratio for each pixel at each location, which is obviously a little rough. Therefore, how to balance these two features in a task-driven adaptive way is also worth studying.

To address these two problems mentioned above, in this paper, we propose an attention-based spatial and spectral network with PCA-guided self-supervised feature extraction for CD in HSIs. The whole framework consists of two parts. In the first part, a PCA-guided self-supervised spatial feature extraction network is devised to extract spatial differential features. Concretely, two HSIs are compared to generate a difference map (DM) first. Then, the principal component analysis is utilized to obtain the transferred image that only contains several principal components. Afterwards, a mapping from the image patch, i.e., a neighborhood with a certain size for each pixel in the DM, to the corresponding principal component vector in the transferred image is established, where the spatial targeted differential features can be extracted. Finally, the extracted spatial features can be used in the subsequently joint analysis combined with the spectral features. In the whole process, no additional supervisory information is involved, and the training data used in the training only comes from the processing of the data itself, which is categorized into the self-supervised learning task recently [21–23]. These methods mine useful supervisory information from the data itself and can obtain performance not weaker than external supervised learning. Besides, the designed mapping relationship can make the extracted spatial features more distinctive. In the second part, we propose an attention-based spatial and spectral CD network. Different from the above-mentioned methods, the attention mechanism [24–26] is introduced to balance the spatial and spectral features adaptively. Specifically, the spatial and spectral features are first combined directly to calculate a weight factor for the corresponding pixel via several fully-connected layers. After that, the calculated factor is applied to weight the two features. Finally, by combining the weighted spatial and spectral features, the final change status for each pixel can be inferred. The introduction of attention mechanism enables the network to calculate its own weight factor for the spatial and spectral features of each pixel, which avoids multiple trials to select the optimal factor and allows for more detailed detection of changes. In order to improve the network performance and the detection effect, a few ground truth labels are used for semi-supervised training detection network. Experiments on several real data sets show the effectiveness and advance of our algorithm. The main contributions of our work are summarized below:

(1) A novel PCA-guided self-supervised spatial feature extraction network, which establishes the mapping relationship from the difference to the principal components of the difference, so as to extract more specific difference representation.
(2) The attention mechanism is introduced, which adaptively balances the proportion of spatial and spectral features, avoiding rough combination with global uniform ratio, making the model more adaptable.
(3) We propose an innovative framework for hyperspectral image change detection, which involves a novel PCA-guided self-supervised spatial feature extraction network and an attention-based spatial-spectral fusion network. Moreover, the proposed ASSCDN can achieve the superior performance using only a small number of training samples on three widely used HSI CD datasets.

The rest of this paper is organized as follows. Related works are presented in Section 2. Section 3 describes the proposed ASSCDN in detail. In Section 4, experiments and analysis based on three pairs of HSI dataset are presented and discussed. Finally, the conclusion is provided in Section 6.

2. Related Works

2.1. Traditional CD Methods

During past few decades, many CD methods have been proposed and applied in practical applications [27,28]. In the early development of CD, two main steps are usually required to realize CD: measuring the difference image (DI) and obtaining the change detection map (CDM). Many techniques are commonly used to measure DI, such as image difference [29], image log-ratio [30], change vector analysis (CVA) [29,31], etc. Generally, these approaches calculate the change magnitude of bi-temporal images by the distance between two pixels. Afterwards, the methods widely used to generate CDM are threshold segmentation techniques (OTSU [32], expectation maximum [33]) or clustering algorithms (k-means [34], fuzzy c-means [35], k-nearest neighbors (KNN) [36], and support vector machines (SVM) [37]). With the development of CD technology, some methods are further promoted to improve the detection performance. For example, Zhuang et al. combined spectral angle mapper and change vector analysis for CD of multispectral images [38]. Thonfeld et al. proposed a robust change vector analysis (RCVA) [39] approach for multi-sensor satellite images CD. In addition to the above methods, some techniques are also helpful to improve the performance of CD, such as principal component analysis (PCA) [34,40], level set [41,42], Markov field [43,44], etc. However, these approaches rely significantly on the quality of hand-crafted features in order to measure the similarity between bi-temporal images.

2.2. Deep Learning-Based CD Methods

In recent years, with the booming development and wide application of deep learning technology in the field of computer vision, many scholars have extended this technology to remote sensing image CD. According to different manners of supervision, we place these deep learning-based CD approaches into three groups [28,45]: supervised CD, unsupervised CD, and semi-supervised CD.

(1) Supervised CD. This kind of method is commonly used in CD, which refers to the method of using artificially labeled samples in model training to realize supervised learning. For instance, in the early stage, Gong et al. designed a deep neural network for synthetic aperture radar (SAR) images CD, which can perform feature learning and generate CDM by supervised learning [46]. Zhang et al. recently promoted a deeply supervised image fusion network for CD, which devises a difference discrimination network to obtain CDM of bi-temporal images through deeply supervised learning [47]. Other methods are available in [48,49]. Although these supervised CD approaches can achieve acceptable performance for CD, manually labeled data is expensive and time consuming, and the quality of the manually labeled data has a significant impact on the performance of the model.

(2) Unsupervised CD. In addition to supervised learning-based CD approaches, unsupervised CD approaches have received much attention, which can acquire CDM directly without the need for manually labeled data. In recent years, many studies have been proposed for unsupervised CD, for example, Saha et al. designed an unsupervised deep change vector analysis (DCVA) method based on pretrained CNN for multiple CD [50]; an unsupervised deep slow feature analysis (DSFA) was proposed based on two symmetric deep networks for multitemporal remote sensing images in [51], which can effectively enhance the separability of changed and unchanged pixels by slow feature analysis. Moreover, other unsupervised change detection methods are available in [52–55]. However, at present, the unsupervised CD method is difficult to promote for practical application, this is because unsupervised CD approaches rely heavily on migrating features from data sources with different distribution, resulting in poor robustness and unreliable results.

(3) Semi-supervised CD. To overcome the limitation of supervised and unsupervised CD methods to a certain extent, semi-supervised learning approaches have been further developed for CD. In semi-supervised CD, in addition to a small amount of labeled data, unlabeled data are also effectively used to achieve the semi-supervised learning, and thus obtaining CDM. For example, Jiang et al. proposed a semi-supervised CD method, which extracts discriminative features by using unlabeled data and limited labeled samples [56]. In [57], a semi-supervised CNN based on a generative adversarial network was proposed, which can employ two discriminators to enhance the feature distribution consistency between the labeled and unlabeled data for CD. These semi-supervised CD methods significantly reduce the dependence on a large number of labeled data, and meanwhile maintain the performance of the model to a certain extent. However, unlabeled data may cause some interference to network training due to its unreliability, so developing reliable methods to apply unlabeled data is a crucial procedure in semi-supervised learning.

3. Proposed Method

In order to effectively detect changes based on the joint spatial and spectral features of HSIs, in this paper, we propose a novel self-supervised feature extraction and attention based CD framework, as shown in Figure 1. From the figure, it can be seen that the entire framework is divided into two steps. In the first step, the PCA-guided self-supervised spatial feature extraction network is designed, which can extract the most important change feature representation in each difference patch. In the second step, in order to effectively combine the extracted spatial and spectral features, the attention mechanism is introduced into the spatial and spectral CD network, which can adaptively learn a matching ratio for the spatial and spectral features of each patch, highlighting where is the most conducive for detecting changes. Below, we will introduce the proposed framework in detail.

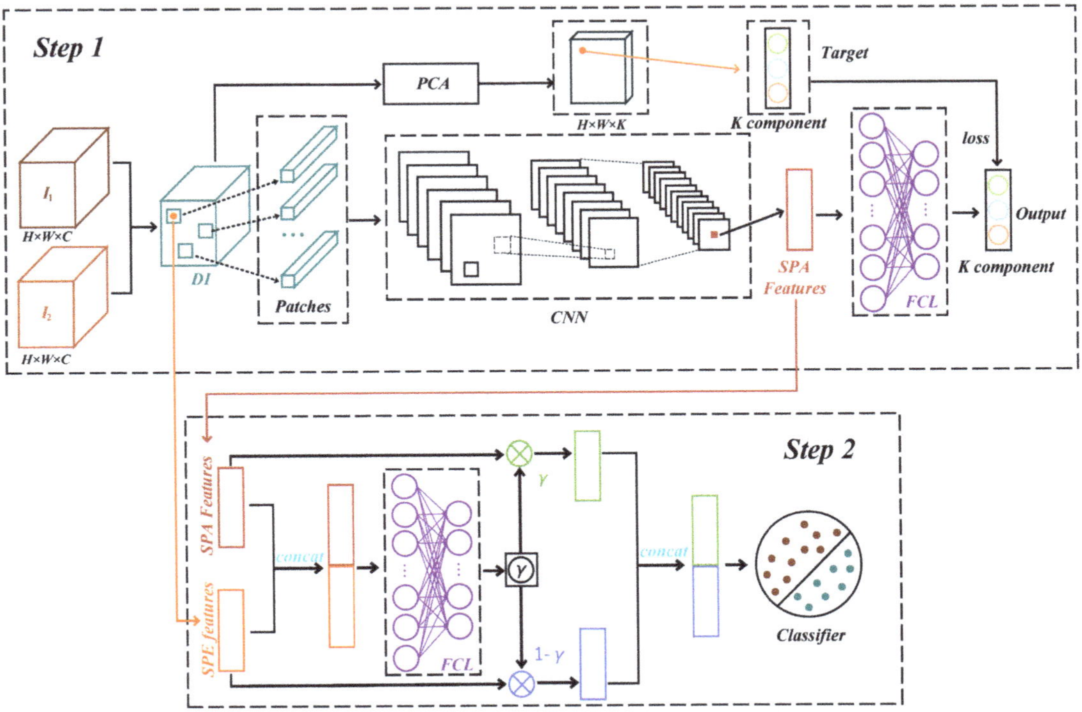

Figure 1. Framework of the proposed ASSCDN. The first step is PCA-guided self-supervised spatial feature extraction network. The second step is to combine the spectral and spatial features by introducing a attention mechanism and obtain the final class.

3.1. Data Preparation

3.1.1. Data Preprocessing

Before comparing and analyzing the target HSIs, as the original HSIs usually contain noise and interference channels caused by atmospheric and water vapor scattering, it is often necessary to perform preprocessing such as dead pixel repair, strip removal, atmospheric correction, etc., on the original images. In addition, as change detection requires joint analysis of these two images, unaligned pixels will cause higher false detection, so joint registration of these two images is also essential.

3.1.2. Training Data Generation

It is a common method to directly analyze the difference image and obtain the final change map, since it can analyze the difference more directly and specifically. In addition, considering the lack of labeled data for HSIs, analysis based on a certain size of neighborhood of each pixel, i.e., a small patch, can often improve the reliability of change detection. After comprehensive consideration, we select the small patch centered on each pixel in the difference map of the two HSIs as the processing unit. Formally, let I_1 and I_2 represent the two HSIs of size $H \times W \times C$ to be detected, where H, W, and C represent the height, width, and the number of spectral bands of the images, respectively. First, by comparing the two images, a difference map DM can be generated, i.e.,

$$\mathrm{DM} = |I_1 - I_2|. \tag{1}$$

Then, by cutting the pixel-by-pixel neighborhood of DM, a total of $H \times W$ patches of size $P \times P \times C$ can be obtained for the input of CD, where P is the patch size.

3.1.3. Principal Component Analysis (PCA) for DM

Principal component analysis (PCA) is a popular dimensional reduction machine learning technique, which has been widely used in change detection due to its simplicity, robustness, and effectiveness. For DM, PCA technique can transform the image into an orthogonal space with larger data variance, where the data can be represented by fewer dimensional features with almost little information loss, consequently finding the most expressive difference representation. Formally, for the DM data matrix $\mathbf{D}$ which has $H \times W \times C$ samples of M-dimensional features, the transformed data can be calculated by

$$\mathbf{D}' = \mathbf{P}\mathbf{D}, \tag{2}$$

where $\mathbf{P}^\top$ is the transposed eigenvector matrix sorted according to the eigenvalue of the eigencovariance matrix $\mathbf{C}$ of $\mathbf{D}$. That is, $\mathbf{P}^\top$ satisfies the following equation:

$$\mathbf{P}^\top\mathbf{C}\mathbf{P} = \begin{bmatrix} \lambda_1 & & & \\ & \lambda_2 & & \\ & & \ddots & \\ & & & \lambda_M \end{bmatrix}, \tag{3}$$

where $\{\lambda_1, \lambda_2, \cdots, \lambda_M\}$ are M eigenvalues of $\mathbf{C}$, which satisfies $\lambda_1 \geq \lambda_2 \geq \cdots \geq \lambda_M$.

In this way, the original data can be transformed into a new feature space, and the former K-dimension features can contain most of the information. The data after dimensionality reduction can be expressed as

$$\tilde{\mathbf{D}} = \mathbf{T}\mathbf{D}, \tag{4}$$

where $\mathbf{T}$ is the matrix of the eigenbasis vectors for the first K rows of $\mathbf{P}$. Then, the obtained $\tilde{\mathbf{D}}$ can be reshaped as the dimension reduced difference map DM_{PCA}.

3.2. PCA-Guided Self-Supervised Spatial Feature Extraction

When the data are ready, it can be fed into the designed framework for change detection. We first extract spatial features based on these patches. As DM_{PCA} contains several major differential features, we expect to establish a mapping relationship from patch to several principal components of its central pixel. In this way, we propose a PCA-guided spatial feature extraction network (PCASFEN) which is supposed learn the spatial features that can express the most dominating features of the central pixel from the neighborhood information. There is no artificially labeled labels involved in the whole learning process; the supervised information can be obtained completely by the transformation of data itself, which is actually a self-supervised task. Specifically, given a patch with of size $P \times P \times C$, several convolutional layers are used to extract deep spatial features. In this process, a pooling layer is not used, mainly considering that the patch size is usually small and pooling may lose more spatial details. In addition, batch normalization is adopted to prevent distributed drift and thus ensure the stability of training. After the feature extraction, in order to ensure the same spatial and spectral dimensions in joint spatial and spectral analysis, the processed features are flattened and processed into a C-dimensional vector with the same feature dimensions as the input via a fully-connected layer. Finally, after several fully connected layers of processing, the output is a vector of K dimensions, which is utilized to regression-fitted with the principal component features of the central pixel of the patch.

3.3. Attention-Based Spatial and Spectral Network

At present, we have obtained spatial and spectral features representing each pixel in the DM. Joint analysis of spatial and spectral features is a common method in change detection tasks, because it can comprehensively analyze data from spatial and spectral perspectives, thus reduce isolated noise points and improve detection robustness. Generally speaking, to better balance these two features, a weighting factor $\gamma \in [0,1]$ is often used. The fusion feature F of a pixel can be represented as

$$F = [\gamma F_{spa}, (1-\gamma)F_{spe}]. \tag{5}$$

It can be seen that γ is a very important parameter, which is used to determine which of the spatial and spectral features contributes more to the final CD result. In most methods, a suitable γ usually requires multiple experiments to obtain, which undoubtedly greatly increases the actual use cost. In addition, for all pixels in the image, γ will eventually be set globally, but in fact, the spatial and spectral features of different pixels contribute differently to their change status. Inspired by the attention mechanism, we propose an attention-based spatial and spectral change detection network (ASSCDN). Concretely, given the spatial feature $F_{spa} \in \mathbb{R}^C$ and a spatial feature $F_{spe} \in \mathbb{R}^C$ of the n-th pixel in DM, first, they are concatenated as $F_n \in \mathbb{R}^{2C}$, where $n = 1, 2, \cdots, H \times W$. Then, F_n is fed into a fully-connected layer to calculate the γ_n only for the corresponding pixel, which can be expressed as

$$\gamma_n = \sigma(wF_n + b) = \frac{1}{1+e^{-(wF_n+b)}}, \tag{6}$$

where σ is the *Sigmoid* activation function which can ensure that γ_n is between 0 and 1, and w and b represent the weight and bias of the fully-connected layer, respectively. Then, F_{spa} and F_{spe} are weighted by multiplying γ_n and $1-\gamma_n$, respectively. At this time, the weighted F_{spa} and F_{spe} can be concatenated into a new feature, represented as

$$F_n{}' = [\gamma_n F_{spa}, (1-\gamma_n)F_{spe}]. \tag{7}$$

Finally, the obtained features can be input into several fully-connected layers for classification to obtain the final change status.

3.4. Training and Testing Process

3.4.1. Training and Testing PCASFEN

As PCASFEN establishes a regression mapping from the patch to the principal component features of the central pixel, the mean square error (*MSE*) function is adopted as the loss of training PCASFEN. Given the input patch and feature pairs, training the PCASFEN can be seen as minimizing the *MSE* loss L_{MSE} between the output K-dimensional vectors $\hat{v}$ and the target principal component features v. L_{MSE} can be represented as

$$L_{MSE} = \frac{1}{N}\sum_{n=1}^{N}(v-\hat{v})^2, \tag{8}$$

where N is the mini-batch size. Here, the Stochastic Gradient Descent (SGD) optimizer is adopted to reduce the loss and update the network parameters. After the training of several epochs, L_{MSE} will converge, and then the C-dimensional spatial features of each pixel neighborhood extracted from the network can be used for subsequent spatial and spectral joint analysis.

3.4.2. Training and Testing ASSCDN

For ASSCDN, it establishes the mapping from the spatial features combined with the spectral features of pixels to the final change status, which is a classification task. Therefore, the cross-entropy loss L_{CE} function is employed to guide parameter updating. L_{CE} can be represented as

$$L_{CE} = -\sum y \log(\hat{y}), \tag{9}$$

where y and $\hat{y}$ are the ground truth label to be fitted and the output of the network, respectively. Similarly, the SGD optimizer is used to optimize the ASSCDN. Due to the effectiveness of the extracted features, only a very small number of labeled samples are enough to satisfy the training. Here, we use random selection from the reference CD map to simulate this process. The number of samples selected will be discussed in detail in the next section. After several rounds of training, the spectral features and the spatial features extracted from PCASFEN of each pixel can be directly input to the well-trained ASSCDN to obtain the change category of this pixel, and thus generate the final change map.

4. Experiments and Analysis

In this section, the experimental datasets are firstly described. Then, the experimental settings, including comparative methods and evaluation metrics are illustrated. Subsequently, the effects of different components in the proposed ASSCDN method on the detection performance are studied and analyzed. Finally, experimental results are presented and discussed in detail.

4.1. Dataset Descriptions

To evaluate the effectiveness of the proposed ASSCDN approach, three groups of HSIs are conducted in the experiments. These datasets are presented as follows.

The first and second datasets are Santa Barbara dataset and Bay Area dataset, which were released in [58]. As shown in Figures 2 and 3, these datasets were captured by AVIRIS sensor, which both have 224 spectral bands. In the Santa Barbara dataset, Figure 2a,b was acquired over the Santa Barbara region, California, in 2013 and 2015, respectively. The images have 30 m/pixel spatial resolution and a size of 984 × 740 pixels. As presented in Figure 3a,b, in the Bay Area dataset, two HSIs were collected over the city of Patterson, California, in 2007 and 2015, respectively. These images are with the size of 600 × 500 pixels and the spatial resolution of 30 m/pixel. Besides, the reference images of two datasets are shown in Figures 2c and 3c, which are obtained by manual interpretation, separately.

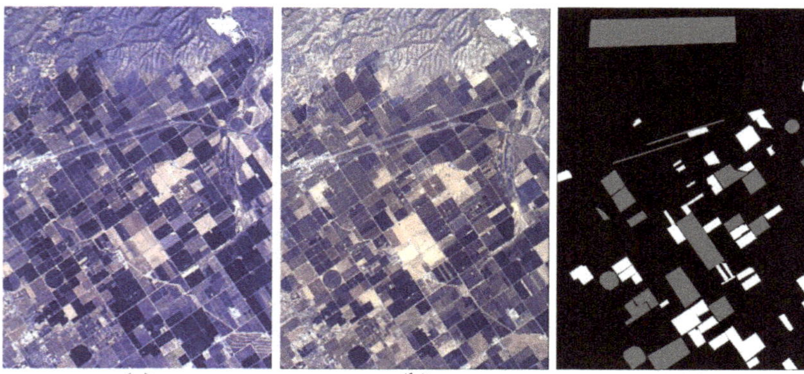

(a) (b) (c)

Figure 2. Barbara dataset: (**a**) T_1-time image, (**b**) T_2-time image, and (**c**) reference image. (Notation: gray color, white color, and black color denote unchanged pixels, changed pixels, and uninteresting pixels, respectively).

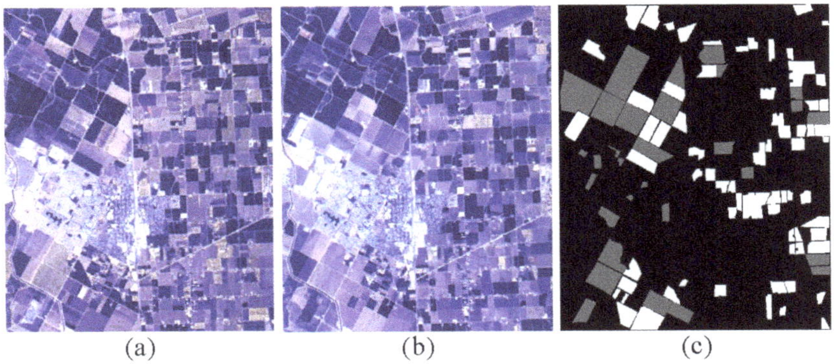

Figure 3. Bay dataset: (**a**) T_1-time image, (**b**) T_2-time image, and (**c**) reference image. (Notation: gray color, white color, and black color denote unchanged pixels, changed pixels, and uninteresting pixels, respectively).

The third dataset is River dataset, which was published in [6], as shown in Figure 4. Figure 4a,b was acquired by Earth Observing-1 (EO-1) Hyperion in 3 May 2013, and 31 December 2013, respectively, which contain total 242 spectral bands, and depict a river area in Jiangsu Province, China. In the River dataset, 198 bands are employed, and these images have a size of 463 × 241 pixels and a spatial resolution of 30 m/pixel. In addition, Figure 4c provides a reference image, which is obtained by manual interpretation.

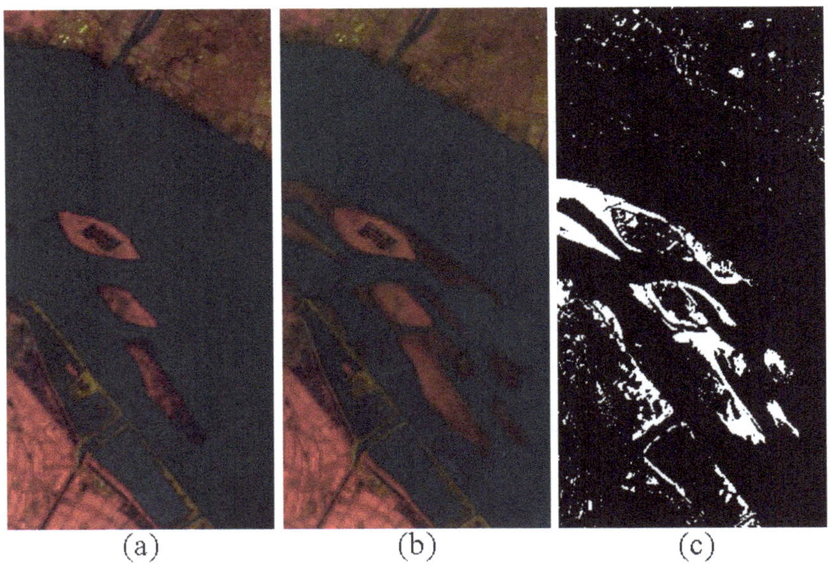

Figure 4. River dataset: (**a**) T_1-time image, (**b**) T_2-time image, and (**c**) reference image. (Notation: white color and black color denote changed pixels and unchanged pixels, respectively).

4.2. Experimental Settings

4.2.1. Evaluation Metrics

To evaluate quantitatively the accuracy of the proposed ASSCDN approach, three commonly used comprehensive evaluation metrics are selected [56,59,60], including overall accuracy (OA), F1-score (F_1), and kappa coefficient (KC). Here, true positive (TP), true negative (TN), false positive (FP), and false negative (FN) are first counted by confusion matrix of the detection results, where TP indicates the number of pixels correctly detected as changed class; TN indicates the number of pixels correctly detected as unchanged class; FP and FN indicate the number of pixels falsely detected as changed and unchanged classes, respectively. On this basis, these evaluation metrics can be computed as follows:

$$\mathrm{OA} = \frac{\mathrm{TP} + \mathrm{TN}}{\mathrm{TP} + \mathrm{TN} + \mathrm{FP} + \mathrm{FN}} \tag{10}$$

$$\mathrm{KC} = \frac{\mathrm{OA} - \mathrm{p_e}}{1 - \mathrm{p_e}} \tag{11}$$

$$\mathrm{p_e} = \frac{(\mathrm{TP} + \mathrm{FP}) \times \mathrm{RC} + (\mathrm{TN} + \mathrm{FN}) \times \mathrm{RU}}{(\mathrm{TP} + \mathrm{TN} + \mathrm{FP} + \mathrm{FN})^2} \tag{12}$$

$$\mathrm{PRE} = \frac{\mathrm{TP}}{\mathrm{TP} + \mathrm{FP}} \tag{13}$$

$$\mathrm{REC} = \frac{\mathrm{TP}}{\mathrm{TP} + \mathrm{FN}} \tag{14}$$

$$\mathrm{F_1} = \frac{2 \times \mathrm{PRE} \times \mathrm{REC}}{\mathrm{PRE} + \mathrm{REC}} \tag{15}$$

where RC and RU represent the number of pixels that are changed and unchanged classes in the reference image, respectively. The larger values of these evaluation metrics indicate better detection performance.

4.2.2. Comparative Methods

In the experiments, eight widely used or state-of-the-art methods are selected to validate the superiority of the proposed ASSCDN approach. These methods are summarized as follows:

(1) CVA, which is a classic method for CD, is a comprehensive measure for the differences in each spectral band [61]. Therefore, CVA is suitable for HSI CD.
(2) KNN, aims to acquire the prediction labels of new data through the labels of the nearest K samples, which is used to acquire CDM.
(3) SVM, a commonly applied supervised classifier, which is exploited to classify a difference image into a binary change detection map.
(4) RCVA, was proposed by Thonfeld et al. for multi-sensor satellite images CD to improve the detection performance [39].
(5) DCVA, can achieve an unsupervised CD based on deep change vector analysis, which implemented a pretrained CNN to extract features of bitemporal images [50].
(6) DSFA, which employs two symmetric deep networks for multitemporal remote sensing images in [51]. This approach can effectively enhance the separability of changed and unchanged pixels by slow feature analysis.
(7) GETNET, which is a benchmark method on River dataset [6]. This method introduces a unmixing-based subpixel representation to fuse multi-source information for HSI CD.
(8) TDSSC, which can capture representative spectral–spatial features by concatenating the feature of spectral direction and two spatial directions, and thus improving detection performance [20].

4.2.3. Implementation Details

In the experiments, the proposed ASSCDN approach and other comparative methods were deployed on Pycharm platform with Pytorch or TensorFlow framework by using a single NVIDIA RTX 3090 or NVIDIA Tesla P40. During the training stage, the parameters of the model were optimized by a SGD optimizer with the momentum of 0.5 and the weight decay of 0.001. In all the experiments, the batch size is set as 32.

4.3. Ablation Study and Parameter Analysis on River Dataset

In this section, to investigate the effectiveness of the proposed ASSCDN, we conduct a series of ablation studies on the River dataset. These ablation studies mainly contain three aspects as follows: (1) In the proposed ASSCDN, we devise a novel PCA-guided self-supervised feature extraction network (PCASFEN) and attention-based CD framework to combine effectively the spatial and spectral features. Therefore, we first test the influence of different components on the performance of CD in the proposed ASSCDN. (2) As the patch size is an inevitable parameter in the proposed self-supervised spatial feature extraction framework, the sensitivity of patch size for network performance is investigated subsequently. (3) In addition, the relationship between the number of training samples and performance is also analyzed to validate the effectiveness of the proposed ASSCDN when only a small number of training samples are available.

4.3.1. Ablation Study for Different Components

In the ablation study, to investigate the contribution of different components in the proposed ASSCDN, three comprehensive evaluation metrics, including OA, KC, and F1, are selected to evaluate quantitatively the results of these ablation studies. Besides, to ensure the fairness of the experiment, we set the same parameter for each experiment, that is, the patch size was set as 15, the number of training samples of each class was 250, and other hyperparameter settings were the same.

In this ablation study, four major components are adopted in the our ASSCDN, i.e., "*spe*", "*spa*", "*spe* + *spa*", and "*spe* + *spa* + Attention", where "*spe*" denotes that only spectral features are used, "*spa*" denotes that only spatial features are exploited, "*spe* + *spa*" indicates that spectral features and spatial features are combined in equal proportions, and "*spe* + *spa* + Attention" indicates that spectral features and spatial features are combined through the application of the proposed attention mechanism. According to the aforementioned settings, the results were obtained on River dataset, as shown in Table 1 and Figure 5. From the quantitative results, compared with "*spe*", "*spa*" can improve the detection performance to a certain extent, which indicates that the most important change feature representation is extracted by our proposed self-supervised spatial feature extraction framework. In addition, "*spe* + *spa*" can achieve better accuracy due to the improved discriminable feature expression by fusing spectral and spatial features, thus ameliorating the detection performance. Note that "*spe* + *spa* + Attention" reached the best accuracy (95.82%, 0.7609, and 78.37%) in terms of OA, KC, and F1. Compared with "*spe* + *spa*", "*spe* + *spa* + Attention" was significantly improved in all three evaluation criteria (1.21%, 0.0575, and 5.10%). From the visual results, the same conclusion can be obtained. Besides, as shown in Figure 6, we also tested the performance of different components with different patch sizes, and the results further verified the contribution of the components of our proposed ASSCDN.

In summary, two aspects can be obtained by the comparison results of the above ablation study: (1) The most useful change feature representation can be captured by our proposed PCASFEN, which can help to enhance the separability between changed and unchanged classes. (2) As it is unreasonable to combine spectral and spatial features by equal proportions for different patches, a novel attention mechanism is designed to adaptively adjust the proportion of spectral and spatial features for different patches to achieve effective and reasonable fusion of spectral and spatial features, thus significantly improving the accuracy of CD. Therefore, the effectiveness of each component of the proposed ASSCDN can be validated, it can join effectively spectral and spatial features by our proposed self-supervised spatial feature extraction network and attention mechanism, thereby elevating the performance of CD for HSI.

Table 1. Quantitative comparison for ablation study of the combination of different features on the River dataset.

Methods	OA(%)	KC	F1 (%)
spe	92.32	0.6441	68.38
spa	93.60	0.6661	70.06
spe + *spa*	94.61	0.7034	73.27
spe + *spa* + Attention	95.82	0.7609	78.37

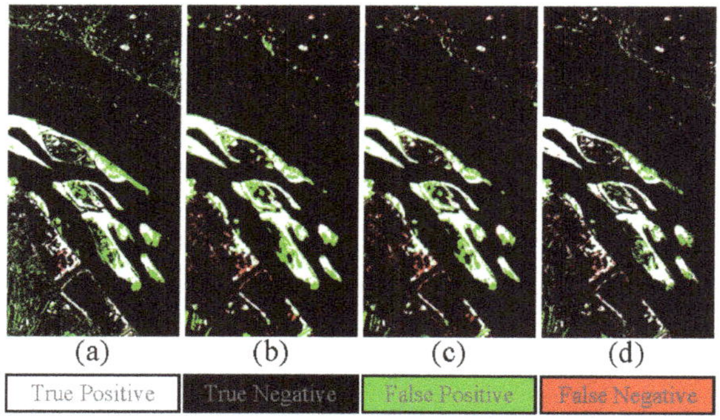

Figure 5. Visual results for ablation study of the combination of different features on the River dataset: (**a**) *spe*, (**b**) *spa*, (**c**) *spe* + *spa*, (**d**) *spe* + *spa* + Attention.

4.3.2. Sensitivity Analysis of Patch Size

In the proposed ASSCDN framework, patch size is an inevitable parameter in our PCASFEN step, which provides the spatial neighborhood information of a central pixel. Therefore, to comprehensively investigate the relationship between the patch size and accuracy, each component of our proposed ASSCDN, including "*spe*", "*spa*", "*spe* + *spa*", and "*spe* + *spa* + Attention", is employed in this experiment. Here, KC is selected to evaluate the results for each component of our proposed ASSCDN. In addition, to ensure the fairness of the comparison, in all experiments, the number of the training samples of each class was fixed to 250, and the other hyperparameter settings were the same.

Based on the above settings, the results of patch sizes ranging from 7 to 17 for each element were acquired, as presented in Figure 6. Notably, "*spe*" does not actually involve patch size as "*spe*" denotes that only spectral features are used to obtain detection results. Therefore, to facilitate comparison with the results of other components, the results of each patch size for the "*spe*" are the same, as the red line shown in Figure 6. By observing Figure 6, we can find that the results of "*spa*" present unstable fluctuation at different patch sizes. That is because different patch sizes may contain different information with various scales. Small patch sizes are more suitable for the different information of the small scale, but the extraction of the difference information of large scale is insufficient, which limits the accuracy. Similarly, larger patch size is more suitable for large-scale difference information, but for small-scale difference information, the noise may be introduced and the performance may is damaged in turn. Moreover, the relationship between the results of "*spe* + *spa*" and "*spe* + *spa* + Attention" and the patch size is similar to that of "*spa*". Overall, compared with "*spa*" and "*spe* + *spa*", the performance of "*spe* + *spa* + Attention" is relatively stable, and can achieve good performance in each patch size.

4.3.3. Analysis of the Relationship between the Number of Training Samples and Accuracy

In this subsection, to further promote the proposed ASSCDN (i.e., "*spe* + *spa* + Attention") in practical application, we conducted an experiment to explore the relationship between the number of training samples and the accuracy. Here, when testing the performance of different numbers of training samples, we set the same hyperparameter, and the patch size was fixed at 11. Additionally, KC is employed to evaluate the accuracy of the all the results. On this basis, the results were acquired with the number of training samples ranging from 10 to 1000 (see Figure 7). As can be seen in Figure 7, with the number of training samples increasing, the value of KC increases gradually, and when the number reaches around 200, the value of KC tends to be stable. Figure 7 also reveals that the proposed ASSCDN can acquire convincing performance even with a small number of training samples.

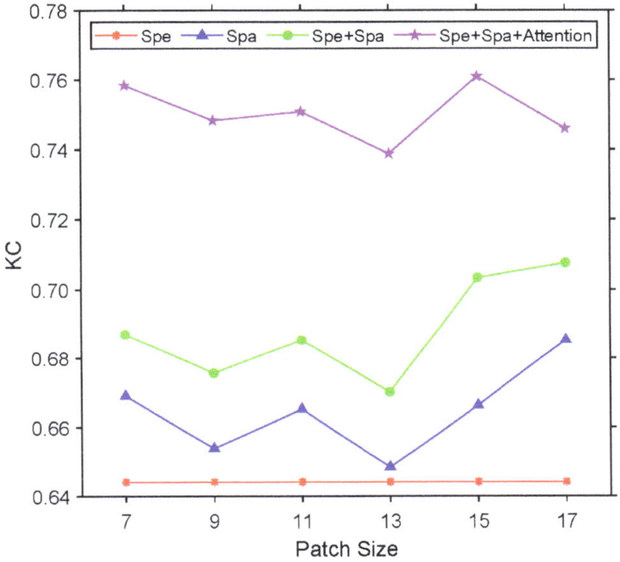

Figure 6. Sensitivity analysis of patch size for each component of the proposed ASSCDN on the River dataset.

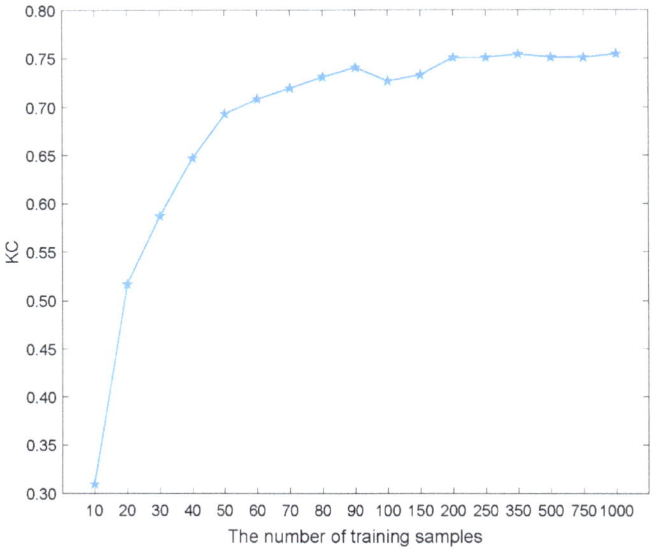

Figure 7. Relationship between the number of training samples and accuracy for the proposed ASSCDN on the River dataset.

4.4. Comparison Results and Analysis

In this section, we tested the performance of the proposed ASSCDN on three real public available HSI datasets. Moreover, to verify the superiority of the proposed ASSCDN, eight approaches are selected for comparison, including four widely used methods: CVA [61], KNN, SVM, and RCVA [39], and four deep learning-based methods: DCVA [50], DSFA [51], GETNET [6], and TDSSC [20]. Furthermore, five metrics (OA, KC, F1, PRE, and REC) are exploited to evaluate the accuracy of the proposed ASSCDN and the compared methods. Moreover, we employed a patch size of 15, and the number of the training samples of 250 to perform the proposed ASSCDN on these three datasets. In addition, to ensure the fairness of comparison, GETNET [6], and TDSSC [20] are deployed under the same semi-supervised learning framework as the proposed ASSCDN.

4.4.1. Results and Comparison on Barbara and Bay Datasets

The CD results were acquired by different approaches on Barbara and Bay datasets, as shown in Figures 8 and 9, and the results of the quantitative evaluation are listed in Tables 2 and 3. From Figures 8a and 9a, the traditional CVA method shows more pixels of false positive due to its lack of effective use of spatial features. Different from CVA, as shown in Figures 8d and 9d, although RCVA introduces neighborhood information, it is unreliable as changed targets of various scales are inevitable. Besides, KNN and SVM present fewer pixels of false positive and false negative for both Barbara and Bay datasets, especially, SVM achieved the highest PRE (93.01%), as listed in Table 2. Notably, unsupervised-based deep learning methods, i.e., DCVA and DSFA, did not reach satisfactory performance on Barbara and Bay datasets, respectively. Among them, DCVA aims to acquire CD results by comparing differences between transferred deep features, but the generalization ability of the transfer model is unreliable, while DSFA may be limited by the results of the pre-detection. GETNET [6] can obtain the second best performance on Barbara dataset, but it cannot get satisfactory accuracy on Bay data. By contrast, TDSSC [20] can achieve relatively stable accuracy on these two datasets as it captures more robust feature representation by fusing the features of spectral direction and two spatial directions. For the proposed ASSCDN, spectral and spatial features are fused adaptively for different patches, which is helpful to obtain more reliable detection results. As listed in Tables 2 and 3, compared with the above methods, our proposed ASSCDN can achieve the best accuracy for both Barbara and Bay datasets in terms of OA, KC, and F1. From the visual results of Barbara and Bay datasets (Figures 8i and 9i see), the proposed ASSCDN acquires very few pixels of false positive and false negative, and it obtains the results closest to the reference image.

Table 2. Quantitative comparison results of various methods applied on the Barbara dataset.

Methods	OA (%)	KC	F1 (%)	PRE (%)	REC (%)
CVA [61]	87.12	0.7320	83.96	82.26	85.72
KNN	91.02	0.8122	88.64	88.24	89.05
SVM	93.21	0.8568	91.20	**93.01**	89.46
RCVA [39]	86.74	0.7226	83.22	82.83	83.62
DCVA [50]	79.21	0.5313	66.96	89.24	53.59
DSFA [51]	86.76	0.7174	69.83	87.06	77.92
GETNET [6]	95.01	0.8962	93.80	91.62	96.09
TDSSC [20]	94.22	0.8789	92.67	92.39	92.95
ASSCDN	**95.39**	**0.9046**	**94.33**	91.45	**97.39**

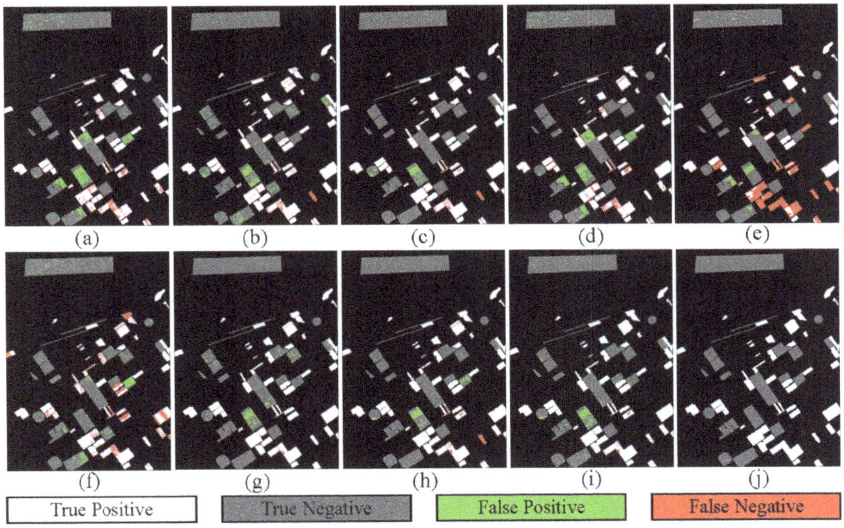

Figure 8. The visual results of different methods on the Barbara dataset: (**a**) CVA [61], (**b**) KNN, (**c**) SVM, (**d**) RCVA [39], (**e**) DCVA [50], (**f**) DSFA [51], (**g**) GETNET [6], (**h**) TDSSC [20], (**i**) our ASSCDN, and (**j**) Reference image.

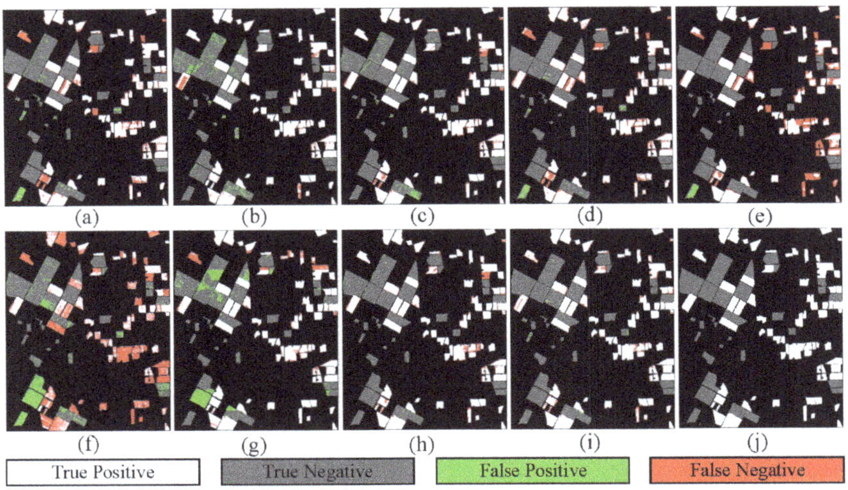

Figure 9. The visual results of different methods on the Bay dataset: (**a**) CVA [61], (**b**) KNN, (**c**) SVM, (**d**) RCVA [39], (**e**) DCVA [50], (**f**) DSFA [51], (**g**) GETNET [6], (**h**) TDSSC [20], (**i**) our ASSCDN, and (**j**) Reference image.

Table 3. Quantitative comparison results of various methods applied on the Bay dataset.

Methods	OA (%)	KC	F1 (%)	PRE (%)	REC (%)
CVA [61]	87.61	0.7534	87.45	94.16	81.64
KNN	91.37	0.8268	91.87	91.58	92.16
SVM	92.58	0.8516	92.80	95.35	90.38
RCVA [39]	87.90	0.7598	87.46	96.77	79.79
DCVA [50]	82.48	0.6546	80.62	97.19	68.87
DSFA [51]	63.37	0.2800	58.34	73.24	48.48
GETNET [6]	85.50	0.7076	86.80	83.73	90.10
TDSSC [20]	94.63	0.8927	94.73	**98.50**	91.19
ASSCDN	**95.53**	**0.9107**	**95.66**	98.45	**93.02**

4.4.2. Results and Comparison on River Dataset

For the River dataset, as presented in Figure 4, more fine changed ground targets exist in this dataset, which increases the difficulty of obtaining fine CD results. As shown in Figure 10, the CD results were obtained by various approaches on the River dataset. From the Figure 4a–c, although typical CVA, KNN, and SVM display a few pixels of false negative, many unchanged pixels are misclassified as changed pixels as spatial information is not considered. Compared with CVA, KNN, and SVM, the result of the RCVA (see Figure 10d) shows fewer noises by introducing spatial contextual information for each pixel. By contrast, DCVA performs poorly performance, as presented in Figure 10e; this is because DCVA depends heavily on transferred deep features. For the DSFA, it generated CD result with relatively few false positive pixels but many missed detection. Both GETNET [6] and TDSSC [20] exhibit fewer false negative pixels, and compared to TDSSC [20], GETNET [6] reaches fewer false positive pixels. From the visual observations, compared with the other methods, our proposed ASSCDN presents the fewest false positive pixels, thus realizing the best visual performance. Although the proposed ASSCDN shows relatively more false negative pixels for GETNET [6] and TDSSC [20], our ASSCDN can obtain a good trade-off between false positive pixels and false negative pixels. In addition to visual comparison, quantitative comparison results have further demonstrated that the proposed ASSCDN can reach the improvements of 0.4%, 0.0113, 0.92%, and 3.47% of OA, KC, F1, and PRE, respectively, as listed in Table 4.

In summary, in this section, the aforementioned comparative experiments based on three real HSIs have been demonstrated that the proposed ASSCDN outperforms some traditional methods and state-of-the-art methods. The comparison results have further verified that effective spatial features can be captured for CD by introducing a novel PCASFEN, which can present the most significant difference representation. Furthermore, spectral and spatial features are fused in an adaptive proportion manner by exploiting an attention mechanism, which is able to enhance feature representation, and thus improves the separability of difference features.

Table 4. Quantitative comparison results of various methods applied on the River dataset.

Methods	OA (%)	KC	F1 (%)	PRE (%)	REC (%)
CVA [61]	92.16	0.6272	66.81	52.86	90.76
KNN	92.58	0.6532	69.17	54.15	95.72
SVM	92.42	0.6504	68.96	53.52	**96.92**
RCVA [39]	94.65	0.6760	70.54	67.62	73.72
DCVA [50]	88.47	0.2466	30.94	32.27	29.72
DSFA [51]	94.61	0.6645	69.41	68.44	70.41
GETNET [6]	95.42	0.7496	77.45	67.71	90.45
TDSSC [20]	94.29	0.7134	74.38	60.94	95.43
ASSCDN	**95.82**	**0.7609**	**78.37**	**71.18**	87.18

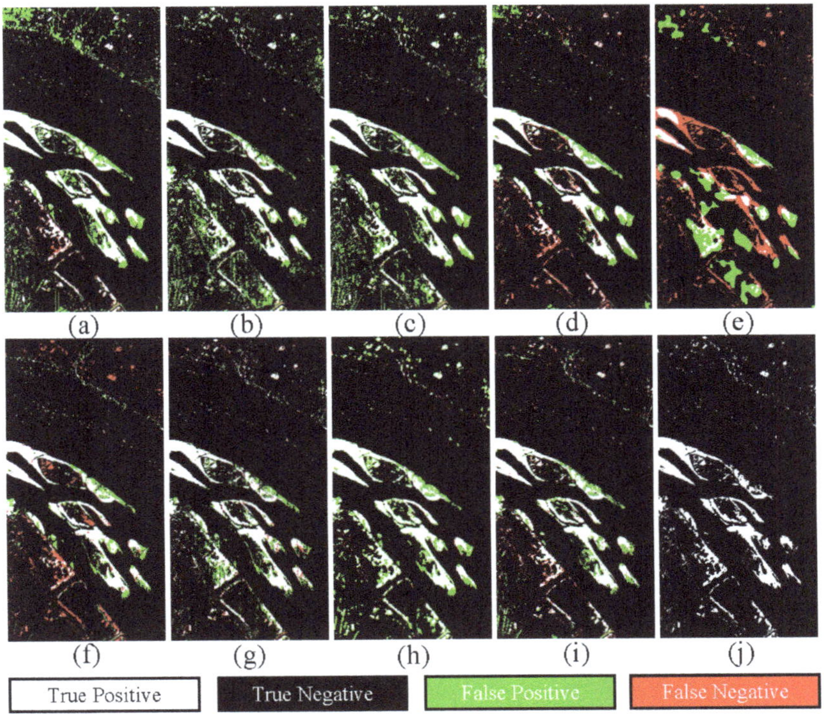

Figure 10. The visual results of different methods on the River dataset: (**a**) CVA [61], (**b**) KNN, (**c**) SVM, (**d**) RCVA [39], (**e**) DCVA [50], (**f**) DSFA [51], (**g**) GETNET [6], (**h**) TDSSC [20], (**i**) our ASSCDN, and (**j**) Reference image.

5. Discussion

In this paper, effective ablation studies and comparison experiments are conducted on three groups of popular benchmark HSI CD datasets. In the ablation studies, three aspects can be observed. First, the effect of different components in our proposed ASSCDN has been proved that the proposed PCA-guided self-supervised feature extraction network and an attention-based CD framework can capture and fuse spatial and spectral features to further improve the performance of HSI CD. Second, although the sensitivity analysis of the patch size reveals that the patch size is more likely to affect the network accuracy (see Figure 6), the proposed ASSCDN significantly improves the accuracy of each patch size. Third, the relationship between the number of training samples and the accuracy has been explored, that is, the results show that the accuracy increases gradually with the increase of the number of training samples. In particular, the proposed ASSCDN can obtain relatively satisfactory performance when fewer training samples are employed. In addition, in the comparison experiments, eight cognate approaches, including four traditional methods (CVA [61], KNN, SVM, and RCVA [39]) and four state-of-the-art methods (DCVA [50], DSFA [51], GETNET [6], and TDSSC [20]), were selected to investigate the performance of the proposed ASSCDN. By observing the quantitative comparison, the proposed ASSCDN is superior to the other eight methods in OA, KC, and F1 for three datasets. Meanwhile, through visual comparison, it can be found that the change detection maps acquired by our ASSCDN can obtain a good trade-off between false detection and missed detection. Despite the proposed ASSCDN can provide a better result for HSI CD, the complexity of performing this method is relatively high, because the training process of our ASSCDN needs to be divided into two stages (i.e., first train the proposed self-supervised spatial feature extraction network, and then train our semi-supervised attention-based spatial and spectral network). Besides, the computational cost of our ASSCDN framework is evaluated by multiply-accumulate operations(MACs), i.e., in the PCA-guided self-supervised spatial feature extraction network step, 0.81 G MACs are needed; in the semi-supervised attention-based spatial and spectral network step, 0.0051 G MACs are needed.

6. Conclusions

In this paper, we propose an attention-based spectral and spatial change detection network (ASSCDN) for hyperspectral images, which mainly contains the following steps as follows. First, the main spatial features of differences can be extracted by our proposed PCASFEN. Second, the attention mechanism is introduced to allocate adaptively the ratio of spectral features and spatial features for fused features. Finally, by the joint analysis of the weighted spatial and spectral features, the change status of each pixel can be obtained. We conducted ablation study and parameter analysis experiment to validate the effectiveness of each component in the proposed ASSCDN. In addition, the experimental comparisons based on three groups of publicly available hyperspectral images have demonstrated that our promoted ASSCDN outperforms the other eight compared methods. In our future work, other HSIs will be collected to further investigate the robustness of this method. Furthermore, there will be a focus on weakly supervised and unsupervised HSI CD.

Author Contributions: Conceptualization, Z.W. and F.J.; methodology, Z.W.; validation, Z.W., F.J. and T.L.; investigation, F.J. and T.L.; writing—original draft preparation, Z.W., F.J. and F.X.; writing—review and editing, F.X. and P.L. All authors have read and agreed to the published version of the manuscript.

Funding: This work was supported by the National Natural Science Foundation of Shaanxi Province under Grant 2021JQ-210, the Fundamental Research Funds for the Central Universities under Grant XJS200216, and the Fundamental Research Funds for the Central Universities and the Innovation Fund of Xidian University.

Acknowledgments: We are grateful to Wang Qi and Javier López-Fandiño who provided the data for this research.

Conflicts of Interest: The authors declare no conflict of interest. The funders had no role in the design of the study; in the collection, analyses, or interpretation of data; in the writing of the manuscript, or in the decision to publish the results.

References

1. Lu, D.; Mausel, P.; Brondizio, E.; Moran, E. Change detection techniques. *Int. J. Remote Sens.* **2004**, *25*, 2365–2401. [CrossRef]
2. Singh, A. Review article digital change detection techniques using remotely-sensed data. *Int. J. Remote Sens.* **1989**, *10*, 989–1003. [CrossRef]
3. Coppin, P.; Jonckheere, I.; Nackaerts, K.; Muys, B.; Lambin, E. Review ArticleDigital change detection methods in ecosystem monitoring: A review. *Int. J. Remote Sens.* **2004**, *25*, 1565–1596. [CrossRef]
4. Liu, S.; Marinelli, D.; Bruzzone, L.; Bovolo, F. A review of change detection in multitemporal hyperspectral images: Current techniques, applications, and challenges. *IEEE Geosci. Remote Sens. Mag.* **2019**, *7*, 140–158. [CrossRef]
5. ZhiYong, L.; Liu, T.; Benediktsson, J.A.; Falco, N. Land Cover Change Detection Techniques: Very-High-Resolution Optical Images: A Review. *IEEE Geosci. Remote Sens. Mag.* **2021**, 2–21. doi: 10.1109/MGRS.2021.3088865. [CrossRef]
6. Wang, Q.; Yuan, Z.; Du, Q.; Li, X. GETNET: A general end-to-end 2-D CNN framework for hyperspectral image change detection. *IEEE Trans. Geosci. Remote Sens.* **2018**, *57*, 3–13. [CrossRef]
7. Liu, S.; Du, Q.; Tong, X.; Samat, A.; Pan, H.; Ma, X. Band selection-based dimensionality reduction for change detection in multi-temporal hyperspectral images. *Remote Sens.* **2017**, *9*, 1008. [CrossRef]
8. Jiang, X.; Gong, M.; Li, H.; Zhang, M.; Li, J. A two-phase multiobjective sparse unmixing approach for hyperspectral data. *IEEE Trans. Geosci. Remote Sens.* **2017**, *56*, 508–523. [CrossRef]
9. Liu, S.; Bruzzone, L.; Bovolo, F.; Zanetti, M.; Du, P. Sequential spectral change vector analysis for iteratively discovering and detecting multiple changes in hyperspectral images. *IEEE Trans. Geosci. Remote Sens.* **2015**, *53*, 4363–4378. [CrossRef]
10. Liu, S.; Bruzzone, L.; Bovolo, F.; Du, P. Hierarchical unsupervised change detection in multitemporal hyperspectral images. *IEEE Trans. Geosci. Remote Sens.* **2014**, *53*, 244–260.
11. Marinelli, D.; Bovolo, F.; Bruzzone, L. A novel change detection method for multitemporal hyperspectral images based on binary hyperspectral change vectors. *IEEE Trans. Geosci. Remote Sens.* **2019**, 57, 4913–4928. [CrossRef]

12. Song, A.; Choi, J.; Han, Y.; Kim, Y. Change detection in hyperspectral images using recurrent 3D fully convolutional networks. *Remote Sens.* **2018**, *10*, 1827. [CrossRef]
13. Zhan, T.; Gong, M.; Jiang, X.; Zhang, M. Unsupervised Scale-Driven Change Detection With Deep Spatial–Spectral Features for VHR Images. *IEEE Trans. Geosci. Remote Sens.* **2020**, *58*, 5653–5665. [CrossRef]
14. Jiao, L.; Liang, M.; Chen, H.; Yang, S.; Liu, H.; Cao, X. Deep fully convolutional network-based spatial distribution prediction for hyperspectral image classification. *IEEE Trans. Geosci. Remote Sens.* **2017**, *55*, 5585–5599. [CrossRef]
15. Yu, C.; Han, R.; Song, M.; Liu, C.; Chang, C.I. A simplified 2D-3D CNN architecture for hyperspectral image classification based on spatial–spectral fusion. *IEEE J. Sel. Top. Appl. Earth Obs. Remote Sens.* **2020**, *13*, 2485–2501. [CrossRef]
16. Wang, D.; Du, B.; Zhang, L.; Xu, Y. Adaptive Spectral–Spatial Multiscale Contextual Feature Extraction for Hyperspectral Image Classification. *IEEE Trans. Geosci. Remote Sens.* **2020**, *59*, 2461–2477. [CrossRef]
17. Hong, D.; Gao, L.; Yao, J.; Zhang, B.; Plaza, A.; Chanussot, J. Graph convolutional networks for hyperspectral image classification. *IEEE Trans. Geosci. Remote Sens.* **2020**, *59*, 5966–5978. [CrossRef]
18. Wu, C.; Du, B.; Zhang, L. Hyperspectral anomalous change detection based on joint sparse representation. *ISPRS J. Photogramm. Remote Sens.* **2018**, *146*, 137–150. [CrossRef]
19. Hou, Z.; Li, W.; Li, L.; Tao, R.; Du, Q. Hyperspectral change detection based on multiple morphological profiles. *IEEE Trans. Geosci. Remote. Sens.* **2021**, 1–12. doi: 10.1109/TGRS.2021.3090802. [CrossRef]
20. Zhan, T.; Song, B.; Sun, L.; Jia, X.; Wan, M.; Yang, G.; Wu, Z. TDSSC: A Three-Directions Spectral–Spatial Convolution Neural Network for Hyperspectral Image Change Detection. *IEEE J. Sel. Top. Appl. Earth Obs. Remote Sens.* **2020**, *14*, 377–388. [CrossRef]
21. Jing, L.; Tian, Y. Self-supervised visual feature learning with deep neural networks: A survey. *IEEE Trans. Pattern Anal. Mach. Intell.* **2020**, *43*, 4037–4058. [CrossRef]
22. Misra, I.; Maaten, L.V.d. Self-supervised learning of pretext-invariant representations. In Proceedings of the IEEE Conference on Computer Vision Recognition, CVPR, Seattle, WA, USA, 14–19 June 2020; pp. 6707–6717.
23. Jaiswal, A.; Babu, A.R.; Zadeh, M.Z.; Banerjee, D.; Makedon, F. A survey on contrastive self-supervised learning. *Technologies* **2021**, *9*, 2. [CrossRef]
24. Chen, H.; Shi, Z. A spatial-temporal attention-based method and a new dataset for remote sensing image change detection. *Remote Sensing* **2020**, *12*, 1662. [CrossRef]
25. Ghaffarian, S.; Valente, J.; Van Der Voort, M.; Tekinerdogan, B. Effect of Attention Mechanism in Deep Learning-Based Remote Sensing Image Processing: A Systematic Literature Review. *Remote Sens.* **2021**, *13*, 2965. [CrossRef]
26. Jiang, H.; Hu, X.; Li, K.; Zhang, J.; Gong, J.; Zhang, M. Pga-siamnet: Pyramid feature-based attention-guided siamese network for remote sensing orthoimagery building change detection. *Remote Sens.* **2020**, *12*, 484. [CrossRef]
27. Liu, T.; Gong, M.; Jiang, F.; Zhang, Y.; Li, H. Landslide Inventory Mapping Method Based on Adaptive Histogram-Mean Distance with Bitemporal VHR Aerial Images. *IEEE Geosci. Remote Sens. Lett.* **2021**. [CrossRef]
28. You, Y.; Cao, J.; Zhou, W. A survey of change detection methods based on remote sensing images for multi-source and multi-objective scenarios. *Remote Sens.* **2020**, *12*, 2460. [CrossRef]
29. Bruzzone, L.; Prieto, D.F. Automatic analysis of the difference image for unsupervised change detection. *IEEE Trans. Geosci. Remote Sens.* **2000**, *38*, 1171–1182. [CrossRef]
30. Bazi, Y.; Bruzzone, L.; Melgani, F. An unsupervised approach based on the generalized Gaussian model to automatic change detection in multitemporal SAR images. *IEEE Trans. Geosci. Remote Sens.* **2005**, *43*, 874–887. [CrossRef]
31. Chen, Q.; Chen, Y. Multi-feature object-based change detection using self-adaptive weight change vector analysis. *Remote Sens.* **2016**, *8*, 549. [CrossRef]
32. Otsu, N. A threshold selection method from gray-level histograms. *IEEE Trans. Syst. Man Cybern.* **1979**, *9*, 62–66. [CrossRef]
33. Bazi, Y.; Melgani, F.; Bruzzone, L.; Vernazza, G. A genetic expectation-maximization method for unsupervised change detection in multitemporal SAR imagery. *Int. J. Remote Sens.* **2009**, *30*, 6591–6610. [CrossRef]
34. Celik, T. Unsupervised change detection in satellite images using principal component analysis and *k*-means clustering. *IEEE Geosci. Remote Sens. Lett.* **2009**, *6*, 772–776. [CrossRef]
35. Shao, P.; Shi, W.; He, P.; Hao, M.; Zhang, X. Novel approach to unsupervised change detection based on a robust semi-supervised FCM clustering algorithm. *Remote Sens.* **2016**, *8*, 264. [CrossRef]
36. Zhan, Y.; Fu, K.; Yan, M.; Sun, X.; Wang, H.; Qiu, X. Change detection based on deep siamese convolutional network for optical aerial images. *IEEE Geosci. Remote Sens. Lett.* **2017**, *14*, 1845–1849. [CrossRef]
37. Migas-Mazur, R.; Kycko, M.; Zwijacz-Kozica, T.; Zagajewski, B. Assessment of Sentinel-2 Images, Support Vector Machines and Change Detection Algorithms for Bark Beetle Outbreaks Mapping in the Tatra Mountains. *Remote Sens.* **2021**, *13*, 3314. [CrossRef]
38. Zhuang, H.; Deng, K.; Fan, H.; Yu, M. Strategies combining spectral angle mapper and change vector analysis to unsupervised change detection in multispectral images. *IEEE Geosci. Remote Sens. Lett.* **2016**, *13*, 681–685. [CrossRef]
39. Thonfeld, F.; Feilhauer, H.; Braun, M.; Menz, G. Robust Change Vector Analysis (RCVA) for multi-sensor very high resolution optical satellite data. *Int. J. Appl. Earth Obs. Geoinf.* **2016**, *50*, 131–140. [CrossRef]
40. Kuncheva, L.I.; Faithfull, W.J. PCA feature extraction for change detection in multidimensional unlabeled data. *IEEE Trans. Neural Netw. Learn. Syst.* **2013**, *25*, 69–80. [CrossRef]
41. Bazi, Y.; Melgani, F.; Al-Sharari, H.D. Unsupervised change detection in multispectral remotely sensed imagery with level set methods. *IEEE Trans. Geosci. Remote Sens.* **2010**, *48*, 3178–3187. [CrossRef]

42. Li, Z.; Shi, W.; Myint, S.W.; Lu, P.; Wang, Q. Semi-automated landslide inventory mapping from bitemporal aerial photographs using change detection and level set method. *Remote Sens. Environ.* **2016**, *175*, 215–230. [CrossRef]
43. Gong, M.; Su, L.; Jia, M.; Chen, W. Fuzzy clustering with a modified MRF energy function for change detection in synthetic aperture radar images. *IEEE Trans. Fuzzy Syst.* **2013**, *22*, 98–109. [CrossRef]
44. Yu, H.; Yang, W.; Hua, G.; Ru, H.; Huang, P. Change detection using high resolution remote sensing images based on active learning and Markov random fields. *Remote Sens.* **2017**, *9*, 1233. [CrossRef]
45. Shi, W.; Zhang, M.; Zhang, R.; Chen, S.; Zhan, Z. Change detection based on artificial intelligence: State-of-the-art and challenges. *Remote Sens.* **2020**, *12*, 1688. [CrossRef]
46. Gong, M.; Zhao, J.; Liu, J.; Miao, Q.; Jiao, L. Change detection in synthetic aperture radar images based on deep neural networks. *IEEE Trans. Neural Netw. Learn. Syst.* **2015**, *27*, 125–138. [CrossRef]
47. Zhang, C.; Yue, P.; Tapete, D.; Jiang, L.; Shangguan, B.; Huang, L. A deeply supervised image fusion network for change detection in high resolution bi-temporal remote sensing images. *ISPRS J. Photogramm. Remote Sens.* **2020**, *166*, 183–200. [CrossRef]
48. Wang, M.; Tan, K.; Jia, X.; Wang, X.; Chen, Y. A deep siamese network with hybrid convolutional feature extraction module for change detection based on multi-sensor remote sensing images. *Remote Sens.* **2020**, *12*, 205. [CrossRef]
49. Lv, Z.; Liu, T.; Kong, X.; Shi, C.; Benediktsson, J.A. Landslide Inventory Mapping With Bitemporal Aerial Remote Sensing Images Based on the Dual-Path Fully Convolutional Network. *IEEE J. Sel. Top. Appl. Earth Obs. Remote Sens.* **2020**, *13*, 4575–4584. [CrossRef]
50. Saha, S.; Bovolo, F.; Bruzzone, L. Unsupervised deep change vector analysis for multiple-change detection in VHR images. *IEEE Trans. Geosci. Remote Sens.* **2019**, *57*, 3677–3693. [CrossRef]
51. Du, B.; Ru, L.; Wu, C.; Zhang, L. Unsupervised deep slow feature analysis for change detection in multi-temporal remote sensing images. *IEEE Trans. Geosci. Remote Sens.* **2019**, *57*, 9976–9992. [CrossRef]
52. Li, X.; Yuan, Z.; Wang, Q. Unsupervised deep noise modeling for hyperspectral image change detection. *Remote Sens.* **2019**, *11*, 258. [CrossRef]
53. Saha, S.; Bovolo, F.; Bruzzone, L. Building change detection in VHR SAR images via unsupervised deep transcoding. *IEEE Trans. Geosci. Remote Sens.* **2020**, *59*, 1917–1929. [CrossRef]
54. Wu, C.; Chen, H.; Du, B.; Zhang, L. Unsupervised Change Detection in Multitemporal VHR Images Based on Deep Kernel PCA Convolutional Mapping Network. *IEEE Trans. Cybern.* **2021**, 1–15. doi: 10.1109/TCYB.2021.3086884 [CrossRef]
55. Shao, P.; Shi, W.; Liu, Z.; Dong, T. Unsupervised change detection using fuzzy topology-based majority voting. *Remote Sens.* **2021**, *13*, 3171. [CrossRef]
56. Jiang, F.; Gong, M.; Zhan, T.; Fan, X. A semisupervised GAN-based multiple change detection framework in multi-spectral images. *IEEE Geosci. Remote Sens. Lett.* **2019**, *17*, 1223–1227. [CrossRef]
57. Peng, D.; Bruzzone, L.; Zhang, Y.; Guan, H.; Ding, H.; Huang, X. SemiCDNet: A semisupervised convolutional neural network for change detection in high resolution remote-sensing images. *IEEE Trans. Geosci. Remote Sens.* **2020**, *59*, 5891–5906. [CrossRef]
58. López-Fandiño, J.; Garea, A.S.; Heras, D.B.; Argüello, F. Stacked autoencoders for multiclass change detection in hyperspectral images. In Proceedings of the 2018 IEEE International Geoscience & Remote Sensing Symposium (IGARSS), Valencia, Spain, 22–27 July 2018; IEEE: New York, NY, USA; pp. 1906–1909.
59. Lv, Z.; Li, G.; Jin, Z.; Benediktsson, J.A.; Foody, G.M. Iterative training sample expansion to increase and balance the accuracy of land classification from VHR imagery. *IEEE Trans. Geosci. Remote Sens.* **2020**, *59*, 139–150. [CrossRef]
60. Lv, Z.; Liu, T.; Benediktsson, J.A. Object-oriented key point vector distance for binary land cover change detection using VHR remote sensing images. *IEEE Trans. Geosci. Remote Sens.* **2020**, *58*, 6524–6533. [CrossRef]
61. Bovolo, F.; Bruzzone, L. A theoretical framework for unsupervised change detection based on change vector analysis in the polar domain. *IEEE Trans. Geosci. Remote Sens.* **2006**, *45*, 218–236. [CrossRef]

remote sensing

MDPI

Article

A Constrained Sparse-Representation-Based Spatio-Temporal Anomaly Detector for Moving Targets in Hyperspectral Imagery Sequences

Zhaoxu Li †, Qiang Ling †, Jing Wu, Zhengyan Wang and Zaiping Lin *

College of Electronic Science and Technology, National University of Defense Technology, Changsha 410073, China; lizhaoxu@nudt.edu.cn (Z.L.); lingqiang16@nudt.edu.cn (Q.L.); jingwu@nudt.edu.cn (J.W.); wangzhengyan@nudt.edu.cn (Z.W.)

* Correspondence: linzaiping@nudt.edu.cn

† These authors contributed equally to this work.

Received: 5 August 2020; Accepted: 25 August 2020; Published: 27 August 2020

Abstract: At present, small dim moving target detection in hyperspectral imagery sequences is mainly based on anomaly detection (AD). However, most conventional detection algorithms only utilize the spatial spectral information and rarely employ the temporal spectral information. Besides, multiple targets in complex motion situations, such as multiple targets at different velocities and dense targets on the same trajectory, are still challenges for moving target detection. To address these problems, we propose a novel constrained sparse representation-based spatio-temporal anomaly detection algorithm that extends AD from the spatial domain to the spatio-temporal domain. Our algorithm includes a spatial detector and a temporal detector, which play different roles in moving target detection. The former can suppress moving background regions, and the latter can suppress non-homogeneous background and stationary objects. Two temporal background purification procedures maintain the effectiveness of the temporal detector for multiple targets in complex motion situations. Moreover, the smoothing and fusion of the spatial and temporal detection maps can adequately suppress background clutter and false alarms on the maps. Experiments conducted on a real dataset and a synthetic dataset show that the proposed algorithm can accurately detect multiple targets with different velocities and dense targets with the same trajectory and outperforms other state-of-the-art algorithms in high-noise scenarios.

Keywords: anomaly detection; constrained sparse representation; hyperspectral imagery; moving target detection; spatio-temporal processing

---

1. Introduction

With the development of optical sensor technology, hyperspectral imagery (HSI) has been dramatically improved in recent years, and HSI sequences are more available in the real world. Because of adequate spectral information with dozens or hundreds of spectrum bands, the HSI detection technique can find and distinguish dim targets, which are unobservable in the visible or infrared images, and has promising prospects in military, security, satellite surveillance, disaster monitoring, and other applications [1]. According to whether prior target spectral information is utilized, the HSI detection technique can be mainly divided into target detection [2–4] and anomaly detection. Due to factors such as camera angle, illumination, atmosphere, and sensor spatial resolution, it is common in HSI that the same object has different spectra. Besides, no prior target spectrum is available for most of the moving target detection scenes. Therefore, current hyperspectral moving target detection technologies [5–12] are mainly based on anomaly detection.

Traditional single-frame anomaly detection is usually accomplished by detecting irregular deviations between the test pixel and background pixels in a hyperspectral image. Designed to detect the presence of a dim target in a multi-band image, the Reed–Xiaoli (RX) algorithm [13] assumes that the global background spectra obey a multivariate Gaussian distribution and applies the Mahalanobis distance to identify anomaly spectra. To solve the problem that the Gaussian distribution is not applicable to the non-stationary global background, the local version of RX [14] divides the local neighborhood of the test pixel into potential regions and background regions by dual-windows and replaces global statistics with local statistics. The Quasi-local-RX (QLRX) algorithm [15] improves point-target detection by utilizing local and global statistics simultaneously. The kernel RX (KRX) algorithm [16], a nonlinear version of RX, maps spectra into a more high-dimensional characteristic space through a kernel function and outperforms the original RX detector in military target and mine detection. The cluster KRX (CKRX) algorithm [17] improves the performance of KRX by replacing background pixels with cluster centers. Support vector data description (SVDD) algorithms [18,19] also determine anomalies in a high-dimensional characteristic space by building a minimal enclosing hypersphere around local background pixels. Sparse representation (SR)-based algorithms [20–27] have made significant progress in anomaly detection in recent years. These algorithms usually assume that background pixels can be presented as linear combinations of the surrounding background, and anomaly pixels cannot. The collaborative representation (CR)-based algorithm [22] adopts l_2-norm minimization to reinforce the collaboration of background representation and is superior to RX and its improved algorithms. To realize the detection of dense small targets, the constrained sparse representation (CSR)-based algorithm [23] imposes two constraints on abundance vectors and can remove anomalous atoms from the local background dictionary. Because background pixels and target pixels are considered low rank and sparse, respectively, low-rank and sparse matrix decomposition-based algorithms [28–30] have also received widespread attention in anomaly detection.

When a hyperspectral staring camera is continuously imaging at short intervals, anomaly detectors can output detection maps in succession. Usually, anomaly detection maps of a hyperspectral imagery sequence can be regarded as an infrared image sequence. Therefore, multi-frame infrared detection or tracking algorithms can be used to detect or track dim moving targets on these maps. Rotman et al. combined hyperspectral target detection and infrared target tracking for the first time [5–7]. They transformed each HSI into a two-dimensional anomaly detection map and then utilized a variance filter (VF) [31] to detect targets moving at subpixel velocity. Besides, Duran et al. focused on tracking small dense objects, such as pedestrians or vehicles, from airborne platforms [8–10]. They adopted endmember techniques to detect subpixel targets and estimated the motion parameters of targets under the framework of the Bayesian filter. Wang et al. proposed a novel temporal anomaly detector in dim moving target detection, which extracts the local spatial background in the previous frame to mine the singularity of the test pixel [11]. Combining the traditional single-frame detection with their proposed temporal detection can effectively reduce temporal noise clutter. Then, Wang et al. introduced a simplified VF to calculate a trajectory history map in the literature [12]. The fusion of the spatial detection map, the temporal detection map, and the trajectory history map (STH) is superior to previous moving target detection algorithms in hyperspectral imagery sequences.

In summary, current anomaly detection algorithms for moving targets still only utilize the spatial neighborhood background of the current frame or the previous frame. However, static or non-moving objects for which the spectra are different from neighborhoods can be regarded as anomaly targets by these detection algorithms. Temporal profile filtering algorithms can detect moving targets, but ask for prior information about speed. Besides, detecting targets in complex motion, such as multiple targets at different velocities and dense targets on the same trajectory, is still a challenge for temporal profile filtering-based algorithms [5–7,11]. To solve these problems, we propose a CSR-based spatio-temporal anomaly detector (CSR-ST), sufficiently employing temporal spectral information in HSI sequences. Unlike hyperspectral change detection (CD) [32,33], which detects anomaly regions under diurnal and seasonal changes, moving target detection asks for a very short interval between frames. This means that camera angle, illumination, weather, and other imaging conditions are almost unchanged in adjacent frames. After frame registration, the spectrum of the same pixel can be regarded as a mixture of spectra in a small local region, only affected by the temporal clutter in different frames. Based on this assumption, we propose a novel temporal anomaly detection framework that calculates the anomaly score of the test pixel employing its former spectra. In our previous work [23], the CSR detector was based on the assumption that a background pixel can be linearly represented by the endmembers present in its spatial neighborhood while an anomaly pixel cannot. Compared to background spectra in the spatial neighborhood, the former spectra of the test pixel in previous frames can provide more pure background endmembers to represent the current spectrum. Therefore, the CSR-based temporal detector has a better ability to recover the test background pixel than the CSR-based spatial detector. Besides, the temporal detector has two insurances to construct a pure temporal background dictionary for the test pixel. The first insurance is to remove potential target spectra from the candidate set of the temporal background dictionary based on spatial detection results. The other insurance is to automatically remove anomaly atoms from the background dictionary when the corresponding abundances are higher than a given upper bound and then solve the model with the new background dictionary. Non-homogeneous background pixels or stationary objects can turn into false alarms in the single-frame detection, while the temporal detector is mainly sensitive to moving targets. However, when some background regions move in the imaging scene, the temporal detector can regard them as targets and be inferior to the spatial detector. The fusion of the spatial detection map and the temporal detection map combines the advantages of the two detectors and can suppress the background and stationary objects. The main contributions of this article are summarized as follows.

1. A novel hyperspectral spatio-temporal anomaly detection algorithm is proposed. Compared to traditional anomaly detection algorithms, the proposed algorithm utilizes the temporal spectral information and extends the CSR algorithm from the spatial domain to the spatio-temporal domain. The spatial detector and the temporal detector play different roles in moving target detection. The former can suppress moving background regions, and the latter can suppress non-homogeneous background and stationary objects. To the best of our knowledge, no literature has introduced the historical spectra of the test pixel to construct the temporal background set in anomaly detection yet.
2. In the CSR-based temporal detection, there are two procedures to purify the background dictionary. The purification procedures can improve the ability of the temporal detector to detect multiple targets in complex motion situations, such as multiple targets with different velocities and dense targets with the same trajectory.
3. An iterative smoothing filter is executed on both spatial and temporal detection maps to suppress the background clutter. Furthermore, the filter can strengthen the detection performance for slow-moving area targets.

The rest of this article is organized as follows. The CSR detector and its kernel version are introduced in Section 2. The proposed CSR-ST algorithm is described in Section 3. The experiments conducted on a real dataset and a synthetic dataset are presented in Section 4, followed by the conclusions in Section 5.

2. Related Work

SR-based anomaly detection algorithms usually assume that a background pixel can exist in a low-dimensional subspace spanned by surrounding background pixels. Meanwhile, anomaly pixels cannot be represented as a sparse linear mixture of background spectra. Suppose $\boldsymbol{y}$ is the test pixel, which has N spectral bands, and $\boldsymbol{A}$ is the background dictionary, which has M atoms; the competing hypotheses for the SR-based algorithms are:

$$\begin{aligned} H_0 : \boldsymbol{y} &= \boldsymbol{A}\boldsymbol{\alpha} + \boldsymbol{n}, \text{ background pixel} \\ H_1 : \boldsymbol{y} &\neq \boldsymbol{A}\boldsymbol{\alpha} + \boldsymbol{n}, \text{ anomaly pixel} \end{aligned} \quad (1)$$

where $A \in \mathbb{R}^{N\times M}$, α is defined as a sparse vector for which each item is the abundance of the correlated atom in A and n is defined as a random noise item.

Usually, the sparse vector α has a sparsity constraint $\|\alpha\|_0 \le K$ imposed in SR-based detection, where K is a sparsity parameter. However, if there is no constraint on each abundance item in α, anomaly pixels can also be linear mixtures of the background dictionary on account of abundance items less than zero. The linear spectral mixture model (LMM) [34] supposes that the abundance vector α of a mixed pixel should satisfy a sum-to-one constraint:

$$\sum_{l=1}^{M} \alpha_l = 1 \tag{2}$$

and a non-negativity constraint:

$$\alpha_l \ge 0,\ l = 1, \cdots, M. \tag{3}$$

The CSR algorithm introduces Equations (2) and (3) into the SR model, and the minimizing problem of CSR can be expressed as:

$$\begin{aligned} \min_{\alpha} \|y - A\alpha\|_2 \quad \text{s.t.} \quad & \|\alpha\|_0 \le K \\ & e^T\alpha = 1 \\ & \alpha_l \ge 0,\ l = 1, \cdots, M \end{aligned} \tag{4}$$

where e represents an $M \times 1$ vector for which each item is one. The objective function can be converted to:

$$\|y - A\alpha\|_2 = \sqrt{\alpha^T A^T A\alpha - 2y^T A\alpha + y^T y} \tag{5}$$

Note that $y^T y$ is a constant and can be removed. If the test pixel is anomalous and the background dictionary contains a few anomaly pixels, the corresponding entries of α can be enormous, resulting in a small reconstitution residual. To avoid missing alarms, an adequately tiny constant C is introduced as an upper limit of α, and Equation (4) can be transformed as:

$$\begin{aligned} \min_{\alpha} \alpha^T A^T A\alpha - 2y^T A\alpha \quad \text{s.t.} \quad & e^T\alpha = 1 \\ & 0 \le \alpha_l \le C,\ l = 1, \cdots, M \end{aligned} \tag{6}$$

where $C \in [1/M, 1]$. According to the Karush–Kuhn–Tucker conditions [35], the constraint $\|\alpha\|_0 \le K$ in Equation (4) can be removed in Equation (6).

When abnormal pixels are tested, the abundances correlated with similar anomalous atoms can reach the maximum. Accordingly, the atoms for which the abundances are C have a significant possibility of being anomalies and should be eliminated from the background dictionary. A pure dictionary $\tilde{A}$ can be built by the remaining atom. With the constraint $0 \le \tilde{\alpha}_i \le 1$ and $\tilde{A}$, reconstruction residuals of anomalous test pixels will be significantly higher than those in the first reconstruction and can be regraded as anomaly scores.

$$r = \sqrt{\tilde{\alpha}^{*T} \tilde{A}^T \tilde{A}\tilde{\alpha}^* - 2y^T \tilde{A}\tilde{\alpha}^* + y^T y} \tag{7}$$

where $\tilde{\alpha}^*$ is the approximately calculated sparse vector without anomalous atoms in the background dictionary $\tilde{A}$.

Given secondary or multiple scattering in the atmosphere, spectrum mixing usually is a nonlinear process [36]. The kernel methods map the original data into a more high-dimensional characteristic space via nonlinear functions and then achieve linear partition of the linearly inseparable data [37]. Skillfully, the inner product in the characteristic space can be replaced by:

$$\langle \phi(x_i), \phi(x_j) \rangle = k(x_i, x_j) \tag{8}$$

where ϕ is a nonlinear function, x_i and x_j are the original data, and k is the kernel function. The kernel CSR (KCSR) algorithm introduces the kernel method and adopts the Gaussian radial basis function kernel:

$$k(x_i, x_j) = e^{-\gamma\|x_i - x_j\|_2^2}. \tag{9}$$

The optimal problem is replaced by:

$$\begin{aligned} \min_{\alpha} \alpha^T K \alpha - 2K_y \alpha \quad \text{s.t.} \quad & e^T \alpha = 1 \\ & 0 \leq \alpha_l \leq C,\ l = 1, \cdots, M \end{aligned} \tag{10}$$

where K is an $M \times M$ Gram matrix for which the i-th row and j-th column item $K_{i,j} = k(a_i, a_j)$. $K_y = \phi(y)^T \phi(A)$ and can also be replaced by:

$$\begin{aligned} K_y &= k(A, y) \\ &= \left[k(a_1, y) \quad k(a_2, y) \quad \cdots \quad k(a_M, y) \right]. \end{aligned} \tag{11}$$

Likewise, the atoms for which abundances are C are removed, and then, a pure background dictionary $\tilde{A}$ is used to solve Equation (10). Therefore, the anomaly score can be replaced by:

$$r = \sqrt{\tilde{\alpha}^{*T} \tilde{K} \tilde{\alpha}^* - 2\tilde{K}_y^T \tilde{\alpha}^* + k(y, y)} \tag{12}$$

where r is the approximate error and $\tilde{K}$ and $\tilde{K}_y$ are both solved by $\tilde{A}$.

3. Spatio-Temporal Anomaly Detection for Moving Targets

In this section, a novel CSR-based spatio-temporal anomaly algorithm is proposed to detect dim moving targets accurately in HSI sequences. Our algorithm is divided into four steps, namely spatial anomaly detection, iterative smoothing filter, temporal anomaly detection, and spatial-temporal fusion. The spatial anomaly detection finds abnormal targets by utilizing the spectral information of the current frame. An iterative smoothing filter can reduce noise and false alarms in the time and space domains. Different from AD, CD, and the temporal detection [12] using the information between two adjacent frames, our proposed temporal anomaly detection constructs background dictionaries with the historical spectral curves of the test pixels. The proposed temporal anomaly detection explores anomaly characteristics in the time dimension and provides anomaly information different from that in the spatial detection. The fusion of spatial and temporal anomaly detection can explore the target information more comprehensively. The framework of the proposed CSR-ST algorithm is displayed in Figure 1.

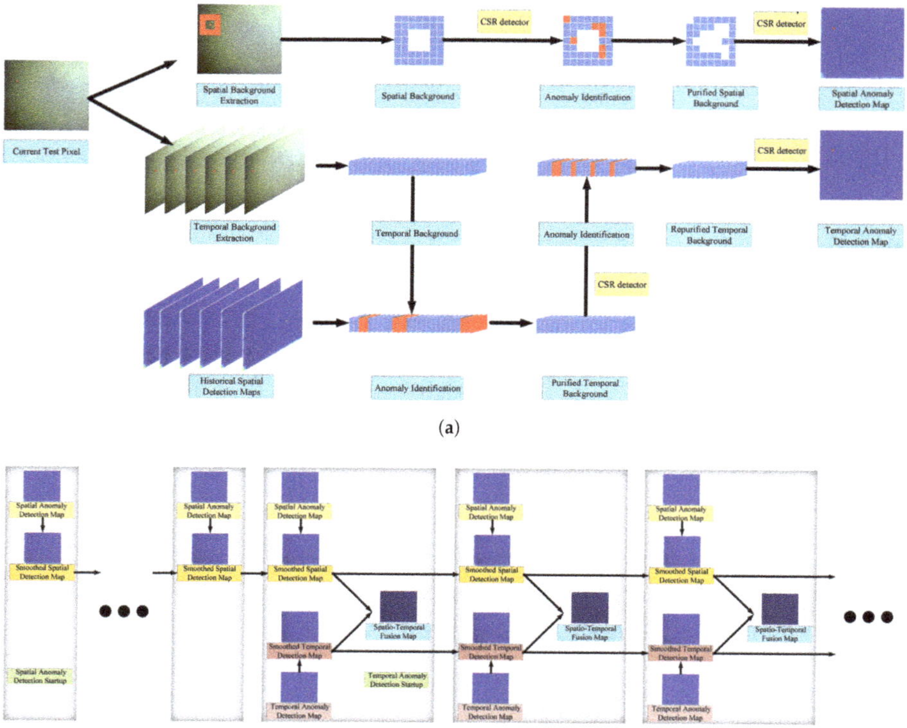

(a)

(b)

Figure 1. The framework of the proposed CSR-ST algorithm. (**a**) The schematic diagram of the CSR-based spatial and temporal detectors. (**b**) The program flowchart of the smoothing filter and fusion on the spatial and temporal detection maps.

3.1. Spatial Anomaly Detection

Let $\boldsymbol{X}_i = \left\{\boldsymbol{x}_i^1, \boldsymbol{x}_i^2, \cdots, \boldsymbol{x}_i^{d_1 \times d_2}\right\} \in \mathbb{R}^N$ denote a hyperspectral cube collected in the current frame, where i is the current sequence number, d_1 and d_2 are defined as the space sizes of the cube, and N is the quantity of spectral bands. Dual concentric windows [38] are used to extract a spatial background dictionary for each pixel $\boldsymbol{x}_i^j$. The dual-windows are centered at each test pixel and divide the neighborhood into a potential target region and a background region. Pixels in the background region are selected as atoms to form a background dictionary $\boldsymbol{A}_i^j$. Then, the spatial anomaly score s_i^j of the test pixel $\boldsymbol{x}_i^j$ is solved by the CSR detector with the corresponding background dictionary $\boldsymbol{A}_i^j$. After all pixels on $\boldsymbol{X}_i$ are detected in sequence, a two-dimensional spatial detection map $\boldsymbol{S}_i$ is obtained:

$$\boldsymbol{S}_i = \begin{bmatrix} s_i^1 & \cdots & s_i^{d_2} \\ \vdots & \ddots & \vdots \\ s_i^{d_1 \times (d_2 - 1) + 1} & \cdots & s_i^{d_1 \times d_2} \end{bmatrix}. \tag{13}$$

3.2. Iterative Smoothing Filter

The spectra change with time due to the measurement noise, resulting in temporal fluctuation of anomaly scores. Meanwhile, spatial background clutter is also generated in the detection maps due to the fluctuation. The literature [17] has used a simple smoothing filter as a post-processing procedure to decrease false alarms and noise in detection maps. Inspired by [17], an iterative smoothing filter is adopted to reduce noise both in the spatial and temporal domains simultaneously.

To avoid the overall drift of anomaly scores on the spatial detection map $\mathbf{S}_i$ caused by sudden changes in imaging conditions, Z-score normalization should be first performed:

$$\bar{\mathbf{S}}_i = \frac{\mathbf{S}_i - \mu}{\sigma}. \quad (14)$$

In typical image preprocessing, μ and σ are the mean value and standard deviation of pixels in the whole image, respectively. However, because anomaly scores of anomalous pixels are much higher than those of background pixels on $\mathbf{S}_i$, it is more accurate to describe the distribution of s_i^j by a truncated normal distribution or a half-normal distribution [39] rather than a normal distribution. Therefore, it is more reasonable to set μ and σ to the mean value and standard deviation of the collection of $\mathbf{S}_i$ and its symmetric set about zero.

Then, an iterative smoothing operation is performed on $\bar{\mathbf{S}}_i$ to reduce spatial and temporal clutter:

$$\tilde{s}_i^j = (1-\rho)\,\tilde{s}_{i-1}^j + \rho \sum_{l \in \mathbf{L}(j)} \varepsilon_l \bar{s}_i^l \quad (15)$$

where $\bar{s}_i^l$ is the normalized spatial anomaly scores of $\boldsymbol{x}_i^l$, $\tilde{s}_i^j$ and $\tilde{s}_{i-1}^j$ are the smoothed spatial anomaly scores of $\boldsymbol{x}_i^j$ and $\boldsymbol{x}_{i-1}^j$, respectively, $\mathbf{L}$ denotes the spatial neighborhood used for smoothing, and ρ and ε_l denote filter weights. When the first spatial detection map is smoothed, let $\rho = 1$. The latter part of Equation (15) is actually a spatial smoothing filter such as the mean filter or the Gaussian filter. Furthermore, one-dimensional denoising algorithms can also replace the temporal iterative smoothing part of Equation (15) to reduce temporal clutter. Compared to the original spatial detection map $\mathbf{S}_i$, background clutter and noise on $\tilde{\mathbf{S}}_i$ are suppressed, and detection performance can be improved.

3.3. Temporal Anomaly Detection

Note that, using the dual-window strategy to select a background dictionary has several disadvantages. Firstly, the selection of an inappropriate dual-window size can cause the local background to be contaminated by target pixels in spatial anomaly detection. If the inner window of dual-windows is too small, the chosen local background of the test target pixel can contain some target pixels. Moreover, the contamination problem can also occur when multiple targets are densely distributed. Secondly, the spatial distributions of moving targets are usually unknown and change in the real world. Therefore, it is difficult to determine the optimal dual-window size to detect moving targets in advance. Thirdly, the performance of these algorithms still varies with the dual-window size, and the best performance of the dual-window-based AD algorithms is a local optimum. For instance, detection results can be further improved after combining with a weight matrix obtained by segmentation or clustering in the literature [40,41], where background pixels are assigned lower weight values. An interesting phenomenon is that the best local background of some detection algorithms for subpixel targets are eight neighborhoods [42], and large dual-windows are harmful to these algorithms. Fourthly, the dual-window-based spatial detection cannot eliminate motionless objects, the spectra of which are also different from the background spectra.

To accurately detect moving targets in HSI sequences, we propose a new approach for constructing background dictionaries of test pixels. Compared to hyperspectral CD, the interval between two contiguous frames in moving target detection is short; thus, the camera angle, illumination, weather, and other imaging conditions are almost unchanged. In this case, the spectrum of the same object in short HSI sequences can only be affected by the measured noise. Moreover, due to camera shake and the error of frame registration, the imaging space corresponding to the same pixel in the HSI moves back and forth in a local background region. Therefore, it can be assumed that the spectra of the same pixel in adjacent frames, $x_i^j, x_{i-1}^j, x_{i-2}^j, ..., x_{i-P}^j$, are a linear combination of the same set of endmembers. According to the LMM, the current pixel x_i^j can be expressed as a linear combination of its former spectra $x_{i-1}^j, x_{i-2}^j, ..., x_{i-P}^j$:

$$x_i^j = \sum_{l=1}^{P} x_{i-l}^j \beta_l + n = B_i^j \beta + n \quad \text{s.t.} \quad \begin{array}{l} \sum_{l=1}^{P} \beta_l = 1 \\ \beta_l \geq 0,\ l = 1, \cdots, P \end{array} \tag{16}$$

where B_i^j is defined as the former spectra matrix, β is defined as the abundance vector, P is defined as the number of former spectra, and n is defined as the noise item.

Equation (16) means that the test pixel x_i^j and its former spectra $x_{i-1}^j, x_{i-2}^j, ..., x_{i-P}^j$ can be also applied to the CSR detector. B_i^j and x_i^j can be considered to consist of the same set of background endmembers. In the spatial anomaly detection, the background dictionary A_i^j constructed by the dual-window strategy contains some endmembers independent of x_i^j. Compared to A_i^j, B_i^j is more suitable as a background dictionary for the CSR and KCSR detectors. In this subsection, temporal anomaly detection is defined as a method to calculate the anomaly scores of the test pixel x_i^j in the current frame by using its former spectra B_i^j. Because the positions of non-homogeneous background pixels or motionless objects are almost unchanged in the HSI after inter-frame registration, the temporal anomaly detection can avoid false alarms caused by these pixels.

However, B_i^j is not a pure background dictionary sometimes. When the target is moving slowly, it takes more than one frame to pass through a pixel. In this case, if x_i^j is a target pixel, it is possible that its former spectra are also target spectra. Besides, if the trajectories of moving targets intersect, the former spectra of pixels at the intersection can also be contaminated by targets. Therefore, we delete the abnormal atoms in B_i^j based on the spatial anomaly detection results. N_D and N_C are defined as the number of atoms in the background dictionary and its candidate set, respectively. Specifically, for the test pixel x_i^j in the current frame, smoothed spatial anomaly scores $\tilde{s}_{i-1}^j, \tilde{s}_{i-2}^j, ..., \tilde{s}_{i-N_C}^j$ of its former spectra are sorted at first. In order from smallest to largest, the sort result is $\tilde{s}_{m_1}^j, \tilde{s}_{m_2}^j, ..., \tilde{s}_{m_{N_C}}^j$, where the subscripts $m_1, m_2, ..., m_{N_C}$ are the sequence numbers. The smaller the spatial anomaly score is, the higher the probability that the corresponding former spectrum belongs to the background. Therefore, N_D former spectra $x_{m_1}^j, x_{m_2}^j, ..., x_{m_{N_D}}^j$ are selected to construct a pure background dictionary $\bar{B}_i^j$ for the test pixel x_i^j. Then, the minimizing problem in the CSR algorithm can be transformed as:

$$\min_{\alpha} \alpha^T \bar{B}_i^{j\,T} \bar{B}_i^j \alpha - 2 x_i^{j\,T} A \alpha \quad \text{s.t.} \quad \begin{array}{l} e^T \alpha = 1 \\ 0 \leq \alpha_l \leq C,\ l = 1, \cdots, N_D \end{array} \tag{17}$$

where $C \in [1/N_D, 1]$. The background dictionary $\bar{B}_i^j$ can be further purified by removing the atoms with $\alpha = C$. The temporal anomaly detection result t_i^j of x_i^j is transformed as:

$$t_i^j = \sqrt{\tilde{\alpha}^{*T} \tilde{B}_i^{j\,T} \tilde{B}_i^j \tilde{\alpha}^* - 2 x_i^{j\,T} \tilde{B}_i^j \tilde{\alpha}^* + x_i^{j\,T} x_i^j} \tag{18}$$

where $\tilde{\alpha}^*$ is the approximately calculated sparse vector without anomalous atoms in the background dictionary $\tilde{B}_i^j$ and t_i^j is the l_2-norm of the approximate error. Similarly, the KCSR algorithm can also be applied to the temporal anomaly detection. After all pixels on X_i are detected in sequence, a two-dimensional temporal detection map T_i is obtained.

The lower limit of the constraint parameter C is connected with the number of anomalous atoms in the background dictionary. To obtain a convenient setting of C in the spatial and temporal anomaly detection, C can be represented as:

$$C = \frac{1}{\nu N_D} \tag{19}$$

where $\nu \in [1/N_D, 1]$. If $\nu < 1/N_D$ and $C > 1$, then the inequality constraint $\alpha_l \leq C$ is invalid. To further explore the meaning of ν, two definitions are given as follows:

$$\eta_1 = \frac{N_a}{N_D} \tag{20}$$

$$\eta_2 = \frac{\sum_{l=1}^{N_a} \alpha_a^l}{N_D} \tag{21}$$

where N_a is defined as the number of anomalous atoms and α_a^l is defined as the abundance relevant to the anomaly endmember in the LMM of the l-th anomalous atom. In the hyperspectral AD, $0 \leq \eta_2 \leq \eta_1 \ll 1$. We proofed a proposition of the parameter ν in the article [23]:

Proposition 1. *To delete all anomalous atoms from the background dictionary, v must satisfy:*

$$v \geq \max\left(\eta_1, \eta_2/\alpha_a\right) \tag{22}$$

where α_a is defined as the abundance relevant to the anomaly endmember in the LMM of the test pixel.

The proposition gives an intuitive interpretation of ν. When ν is larger than $\max(\eta_1, \eta_2/\alpha_a)$, all anomalous atoms can be deleted. Regardless of spatial detection or temporal detection, α_a of the same test pixel is constant. Therefore, it is practicable to set ν to the same value in both detections. η_1 and η_2/α_a in temporal detection can be set to values smaller than those in spatial detection by reducing the proportion of anomalous atoms in $\bar{\boldsymbol{B}}_i^j$. One method is to enlarge N_D, the size of $\bar{\boldsymbol{B}}_i^j$. Another method is to decrease N_a, the number of anomalous atoms, by enlarging the size of the candidate set $\boldsymbol{B}_i^j$ or sample the former spectra at intervals before constructing $\boldsymbol{B}_i^j$. Through the above operations, the lower limit of ν in temporal detection is less than that in spatial detection. When ν is set to an excessively large value, numerous background atoms are exorbitantly deleted, resulting in slight degeneration in the ability of the CSR and KCSR algorithms to represent test background pixels. Therefore, ν should be a trade-off value between the inadequate deletion of anomalous atoms and unnecessary deletion in spatial detection. The same ν can cause the excessive deletion of atoms in temporal detection, but a large N_D can avoid this situation.

3.4. Spatio-Temporal Fusion

Compared to the spatial anomaly detection, the temporal anomaly detection can suppress spatially non-homogeneous background pixels and stationary objects. Furthermore, compared to the temporal profile filtering algorithms, the proposed temporal anomaly detection can identify moving targets with different speeds simultaneously and is robust to the situation where multiple targets pass through the same trajectory one after the other. However, the temporal detection is inferior to the spatial detection in some situations. If there are some moving background pixels in the scene, such as clouds, temporal anomaly detection can judge them as targets. Besides, if the frame registration error is too large, the temporal background dictionary cannot describe the background accurately. To improve the stability and robustness of the detection algorithm, it is necessary to combine spatial and temporal detection results.

Before fusion, the filtering operation in Section 3.2 can also be performed on the temporal detection map $\boldsymbol{T}_i$. First, perform Z-score normalization on $\boldsymbol{T}_i$:

$$\bar{T}_i = \frac{T_i - \mu}{\sigma}. \tag{23}$$

where μ and σ are set to the mean value and standard deviation of the collection of S_i and its symmetric set about zero. Then, the same iterative smoothing operation as Equation (15) is performed on $\bar{T}_i$ to reduce temporal clutter:

$$\tilde{t}_i^j = (1-\rho)\,\tilde{t}_{i-1}^j + \rho \sum_{l \in L(j)} \varepsilon_l \bar{t}_i^l \tag{24}$$

where $\bar{t}_i^l$ is the normalized temporal anomaly scores of x_i^l and $\tilde{t}_i^j$ and $\tilde{t}_{i-1}^j$ are the temporal spatial anomaly scores of x_i^j and x_{i-1}^j, respectively. The smoothed detection maps can be combined by the multiplication fusion strategy:

$$ST_i = \frac{\tilde{S}_i - \min(\tilde{S}_i)}{\max(\tilde{S}_i) - \min(\tilde{S}_i)} \circ \frac{\tilde{T}_i - \min(\tilde{T}_i)}{\max(\tilde{T}_i) - \min(\tilde{T}_i)} \tag{25}$$

where $\max(\tilde{S}_i)$ and $\max(\tilde{T}_i)$ are the maximum values in $\tilde{S}_i$ and $\tilde{T}_i$, $\min(\tilde{S}_i)$ and $\min(\tilde{T}_i)$ are the minimum values in $\tilde{S}_i$ and $\tilde{T}_i$, the symbol $\circ$ denotes the Hadamard product, and ST_i is the fusion spatio-temporal detection map. The overall description of the proposed spatio-temporal anomaly detection is presented in Algorithm 1.

Algorithm 1 CSR-based spatio-temporal anomaly detection for moving targets

Input: Hyperspectral sequences, dual-window size (w_{in}, w_{out}), temporal background dictionary size N_D, candidate set size N_C, parameter ν, and kernel parameter γ for KCSR.

for each frame X_i in the hyperspectral sequences **do**

1. **for** each pixel x_i^j in X_i **do**
   (a) Collect the spatial background dictionary based on the hollow window;
   (b) Calculate the spatial anomaly score s_i^j by the CSR or KCSR detector;
2. Smooth the spatial detection map S_i by Equations (14) and (15);
3. **if** $i > N_C$ **then**
   (a) **for** each pixel x_i^j in X_i **do**
      i. According to the sorting of smoothed spatial detection results $\hat{s}_{i-1}^j, \hat{s}_{i-2}^j, \ldots, \hat{s}_{i-N_C}^j$, select N_D dictionary atoms from former spectra $x_{i-1}^j, x_{i-2}^j, \ldots, x_{i-N_C}^j$ to construct the temporal background dictionary $\bar{B}_i^j$;
      ii. Calculate the temporal anomaly score t_i^j by the CSR or KCSR detector;
   (b) Smooth the spatial detection map T_i by Equations (23) and (24);
   (c) Calculate the spatio-temporal fusion map ST_i by Equation (25);

   end if

Output: Spatio-temporal anomaly detection map ST_i when $i > N_C$.

4. Experimental Results and Discussion

In the beginning of this section, a real HSI sequence dataset and a synthetic dataset are introduced. Subsequently, the capability of the proposed temporal anomaly detection with different background dictionary sizes and different spatial detection results is demonstrated in detail. Additionally, the proposed spatio-temporal anomaly detection is compared to several existing algorithms in the detection performance.

4.1. Datasets and Evaluation Metrics

The Cloud dataset is an HSI sequence under a complex cloudy background and was collected by the Interuniversity Microelectronics Centre of Beihang University with the xiSpec snapshot mosaic hyperspectral cameras [12]. The dataset has a spatial size of 409 × 216 pixels and 25 spectral bands including the 682–957 nm spectral region. The HSI sequence consists of 500 frames, where an aircraft (Target A) rises from the bottom of the imagery. Since the distance between the camera and the aircraft increases with the frames, the size of the aircraft decreases over time, 53 pixels in the 1st frame and 21 pixels in the 500th frame, resulting in a descending spectral difference from the background. However, because of the aircraft's speed on HSIs also decreases, the number of frames that the aircraft needs to pass through a pixel increases. Three small flying targets (Target B, Target C, and Target D) with no more than 10 pixels exist in the 250th–393rd, 256th–363rd, and 417th–466th frames, respectively, and their velocities are all greater than 5 pixels per frame. As shown in Figure 2, there is a noise clutter in the cloudy background.

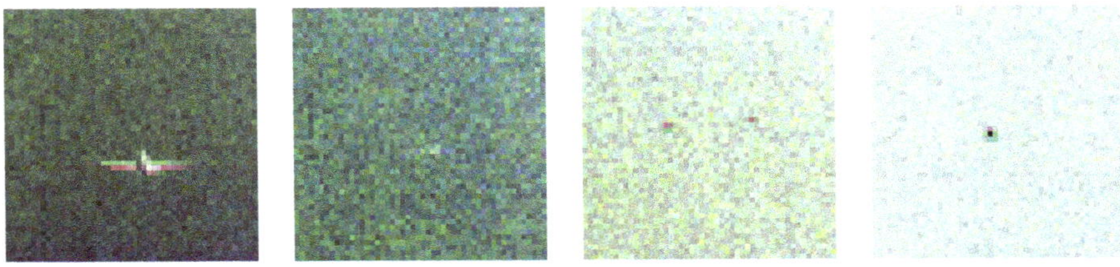

(a) (b) (c) (d)

Figure 2. False color local image around targets in the Cloud dataset. (**a**) Target A in the 50th frame. (**b**) Target A in the 500th frame. (**c**) Target B and Target C. (**d**) Target D.

The synthetic dataset is based on the Terrain dataset acquired by the Hyperspectral Digital Image Collection Experiment sensor. The dataset has a spatial size of 180 × 180 pixels and 210 spectral bands including the 400–2500 nm spectral region, as shown in Figure 3a. The spatial resolution is 1 m, and the spectral resolution is 10 nm. The water absorption and high noise bands are deleted, and one-hundred sixty-two spectral bands are usable in the experiments. According to the LMM, synthetic targets can be added to the Terrain dataset by:

$$\tilde{\boldsymbol{a}} = (1 - \lambda)\,\boldsymbol{b} + \lambda \boldsymbol{a} + \boldsymbol{n} \tag{26}$$

where $\boldsymbol{a}$ is a pure target spectrum, $\boldsymbol{b}$ is an original background spectrum, $\tilde{\boldsymbol{a}}$ is a mixed target spectrum, $\boldsymbol{n}$ is the added zero mean Gaussian noise vector, and λ is the target abundance to be set. Considering that the radiation response interval of the background varies with bands, Gaussian noise with different variance is added to each band of a hyperspectral cube. Noise intensity is adjusted by the signal-to-noise ratio (SNR), expressed in this dataset by:

$$\mathrm{SNR}_{\mathrm{dB}} = 10 \log_{10}\left(\frac{\sigma_{b,l}^2}{\sigma_{n,l}^2}\right) \tag{27}$$

where $\sigma_{b,l}^2$ and $\sigma_{n,l}^2$ are the variances of the background and noise in the l-th band. Three targets with a size of 5 × 5 pixels and a speed of 2 pixels per frame are added to the Terrain dataset and move 100 frames. The plane trajectories of targets are the same, and the distance between the two targets is ten frames. Considering that the boundaries between neighboring objects are often accompanied by severe spectral mixing in the real data, λ of 16 pixels on the periphery of targets is set to 10%, while that of 9 pixels in the center of targets is set to 40%. To explore the noise immunity of CSR-ST, the SNR is set to 20 dB, 10 dB, 5 dB, and 0 dB in turn. Figure 3b–f shows background spectra and mixed target spectra in different noise environments. With the decrease of SNR, the discriminability between background and mixed targets also decreases. When the SNR is 0 dB, background spectra and mixed target spectra are almost indistinguishable.

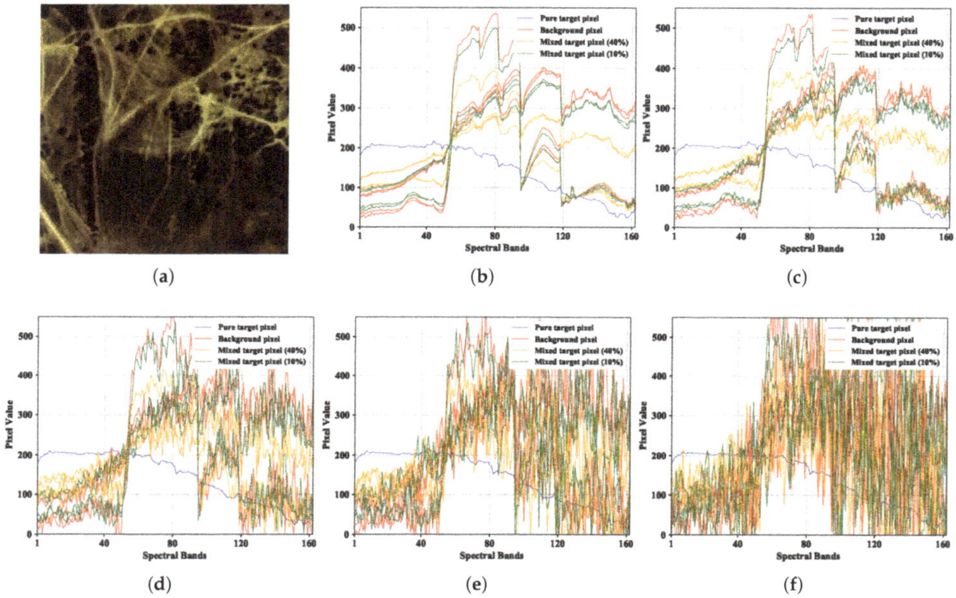

Figure 3. The synthetic Terrain dataset. (**a**) Original false color image. (**b**–**f**) Spectral curves of background and target pixels with different noise. The blue curve is a pure target pixel; the four red curves are background pixels; orange curves are mixed target pixels with a target abundance of 40%; and green curves are mixed target pixels with a target abundance of 10%. (**b**) No noise. (**c**) SNR = 20 dB. (**d**) SNR = 10 dB. (**e**) SNR = 5 dB. (**f**) SNR = 0 dB.

To evaluate anomaly detection performance, this article adopts the receiver operating characteristic (ROC) curve and the area under the ROC curve (AUC). The detection probability (P_d) and false alarm rate (P_f) are computed on a segmentation map, which is obtained on the detection map by a given threshold. After the threshold is iterated over, a set of P_d and P_f can be used to plot the ROC curve. An excellent detector has an upper left ROC curve [43]. However, the ROC curve can only qualitatively analyze detection performance. AUC [44] can give an intuitive and quantitative description and is calculated by several trapezoids:

$$\mathrm{AUC} = \frac{1}{2}\sum_{l=1}^{n-1}(P_f^{l+1} - P_f^{i})(P_d^{l+1} + P_d^{l}) \tag{28}$$

where (P_f^l, P_d^l) is defined as the l-th coordinate point and n is defined as the number of coordinate points constituting the curve. The closer to 1 AUC values are, the better the detection algorithms are. For the anomaly detector in an HSI sequence, the mean ROC of all frames can describe the performance.

Considering that kernel space can represent hyperspectral data better, the proposed spatio-temporal anomaly detection algorithm is based on the KCSR model in the following experiments. KCSR-S, KCSR-SF, KCSR-T, and KCSR-ST denote spatial detection, smoothed spatial detection, temporal detection, and spatio-temporal fusion detection, respectively. All the experiments were implemented on a machine that was equipped with an Intel Core i9-9980XE CPU and 128-GB RAM, and the programs were written in Python.

4.2. Temporal Detection Performance under Different Settings of the Temporal Background Dictionary

For the KCSR-based temporal detection, the parameter ν can be set to the same value in spatial detection, which was analyzed in Section 3.3. Moreover, because of the same background spectra for the spatial and temporal detection, the kernel space in spatial detection is also suitable for temporal detection. Therefore, after the parameters of spatial detection are adjusted, the settings to be adjusted in the temporal detection are N_D and N_C, denoting the sizes of the temporal background dictionary and its candidate set, respectively. To further explain N_D and N_C, we define the number of removed atoms as:

$$N_R = N_C - N_D. \tag{29}$$

The meaning of the candidate set is to prevent the background dictionary from the target contamination, and N_R should ensure than most of the abnormal spectra can be removed from the candidate set.

4.2.1. Experiments on the Cloud Dataset

Traditional temporal profile filtering algorithms ask for strong prior information about the target velocity. We count the number of frames that targets take to pass through a single pixel in the Cloud dataset and draw the histogram. As shown in Figure 4, three-thousand two-hundred forty-one pixels are passed through by targets in 20 frames, while only 130 pixels are passed through by targets more than 20 frames. The latter occurs mainly in the latter half of the sequence because the airport is far away from the camera and becomes slower in the imagery.

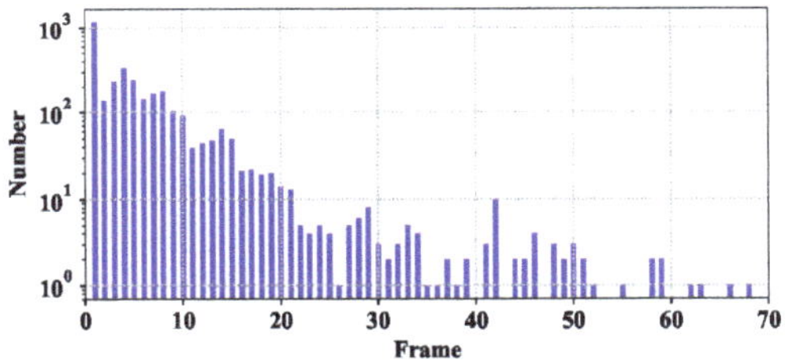

Figure 4. Histogram of the number of frames that targets take to pass through a single pixel in the Cloud dataset.

To explore the impact of the temporal background dictionary on the temporal detection performance, we set N_C to 20, 30, 50, 80, and 100, respectively. N_R was set to 10, 20, 30, and 40, respectively. Because the first 100 frames in the Cloud dataset are selected as the temporal background candidate set, the temporal anomaly detection starts at the 101st frame. The parameters ν, γ and the dual-window size of KCSR-S are empirically tuned to acquire the best detection capability in the first frame. The mean AUCs of KCSR-T in the Cloud dataset are shown in Table 1. When N_C is set to 20, the mean AUC of KCSR-T becomes the worst value of 0.970966 in the table. That is because if the dictionary candidate set size is too small, temporal background dictionaries of some target pixels can consist mainly of target spectra. When N_C is set to 50, 80, and 100, the mean AUCs of KCSR-T when N_R is 20 are better than those when N_R is 10. That is because the former can remove more target spectra in the dictionary candidate set than the latter. When N_C is set to 30 and 50, the mean AUCs of KCSR-T when N_D is 10 are worse than those when N_D is 20. It is indicated that a small temporal background dictionary size is not conducive to the representation of spectral features. Moreover, the best mean AUC in Table 1 is 0.980302 and achieved when N_C is 50 and N_R is 20.

Table 1. The mean AUC performance achieved by KCSR-T on the Cloud dataset with different settings of the temporal background dictionary.

N_R \ N_C	20	30	50	80	100
10	0.970966	0.978610	0.980203	0.978596	0.978292
20	–	0.976092	0.980302	0.979089	0.978912
30	–	–	0.979407	0.979447	0.979326
40	–	–	0.978459	0.979405	0.979726

4.2.2. Experiments on the Synthetic Terrain Dataset

To explore how to set the temporal background dictionary on the synthetic Terrain dataset, we set N_C to 20, 30, 40, and 50 in turn. N_R was set to 10, 20, 30, and 40, respectively. The first 50 frames in the Cloud dataset are selected as the temporal background candidate set, and the KCSR-T is performed on the last 50 frames. The parameters ν, γ and the dual-window size of KCSR-S are empirically tuned to acquire the best detection capability in the first frame. As shown in Table 2, when the background dictionary size N_D is fixed to 10, the worst mean AUC is achieved by $N_R = 20$. That is because the former spectra of target pixels contain at most 8 target spectra, and a small N_R is not conducive to removing them from the background dictionary candidate set. With SNR decreasing, the distinction between background and target spectra decreases, and the gaps of mean AUCs between $N_R = 10$ and other settings become larger. In addition, the best mean AUCs in the four noise conditions are achieved by $N_C = 50$ and $N_R = 20$, which are 0.999258, 0.932968, 0.819948, and 0.685078, respectively.

Table 2. The mean AUC performance achieved by KCSR-T on the synthetic Terrain dataset in four noise conditions with different settings of the temporal background dictionary. (**a**) SNR = 20 dB. (**b**) SNR = 10 dB. (**c**) SNR = 5 dB. (**d**) SNR = 0 dB.

a

N_R \ N_C	20	30	40	50
10	0.997661	0.998487	0.998910	0.999168
20	–	0.998424	0.998939	0.999258
30	–	–	0.998367	0.999095
40	–	–	–	0.998408

b

N_R \ N_C	20	30	40	50
10	0.922512	0.926485	0.929855	0.931893
20	–	0.925626	0.930665	0.932968
30	–	–	0.926307	0.930971
40	–	–	–	0.927911

c

N_R \ N_C	20	30	40	50
10	0.807302	0.810852	0.814612	0.816796
20	–	0.811735	0.817965	0.819948
30	–	–	0.814413	0.818968
40	–	–	–	0.815303

d

N_R \ N_C	20	30	40	50
10	0.670573	0.681748	0.679998	0.682716
20	–	0.678515	0.683091	0.685078
30	–	–	0.681748	0.685298
40	–	–	–	0.684125

4.3. Detection Performance under Different Settings of the Dual-Window

As mentioned in Section 3.3, it is different for moving target detection to set the optimal dual-window size in advance in the spatial anomaly detection. One important reason for this is that the sizes of moving targets can change. For the Cloud dataset, as the airplane moves away from the camera, the aircraft size in the HSI becomes smaller. In Section 4.2.1, the dual-window size (w_{in}, w_{out}) of KCSR-S was set to (29, 31), which is the optimal size in the first frame. However, (29, 31) is too large for the aircraft in the 500th frame, which only has 21 pixels. To explore the impact of different settings of the dual-window on KCSR-S and KCSR-T, we set the dual-window size to (3, 5), (9, 11), (13, 15), (19, 21), (23, 25), and (29, 31), respectively. Considering that the iterative smoothing filter can improve the spatial detection map of KCSR-S, KCSR-T uses the original spatial detection results instead of smoothed results to select the temporal background dictionary in this subsection.

As shown in Table 3, the dual-window size has a significant influence on the detection capability of KCSR-S. The best mean AUC of KCSR-S in the 101st–500th frames is 0.970251, while the worst mean AUC is 0.961412. However, the mean AUC of KCSR-T is better than that of KCSR-S under different dual-window sizes and fluctuates in a small range from 0.979962 to 0.980175. To give a more intuitive representation, we fit the variation curves of AUC with time in the 101st–500th frames by a power function with the highest power of 15. As shown in Figure 5a, with the change of the aircraft size, the optimal dual-window size also changes at different times. The optimal size around the 200th frame is (23, 25) and then becomes (19, 21) in the 300th frame. When it reaches the 200th frame, the 300th frame, the 450th frame, and the 480th frame, respectively, the optimal size is (23, 25), (19, 21), (13, 15), and (9, 11), respectively. Although the fitted curve with a dual-window size of (29, 31) performs well at the beginning of the sequence, the gap between the curve and the best AUC increases over time. However, the AUC of KCSR-T is almost impervious to the dual-window size of KCSR-S. Compared to Figure 5a, the curves with different dual-window size in Figure 5b are almost the same. There are two reasons why KCSR-T is robust to the dual-window size of KCST-S. On the one hand, different dual-window sizes can result in different anomaly scores of target pixels in the spatial detection, and an unsuitable size can lead to lower anomaly scores. However, for the same pixel, even though under unsuitable dual-window sizes, the gap between anomaly scores within and without targets is still large enough to remove anomalous spectra in the candidate set of the temporal background dictionary. On the other hand, KCSR-T can also automatically remove anomalous atoms from the background dictionary during the temporal detection process. In conclusion, the proposed temporal anomaly detection is remarkably robust to the dual-window size in the spatial detection, and the combination of the spatial and temporal detection can overcome the disadvantages of the dual-window strategy.

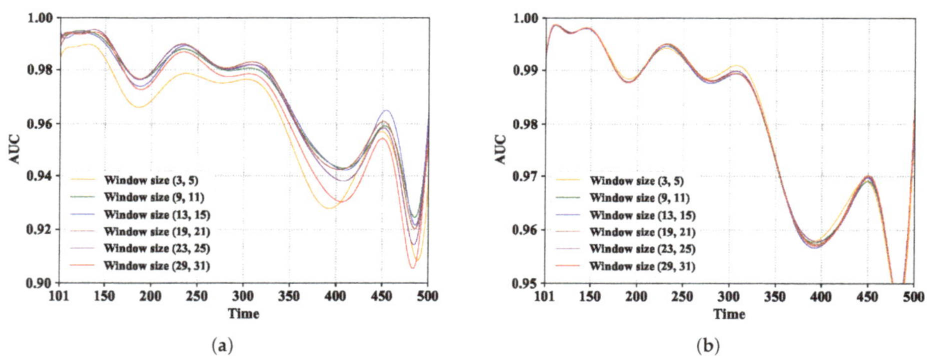

Figure 5. The fitted variation curves of AUC with time in the 101st–500th frames under different settings of the dual-window. (**a**) The curves of KCSR-S. (**b**) The curves of KCSR-T.

Table 3. The mean AUC performance achieved by KCSR-S and KCSR-T on the Cloud dataset with different settings of the temporal background dictionary.

(w_{in}, w_{out})	(3, 5)	(9, 11)	(13, 15)	(19, 21)	(23, 25)	(29, 31)
KCSR-S	0.961412	0.969947	0.970367	0.970251	0.968993	0.964927
KCSR-T	0.980175	0.979962	0.979918	0.980145	0.980115	0.980010

4.4. Comparison to the State-of-the-Art

In the subsection, the KCSR-ST algorithm is contrasted with several single-frame HSI anomaly detection algorithms, including RX [13], QLRX [15], KSVDD [19], KRX [16], CR [22], KCR [22], and CSR. Meanwhile, the proposed algorithm is also contrasted with two detection algorithms for moving targets, including VF [5] and STH [12]. In fairness, both VF and STH are based on KCSR in the following experiments, denoted by KCSR-VF and KCSR-STH, respectively. All parameters of these algorithms are empirically tuned to acquire the best detection capability at the beginning of the sequences. The dual-window sizes are set to (29, 31) and (9, 15) for the Could and Terrain dataset, respectively. The N_C and N_D on the two datasets are set to the optimal values obtained in Section 4.2. The temporal filter weight ρ is set to 0.5, and the spatial smooth filter adopts a simple 3×3 mean smoothing filter. The AUC performances and detection maps of KCSR-S, KCSR-SF, and KCSR-T are also shown to explore the role of each step in the proposed KCSR-ST algorithm.

4.4.1. Experiments on the Cloud Dataset

The ROC curves obtained on the Cloud dataset are shown in Figure 6; the AUC values are shown in Table 4; and the color detection maps are shown in Figure 7. These all illustrate that KCSR-ST is superior to all single-frame and multiple-frame anomaly detection algorithms.

Table 4. The mean AUC performance obtained on the Cloud dataset.

RX	QLRX	KRX	KSVDD	CR	KCR	CSR	KCSR-S	KCSR-VF	KCSR-STH	KCSR-SF	KCSR-T	KCSR-ST
0.9371	0.9440	0.9644	0.9617	0.9360	0.9442	0.9620	0.9649	0.9784	0.9892	0.9901	0.9803	**0.9976**

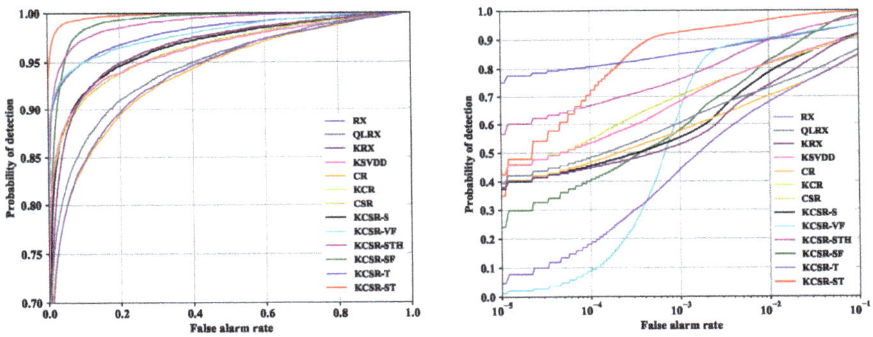

(**a**) (**b**)

Figure 6. ROC curves obtained on the Cloud dataset. (**a**) Logarithmic abscissa; (**b**) linear abscissa.

As shown in Table 4, the best AUC value among single-frame anomaly detection algorithms is 0.9649 and achieved by KCSR. Taking advantage of temporal information, the AUC values of KCSR-VF, KCSR-STH, and KCSR-T are all higher than single-frame algorithms. The reason for this phenomenon can be explained in Figure 7. As shown in Figure 7a, obvious vignetting exists at the edges of false color images. Vignetting is a common phenomenon in photography, but turns edges of HSIs into heterogeneous background pixels. Therefore, there always exists a relatively large number of false alarms at the edges of detection maps obtained by single-frame algorithms, which is shown in Figure 7c–j. Because KCSR-VF and KCSR-T make use of the historical spatial detection results and the former spectra of test pixels, respectively, the heterogeneous background pixels rarely lead to false alarms in the corresponding detection maps, which is shown in Figure 7k,n.

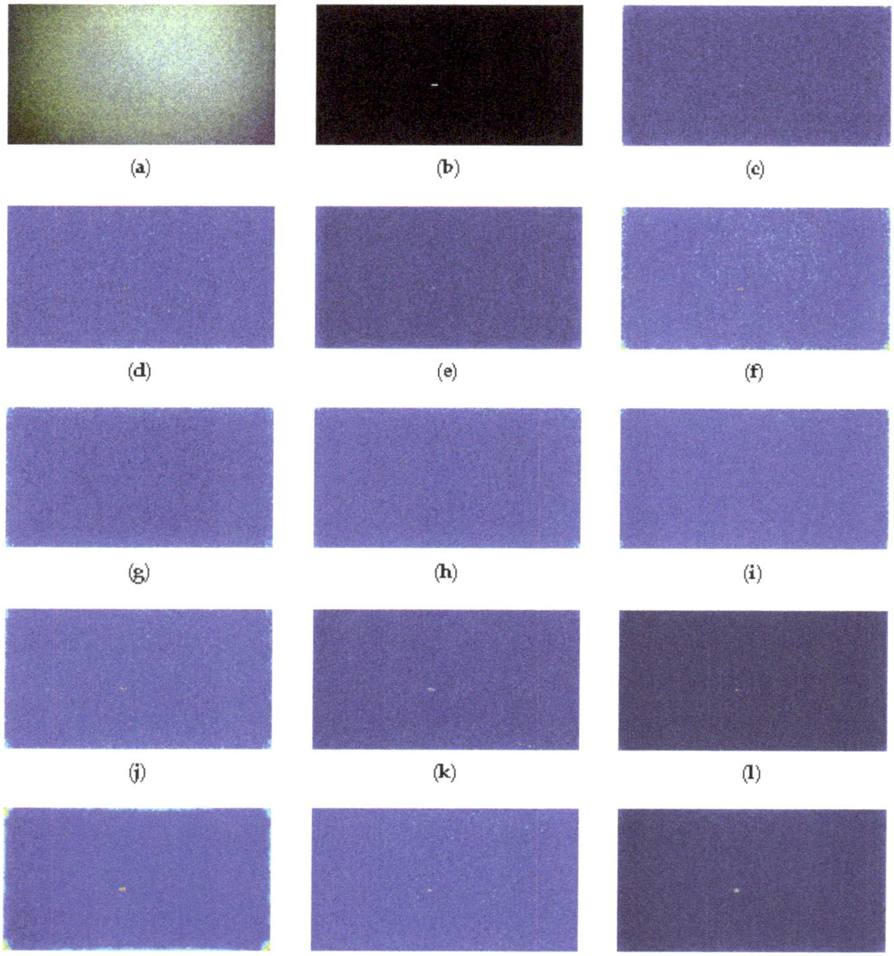

(**a**) (**b**) (**c**) (**d**) (**e**) (**f**) (**g**) (**h**) (**i**) (**j**) (**k**) (**l**) (**m**) (**n**) (**o**)

Figure 7. Color detection maps obtained in the 400th frame of the Cloud dataset. (**a**) False color image; (**b**) ground-truth map; (**c**) RX; (**d**) QLRX; (**e**) KRX; (**f**) KSVDD; (**g**) CR; (**h**) KCR; (**i**) CSR; (**j**) KCSR-S; (**k**) KCSR-VF; (**l**) KCSR-STH; (**m**) KCSR-SF; (**n**) KCSR-T; (**o**) KCSR-ST.

However, the historical trajectory of Target B turns into false alarms in the detection map of KCSR-VF in the 400th frame. That is because the VF algorithm is mainly designed to detect slow targets, and the parameter setting of VF depends on the speed of targets. Because velocities of Targets B, C, and D are all greater than 5 pixels per frame and go through a pixel in a frame, the temporal variance-calculation window suitable for Target A is too long for them. As long as the temporal variance-calculation window contains the trajectory of Targets B, C, and D, the detection results can have high values and become false alarms in the VF detection map. Moreover, KCSR-STH combines KCSR-VF with other spatial detection maps and is slightly affected by these false alarms, shown in Figure 7l.

As shown in Figure 7j,m, there is much background clutter on the detection maps of KCSR-S and KCSR-T. Compared to KCSR-S, the spatial detection map after the iterative smoothing filter, KCSR-SF, suppresses the background clutter and enhances the target. However, false alarms resulting from the heterogeneous background are also enhanced in Figure 7m. KCSR-ST combines the smoothed spatial detection map (KCSR-SF) with the smoothed temporal detection map, and the heterogeneous background and the background clutter are entirely suppressed in Figure 7o. As shown in Figure 6a, the ROC curve of KCSR-ST is on the upper left of those of other algorithms, which indicates that KCSR-ST is superior to the single-frame and multi-frame anomaly detection algorithms. However, when P_f is limited to an extremely low value range, the detection performance of KCSR-ST is inferior to KCSR-T. As shown in Figure 6b, when P_f is 10^{-5}, the P_d of KCSR-T is about 0.75, while that of KCSR-ST is only about 0.35. Furthermore, when P_f is smaller than 10^{-5}, the ROC curve of KCSR-S outperforms KCSR-SF. This is because the iterative smoothing filter enhances target pixels and pixels around targets. Compared to KCSR-S and KCSR-T, KCSR-SF and KCSR-ST blurred the boundary between target and background in the detection maps. However, the iterative smoothing filter can still be regarded as a useful strategy. Although reducing P_d when P_f is low, the enhancement improves the ability to detect slow targets and the robustness to the different moving speeds of the targets. For most hyperspectral anomaly detection scenarios, the focus is on whether the target exists rather than the shape of the target. The false alarms that result from the enhancement of pixels around the target have little influence on the judgment of whether the target exists. Besides, the enhancement from the iterative smoothing filter can be optimized by adjusting the filter weights or changing the smoothing strategy.

4.4.2. Experiments on the Terrain Dataset

The ROC curves achieved on the synthetic Terrain dataset under different noise environments are shown in Figure 8; the color detection maps are shown in Figures 9–11; and the AUC results are shown in Table 5. Our proposed KCSR-ST algorithm is considerably robust to noise and superior to all single-frame anomaly detection algorithms. When the SNR is set to 20 dB, 10 dB, 5 dB, and 0 dB, respectively, the corresponding mean AUC of KCSR-ST is 0.9996, 0.9959, 0.9461, and 0.7516, respectively; whereas, the best mean AUC among single-frame algorithms is 0.8402, 0.7438, 0.7057, and 0.6205, respectively. As shown in Figure 9c–j, Figure 10c–j, and Figure 11c–j, there are a large number of false alarms on the detection maps of single-frame algorithms because some trees are sparsely distributed in the scene.

Table 5. The mean AUC performance obtained on the synthetic Terrain dataset.

SNR	RX	QLRX	KRX	KSVDD	CR	KCR	CSR	KCSR-S	KCSR-VF	KCSR-STH	KCSR-SF	KCSR-T	KCSR-ST
20	0.7696	0.8400	0.6407	0.6166	0.8199	0.8371	0.7889	0.8402	0.8829	0.8852	0.9228	0.9993	**0.9996**
10	0.7047	0.6778	0.6204	0.6128	0.7323	0.7438	0.6970	0.7325	0.7664	0.7769	0.8888	0.9330	**0.9959**
5	0.6465	0.6383	0.5881	0.6051	0.6905	0.7057	0.6591	0.6858	0.7136	0.7265	0.8162	0.8199	**0.9461**
0	0.5155	0.6219	0.4960	0.5705	0.6007	0.6205	0.5769	0.5943	0.5671	0.6005	0.6390	0.6851	**0.7516**

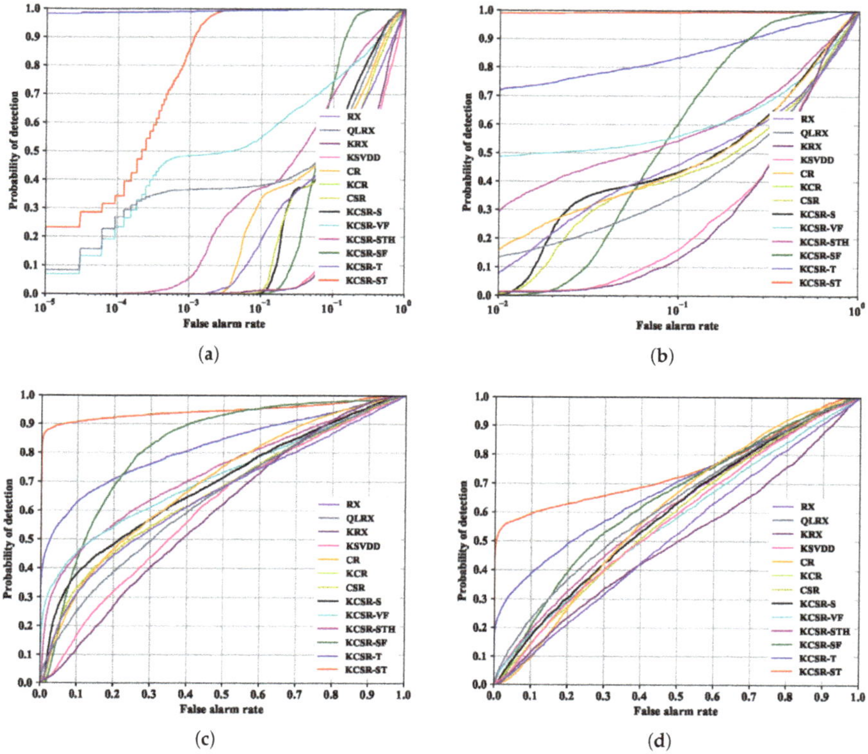

Figure 8. ROC curves obtained on the synthetic Terrain dataset. (**a**,**b**) adopt logarithmic abscissas and (**c**,**d**) adopt linear abscissas. (**a**) SNR = 20 dB; (**b**) SNR = 10 dB; (**c**) SNR = 5 dB; (**d**) SNR = 0 dB.

Although KCSR-VF and KCSR-STH are also superior to single-frame algorithms, their detection performance is far inferior to that of KCSR-ST on the Terrain dataset. As shown in Figure 9k, the trajectory of targets results in false alarms on the detection map of KCSR-VF. That is because the targets share the same trajectory, and the baseline background of VF cannot be estimated accurately. KCSR-STH combines KCSR-VF, KCSR-S, and its temporal detection and suppresses the background and false alarms. However, because the temporal detection of STH extracts the background dictionary of the test pixel in the forward frame by the same dual-window as KCSR-S, the false alarms resulting from sparse trees are still on the temporal detection map of KCSR-STH and then appear in the final fusion map, i.e., Figure 9l.

As shown in Figure 9k, when the SNR is 20 dB, KCSR-T has an excellent ability to detect moving targets. Although the mean AUC of KCSR-T is slightly lower than KCSR-ST, the ROC performance of KCSR-T outperforms KCSR-ST when P_f is smaller than 10^{-3}. As shown in Figure 8a, when P_f is 10^{-5}, the P_d of KCSR-T is about 0.98, while the P_ds of all the other algorithm are higher than 0.25. When the SNR is 10 dB, the mean AUC of KCSR-T is much lower that of KCSR-ST, and the ROC performance of KCSR-T is also inferior to that of KCSR-ST. That is because the target abundances of target peripheral pixels are lower, and then, these pixels cannot be detected by KCSR-T, which is shown in Figure 10n. Employing the iterative smoothing filter, KCSR-ST enhances the anomaly scores of pixels around targets and then performs prominently under the ROC and AUC evaluation metrics. When SNR comes to 5 dB, there is much noise clutter on the detection map of KCSR-T, i.e., Figure 11n, and the mean AUC of KCSR-T descends to 0.8199. As shown in Figure 8c, the ROC curves of single-frame algorithms are close to the diagonal, which means that the detection abilities of single-frame algorithms of moving targets are incredibly inferior. KCSR-ST can effectively suppress the noise clutter and false alarms on the detection map, which is shown in Figure 11n, and its ROC performances are much better than the other curves in Figure 8c. Even though the SNR descends to 0 dB, KCSR-ST still can detect targets. As shown in Figure 8d, the ROC curves of other algorithms are around the diagonal, while the P_d of KCSR-ST can reach 0.6 when the P_f is 0.1.

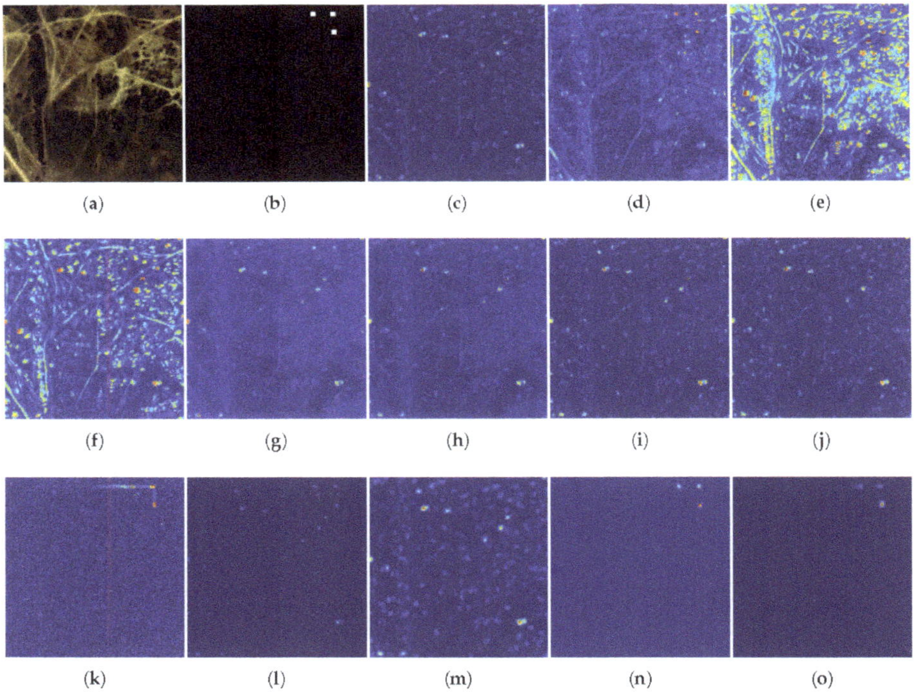

Figure 9. Color detection maps obtained in the 60th frame of the synthetic Terrain dataset when SNR = 20 dB. (**a**) False color image; (**b**) ground-truth map; (**c**) RX; (**d**) QLRX; (**e**) KRX; (**f**) KSVDD; (**g**) CR; (**h**) KCR; (**i**) CSR; (**j**) KCSR-S; (**k**) KCSR-VF; (**l**) KCSR-STH; (**m**) KCSR-SF; (**n**) KCSR-T; (**o**) KCSR-ST.

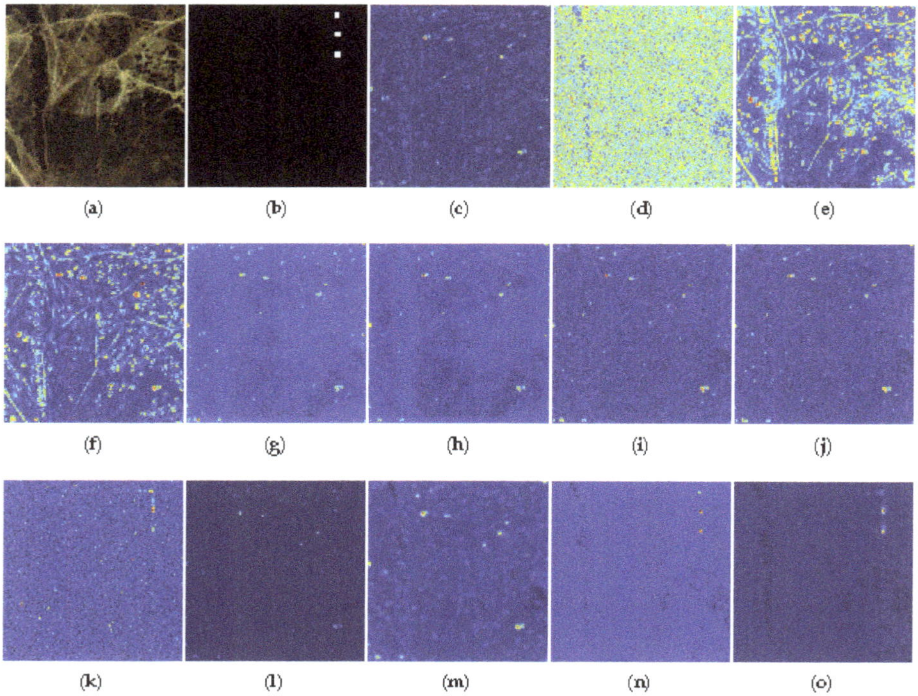

Figure 10. Color detection maps obtained in the 70th frame of the synthetic Terrain dataset when SNR = 10 dB. (**a**) False color image; (**b**) ground-truth map; (**c**) RX; (**d**) QLRX; (**e**) KRX; (**f**) KSVDD; (**g**) CR; (**h**) KCR; (**i**) CSR; (**j**) KCSR-S; (**k**) KCSR-VF; (**l**) KCSR-STH; (**m**) KCSR-SF; (**n**) KCSR-T; (**o**) KCSR-ST.

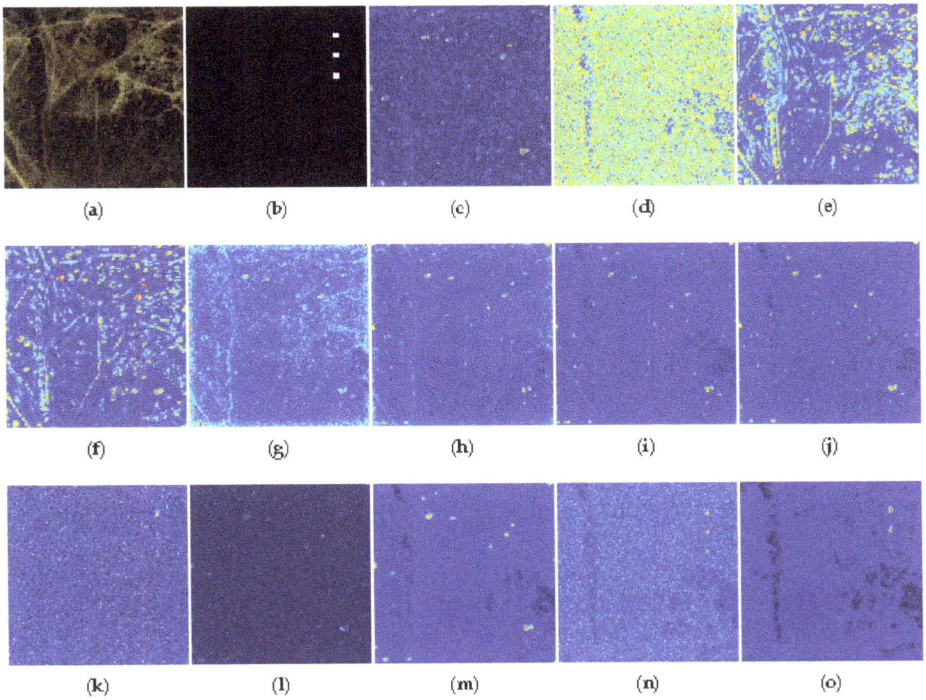

Figure 11. Color detection maps obtained in the 80th frame of the synthetic Terrain dataset when SNR = 5 dB. (**a**) False color image; (**b**) ground-truth map; (**c**) RX; (**d**) QLRX; (**e**) KRX; (**f**) KSVDD; (**g**) CR; (**h**) KCR; (**i**) CSR; (**j**) KCSR-S; (**k**) KCSR-VF; (**l**) KCSR-STH; (**m**) KCSR-SF; (**n**) KCSR-T; (**o**) KCSR-ST.

5. Conclusions

In the traditional single-frame anomaly detection, false alarms on stationary targets and non-homogeneous backgrounds are unavoidable. Besides, detecting targets in complex motion is still a challenge for multi-frame algorithms. In this article, a constrained sparse representation- based spatio-temporal AD algorithm is proposed to identify small and dim moving targets in hyperspectral sequences and overcomes the aforementioned drawbacks. Our algorithm includes a spatial detector and a temporal detector. The former can suppress moving background regions, and the latter can suppress non-homogeneous background and stationary objects. Moreover, two temporal background purification procedures ensure the effectiveness of the temporal detector for targets in complex motion. Experiments accomplished on the Cloud dataset and the synthetic Terrain dataset indicate that our algorithm is superior to other classic detection algorithms. Even though the noise clutter is extreme, our algorithm can also suppress the clutter and effectively detect small and dim moving targets.

Our algorithm provides a novel spatio-temporal anomaly detection framework for hyperspectral remote sensing. In addition, adaptive anomaly elimination in the temporal background is a good idea for detecting targets in complex motion. However, the proposed algorithm needs accurate frame registration and has enormous demand for data storage equipment. Besides, the iterative smoothing filter can effectively suppress background clutter, but blurs the boundary between the target and the background. In future work, we will focus on reducing the algorithm's need for inter-frame matching and data storage and improve the iterative smoothing filter by introducing edge-preserving filters. Furthermore, the proposed algorithm can be combined with target tracking, state estimation, and trajectory prediction and then provide motion information about targets.

Author Contributions: Conceptualization, methodology, and software, Q.L. and Z.L. (Zhaoxu Li); writing, original draft preparation, Z.L. (Zhaoxu Li) and Z.W.; writing, review and editing, Q.L., Z.L. (Zaiping Lin), and J.W.; visualization, Z.W.; project administration, Z.L. (Zaiping Lin). All authors read and agreed to the published version of the manuscript.

Funding: This research was funded by the National Natural Science Foundation of China under Grant 61605242, Grant 61602499, and Grant 61471371.

Acknowledgments: Thanks to Wang of Beihang University for providing the Cloud dataset.

Conflicts of Interest: The authors declare no conflict of interest.

Abbreviations

The following abbreviations are used in this manuscript:

AD	Anomaly detection
CD	Change detection
HSI	Hyperspectral imagery
RX	Reed–Xiaoli algorithm
KRX	Kernel version of RX
QLRX	Quasi-local-RX
SVDD	Support vector data description
KSVDD	Kernel version of SVDD
SR	Sparse representation
CR	Collaborative representation
KCR	Kernel version of CR
CSR	Constrained sparse representation
KCSR	Kernel version of KCR
SMO	sequential minimal optimization
VF	Variance filter
KCSR-VF	VF based on KCSR
STH	Fusion of the spatial detection map, temporal detection map, and trajectory history map
KCSR-STH	STH based on KCSR
CSR-ST	CSR-based spatio-temporal anomaly detector
KCSR-ST	Kernel version of KCSR-ST
KCSR-S	Spatial detection of CSR-ST
KCSR-SF	Smoothed spatial detection of KCSR-ST
KCSR-T	Temporal detection of KCSR-ST
ROC	Receiver operating characteristic
SNR	Signal to noise ratio
AUC	Area under the ROC curve

References

1. Borengasser, M.; Hungate, W.S.; Watkins, R. *Hyperspectral Remote Sensing—Principles and Applications*; CRC Press: Boca Raton, FL, USA, 2008.
2. Ling, Q.; Guo, Y.; Lin, Z.; Liu, L.; An, W. A Constrained Sparse-Representation-Based Binary Hypothesis Model for Target Detection in Hyperspectral Imagery. *IEEE J. Sel. Top. Appl. Earth Obs. Remote Sens.* **2019**, *12*, 1933–1947. [CrossRef]
3. Gao, L.; Yang, B.; Du, Q.; Zhang, B. Adjusted spectral matched filter for target detection in hyperspectral imagery. *Remote Sens.* **2015**, *7*, 6611–6634. [CrossRef]
4. Zhang, Y.; Wu, K.; Du, B.; Zhang, L.; Hu, X. Hyperspectral target detection via adaptive joint sparse representation and multi-task learning with locality information. *Remote Sens.* **2017**, *9*, 482. [CrossRef]
5. Varsano, L.; Rotman, S.R. Point target tracking in hyperspectral images. *Opt. Eng.* **2005**, *5806*, 1269–1278.
6. Varsano, L.; Yatskaer, I.; Rotman, S.R. Temporal target tracking in hyperspectral images. *Opt. Eng.* **2006**, *45*, 126201. [CrossRef]

7. Aminov, B.; Rotman, S.R. Spatial and temporal point tracking in real hyperspectral images. In Proceedings of the 2006 IEEE 24th Convention of Electrical Electronics Engineers in Israel, Eilat, Israel, 15–17 November 2006; pp. 16–20.
8. Duran, O.; Onasoglou, E.; Petrou, M. Fusion of Kalman Filter and anomaly detection for multispectral and hyperspectral target tracking. In Proceedings of the 2009 IEEE International Geoscience and Remote Sensing Symposium, Cape Town, South Africa, 12–17 July 2009; Volume 4, pp. IV-753–IV-759.
9. Duran, O. Subpixel tracking using spectral data and Kalman filter. *Energy Policy* **2014**, *17*, 429–430.
10. Duran, O.; Petrou, M. Subpixel temporal spectral imaging. *Pattern Recognit. Lett.* **2014**, *48*, 15–23. [CrossRef]
11. Li, Y.; Wang, J.; Liu, X.; Xian, N.; Xie, C. DIM moving target detection using spatio-temporal anomaly detection for hyperspectral image sequences. In Proceedings of the IGARSS 2018—2018 IEEE International Geoscience and Remote Sensing Symposium, Valencia, Spain, 22–27 July 2018; pp. 7086–7089.
12. Wang, J.; Li, Y. A rapid detection method for dim moving target in hyperspectral image sequences. *Infrared Phys. Technol.* **2019**, *102*, 102967. [CrossRef]
13. Reed, I.; Yu, X. Adaptive multiple-band CFAR detection of an optical pattern with unknown spectral distribution. *IEEE Trans. Acoust. Speech Signal Process.* **1990**, *38*, 1760–1770. [CrossRef]
14. Yu, X.; Reed, I.; Stocker, A.D. Comparative performance analysis of adaptive multispectral detectors. *IEEE Trans. Signal Process.* **1993**, *41*, 2639–2656. [CrossRef]
15. Caefer, C.E.; Silverman, J.; Orthal, O.; Antonelli, D.; Sharoni, Y.; Rotman, S.R. Improved covariance matrices for point target detection in hyperspectral data. *Opt. Eng.* **2008**, *47*, 1–13.
16. Heesung K.; Nasrabadi, N.M. Kernel RX-algorithm: A nonlinear anomaly detector for hyperspectral imagery. *IEEE Trans. Geosci. Remote Sens.* **2005**, *43*, 388–397. [CrossRef]
17. Zhou, J.; Kwan, C.; Ayhan, B.; Eismann, M.T. A Novel Cluster Kernel RX Algorithm for Anomaly and Change Detection Using Hyperspectral Images. *IEEE Trans. Geosci. Remote Sens.* **2016**, *54*, 6497–6504. [CrossRef]
18. Banerjee, A.; Burlina, P.; Diehl, C. A Support Vector Method for Anomaly Detection in Hyperspectral Imagery. *IEEE Trans. Geosci. Remote Sens.* **2006**, *44*, 2282–2291. [CrossRef]
19. Gurram, P.; Kwon, H. Support-Vector-Based Hyperspectral Anomaly Detection Using Optimized Kernel Parameters. *IEEE Geosci. Remote Sens. Lett.* **2011**, *2*, 1060–1064. [CrossRef]
20. Xu, Y.; Wu, Z.; Li, J.; Plaza, A.; Wei, Z. Anomaly Detection in Hyperspectral Images Based on Low-Rank and Sparse Representation. *IEEE Trans. Geosci. Remote Sens.* **2016**, *54*, 1990–2000. [CrossRef]
21. Li, J.; Zhang, H.; Zhang, L.; Ma, L. Hyperspectral Anomaly Detection by the Use of Background Joint Sparse Representation. *IEEE J. Sel. Top. Appl. Earth Obs. Remote Sens.* **2015**, *8*, 2523–2533. [CrossRef]
22. Li, W.; Du, Q. Collaborative Representation for Hyperspectral Anomaly Detection. *IEEE Trans. Geosci. Remote Sens.* **2015**, *53*, 1463–1474. [CrossRef]
23. Ling, Q.; Guo, Y.; Lin, Z.; An, W. A Constrained Sparse Representation Model for Hyperspectral Anomaly Detection. *IEEE Trans. Geosci. Remote Sens.* **2019**, *57*, 2358–2371. [CrossRef]
24. Sun, W.; Tian, L.; Xu, Y.; Du, B.; Du, Q. A randomized subspace learning based anomaly detector for hyperspectral imagery. *Remote Sens.* **2018**, *10*, 417. [CrossRef]
25. Soofbaf, S.R.; Sahebi, M.R.; Mojaradi, B. A sliding window-based joint sparse representation (swjsr) method for hyperspectral anomaly detection. *Remote Sens.* **2018**, *10*, 434. [CrossRef]
26. Ma, D.; Yuan, Y.; Wang, Q. Hyperspectral anomaly detection via discriminative feature learning with multiple-dictionary sparse representation. *Remote Sens.* **2018**, *10*, 745. [CrossRef]
27. Zhu, L.; Wen, G. Hyperspectral anomaly detection via background estimation and adaptive weighted sparse representation. *Remote Sens.* **2018**, *10*, 272.
28. Niu, Y.; Wang, B. Hyperspectral anomaly detection based on low-rank representation and learned dictionary. *Remote Sens.* **2016**, *8*, 289. [CrossRef]
29. Yang, Y.; Zhang, J.; Song, S.; Liu, D. Hyperspectral anomaly detection via dictionary construction-based low-rank representation and adaptive weighting. *Remote Sens.* **2019**, *11*, 192. [CrossRef]
30. Zhu, L.; Wen, G.; Qiu, S. Low-rank and sparse matrix decomposition with cluster weighting for hyperspectral anomaly detection. *Remote Sens.* **2018**, *10*, 707. [CrossRef]
31. Silverman, J.; Caefer, C.E.; DiSalvo, S.; Vickers, V.E. Temporal filtering for point target detection in staring IR imagery: II. Recursive variance filter. In *Signal and Data Processing of Small Targets 1998*; Drummond, O.E., Ed.; International Society for Optics and Photonics SPIE: Washington, DC, USA, 1998; Volume 3373, pp. 44–53.

32. Liu, S.; Du, Q.; Tong, X.; Samat, A.; Pan, H.; Ma, X. Band selection-based dimensionality reduction for change detection in multi-temporal hyperspectral images. *Remote Sens.* **2017**, *9*, 1008. [CrossRef]
33. Song, A.; Choi, J.; Han, Y.; Kim, Y. Change detection in hyperspectral images using recurrent 3D fully convolutional networks. *Remote Sens.* **2018**, *10*, 1827. [CrossRef]
34. Liu, J.; Zhang, J. Spectral Unmixing via Compressive Sensing. *IEEE Trans. Geosci. Remote Sens.* **2014**, *52*, 7099–7110.
35. Boyd, S.; Vandenberghe, L. *Convex Optimization*; Cambridge University Press: Cambridge, UK, 2004; p. 244.
36. Wang, T.; Du, B.; Zhang, L. A Kernel-Based Target-Constrained Interference-Minimized Filter for Hyperspectral Sub-Pixel Target Detection. *IEEE J. Sel. Top. Appl. Earth Obs. Remote Sens.* **2013**, *6*, 626–637. [CrossRef]
37. Müller, K.R.; Mika, S.; Rätsch, G.; Tsuda, K.; Schölkopf, B. An introduction to kernel-based learning algorithms. *IEEE Trans. Neural Netw.* **2001**, *12*, 181–201. [CrossRef] [PubMed]
38. Bioucas-Dias, J.M.; Plaza, A.; Camps-Valls, G.; Scheunders, P.; Nasrabadi, N.; Chanussot, J. Hyperspectral Remote Sensing Data Analysis and Future Challenges. *IEEE Geosci. Remote Sens. Mag.* **2013**, *1*, 6–36. [CrossRef]
39. Gómez, H.J.; Olmos, N.M.; Varela, H.; Bolfarine, H. Inference for a truncated positive normal distribution. *Appl. Math. A J. Chin. Univ.* **2018**, *33*, 163–176. [CrossRef]
40. Zhu, L.; Wen, G.; Qiu, S.; Zhang, X. Improving Hyperspectral Anomaly Detection With a Simple Weighting Strategy. *IEEE Geosci. Remote Sens. Lett.* **2019**, *16*, 95–99. [CrossRef]
41. Li, Z.; Ling, Q.; Lin, Z.; Wu, J. Segmentation-Based Weighting Strategy for Hyperspectral Anomaly Detection. *IEEE Geosci. Remote Sens. Lett.* **2020**, 1–5. [CrossRef]
42. Cohen, Y.; August, Y.; Dan, G.B.; Rotman, S.R. Evaluating Subpixel Target Detection Algorithms in Hyperspectral Imagery. *J. Electr. Comput. Eng.* **2012**, *2012*, 103286. [CrossRef]
43. Manolakis, D.; Shaw, G. Detection algorithms for hyperspectral imaging applications. *IEEE Signal Process. Mag.* **2002**, *19*, 29–43. [CrossRef]
44. Davis, J.; Goadrich, M. The relationship between Precision-Recall and ROC curves. In Proceedings of the 23rd International Conference on Machine Learning, Pittsburgh, PA, USA, 25–29 June 2006; pp. 233–240.

remote sensing

MDPI

Article

Toward Super-Resolution Image Construction Based on Joint Tensor Decomposition

Xiaoxu Ren [1], Liangfu Lu [2,*] and Jocelyn Chanussot [3]

1 College of Intelligence and Computing, Tianjin University, Tianjin 300350, China; xiaoxuren@tju.edu.cn
2 School of Mathematics, Tianjin University, Tianjin 300350, China
3 LJK, CNRS, Inria, Grenoble INP, Université Grenoble Alpes, 38000 Grenoble, France; jocelyn.chanussot@grenoble-inp.fr
* Correspondence: liangfulv@tju.edu.cn

Received: 30 June 2020; Accepted: 3 August 2020; Published: 6 August 2020

Abstract: In recent years, fusing hyperspectral images (HSIs) and multispectral images (MSIs) to acquire super-resolution images (SRIs) has been in the spotlight and gained tremendous attention. However, some current methods, such as those based on low rank matrix decomposition, also have a fair share of challenges. These algorithms carry out the matrixing process for the original image tensor, which will lose the structure information of the original image. In addition, there is no corresponding theory to prove whether the algorithm can guarantee the accurate restoration of the fused image due to the non-uniqueness of matrix decomposition. Moreover, degenerate operators are usually unknown or difficult to estimate in some practical applications. In this paper, an image fusion method based on joint tensor decomposition (JTF) is proposed, which is more effective and more applicable to the circumstance that degenerate operators are unknown or tough to gauge. Specifically, in the proposed JTF method, we consider SRI as a three-dimensional tensor and redefine the fusion problem with the decomposition issue of joint tensors. We then formulate the JTF algorithm, and the experimental results certify the superior performance of the proposed method in comparison to the current popular schemes.

Keywords: hyperspectral image; multispectral image; image fusion; joint tensor decomposition

1. Introduction

With the flourishing of both artificial intelligence (AI) and mathematical theory, image fusion has always been the focus and hotspot in neuroscience, metabonomics, remote sensing, and many other fields [1–3]. Generally, image fusion refers to synthesizing images' data obtained from the diverse image acquisition equipment. It is aimed at achieving complementary information from different information sources to further acquire clearer, more informative, and higher quality reconstructed images. In 1994, Genderen and Phol proposed a simple and intuitive definition of image fusion [4]: image fusion is merging two or more images into a new image using some algorithms. Therein, hyperspectral images (HSIs), playing a catalytic role in image fusion, have been widely leveraged in geophysical exploration [5], agricultural remote sensing [6], marine remote sensing [7], environmental monitoring [8], and other fields [9–13] because of their rich spectral information. However, the spatial resolution of HSIs is still relatively low, subjected to the imaging equipment of HSIs and the complex imaging environment, which cannot meet the application requirements of mixing, classification, detection, etc., while this further limits the prospect of HSIs. Therefore, how to improve the resolution of hyperspectral images has become the hot issue in the field of image processing.

Specially, panchromatic fusion and hyperspectral-multispectral fusion are two forms of hyperspectral super-resolution image reconstruction. Panchromatic fusion refers to fusing multispectral images (or hyperspectral images) and panchromatic images to obtain images with more spatial information. In [14], the authors classified the pansharpening techniques into component substitution (CS) [15] and multi-resolution analysis (MRA) [16]. Meanwhile, they proposed a hybrid method, combining the better spatial information of CS and the more accurate spectral information of MRA techniques, to improve the spatial resolution while preserving the original spectral information as much as possible.

Concretely, a multispectral image (MSI) generally consists of dozens of bands, and most of them are in the range of the visible region. Given a low-spatial resolution HSI, the operation of spatial resolution enhancement using MSI under the same scene is termed hyperspectral image fusion or super-resolution image (SRI) reconstruction. Generally, the MSI has a higher spatial resolution than the HSI, which is complementary to the HSI. In this paper, we mainly study the method of SRI reconstruction by combining the spectral information of HSI with the spatial information of MSI under the same scene.

Nevertheless, the existing technologies can neither avoid the distortion of image spectral characteristics, nor the complex and time-consuming frequency decomposition and reconstruction. Therefore, Yokoya proposed a simple spectral preservation fusion technique: the smoothing filter based intensity modulation (**SFIM**) [17], which is also called the generalized Laplace pyramid (**GLP**) [18–21] based on the modulation transfer function (**MTF**) used to fuse HSI and MSI by hypersharpening [22]. Then, utilizing the ratio between the high-resolution image and its low-pass filter (with a smoothing filter) image, spatial details could be modulated into co-registered low-resolution MSIs without changing their spectral characteristics and contrast. Compared with Brovey transform [23], **SFIM** is an advanced fusion technology to improve the spatial details of MSIs, and its spectral characteristics are reliably preserved.

Beyond that, Eismann introduced a maximum a posteriori estimation method [24]. It combined the stochastic mixing model of the content under the spectral scene and developed a cost function that could optimize the estimation of the hyperspectral scene related to the observed hyperspectral and auxiliary images. Moreover, this method can generally reconstruct sub-pixel information of several main components in SRI estimation. Furthermore, sparse representation has often been employed to deal with various types of image processing problems, especially in the inverse problem. In 2006, Elad denoised the image with the sparse representation method [25]. This not only achieved the state-of-the-art effect, but introduced the K-SVD dictionary training method [26]. In 2010, the authors of [27] proposed a single-frame super-resolution image reconstruction method based on sparse representation.

In contrast, traditional methods, such as principal component substitution and enhancement of least squares estimation, are primarily limited to the first principal component. For remote sensing images, the spectral characteristics of pixels are denoted as endmembers, including mixed endmembers and pure endmembers. Since each pixel has mixed endmembers, unmixing is a technique for estimating the number of pure endmembers in each pixel, the spectral characteristics, and the abundance of the endmembers [28]. The SRI reconstruction method based on unmixing usually decomposes the HSI and the MSI of the same scene. The endmember matrix of the decomposed HSI and the abundance matrix of the MSI are combined to obtain the reconstructed HSI with high spatial resolution.

In the effort of [29], the authors proposed the method of enhancing the spatial resolution of the HSI in terms of unmixing technology: coupled nonnegative matrix factorization (**CNMF**). They exploited nonnegative matrix factorization to unmix the HSI and MSI sequentially and iteratively obtained the endmember matrix and abundance matrix. Ultimately, the SRI was obtained by combining the two matrices. The nonnegative matrix decomposition usually cannot guarantee the unique solution, although **CNMF** could achieve good reconstruction results. To address this matter, Eliot Wycoff presented a nonnegative sparse enhancement model for SRI reconstruction [30], which testified that the solution was not unique in the **CNMF** method, and it had high computational complexity and a high requirement for CPU operation ability. To further boost the effect of super-resolution reconstruction, the authors of [31] came up with a method to resolve the problem of super-resolution and hyperspectral unmixing simultaneously. Unlike the measures in [29], they took advantage of the nearest alternating linear minimization (PALM) [32] to update them simultaneously, while the initialization of the endmember matrix applied SISAL [33] for endmember extraction.

Simultaneously, some researchers studied the fusion of the MSI and HSI based on the tensor [34–40], mainly considering the natural tensor structure of spectral images, so as to reduce the information lost in matricization and increase the performance. Generally, multi-channel images and other data have their own natural tensor structure. In addition, since the tensor has good expressive ability and computational characteristics, it is very meaningful to study the tensor analysis of images. Moreover, tensor decomposition can preserve the structural characteristics of the original image data. For HSIs, tensor decomposition makes full use of spatial and spectral redundancy between images and compresses and extracts relevant feature information with high quality. Based on HSIs, a nonnegative tensor canonical polyadic decomposition (CP) algorithm was raised that was applied to dispose of the blind source separation [41]. Shashua utilized CP decomposition for image compression and classification [42], while Bauckhage introduced discriminant analysis to high-order data such as color images for classification [43]. Xiao Fu proposed a coupled tensor decomposition framework [44], which could guarantee the identifiability of SRIs under mild and realistic conditions. Meanwhile, Shutao Li and Renwei Dian put forward a coupled sparse tensor factorization (**CSTF**) [45]; they regarded the SRI as a three-dimensional tensor and redefined the fusion problem as the core tensor and dictionary estimation of three modes. The high-spatial spectral correlation in the SRI was modeled by a regularizer, which could promote the generation of sparse core tensors. However, **CSTF** is an optimization model based on tensor Tucker decomposition, which is not unique. Moreover, most existing methods assume that known (or easily estimated) degenerate operators are applied to SRI to form the corresponding HSI and MSI, which is practically absent. In this paper, we deal with the super-resolution problem under the condition that the degenerate operators are seldom known and contain noise. A joint tensor decomposition model is proposed by taking advantage of the multi-dimensional tensor structure of the HSI and MSI.

The main content of this paper is to utilize the joint tensor decomposition (**JTF**) algorithm for the fusion of HSI and MSI, so as to explore the problem of SRI reconstruction. The contributions are listed as follows:

- In the proposed method, the high-spatial resolution HSI is regarded as a three-dimensional tensor, while the fusion issue is redefined as the joint estimation of the coupling factor matrix, which is also expressed as the joint tensor decomposition problem for the hyperspectral image tensor, multispectral image tensor, and noise regularization term.
- In order to observe the reconstruction effect of this method, the performance of this algorithm is compared with the five algorithms. Experiments show that the **JTF** method provides clearer spatial details than the real SRI, while the running time of **JTF** is acceptable compared with the excellent performance.
- Besides, we conduct experiments under the incorrect Gaussian kernel ($3 \times 3, 5 \times 5, 7 \times 7$), correct Gaussian kernel (9×9), and different noises, while showing the fusion effect of the six test methods with the Pavia University data captured by the ROSISsensor as well. The results reveal that the **JTF** method performs best in comparison with the other methods in terms of reconstruction accuracy regardless of whether the Gaussian kernel is correct and the level of added noise. This indicates that the **JTF** algorithm is more suitable for degradation operators that are unknown or contain noise.

The outline of this paper is organized as follows. In Section 1, we mainly introduce some basic notations and definitions for tensors. In Section 2, we give a basic overview of tensors. The proposed coupled image fusion algorithms are introduced in Section 3. In Section 4, experimental results on the algorithms are presented. Conclusions and future research directions are given in Section 6.

2. Preliminaries on Tensors

2.1. Definition and Notations

In this section, we first briefly introduce some necessary notions and preliminaries. The general tensor is denoted as $\mathcal{X}$, while the element (i,j,k) of a third-order tensor $\mathcal{X}$ is signed by x_{ijk}. The matrix is denoted as **X**, and the scalar (or the vector) is represented by x. A fiber is defined by fixing every index but one. Third-order tensors have column, row, and tube fibers, denoted by $x_{:jk}$, $x_{i:k}$, and $x_{ij:}$, respectively; see Figure 1. Fibers are always assumed to be column vectors. The mode-n matricization of a tensor $\mathcal{X} \in \mathbb{R}^{I_1 \times I_2 \times \cdots \times I_N}$ is signed by $\mathbf{X}_{(n)}$, which arranges the mode-n fibers to be the columns of the matrix and can reduce the dimension of the tensor.

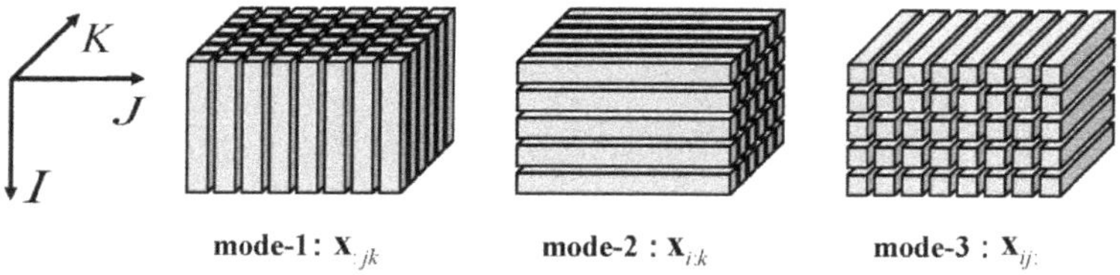

Figure 1. Fibers of a third-order tensor .

Definition 1. *The Kronecker product of matrices* $\mathbf{A} \in \mathbb{R}^{I \times J}$ *and* $\mathbf{B} \in \mathbb{R}^{K \times L}$ *is defined by Equation (1), which is denoted as* $\mathbf{A} \otimes \mathbf{B}$*, and the calculation result is a matrix of size* $IK \times JL$*, i.e.,*

$$\mathbf{A} \otimes \mathbf{B} = \begin{bmatrix} a_{11}\mathbf{B} & a_{12}\mathbf{B} & \cdots & a_{1J}\mathbf{B} \\ a_{21}\mathbf{B} & a_{22}\mathbf{B} & \cdots & a_{2J}\mathbf{B} \\ \vdots & \vdots & \ddots & \vdots \\ a_{I1}\mathbf{B} & a_{I2}\mathbf{B} & \cdots & a_{IJ}\mathbf{B} \end{bmatrix} \tag{1}$$
$$= [a_1 \otimes b_1 \quad a_1 \otimes b_2 \quad a_1 \otimes b_3 \cdots a_J \otimes b_{L-1} \quad a_J \otimes b_L].$$

Then, we have a new matrix-matrix production termed as the Khatri–Rao product.

Definition 2. *Let* $\mathbf{A} \in \mathbb{R}^{I \times K}$ *and* $\mathbf{B} \in \mathbb{R}^{J \times K}$*. Then, the Khatri–Rao product is a matrix of size* $IJ \times K$ *defined as:*

$$\mathbf{A} \odot \mathbf{B} = [a_1 \otimes b_1 \quad a_2 \otimes b_2 \quad \cdots \quad a_K \otimes b_K], \tag{2}$$

where $\otimes$ *is the Kronecker product.*

Next, we discuss some properties of the Khatri–Rao product, which will be useful in our later discussion.

$$\begin{aligned} \mathbf{A} \odot \mathbf{B} \odot \mathbf{C} &= (\mathbf{A} \odot \mathbf{B}) \odot \mathbf{C} = \mathbf{A} \odot (\mathbf{B} \odot \mathbf{C}), \\ (\mathbf{A} \odot \mathbf{B})^{\mathrm{T}}(\mathbf{A} \odot \mathbf{B}) &= \mathbf{A}^{\mathrm{T}}\mathbf{A} * \mathbf{B}^{\mathrm{T}}\mathbf{B}, \\ (\mathbf{A} \odot \mathbf{B})^{\dagger} &= ((\mathbf{A}^{\mathrm{T}}\mathbf{A}) * (\mathbf{B}^{\mathrm{T}}\mathbf{B}))^{\dagger}(\mathbf{A} \odot \mathbf{B})^{\mathrm{T}}. \end{aligned} \tag{3}$$

Definition 3. *The n-mode (matrix) product of a tensor* $\mathcal{X} \in \mathbb{R}^{I_1 \times I_2 \times \cdots \times I_N}$ *with a matrix* $\mathbf{M} \in \mathbb{R}^{J \times I_n}$ *is represented as:*

$$(\mathcal{X} \times_n \mathbf{M})_{i_1 \cdots i_{n-1} j i_{n+1} \cdots i_N} = \sum_{i_n=1}^{i_N} x_{i_1 i_2 \cdots i_N} m_{j i_n}, \tag{4}$$

which can be denoted by $\mathcal{X} \times_n \mathbf{M}$ *and is a tensor with a size of* $I_1 \times \cdots \times I_{n-1} \times J \times I_{n+1} \times \cdots \times I_N$.

Definition 4. *The Frobenius norm of a tensor* $\mathcal{X} \in \mathbb{R}^{I_1 \times I_2 \times \cdots \times I_N}$ *is represented as:*

$$\| \mathcal{X} \| = \sqrt{\sum_{i_1=1}^{I_1} \sum_{i_2=1}^{I_2} \cdots \sum_{i_N=1}^{I_N} x^2_{i_1,i_2 \cdots iN}}. \tag{5}$$

2.2. Tensor Decomposition

The general tensor decomposition models involve CP decomposition and Tucker decomposition. Specifically, the CP decomposition is a special case of the Tucker decomposition. Due to the special structure of tensors, these tensor decomposition methods are leveraged in hyperspectral image processing. For a tensor $\mathcal{X} \in \mathbb{R}^{I_1 \times I_2 \times \cdots \times I_N}$, the CP decomposition could be expressed as:

$$\mathcal{X} \approx \sum_{r=1}^{R} \lambda_r a_r^{(1)} \circ a_r^{(2)} \circ \cdots \circ a_r^{(N)} = [\![\lambda; \mathbf{A}^{(1)}, \mathbf{A}^{(2)}, \cdots, \mathbf{A}^{(N)}]\!], \tag{6}$$

where "$\circ$" is the outer product of the vectors, R is a positive integer, and $\mathbf{A}^{(n)}$ is the factor matrix. For $n = 1, 2, \cdots, N$, $\lambda \in \mathbb{R}^R$, $a_r^{(n)} \in \mathbb{R}^{I_n}$, $\mathbf{A}^{(n)} \in \mathbb{R}^{I_n \times R}$, the factor matrix is a combination of the rank one vector $a_r^{(n)}$ and denoted as:

$$\mathbf{A}^{(n)} = [a_1^{(n)}, a_2^{(n)}, \cdots, a_R^{(n)}], \tag{7}$$

Let the three-order tensor $\mathcal{X} \in \mathbb{R}^{I \times J \times K}$ be a hyperspectral image, the CP decomposition could be formulated as:

$$\mathcal{X} \approx \sum_{r=1}^{R} \lambda_r a_r \circ b_r \circ c_r = [\![\lambda; \mathbf{A}, \mathbf{B}, \mathbf{C}]\!], \tag{8}$$

where I, J, and K are the numbers of the row, column, and spectral dimensions, respectively, while $r = 1, 2, \cdots, R$, $\lambda \in \mathbb{R}^R$, $a_r \in R^I$, $b_r \in R^J$, $c_r \in R^K$.

Each column of the above factor matrices **A**, **B**, and **C** is normalized, and λ_r is the weight. If there is no requirement to standardize the factor matrix, the CP decomposition can also be reformulated as:

$$\mathcal{X} \approx (\mathbf{A}', \mathbf{B}', \mathbf{C}'). \tag{9}$$

where $\mathbf{A}', \mathbf{B}', \mathbf{C}'$ mean the general factor matrices, which are constructed by assigning the weight to the factor matrices $\mathbf{A}, \mathbf{B}, \mathbf{C}$.

The schematic diagram of CP decomposition is shown in Figure 2. If R denotes the minimum number of outer products needed to express $\mathcal{X}$, then the tensor rank is R, i.e., $rank(\mathcal{X}) = R$, and the decomposition is known as rank decomposition, which is a particular case of CP decomposition. At present, there is no specific method to directly solve the rank of any given tensor, which has been proven to be an NP-hard problem. Through the factor matrix, the CP decomposition of a third-order tensor can be written in expansion form.

$$\begin{aligned} \mathbf{X}_{(1)} &= \mathbf{A}'(\mathbf{C}' \odot \mathbf{B}')^{\mathrm{T}}, \\ \mathbf{X}_{(2)} &= \mathbf{B}'(\mathbf{C}' \odot \mathbf{A}')^{\mathrm{T}}, \\ \mathbf{X}_{(3)} &= \mathbf{C}'(\mathbf{B}' \odot \mathbf{A}')^{\mathrm{T}}. \end{aligned} \tag{10}$$

A third-order tensor can be denoted as follows by applying mode-n products:

$$\mathcal{X}' = \mathcal{X} \times_1 \mathbf{D_1} \times_2 \mathbf{D_2} \times_3 \mathbf{D_3}, \tag{11}$$

The above formula can be expressed in the form of factor matrices:

$$\mathcal{X}' \approx (\mathbf{D_1 A}', \mathbf{D_2 B}', \mathbf{D_3 C}'). \tag{12}$$

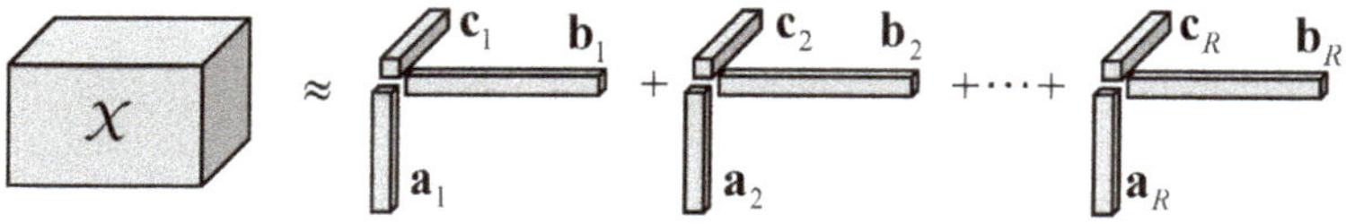

Figure 2. Canonical polyadic (CP) decomposition of third-order tensors .

Theorem 1. *Correspondingly, we consider how many rank-one tensors (components) of the decomposition of the CP model are added to minimize the error. The usual practice is to start with* $R = 1$ *until you encounter a "good" result. Of course, if you have a strong application background and prior information, you can also specify it in advance. For a given number of components, there is still no universal solution for CP decomposition. Specifically, the alternating least squares (ALS) algorithm is a more popular method in the case that the number of components is pre-given [46]. For the CP decomposition of tensors, even if R is much larger than* $max\{i, j, k\}$*, the CP decomposition model is essentially unique. The lower order decomposition of matrices and Tucker decomposition of tensors are generally not unique, which is a significant difference between them. The most famous result about the uniqueness of tensor decomposition is due to Kruskal [47]. One result of the Kruskal criteria is the following statement, which applies to general tensors, which provides the uniqueness proof of the CP decomposition model.*

Suppose $\mathcal{X} = [\![\mathbf{A}, \mathbf{B}, \mathbf{C}]\!]$ *and tensor* $\mathcal{X} \in \mathbb{R}^{I \times J \times K}$ *of rank R has a unique decomposition if:*

$$R \leq \frac{1}{2}[min(I, R) + min(J, R) + min(K, R) - 2].$$

where $\mathbf{A} \in \mathbb{R}^{I \times R}$, $\mathbf{B} \in \mathbb{R}^{J \times R}$, *and* $\mathbf{C} \in \mathbb{R}^{K \times R}$.

The uniqueness condition of the tensor CP decomposition model is relatively relaxed compared with that of the matrix decomposition model. Since the rank of matrix decomposition must be lower than the dimension of the matrix and needs nonnegative, sparse, and geometric conditions, certainly, we can also judge whether the CP decomposition model is unique according to the rank of the given tensor.

3. Problem Formulation

The purpose of HSI and MSI fusion is to estimate the unobservable SRI ($\mathcal{S} \in \mathbb{R}^{I \times J \times K}$) from the observable low-spatial resolution HSI ($\mathcal{H} \in \mathbb{R}^{i \times j \times K}$) and the high-spatial resolution MSI ($\mathcal{M} \in \mathbb{R}^{I \times J \times k}$), where $I(i)$ and $J(j)$ denote the spatial dimensions and $K(k)$ denotes the number of spectral bands. The tensor $\mathcal{H}$ is spatially downsampled with respect to (w.r.t.) $\mathcal{S}$, that is $I > i$ and $J > j$, while the tensor $\mathcal{M}$ is spectrally downsampled w.r.t. $\mathcal{S}$, that is $K > k$. We assume that the two observed data are obtained under the same atmospheric and illumination conditions and are geometrically combined with radiation correction.

3.1. Image Fusion Based on Matrix Decomposition

The fusion method based on matrix factorization assumes that each spectral vector of the target SRI can be written as a small number of linear combinations of different spectral characteristics [48], which can be represented as:

$$\mathbf{S}_{(3)} = \mathbf{WH}, \tag{13}$$

where $\mathbf{S}_{(3)} \in \mathbb{R}^{IJ \times K}$ is the three-mode unfolding matrix of the tensor $\mathcal{S}$. Matrices $\mathbf{W} \in \mathbb{R}^{IJ \times R}$ and $\mathbf{H} \in \mathbb{R}^{R \times K}$ represent the spectral basis and the corresponding coefficient matrix, respectively, where $R \ll min\{IJ, K\}$.

The spatial domain of low-spatial resolution hyperspectral data is degraded from the spatial domain of multispectral data. On the other hand, multispectral data are a form of spectral degradation of high-spatial resolution hyperspectral data. Therefore, $\mathcal{H}$ and $\mathcal{M}$ are modeled as:

$$\mathbf{H}_{(3)} = \mathbf{W}\mathbf{H}_h, \mathbf{M}_{(3)} = \mathbf{W}_m\mathbf{H}, \tag{14}$$

where $\mathbf{H}_h = \mathbf{H}\mathbf{D}_\mathbf{H}$, $\mathbf{W}_m = \mathbf{D}_\mathbf{M}\mathbf{W}$, $\mathbf{H}_{(3)} \in \mathbb{R}^{ij\times K}$, and $\mathbf{M}_{(3)} \in \mathbb{R}^{IJ\times k}$ are the three-mode unfolding matrices of the HSI (tensor $\mathcal{H}$) and MSI (tensor $\mathcal{M}$), respectively. $\mathbf{D}_\mathbf{H} \in \mathbb{R}^{IJ\times ij}$ is a matrix modeling the point spread function (PSF) and the spatial subsampling process in the hyperspectral sensor. $\mathbf{D}_\mathbf{M} \in \mathbb{R}^{K\times k}$ is a matrix modeling spectral downsampling in the multispectral sensor, whose rows contain the spectral response of the multispectral sensor. Therefore, the matricized HSI and MSI are modeled as:

$$\mathbf{H}_{(3)} = \mathbf{W}\mathbf{H}\mathbf{D}_\mathbf{H}, \mathbf{M}_{(3)} = \mathbf{D}_\mathbf{M}\mathbf{W}\mathbf{H}, \tag{15}$$

In the matrix decomposition based fusion approaches, if the spectral basis $\mathbf{D}_\mathbf{H}$ and coefficient matrix $\mathbf{D}_\mathbf{M}$ can be estimated by jointly factoring from $\mathbf{H}_{(3)}$ and $\mathbf{M}_{(3)}$, the SRI can be restored according to Equation (13), which is the main idea based on matrix decomposition.

3.2. Image Fusion Based on Tensor Decomposition

Matrix based methods usually assume that degradation operators are known or easily estimated, but in practice, it is difficult to determine. By comparing the spectral properties of hyperspectral and multispectral sensors, the degradation operator $\mathbf{D}_\mathbf{M}$ can be modeled and estimated relatively easily. However, the spatial operator becomes a bit difficult. A common model assumption from SRI to HSI conversion is a combination of the blurring by a Gaussian kernel and a downsampling process. Of course, this is a rough approximation and may be far from accurate. Even if this assumption is approximately correct, there are still many uncertainties.

In order to solve the non-uniqueness of matrix decomposition and under the condition of little knowledge of degenerate operators and noise, we propose a method based on joint tensor decomposition to fuse the HSI and MSI in this section. Tensor based models have many advantages. For example, it is a very efficient strategy to abstract image data into tensor representation and then input them in the image fusion model. For the output data, we can choose the desired format to save them conveniently. Formally, we represent the SRI as the following equation via CP decomposition:

$$\mathcal{S} = [\![\mathbf{A}, \mathbf{B}, \mathbf{C}]\!] \tag{16}$$

where $\mathcal{S} \in \mathbb{R}^{I\times J\times K}, \mathbf{A} \in \mathbb{R}^{I\times R}, \mathbf{B} \in \mathbb{R}^{J\times R}, \mathbf{C} \in \mathbb{R}^{K\times R}$, and R is the the number of components.

The HSI is the spatial downsampling version of the SRI. Assuming that the point spread function (PSF) of the hyperspectral sensor is separable from the downsampling matrix of the wide mode and the high mode, we can have:

$$\mathcal{H} = \mathcal{S} \times_1 \mathbf{D}_1 \times_2 \mathbf{D}_2 \tag{17}$$

where $\mathbf{D}_1 \in \mathbb{R}^{I\times i}, \mathbf{D}_2 \in \mathbb{R}^{J\times j}$ are the spatial degradation along the width and height modes, respectively. For subsampling, the separability hypothesis implies that the function of spatial subsampling matrix $\mathbf{D}_\mathbf{H}$ is decoupled from the two spatial patterns of $\mathcal{S}$, and thus, the degenerate operator $\mathbf{D}_\mathbf{H} = \mathbf{D}_2 \otimes \mathbf{D}_1$ in the matricized form. Under the separability assumption, the HSI ($\mathcal{H}$) can be represented as:

$$\mathcal{H} = [\![\mathbf{A}', \mathbf{B}', \mathbf{C}]\!] \tag{18}$$

where $\mathbf{A}' = \mathbf{D}_1\mathbf{A} \in \mathbb{R}^{I\times R}$, $\mathbf{B}' = \mathbf{D}_2\mathbf{B} \in \mathbb{R}^{J\times R}$, and $\mathbf{C} \in \mathbb{R}^{k\times R}$. In this paper, we assume that spectral response $\mathbf{D}_\mathbf{M}$ has noise, i.e., rough sampling in the process of conversion from the SRI to the MSI. Formally, we represent it as:

$$\mathbf{D}'_\mathbf{M} = \mathbf{D}_\mathbf{M} + \Gamma \tag{19}$$

where Γ is Gaussian random noise. Analogously, the MSI ($\mathcal{M}$) can be represented as:

$$\mathcal{M} = \mathcal{S} \times_3 \mathbf{D}_{\mathbf{M}}^{'} \tag{20}$$

where $\mathbf{D}_{\mathbf{M}}^{'} \in \mathbb{R}^{K \times k}$ is the downsampling matrix of the spectral mode. We substitute Formula (16) into (20) to obtain:

$$\mathcal{M} = [\![\mathbf{A}, \mathbf{B}, \mathbf{C}^{'}]\!] \tag{21}$$

where $\mathbf{A} \in \mathbb{R}^{i \times R}$, $\mathbf{B} \in \mathbb{R}^{j \times R}$, and $\mathbf{C}^{'} = \mathbf{D}_{\mathbf{M}}^{'}\mathbf{C} \in \mathbb{R}^{K \times R}$. In order to reconstruct the SRI, we need to estimate the factor matrices $\mathbf{A}, \mathbf{B}, \mathbf{C}$.

4. The Joint Tensor Decomposition Method

The Joint Tensor Decomposition Method

In this section, we consider that when $\mathbf{D_M}$ contains noise and the spatial degradation operator $\mathbf{D_H} = \mathbf{D_2} \otimes \mathbf{D_1}$ is completely unknown, even though this type of operation is called a combination of blurring and downsampling, in practice, hyperparameters such as the blurring kernel type, kernel size, and downsampling offset are barely known. Therefore, the joint tensor decomposition model can be generalized to the following model:

$$\min_{\mathbf{A},\mathbf{B},\mathbf{C}} \parallel \mathcal{H} - (\mathbf{A}^{'}, \mathbf{B}^{'}, \mathbf{C}) \parallel_{\mathrm{F}}^{2} + \parallel \mathcal{M} - (\mathbf{A}, \mathbf{B}, \mathbf{C}^{'}) \parallel_{\mathrm{F}}^{2} + \beta \parallel \mathbf{C}^{'} - \mathbf{D}_{\mathbf{M}}^{'}\mathbf{C} \parallel_{\mathrm{F}}^{2}, \tag{22}$$

We use the following optimization models to obtain factor matrices $\mathbf{A}, \mathbf{B}$ and $\mathbf{C}$, where β is the regularization parameter. The above optimization problem is non-convex, and the solutions of the factor matrices $\mathbf{A}, \mathbf{B}$, and $\mathbf{C}$ are not unique. However, the objective function in (22) is convex for each variable block, remaining unchanged with other variables. Therefore, we choose the proximal alternate optimization (PAO) scheme to solve the above optimization problem, which guarantees that the optimization problem converges to the critical point under certain conditions. Then, each step of the iterative update of the factor matrix is reduced to solving an easy-to-handle Sylvester equation by matricization of tensor HSI and MSI. Specifically, the $\mathbf{A}, \mathbf{B}$, and $\mathbf{C}$ iterations are updated as follows:

- Optimization with respect to $\mathbf{C}$:

  When $\mathbf{A}$, $\mathbf{B}$, $\mathbf{A}^{'}$, $\mathbf{B}^{'}$, and $\mathbf{C}^{'}$ are fixed, the optimization w.r.t. $\mathbf{C}$ in (22) can be written as:

$$\min_{\mathbf{C}} \parallel \mathcal{H} - (\mathbf{A}^{'}, \mathbf{B}^{'}, \mathbf{C}) \parallel_{\mathrm{F}}^{2} + \parallel \mathcal{M} - (\mathbf{A}, \mathbf{B}, \mathbf{C}^{'}) \parallel_{\mathrm{F}}^{2} + \beta \parallel \mathbf{C}^{'} - \mathbf{D}_{\mathbf{M}}^{'}\mathbf{C} \parallel_{\mathrm{F}}^{2},$$

The above optimization problem can be transformed into the following one by using the properties of n-mode matrix unfolding.

$$\min_{\mathbf{C}} \|\|\ \mathbf{H}_{(3)} - \mathbf{C}(\mathbf{B}^{'} \odot \mathbf{A}^{'})^{\mathrm{T}}) \parallel_{\mathrm{F}}^{2} + \beta \parallel \mathbf{C}^{'} - \mathbf{D}_{\mathbf{M}}^{'}\mathbf{C} \parallel_{\mathrm{F}}^{2} \tag{23}$$

where $\mathbf{H}_{(3)}$ is the three-mode unfolding matrix of tensors $\mathcal{H}$. The optimization problem (23) is quadratic, and its unique solution is equal to the calculation of the general Sylvester matrix equation.

$$\beta\mathbf{D}_{\mathbf{M}}^{'\mathrm{T}}\mathbf{D}_{\mathbf{M}}^{'}\mathbf{C} + \mathbf{C}\mathbf{E} - \beta\mathbf{D}_{\mathbf{M}}^{'\mathrm{T}}\mathbf{C}^{'} = \mathbf{H}_{(3)}\mathbf{E} \tag{24}$$

where $\mathbf{E} = (\mathbf{B}^{'\mathrm{T}}\mathbf{B}^{'}) * (\mathbf{A}^{'\mathrm{T}}\mathbf{A}^{'})$.

We use the Sylvester function in the MATLAB toolbox to solve the above equation.

- Optimization with respect to $\mathbf{A}^{'}$:

When $\mathbf{A}$, $\mathbf{B}$, $\mathbf{C}$, $\mathbf{B}'$, and $\mathbf{C}'$ are fixed, the optimization w.r.t. $\mathbf{A}'$ in (22) can be written as:

$$\min_{\mathbf{A}'} \parallel \mathcal{H} - (\mathbf{A}', \mathbf{B}', \mathbf{C}) \parallel_F^2, \tag{25}$$

The above optimization problem can be transformed into the following one by using the properties of n-mode matrix unfolding.

$$\min_{\mathbf{A}'} \parallel \mathbf{H}_{(1)} - \mathbf{A}'(\mathbf{C} \odot \mathbf{B}')^{\mathrm{T}}) \parallel_F^2, \tag{26}$$

where $\mathbf{H}_{(1)}$ is the one-mode unfolding matrix of tensors $\mathcal{H}$. The optimization problem (26) is convex, and the optimal solution is then given by:

$$\mathbf{A}' = \mathbf{H}_{(1)}[(\mathbf{C} \odot \mathbf{B}')^{\mathrm{T}}]^{\dagger}, \tag{27}$$

According to the property of the Khatri–Rao product pseudo-inverse, we can rewrite the solution as:

$$\mathbf{A}' = \mathbf{H}_{(1)}(\mathbf{C} \odot \mathbf{B}')(\mathbf{C}^{\mathrm{T}}\mathbf{C} * \mathbf{B}'^{\mathrm{T}}\mathbf{B}')^{\dagger}, \tag{28}$$

The advantage of solving the above equation is that we only need to compute the pseudo-inverse matrix of the $R \times R$ matrix, but not the $jK \times R$ matrix. The solving process of factor matrix $\mathbf{B}'$ is similar to that of $\mathbf{A}'$, and we can rewrite the solution as:

$$\mathbf{B}' = \mathbf{H}_{(2)}(\mathbf{C} \odot \mathbf{A}')(\mathbf{C}^{\mathrm{T}}\mathbf{C} * \mathbf{A}'^{\mathrm{T}}\mathbf{A}')^{\dagger}. \tag{29}$$

- Optimization with respect to $\mathbf{C}'$:

When $\mathbf{A}$, $\mathbf{B}$, $\mathbf{C}$, $\mathbf{A}'$, and $\mathbf{B}'$ are fixed, the optimization w.r.t. $\mathbf{C}'$ in (22) can be written as the following one by using the properties of n-mode matrix unfolding.

$$\min_{\mathbf{C}'} \parallel \mathbf{M}_{(3)} - \mathbf{C}'(\mathbf{B} \odot \mathbf{A})^{\mathrm{T}}) \parallel_F^2 + \beta \parallel \mathbf{C}' - \mathbf{D}'_{\mathbf{M}}\mathbf{C} \parallel_F^2 \tag{30}$$

where $\mathbf{M}_{(3)}$ is the three-mode unfolding matrix of tensors $\mathcal{M}$. The optimization problem (30) is quadratic, and its unique solution is equal to the calculation of the general Sylvester matrix equation.

$$\beta \mathbf{I}^{\mathrm{T}}\mathbf{I}\mathbf{C}' + \mathbf{C}'\mathbf{F} - \beta \mathbf{D}'_{\mathbf{M}}\mathbf{C} = \mathbf{H}_{(3)}\mathbf{F} \tag{31}$$

where $\mathbf{F} = (\mathbf{B}^{\mathrm{T}}\mathbf{B}) * (\mathbf{A}^{\mathrm{T}}\mathbf{A})$, and $\mathbf{I}$ is the unit matrix of 4×4.

We use the Sylvester function in the MATLAB toolbox to solve the above equation.

- Optimization with respect to $\mathbf{A}$:

When $\mathbf{B}$, $\mathbf{C}$, $\mathbf{A}'$, $\mathbf{B}'$, and $\mathbf{C}'$ are fixed, the optimization w.r.t. $\mathbf{A}$ in (22) can be written as:

$$\min_{\mathbf{A}} \parallel \mathcal{M} - (\mathbf{A}, \mathbf{B}, \mathbf{C}') \parallel_F^2, \tag{32}$$

The above optimization problem can be transformed into the following one by using the properties of n-mode matrix unfolding.

$$\min_{\mathbf{A}} \parallel \mathbf{M}_{(1)} - \mathbf{A}(\mathbf{C}' \odot \mathbf{B})^{\mathrm{T}} \parallel_F^2, \tag{33}$$

where $\mathbf{M}_{(1)}$ is the one-mode unfolding matrix of tensors $\mathcal{M}$. The optimization problem (33) is convex, and the optimal solution is then given by:

$$\mathbf{A} = \mathbf{M}_{(1)}[(\mathbf{C}' \odot \mathbf{B})^{\mathrm{T}}]^{\dagger}, \tag{34}$$

According to the property of the Khatri–Rao product pseudo-inverse, we can rewrite the solution as:

$$\mathbf{A} = \mathbf{M}_{(1)}(\mathbf{C}' \odot \mathbf{B})(\mathbf{C}'^{\mathrm{T}}\mathbf{C} * \mathbf{B}'^{\mathrm{T}}\mathbf{B})^{\dagger}, \tag{35}$$

Similarly, we only need to compute the pseudo-inverse matrix of the $R \times R$ matrix, but not the $Jk \times R$ matrix. The solving process of factor matrix **B** is similar to that of **A**, and we can rewrite the solution as:

$$\mathbf{B} = \mathbf{M}_{(2)}(\mathbf{C}' \odot \mathbf{A})(\mathbf{C}'^{\mathrm{T}}\mathbf{C}' * \mathbf{A}^{\mathrm{T}}\mathbf{A})^{\dagger}, \tag{36}$$

For each iteration update, we discuss it in detail. The specific algorithm is shown in Algorithm 1. After obtaining the estimated values of **A**, **B**, and **C**, the super-resolution tensor reconstruction is obtained from the following formula:

$$\mathcal{S} \approx (\mathbf{A}, \mathbf{B}, \mathbf{C}). \tag{37}$$

The detailed steps of the proposed method are given in Algorithm 1.

Algorithm 1: Algorithm for coupled images.

Initialization: $\beta, R, \mathbf{A}_0, \mathbf{B}_0, \mathbf{C}_0, \mathbf{A}_0', \mathbf{B}_0', \mathbf{C}_0'$
  Apply blind STEREO Algorithm [44] with random initializations to obtain $\mathbf{A}, \mathbf{B}, \mathbf{C}, \mathbf{A}', \mathbf{B}', \mathbf{C}'$
While not converged, do
  $\mathbf{C} \leftarrow \arg\min_{\mathbf{C}} \| \mathbf{H}_{(3)} - \mathbf{C}(\mathbf{B}' \odot \mathbf{A}')^{\mathrm{T}}) \|_{\mathrm{F}}^2 + \beta \| \mathbf{C}' - \mathbf{D}_{\mathbf{M}}'\mathbf{C} \|_{\mathrm{F}}^2$
  $\mathbf{A}' \leftarrow \arg\min_{\mathbf{A}'} \| \mathbf{H}_{(1)} - \mathbf{A}'(\mathbf{C} \odot \mathbf{B}')^{\mathrm{T}}) \|_{\mathrm{F}}^2,$
  $\mathbf{B}' \leftarrow \arg\min_{\mathbf{B}'} \| \mathbf{H}_{(2)} - \mathbf{B}'(\mathbf{C} \odot \mathbf{A}')^{\mathrm{T}}) \|_{\mathrm{F}}^2,$
  $\mathbf{C}' \leftarrow \arg\min_{\mathbf{C}'} \| \mathbf{M}_{(3)} - \mathbf{C}'(\mathbf{B} \odot \mathbf{A})^{\mathrm{T}}) \|_{\mathrm{F}}^2 + \beta \| \mathbf{C}' - \mathbf{D}_{\mathbf{M}}'\mathbf{C} \|_{\mathrm{F}}^2,$
  $\mathbf{A} \leftarrow \arg\min_{\mathbf{A}} \| \mathbf{M}_{(1)} - \mathbf{A}(\mathbf{C}' \odot \mathbf{B})^{\mathrm{T}}) \|_{\mathrm{F}}^2,$
  $\mathbf{B} \leftarrow \arg\min_{\mathbf{B}} \| \mathbf{M}_{(2)} - \mathbf{B}(\mathbf{C}' \odot \mathbf{A})^{\mathrm{T}}) \|_{\mathrm{F}}^2.$
end while

5. Experiments And Results

5.1. Experimental Data

To obtain an MSI from an SRI, we used the spectral specifications of the multispectral sensor, which were taken from the QuickBird sensor in our experiments [49]. The QuickBird sensor produces four-band MSI in the following spectral bands: blue (430–545 nm), green (466–620 nm), red (590–710 nm), and near-infrared (715–918 nm). Then, the spectral response matrix $\mathbf{D_M}$ is formed by comparing the SRI obtained in the experiment from 400 to 2500 nm with the multi-spectral sensor band, and the MSI image is obtained by assuming that there is random Gaussian noise in $\mathbf{D_M}$. More precisely, $\mathbf{D_M}$ is a selective averaging matrix that acts on the common wavelength of the SRI and MSI, and the experimental data came from [44]. The data selected in this paper were taken from Pavia University in Italy and were captured by the ROSIS sensor. The SRI, HSI, and MSI have sizes of $608 \times 336 \times 103$, $152 \times 84 \times 103$, and $608 \times 336 \times 4$, respectively. Specifically, the MSI is generated by QuickBird simulation, while the HSI is generated by SRI by 9×9 Gaussian blur and downsampling, and the MSI is generated for the Pavia University image according to the QuickBird specification. The degradation process from the SRI to the HSI is a combination of spatial blurring of the 9×9 Gaussian kernel and the $D = 4$ factor along two spatial directions to model the blurred image.

On the 3.6 GHz kernel and 8 GB RAM Windows server, the simulation is carried out by MATLAB. According to the algorithm **JTF**, factors **A**, **B**, and **C** are obtained through the joint tensor decomposition of the MSI and HSI, which mainly solves the least squares problem and preliminarily estimates the potential factors, where the CP decomposition is computed by TensorLab [50]. The maximum number of iterations for tensor decomposition was set to 25 in the initialization, while the number of iteration updates for factor matrix requires continuing numerical simulation. In this paper, we fixed $\beta = 1$. We mainly refer to [44]; this paper proved that the performance of super-resolution tensor reconstruction is best when beta is equal to one. The proposed model adds the noise term in the objective function based on [44]. Therefore, we selected a similar parameter.

To further demonstrate the performance of our proposed algorithm, this method is compared with the following five HSI-MSI fusion methods: **Blind Stereo** [44], **CNMF** (coupled nonnegative matrix factorization) [29], **SFIM** (smoothing filter based intensity modulation) [51], **MTF-GLP** (modulation transfer function based generalized Laplacian pyramid) [19], and **MAPSMM** (maximum a posterior estimation with a stochastic mixing model) [24].

5.2. Evaluation Criterion

In order to evaluate the quality of reconstructed high-spatial resolution HSIs, we introduce several intuitive evaluation indicators. The first index is the reconstruction signal-to-noise ratio (R-SNR) criterion defined as:

$$\mathrm{R-SNR} = 10\log_{10}\left(\frac{\sum_{k=1}^{K}\|\mathbf{S_k}\|_F^2}{\sum_{k=1}^{K}\|\mathbf{S_k'} - \mathbf{S_k}\|_F^2}\right). \tag{38}$$

where $\mathbf{S_k'}$ and $\mathbf{S_k}$ are the frontal slices of reconstructed SRI and ground truth SRI. The higher the R-SNR is, the better the reconstruction quality.

The second index is the root mean squared error (RMSE), i.e.,

$$\mathrm{RMSE} = \sqrt{\frac{\| \mathcal{S}' - \mathcal{S} \|_F^2}{\mathbf{WHS}}}. \tag{39}$$

where $\mathcal{S}'$ and $\mathcal{S}$ are the reconstructed SRI and ground truth SRI, **B** is the number of bands of hyperspectral images, and **W** and **H** are the spatial dimensions of total spectral images. Low RMSE values indicate good reconstruction performance.

The third index is the spectral angle mapper (SAM), which is defined as:

$$\mathrm{SAM} = \frac{1}{IJ}\sum_{n=1}^{IJ}\arccos\left(\frac{\mathbf{S}_{(3)}(n,:)\mathbf{S}'_{(3)}(n,:)^{\mathrm{T}}}{\|\mathbf{S}_{(3)}(n,:)\|_2\|\mathbf{S}'_{(3)}(n,:)\|_2}\right). \tag{40}$$

where $\mathbf{S}'_{(3)}(n,:)$ and $\mathbf{S}_{(3)}(n,:)$ express respectively the fibers of the reconstructed and the ground-truth SRI. SAM measures the angles between the reconstructed and the ground-truth fibers of the SRI, and a small SAM is equivalent to good performance.

The fourth index is the relative dimensionless global error in synthesis (ERGAS), which is represented as:

$$\mathrm{ERGAS} = 100c\sqrt{\frac{1}{IJK}\sum_{k=1}^{K}\frac{\|\mathbf{S_k'} - \mathbf{S_k}\|_F^2}{\mu_k^2}}. \tag{41}$$

where $c = \frac{I}{i} = \frac{J}{j}$ is the spatial downsampling factor and μ_k is the mean values of the elements in $\mathbf{S_k}$. After image reconstruction, we hope to get a smaller ERGAS.

The fifth index is the universal image quality index (UIQI), which is defined as:

$$\mathrm{UIQI} = \frac{1}{S}\sum_{i=1}^{S}\mathrm{UIQI}(\mathbf{S}'^{i}, \mathbf{S}^{i}). \tag{42}$$

where:

$$\mathrm{UIQI}(\mathbf{S}'^{i},\mathbf{S}^{i}) = \frac{1}{P}\sum_{j=1}^{P}\frac{\sigma_{\mathbf{s}_j^i\mathbf{s}_j'^i}}{\sigma_{\mathbf{s}_j^i}\sigma_{\mathbf{s}_j'^i}}\frac{2\mu_{\mathbf{s}_j^i}\mu_{\mathbf{s}_j'^i}}{\mu_{\mathbf{s}_j^i}+\mu_{\mathbf{s}_j'^i}}\frac{2\sigma_{\mathbf{s}_j^i}\sigma_{\mathbf{s}_j'^i}}{\sigma_{\mathbf{s}_j^i}+\sigma_{\mathbf{s}_j'^i}} \tag{43}$$

$\mathbf{S}_j^i$ and $\mathbf{S}_j'^i$ show the jth window of the ith band ground truth image and reconstructed image, respectively. P represents the number of window positions. $\sigma_{\mathbf{s}_j^i\mathbf{s}_j'^i}$ means the sample covariance between $\mathbf{S}_j^i$ and $\mathbf{S}_j'^i$, and $\mu_{\mathbf{s}_j^i}$ and $\sigma_{\mathbf{s}_j^i}$ denote the mean value and standard deviation of $\mathbf{S}_j^i$. The range of the index is $[-1,1]$. The larger the value of UIQI, the better the fusion effect.

The sixth index is the normalized mean squared error (NMSE), which is represented as:

$$\mathrm{NMSE} = \frac{||\mathbf{S}_3' - \mathbf{S}_3||_\mathrm{F}}{||\mathbf{S}_3||_\mathrm{F}}. \tag{44}$$

where $\mathbf{S}_3'$ and $\mathbf{S}_3$ are the three-mode unfolding matrix of the reconstructed and ground-truth SRI. The smaller the NMSE, the closer the effect of image fusion is to the ground truth image.

In addition to the above evaluation indicators, the other simplest performance indicator is the running time of the algorithm. In this paper, the efficiency of several reconstruction algorithms is compared with the computational time.

5.3. Selection Of Parameters

In this section, we select different iterations and the number of components of CP decomposition and experimented under these conditions to obtain the best parameter values, so as to evaluate the sensitivity of the **JTF** algorithm to important parameters in the model. Because the algorithm in this paper is based on the unknown $\mathbf{D_H}$, we first consider experimenting under the condition that the algorithm incorrectly assumes a 7×7 Gaussian blur kernel instead of using the correct 9×9 Gaussian kernel. Certainly, we have also made experimental comparisons under the correct Gaussian kernels.

Considering the **JTF** algorithm under different signal-to-noise ratios and assuming that the signal-to-noise ratio (SNR) of the HSI and MSI is the same, where the SNR here is set 20 db, to evaluate the effect of the number of iterations **Iter** on the image fusion in the algorithm, we run the **JTF** algorithm based on the number of iterations **Iter**. Because the **JTF** algorithm is based on the modification of the **Blind Stereo** algorithm [44], we compare the performance of the algorithms with or without noise of the degenerate operator. Figure 3 shows the evaluation metrics of SRI after the reconstruction of Pavia University with the change of iteration number **Iter**. In order to reduce the running time of the algorithm, without loss of generality, here $R = 100$, $\beta = 1$, where the black line represents **Blind Stereo** algorithm performance in matrix $\mathbf{D_M}$ with noise, the red is the reconstruction results of the **JTF** algorithm under the same condition, and the blue trend line indicates the fusion performance by the **Blind Stereo** algorithm when $\mathbf{D_M}$ does not contain noise.

As can be seen from Figure 3, when **Iter** changes from one to 20, the R-SNR of Pavia University decreases, while the values of NMSE, ERGAS, and SAM increase. Among them, R-SNR declines sharply and then rises, while the other evaluation metrics show the opposite trend. When the number of iterations is less than five, the reconstruction effect is better. Therefore, the maximum number of iterations of the **JTF** algorithm is set between one and five. In addition, the reconstruction effect of this algorithm is always superior to the Blind stereofusion algorithm with noise.

Then, we change the number of components from $R = 50$ to $R = 600$ to observe the effect of the number of tensor decomposition components on the image fusion, which depicts different evaluation metrics of the recovered HSIs for Pavia University; see Figure 4.

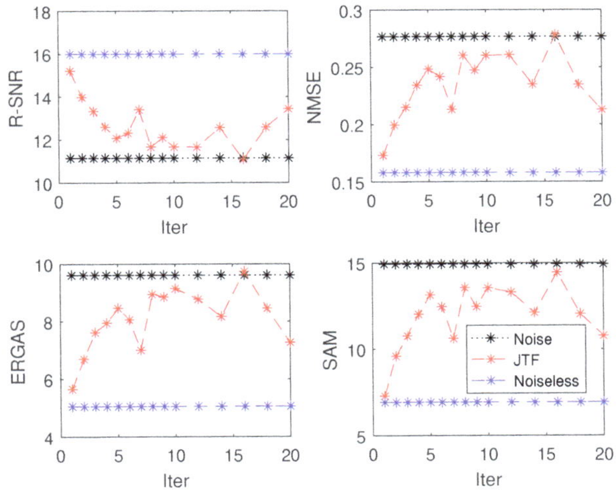

Figure 3. The results of the evaluation criterion as functions of the number of iterations **Iter** for the proposed joint tensor decomposition (JTF) method. SAM, spectral angle mapper; ERGAS, relative dimensionless global error in synthesis; R-SNR, reconstruction signal-to-noise ratio.

Figure 4 shows the effect of image fusion in three cases with a different number of components. When $\mathbf{D_H}$ is unknown and $\mathbf{D_M}$ contains noise, we compare the above three cases. From the six evaluation metrics, it can be seen that when the number of components is less than 100, the performance of the proposed algorithm is almost the same as the case that $\mathbf{D_M}$ is clean. It completely achieves the denoising effect and is always better than the **Blind Stereo** algorithm in the same situation. With the increase of the number of components, all three cases show good performance. However, when the number of components increases to more than 300, the reconstruction effect has a downward trend. According to the above Theorem [47], this is because when the number of components does not satisfy Theorem 1, the algorithm cannot guarantee the uniqueness of CP decomposition, which affects the initialization of the initial factor matrix in the algorithm, resulting in the poor performance of the algorithm.

Then, consider the selection range of the number of components under the condition of the uniqueness of tensor decomposition. As the **JTF** algorithm only decomposes MSI by CP, we set $I = 608$, $J = 336$, $K = 4$ in Theorem 1. According to the dimension of MSI, we can divide the selection of parameter R into the following four cases:

(1) When $R < K = 4$, bring R into Proposition 1, i.e.,

$$R \leq \frac{1}{2}(R + R + R - 2) \Rightarrow R \geq 1. \tag{45}$$

By synthesizing the formulas and conditions, we can get the range of R in the first case, which is $1 \leq R < 4$.

(2) When $4 = K \leq R < J = 336$, bring R into Proposition 1, i.e.,

$$R \leq \frac{1}{2}(R + R + 4 - 2) \Rightarrow R \leq R + 1. \tag{46}$$

The above derivation is obviously valid, so we only consider the conditions, and we can get the range of R in the second case, which is $4 \leq R < 336$.

(3) When $J = 336 \leq R < I = 608$, bring R into Proposition 1, i.e.,

$$R \leq \frac{1}{2}(R + 336 + 4R - 2) \Rightarrow R \geq 338. \tag{47}$$

By synthesizing the formulas and conditions, we can get the range of R in the third case, which is $336 \leq R < 338$.

(4) When $R \geq 608 = I$, bring R into Proposition 1, i.e.,

$$R \leq \frac{1}{2}(608 + 336 + 4R - 2) \Rightarrow R \geq 475. \tag{48}$$

As such, we can get the range of R in the fourth case, which is $R < 475$ and $R \geq 608$ by synthesizing the formulas and conditions. Therefore, we can conclude that there is a contradiction between the deduced range of R and the range of conditions, so this situation does not exist.

To sum up, combining the above four cases, in order to guarantee the uniqueness of CP decomposition, the range of the number of components is $1 \leq R \leq 338$. In this paper, according to the fusion effect of the algorithm while ensuring the uniqueness of CP decomposition, we fixed $R = 275$.

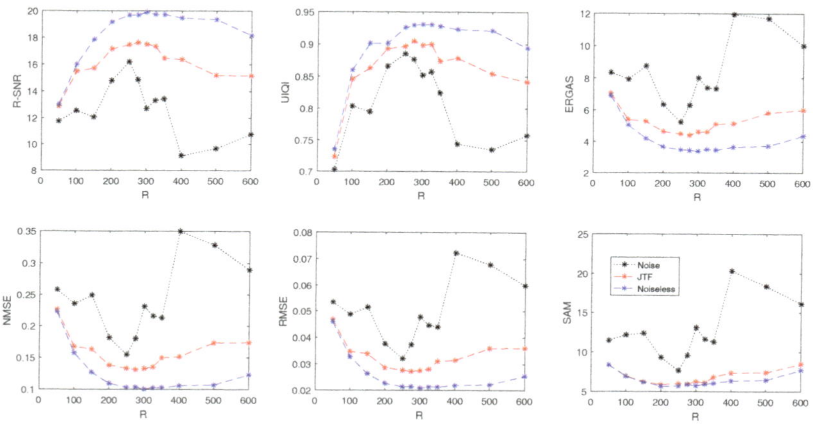

Figure 4. The results of evaluation metrics as functions of the number of components R for the proposed JTF method. UIQI, universal image quality index.

5.4. Experimental Results

To further investigate the performance of the method, we conduct experiments under the incorrect Gaussian kernel (3×3, 5×5, 7×7) and correct Gaussian kernel (9×9) and show the fusion effect of the six test methods on Pavia University. Table 1 shows the R-SNR, NMSE, RMSE, ERGAS, SAM, and UIQI of the HSI recovered from Pavia University, and we present the best of the six algorithms in bold. As can be seen from the table, in the case of incorrect estimation of the Gaussian kernel, the fusion effect of other algorithms excluding **JTF** is worse than that of the correct Gaussian kernel. The closer these five algorithms are to the correct Gaussian kernel, the better the results will be. Nevertheless, the **JTF** method performs best in the comparison of the methods in terms of reconstruction accuracy whether the Gaussian kernel is correctly estimated or not. Overall, the **JTF** and **CNMF** methods are very effective in the reconstruction of Pavia University. On the contrary, the **JTF** algorithm proposed does not degrade the image reconstruction effect due to the incorrect estimation of the Gaussian kernel. More specifically, the property of the proposed algorithm is greater under the hypothetical Gaussian kernel, which also proves that the **JTF** algorithm has more generalization significance and application prospects.

Table 1. Quantitative results of the test methods on Pavia University under the different Gaussian kernels. CNMF, coupled nonnegative matrix factorization; SFIM, smoothing filter based intensity modulation; MTF-GLP, modulation transfer function based generalized Laplacian pyramid; MAPSMM, maximum a posterior estimation with a stochastic mixing model.

Gaussian Kernel	Method	R-SNR	NMSE	RMSE	ERGAS	SAM	UIQI
3 × 3	JTF	**17.8267**	**0.1284**	**0.0266**	**4.3495**	**5.95**	**0.9031**
	Blind STEREO	11.0252	0.281	0.0581	9.9924	15.841	0.7777
	CNMF	15.8798	0.1607	0.0332	5.684	7.8655	0.8922
	SFIM	10.7811	0.289	0.0598	10.3849	11.2828	0.7374
	MTF-GLP	12.8459	0.2279	0.0471	7.608	10.4577	0.7845
	MAPSMM	11.8642	0.2551	0.0528	8.3346	10.4203	0.7449
5 × 5	JTF	**17.6357**	**0.1313**	**0.0272**	**4.4351**	**5.9266**	**0.8983**
	Blind STEREO	11.163	0.2766	0.0572	9.6563	15.6988	0.7938
	CNMF	15.5546	0.1668	0.0345	5.7987	7.9809	0.8623
	SFIM	12.6416	0.2333	0.0483	7.9097	10.2705	0.7725
	MTF-GLP	13.6742	0.2072	0.0428	6.9561	9.7086	0.8038
	MAPSMM	12.8407	0.228	0.0472	7.4954	9.5283	0.7733
7 × 7	JTF	**17.7156**	**0.1301**	**0.0269**	**4.4008**	**5.8517**	**0.9001**
	Blind STEREO	13.4084	0.2136	0.0442	7.421	11.9428	0.8584
	CNMF	16.0028	0.1584	0.0328	5.5489	7.2096	0.8747
	SFIM	13.1113	0.221	0.0457	7.493	9.9524	0.7804
	MTF-GLP	13.9393	0.2009	0.0416	6.7583	9.461	0.8057
	MAPSMM	13.1616	0.2197	0.0454	7.2442	9.3344	0.7777
9 × 9	JTF	**17.603**	**0.1318**	**0.0273**	**4.4091**	**5.7811**	**0.8991**
	Blind STEREO	13.7402	0.2056	0.0425	7.0851	11.3721	0.8621
	CNMF	16.3576	0.1521	0.0315	5.3307	7.1092	0.8792
	SFIM	13.3654	0.2147	0.0444	7.2704	9.7603	0.7875
	MTF-GLP	14.1333	0.1965	0.0406	6.6077	9.3225	0.8109
	MAPSMM	13.2983	0.2163	0.0447	7.1304	9.1704	0.7813

Figure 5 reveals the fusion experimental results for Pavia University under the incorrect Gaussian kernel (7 × 7), which contains the 50th and 100th bands' fused images and the corresponding error images reconstructed by the six algorithms, were Line 1 and Line 2 in Figure 5 denote the fused HSIs of the 50th band and the corresponding error HSIs of each method, respectively. Moreover, Figure 5g shows the reference HSIs, while the third and forth rows show the reconstructed images for the 100th band and corresponding error images, respectively. Except for the last column, each column in Figure 5 shows the experimental results corresponding to each method. The error image reflects the difference between the fusion result and the ground truth. As depicted in Figure 5, this paper uses the red box to show the more obvious areas in order to compare the difference of error images of different algorithms clearly. By visualized comparison of the fused HSIs with the reference HSIs, the fusion result of the **Blind STEREO** method shows slight spectral distortion on the top of the building, while the **MAPSMM** method generates fuzzy spatial details in some areas, and the spatial information of the fused image is well enhanced by the **CNMF** method. A closer inspection reveals that the spectral and spatial differences of fused HSIs obtained by the six methods are not obvious. Therefore, in order to further compare the performance of each fusion method, the second and fourth lines of Figure 5 show the error images of the six methods under two spectral bands. The error image is the difference (absolute value) between the fused HSI and the reference HSI pixel value. We magnify the data element value in the error image by 10 times, so that we can inspect it more carefully. It can be seen that the **Blind STEREO**, **SFIM**, **MTF-GLP**, and **MAPSMM** methods have large differences, while the **CNMF** method generates relatively smaller differences, and the **JTF** method has the smallest differences in most regions, indicating that this method has good fusion ability and provides clearer spatial details than the other five algorithms.

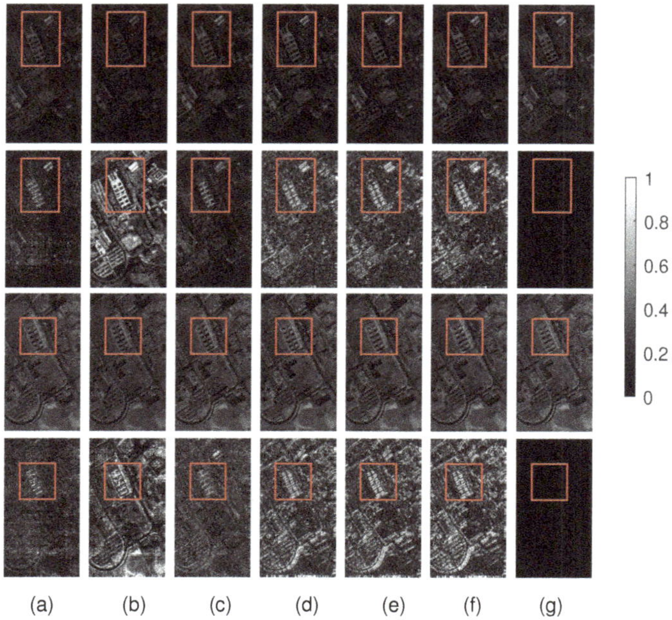

Figure 5. Reconstructed images and corresponding error images of Pavia University for the 50th and 100th bands with unknown $\mathbf{D_H}$ and noisy $\mathbf{D_M}$: (**a**) JTF; (**b**) Blind STEREO; (**c**) CNMF; (**d**) SFIM; (**e**) MTF-GLP; (**f**) MAPSMM; (**g**) ground truth.

Similar to the previous experiments, Figure 6 shows the fusion experimental results for Pavia University under the correct Gaussian kernel (9 × 9), which contain the 50th and 100th bands' fused images and the corresponding error images reconstructed by the six algorithms. Figure 6 shows the fused HSIs of the 50th band and the corresponding error HSIs of each method, which are displayed in Lines 1–2. Moreover, Figure 6g shows the reference HSIs, while the third and forth rows show the reconstructed images for the 100th band and corresponding error images, respectively. Similarly, in order to compare the difference of the error images of different algorithms clearly, the data element values in the error image are magnified 10 times, and the red box is applied to display the region with obvious errors. The spectral distortion caused by the **Blind STEREO** method is very obvious and is affected by the Gaussian kernel changes, as shown in Figure 6b. Compared with the **Blind STEREO** method, other methods can effectively improve the spatial performance while maintaining the spectral information, and the difference between the fused images is not significant. Therefore, in order to further verify the fusion performance of the proposed method, the second and fourth lines of Figure 6 show the error images corresponding to different methods, respectively. It can be seen that the error image obtained by the **JTF** method is the lowest in most regions, and the fusion effect is not affected by the Gaussian kernel, which indicates that the **JTF** method has a superior image reconstruction effect and is more robust. Overall, the **JTF** method has better reconstruction performance and clearer fusion effects than the other five algorithms.

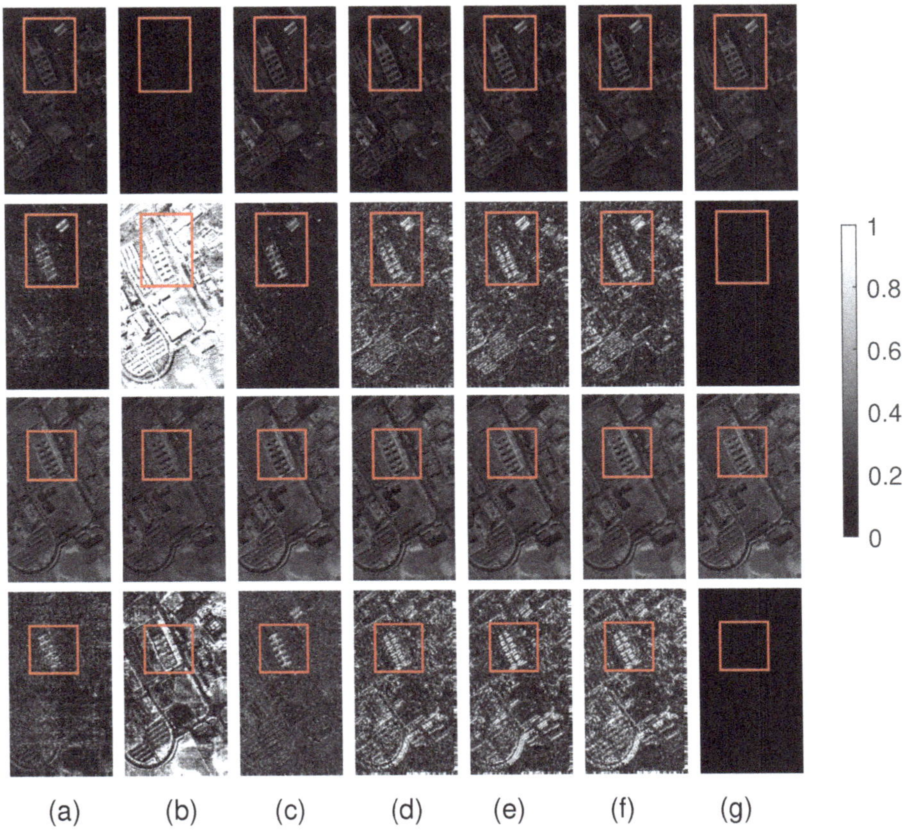

Figure 6. Reconstructed images and corresponding error images of Pavia University for the 20th and 60th bands with unknown $\mathbf{D_H}$ and noisy $\mathbf{D_M}$: (**a**) JTF; (**b**) Blind STEREO; (**c**) CNMF; (**d**) SFIM; (**e**) MTF-GLP; (**f**) MAPSMM; (**g**) ground truth.

5.5. Experimental Results of the Noisy Case

In practice, there exits additive noise in the hyperspectral and multispectral imaging processes. Therefore, to test the robustness of the proposed **JTF** method to the noise, we firstly simulate the tensor images $\mathcal{M}$ and $\mathcal{H}$ in the same way as the previous experiments for Pavia University and then add Gaussian noise to the HSI and MSI. Because the noise level in the HSI is often higher than that of the MSI, we fix the SNR added to the HSI to be 20db and compare the evaluation indicators with the traditional five classical models with the change of noise added to MSI.

Figure 7 presents the quality metric values of the noisy cases on Pavia University. It can be seen that from the reconstruction performance of the six fusion algorithms emerges a trend of enhancement with the increase of MSI image noise. Although the fusion effect of **CNMF** is closer to that of the **JTF** algorithm when the noise is high, the **JTF** method is still better than other test methods in the case of noise as a whole.

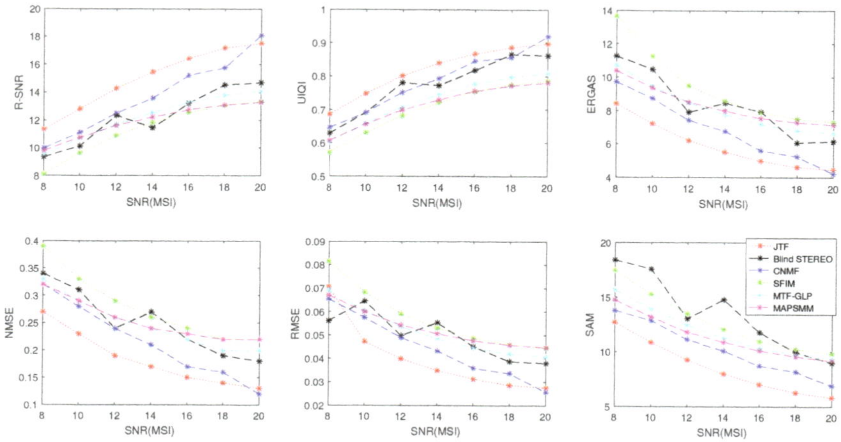

Figure 7. The results of evaluation metrics under different noises. MSI, multispectral image.

5.6. Analysis of Computational Costs

In this section, experiments are carried out on six classical methods to demonstrate the computational efficiency of the proposed method, which are accomplished with MATLAB R2016b on a PC with Intel Core i7-7500 CPU and 8 GB RAM. The mean time (in terms of seconds) of all comparison methods is shown as Table 2.

As can be seen from Table 2, the method based on the filter fusion (**SFIM**) does not need to calculate the optimal factor matrix of each mode, and its running time is shorter than the method based on tensor samples. For tensor based methods, since the iterative strategy is used to obtain the optimal solution of each unknown factor, the time of the two methods (**JTF**, **Blind STEREO**) is almost the same. **MTF-GLP** runs between the first two classes of methods, while **CNMF** and **MAPSMM** have long running times. Compared with the excellent performance, the running time of **JTF** is acceptable. Analyses were conducted based on various noises under different Gaussian kernels to further observe the performance of different algorithms. The results indicate that the running time of most algorithms is shorter under the premise of correct estimation of Gaussian kernels. However, there is little difference with the run time of the **JTF** algorithm under unknown Gaussian kernels, and the running time has little relation with the magnitude of additive noise, which indirectly proves that our algorithm is more robust.

Table 2. Time (s) of the test methods on Pavia University under the different conditions.

Method	Gaussian Kernel (7 × 7) SNR (10 db)	Gaussian Kernel (9 × 9) SNR (10 db)	Gaussian Kernel (7 × 7) SNR (20 db)	Gaussian Kernel (9 × 9) SNR (20 db)
JTF	17.266912	15.883985	15.303149	15.035694
Blind STEREO	15.179482	13.761605	13.057704	12.79495
CNMF	91.838425	89.438305	82.124058	82.466387
SFIM	1.421629	0.821533	0.836197	0.829191
MTF-GLP	24.859017	25.753572	24.809271	31.731484
MAPSMM	301.682105	283.545812	266.800409	299.233003

6. Conclusions

In this paper, a joint tensor decomposition method was proposed to fuse hyperspectral and multispectral images to address the hyperspectral super-resolution issue. The JTF algorithm regards the fusion problem as the joint tensor decomposition, which not only ensures the non-uniqueness of decomposition, but is applicable to the circumstance that degenerate operators are unknown or tough to gauge. In order to observe the reconstruction effect of this method, we compare the performance of the proposed algorithm with that of the five algorithms. Experiments show that the proposed algorithm has great performance advantages and certain simulation prospects. For our future work, we would concentrate on the novel scenario that adds the non-negative constraints for the joint tensor decomposition of super-resolution images.

Author Contributions: Methodology, X.R. and L.L.; resources, J.C.; validation, X.R.; writing, original draft, X.R. and L.L.; writing, review and editing, L.L. All authors read and agreed to the published version of the manuscript.

Funding: The authors would like to thank the Editors and anonymous Reviewers. This work was partially supported by the National Natural Science Foundation of China under No.51877144.

Acknowledgments: The authors would like to thank the Editors and Reviewers of the Remote Sensing journal for their constructive comments and suggestions.

Conflicts of Interest: The authors declare no conflict of interest.

References

1. Zhu, Z.Y.; Dong, S.J.; Yu, C.L.; He, J. A Text Hybrid Clustering Algorithm Based on HowNet Semantics. *Key Eng. Mater.* **2011**, *474*, 2071–2078. [CrossRef]
2. Sui, J.; Adali, T.; Yu, Q.; Chen, J.; Calhoun, V.D. A review of multivariate methods for multimodal fusion of brain imaging data. *J. Neurosci. Methods* **2012**, *204*, 68–81. [CrossRef]
3. Acar, E.; Lawaetz, A.J.; Rasmussen, M.A.; Bro, R. Structure-revealing data fusion model with applications in metabolomics. *IEEE Eng. Med. Biol. Soc.* **2013**, *14*, 6023–6026.
4. Pohl, C.; Van Genderen, J.L. Review article Multisensor image fusion in remote sensing: Concepts, methods and applications. *Int. J. Remote Sens.* **1998**, *19*, 823–854. [CrossRef]
5. Chabrillat, S.; Pinet, P.C.; Ceuleneer, G.; Johnson, P.E.; Mustard, J.F. Ronda peridotite massif: Methodology for its geological mapping and lithological discrimination from airborne hyperspectral data. *Int. J. Remote Sens.* **2000**, *21*, 2363–2388. [CrossRef]
6. Haboudane, D.; Miller, J.R.; Pattey, E.; Zarco-Tejada, P.J.; Strachan, I.B. Hyperspectral vegetation indices and novel algorithms for predicting green LAI of crop canopies: Modeling and validation in the context of precision agriculture. *Remote Sens. Environ.* **2004**, *90*, 337–352. [CrossRef]
7. Adam, E.; Mutanga, O.; Rugege, D. Multispectral and hyperspectral remote sensing for identification and mapping of wetland vegetation: A review. *Wetl. Ecol. Manag.* **2010**, *18*, 281–296. [CrossRef]
8. Ellis, R.J.; Scott, P. Evaluation of hyperspectral remote sensing as a means of environmental monitoring in the St. Austell China clay (kaolin) region, Cornwall, UK. *Remote Sens. Environ* **2004**, *93*, 118–130. [CrossRef]
9. Xu, Y.; Wu, Z.; Li, J.; Plaza, A.; Wei, Z. Anomaly detection in hyperspectral images based on low-rank and sparse representation. *IEEE Trans. Geosci. Remote Sens.* **2016**, *54*, 1990–2000. [CrossRef]
10. Zhang, M.; Li, W.; Du, Q. Diverse region based CNN for hyperspectral image classification. *IEEE Trans. Image Process* **2018**, *27*, 2623–2634. [CrossRef]
11. Xue, B.; Yu, C.; Wang, Y.; Song, M.; Li, S.; Wang, L.; Chen, H.M.; Chang, C.I. A subpixel target detection approach to hyperspectral image classification. *IEEE Trans. Geosci. Remote Sens.* **2017**, *55*, 5093–5114. [CrossRef]
12. Kang, X.; Duan, P.; Xiang, X.; Li, S.; Benediktsson, J.A. Detection and correction of mislabeled training samples for hyperspectral image classification. *IEEE Trans. Geosci. Remote Sens.* **2018**, *56*, 5673–5686. [CrossRef]
13. Lv, Z.; Shi, W.; Zhou, X.; Benediktsson, J.A. Semi-automatic system for land cover change detection using bi-temporal remote sensing images. *Remote Sens.* **2017**, *9*, 1112. [CrossRef]
14. Licciardi, G.; Vivone, G.; Dalla Mura, M.; Restaino, R.; Chanussot, J. Multi-resolution analysis techniques and nonlinear PCA for hybrid pansharpening applications. *Multidimens. Syst. Signal Process.* **2015**, *27*, 807–830. [CrossRef]
15. Shah, V.P.; Younan, N.H.; King, R.L. An efficient pan-sharpening method via a combined adaptive PCA approach and contourlets. *IEEE Trans. Geosci. Remote Sens.* **2008**, *46*, 1323–1335. [CrossRef]
16. Nason, G.P.; Silverman, B.W. The Stationary Wavelet Transform and Some Statistical Applications. In *Wavelets and Statistics*; Antoniadis, A., Oppenheim, G., Eds.; Lecture Notes in Statistics; Springer: New York, NY, USA, 1995; Volume 103.

17. Yokoya, N.; Grohnfeldt, C.; Chanussot, J. Hyperspectral and multispectral data fusion: A comparative review of the recent literature. *IEEE Geosci. Remote Sens. Mag.* **2017**, *5*, 29–56. [CrossRef]
18. Aiazzi, B.; Alparone, L.; Baronti, S.; Garzelli, A. Context-driven fusion of high spatial and spectral resolution images based on oversampled multiresolution analysis. *IEEE Trans. Geosci. Remote Sens.* **2002**, *40*, 2300–2312. [CrossRef]
19. Aiazzi, B.; Alparone, L.; Baronti, S.; Garzelli, A.; Selva, M. MTF-tailored multiscale fusion of high-resolution MS and pan imagery. *Photogramm. Eng. Remote Sens.* **2006**, *72*, 591–596. [CrossRef]
20. Vivone, G.; Restaino, R.; Dalla Mura, M.; Licciardi, G.; Chanussot, J. Contrast and error based fusion schemes for multispectral image pansharpening. *IEEE Geosci. Remote Sens. Lett.* **2013**, *11*, 930–934. [CrossRef]
21. Vivone, G.; Alparone, L.; Chanussot, J.; Dalla Mura, M.; Garzelli, A.; Licciardi, G.A.; Restaino, R.; Wald, L. A critical comparison among pansharpening algorithms. *IEEE Trans. Geosci. Remote Sens.* **2015**, *53*, 2565–2586. [CrossRef]
22. Selva, M.; Aiazzi, B.; Butera, F.; Chiarantini, L.; Baronti, S. Hyper-sharpening: A first approach on SIM-GA data. *IEEE J. Sel. Top. Appl. Earth Obs. Remote Sens.* **2015**, *8*, 3008–3024. [CrossRef]
23. Wang, Z.; Ziou, D.; Armenakis, C.; Li, D.; Li, Q. A comparative analysis of image fusion methods. *IEEE Trans. Geosci. Remote Sens.* **2005**, *43*, 1391–1402. [CrossRef]
24. Eismann, M.T. Resolution enhancement of hyperspectral imagery using maximum a posteriori estimation with a stochastic mixing model. *Diss. Abstr. Int.* **2004**, *65*, 1385.
25. Elad, M.; Aharon, M. Image denoising via learned dictionaries and sparse representation. In Proceedings of the IEEE Computer Society Conference on Computer Vision and Pattern Recognition, New York, NY, USA, 17–22 June 2006; pp. 895–900.
26. Aharon, M.; Elad, M.; Bruckstein, A. R_K-SVD: An algorithm for designing overcomplete dictionaries for sparse representation. *IEEE Trans. Signal Process.* **2006**, *54*, 4311–4322. [CrossRef]
27. Yang, J.; Wright, J.; Huang, T.S.; Ma, Y. Image super-resolution via sparse representation. *IEEE Trans. Image Process.* **2010**, *19*, 2861–2873. [CrossRef]
28. Bioucas-Dias, J.M.; Plaza, A.; Dobigeon, N.; Parente, M.; Du, Q.; Gader, P.; Chanussot, J. Hyperspectral unmixing overview: Geometrical, statistical, and sparse regression based approaches. *IEEE J. Sel. Top. Appl. Earth Obs. Remote Sens.* **2012**, *5*, 354–379. [CrossRef]
29. Yokoya, N.; Yairi, T.; Iwasaki, A. Coupled nonnegative matrix factorization unmixing for hyperspectral and multispectral data fusion. *IEEE Trans. Geosci. Remote Sens.* **2012**, *50*, 528–537. [CrossRef]
30. Wycoff, E.; Chan, T.H.; Jia, K.; Ma, W.K.; Ma, Y. A non-negative sparse promoting algorithm for high resolution hyperspectral imaging. In Proceedings of the IEEE International Conference on Acoustics, Speech, and Signal Processing, Vancouver, BC, Canada, 26–31 May 2013; pp. 1409–1413.
31. Lanaras, C.; Baltsavias, E.; Schindler, K. Hyperspectral super-resolution by coupled spectral unmixing. In Proceedings of the IEEE International Conference on Computer Vision, Santiago, Chile, 7–13 December 2015; pp. 3586–3594.
32. Bolte, J.; Sabach, S.; Teboulle, M. Proximal alternating linearized minimization for nonconvex and nonsmooth problems. *Math. Program.* **2014**, *146*, 459–494. [CrossRef]
33. Bioucas-Dias, J.M. A variable splitting augmented lagrangian approach to linear spectral unmixing. In Proceedings of the First Workshop on Hyperspectal Image and Signal Processing, Grenoble, France, 26–28 August 2009; pp. 1–4.
34. Li, W.; Du, Q. MLaplacian Regularized Collaborative Graph for Discriminant Analysis of Hyperspectral Imagery. *IEEE Trans. Geosci. Remote Sens.* **2016**, *54*, 7066–7076. [CrossRef]
35. Guo, X.; Huang, X.; Zhang, L.; Zhang, L.; Plaza, A.; Benediktsson, J.A. Support Tensor Machines for Classification of Hyperspectral Remote Sensing Imagery. *IEEE Trans. Geosci. Remote Sens.* **2016**, *54*, 3248–3264. [CrossRef]
36. An J.; Zhang, X.; Zhou, H.; Jiao, L. Tensor-Based Low-Rank Graph with Multimanifold Regularization for Dimensionality Reduction of Hyperspectral Images. *IEEE Trans. Geosci. Remote Sens.* **2018**, *56*, 4731–4746. [CrossRef]
37. Renard, N.; Bourennane, S.; Blanc-Talon, J. Denoising and Dimensionality Reduction Using Multilinear Tools for Hyperspectral Images. *IEEE Geosci. Remote Sens. Lett.* **2008**, *5*, 138–142. [CrossRef]

38. Zhong, Z.; Fan, B.; Duan, J.; Wang, L.; Ding, K.; Xiang, S.; Pan, C. Discriminant Tensor Spectral Spatial Feature Extraction for Hyperspectral Image Classification. *IEEE Geosci. Remote Sens. Lett.* **2017**, *12*, 1028–1032. [CrossRef]
39. Makantasis, K.; Doulamis, A.D.; Doulamis, N.D.; Nikitakis, A. Tensor-Based Classification Models for Hyperspectral Data Analysis. *IEEE Trans. Geosci. Remote Sens.* **2018**, *56*, 6884–6898. [CrossRef]
40. Xu, Y.; Wu, Z.; Chanussot, J.; Comon, P.; Wei, Z. Nonlocal Coupled Tensor CP Decomposition for Hyperspectral and Multispectral Image Fusion. *IEEE Trans. Geosci. Remote Sens.* **2020**, *58*, 348–362, doi:10.1109/TGRS.2019.2936486. [CrossRef]
41. Cohen, J.; Farias, R.C.; Comon, P. Fast decomposition of large nonnegative tensors. *IEEE Signal Process. Lett.* **2015**, *22*, 862–866. [CrossRef]
42. Shashua, A.; Levin, A. Linear image coding for regression and classification using the tensor-rank principle. In Proceedings of the 2001 IEEE Computer Society Conference on Computer Vision and Pattern Recognition, Kauai, HI, USA, 8–14 December 2001; pp. 42–49.
43. Bauckhage, C. Robust tensor classifiers for color object recognition. *Int. Conf. Image Anal. Recognit.* **2007**, *4633*, 352–363.
44. Kanatsoulis, C.I.; Fu, X.; Sidiropoulos, N.D.; Ma, W.K. Hyperspectral super-resolution: A coupled tensor factorization approach. *IEEE Trans. Signal Process.* **2018**, *66*, 6503–6517. [CrossRef]
45. Li, S.; Dian, R.; Fang, L.; Bioucas-Dias, J.M. Fusing hyperspectral and multispectral images via coupled sparse tensor factorization. *IEEE Trans. Image Process.* **2018**, *27*, 4118–4130. [CrossRef]
46. Kolda, T.G.; Bader, B.W. Tensor Decompositions and Applications. *Siam Rev.* **2009**, *51*, 455–500. [CrossRef]
47. Chiantini, L.; Ottaviani, G. On generic identifiability of 3-tensors of small rank. *SIAM J. Matrix Anal. Appl.* **2012**, *33*, 1018–1037. [CrossRef]
48. Iordache, M.D.; Bioucas-Dias, J.M.; Plaza, A. Sparse unmixing of hyperspectral data. *IEEE Trans. Geosci. Remote Sens.* **2011**, *49*, 2014–2039. [CrossRef]
49. Quickbird Satellite Sensor. Available online: http://www.satimagingcorp.com/satellite-sensors/quickbird (accessed on 4 April 2018).
50. Vervliet, N.; Debals, O.; Sorber, L.; Barel, M.V.; Lathauwer, L.D. Tensorlab v3.0, March 2016. Available online: http://www.tensorlab.net (accessed on 4 March 2018).
51. Liu, J.G. Smoothing filter based intensity modulation: A spectral preserve image fusion technique for improving spatial details. *Int. J. Remote Sens.* **2000**, *21*, 3461–3472. [CrossRef]

remote sensing MDPI

Article

Hyperspectral Image Super-Resolution with Self-Supervised Spectral-Spatial Residual Network

Wenjing Chen [1,2], Xiangtao Zheng [1,*] and Xiaoqiang Lu [1]

1 Key Laboratory of Spectral Imaging Technology CAS, Xi'an Institute of Optics and Precision Mechanics, Chinese Academy of Sciences, Xi'an 710119, China; chenwenjing2017@opt.cn (W.C.); luxiaoqiang@opt.ac.cn (X.L.)
2 University of Chinese Academy of Sciences, Beijing 100049, China
* Correspondence: zhengxiangtao@opt.cn

Abstract: Recently, many convolutional networks have been built to fuse a low spatial resolution (LR) hyperspectral image (HSI) and a high spatial resolution (HR) multispectral image (MSI) to obtain HR HSIs. However, most deep learning-based methods are supervised methods, which require sufficient HR HSIs for supervised training. Collecting plenty of HR HSIs is laborious and time-consuming. In this paper, a self-supervised spectral-spatial residual network (SSRN) is proposed to alleviate dependence on a mass of HR HSIs. In SSRN, the fusion of HR MSIs and LR HSIs is considered a pixel-wise spectral mapping problem. Firstly, this paper assumes that the spectral mapping between HR MSIs and HR HSIs can be approximated by the spectral mapping between LR MSIs (derived from HR MSIs) and LR HSIs. Secondly, the spectral mapping between LR MSIs and LR HSIs is explored by SSRN. Finally, a self-supervised fine-tuning strategy is proposed to transfer the learned spectral mapping to generate HR HSIs. SSRN does not require HR HSIs as the supervised information in training. Simulated and real hyperspectral databases are utilized to verify the performance of SSRN.

Keywords: hyperspectral image super-resolution; data fusion; spectral-spatial residual network; multispectral image; self-supervised training

Citation: Chen, W.; Zheng, X.; Lu, X. Hyperspectral Image Super-Resolution with Self-Supervised Spectral-Spatial Residual Network. *Remote Sens.* **2021**, *13*, 1260. https://doi.org/10.3390/rs13071260

Academic Editor: Chein-I Chang

Received: 16 February 2021
Accepted: 23 March 2021
Published: 26 March 2021

1. Introduction

Hyperspectral imaging sensors collect hyperspectral images (HSIs) across many narrow spectral wavelengths, which contain rich physical properties of observed scenes [1]. HSIs with high spectral resolution are beneficial for various tasks, e.g., classification [2] and detection [3]. However, as the amount of incident energy is limited, observed HSIs usually have low spatial resolution (LR) [4]. Contrary to HSIs, observed multispectral images (MSIs) have high spatial resolution (HR) but low spectral resolution [5,6]. Exploring both MSIs and HSIs captured in the same scene is a feasible and effective way for improving the spatial resolution of HSIs [7].

Over decades, many methods [8,9] have been proposed to reconstruct the desired HR HSI by fusing HR MSIs and LR HSIs, including sparse representation-based methods [10,11], Bayesian-based methods [12,13], spectral unmixing-based methods [1,14], and tensor factorization-based methods [15,16]. Sparse representation-based, Bayesian-based, and spectral unmixing-based methods usually first learn spectral bases (or endmembers) from the LR HSI [9,10]. Then, the learned spectral bases are transformed to extract the sparse codes (or abundances) from the HR MSI. Finally, the desired HR HSI is reconstructed using the learned spectral bases and sparse codes. These methods usually treat the HR MSI and LR HSI as 2-D matrices, which result in the spatial structure information of HR MSIs and LR HSIs not being effectively exploited [15]. Tensor factorization-based methods [15,16] consider HR MSIs and LR HSIs as 3-D tensors to fully explore the spatial structure information of HR MSIs and LR HSIs. In general, previous methods mainly focus on exploiting various handcrafted priors (e.g., sparsity and low-rankness) to improve the quality of the reconstructed HR HSI [9]. However, sparsity and low-rankness priors may not hold in real complicated scenarios [17], which can result in unsatisfactory super-resolved results [18].

Recent works [19–22] usually build various deep learning (DL) architectures to learn deep priors for fusing HR MSIs and LR HSIs. Due to powerful feature learning capabilities, DL-based methods have shown superior performance. In most DL-based methods, deep networks are usually utilized to learn the deep priors between LR HSIs and HR HSIs [22–24]. For example, Li et al. [25] employed a Laplacian pyramid network instead of the bicubic interpolation to upsample HSIs for the guided filtering-based MSI and HSI fusion. Dian et al. [21] proposed to utilize a residual network to learn deep priors of HR HSIs. However, these methods are supervised methods, which require plentiful HR HSIs as the supervised information to optimize weight parameters of deep networks. It is an intractable problem to collect a mass of HR HSIs for supervised training [26].

To mitigate the dependence on HR HSIs as the supervised information, several works [26,27] have designed unsupervised deep networks. Yuan et al. [26] transferred the deep priors between LR and HR nature images to HSIs. Sidorov et al. [27] utilized a fully convolutional encoder–decoder network to explore deep hyperspectral priors. However, these methods [26,27] cannot exploit HR MSIs for reconstructing HR HSIs. To leverage both LR HSIs and the corresponding HR MSI, several works [17,28] attempted to build two-branch deep networks. Qu et al. [28] designed two sparse Dirichlet autoencoder networks: one for extracting spectral bases from LR HSI and the other for extracting spatial representations from HR MSIs. Ma et al. [17] proposed a generative adversarial network with two discriminators to reconstruct HR HSIs. One discriminator is utilized to preserve the spectral information of HR HSIs consistent with that of LR HSIs, and the other discriminator is designed to preserve the spatial structures of HR HSIs consistent with that of HR MSIs. However, these methods [17,28] ignore the potential spectral mapping relationship between the observed MSI and HSI.

In this paper, the fusion problem of HR MSIs and LR HSIs is considered a problem of learning the pixel-wise spectral mapping from MSIs to HSIs. The pixel-wise spectral mapping can be utilized to reconstruct hyperspectral pixels directly from multispectral pixels. Since the LR HSI and the reconstructed HR HSI contain the same observed scene, the spectral mapping between the HR MSI and HR HSI is assumed to be approximately equal to that between the corresponding LR MSI and LR HSI. In this paper, as shown in Figure 1, a self-supervised spectral-spatial residual network (SSRN) is proposed to learn the pixel-wise spectral mapping between LR MSIs and LR HSIs. Then, the learned spectral mapping is transferred to reconstruct the desired HR HSI from HR MSIs. In the proposed SSRN, the LR MSI utilized for training is the spatial degradation of HR MSIs. Additionally, SSRN takes the observed LR HSI instead of the HR HSI as supervised information in training. There are two advantages to consider the fusion problem of HR MSIs and LR HSIs as the problem of learning the pixel-wise spectral mapping. The first advantage is that reconstructing HR HSIs directly from HR MSIs, which contains the desired HR spatial structure information, can mitigate the distortion of spatial structures in HR HSIs. The second advantage is that there are plentiful multispectral and hyperspectral pixel pairs naturally between MSIs and HSIs, which are sufficient for training deep networks without the need to introduce other supervised information.

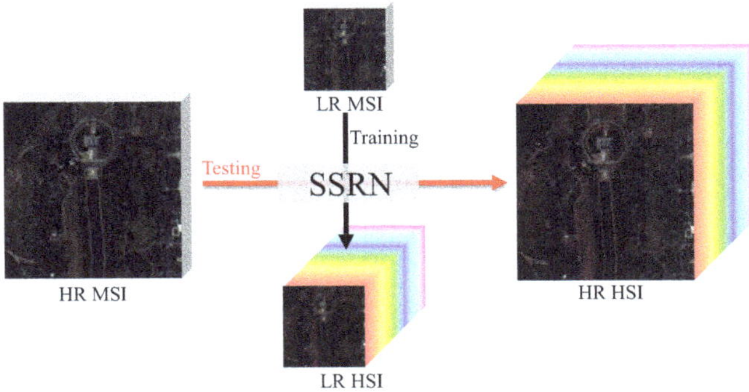

Figure 1. Illustration of the proposed spectral-spatial residual network (SSRN) framework. Firstly, a low spatial resolution (LR) multispectral image (MSI) and an LR hyperspectral image (HSI) are utilized to train the proposed SSRN to learn the pixel-wise spectral mapping. Then, the learned pixel-wise spectral mapping is exploited to estimate high spatial resolution (HR) HSIs from HR MSIs.

The proposed SSRN includes two modules: the spectral module and the spatial module. First, the spectral module is proposed to extract spectral features from MSIs. In the spectral module, the concatenation operation is employed to explore the complementarity among multi-layer features. Second, the spatial module is added following the spectral module to capture spectral-spatial features for facilitating learning of the spectral mapping. Especially, an attention mechanism is employed in the spatial module to make SSRN extract spectral-spatial features from homogeneous adjacent pixels, since homogeneous adjacent pixels in HSIs usually share similar spectral signatures. Finally, a self-supervised fine-tuning strategy is employed to further improve the performance of SSRN. In fact, the spatial degradation from the HR image to the LR image usually interferes with the spectral signatures of the LR image, which makes the spectral mapping between LR MSIs and LR HSIs slightly different from the spectral mapping between HR MSIs and HR HSIs. The self-supervised fine-tuning strategy is utilized to obtain the spectral mapping between HR MSIs and HR HSIs from the spectral mapping between LR MSIs and LR HSIs. The experimental results demonstrate that SSRN performs better than the state-of-the-art methods.

The major contributions of this paper are as follows:

- A spectral-spatial residual network is proposed to consider the fusion of HR MSIs and LR HSIs as a pixel-wise spectral mapping problem. In SSRN, the HR HSI is estimated from the HR MSI at the desired spatial resolution, which can effectively preserve spatial structures of HR HSIs.
- A self-supervised fine-tuning strategy is proposed to promote SSRN learning optimal spectral mapping. The self-supervised fine-tuning does not require HR HSIs as the supervised information.
- A spatial module configured with the attention mechanism is proposed to explore the complementarity of adjacent pixels. The attention mechanism can explore the spectral-spatial features from homogeneous adjacent pixels, which is beneficial to the learning of pixel-wise spectral mapping.

The remaining sections are as follows. In Section 2, recent HSI super-resolution methods are reviewed. In Section 3, the proposed SSRN is introduced. The experimental results of SSRN and the compared methods are reported in Section 4. The performance of SSRN is discussed in Section 5. Finally, Section 6 concludes this paper.

2. Related Work

Many methods have been proposed to reconstruct HR HSIs by fusing the observed LR HSI and HR MSI. In light of whether deep networks are utilized, the existing methods are roughly categorized into traditional methods and DL-based methods.

2.1. Traditional Methods

According to different technique frameworks, traditional methods can be further divided into sparse representation-based methods, Bayesian-based methods, spectral unmixing-based methods, and tensor factorization-based methods.

Sparse representation-based methods [29] learn a dictionary from the observed LR HSI. The dictionary represents the reflectance spectrum of the scene and is then employed to learn the sparse code of HR MSIs. Akhtar et al. [30] proposed a generalization of the simultaneous orthogonal matching pursuit (GSOMP) method. Wei et al. [31] proposed a variational-based fusion method and designed a sparse regularization term.

Bayesian-based methods [32] intuitively interpret the process of fusion through the posterior distribution. Eismann et al. [33] proposed a maximum a posteriori probability (MAP) estimation method. Wei et al. [34] proposed a hierarchical Bayesian fusion method to fuse spectral images. Irmak et al. [35] proposed a MAP-based energy function to enhance the spatial resolution of HSI.

Spectral unmixing-based methods usually employ nonnegative matrix factorization to decompose HR MSIs and LR HSIs [36,37]. A classic method is coupled nonnegative matrix factorization (CNMF) [36]. In CNMF, HR MSIs and LR HSIs are alternately decomposed. Then, the estimated endmember matrix of the LR HSI and the estimated abundance matrix of the HR MSI are multiplied to reconstruct the HR HSI. Borsoi et al. [1] embedded an explicit parameter into a spectral unmixing-based method to model the spectral variability between the HR MSI and LR HSI.

Tensor factorization-based methods treat HSIs as a 3-D tensor to estimate a core tensor and the dictionaries of the width, height, and spectral modes [15,16]. Dian et al. [16] introduced the sparsity prior into tensor factorization to extract non-local spatial information from HR MSIs and spectral information from LR HSIs, respectively. Li et al. [38] proposed a coupled sparse tensor factorization to estimate the core tensor.

Traditional methods have achieved favorable performances by exploiting the priors (e.g., sparsity and low-rankness), but such priors may not hold in some complicated scenarios [9,17,18].

2.2. Deep Learning-Based Methods

Recently, many works have designed various deep networks for fusing HR MSIs and LR HSIs, which can be divided into supervised DL-based methods [39] and unsupervised DL-based methods [40].

Supervised DL-based methods usually exploit massive HR HSIs as training images to learn potential HSI priors or the mapping relationship between LR and HR HSIs [20,25]. Xie et al. [20] exploited the low-rankness prior of HSIs to construct an MSI and HSI fusion model, which can be optimized iteratively with the proximal gradient. Subsequently, the iterative optimization is unfolded into a convolutional network structure for end-to-end training. Wei et al. [23] proposed a residual convolutional network to learn the mapping relationship between LR MSIs and HR MSIs. To mitigate dependence on the point spread function and spectral response function, Wang et al. [24] proposed a blind iterative fusion network to iteratively optimize the observation model. Li et al. [39] proposed a two-stream network to reconstruct HR HSIs, where one is a 1-D convolutional stream to extract spectral features and the other is a 2-D convolutional stream to extract spatial features. However, in practice, collecting plenty of HR HSIs as supervised information for training is time-consuming and laborious [26,27].

Unsupervised DL-based methods are dedicated to leveraging spectral and spatial ingredients from the given HR MSI and LR HSI to reconstruct the desired HR HSI [17,28,41]. Huang et al. [42] utilized a sparse denoising autoencoder to learn the spatial mapping relationship between LR and HR panchromatic images, where LR panchromatic images are obtained from the spectral degradation of LR MSIs. Then, the learned spatial mapping relationship was exploited to improve the spatial resolution of each spectral band of LR MSIs. Fu et al. [40] proposed a plain network simply composed of five convolution layers to fuse HR MSIs and LR HSIs. The HR MSI was concatenated with the feature maps of every convolution layer to guide the spatial structure reconstruction of HR HSIs. Although recent methods have achieved superior performance [17,28], designing deep networks suitable for HSI super-resolution that do not require additional supervision information for training is still an open problem.

3. Materials and Methods

3.1. Proposed Method

3.1.1. Problem Formulation

The goal of the proposed SSRN is to estimate the HR HSI by fusing the observed HR MSI and LR HSI of the same scene. Let the HR HSI be $\mathbf{X}_H \in \mathbb{R}^{B\times W\times H}$, the observed LR HSI be $\mathbf{X}_L \in \mathbb{R}^{B\times w\times h}$, and the observed HR MSI be $\mathbf{Y}_H \in \mathbb{R}^{b\times W\times H}$, where B and b represent spectral band numbers, W and w represent the width, and H and h represent the height. The observed LR HSI has a higher spectral resolution and a lower spatial resolution than the observed HR MSI, *i.e.*, $W = D \times w$, $H = D \times h$, and $B \gg b$ (D is the scaling factor). In fact, one pixel $\mathbf{Y}_H(i,j) \in \mathbb{R}^b$ of $\mathbf{Y}_H$ uniquely corresponds to one pixel $\mathbf{X}_H(i,j) \in \mathbb{R}^B$ of $\mathbf{X}_H$, where (i,j) represents the spatial location in the ith row and the jth column. This paper exploits a convolutional network to learn a nonlinear pixel-wise spectral mapping $F : \mathbb{R}^b \rightarrow \mathbb{R}^B$ that maps $\mathbf{Y}_H(i,j)$ to $\mathbf{X}_H(i,j)$. The pixel-wise spectral mapping can be formulated as

$$\mathbf{X}_H = F(\mathbf{Y}_H). \tag{1}$$

Since HR HSIs are difficult to obtain in practice [26,28], the proposed SSRN does not use HR HSIs as supervised information. In this paper, the spectral signatures of LR HSI $\mathbf{X}_L$ are first used as the supervised information to learn the spectral mapping $\hat{F} : \mathbb{R}^b \rightarrow \mathbb{R}^B$ between LR MSIs and LR HSIs. During the training phase, the input of SSRN is the LR MSI $\mathbf{Y}_L \in \mathbb{R}^{b\times w\times h}$ and the output is the LR HSI $\mathbf{X}_L$. $\mathbf{Y}_L$ is obtained by spatially blurring and then downsampling $\mathbf{Y}_H$.

$$\mathbf{Y}_L = E(\mathbf{Y}_H), \tag{2}$$

where $E(\cdot)$ represents the spatially blurring and downsampling operations [12]. Then, the learned spectral mapping $\hat{F}$ between the LR MSI $\mathbf{Y}_L$ and LR HSI $\mathbf{X}_L$ is transformed to the spectral mapping F with a self-supervised fine-tuning strategy, which can reconstruct the HR HSI $\mathbf{X}_H$ from the HR MSI $\mathbf{Y}_H$.

Previous methods [10,28,30,36] usually focus on extracting spectral ingredients (spectral bases or endmembers) from the LR HSI $\mathbf{X}_L$ and extracting spatial ingredients (sparse codes or abundances) from the HR MSI $\mathbf{Y}_H$. Then, the spectral ingredients of $\mathbf{X}_L$ and the spatial ingredients of $\mathbf{Y}_H$ are utilized to reconstruct the HR HSI $\mathbf{X}_H$. However, the observed scene in the HR MSI $\mathbf{Y}_H$ usually contains complex spatial distributions of land-covers; hence, there are still many challenges in accurately extracting spatial ingredients from HR MSI $\mathbf{Y}_H$ [8,9]. In previous methods, inaccurate spatial ingredients extracted from the HR MSI $\mathbf{Y}_H$ can cause spatial distortion of the reconstructed HR HSI. Different from previous methods [10,28,30,36], the proposed SSRN avoids the process of spatial ingredient extraction from HR MSI $\mathbf{Y}_H$. The proposed SSRN considers the fusion problem of HR MSI and LR HSI as a problem of spectral mapping learning. Based on the learned spectral mapping F, HR HSI $\mathbf{X}_H$ is directly reconstructed from HR MSI $\mathbf{Y}_H$. All the spatial ingredients of HR MSI $\mathbf{Y}_H$ can be used to reconstruct the HR HSI $\mathbf{X}_H$. Therefore, compared with previous methods, the proposed SSRN can better preserve the spatial structures of the reconstructed HR HSI.

The proposed method is similar to recent spectral resolution enhancement methods [43,44] that focus on learning the spectral mapping between MSIs and HSIs. However, the methods for spectral resolution enhancement are usually supervised training methods [45,46], which learn the spectral mapping from plentiful MSI and HSI pairs that are collected in other observed scenes. In contrast, in the HR MSI and LR HSI fusion task, the HR MSI and LR HSI are captured in the same observed scene. Our proposed method is a self-supervised training method specially designed for the HR MSI and LR HSI fusion task. The details of SSRN are introduced in the following subsections.

3.1.2. Architecture of SSRN

A detailed architecture of SSRN is shown in Figure 2. SSRN consists of two modules: the spectral module and the spatial module. In SSRN, the input is an MSI patch $\hat{\mathbf{Y}} \in \mathbb{R}^{b \times K \times K}$ and the output is an HSI patch $\hat{\mathbf{X}} \in \mathbb{R}^{B \times K \times K}$, where b and B represent spectral band numbers and $K \times K$ represents the spatial size. First, a 1×1 convolution layer is used to generate initial shallow spectral features from the MSI patch $\hat{\mathbf{Y}}$. Then, the spectral module is utilized to extract spectral features from the initial shallow spectral features, and the spatial module is added following the spectral module to extract spectral-spatial features to facilitate learning of spectral mapping.

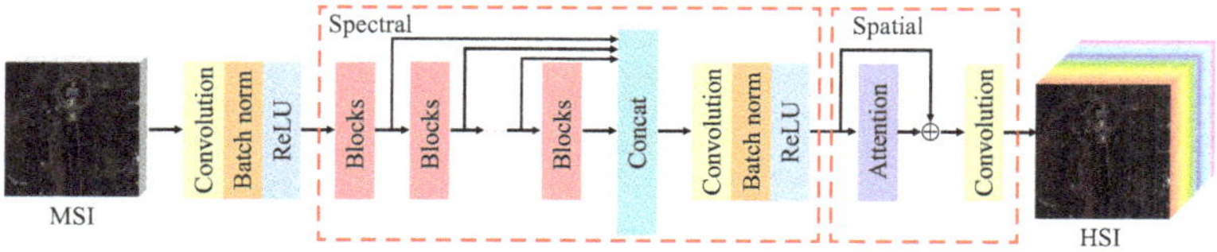

Figure 2. Architecture of the proposed SSRN. A spectral module and a spatial module are utilized to learn the pixel-wise spectral mapping between MSIs and HSIs. ⊕ represents the residual connection.

In this paper, spectral features refer to the features of multispectral pixels in the spectral dimension, which do not involve any information of the spatially adjacent pixels. The spectral module mainly consists of several residual blocks and a multi-layer feature aggregation (MLFA) component. As shown in Figure 3, the setting of the residual blocks is similar to that in the literature [47], where the residual connection can facilitate the convergence of SSRN. In residual blocks, the kernel size of all convolution layers is set to 1×1 to ensure that spectral feature extraction is only performed in the spectral dimension of MSI. The different residual blocks can extract different spectral features, which are beneficial for learning spectral mapping [48,49]. To explore the complementarity among the features of different residual blocks, an MLFA component is employed to integrate these features into the final spectral feature. The MLFA component is composed of a concatenation layer and 1×1 convolution, which do not introduce any information from the spatially adjacent multispectral pixels.

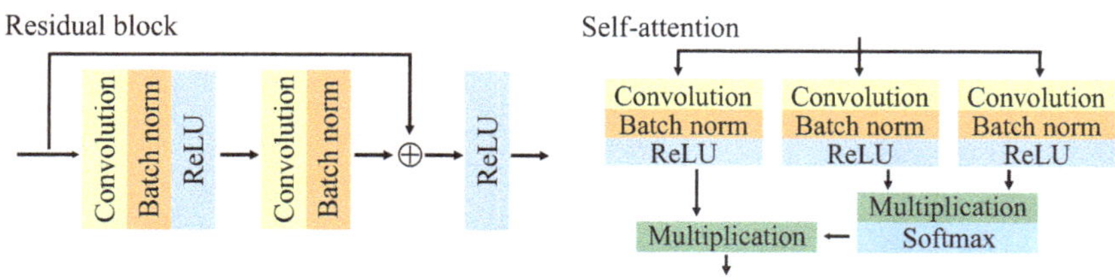

Figure 3. Structures of the residual block and the self-attention module. ⊕ represents the residual connection. ReLU represents the rectified linear unit.

The spatial module aims to extract complementary spatial information from adjacent pixels to learn spectral mapping. In this paper, the spatial information from adjacent pixels refers to the spatial structure information and spectrums contained in adjacent pixels. In practice, due to that adjacent pixels in real MSIs or HSIs potentially corresponding to the same object, adjacent pixels may have similar spectral signatures [50–52], which can be used as a prior to refine the reconstruct HR HSI. The adjacent pixels with similar spectral signatures are called homogeneous adjacent pixels. The spatial information of homogeneous adjacent pixels in HR MSI is beneficial in the learning of the pixel-wise spectral mapping between MSIs and HSIs [53]. Previous methods [43,44] usually use 3 × 3 convolution to extract spatial information. However, the 3 × 3 convolution can introduce the interference information from inhomogeneous adjacent pixels [2]. In this paper, the spatial module employs a self-attention module [54] to extract spectral-spatial features from the homogeneous adjacent pixels. The self-attention module can capture homogeneous adjacent pixels based on the correlation between different pixels [54] and then aggregate the information from these homogeneous adjacent pixels to generate spectral-spatial features. The details of the self-attention module are shown in Figure 3. The self-attention module takes the final spectral feature from the spectral module as the input and outputs the spectral-spatial features. The final spectral feature is denoted as $\mathbf{S} \in \mathbb{R}^{C \times K \times K}$, where C is the channel number and $K \times K$ is the spatial size. First, $\mathbf{S}$ is fed into three 1 × 1 convolution layers to generate abstract features $f(\mathbf{S}) \in \mathbb{R}^{C \times K \times K}$, $g(\mathbf{S}) \in \mathbb{R}^{C \times K \times K}$, and $n(\mathbf{S}) \in \mathbb{R}^{C \times K \times K}$, respectively. Second, $f(\mathbf{S}), g(\mathbf{S})$, and $n(\mathbf{S})$ are reshaped to $\bar{f}(\mathbf{S}), \bar{g}(\mathbf{S})$, and $\bar{n}(\mathbf{S}) \in \mathbb{R}^{C \times M}$, where $M = K \times K$. Each column of the reshaped $\bar{f}(\mathbf{S}), \bar{g}(\mathbf{S})$, and $\bar{n}(\mathbf{S})$ represents the spectral feature of a certain pixel. The correlation among pixels in spectral features is calculated as follows

$$\mathbf{N} = \bar{f}(\mathbf{S})^T \bar{g}(\mathbf{S}), \tag{3}$$

where $(\cdot)^T$ denotes the transpose and $\mathbf{N} \in \mathbb{R}^{M \times M}$. A softmax function is employed to normalize the value of all elements in $\mathbf{N}$ to the range $[0, 1]$. Then, the spectral-spatial features are generated by multiplying $\bar{n}(\mathbf{S})$ with the normalized correlation $\mathbf{N}$. With the normalized correlation $\mathbf{N}$, the homogeneous pixels from adjacent regions can be aggregated to facilitate the learning of the spectral mapping. Finally, the shape of spectral-spatial features is reshaped to $\mathbb{R}^{C \times K \times K}$ for the following operations. After the self-attention module, a 1 × 1 convolution layer with B kernels is utilized to reconstruct the desired HR HSI from the spectral-spatial features.

In the proposed SSRN, the kernel size of all convolution layers is set to 1 × 1, which can mitigate the difficulty of training SSRN caused by too many weight parameters.

3.1.3. Loss Function

In the proposed SSRN, a reconstruction loss L_{rec} and a cosine similarity loss L_{cos} are employed as the loss functions. Let $\mathbf{U}$ represent the generated HSI and $\mathbf{V}$ represent the ground truth. For convenience, $\mathbf{U}$ and $\mathbf{V}$ are reshaped to $\mathbb{R}^{P \times Q}$, where P is the number of spectral bands and Q is the number of pixels. Each column of $\mathbf{U}$ and $\mathbf{V}$ represents the spectral vector of a hyperspectral pixel. The reconstruction loss L_{rec} is a classic metric function that measures the numerical differences between two HSIs. L_{rec} is defined as

$$L_{rec}(\mathbf{U}, \mathbf{V}) = \|\mathbf{U} - \mathbf{V}\|_F^2, \tag{4}$$

where $\|\cdot\|_F$ represents the Frobenius norm. The cosine similarity loss L_{cos} measures the spectral distortion based on the angle between two spectral signatures. L_{cos} is defined as

$$L_{cos}(\mathbf{U}, \mathbf{V}) = 1 - \frac{1}{Q} \sum_{i=1}^{Q} \frac{\mathbf{U}^{(i)} \cdot \mathbf{V}^{(i)}}{\|\mathbf{U}^{(i)}\|_2 \|\mathbf{V}^{(i)}\|_2}, \tag{5}$$

where $\mathbf{U}^{(i)}$ is the ith column of $\mathbf{U}$ that denotes the spectral vector of the ith pixel of $\mathbf{U}$ and $\mathbf{V}^{(i)}$ is the ith column of $\mathbf{V}$ that denotes the spectral vector of the ith pixel of $\mathbf{V}$ $(1 \leqslant i \leqslant Q)$.

In the training phase, the LR MSI $\mathbf{Y}_L$ and the LR HSI $\mathbf{X}_L$ are cropped into small patches for training. Let $\hat{\mathbf{Y}}_L \in \mathbb{R}^{b \times K \times K}$ be the LR MSI patch cropped from $\mathbf{Y}_L$, $\hat{\mathbf{X}}_L \in \mathbb{R}^{B \times K \times K}$ be the corresponding LR HSI patch cropped from $\mathbf{X}_L$, and $\bar{\mathbf{X}}_L = \hat{F}(\hat{\mathbf{Y}}_L) \in \mathbb{R}^{B \times K \times K}$ be the reconstructed LR HSI patch. To facilitate calculation of the loss, $\hat{\mathbf{Y}}_L$ is reshaped to $\mathbb{R}^{b \times M}$, and $\hat{\mathbf{X}}_L$ and $\bar{\mathbf{X}}_L$ are reshaped to $\mathbb{R}^{B \times M}$, where $M = K \times K$. The details of the loss function in SSRN are as follows.

First, the loss $Loss_{HSI}$ between the reconstructed $\bar{\mathbf{X}}_L$ and the ground truth $\hat{\mathbf{X}}_L$ are measured by the reconstruction loss L_{rec} and the cosine similarity loss L_{cos}.

$$Loss_{HSI}(\bar{\mathbf{X}}_L, \hat{\mathbf{X}}_L) = L_{rec}(\bar{\mathbf{X}}_L, \hat{\mathbf{X}}_L) + \lambda L_{cos}(\bar{\mathbf{X}}_L, \hat{\mathbf{X}}_L), \tag{6}$$

where λ is the balancing parameter to control the tradeoff between L_{rec} and L_{cos}.

Second, according to the observation model, the LR MSI patch $\hat{\mathbf{Y}}_L$ is the spectral degradation of the LR HSI patch $\hat{\mathbf{X}}_L$ [14], which can be formulated as

$$\hat{\mathbf{Y}}_L = R(\hat{\mathbf{X}}_L), \tag{7}$$

where $R(\cdot)$ represents the spectral degradation. This means that the spectral degradation of the reconstructed HSI patch $\bar{\mathbf{X}}_L$ should also be consistent with the input MSI patch $\hat{\mathbf{Y}}_L$. To maintain the consistency between $R(\bar{\mathbf{X}}_L)$ and $\hat{\mathbf{Y}}_L$, another loss function $Loss_{MSI}$ is established in this paper. Similar to $Loss_{HSI}$, $Loss_{MSI}$ is formulated as

$$Loss_{MSI}(R(\bar{\mathbf{X}}_L), \hat{\mathbf{Y}}_L) = L_{rec}(R(\bar{\mathbf{X}}_L), \hat{\mathbf{Y}}_L) + \beta L_{cos}(R(\bar{\mathbf{X}}_L), \hat{\mathbf{Y}}_L), \tag{8}$$

where β is simply set to the same value as λ of Equation (6), since the second terms in Equations (6) and (8) are all the cosine similarity loss. Overall, the loss function of SSRN is set as

$$Loss_{train} = Loss_{HSI}(\bar{\mathbf{X}}_L, \hat{\mathbf{X}}_L) + \phi Loss_{MSI}(R(\bar{\mathbf{X}}_L), \hat{\mathbf{Y}}_L). \tag{9}$$

In the proposed SSRN, $Loss_{HSI}$ and $Loss_{MSI}$ are equally important for reconstructing HR HSI. Therefore, ϕ is simply set to 1 in the following experiments.

3.1.4. Self-Supervised Fine-Tuning

This paper assumes that the pixel-wise spectral mapping F between HR MSI and HR HSI can be estimated on the basis of the pixel-wise spectral mapping $\hat{F}$ between LR MSI and LR HSI. The training process of the proposed SSRN includes two stages: the pretraining stage and the fine-tuning stage. In the pretraining stage, the pixel-wise spectral mapping $\hat{F}$ can be easily learned from the paired LR MSI patches and LR HSI patches using the proposed SSRN. In this stage, the proposed SSRN is supervised by LR MSIs and LR HSIs simultaneously. In fact, the spectral signatures of LR MSIs and LR HSIs are usually influenced by spatial degradation. The spectral mapping $\hat{F}$ is not exactly equal to the spectral mapping F. Hence, in this paper, a fine-tuning strategy is proposed to further estimate the spectral mapping F from the spectral mapping $\hat{F}$. The SSRN trained with LR MSIs and LR HSIs serves as a pretrained network. Then, in the fine-tuning stage, the pretrained SSRN is further fine-tuned with the HR MSI. Since the HR HSI is hard to be obtained in practice, SSRN does not utilize the HR HSI as supervised information in training. Therefore, Equation (6) cannot be employed as the loss function in the fine-tuning stage. Equation (8) is employed as the fine-tuning loss $Loss_{FT}$ to maintain the consistency between $R(\bar{\mathbf{X}}_H)$ and $\hat{\mathbf{Y}}_H$, where $R(\cdot)$ has the same definition as that in Equation (7), $\bar{\mathbf{X}}_H$ is the reconstructed HR HSI patch, and $\hat{\mathbf{Y}}_H$ is the input HR MSI patch. $Loss_{FT}$ can be expressed as

$$Loss_{FT} = Loss_{MSI}(R(\bar{\mathbf{X}}_H), \hat{\mathbf{Y}}_H). \tag{10}$$

In the fine-tuning stage, the proposed SSRN is only supervised by HR MSI. Therefore, the fine-tuning stage is a self-supervised training style. After fine-tuning, the spectral mapping F between HR MSI and HR HSI is obtained. The desired HR HSI can be reconstructed with Equation (1).

3.2. *Software and Package*

The proposed SSRN is implemented in a computer workstation that is configured with the Ubuntu 14.04 system, 64G RAM, Intel Core i7-5930K, and NVIDIA TITAN X. The software used in the experiments is PyCharm. The packages used in the experiments include Python, TensorFlow, NumPy, and SciPy.

3.3. *Databases*

To evaluate the performance of SSRN, the experiments are conducted on simulated databases and real databases, respectively. First, the Pavia University (PU) database (http://www.ehu.eus/ccwintco/index.php?title=Hyperspectral_Remote_Sensing_Scenes#Pavia_University_scene, accessed on 16 December 2020) and the Washington DC Mall (WDCM) database (https://engineering.purdue.edu/~biehl/MultiSpec/hyperspectral.html, accessed on 16 December 2020) are utilized to simulate MSIs for experiments. Then, the Paris database (https://github.com/alfaiate/HySure/tree/master/data, accessed on 17 December 2020) [12] and the Ivanpah Playa database (https://github.com/ricardoborsoi/FuVar Release/tree/master/DATA, accessed on 17 December 2020) [1], which contain both real MSIs and HSIs, are employed to conduct experiments. Finally, the CAVE database (https://www.cs.columbia.edu/CAVE/databases/multispectral/, accessed on 19 December 2020) [55] is employed to further explore the performance of SSRN.

The PU database is captured by the ROSIS sensor over Pavia University. The PU database contains an HSI with 103 spectral bands and 610 × 340 pixels. The WDCM database is collected by the HYDICE sensor over the National Mall. The WDCM database consists of an HSI with 191 spectral bands and 1280 × 307 pixels. Similar to the literature [37], a 200 × 200 subimage of the PU database and a 240 × 240 subimage of the WDCM database are utilized for experiments. The original HSIs in the PU and WDCM databases are regarded as the ground truth. The ground truth is blurred and then spatially downsampled with the scaling factor of 4 to simulate the observed LR HSI. The observed HR MSI is obtained by spectrally downsampling the ground truth. The setting of the spectral response function is the same as that in the literature [37].

The Paris database contains an HSI captured by the hyperion instrument and an MSI collected by the ALI instrument [12]. The HSI contains 128 spectral bands. The MSI contains 9 spectral bands. Both the HSI and the MSI have 72 × 72 pixels. The Ivanpah Playa database consists of an HSI with 173 spectral bands and an MSI with 10 spectral bands. The HSI and the MSI on the Ivanpah Playa database contain 80 × 128 pixels. According to the literature [1], the HSIs on the Paris and Ivanpah Playa databases are treated as the ground truth, which are blurred and spatially downsampled with the scaling factor of 4 to generate the observed LR HSI. The MSIs on the Paris and Ivanpah Playa databases are treated as the observed HR MSI.

The CAVE database contains 32 HSIs, which are captured by the cooled charge-coupled device (CCD) camera on the ground [55]. On the CAVE database, each HSI consists of 512 × 512 pixels, where each pixel is composed of 31 spectral bands ranging from 400 nm to 700 nm. Following the literature [56], the original HSIs on the CAVE database are treated as the ground truth. Then, the ground truth is blurred and spatially downsampled with the scaling factor of 4 to obtain the observed LR HSI. The ground truth is spectrally downsampled by the spectral response function of Nikon D700 (https://maxmax.com/spectral_response.htm, accessed on 19 December 2020) to obtain the observed HR MSI.

3.4. *Evaluation Metrics*

Five quantitative quality metrics are employed for performance evaluation, including peak signal-to-noise ratio (PSNR), spectral angle mapper (SAM), universal image quality index (UIQI), erreur relative globale adimensionnelle de synthèse (ERGAS), and root mean squared error (RMSE). PSNR measures the spatial reconstruction quality of each spectral band in the reconstructed HR HSI. SAM measures the spectral distortions of each hyperspectral pixel in the reconstructed HR HSI. UIQI measures the spatial structural similarity between the reconstructed HR HSI and the ground truth based on the combination of luminance, contrast, and correlation comparisons. ERGAS takes into account the ratio of ground sample distances between HR MSI and LR HSI to measure the global statistical quality of the reconstructed HR HSI. RMSE measures the global statistical error between the reconstructed HR HSI and the ground truth. The larger values of PSNR and UIQI indicate the better quality of the reconstructed HR HSI. When the values of SAM, ERGAS, and RMSE are smaller, the quality of the reconstructed HR HSI is better. The best value of PSNR is $+\infty$. The best value of SAM is 0. The best value of UIQI is 1. The best values of ERGAS and RMSE are 0.

In this paper, the ground truth $\tilde{\mathbf{X}}_H \in \mathbb{R}^{B\times W\times H}$ and the reconstructed HR HSI $\mathbf{X}_H \in \mathbb{R}^{B\times W\times H}$ are converted into 8-bit images to calculate quantitative performance, where B, W, and H are the numbers of the band, width, and height, respectively. The formulations of the above quality metrics for the ground truth $\tilde{\mathbf{X}}_H$ and the reconstructed HR HSI $\mathbf{X}_H$ are given below.

PSNR is formulated as

$$\mathrm{PSNR} = \frac{1}{B}\sum_{i=1}^{B} 10\log_{10}\left(\frac{max(\tilde{\mathbf{X}}_{H_i})^2}{\frac{1}{W\times H}\sum_{j=1}^{W\times H}\left(\tilde{\mathbf{X}}_{H_{ij}} - \mathbf{X}_{H_{ij}}\right)^2}\right), \tag{11}$$

where $max(\tilde{\mathbf{X}}_{H_i})$ represents the maximum pixel value in the ith band of $\tilde{\mathbf{X}}_H$. $\tilde{\mathbf{X}}_{H_{ij}}$ and $\mathbf{X}_{H_{ij}}$ $(1 \leqslant i \leqslant B, 1 \leqslant j \leqslant W\times H)$ represent the jth pixel in the ith band of $\tilde{\mathbf{X}}_H$ and $\mathbf{X}_H$, respectively.

SAM is formulated as

$$\mathrm{SAM} = \frac{1}{W\times H}\sum_{j=1}^{W\times H} \arccos\left(\frac{(\tilde{\mathbf{X}}_H[j])^T\mathbf{X}_H[j]}{\|\tilde{\mathbf{X}}_H[j]\|_2\|\mathbf{X}_H[j]\|_2}\right), \tag{12}$$

where $\tilde{\mathbf{X}}_H[j] \in \mathbb{R}^{B\times 1}$ and $\mathbf{X}_H[j] \in \mathbb{R}^{B\times 1}$ $(1 \leqslant j \leqslant W\times H)$ denote the spectra of the jth pixel of $\tilde{\mathbf{X}}_H$ and $\mathbf{X}_H$, respectively. $(\cdot)^T$ denotes the transpose, and $\|\cdot\|_2$ denotes the ℓ_2 vector norm.

UIQI is formulated as

$$\mathrm{UIQI} = \frac{1}{B}\sum_{i=1}^{B}\left(\frac{1}{Z}\sum_{q=1}^{Z}\frac{4\mu_{\tilde{z}_{iq}}\mu_{z_{iq}}\sigma_{\tilde{z}_{iq}z_{iq}}}{\left({\mu_{\tilde{z}_{iq}}}^2 + {\mu_{z_{iq}}}^2\right)\left({\sigma_{\tilde{z}_{iq}}}^2 + {\sigma_{z_{iq}}}^2\right)}\right), \tag{13}$$

where a sliding window moving pixel by pixel is used to divide the ith band of $\tilde{\mathbf{X}}_H$ and $\mathbf{X}_H$ at the same position into Z image patch pairs $\tilde{z}_{iq}$ and z_{iq} $(1 \leqslant i \leqslant B, 1 \leqslant q \leqslant Z)$, respectively. Z is the image patch pair number. $\mu_{\tilde{z}_{iq}}$ and $\mu_{z_{iq}}$ are mean pixel values of image patches $\tilde{z}_{iq}$ and z_{iq}, respectively. $\sigma_{\tilde{z}_{iq}}$ and $\sigma_{z_{iq}}$ are the corresponding variance. $\sigma_{\tilde{z}_{iq}z_{iq}}$ is the covariance.

ERGAS is formulated as

$$\mathrm{ERGAS} = 100d\sqrt{\frac{1}{B}\sum_{i=1}^{B}\frac{\frac{1}{W\times H}\sum_{j=1}^{W\times H}\left(\tilde{\mathbf{X}}_{H_{ij}} - \mathbf{X}_{H_{ij}}\right)^2}{\left(\mu_{\tilde{\mathbf{X}}_{H_i}}\right)^2}}, \tag{14}$$

where d is the ratio of ground sample distances between HR MSI and LR HSI. $\mu_{\tilde{\mathbf{X}}_{H_i}}$ $(1 \leqslant i \leqslant B)$ denotes the mean pixel value in the ith band of the ground truth HSI $\tilde{\mathbf{X}}_H$.

RMSE is formulated as

$$\text{RMSE} = \sqrt{\frac{1}{B}\sum_{i=1}^{B}\left(\frac{1}{W \times H}\sum_{j=1}^{W \times H}\left(\tilde{\mathbf{X}}_{H_{ij}} - \mathbf{X}_{H_{ij}}\right)^2\right)}, \tag{15}$$

where $\tilde{\mathbf{X}}_{H_{ij}}$ and $\mathbf{X}_{H_{ij}}$ $(1 \leqslant i \leqslant B, 1 \leqslant j \leqslant W \times H)$ represent the jth pixel in the ith band of $\tilde{\mathbf{X}}_H$ and $\mathbf{X}_H$, respectively.

4. Results

4.1. Parameter Settings of SSRN

This subsection explores the parameter settings of SSRN. The WDCM database has plenty of spectral bands and contains complicated land-cover distributions, making the fusion task challenging [16]. Therefore, the WDCM database is utilized for parameter setting experiments. For convenience, this subsection directly uses PSNR and SAM to measure the quality of the reconstructed HR HSI. Moreover, the fine-tuning strategy is not employed in the parameter setting experiments.

4.1.1. Number of Convolutional Kernels

In the experiments, the spatial size of input image patches is set as 4×4. The number of training epochs is set as 200. The learning rate is initially set as 0.01, which then drops by a factor of 10 after 100 epochs. The balancing parameter λ in the loss function is initially set as 0.1. The number of residual blocks is set as 3. For convenience, all convolution layers of SSRN (except the last convolution layer) are configured with the same number of convolutional kernels, which is set as 16, 32, 64, 128, 256m and 512 for the experiments. The PSNR and SAM of SSRN with different numbers of convolutional kernels on the WDCM database are shown in Table 1. As the kernel number increases from 16 to 256, the performance of SSRN increases. As the kernel number increases from 256 to 512, the performance of SSRN decreases due to too many weight parameters, making SSRN training difficult. As shown in Table 1, the number of convolutional kernels in SSRN except the last convolution layer is set as 256 in the following experiments.

Table 1. Peak signal-to-noise ratio (PSNR) and spectral angle mapper (SAM) of SSRN with different numbers of convolutional kernels on the Washington DC Mall (WDCM) database.

Number	16	32	64	128	256	512
PSNR	31.772	32.477	32.926	33.044	33.123	33.048
SAM	1.485	1.385	1.287	1.245	1.228	1.245

4.1.2. Number of Residual Blocks

SSRN utilizes several residual blocks to extract spectral features from the MSI. To explore the effects of different numbers of residual blocks on the performance of SSRN, the number of residual blocks is set as 1, 2, 3, 4, 5, and 6 for the experiments. The PSNR and SAM of SSRN on the WDCM database are shown in Table 2. SSRN with 4 residual blocks achieves the best performance, where the PSNR and SAM are 33.167 and 1.213, respectively. In following experiments, the residual block number of SSRN is set as 4.

Table 2. PSNR and SAM of SSRN with different numbers of residual blocks on the WDCM database.

Number	1	2	3	4	5	6
PSNR	32.850	33.149	33.123	33.167	32.978	32.941
SAM	1.232	1.215	1.228	1.213	1.219	1.240

4.1.3. Balancing Parameter λ

The balancing parameter λ is a key parameter that controls the tradeoff between the reconstruction loss and the cosine similarity loss in SSRN. If the value of the balancing parameter λ is too small, the cosine similarity loss in SSRN will be invalidated, resulting in a large SAM value of the reconstructed HR HSI. If the value of the balancing parameter λ is too large, the reconstruction loss will be invalidated, resulting in a decrease in the quality of the reconstructed HR HSI. To explore the impacts of the balancing parameter λ on the performance of SSRN, λ is set as 0.001, 0.01, 0.1, 1, 5, and 10 for the experiments. The PSNR and SAM of SSRN with different balancing parameter λ are shown in Table 3. As λ increases from 0.001 to 0.1, the performance of SSRN increases. However, as λ increases from 0.1 to 10, the performance of SSRN decreases. The balancing parameter λ of SSRN is set as 0.1 in the following experiments.

Table 3. PSNR and SAM of SSRN with different λ on the WDCM database.

λ	**0.001**	**0.01**	**0.1**	**1**	**5**	**10**
PSNR	31.959	32.473	33.167	33.060	31.884	29.940
SAM	1.436	1.411	1.213	1.244	1.254	1.267

4.2. Ablation Study

The proposed SSRN can be specifically decomposed into five components, including the basic network, the MLFA component, the spatial module, the cosine similarity loss, and the fine-tuning. The basic network refers to the proposed spectral module without the MLFA, which can be utilized to coarsely learn the pixel-wise spectral mapping. The loss function of the basic network is a reconstruction loss. The other four components are utilized to improve the performance of this basic network. In this subsection, the ablation experiments for these four components are conducted on the WDCM database. The experimental results are shown in Table 4. The basic network achieves the worst performance. It is indicated that spatial features are not adequately exploited by the basic network. The MLFA component is added to the basic network to demonstrate that aggregating features of different convolution layers can improve the performance of the basic network. After further introducing the spatial module in the basic network and MLFA, the PSNR of the estimated HSI improved. Although the spatial module can improve the spatial quality of the estimated HSI, it cannot significantly reduce the spectral distortion. Then, the cosine similarity loss is further added into the basic network combined with the MLFA and the spatial module. As shown in Table 4, the cosine similarity loss can effectively alleviate the problem of spectral distortion in the estimated HSI. Finally, the fine-tuning strategy is added into the basic network combined with other three components. The proposed SSRN shows superior performance, which demonstrates the effectiveness of the fine-tuning strategy. Therefore, the MLFA, the spatial module, the cosine similarity loss, and the fine-tuning are all crucial components for the proposed SSRN.

Table 4. Ablation experiments of SSRN on the WDCM databases.

	Ablation Study				
MLFA	×	√	√	√	√
Spatial module	×	×	√	√	√
Cosine similarity loss	×	×	×	√	√
Fine-tuning	×	×	×	×	√
PSNR	31.902	32.061	32.150	33.167	33.232
SAM	1.514	1.482	1.494	1.213	1.211

1. × represents that the basic network is configured with the component. 2. √ represents that the basic network is not configured with the component.

4.3. Comparisons with Other Methods on Simulated Databases

In this subsection, the PU and WDCM databases are employed to simulate HR MSIs to evaluate the proposed SSRN. The proposed SSRN is compared with several state-of-the-art HSI super-resolution methods, including coupled nonnegative matrix factorization (CNMF) (http://naotoyokoya.com/assets/zip/CNMF_MATLAB.zip, accessed on 12 October 2020) [36], generalization of simultaneous orthogonal matching pursuit (GSOMP) (http://www.csse.uwa.edu.au/~ajmal/code/HSISuperRes.zip, accessed on 12 October 2020) [30], hyperspectral image super-resolution via subspace-based regularization (HySure) (https://github.com/alfaiate/HySure, accessed on 12 October 2020) [12], transfer learning-based super-resolution (TLSR) [26], unsupervised sparse Dirichlet-net (USDN) (https://github.com/aicip/uSDN, accessed on 20 October 2020) [28], and deep hyperspectral prior (DHSP) (https://github.com/acecreamu/deep-hs-prior, accessed on 20 October 2020) [27]. CNMF, GSOMP, and HySure are traditional methods. TLSR, USDN, and DHSP are recent unsupervised DL-based methods. On the PU and WDCM databases, the number of training epochs is set as 200 for SSRN. The learning rate of SSRN is initially set as 0.01, which then drops by a factor of 10 after every 100 epochs. Compared methods use the parameter settings from the original literature. All experiments are implemented 5 times, and then, the average results are reported.

4.3.1. PU Database

The quantitative results of SSRN and the compared methods on the PU database are reported in Table 5. TLSR and DHSP perform worse than traditional methods, since TLSR and DHSP only employ a single hyperspectral image to reconstruct HR HSIs. TLSR and DHSP cannot utilize the spatial information of the MSI to estimate the HR HSI. USDN utilizes two autoencoder networks to extract spatial information from HR MSIs and spectral information from LR HSIs, respectively. USDN shows superior performance to the traditional methods. Different from CNMF, GSOMP, HySure, and USDN, the proposed SSRN learns a pixel-wise spectral mapping between MSIs and HSIs. In SSRN, the desired HSI is directly estimated from MSIs with the desired high spatial resolution, which can preserve the spatial structures. In addition, the proposed SSRN employs cosine similarity loss for training, which can reduce the distortion of spectral signatures. As shown in Table 5, the proposed SSRN outperforms other methods on the PU database.

Table 5. Quantitative experimental results on the Pavia University (PU) database.

Methods	PSNR	UIQI	RMSE	ERGAS	SAM
CNMF [36]	33.072	0.963	5.828	3.654	3.710
GSOMP [30]	35.117	0.971	4.819	3.230	4.050
HySure [12]	38.710	0.983	3.226	2.037	3.453
TLSR [26]	25.349	0.783	14.093	8.625	6.815
USDN [28]	36.944	0.977	3.835	2.620	3.340
DHSP [27]	25.702	0.799	13.504	8.282	6.606
SSRN	39.741	0.985	2.886	1.980	2.781

To visualize the experimental results, the visual images and error maps of SSRN and the compared methods are displayed in Figure 4. The HSIs estimated by TLSR and DHSP are blurry, since TLSR and DHSP cannot utilize the spatial information of the MSI. In the estimated HSIs of TLSR and DHSP, the small targets that only cover one or two pixels are missing. As shown in the error maps, the proposed SSRN effectively preserves the spatial structures of the estimated HSI.

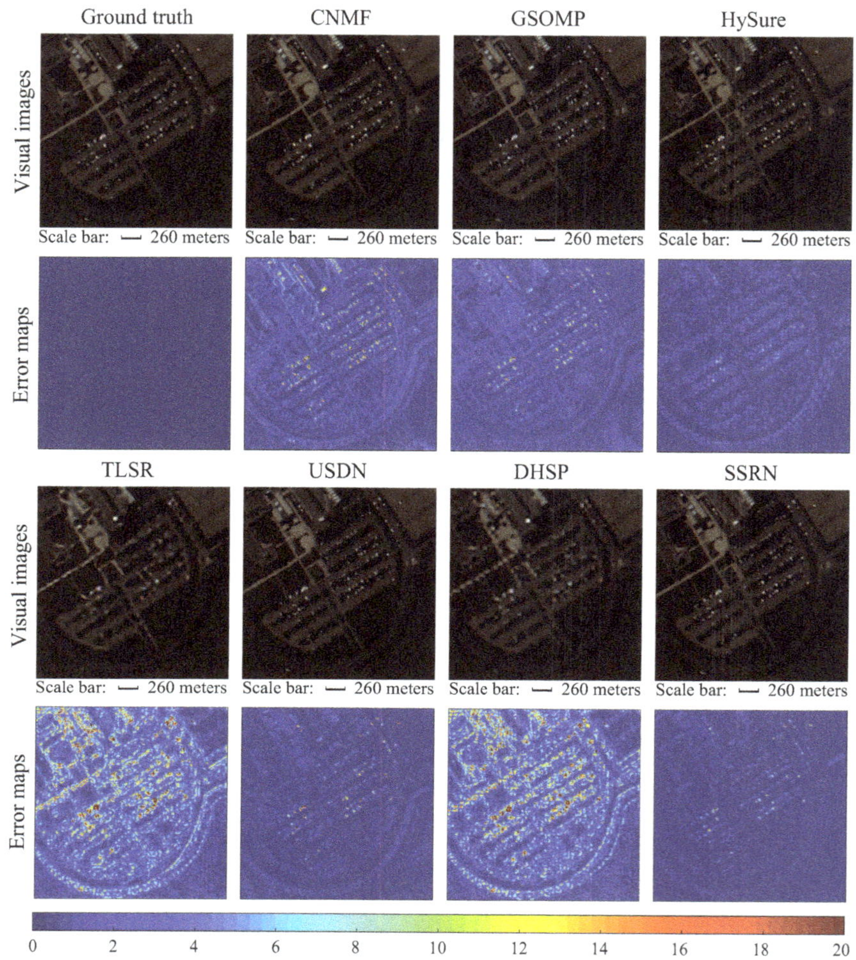

Figure 4. Visual images (R: 60, G: 30, and B: 10) and error maps of SSRN and the compared methods on the PU database. The error maps are the sum of absolute differences in all spectral bands between the estimated HSI and the ground truth.

4.3.2. WDCM Database

The quantitative results of SSRN and the compared methods on the WDCM database are reported in Table 6. CNMF, GSOMP, and HySure show better performance than TLSR and DHSP, since the spatial information of the MSI is utilized. The performance of USDN can compete with CNMF, GSOMP, and HySure. As shown in Table 6, the performance of the proposed SSRN is better than the compared methods in terms of PSNR, RMSE, and SAM. In terms of UIQI and ERGAS, the proposed SSRN shows favorable performance, which is close to the results of GSOMP.

Visual images and error maps of SSRN and the compared methods on the WDCM database are shown in Figure 5. The visual images of CNMF, GSOMP, HySure, USDN, and the proposed SSRN have good visualization results, owing to the reliable spatial information provided by the HR MSI. As shown in the error maps, the errors of TLSR and DHSP are mainly concentrated on the edges of complicated land-covers. The proposed SSRN shows superior performance in complicated land-covers.

Table 6. Quantitative experimental results on the WDCM database.

Methods	PSNR	UIQI	RMSE	ERGAS	SAM
CNMF [36]	32.217	0.948	1.520	74.197	1.944
GSOMP [30]	31.979	0.956	1.729	57.587	1.877
HySure [12]	30.484	0.940	2.316	59.799	2.518
TLSR [26]	21.663	0.712	8.595	61.663	6.095
USDN [28]	31.355	0.935	1.805	122.336	2.264
DHSP [27]	21.917	0.749	8.566	122.069	5.967
SSRN	33.232	0.954	1.448	61.216	1.211

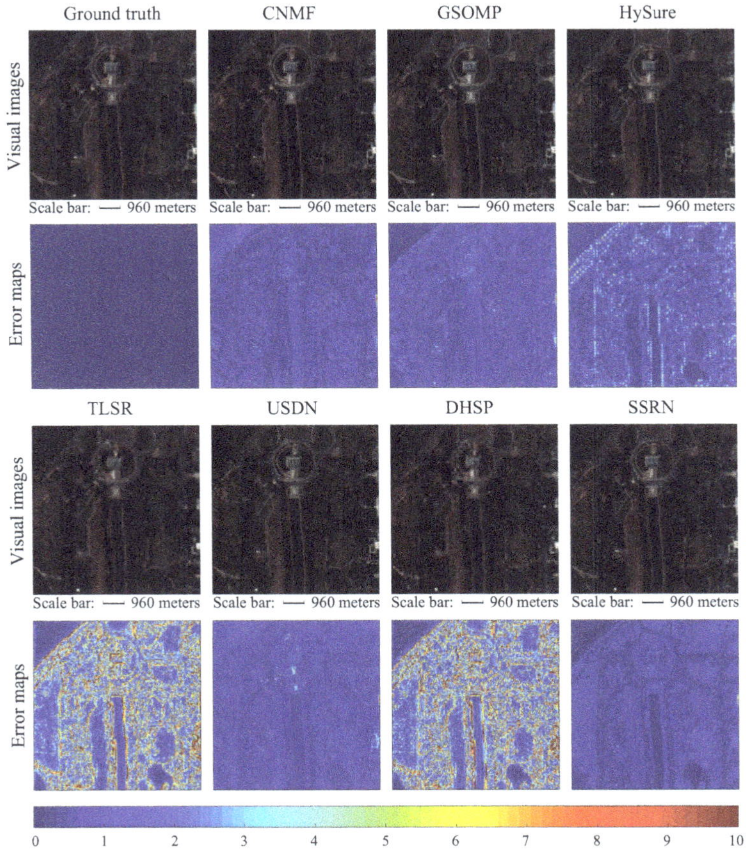

Figure 5. Visual images (R: 50, G: 30, and B: 20) and error maps of SSRN and the compared methods on the WDCM database.

4.4. Comparisons with Other Methods on Real Databases

In this subsection, SSRN and the compared methods are evaluated on two real databases. On the Paris and Ivanpah Playa databases, the number of training epochs is set as 400 for SSRN. The learning rate of SSRN is initially set as 0.01, which then drops by a factor of 10 after 200 epochs.

4.4.1. Paris Database

On the Paris database, HR MSIs and LR HSIs are captured at the same time instant. On this database, the LR HSI is generated from the original HSI for training. After spatially downsampling with the scaling factor of 4, the LR HSI contains only 18×18 pixels, which is far less than the number of pixels on simulated databases. Insufficient pixels in the LR HSI can make the proposed SSRN difficult to train. To alleviate the problem of insufficient pixels, the training samples are flipped left and right. Furthermore, the training samples are rotated 90, 180, and 270 degrees. On the Paris database, the spectral response function is estimated with the method proposed in the literature [12]. The performance of SSRN and the compared methods on the Paris database is shown in Table 7. In comparison with the PU and WDCM databases, the performance of SSRN and the compared methods decreased due to the too complicated land-cover distributions on the Paris database. As shown in Table 7, the proposed SSRN still shows better performance than the compared methods.

Table 7. Quantitative experimental results on the Paris database.

Methods	PSNR	UIQI	RMSE	ERGAS	SAM
CNMF [36]	27.879	0.819	7.564	3.601	3.534
GSOMP [30]	28.235	0.817	7.299	3.517	3.381
HySure [12]	27.621	0.824	7.886	3.763	3.759
TLSR [26]	24.671	0.520	10.985	5.130	4.806
USDN [28]	27.975	0.803	7.509	3.622	3.435
DHSP [27]	24.569	0.516	11.106	5.185	4.935
SSRN	28.350	0.829	7.185	3.434	3.334

Visual images and error maps of SSRN and the compared methods are shown in Figure 6. Since the proposed SSRN estimates the HSI directly from the MSI, the spatial information of the MSI can be fully utilized. As shown in Figure 6, compared to the error maps of other methods, the proposed SSRN effectively mitigates the spatial distortion.

4.4.2. Ivanpah Playa Database

The Ivanpah Playa database is a real database that consists of a LR HSI collected on 26 October 2015 and a HR MSI captured on 17 December 2017. On the Ivanpah Playa database, the HR MSI and LR HSI are collected during different seasons. In practice, seasonal changes may result in that the same land-cover material having different intrinsic spectral signatures [1]. Therefore, the intrinsic spectral signatures of the same land-cover may be different in LR MSIs and HR HSIs on the Ivanpah Playa database. It is challenging to perform HR MSI and LR HSI fusion on the Ivanpah Playa database. On this database, similar to the literature [1], the spectral response function from calibration measurements (https://earth.esa.int/web/sentinel/user-guides/sentinel-2-msi/document-library/-/asset_publisher/Wk0TKajiISaR/content/sentinel-2aspectral-responses, accessed on 17 December 2020) is employed for the compared methods.

Experimental results of SSRN and the compared methods on the Ivanpah Playa database are reported in Table 8. Different from that on the PU, WDCM, and Paris databases, TLSR and DHSP perform better than traditional methods and USDN on the Ivanpah Playa database. CNMF, GSOMP, HySure, and USDN usually rely on the assumption that the intrinsic spectral signatures of the same land-cover in HR MSIs and LR HSIs are the same [28,36]. In these methods, the spectral response function is usually directly used to obtain the spectral ingredients of HR MSIs from the spectral ingredients of LR HSIs. However, this assumption is not satisfied on the Ivanpah Playa database, which results in the performance degradations of CNMF, GSOMP, HySure, and USDN.

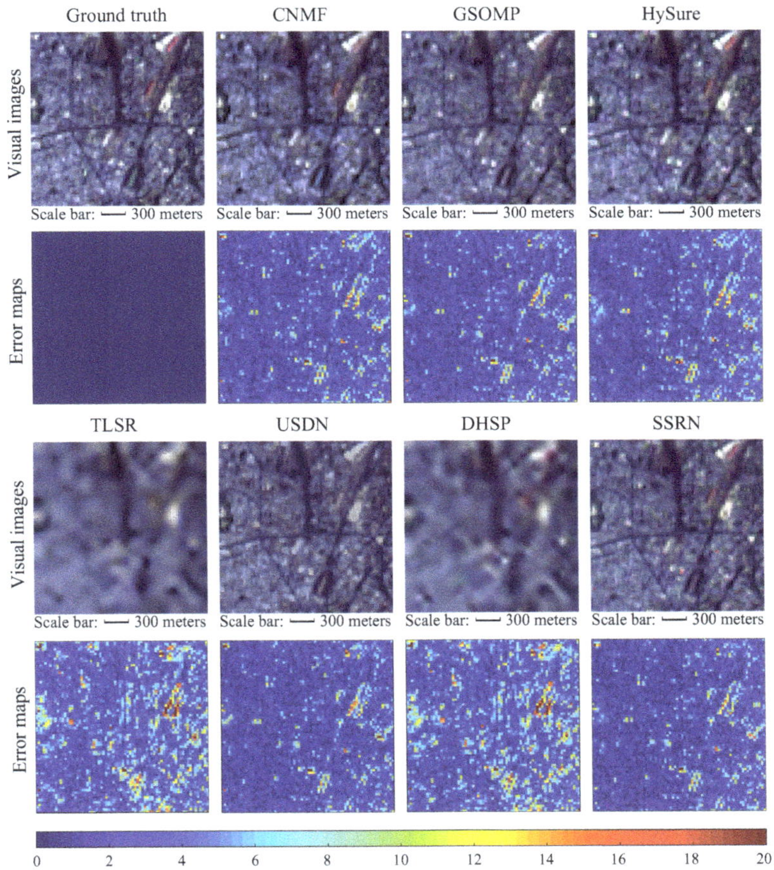

Figure 6. Visual images (R: 24, G: 14, and B: 3) and error maps of the proposed SSRN and the compared methods on the Paris database.

Different from CNMF, GSOMP, HySure, and USDN, the proposed SSRN does not rely on the assumption that the intrinsic spectral signatures of the same land-cover in HR MSIs and LR HSIs are the same. In the proposed SSRN, the fusion problem of the HR MSI and the LR HSI is considered a problem of spectral mapping learning. The proposed SSRN is utilized to directly learn spectral mapping from the multispectral pixels to the hyperspectral pixels. Owing to the powerful nonlinear representation ability of deep convolutional networks, SSRN can model the spectral variability increased by the seasonal changes between multispectral and hyperspectral pixels. In addition, due to HR MSI and LR HSI on the Ivanpah Playa database being collected at different time instants, the imaging environments (e.g., illumination, atmospheric, and weather) of HR MSIs and LR HSIs are different. Different imaging environments may result in it being difficult to accurately obtain real spectral response function [12]. In the proposed SSRN, only the loss function requires a spectral response function. To reduce the errors caused by the estimated spectral response function, the second term in the loss function Equation (9) and the fine-tuning strategy of SSRN are removed in the experiments on the Ivanpah Playa database. Data augmentation that is same as that on the Paris database is also employed on the Ivanpah Playa database to increase the number of training samples. As shown in Table 8, in terms of PSNR, UIQI, RMSE, and ERGAS, the proposed SSRN shows superior performance.

Table 8. Quantitative experimental results on the Ivanpah Playa database.

Methods	PSNR	UIQI	RMSE	ERGAS	SAM
CNMF [36]	23.399	0.721	15.600	2.395	1.456
GSOMP [30]	20.855	0.481	21.703	3.295	3.575
HySure [12]	21.658	0.531	19.126	2.939	2.221
TLSR [26]	23.702	0.786	15.149	2.330	1.440
USDN [28]	22.143	0.487	18.048	2.769	2.169
DHSP [27]	23.963	0.792	14.672	2.257	1.418
SSRN	27.770	0.807	9.447	1.451	1.451

Visual images and error maps are shown in Figure 7. The spatial structures on the Ivanpah Playa database are relatively smooth. Although TLSR and DHSP cause the high-frequency information of the reconstructed image to be blurred, the experimental results of TLSR and DHSP in the smooth land-cover regions are favorable. According to the visual image and the error map of SSRN, the proposed SSRN effectively preserves the spatial structures of the HR MSI in the estimated HSI.

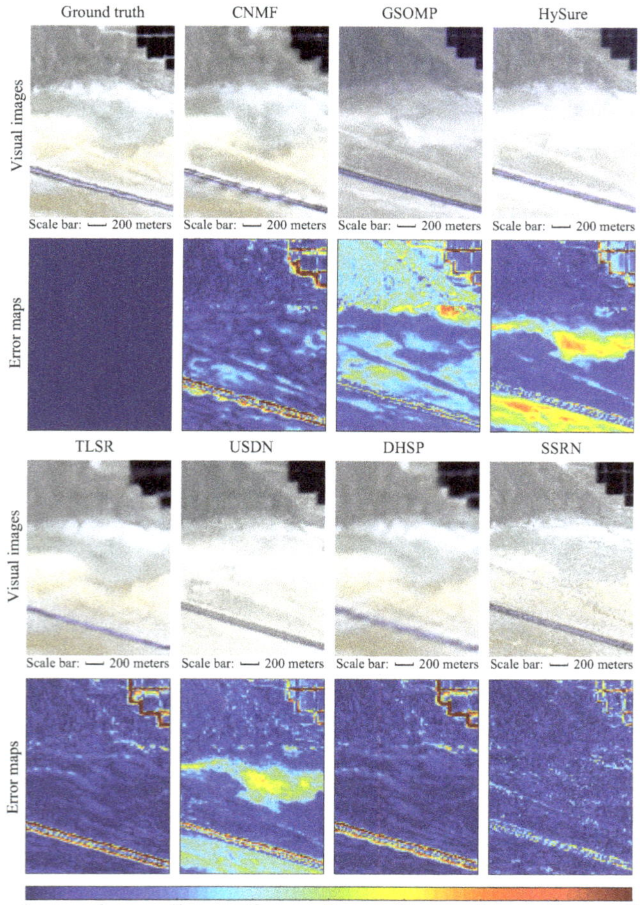

Figure 7. Visual images (R: 32, G: 20, and B: 8) and error maps of the proposed SSRN and the compared methods on the Ivanpah Playa database.

4.5. Time Cost

The total time cost of SSRN and the compared methods on the PU, WDCM, Paris, and Playa databases is shown in Table 9. In this paper, all experiments are conducted on the Ubuntu 14.04 system, 64G RAM, Intel Core i7-5930K, and NVIDIA TITAN X. CNMF, GSOMP, HySure, and TLSR are implemented with MATLAB. DHSP is implemented with PyTorch. USDN and the proposed SSRN are implemented with TensorFlow. The codes ofthe traditional methods (CNMF, GSOMP, and HySure) are implemented with the CPU. In the compared methods, the DL-based methods include TLSR, USDN, and DHSP. However, the code of TLSR provided by the original literature [26] is implemented with the CPU rather than the GPU. The codes of other deep learning-based methods (USDN, DHSP, and the proposed SSRN) are implemented with the GPU. As shown in Table 9, CNMF has superior computational efficiency. In general, DL-based methods usually take more time than traditional methods due to plenty of weight parameters. In the training process, the inputs of the proposed SSRN are image patches and the inputs of TLSR, USDN, and DHSP are entire images. Therefore, the proposed SSRN has less time cost than TLSR, USDN, and DHSP.

Table 9. Time cost of different methods on different databases (seconds).

Methods	CPU/GPU	PU	WDCM	Paris	Playa
CNMF [36]	CPU	12.37	14.26	1.56	3.64
GSOMP [30]	CPU	77.60	160.70	10.59	20.52
HySure [12]	CPU	40.96	58.73	6.85	26.21
TLSR [26]	CPU	1130.83	541.05	251.68	492.56
USDN [28]	GPU	782.65	198.68	151.53	134.43
DHSP [27]	GPU	796.92	1885.74	259.89	470.56
SSRN	GPU	74.09	117.21	82.88	181.46

5. Discussion

The performance of the proposed SSRN heavily depends on the learning of the spectral mapping. When the spectral information contained in MSI is too little, it becomea difficult to learn effective spectral mapping, which may weaken the performance of the proposed SSRN. For instance, RGB images (special MSIs), containing only three spectral bands, have little spectral information. Similar colors in RGB images may represent different objects. In other words, similar RGB image pixels may correspond to different HSI pixels, which makes it challenging to learn the spectral mapping between MSIs and HSIs. In this subsection, to explore the performance of SSRN when the MSI contains little spectral information, the CAVE database [55] is employed to conducted experiments. The average quantitative results are reported in Table 10. On the CAVE database, the MSI only contains three spectral bands, making it challenging to learn the spectral mapping between MSIs and HSIs. In terms of PSNR, UIQI, RMSE, and ERGAS, the performance of SSRN is weaker than that of CNMF and HySure. In terms of SAM, the proposed SSRN outperforms the compared methods.

Table 10. Quantitative experimental results on the CAVE database.

Methods	PSNR	UIQI	RMSE	ERGAS	SAM
CNMF [36]	42.403	0.845	2.273	2.337	6.629
GSOMP [30]	37.204	0.824	5.122	5.439	12.556
HySure [12]	41.331	0.814	2.130	2.530	6.645
TLSR [26]	34.148	0.744	5.206	5.879	6.221
USDN [28]	37.711	0.825	3.769	3.847	11.493
DHSP [27]	34.205	0.703	5.190	5.840	7.096
SSRN	40.558	0.816	2.520	3.031	5.523

6. Conclusions

In this paper, a spectral-spatial residual network is proposed to estimate HR HSI based on the observed HR MSI and LR HSI. Different from previous methods that focus on extracting spectral ingredients from LR HSI and extracting spatial ingredients from HR MSI, the proposed SSRN directly learns pixel-wise spectral mapping between MSIs and HSIs. In SSRN, a spectral module is proposed to extract spectral features from MSIs and a spatial module is proposed to explore the complementarity of homogeneous adjacent pixels to facilitate learning of spectral mapping. Finally, a self-supervised fine-tuning strategy is proposed to estimate the spectral mapping between HR MSIs and HR HSIs on the basis of the learned pixel-wise spectral mapping between LR MSIs and LR HSIs. Experiments on simulated and real databases show that SSRN can effectively reduce spatial and spectral distortions and can achieve superior performance. In the future, we will study more efficient deep networks for learning spectral mapping between MSIs and HSIs.

Author Contributions: Conceptualization, W.C. and X.Z.; methodology, W.C.; software, W.C.; validation, W.C., X.Z., and X.L.; formal analysis, X.Z.; investigation, W.C.; resources, X.Z.; data curation, W.C.; writing—original draft preparation, W.C.; writing—review and editing, X.Z.; visualization, X.Z.; supervision, X.L.; project administration, X.L.; funding acquisition, X.L. All authors have read and agreed to the published version of the manuscript.

Funding: This work was funded in part by the National Science Fund for Distinguished Young Scholars under grant 61925112, in part by the National Natural Science Foundation of China under grant 61806193 and grant 61772510, in part by the Innovation Capability Support Program of Shaanxi under grant 2020KJXX-091 and grant 2020TD-015, and in part by the Natural Science Basic Research Program of Shaanxi under grants 2019JQ-340 and 2019JC-23.

Institutional Review Board Statement: Not applicable.

Informed Consent Statement: Not applicable.

Data Availability Statement: Not applicable.

Acknowledgments: The authors would like to thank the editors and reviewers for their insightful suggestions.

Conflicts of Interest: The authors declare no conflicts of interest.

References

1. Borsoi, R.A.; Imbiriba, T.; Bermudez, J.C.M. Super-Resolution for Hyperspectral and Multispectral Image Fusion Accounting for Seasonal Spectral Variability. *IEEE Trans. Image Process.* **2020**, *29*, 116–127. [CrossRef] [PubMed]
2. Sun, H.; Zheng, X.; Lu, X.; Wu, S. Spectral-Spatial Attention Network for Hyperspectral Image Classification. *IEEE Trans. Geosci. Remote Sens.* **2020**, *58*, 3232–3245. [CrossRef]
3. Zhao, X.; Li, W.; Zhang, M.; Tao, R.; Ma, P. Adaptive Iterated Shrinkage Thresholding-Based Lp-Norm Sparse Representation for Hyperspectral Imagery Target Detection. *Remote Sens.* **2020**, *12*, 3991. [CrossRef]
4. Ren, X.; Lu, L.; Chanussot, J. Toward Super-Resolution Image Construction Based on Joint Tensor Decomposition. *Remote Sens.* **2020**, *12*, 2535. [CrossRef]
5. Zhang, K.; Wang, M.; Yang, S. Multispectral and Hyperspectral Image Fusion Based on Group Spectral Embedding and Low-Rank Factorization. *IEEE Trans. Geosci. Remote Sens.* **2017**, *55*, 1363–1371. [CrossRef]
6. Chen, W.; Lu, X. Unregistered Hyperspectral and Multispectral Image Fusion with Synchronous Nonnegative Matrix Factorization. In *Chinese Conference on Pattern Recognition and Computer Vision, Proceedings of the Third Chinese Conference, PRCV 2020, Nanjing, China, 16–18 October 2020*; Springer: Berlin/Heidelberg, Germany, 2020; pp. 602–614.
7. Liu, W.; Lee, J. An Efficient Residual Learning Neural Network for Hyperspectral Image Superresolution. *IEEE J. Sel. Top. Appl. Earth Obs. Remote Sens.* **2019**, *12*, 1240–1253. [CrossRef]
8. Loncan, L.; de Almeida, L.B.; Bioucas-Dias, J.M.; Briottet, X.; Chanussot, J.; Dobigeon, N.; Fabre, S.; Liao, W.; Licciardi, G.A.; Simões, M.; et al. Hyperspectral Pansharpening: A Review. *IEEE Geosci. Remote Sens. Mag.* **2015**, *3*, 27–46. [CrossRef]
9. Yokoya, N.; Grohnfeldt, C.; Chanussot, J. Hyperspectral and Multispectral Data Fusion: A comparative review of the recent literature. *IEEE Geosci. Remote Sens. Mag.* **2017**, *5*, 29–56. [CrossRef]
10. Han, X.; Shi, B.; Zheng, Y. Self-Similarity Constrained Sparse Representation for Hyperspectral Image Super-Resolution. *IEEE Trans. Image Process.* **2018**, *27*, 5625–5637. [CrossRef]

11. Feng, X.; He, L.; Cheng, Q.; Long, X.; Yuan, Y. Hyperspectral and Multispectral Remote Sensing Image Fusion Based on Endmember Spatial Information. *Remote Sens.* **2020**, *12*, 1009. [CrossRef]
12. Simões, M.; Bioucas-Dias, J.; Almeida, L.B.; Chanussot, J. A Convex Formulation for Hyperspectral Image Superresolution via Subspace-Based Regularization. *IEEE Trans. Geosci. Remote Sens.* **2015**, *53*, 3373–3388. [CrossRef]
13. Wei, Q.; Dobigeon, N.; Tourneret, J. Fast Fusion of Multi-Band Images Based on Solving a Sylvester Equation. *IEEE Trans. Image Process.* **2015**, *24*, 4109–4121. [CrossRef]
14. Zhou, Y.; Feng, L.; Hou, C.; Kung, S. Hyperspectral and Multispectral Image Fusion Based on Local Low Rank and Coupled Spectral Unmixing. *IEEE Trans. Geosci. Remote Sens.* **2017**, *55*, 5997–6009. [CrossRef]
15. Zhang, K.; Wang, M.; Yang, S.; Jiao, L. Spatial-Spectral-Graph-Regularized Low-Rank Tensor Decomposition for Multispectral and Hyperspectral Image Fusion. *IEEE J. Sel. Top. Appl. Earth Obs. Remote Sens.* **2018**, *11*, 1030–1040. [CrossRef]
16. Dian, R.; Li, S.; Fang, L.; Lu, T.; Bioucas-Dias, J.M. Nonlocal Sparse Tensor Factorization for Semiblind Hyperspectral and Multispectral Image Fusion. *IEEE Trans. Cybern.* **2020**, *50*, 4469–4480. [CrossRef]
17. Ma, J.; Yu, W.; Chen, C.; Liang, P.; Guo, X.; Jiang, J. Pan-GAN: An unsupervised pan-sharpening method for remote sensing image fusion. *Inf. Fusion* **2020**, *62*, 110–120. [CrossRef]
18. Zhang, L.; Nie, J.; Wei, W.; Zhang, Y.; Liao, S.; Shao, L. Unsupervised Adaptation Learning for Hyperspectral Imagery Super-Resolution. In Proceedings of the IEEE/CVF Conference on Computer Vision and Pattern Recognition (CVPR), Seattle, WA, USA, 14–19 June 2020; pp. 3070–3079. [CrossRef]
19. Zhang, X.; Huang, W.; Wang, Q.; Li, X. SSR-NET: Spatial-Spectral Reconstruction Network for Hyperspectral and Multispectral Image Fusion. *IEEE Trans. Geosci. Remote Sens.* **2020**, 1–13. [CrossRef]
20. Xie, Q.; Zhou, M.; Zhao, Q.; Meng, D.; Zuo, W.; Xu, Z. Multispectral and Hyperspectral Image Fusion by MS/HS Fusion Net. In Proceedings of the IEEE/CVF Conference on Computer Vision and Pattern Recognition (CVPR), Long Beach, CA, USA, 16–20 June 2019; pp. 1585–1594. [CrossRef]
21. Dian, R.; Li, S.; Guo, A.; Fang, L. Deep Hyperspectral Image Sharpening. *IEEE Trans. Neural Netw. Learn. Syst.* **2018**, *29*, 5345–5355. [CrossRef]
22. Xie, W.; Jia, X.; Li, Y.; Lei, J. Hyperspectral Image Super-Resolution Using Deep Feature Matrix Factorization. *IEEE Trans. Geosci. Remote Sens.* **2019**, *57*, 6055–6067. [CrossRef]
23. Wei, Y.; Yuan, Q.; Shen, H.; Zhang, L. Boosting the Accuracy of Multispectral Image Pansharpening by Learning a Deep Residual Network. *IEEE Geosci. Remote Sens. Lett.* **2017**, *14*, 1795–1799. [CrossRef]
24. Wang, W.; Zeng, W.; Huang, Y.; Ding, X.; Paisley, J. Deep Blind Hyperspectral Image Fusion. In Proceedings of the IEEE/CVF International Conference on Computer Vision (ICCV), Long Beach, CA, USA, 16–20 June 2019; pp. 4149–4158. [CrossRef]
25. Li, K.; Xie, W.; Du, Q.; Li, Y. DDLPS: Detail-Based Deep Laplacian Pansharpening for Hyperspectral Imagery. *IEEE Trans. Geosci. Remote Sens.* **2019**, *57*, 8011–8025. [CrossRef]
26. Yuan, Y.; Zheng, X.; Lu, X. Hyperspectral Image Superresolution by Transfer Learning. *IEEE J. Sel. Top. Appl. Earth Obs. Remote Sens.* **2017**, *10*, 1963–1974. [CrossRef]
27. Sidorov, O.; Hardeberg, J.Y. Deep Hyperspectral Prior: Single-Image Denoising, Inpainting, Super-Resolution. In Proceedings of the IEEE/CVF International Conference on Computer Vision (ICCV), Long Beach, CA, USA, 16–20 June 2019; pp. 3844–3851. [CrossRef]
28. Qu, Y.; Qi, H.; Kwan, C. Unsupervised Sparse Dirichlet-Net for Hyperspectral Image Super-Resolution. In Proceedings of the IEEE/CVF Conference on Computer Vision and Pattern Recognition, Salt Lake City, UT, USA, 18–23 June 2018; pp. 2511–2520. [CrossRef]
29. Fang, L.; Zhuo, H.; Li, S. Super-resolution of hyperspectral image via superpixel-based sparse representation. *Neurocomputing* **2018**, *273*, 171–177. [CrossRef]
30. Akhtar, N.; Shafait, F.; Mian, A. Sparse Spatio-spectral Representation for Hyperspectral Image Super-resolution. In *European Conference on Computer Vision, Proceedings of the Computer Vision—ECCV 2014, Zurich, Switzerland, 6–12 September 2014*; Springer International Publishing: Cham, Switzerland, 2014; pp. 63–78.
31. Wei, Q.; Bioucas-Dias, J.; Dobigeon, N.; Tourneret, J. Hyperspectral and Multispectral Image Fusion Based on a Sparse Representation. *IEEE Trans. Geosci. Remote Sens.* **2015**, *53*, 3658–3668. [CrossRef]
32. Wei, Q.; Dobigeon, N.; Tourneret, J. Bayesian fusion of hyperspectral and multispectral images. In Proceedings of the IEEE International Conference on Acoustics, Speech and Signal Processing (ICASSP), Florence, Italy, 4–9 May 2014; pp. 3176–3180. [CrossRef]
33. Eismann, M.T.; Hardie, R.C. Hyperspectral resolution enhancement using high-resolution multispectral imagery with arbitrary response functions. *IEEE Trans. Geosci. Remote Sens.* **2005**, *43*, 455–465. [CrossRef]
34. Wei, Q.; Dobigeon, N.; Tourneret, J. Bayesian Fusion of Multi-Band Images. *IEEE J. Sel. Top. Signal Process.* **2015**, *9*, 1117–1127. [CrossRef]
35. Irmak, H.; Akar, G.B.; Yuksel, S.E. A MAP-Based Approach for Hyperspectral Imagery Super-Resolution. *IEEE Trans. Image Process.* **2018**, *27*, 2942–2951. [CrossRef] [PubMed]
36. Yokoya, N.; Yairi, T.; Iwasaki, A. Coupled Nonnegative Matrix Factorization Unmixing for Hyperspectral and Multispectral Data Fusion. *IEEE Trans. Geosci. Remote Sens.* **2012**, *50*, 528–537. [CrossRef]

37. Lin, C.; Ma, F.; Chi, C.; Hsieh, C. A Convex Optimization-Based Coupled Nonnegative Matrix Factorization Algorithm for Hyperspectral and Multispectral Data Fusion. *IEEE Trans. Geosci. Remote Sens.* **2018**, *56*, 1652–1667. [CrossRef]
38. Li, S.; Dian, R.; Fang, L.; Bioucas-Dias, J.M. Fusing Hyperspectral and Multispectral Images via Coupled Sparse Tensor Factorization. *IEEE Trans. Image Process.* **2018**, *27*, 4118–4130. [CrossRef] [PubMed]
39. Li, J.; Cui, R.; Li, B.; Song, R.; Li, Y.; Du, Q. Hyperspectral Image Super-Resolution with 1D-2D Attentional Convolutional Neural Network. *Remote Sens.* **2019**, *11*. [CrossRef]
40. Fu, Y.; Zhang, T.; Zheng, Y.; Zhang, D.; Huang, H. Hyperspectral Image Super-Resolution with Optimized RGB Guidance. In Proceedings of the IEEE/CVF Conference on Computer Vision and Pattern Recognition (CVPR), Long Beach, CA, USA, 16–20 June 2019; pp. 11653–11662. [CrossRef]
41. Wang, Z.; Chen, B.; Lu, R.; Zhang, H.; Liu, H.; Varshney, P.K. FusionNet: An Unsupervised Convolutional Variational Network for Hyperspectral and Multispectral Image Fusion. *IEEE Trans. Image Process.* **2020**, *29*, 7565–7577. [CrossRef]
42. Huang, W.; Xiao, L.; Wei, Z.; Liu, H.; Tang, S. A New Pan-Sharpening Method with Deep Neural Networks. *IEEE Geosci. Remote Sens. Lett.* **2015**, *12*, 1037–1041. [CrossRef]
43. Li, J.; Wu, C.; Song, R.; Xie, W.; Ge, C.; Li, B.; Li, Y. Hybrid 2-D-3-D Deep Residual Attentional Network with Structure Tensor Constraints for Spectral Super-Resolution of RGB Images. *IEEE Trans. Geosci. Remote Sens.* **2020**, 1–15. [CrossRef]
44. Fu, Y.; Zhang, T.; Zheng, Y.; Zhang, D.; Huang, H. Joint Camera Spectral Response Selection and Hyperspectral Image Recovery. *IEEE Trans. Pattern Anal. Mach. Intell.* **2020**. [CrossRef]
45. Akhtar, N.; Mian, A. Hyperspectral Recovery from RGB Images using Gaussian Processes. *IEEE Trans. Pattern Anal. Mach. Intell.* **2020**, *42*, 100–113. [CrossRef]
46. Yan, L.; Wang, X.; Zhao, M.; Kaloorazi, M.; Chen, J.; Rahardja, S. Reconstruction of Hyperspectral Data from RGB Images with Prior Category Information. *IEEE Trans. Comput. Imaging* **2020**, *6*, 1070–1081. [CrossRef]
47. He, K.; Zhang, X.; Ren, S.; Sun, J. Deep Residual Learning for Image Recognition. In Proceedings of the IEEE Conference on Computer Vision and Pattern Recognition (CVPR), Las Vegas, NV, USA, 27–30 June 2016; pp. 770–778. [CrossRef]
48. Song, W.; Li, S.; Fang, L.; Lu, T. Hyperspectral Image Classification with Deep Feature Fusion Network. *IEEE Trans. Geosci. Remote Sens.* **2018**, *56*, 3173–3184. [CrossRef]
49. Sun, H.; Li, S.; Zheng, X.; Lu, X. Remote Sensing Scene Classification by Gated Bidirectional Network. *IEEE Trans. Geosci. Remote Sens.* **2020**, *58*, 82–96. [CrossRef]
50. Sun, H.; Zheng, X.; Lu, X. A Supervised Segmentation Network for Hyperspectral Image Classification. *IEEE Trans. Image Process.* **2021**, *30*, 2810–2825. [CrossRef]
51. Lu, X.; Dong, L.; Yuan, Y. Subspace Clustering Constrained Sparse NMF for Hyperspectral Unmixing. *IEEE Trans. Geosci. Remote Sens.* **2020**, *58*, 3007–3019. [CrossRef]
52. Du, X.; Zheng, X.; Lu, X.; Doudkin, A.A. Multisource Remote Sensing Data Classification with Graph Fusion Network. *IEEE Trans. Geosci. Remote Sens.* **2021**, 1–11. [CrossRef]
53. Dian, R.; Li, S. Hyperspectral Image Super-Resolution via Subspace-Based Low Tensor Multi-Rank Regularization. *IEEE Trans. Image Process.* **2019**, *28*, 5135–5146. [CrossRef] [PubMed]
54. Wang, X.; Girshick, R.; Gupta, A.; He, K. Non-local Neural Networks. In Proceedings of the IEEE/CVF Conference on Computer Vision and Pattern Recognition, Salt Lake City, UT, USA, 18–23 June 2018; pp. 7794–7803. [CrossRef]
55. Yasuma, F.; Mitsunaga, T.; Iso, D.; Nayar, S.K. Generalized Assorted Pixel Camera: Postcapture Control of Resolution, Dynamic Range, and Spectrum. *IEEE Trans. Image Process.* **2010**, *19*, 2241–2253. [CrossRef]
56. Akhtar, N.; Shafait, F.; Mian, A. Bayesian sparse representation for hyperspectral image super resolution. In Proceedings of the IEEE Conference on Computer Vision and Pattern Recognition (CVPR), Boston, MA, USA, 7–12 June 2015; pp. 3631–3640. [CrossRef]

remote sensing MDPI

Article

A Spatial-Enhanced LSE-SFIM Algorithm for Hyperspectral and Multispectral Images Fusion

Yulei Wang [1,2], Qingyu Zhu [1], Yao Shi [1], Meiping Song [1,*] and Chunyan Yu [1]

1 Center of Hyperspectral Imaging in Remote Sensing, Information Science and Technology College, Dalian Maritime University, Dalian 116026, China; wangyulei@dlmu.edu.cn (Y.W.); zhuqingyu@dlmu.edu.cn (Q.Z.); 1120180233@dlmu.edu.cn (Y.S.); yucy@dlmu.edu.cn (C.Y.)
2 State Key Laboratory of Integrated Services Networks, Xidian University, Xi'an 710000, China
* Correspondence: smping@dlmu.edu.cn

Abstract: The fusion of a hyperspectral image (HSI) and multispectral image (MSI) can significantly improve the ability of ground target recognition and identification. The quality of spatial information and the fidelity of spectral information are normally contradictory. However, these two properties are non-negligible indicators for multi-source remote-sensing images fusion. The smoothing filter-based intensity modulation (SFIM) method is a simple yet effective model for image fusion, which can improve the spatial texture details of the image well, and maintain the spectral characteristics of the image significantly. However, traditional SFIM has a poor effect for edge information sharpening, leading to a bad overall fusion result. In order to obtain better spatial information, a spatial filter-based improved LSE-SFIM algorithm is proposed in this paper. Firstly, the least square estimation (LSE) algorithm is combined with SFIM, which can effectively improve the spatial information quality of the fused image. At the same time, in order to better maintain the spatial information, four spatial filters (mean, median, nearest and bilinear) are used for the simulated MSI image to extract fine spatial information. Six quality indexes are used to compare the performance of different algorithms, and the experimental results demonstrate that the LSE-SFIM based on bilinear (LES-SFIM-B) performs significantly better than the traditional SFIM algorithm and other spatially enhanced LSE-SFIM algorithms proposed in this paper. Furthermore, LSE-SFIM-B could also obtain similar performance compared with three state-of-the-art HSI-MSI fusion algorithms (CNMF, HySure, and FUSE), while the computing time is much shorter.

Keywords: hyperspectral image; multi-source image fusion; SFIM; least square estimation; spatial filter

Citation: Wang, Y.; Zhu, Q.; Shi, Y.; Song, M.; Yu, C. A Spatial-Enhanced LSE-SFIM Algorithm for Hyperspectral and Multispectral Images Fusion. *Remote Sens.* **2021**, *13*, 4967. https://doi.org/10.3390/rs13244967

Academic Editors: Chein-I Chang, Haoyang Yu, Jiaojiao Yu, Lin Wang, Hsiao-Ch Li and Xiaorun Li

Received: 14 October 2021
Accepted: 3 December 2021
Published: 7 December 2021

1. Introduction

In recent years, a large number of remote-sensing satellites have been launched continuously with the development of Earth observation technology [1,2]. Modern remote-sensing technology has reached a new developmental stage of multi-platform, multi-sensor, and multi-angle observation [3–5]. The continuous development of remote-sensing applications such as geological exploration [6], resource and environmental investigation [7–9], agricultural monitoring [10–12], urban planning [13–16], etc., has greatly promoted the demand for remote-sensing data and the improvement of the performance of satellite sensors. However, due to the limitations of optical diffraction, modulation transfer function, signal-to-noise ratio, and the sensor hardware conditions, a single sensor normally cannot obtain data with both high-spatial and high-spectral resolutions at the same time. Multi-sensor data fusion has arisen at an historic moment, which can effectively explore the complementary information from multi-platform observations, making land surface monitoring more accurate and comprehensive. Multi-source remote-sensing data fusion refers to the processing of multi-source data with complementary information in time or space according to certain rules, so as to obtain a more accurate and informative composite images than any single data source.

A variety of multi-source remote-sensing fusion techniques have been developed in the last decade to enhance the spatial resolution of hyperspectral images and obtain information-rich HSI data with both high-spectral and high-spatial resolutions. The HSI-MSI fusion algorithms can be divided into the following four categories: component substitution (CS), multi-resolution analysis (MRA), spectral unmixing (SU), and Bayesian representation (BR). The idea of CS-based fusion algorithms is straightforward, it transforms the original HSI data, replaces the spatial information in low-spatial HSI data set with the spatial information in the high-spatial MSI data set, and finally inverts the reconstructed data to obtain the fused hyperspectral image. The typical CS-based methods are proposed toward generalizing existing pansharpening methods for HSI-MSI fusion, including the IHS [17] transform method proposed by W. J. Carper in 1990, PCA [18] transform proposed by P. S. Chavez in 1991, Gram–Schmidt (GS) [19] transform proposed by B. Aiazzi in 2007, and their variants [20–22], etc. These methods are simple and easy to implement, but have serious spectral distortion and cannot be used well for the fusion of hyperspectral images. The MRA-based methods obtain the fusion result by filtering the high-resolution image and adding the high-frequency detailed information to the hyperspectral image. The earliest MRA-based methods realized multi-scale image decomposition through pyramid transform, while the most representative and most widely used multi-resolution analysis methods include the fusion method based on various wavelet transforms [23], the smoothing filter-based intensity modulation (SFIM) [24] proposed by Liu and the generalized Laplacian pyramid (GLP) method proposed by Aiazzi [25]. The advantages of these MRA-based methods are less spectral distortion and anti-aliasing, but the algorithm is complex and the spatial feature loss often occurs in the fusion results. The SU-based methods utilize the hyperspectral linear unmixing model and apply it to the fusion optimization model. The advantages of these methods are less spatial and spectral information loss, but they always have higher computational complexity. Typical methods include the coupled non-negative matrix factorization (CNMF) [26] method proposed by Yokoya in 2012, the subspace-based regularization (HySure) [27] method proposed by Simoes in 2015, and the coupled sparse tensor factorization (CSTF) [28] method proposed by Li in 2018. The BR-based methods transform the problem of high-resolution image and hyperspectral image fusion into the problem of solving the Bayesian optimization model, and obtain the fusion result through solving the optimization. Typical BR-based methods include a maximum posteriori-stochastic mixing model (MAP-SMM) [29] proposed by Eismann in 2004, Bayesian sparse method [30] proposed by Wei in 2015, and fast fusion based on the Sylvester equation (FUSE) [31] method proposed by Wei in 2015. The BR-based methods have the advantage of less spatial and spectral information loss, but also result in a disadvantage of high computational complexity.

Recently, an increasing number of HSI-MSI fusion algorithms have been proposed [32–34]. These algorithms have been proved to be effective with good fusion performance. However, most of the researchers focus too much on performance improvements using modern technologies such as sparse representation, deep learning processing, etc., ignoring the computing time. In other words, these algorithms improve the fusion performance at the cost of increased computational complexity. As one of the effective MRA-based fusion methods, SFIM is proposed by Liu [24] for image fusion as mentioned above. Compared with traditional methods, SFIM is simple to calculate, easy to implement, and the spectral information is normally retained well, but there are problems such as fuzzy edge information of the image and insufficient improvement of detailed spatial information. In recent years, many improved SFIM algorithms have been studied, most of which focused on how to obtain simulated multispectral images with spatial information characteristics consistent with hyperspectral images and spectral features consistent with multispectral images. This paper combines the least square estimation LSE algorithm with SFIM, which can effectively improve the spatial information quality of the fused image. This paper also compares several spatial filters to extract spatial information to enhance the simulated MSI's boundary spatial information, and proposes an improved LSE-SFIM fusion algorithm based on spatial information promotion to obtain an optimal fusion result. Experimental results on three HSI-MSI data sets show the effectiveness of the proposed algorithm using six image quality indexes.

The remainder of this article is organized as follows. Section 2 gives a detailed description of the proposed method. In Section 3, experimental results and analysis of different data sets are presented. Finally, conclusions are drawn in Section 4.

2. Proposed Method

2.1. Basic Smoothing Filter-Based Intensity Modulation (SFIM) Algorithm

The SFIM algorithm was proposed by Liu for image fusion in 2000, which is based on the simplified solar radiation and surface reflection model. Even though it was proposed some time ago, this algorithm is still in use due to its simplicity with good spectral preservation. The basic principle of the traditional SFIM is expressed as follows [24]:

$$DN_{SFIM} = \frac{DN_{low} DN_{high}}{MeanDN_{high}} \quad (1)$$

where DN_{low}, DN_{high} represent the gray values of low-resolution and high-resolution images respectively, $MeanDN_{high}$ represents the simulated low-resolution image obtained by the local mean value of DN_{high}.

2.2. The Proposed Spatial Filter-Based Least Square Estimation (LSE)-SFIM

For HSI-MSI image fusion, the formula (1) can be expressed as:

$$\text{Fusion} = \frac{HSI' \times MSI}{MSI''} \quad (2)$$

where HSI' is the up-sampling of original low-resolution hyperspectral data HSI, MSI represents the original high-resolution multispectral data MSI, and MSI'' is the up-sampling of MSI' where MSI' represents the simulated low-resolution image obtained by MSI. The algorithm performance is influenced by two factors: (1) how to obtain the simulated low-resolution image MSI'; and (2) how to get the up-sampled HIS' and MS''. The traditional SFIM uses mean filter to obtain the simulated low-resolution MSI' (down-sampling) and uses the same filter to obtain up-sampled HSI' and MSI''. The edge information is lost by the mean filters, and this paper takes two steps to solve the problem: (1) least squares estimation (LSE) is used to adjust the coefficient and obtain MSI' with as similar spatial information as the original HSI image, with the details discussed in Section 2.2.1; (2) filtering and interpolation methods are compared in the up-sampling stage to obtain the best enhanced spatial information, with the details discussed in Section 2.2.2. A bilinear approach proved to be the best in the experiments for this paper. The flow chart of the proposed algorithm is shown in Figure 1.

In order to make it clear how we obtain MSI, MSI', MSI'', HSI, and HSI', Figure 2 gives a graphic abstract with detailed steps of the proposed algorithm. It can easily be seen from Figure 2 that, MSI is the original high-spatial multispectral data set, MSI' is the down-sampling of MSI where the spatial size can be shrunk into the same as the original HIS (LSE is used here to adjust MSI' for preserving better spatial information), and MSI'' is the up-sampling of MSI' with the same size as the original high-spatial resolution MSI. HSI is the original low-spatial resolution hyperspectral data set, and HSI' is the up-sampling of HSI to the same size as the high-spatial resolution.

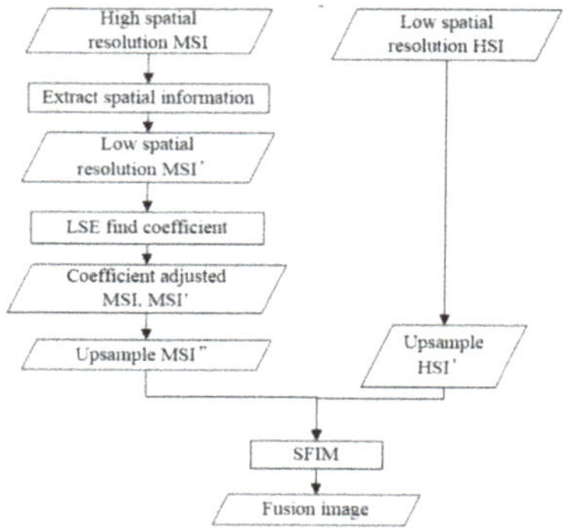

Figure 1. Flowchart of the proposed fusion algorithm.

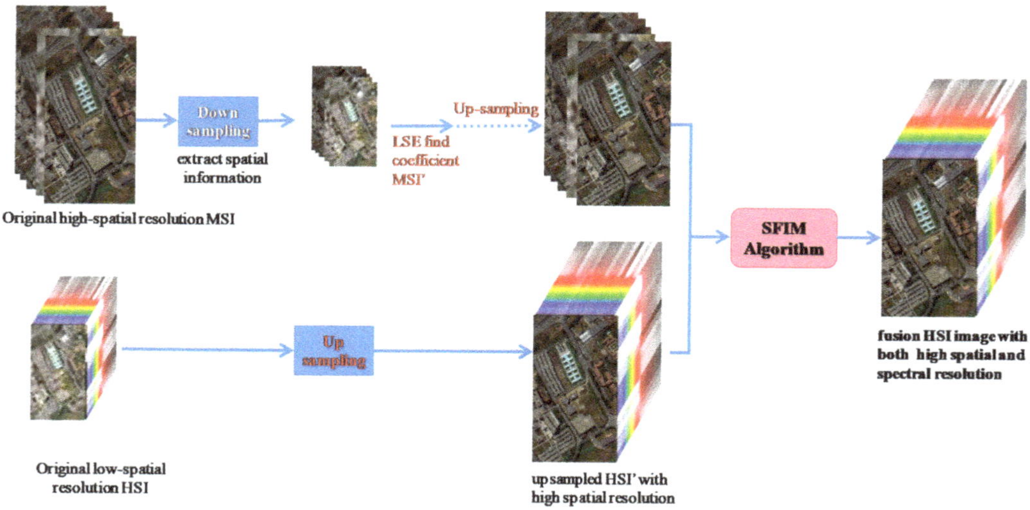

Figure 2. Graphic abstract with detailed steps of the proposed algorithm.

2.2.1. Least Square Estimation Based SFIM Algorithm (LSE-SFIM)

Assuming that there is an ideal simulated multispectral image MSI′, it should have two characteristics: (1) the spatial information characteristics are consistent with the original hyperspectral image, which can ensure that the spatial information of the hyperspectral image is counteracted; (2) the spectral information characteristics should be consistent with the original multispectral image, which ensures that the spectral characteristics of the multispectral image can be counteracted. The least square estimation algorithm solves these two problems well.

The LSE algorithm finds the best matching function by minimizing the sum of squares of errors. It is often used to solve linear regression coefficients in the processing of remote-sensing images. The LSE-SFIM algorithm uses LSE to solve the linear regression coefficient that can minimize the spatial information error between the hyperspectral image and the simulated multispectral image MSI′, so that the latter can have as much spatial information as possible as with the hyperspectral image.

The LSE-SFIM fusion algorithm first down-samples and extracts the spatial information of the high-resolution multispectral image MSI to obtain the simulated MSI′, and then uses the least square estimation algorithm to solve the problem that can minimize the linear regression coefficient of the spatial information error between the multispectral image MSI′ and the hyperspectral image, and use this linear regression coefficient to update the MSI and MSI′, so as to ensure that the spectral information of the MSI is close to the MSI′, and it can ensure that both MSI and MSI′ can have the same spatial information as HSI as possible. Finally, the HSI′ and MSI″ are obtained by up-sampling, which will be introduced in the next section, and fused separately according to the bands to obtain the fused image.

2.2.2. Spatial Information Enhanced LSE-SFIM

When using the LSE-SFIM fusion image, the most critical step concerns how to obtain the simulated multispectral image MSI″ whose spatial information features are consistent with the hyperspectral image and spectral features are consistent with the multispectral image, so as to effectively improve the spatial resolution of the hyperspectral image and achieve the purpose of fusion. This step is the up-sampling process to obtain MSI″ and HSI′ in Figure 2. This paper compares several methods of extracting boundary spatial information from low-spatial resolution multispectral images, and obtains the best fusion result.

Filtering Method

Mean filtering and median filtering are two commonly used filtering methods. The main idea of mean filtering is to replace the gray value of the central pixel with the mean value of the gray value of the pixel to be found and the surrounding pixels in the middle, so as to achieve the purpose of filtering. Mean filtering can be simplified as Equation (3):

$$g(x,y) = \frac{1}{M}\sum_{f\in W} f(x,y) \tag{3}$$

where M is the filtering window size (pixels within the current window), and W is the current window.

Median filtering, as the name implies, is to replace the value of the pixel with the median value of the gray-scale values in the neighborhood window of a pixel. Median filtering can be simplified as Equation (4):

$$g(x,y) = \text{med}\{f(x-k, y-l), (k,l \in W)\} \tag{4}$$

Taking 3×3 window size as the example, mean and median filtering are shown in Figure 3, where (a) represents gray values before filtering, (b) represents gray values after mean filtering, and (c) represents gray values after median filtering.

1	2	1	4	3
1	2	2	4	3
5	7	6	8	5
4	5	4	7	6
5	9	9	8	1

(**a**)

1	2	1	4	3
1	3	4	4	3
5	4	5	5	5
4	6	7	6	6
5	9	9	8	1

(**b**)

1	2	1	4	3
1	2	4	4	3
5	4	5	5	5
4	5	7	6	6
5	9	9	8	1

(**c**)

Figure 3. 3×3 mean and median filtering, (**a**) gray values before filtering in green color, (**b**) gray values after mean filtering in red color, and (**c**) gray values after median filtering in purple color.

Interpolation Method

Image interpolation algorithm is a basic technology in image processing. Nearest neighbor interpolation and bilinear interpolation are two commonly used image interpolation algorithms. The nearest neighbor interpolation algorithm has the least amount of calculation and the simplest principle. The gray value of the nearest point among the neighboring pixels around the point to be sampled is used as the gray value of the point. There is a linear relationship between the pixel values of different points in the image. According to this idea, the bilinear interpolation algorithm considers the pixel values in the horizontal and vertical directions at the same time, so that the problem of grayscale discontinuity in the image is improved, and the overall effect of the image can also be improved.

3. Experimental Results and Analysis

In order to verify the effectiveness of the algorithm proposed in this paper, three sets of simulation experiment data sources are selected, namely Pavia University, Chikusei and HyMap Rodalquilar. This paper uses MATLAB R2018b software platform to program on Windows 10 64-bit system, and the processor is Intel (R) Core (TM) i5-8250U, 8G memory.

3.1. Hyperspectral Datasets

In order to evaluate the performance of the fusion method objectively and quantitatively, we use low-spatial resolution hyperspectral images obtained from real data resampling in the spatial domains, and high-spatial resolution multispectral images obtained in the spectral domains to carry out simulation data experiments. Table 1 shows the parameters of the three datasets used in this paper for verification experiments.

Table 1. Parameters of three hyperspectral datasets.

Dataset	Year	Original Sensor	Spectral Range (μm)	Spatial Resolution (m)	Bands
Pavia University	2003	ROSIS-3	0.43–0.84	1.3	103
Chikusei	2014	Hyperspec	0.36–1.02	2.5	128
HyMap Rodalquilar	2003	HyMap	0.4–2.5	10	167

3.1.1. Pavia University

Pavia University acquired a ROSIS sensor in 2001. The image size is 610 × 340 with a spatial resolution of 1.3 m per pixel, and the experimental data selected in this section contains 560 × 320 pixels. Its spectral range is 0.43–0.86 μm with a total of 115 bands, and 103 bands are remaining used after removing 12 noise bands. The low-spatial resolution HSI image was obtained from the original HSI data through the isotropic Gaussian point spread function down-sampling eight times with a total of 103 bands and 70 × 40 pixels, the pseudo-color image is shown in Figure 4a. The high-spatial resolution MSI image data was synthesized from the original HSI data according to the SRF down-sampling of the ROSIS sensor. There were four bands in total with an image size of 560 × 320, as shown in Figure 4b. The reference image of the original HIS is shown in Figure 4c.

(a) HSI (b) MSI (c) REF

Figure 4. Pavia University datasets with (**a**) low spatial resolution hyperspectral image (HSI), (**b**) high spatial resolution multispectral image (MSI) and (**c**) high spatial resolution HSI as the reference.

3.1.2. Chikusei

Chikusei dataset was collected by a Headwall Hyperspec-VNIR-C sensor on 29 July 2014 in Chikusei City, Japan. It was then produced and published by Naoto Yokoya and Akira Iwasaki of the University of Tokyo [26]. The spatial resolution is 2.5 m and the scene is 2517 × 2335. It consists of several pixels, mainly including agricultural and urban areas. In the experiment, a size of 540 × 420 pixels image is selected for experiments. The data spectrum range is 0.36–1.02 μm, including 128 bands in total. Among them, the low-spatial resolution HSI image was obtained from the original HSI data through the isotropic Gaussian point spread function down-sampling six times, with a total of 128 bands and 90 × 70 pixels, the pseudo-color image is shown in Figure 5a. The high-spatial resolution MSI data were synthesized from the original HSI data according to the SRF of the WV-2 sensor, with eight bands and 540 × 420 pixels, and the pseudo-color image of MSI was shown in Figure 5b. The reference image of the original HSI is shown in Figure 5c.

(**a**) HSI (**b**) MSI (**c**) REF

Figure 5. Chikusei datasets with (**a**) low-spatial resolution HSI, (**b**) high-spatial resolution MSI and (**c**) high-spatial resolution HSI as the reference.

3.1.3. HyMap Rodalquilar

The HyMap image was taken in Rodalquilar, Spain in June 2003 [35], covering a gold mining area in the Cabo de Gata Mountains. The spatial resolution of the data was 10 m. The experimental data selected in this paper contains 867 × 261 pixels. After removing the water absorption band, 167 bands are selected for experimentation, and the spectral range is 0.4 μm–2.5 μm. Among them, the low-spatial resolution HSI image was obtained from the original HSI data through the isotropic Gaussian point spread function down-sampling three times, with a total of 167 bands and 289 × 87 pixels, and the resulting pseudo-color image is shown in Figure 6a. The high-spatial resolution MSI image data were synthesized from the original HSI data according to the SRF down-sampling of the HyMap sensor. There were four bands in total with the image size of 867 × 261 pixels, as shown in Figure 6b. The reference image is shown in Figure 6c.

3.2. Comparative Analysis of the Proposed Spatial Enhanced LSE-SFIM Using Different Spatial Filters

In this section, different spatial enhanced methods in Section 2.2.2 are used to extract boundary information and obtain better fusion results, and the performance discussed to find the best method. Six methods are discussed in this section, the traditional SFIM (named as SFIM), LSE-based SFIM (named LSE-SFIM), mean filtering LSE-SFIM (named LSE-SFIM-M), median filtering LSE-SFIM (named LSE-SFIM-Med), neighboring interpolation LSE-SFIM (named LSE-SFIM-N), and bilinear interpolation LSE-SFIM (named LSE-SFIM-B). Both subjective and objective evaluations are discussed, and spectral distortion are compared among all six algorithms.

(**a**) HSI (**b**) MSI (**c**) REF

Figure 6. HyMap Rodalquilar datasets with (**a**) low spatial resolution HSI, (**b**) high spatial resolution MSI and (**c**) high spatial resolution HSI as the reference.

3.2.1. Subjective Evaluation

The subjective evaluation mainly uses human eyes to observe the fusion results. The comparison chart of the fusion results of the three groups of experiments is shown in Figures 7–9. Observing the fusion result from a subjective point of view, it can be known that the method of using bilinear interpolation to obtain simulated MSI′ has better visibility, and the fusion result obtained has a clearer texture and better spectrum retention performance.

(**a**) SFIM (**b**) LSE-SFIM (**c**) LSE-SFIM-M (**d**) LSE-SFIM-Med (**e**) LSE-SFIM-N (**f**) LSE-SFIM-B

Figure 7. Fusion results of Pavia University using six smoothing filter-based intensity modulation (SFIM)-based algorithms.

(**a**) SFIM (**b**) LSE-SFIM (**c**) LSE-SFIM-M (**d**) LSE-SFIM-Med (**e**) LSE-SFIM-N (**f**) LSE-SFIM-B

Figure 8. Fusion results of Chikusei using six SFIM-based algorithms.

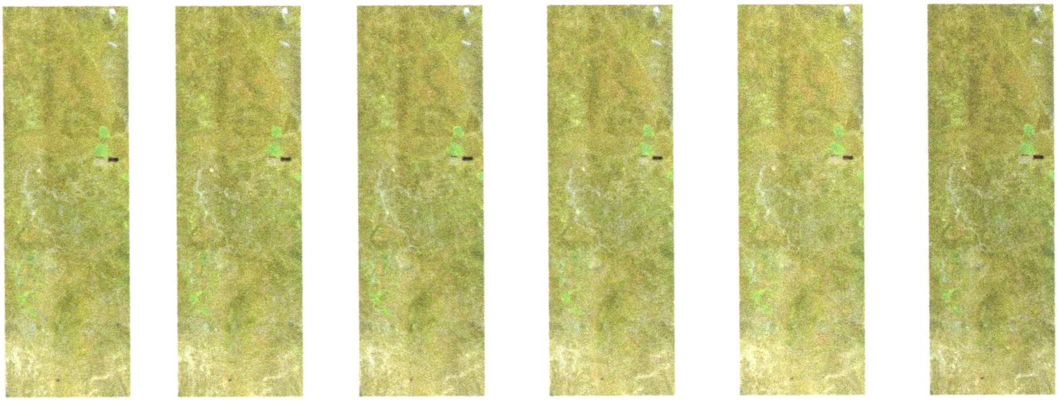

(**a**) SFIM (**b**) LSE-SFIM (**c**) LSE-SFIM-M (**d**) LSE-SFIM-Med (**e**) LSE-SFIM-N (**f**) LSE-SFIM-B

Figure 9. Fusion results of Chikusei using six SFIM-based algorithms.

Figure 7a–f shows that the Pavia University dataset has been subjected to SFIM, LSE-SFIM, LSE-SFIM-M, LSE-SFIM-Med, LSE-SFIM-N and LSE-SFIM-B. It can be seen from the figures that the spectral characteristics of the fusion results by all methods are maintained well. In terms of spatial geometric features, the fusion result of LSE-SFIM-N is visually blurred, and the edge details are not highlighted. In addition, the fusion results obtained by other algorithms have higher clarity, and the LSE-SFIM-B algorithm, whether in terms of spectral characteristics or spatial characteristics, has the closest visual effect to the reference image.

Figure 8a–f are the fusion results of Chikusei dataset through 6 algorithms, namely SFIM, LSE-SFIM, LSE-SFIM-M, LSE-SFIM-Med, LSE-SFIM-N and LSE-SFIM-B. It is obvious that, in terms of spectral characteristics, the fusion results of each method have no more spectral distortions of ground objects, and the color information performs well. In terms of spatial features, the fusion result of LSE-SFIM-N has unclear ground textures and non-obvious edge details. In addition, the fusion results obtained by other algorithms maintain both texture features and edge details of the ground features, especially the LSE-SFIM-B algorithm, which retains more spatial features and the edges of the ground features are more obvious.

Figure 9a–f are the fusion results of the HyMap Rodalquilar data set by 6 algorithms, namely SFIM, LSE-SFIM, LSE-SFIM-M, LSE-SFIM-Med, LSE-SFIM-N and LSE-SFIM-B. In terms of spectral characteristics, the fusion results of each method are not too distorted in the maintenance of the ground object spectrum, and the color information is maintained well. In terms of spatial features, the LSE-SFIM-N fusion result has a poor spatial information enhancement effect. In addition, the fusion results obtained by other algorithms maintain the texture features of hyperspectral images and multispectral images well, especially the LSE-SFIM-B algorithm, which maintains the spatial characteristics better, and the edges of the features are also the most obvious.

In general, through the subjective evaluation of the fusion results by human eyes, it can be found that the proposed LSE-SFIM-B fusion algorithm has the best performance, and the fusion image with clearer boundaries can be obtained from the visual results, especially in Figures 7 and 8. It is the LSE-SFIM-B algorithm makes full use of the complementary characteristics of HIS and MSI images, which realizes the fusion of spectral and spatial features of multiple source images, improves the geometric features of ground objects, and verifies the effectiveness of this algorithm. As for Figure 9, due to image abbreviation, the spatial information enhancement effect of some images is not easy to see, and it is difficult to subjectively judge which method is better. Therefore, objective evaluation is particularly important.

3.2.2. Objective Evaluation

By observing the fusion results in Figures 7–9, it can be seen that the method of obtaining simulated multispectral images by using the LSE-SFIM-B method has better visibility, and the fusion results obtained have clearer texture and better spectrum retention capabilities. To further objectively evaluate the quality of the fusion images by different algorithms, this paper also calculates and analyzes the fusion results from a quantitative perspective by comparing the six objective evaluation indicators of PSNR (peak signal-to-noise ratio), SAM (spectral angle mapping), CC (cross correlation), $Q2^n$ (quality 2^n), RMSE (root mean square error) and ERGAS (error relative global adimensionnelle synthesizer). The quantitative comparisons are shown in Tables 2–4, with the histogram comparison of evaluation indicators shown in Figures 10–12.

Table 2. Quantitative comparisons of fusion performance by six algorithms for Pavia University.

Evaluation Index	SFIM	LSE-SFIM	LSE-SFIM-M	LSE-SFIM-Med	LSE-SFIM-N	LSE-SFIM-B
PSNR	25.8772	36.1762	37.9836	33.4864	28.7486	**42.1976**
SAM	9.3271	3.4442	3.0933	4.1955	5.7274	**2.6762**
CC	0.82418	0.98594	0.99101	0.97659	0.92191	**0.99362**
$Q2^n$	0.46532	0.72695	0.75624	0.7093	0.56614	**0.8975**
RMSE	0.4142	0.01298	0.01075	0.01767	0.030804	**0.007333**
ERGAS	6.1140	1.2945	1.0702	1.6944	2.8878	**0.76253**

Table 3. Quantitative comparison of fusion results by six algorithms for Chikusei.

Evaluation Index	SFIM	LSE-SFIM	LSE-SFIM-M	LSE-SFIM-Med	LSE-SFIM-N	LSE-SFIM-B
PSNR	24.0379	37.8579	40.1123	35.3821	31.0431	**46.6653**
SAM	7.4477	1.8777	1.5704	2.0696	2.8164	**1.3432**
CC	0.76329	0.9873	0.99117	0.98013	0.94795	**0.99341**
$Q2^n$	0.35951	0.87498	0.87572	0.85253	0.83137	**0.91992**
RMSE	0.4142	0.0078549	0.0061514	0.010291	0.017394	**0.0037586**
ERGAS	6.1140	1.7005	1.4777	2.0937	3.0949	**1.2483**

Table 4. Quantitative comparison of fusion results by six algorithms for HyMap Rodalquilar.

Evaluation Index	SFIM	LSE-SFIM	LSE-SFIM-M	LSE-SFIM-Med	LSE-SFIM-N	LSE-SFIM-B
PSNR	36.9969	36.3762	37.8249	36.518	35.2463	**39.6276**
SAM	2.9165	2.7045	2.6616	2.6922	2.7101	**2.6475**
CC	0.95908	0.96912	0.97943	0.97128	0.96045	**0.9855**
$Q2^n$	**0.67894**	0.51703	0.54971	0.53345	0.49201	0.6217
RMSE	0.018907	0.016785	0.015491	0.01656	0.018003	**0.014377**
ERGAS	4.2584	2.3641	2.1552	2.3246	2.5765	**1.9779**

According to Table 2 and Figure 10, it can be seen that for Pavia University data, the method based on the LSE-SFIM algorithm has better fusion effect than traditional SFIM, and the algorithms LSE-SFIM-M and LSE-SFIM-B are better than the original LSE-SFIM algorithm, and the effect of LSE-SFIM-B is even better than that of LSE-SFIM-M, which is the best among several comparison methods. In terms of PSNR, CC and $Q2^n$, the LSE-SFIM-B fusion algorithm is significantly higher than the results of the other algorithms, indicating that the spatial quality information of the fusion image is better, the fusion result has more detailed spatial information, and it is correlated with the reference image. In terms of SAM, RMSE and ERGAS, the LSE-SFIM-B fusion algorithm is still superior to other algorithms, indicating that the fusion result can better maintain the spectrum, and the error with the reference image is the smallest, and the fusion result is the closest to the reference image.

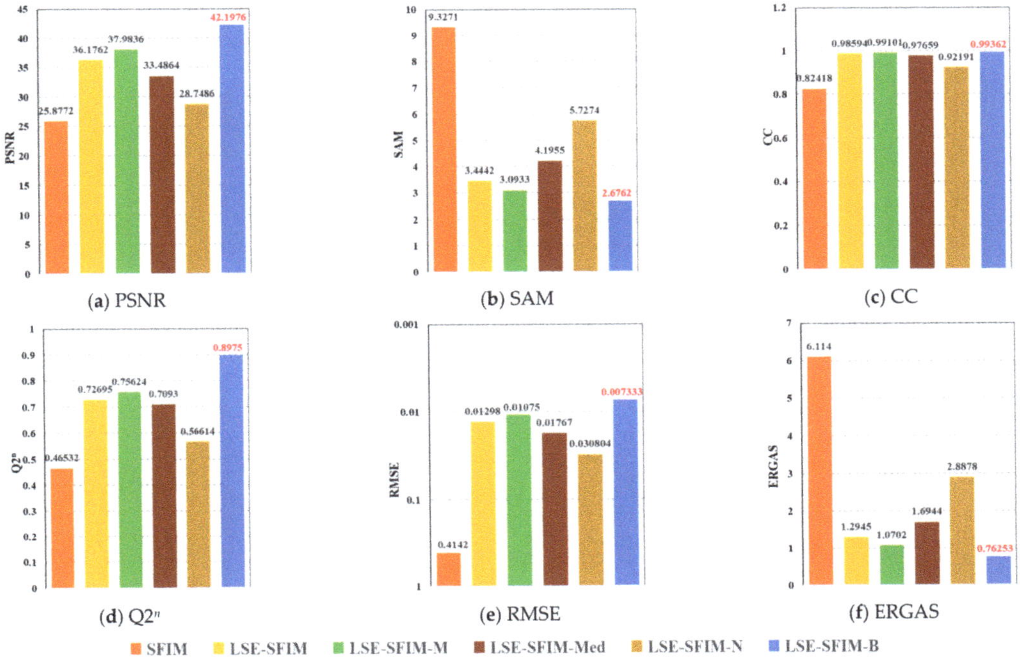

Figure 10. Histograms comparison of evaluation indicators by six algorithms for Pavia University.

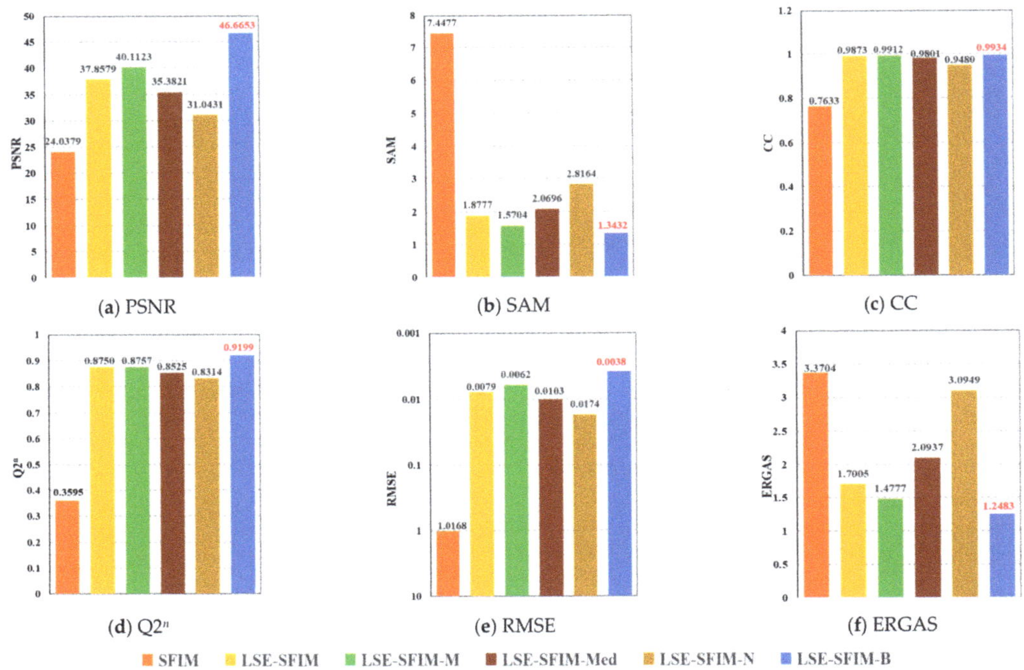

Figure 11. Histogram comparison of evaluation indicators by six algorithms for Chikusei.

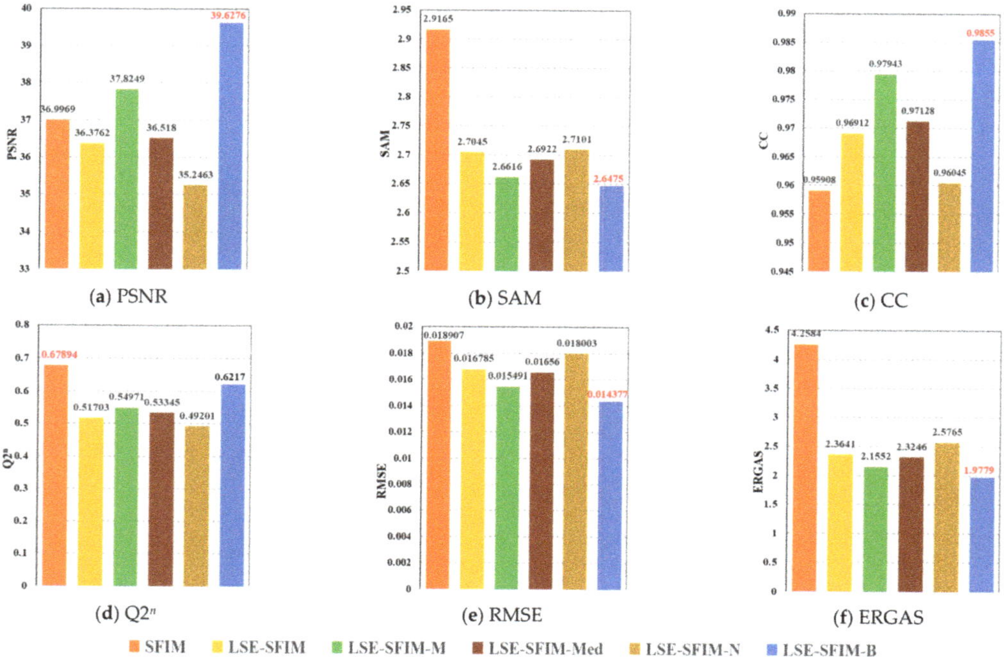

Figure 12. Histogram comparison of evaluation indicators by six algorithms for HyMap Rodalquilar.

In order to further illustrate that the algorithms proposed in this paper have a good performance using the data of different images, Table 3 and Figure 11 give the objective evaluation of Chikusei data, and Table 4 and Figure 12 give the objective evaluation of HyMap Rodalquilar data. It can be seen that for all the data sets, the fusion algorithm using LSE-SFIM-B is also the most outstanding in terms of spatial resolution enhancement and spectral characteristic maintenance. Specifically, the LSE-SFIM-B algorithm is significantly higher than other algorithms in terms of PSNR, CC and $Q2^n$, indicating that the fusion image has good ground quality information, the fusion result is more detailed, and the correlation with the reference image is relatively high. In terms of SAM, RMSE and ERGAS, the LSE-SFIM-B fusion algorithm has the best performance, indicating that the error between the fusion result and the reference image is the smallest, and the spectrum can be better maintained.

3.2.3. Spectral Distortion Comparison

A good fusion method should minimize spectral distortion as much as possible while improving the spatial resolution. In this section, to further analyze the spectral distortion for different SFIM-based algorithms, Figures 13–15 show SAM plots of the experimental results of three hyperspectral data sets. The SAM plot is conducted for every pixel to compute the SAM value between the fusion result and the reference image. In the figures, each pixel uses the change from cold to warm to indicate the level of spectral similarity at that pixel. The closer the color of the pixel point is to the warm color, that is, the closer to dark red, the lower the spectral similarity and the worse the spectral quality relative to other pixels; the closer the color of the pixel point is to the cool color, that is, the closer to dark blue, the higher the spectral similarity and the higher the spectral quality relative to other pixels. The larger the area occupied by the blue part in the figure, the better the overall spectral quality. Compared with other algorithms, it can be seen from the SAM graph of the algorithm experiment results that the spectral performance of LSE-SFIM-B on the three data sets is relatively better.

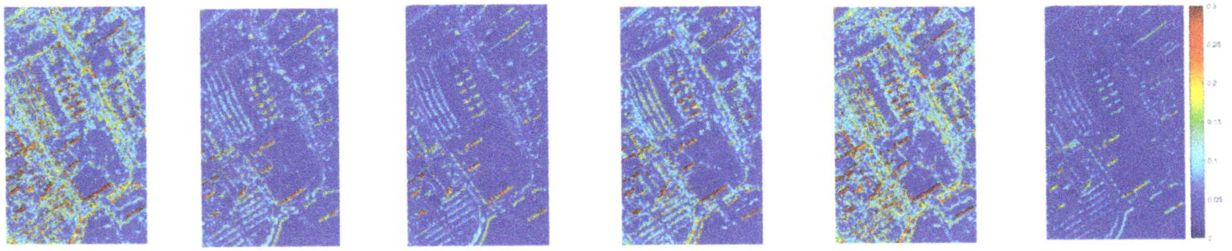

(**a**) SFIM (**b**) LSE-SFIM (**c**) LSE-SFIM-M (**d**) LSE-SFIM-Med (**e**) LSE-SFIM-N (**f**) LSE-SFIM-B

Figure 13. Spectral angle mapping (SAM) map of six algorithms for Pavia University data experiment results.

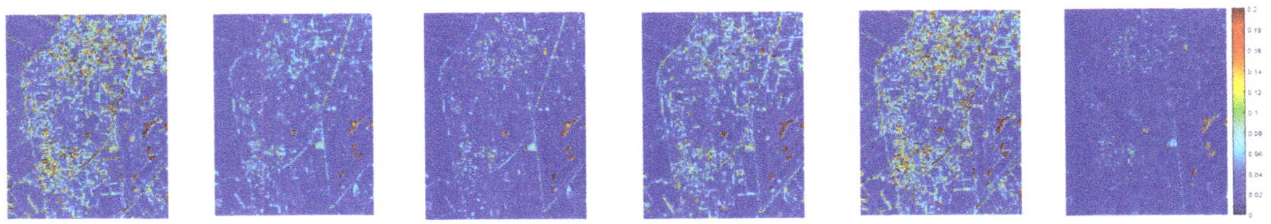

(**a**) SFIM (**b**) LSE-SFIM (**c**) LSE-SFIM-M (**d**) LSE-SFIM-Med (**e**) LSE-SFIM-N (**f**) LSE-SFIM-B

Figure 14. SAM map of six algorithms for Chikusei data experiment results.

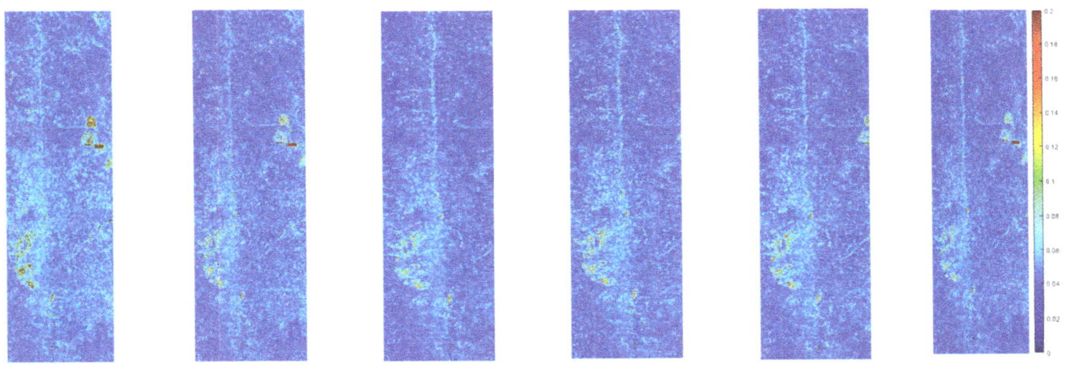

(**a**) SFIM (**b**) LSE-SFIM (**c**) LSE-SFIM-M (**d**) LSE-SFIM-Med (**e**) LSE-SFIM-N (**f**) LSE-SFIM-B

Figure 15. SAM map of six algorithms for HyMap Rodalquilar data experiment results.

3.2.4. Influence of Spatial Scale Factor between MSI and HSI

The above experiments have proved the LSE-SFIM-B algorithm to be effective among all SFIM-based algorithms. It would be interesting to know the performance of the proposed LSE-SFIM-B algorithm according to the spatial scale factor between the high-resolution MSI and low-resolution HSI images. In order to see how the algorithm performs on different scale factors, the Pavia University data are used in this section, where the spatial scale factors (SF) are set to SF = 2, 4, 8, respectively. The performance comparison is given in Table 5, where the spatial scale value is the down-sampling rate of the HSI data.

Table 5. Performance of different spatial scale factors of MSI and HSI data using LSE-SFIM-B algorithm for Pavia University.

Indexes/Spatial Scales	2	4	8
PSNR	44.1518	43.2708	42.1976
SAM	2.1803	2.4104	2.6762
CC	0.99553	0.99458	0.99362
$Q2^n$	0.93804	0.93213	0.8975
RMSE	0.005912	0.006572	0.007333
ERGAS	2.5732	1.4025	0.76253

The results in Table 5 are very interesting and show that the PSNR, SAM, CC, $Q2^n$ and RMSE values tend to worsen as the spatial scale factor increases, while the ERGAS value becomes smaller (better) as the spatial scale factor increases.

3.3. Performance Analysis of the Proposed SFIM-Based Algorithm and Other Commonly Used Algorithms

In order compare the fusion performance of the proposed SFIM-based algorithm with the existing representative algorithms, this section chooses some state-of-the-art algorithms for comparison. Three state-of-the-art algorithms have been used in this section, which are CNMF proposed in 2012, HySure proposed in 2015, and FUSE proposed in 2018. Since the LSE-SFIM-B method has been proved to be most effective in the previous section, we will use this one for comparison. Since the robustness of the algorithm for different data sets has been proved in the previous section, in this section only the Chikusei data set is use to reduce repetition.

The experimental settings are as follows: (1) for the Chikusei data set, the number of endmembers (D) is set to $D = 30$ for any algorithm needed. (2) In CNMF algorithm, the maximum number of iterations for inner loops (I_{in}) and the maximum number of iterations for outer loops (I_{out}) are $I_{in} = 200$, $I_{out} = 2$. (3) In the HySure algorithm, the parameters are set to $\lambda_\varphi = 10^{-3}, \lambda_B = \lambda_R = 10$.

Figure 16a–e are the fusion results of the Chikusei dataset through five algorithms, which are conventional SFIM, the proposed LSE-SFIM-B, CNMF, HySure, and FUSE. The visual effects seem to be similar to the last four algorithms, in which the SFIM seems to have the worst performance. In order to further evaluate the performance of the proposed LSE-SFIM-B algorithm and the other three state-of-the-art algorithms, Table 6 gives the comparison for objective evaluation indicators PSNR, SAM, CC, $Q2^n$, RMSE and ERGAS.

(**a**) SFIM (**b**) LSE-SFIM-B (**c**) CNMF (**d**) HySure (**e**) FUSE

Figure 16. Fusion results of Chikusei dataset by five different algorithms.

Table 6. Quantitative comparison of fusion results by five different algorithms for Chikusei.

Evaluation Index	SFIM	LSE-SFIM-B	CNMF	HySure	FUSE
PSNR	24.0379	46.6653	46.1716	**47.3149**	45.4159
SAM	7.4477	1.3432	1.2497	**1.1544**	1.4699
CC	0.76329	**0.99341**	0.98988	0.99093	0.98855
$Q2^n$	0.35951	0.91992	0.9485	**0.9606**	0.91975
RMSE	0.4142	0.0037586	0.0035002	**0.0032553**	0.0044402
ERGAS	6.1140	**1.2483**	1.5456	1.4725	1.6222

It can be seen that the conventional SFIM algorithm has the worst performance of all six indicators. The other four algorithms, including the proposed LSE-SFIM-B, CNMF, HySure, and FUSE, have similar performance. In the HySure algorithm four of the six indicators are optimal, and the other two optimal indicators are obtained by the proposed LSE-SFIM-B algorithm. However, as mentioned in Section 1, the SFIM-based algorithm is simple to calculate and easy to implement. In order to verify the computational complexity of these different algorithms, Table 7 shows the computing time comparisons for different algorithms, and the proposed LSE-SFIM-B algorithm has the least computing time. To further show the time-efficient superiority of LSE-SFIM-B, the speed-up ratio is also provided in Table 7 which is calculated using the computing time of the other three algorithms (CNMF, HySure, and FUSE) divided by the computing time of the proposed LSE-SFIM-B algorithm. As a result, even though the HySure algorithm has four performance indicators that are better than LSE-SFIM-B algorithm, it is time consuming, especially when the data set is very large, the LSE-SFIM-B could demonstrate its excellent time efficiency while maintaining the performance.

Table 7. Computational complexity analysis by five different algorithms for Chikusei.

Algorithms	LSE-SFIM-B	CNMF	HySure	FUSE
Computing time/second	**0.77**	32.36	317.43	3.00
speed-up ratio/times	–	42.03	412.25	3.40

4. Discussion and Conclusions

This paper proposes a spatial-enhanced LSE-SFIM algorithm for HSI-MSI images fusion. The contributions of the proposed algorithm can be summarized as follows:

1. Improving the performance of tradition SFIM algorithm. The traditional SFIM fusion algorithm has problems such as blurred image edge information and insufficient spatial detail information. In this paper, two steps are taken to solve the above problems: (1) LSE is used to adjust the obtained simulated low-spatial MSI′ so that the linear regression can minimize the spatial information error, and the simulated MSI′ can have as much as possible the same spatial information as HSI; (2) four different spatial filters are then used to further improve the detailed spatial information in the process of up-sampling, and the experimental results show that the use of bilinear interpolation in LSE-SFIM-B fusion algorithm has the best performance among all SFIM based algorithms.
2. Achieving similar performance with much less computing time. This paper also employs three state-of-the art algorithms (CNMF, HySure and FUSE) to compare the performance, and the experimental results show that the proposed LSE-SFIM-B algorithm can achieve similar performance as these, while the computing time is much less. As a result, in the case of high time requirements or in the case of processing a very large data set, the proposed LSE-SFIM-B algorithm can show a good ability in both processing performance and time effect with practical significance.

The proposed algorithm can achieve a good performance in most cases, performs better than traditional SFIM algorithm with better spatial preserving and less spectral distortion, and also has less computational complexity than the state-of-the-art fusion algorithms. However, the spectral fidelity is not good enough, since the SFIM-based model performs the fusion band by band, without considering the spectral correlations. Adding spectral constraints to the model can be considered in a future study.

Author Contributions: Conceptualization, Y.W.; Methodology, Y.W., Q.Z. and Y.S.; Experiments: Y.S.; Data Curation, Y.W. and M.S.; Formal Analysis, M.S. and C.Y.; Writing—Original Draft, Y.W. and Y.S.; Writing—Review & Editing: Y.W. All authors have read and agreed to the published version of the manuscript.

Funding: The work of Y. Wang was supported in part by the National Nature Science Foundation of China (61801075), China Postdoctoral Science Foundation (2020M670723), the Fundamental Research Funds for the Central Universities (3132019341), and Open Research Funds of State Key Laboratory of Integrated Services Networks, Xidian University (ISN20-15). The work of Meiping Song was supported by the National Nature Science Foundation of China (61971082).

Conflicts of Interest: The authors declare no conflict of interest.

References

1. McCabe, M.F.; Rodell, M.; Alsdorf, D.E.; Miralles, D.G.; Uijlenhoet, R.; Wagner, W.; Lucieer, A.; Houborg, R.; Verhoest, N.E.C.; Franz, T.E.; et al. The future of Earth observation in hydrology. *Hydrol. Earth Syst. Sci.* **2017**, *21*, 3879–3914. [CrossRef]
2. Selva, D.; Krejci, D. A survey and assessment of the capabilities of Cubesats for Earth observation. *Acta Astronaut.* **2012**, *74*, 50–68. [CrossRef]
3. Xiang, S.; Wang, L.; Xing, L.; Du, Y.; Zhang, Z. Knowledge-based memetic algorithm for joint task planning of multi-platform earth observation system. *Comput. Ind. Eng.* **2021**, *160*, 107559. [CrossRef]
4. Shayeganpour, S.; Tangestani, M.H.; Gorsevski, P.V. Machine learning and multi-sensor data fusion for mapping lithology: A case study of Kowli-kosh area, SW Iran. *Adv. Space Res.* **2021**, *68*, 3992–4015. [CrossRef]
5. Si, Y.; Lu, Q.; Zhang, X.; Hu, X.; Wang, F.; Li, L.; Gu, S. A review of advances in the retrieval of aerosol properties by remote sensing multi-angle technology. *Atmos. Environ.* **2021**, *244*, 117928. [CrossRef]
6. Zhang, L.P.; Shen, H.F. Progress and future of remote sensing data fusion. *J. Remote Sens.* **2016**, *20*, 1050–1061.
7. Tohid, Y.; Farhang, A.; Ali, A.; Calagari, A.A. Integrating geologic and Landsat-8 and ASTER remote sensing data for gold exploration: A case study from Zarshuran Carlin-type gold deposit, NW Iran. *Arab. J. Geosci.* **2018**, *11*, 482.
8. Erdelj, M.; Natalizio, E.; Chowdhury, K.R.; Akyildiz, I.F. Help from the Sky: Leveraging UAVs for Disaster Management. *IEEE Pervasive Comput.* **2017**, *16*, 24–32. [CrossRef]
9. San Juan, R.F.V.; Domingo-Santos, J.M. The Role of GIS and LIDAR as Tools for Sustainable Forest Management. *Front. Inf. Syst.* **2018**, *1*, 124–148.
10. Zhu, Q.; Zhang, J.; Ding, Y.; Liu, M.; Li, Y.; Feng, B.; Miao, S.; Yang, W.; He, H.; Zhu, J. Semantics-Constrained Advantageous Information Selection of Multimodal Spatiotemporal Data for Landslide Disaster Assessment. *ISPRS Int. J. Geo Inf.* **2019**, *8*, 68. [CrossRef]
11. Khanal, S.; Fulton, J.; Shearer, S.A. An overview of current and potential applications of thermal remote sensing in precision agriculture. *Comput. Electron. Agric.* **2017**, *139*, 22–32. [CrossRef]
12. Chlingaryan, A.; Sukkarieh, S.; Whelan, B. Machine learning approaches for crop yield prediction and nitrogen status estimation in precision agriculture: A review. *Comput. Electron. Agric.* **2018**, *151*, 61–69. [CrossRef]
13. Zhou, L.; Chen, N.; Chen, Z.; Xing, C. ROSCC: An Efficient Remote Sensing Observation-Sharing Method Based on Cloud Computing for Soil Moisture Mapping in Precision Agriculture. *IEEE J. Sel. Top. Appl. Earth Obs. Remote Sens.* **2016**, *9*, 5588–5598. [CrossRef]
14. Wu, B.; Yu, B.; Yao, S.; Wu, Q.; Chen, Z.; Wu, J. A surface network based method for studying urban hierarchies by night time light remote sensing data. *Int. J. Geogr. Inf. Sci.* **2019**, *33*, 1377–1398. [CrossRef]
15. Zhang, Z.; Liu, F.; Zhao, X.; Wang, X.; Shi, L.; Xu, J.; Yu, S.; Wen, Q.; Zuo, L.; Yi, L.; et al. Urban Expansion in China Based on Remote Sensing Technology: A Review. *Chin. Geogr. Sci.* **2018**, *28*, 727–743. [CrossRef]
16. Shen, Q.; Yao, Y.; Li, J.; Zhang, F.; Wang, S.; Wu, Y.; Ye, H.; Zhang, B. A CIE Color Purity Algorithm to Detect Black and Odorous Water in Urban Rivers Using High-Resolution Multispectral Remote Sensing Images. *IEEE Trans. Geosci. Remote Sens.* **2019**, *57*, 6577–6590. [CrossRef]
17. Carper, W.J.; Lillesand, T.M.; Kiefe, R.W. Use of intensity-hue-saturation transformations for merging SPOT panchromatic and multispectral image data. *Photogramm. Eng. Remote Sens.* **1990**, *56*, 459–467.
18. Chavez, P.S.; Sides, S.C.; Anderson, J.A. Comparison of three different methods to merge multiresolution and multispectral data: Landsat TM and SPOT panchromatic. *Photogramm. Eng. Remote Sens.* **1991**, *57*, 265–303.
19. Aiazzi, B.; Baronti, S.; Selva, M. Improving Component Substitution Pansharpening Through Multivariate Regression of MS + Pan Data. *IEEE Trans. Geosci. Remote Sens.* **2007**, *45*, 3230–32399. [CrossRef]

20. Sulaiman, A.G.; Elashmawi, W.H.; Eltaweel, G. IHS-based pan-sharpening technique for visual quality improvement using KPCA and enhanced SML in the NSCT domain. *Int. J. Remote Sens.* **2021**, *42*, 537–566. [CrossRef]
21. Li, Y.; Qu, J.; Dong, W.; Zheng, Y. Hyperspectral pansharpening via improved PCA approach and optimal weighted fusion strategy. *Neurocomputing* **2018**, *315*, 371–380. [CrossRef]
22. Aiazzi, B.; Alparone, L.; Arienzo, A.; Garzelli, A.; Lolli, S. Fast multispectral pansharpening based on a hyper-ellipsoidal color space. In Proceedings of the Conference on Image and Signal Processing for Remote Sensing XXV, Strasbourg, France, 11 September 2019.
23. Singh, R.; Khare, A. Fusion of multimodal medical images using Daubechies complex wavelet transform–A multiresolution approach. *Inf. Fusion* **2014**, *19*, 49–60. [CrossRef]
24. Liu, J.G. Smoothing Filter-based Intensity Modulation: A spectral preserve image fusion technique for improving spatial details. *Int. J. Remote Sens.* **2000**, *21*, 3461–3472. [CrossRef]
25. Aiazzi, B.; Alparone, L.; Baronti, S.; Garzelli, A.; Selva, M. MTF-tailored Multiscale Fusion of High-resolution MS and Pan Imagery. *Photogramm. Eng. Remote Sens.* **2006**, *72*, 591–596. [CrossRef]
26. Yokoya, N.; Mayumi, N.; Iwasaki, A. Coupled Nonnegative Matrix Factorization Unmixing for Hyperspectral and Multispectral Data Fusion. *IEEE Trans. Geosci. Remote Sens.* **2012**, *50*, 528–537. [CrossRef]
27. Simoes, M.; Bioucas-Dias, J.; Almeida, L.B.; Chanussot, J. A Convex Formulation for Hyperspectral Image Superresolution via Subspace-Based Regularization. *IEEE Trans. Geosci. Remote Sens.* **2015**, *53*, 3373–3388. [CrossRef]
28. Li, S.; Dian, R.; Fang, L.; Bioucas-Dias, J. Fusing Hyperspectral and Multispectral Images via Coupled Sparse Tensor Factorization. *IEEE Trans. Image Process.* **2018**, *27*, 4118–4130. [CrossRef]
29. Eismann, M.T. Resolution Enhancement of Hyperspectral Imagery Using Maximum a Posteriori Estimation with a Stochastic Mixing Model. Ph.D. Thesis, University of Dayton, Dayton, OH, USA, 2004.
30. Wei, Q.; Bioucas-Dias, J.; Dobigeon, N.; Tourneret, J.-Y. Hyperspectral and Multispectral Image Fusion Based on a Sparse Representation. *IEEE Trans. Geosci. Remote Sens.* **2015**, *53*, 3658–3668. [CrossRef]
31. Wei, Q.; Dobigeon, N.; Tourneret, J.-Y. Fast Fusion of Multi-Band Images Based on Solving a Sylvester Equation. *IEEE Trans. Image Process.* **2015**, *24*, 4109–4121. [CrossRef]
32. Ren, X.; Lu, L.; Chanussot, J. Toward Super-Resolution ImageConstruction Based on Joint Tensor Decomposition. *Remote Sens.* **2020**, *12*, 2535. [CrossRef]
33. Lu, X.; Yang, D.; Zhang, J.; Jia, F. Hyperspectral Image Super-Resolution Based on Spatial Correlation-Regularized Unmixing Convolutional Neural Network. *Remote Sens.* **2021**, *13*, 4074. [CrossRef]
34. Zhang, L.; Nie, J.; Wei, W.; Li, Y.; Zhang, Y. Deep Blind Hyperspectral Image Super-Resolution. *IEEE Trans. Neural Netw. Learn. Syst.* **2021**, *32*, 2388–2400. [CrossRef] [PubMed]
35. Bedini, E.; Van Der Meer, F.; van Ruitenbeek, F. Use of HyMap imaging spectrometer data to map mineralogy in the Rodalquilar caldera, southeast Spain. *Int. J. Remote Sens.* **2008**, *30*, 327–348. [CrossRef]

remote sensing

MDPI

Article

Novel Air Temperature Measurement Using Midwave Hyperspectral Fourier Transform Infrared Imaging in the Carbon Dioxide Absorption Band

Sungho Kim

Department of Electronic Engineering, Yeungnam University, 280 Daehak-Ro, Gyeongsan, Gyeongbuk 38541, Korea; sunghokim@ynu.ac.kr; Tel.: +82-53-810-3530

Received: 15 May 2020; Accepted: 4 June 2020; Published: 8 June 2020

Abstract: Accurate visualization of air temperature distribution can be useful for various thermal analyses in fields such as human health and heat transfer of local area. This paper presents a novel approach to measuring air temperature from midwave hyperspectral Fourier transform infrared (FTIR) imaging in the carbon dioxide absorption band (between 4.25–4.35 μm). In this study, the proposed visual air temperature (VisualAT) measurement is based on the observation that the carbon dioxide band shows zero transmissivity at short distances. Based on analysis of the radiative transfer equation in this band, only the path radiance by air temperature survives. Brightness temperature of the received radiance can provide the raw air temperature and spectral average, followed by a spatial median-mean filter that can produce final air temperature images. Experiment results tested on a database obtained by a midwave extended FTIR system (Telops, Quebec City, QC, Canada) from February to July 2018 show a mean absolute error of 1.25 °K for temperature range of 2.6−26.4 °C.

Keywords: air temperature; spatial measurement; FTIR; MWIR; carbon dioxide absorption

1. Introduction

How accurately can we measure and visualize air temperature remotely using thermal sensing? Air temperature is an important meteorological factor, which has a wide range of applications in fields like human health [1], virus propagation [2], growth and reproduction of plants [3], climate change [4], and hydrology [5].

Air temperature can be measured in numerous ways, including contact sensors and remote sensors. Contact sensor-based methods include thermistors, thermocouples, and mercury thermometers [6]. Thermistors are metallic devices that undergo predictable changes in resistance in response to changes in temperature. This resistance is measured and converted to a temperature reading in Celsius, Fahrenheit, or Kelvin. A thermocouple consists of two dissimilar electrical conductors forming an electrical junction, which produces a temperature-dependent voltage, and this voltage can be converted to a temperature [7]. A mercury thermometer consists of liquid in a glass rod with a very thin tube in it. Mercury or red-colored alcohol inside the tube expands when the temperature rises. These sensors should be located in the shade to measure air temperature. If the sun shines on the thermometer directly, it heats the liquid and produces an incorrect, higher temperature than the true air temperature. In addition, it needs enough time (at least several minutes for the liquid to expand) to measure outdoor air temperature. Furthermore, it requires hundreds of thousands contact sensors to measure spatial distribution of air temperature.

The thermal remote sensing approaches will use the relationships between land surface temperature (LST) and near surface air temperature. Land surface radiance is measured by thermal infrared sensors mounted on a satellite or an airborne platform, and LST is retrieved via temperature-emissivity separation (TES) [8,9]. Surface air temperature can be estimated by the temperature-vegetation index (TVX), thermodynamics, and the regression method. The TVX is based on the assumption that a thick vegetation canopy can approximate air temperature [10,11]. It can be useful only if there is high vegetation cover. The thermodynamics-based method uses the energy balance between LST and the surface environment, such as air and water [12,13]. This method provides good air temperature measurement, but it requires many parameters as input. The last type is data-based regression between air temperature and LST. There is a linear regression model [14] and a nonlinear model, especially as a deep learning method [15]. Such machine learning models have reported successful air temperature estimation. One recent deep learning-based method, a five-layer deep belief network (DBN), showed promising air temperature estimation by establishing the relationship between ground station air temperature and multi-source data (remote sensing radiance, socioeconomic data, and assimilation data) [15]. Although the deep learning method showed quite accurate air temperature estimation, it requires huge amounts of data to train the multi-layered deep neural architecture.

The above mentioned approaches are not suitable for instantaneously measuring and visualizing air temperature of a viewing area. Using contact sensors requires several minutes and a huge number of sensors to measure spatial air temperature [16]. Non-contact (remote) sensors, such as thermal infrared (TIR) imagers, require LST and regression with huge amounts of training data. Despite being trained correctly, the regression-based approach is sensitive to many spatio-temporal conditions [17]. In addition, this approach is only suitable for aerial-based sensing in satellites or airborne platforms [18].

In this paper, our research focuses on how to measure spatial air temperature of the viewing area and visualize it instantaneously for environment monitoring of the surveillance area. The key idea is to use up-welling information in the carbon dioxide absorption band (4.25−4.35 μm) with a midwave Fourier transform infrared (FTIR) imager. Most temperature estimation research in TES and LST generates up-welling information by using the moderate-resolution atmospheric radiance and transmittance model (MODTRAN) for atmospheric correction (compensation) purposes [19–21]. Harig proposed a passive sensing of pollutant clouds by longwave FTIR to find optimal SNR [22]. His work focused on detecting cloud target in ground background not air temperature. In this paper, it is possible to measure and visualize air temperature accurately through careful analysis of upwelling (path) spectral radiance in the proposed visual air temperature (VisualAT).

The contributions from this paper can be summarized as follows. First, VisualAT can measure spatial air temperature accurately (mean absolute error [MAE]: 1.25 K). Second, VisualAT can visualize the distribution of air temperature with a high spatial resolution. Third, it can measure and visualize air temperature instantaneously. Finally, the proposed VisualAT can be used for various outdoor air temperature measurement applications, such as health monitoring, weather monitoring, and thermal surveillance of local area.

The remainder of this paper is organized as follows. Section 2 introduces the marials for FTIR analysis and Section 3 explains the proposed VisualAT method, including the radiative transfer equation. Section 4 analyzes VisualAT for temperature measurement applications, considering a range of environmental changes. The paper concludes in Section 5.

2. Materials for FTIR Analysis

2.1. Outdoor Hyperspectral Data Acquisition System

Figure 1 presents a measurement scenario in an outdoor environment consisting of a scene and a sensor system. A painted target in front of a sky-and-sea background is 78 m away from the observation laboratory. MWIR hyperspectral images were acquired with the Telops Hyper-Cam MWE model [23]. It can provide calibrated spectral radiance images with a high spatial and spectral resolution from a Michelson interferometer in the short-wave to midwave band (1.5–5.6 μm). The spatial image resolution is 320 × 240, and the spectral resolution is up to 0.25 cm^{-1}. The noise equivalent spectral radiance (NESR) is 7 [$nW/(cm^2 \cdot sr \cdot cm^{-1})$], and the radiometric accuracy is approximately 2 K. The field of view is 6.5 × 5.1 deg.

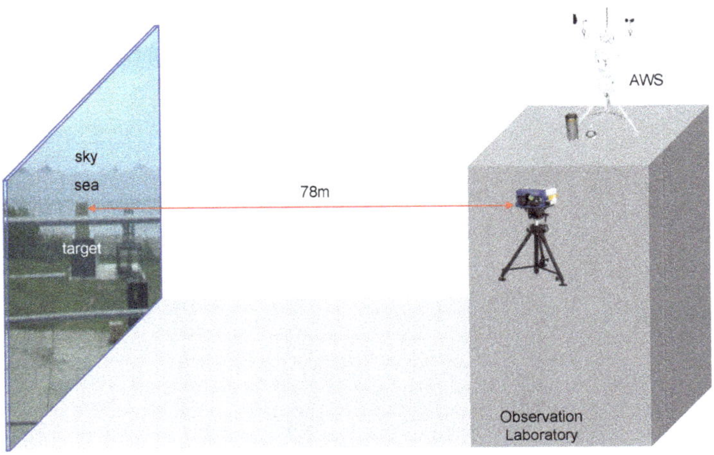

Figure 1. Measurement scenario in an outdoor environment: FTIR is located in an air conditioned room and AWS is installed on the roof. The background scene is 78 m away from the camera.

The objective of this research is to estimate air temperature spatially and to visualize the temperature distribution. A cropped hypercube image provides 128 × 200 × 374 (width × height × bands) data. A hyperspectral spectral imager (HSI) database was recorded daily from February to July 2018. Recording was done three times a day (10:30, 13:30, and 15:30 h) and four times an hour for a total of 12 times per day. In addition, an automatic weather station (AWS) recorded environmental information, such as the date, time, air temperature, humidity, air pressure, and visibility. The measured air temperatures were used to evaluate the temperature estimation accuracy by the proposed VisuaAT method.

2.2. FTIR Data Acquisition Interface

The selection of the CO_2 absorption band is critical in the proposed VisualAT method. An initial baseline band range was selected by the developed GUI shown in Figure 2. The midwave FTIR spectral image software platform was developed for hypercube image display and spectral profile analysis by varying the wavelength parameter and units. The top left image in Figure 2 represents a spectral image in a specific band, and the lower graph shows a spectral profile at a selected point (indicated by +). The top right image in Figure 2 is a broad-band image with a selected spectral range.

2.3. Radiometric Calibration of FTIR Imaging

The wavenumber and radiometric calibration of midwave hyperspectral data should be done accurately to measure air temperature. The technical details are described in [24] and one full set of data including the raw IR spectrum, internal calibration, and temperature extraction is described in the following paragraphs. A Michelson interferometer produces interferomgrams by moving a mirror. Figure 3a presents an interferogram image at optical path difference (OPD) ID 500 (total 1186). Figure 3b gives an example of the whole interferogram at pixel (60, 60). Figure 3c shows the results of spectrum extraction by applying the fast Fourier transform (FFT) to the interferogram (Figure 3b). The unit of the *y-axis* in Figure 3c is just spectral intensity in arbitrary units. The wavelength calibration is performed using the HeNe laser (wavelength $\lambda = 632.8$ nm). Figure 3d shows the wavenumber calibrated spectrum.

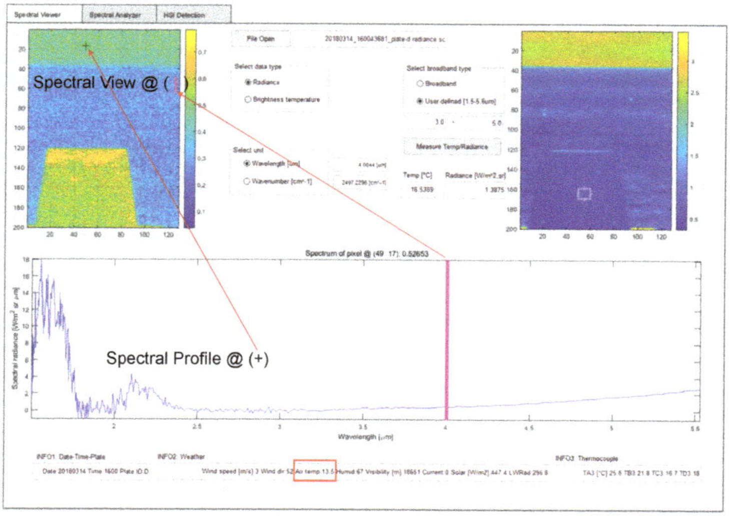

Figure 2. Midwave FTIR image analysis software platform. A spectral image at a selected wavelength can be visualized interactively.

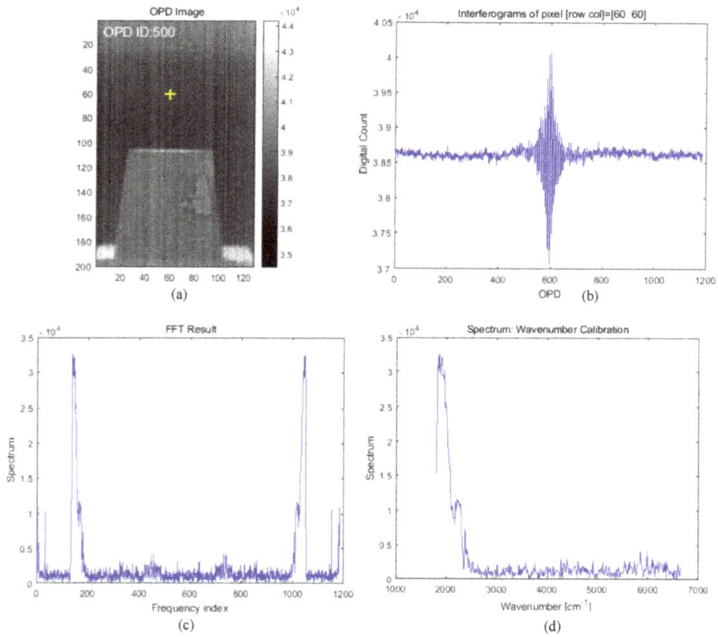

Figure 3. Spectral calibration process: (**a**) interferogram image at optical path difference (OPD) ID = 500 (ZPD); (**b**) interferogram at a pixel ((row, col) = (60, 60)); (**c**) fast Fourier transform (FFT) results; (**d**) wavenumber calibration results.

The next step is to calibrate radiometrically using two blackbodies. The HYPER-CAM MWE can provide the spectral radiance data using two built-in BBs (hot, cold) [25]. Figure 4 shows the measured interferogram image, spectra (arbitrary units), and calculated spectral radiances for the hot (95 °C) and cold (25 °C) blackbodies. Figures 5a,b show the estimated gain and offset magnitude at pixel (60, 60), respectively. Figure 5c,d presents the calculated spectral radiance with the unit of wavenumber and wavelength, respectively.

The amount of spectral radiance energy can be converted into equivalent brightness temperatures [26]. By inverting Equation (7), the temperature T (K) can be obtained as

$$T = \frac{(hc/k)\widetilde{\nu}}{In[2hc^2\widetilde{\nu}^3/L_S(\widetilde{\nu})+1]}. \quad (1)$$

2.4. MODTRAN Simulator

The moderate-resolution atmospheric radiance and transmittance model (MODTRAN, http://modtran.spectral.com/) is used in this paper to simulate atmospheric transmittance and path thermal calculation. Figure 6 shows a GUI interface to simulate spectral path radiance by setting geometric and atmospheric parameters.

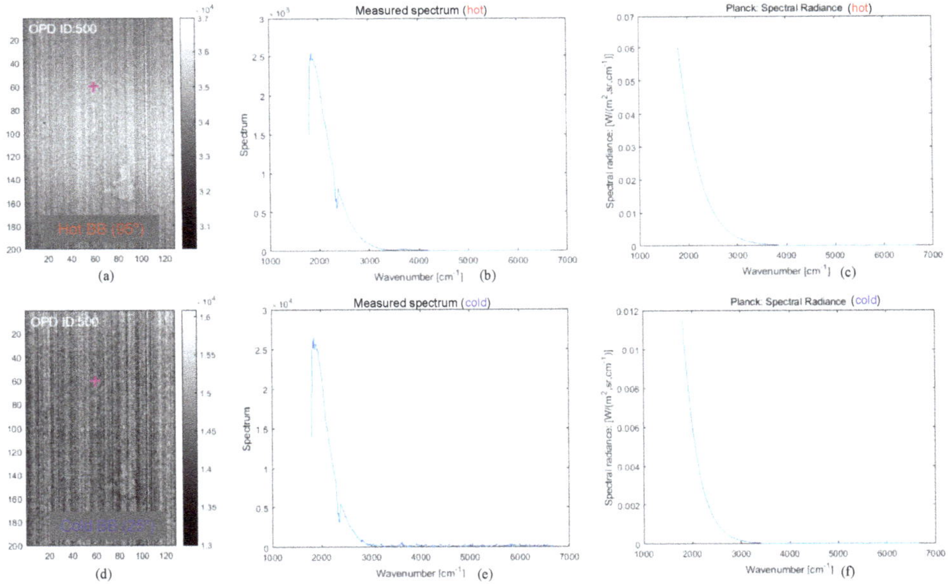

Figure 4. Blackbody spectrum and spectral radiance extraction for a radiometric calibration: (**a**) interferogram image of a hot blackbody (95 °C); (**b**) measured spectrum of a hot blackbody; (**c**) calculated spectral radiance at hot temperature; (**d**) interferogram image of a cold blackbody (25 °C); (**e**) mesured spectrum of a cold blackbody; (**f**) calculated spectral radiance at cold temperature.

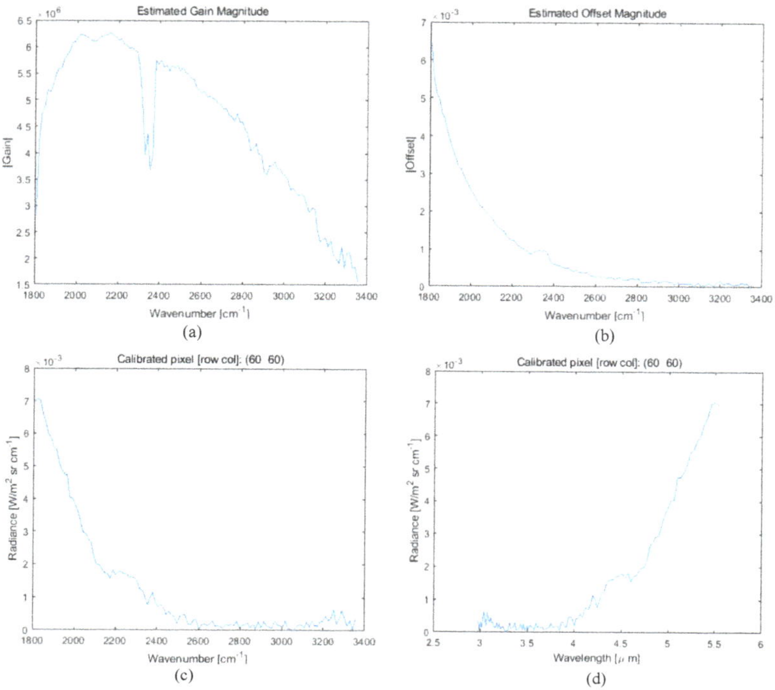

Figure 5. Radiometric calibration and spectral radiance extraction: (**a**) estimated gain magnitude; (**b**) estimated offset magnitude; (**c**) spectral radiance vs. wavenumber; (**d**) spectral radiance vs. wavelength.

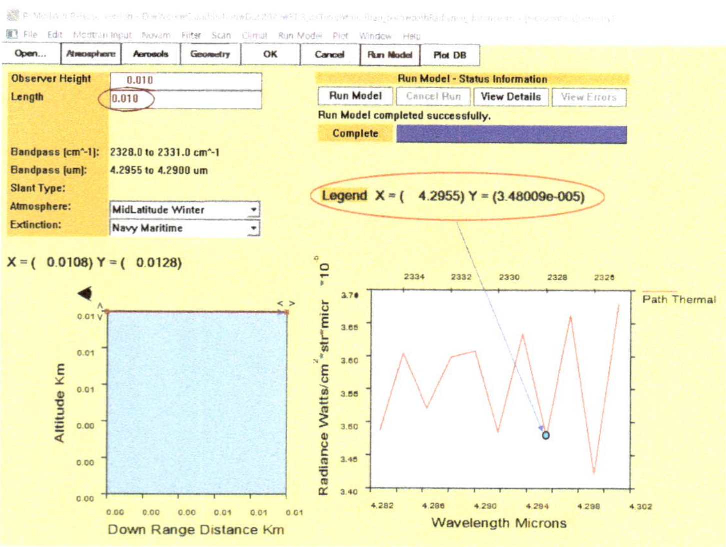

Figure 6. MODTRAN simulation environment for path thermal calculation.

3. Proposed Visual Air Temperature Measurement Method

3.1. Derivation of Radiative Ttransfer Equation

Figure 7 shows the air temperature measurement scenario. Air temperature from VisualAT measurement can be derived from radiative transfer Equation (2). We adopt the radiative transfer equation used in MODTRAN [27]. In general, at-sensor received radiance in the midwave infrared (MWIR) region consists of opaque object-emitted radiance, reflected downwelling radiance, and total atmospheric path radiance (thermal+solar components).

$$L_{obs}(\lambda) = \tau(\lambda)\left[\varepsilon(\lambda)L_{obj}(\lambda, T_{obj}) + (1-\varepsilon(\lambda))(L_s^{\downarrow}(\lambda) + L_t^{\downarrow}(\lambda))\right] + L_s^{\uparrow}(\lambda) + L_t^{\uparrow}(\lambda) \quad (2)$$

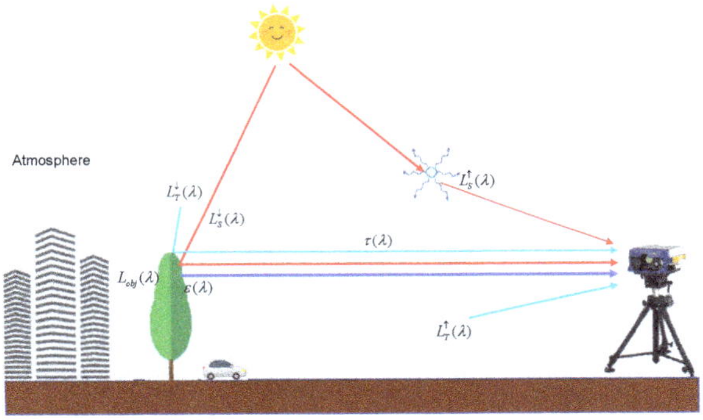

Figure 7. Operational concept of visual air temperature (VisualAT) measurement using the passive open path Fourier transform infrared (FTIR) imaging system.

$L_{obs}(\lambda)$ is the at-sensor radiance; λ is wavelenth; $\varepsilon(\lambda)$ is spectral object surface emissivity; $L_{obj}(\lambda, T_{obj})$ is the spectral radiance of the object, assuming a blackbody in the Planck function with object surface temperature (T_{obj}). $L_s^{\downarrow}(\lambda)$ and $L_t^{\downarrow}(\lambda)$ represent spectral downwelling solar radiance and thermal irradiance, respectively; $\tau(\lambda)$ is the spectral atmospheric transmittance, and $L_s^{\uparrow}(\lambda)$ and $L_t^{\uparrow}(\lambda)$ are the spectral upwelling solar and thermal path radiance, respectively, reaching the sensor.

According to the MODTRAN simulation in the MWIR band, the spectral transmittance of the carbon dioxide (CO_2) band (4.25–4.35 µm) decreases abruptly with distance, as shown in Figure 8. The average transmittance in the CO_2 band is 0.5, 0.13, 0.03, 0.005, 0.0001, and 0 at 1 m, 5 m, 10 m, 20 m, 50 m, and 100 m, respectively. If we consider only the CO_2 band (λ_{CO_2}) with a minimum 20 m object distance, the transmittance ($\tau(\lambda_{CO_2})$) can be regarded as 0, which leads to Equation (3). An MWIR FTIR camera receives only the upwelling of path solar and thermal radiances in the λ_{CO_2} band where the range is normally 4.25–4.35 µm.

$$L_{obs}(\lambda_{CO_2}) = L_s^{\uparrow}(\lambda_{CO_2}) + L_t^{\uparrow}(\lambda_{CO_2}) \quad (3)$$

According to the MWIR radiometric characteristics [28], the contribution of solar radiance ($L_s^{\uparrow}(\lambda_{CO_2})$) from air scattering is very small, even for very dry conditions (less than 2% at 5 µm) [28]. Ignoring the first term, we can simplify Equation (3) into Equation(4):

$$L_{obs}(\lambda_{CO_2}) = L_t^{\uparrow}(\lambda_{CO_2}) \quad (4)$$

The definition of thermal upwelling is defined with Equation (5):

$$L_t^{\uparrow}(\lambda_{CO_2}) = (1 - \tau(\lambda_{CO_2}))B(\lambda_{CO_2}, T_{air}) \quad (5)$$

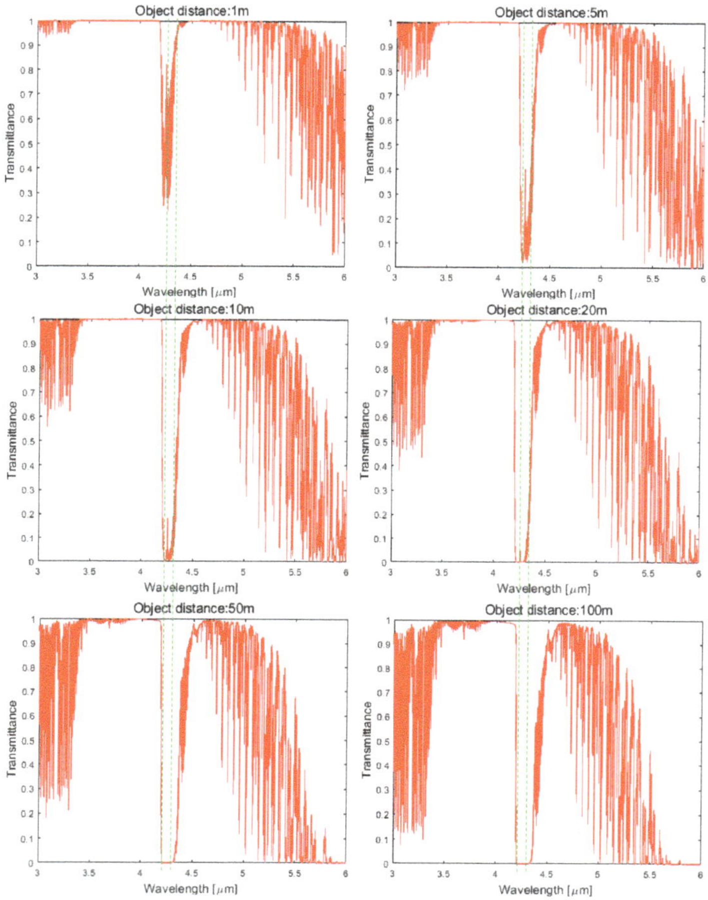

Figure 8. Spectral transmittance according to object-camera distance in the MWIR band. Note the abrupt absorption at 4.25–4.35 μm.

Since the spectral transmittance in the CO_2 band is 0 ($\tau(\lambda_{CO_2}) = 0$), the final form is approximated in Equation (6):

$$L_{obs}(\lambda_{CO_2}) \simeq B(\lambda_{CO_2}, T_{air}) \quad (6)$$

where $B(\lambda_{CO_2}, T_{air})$ denotes the spectral radiance ([W/(m^2·sr·μm)]) of a blackbody (Planck's law [29]), and T_{air} is the air temperature in degrees Kelvin $[K]$ of the atmosphere between the object and the camera sensor. The spectral radiation of atmosphere is modeled as blackbody [30–32]. Atmospheric path radiance can be described in difference methods, but the simplest way is to model the particles as blackbodies [32]. $B(\lambda_{CO_2}, T_{air})$ is defined with Equation (7):

$$B(\lambda_{CO_2}, T_{air}) = \frac{2hc^2}{\lambda^5(e^{hc/\lambda_{CO_2}kT_{air}} - 1)} \quad (7)$$

where h denotes Planck's constant, c is the speed of light, and k is the Boltzmann constant.

The amount of spectral radiance energy can be converted into equivalent brightness temperatures ([33]). By inverting Equation (7), temperature T_{air} $[K]$ can be obtained as follows:

$$T_{air}(\lambda_{CO_2}) = \frac{hc}{\lambda_{CO_2}k \ln\left(2hc^2/\lambda_{CO_2}^5 B(\lambda_{CO_2}, T_{air}) + 1\right)}. \quad (8)$$

3.2. *VisualAT: Proposed Visual Air Temperature Measurement*

Figure 9 summarizes the overall processing flow of the proposed VisualAT method. The first row represents the three steps in spectral brightness air temperature extraction using Equation (8), described as follows.

(1) Given is an MWIR hyperspectral image cube (374 bands, spatial resolution: 200 × 128).
(2) CO_2 band images (4.29, 4.31, 4.34 µm) are selected. The band region between 4.25–4.35 µm is selected based on the MODTRAN-based transmissivity analysis ($\tau(\lambda) = 0$) and visual inspection. A wavelength can be in the CO_2 absorption band if there is no object signature and it looks like a noisy image.
(3) The selected spectral radiance images are converted to spectral brightness images using Equation (8).
The second row of Figure 9 represents the image processing for visual air temperature image generation, explained in the following steps.
(4) A raw temperature image is extracted via pixel-wise temperature mean filter along the spectral axis. It still shows a noise-like image consisting of salt-and-pepper noise and thermal noise. This can be removed by consecutive spatial 2D median filtering and Gaussian filtering [34]. The Gaussian filtering is adopted to reduce spatial thermal noise. The empirically tuned kernel size of the median filter is 10 × 15, and sigma of the Gaussian filter is set to 2.

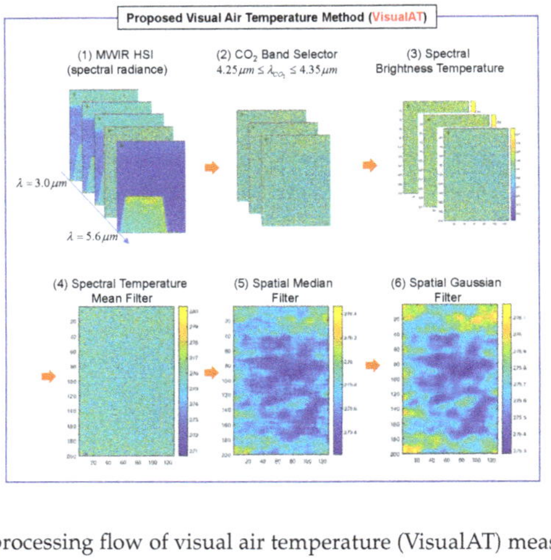

Figure 9. Overall processing flow of visual air temperature (VisualAT) measurement method.

Figure 10a shows the spectral brightness temperature by applying Equation (1) to the calibrated spectral radiance at pixel (60, 60). Figure 10b shows the enlarged brightness temperature around CO_2 band (4.29–4.34 μm). Figure 10c represents the spatial raw temperature distribution of the CO_2 band image. Final visual air temperature image (Figure 10e) is acquired through the median and Gaussian filtering processes.

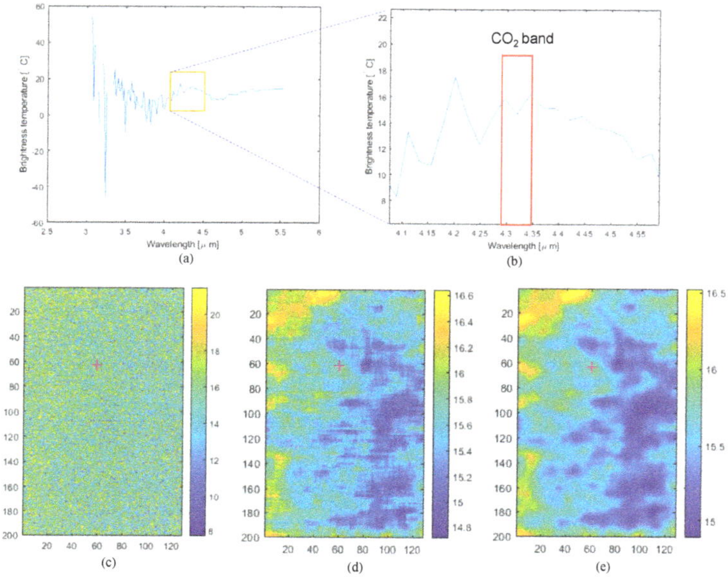

Figure 10. Brightness temperature extraction and VisualAT results: (**a**) brightness temperature; (**b**) enlarged brightness temperature with CO_2 band region; (**c**) CO_2 band image; (**d**) median filtered image; (**e**) Gaussian filtered temperature image.

3.3. *Analysis of Air Temperature Measurement*

The proof of radiometric air temperature measurement and visualization is possible by mathematical derivations as Equations (2)–(8). However, it is a challenging problem to validate the air temperature measurement and visualization experimentally because we need a huge dark room with controllable air temperature. In this subsection, we analyze the properties of the VisualAT method by demonstrating extreme air temperature measurement and by thermal air flow visualization. Figure 11 demonstrates the air temperature measurement and visualization using the hot summer and cold winter data set. Upper row images represent visual temperature extraction process and camera's internal temperature information for hot summer data. The ground truth of air temperature is 26.4 °C and the estimated temperature is 25.6 °C. Likewise, lower row images represent the same process for cold winter data. Note that the ground truth of air temperature is 2.6 °C and the estimated temperature is 2.52 °C. The internal camera temperatures of IR lens, front wall, beam splitter, and etc are approximately 29 °C and they do not affect to the air temperature estimation because hot/cold blackbody-based radiometric calibration can remove the effect of stray light before each measurement. The proposed VisualAT can visualize thermal air flow using consecutive FTIR hyper-cubes acquired in very short period (1 s). Figure 12 shows the air flow directions indicated by the curves and arrows. Because the sea wind is so strong, the thermal air flow changes dynamically in very short time. Therefore, this result can be another indirect proof of imaging variations in air temperature.

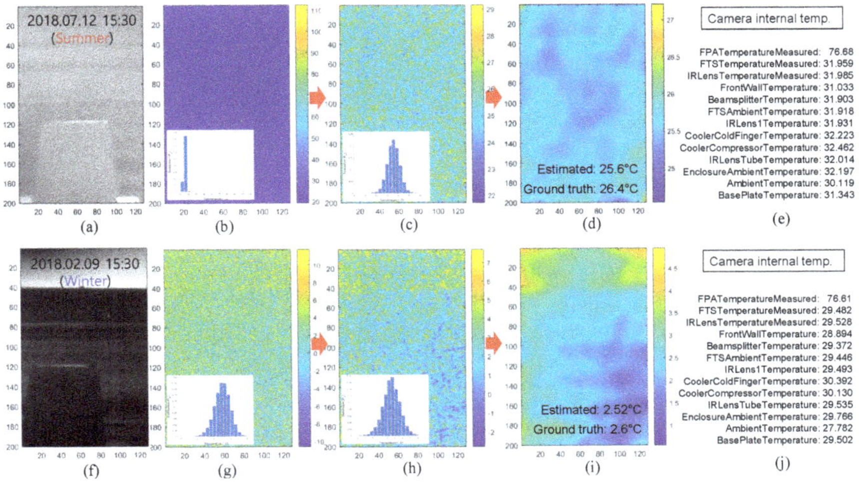

Figure 11. Indirect proof of air temperature measurement by extreme weather conditions: Hot summer-(**a**) broad-band image, (**b**) CO_2 band image, (**c**) median filtered image, (**d**) Gaussian filtered image, (**e**) camera internal temperature information; Cold winder-(**f**) broad-band image, (**g**) CO_2 band image, (**h**) median filtered image, (**i**) Gaussian filtered image, (**j**) camera internal temperature information.

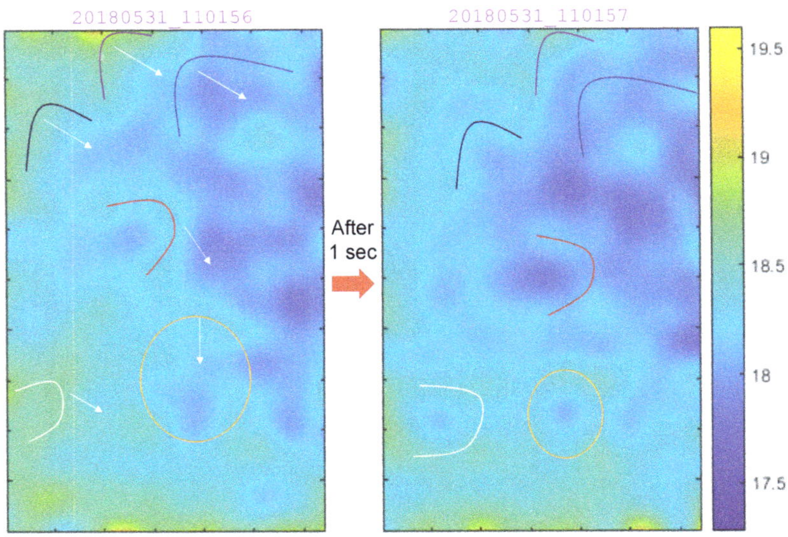

Figure 12. Indirect proof of air temperature measurement by air flow visualization. The arrows indicate the directions of air flow.

3.4. Signal Analysis of Air Temperature Monitoring

It is important to analyze how the IR radiation from different parts of the probed air contributes to the received thermal signal. Figure 13 represents the geometric relationship between atmosphere plane and a pixel. If an atmospheric plane at distance R is considered, the probing area (A) is $a' \times b'$ using the instantaneous field of view (IFOV). The received thermal flux (Φ_λ) at the pixel detector is $\Phi_\lambda = (1 - \tau(\lambda_{CO_2})) \cdot L_\lambda(T_{air}) \cdot A \cdot \Omega \cdot \tau_o$ where Ω is solid angle and τ_o is lens transmittance [35]. Because air particles are regarded as blackbodies [32], the emissivity is regarded as 1. If we use basic geometrical relationships, the final form is changed to $\Phi_\lambda = (1 - \tau(\lambda_{CO_2})) \cdot L_\lambda(T_{air}) \cdot A_{IFOV} \cdot \Omega_{IFOV} \cdot \tau_o$ where A_{IFOV} denotes the pixel area and Ω_{IFOV} represents the solid angle of IFOV. The received thermal flux is strongly related to $(1 - \tau(\lambda_{CO_2}))$. The simulation of atmospheric transmittance is conducted by MODTRAN 4.0 and Figure 14a represents the result. As the distance increases up to 20 m, the thermal contribution of air particles increases. This analysis is confirmed through the MODTRAN-based path thermal simulation as shown in Figure 14b and Figure 6. Therefore, we can conclude that the received air flux is contributed dominantly by the air temperature at 20 m distance.

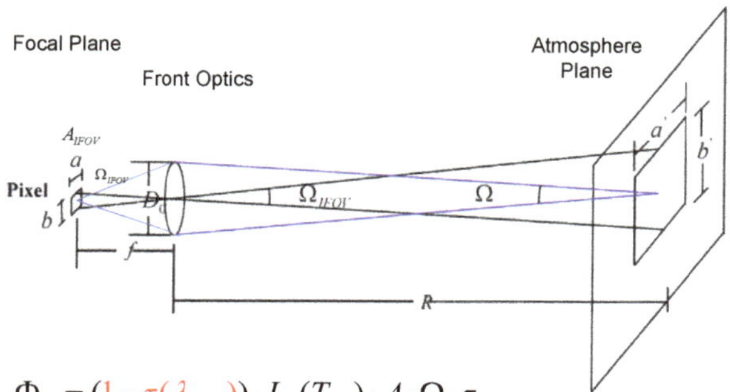

$$\Phi_\lambda = (1 - \tau(\lambda_{CO_2})) \cdot L_\lambda(T_{air}) \cdot A \cdot \Omega \cdot \tau_o$$

$$A \cdot \Omega = \frac{a'^2 \pi D_0^2}{4R^2}, \text{ since } \frac{a'}{R} = \frac{a}{f} \qquad L_\lambda(T_{air}) = \frac{2hc^2}{\lambda^5} \frac{1}{e^{hc/\lambda k T_{air}} - 1} \qquad A = a'b' = a'^2$$

$$\frac{a^2 \pi D_0^2}{4f^2} = A_{IFOV} \cdot \Omega_{IFOV} \qquad \Omega = \frac{\pi D_0^2}{4R^2}$$

$$\Phi_\lambda = (1 - \tau(\lambda_{CO_2})) \cdot L_\lambda(T_{air}) \cdot A_{IFOV} \cdot \Omega_{IFOV} \cdot \tau_o$$

Figure 13. Geometrical model of received thermal flux in a pixel.

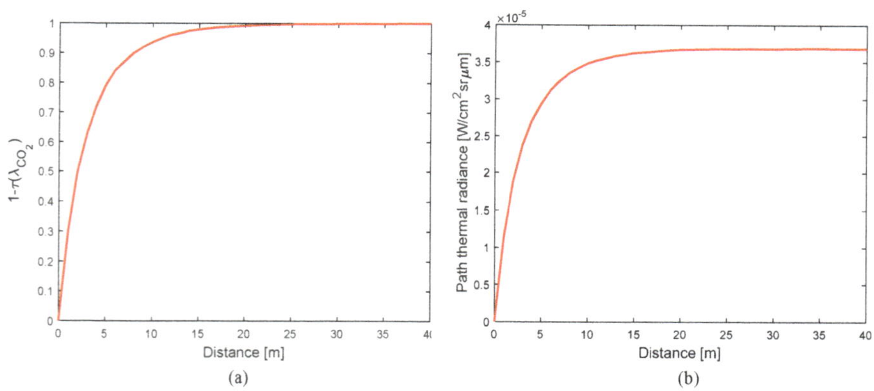

Figure 14. (**a**) (1-atmospheric transmittance) vs. distance (MODTRAN-based simulation at CO_2 absorption band (4.29 μm)), (**b**) path thermal radiance vs. distance (MODTRAN-based simulation).

3.5. Performance Metric

The mean absolute error (MAE) metric was used to compute the performance of the VisualAT method in predicting air temperature. The MAE metric is defined in Equation (9):

$$MAE = \frac{1}{N}\sum_{k=1}^{N}\left|T_{air}^{k} - T_{GT}^{k}\right| \tag{9}$$

where T_{air}^{k} denotes the $k-th$ predicted air temperature by the VisualAT method, and T_{GT}^{k} is the corresponding air temperature measured by the AWS, as shown in Figure 15.

Date-Time	Wind (m/s)	Max Wind (m/s)	Direction (°)	Max Direction (°)	**Air Temperature (°C)**	Humidity (%)	Pressure (hPa)	Visibility (m)
2018-03-14 0:01	0.2	0.5	120	118	**8.4**	84	1017.6	18238
2018-03-14 0:01	0.2	0.6	95	315	**8.4**	84	1017.7	18238
2018-03-14 0:01	0.3	1.7	66	6	**8.4**	84	1017.7	17842
2018-03-14 0:01	0.4	2	40	338	**8.4**	84	1017.7	17454
2018-03-14 0:01	0.5	2	20	338	**8.4**	84	1017.7	17454
2018-03-14 0:01	0.6	2	7	338	**8.4**	84	1017.7	17516
2018-03-14 0:02	0.8	2	1	338	**8.5**	84	1017.7	17165
2018-03-14 0:02	0.9	2	1	338	**8.5**	84	1017.7	17165
2018-03-14 0:02	1.1	2.2	3	22	**8.5**	84	1017.7	16104
2018-03-14 0:02	1.2	2.2	9	22	**8.5**	84	1017.7	15168
2018-03-14 0:02	1.2	2.2	16	22	**8.5**	84	1017.7	15168
2018-03-14 0:02	1.2	2.2	25	22	**8.5**	84	1017.7	14326
2018-03-14 0:03	1.2	2.2	34	22	**8.5**	84	1017.7	13516
2018-03-14 0:03	1.2	2.2	43	22	**8.5**	84	1017.6	13516
2018-03-14 0:03	1.2	2.2	52	22	**8.5**	84	1017.6	13452
2018-03-14 0:03	1.2	2.2	58	22	**8.5**	84	1017.6	13632
2018-03-14 0:03	1.2	2.2	63	22	**8.5**	84	1017.6	13632
2018-03-14 0:03	1.2	2.2	64	22	**8.5**	84	1017.6	13904
2018-03-14 0:04	1.2	2.2	64	22	**8.5**	84	1017.6	13844
2018-03-14 0:04	1.1	2	62	28	**8.5**	84	1017.6	13844
2018-03-14 0:04	1.1	1.6	56	45	**8.5**	84	1017.6	13894
2018-03-14 0:04	1.1	1.6	51	51	**8.5**	84	1017.6	13435
2018-03-14 0:04	1	1.6	46	51	**8.5**	84	1017.6	13435
2018-03-14 0:04	1	1.6	43	51	**8.5**	84	1017.6	13637
2018-03-14 0:05	1	1.6	41	51	**8.5**	84	1017.6	14544
2018-03-14 0:05	0.9	1.6	41	51	**8.5**	84	1017.6	14544
2018-03-14 0:05	0.9	1.6	41	51	**8.5**	84	1017.6	14766
2018-03-14 0:05	0.9	1.2	42	51	**8.5**	84	1017.6	16395
2018-03-14 0:05	0.8	1.2	43	51	**8.5**	84	1017.6	16395
2018-03-14 0:05	0.8	1.2	43	51	**8.5**	84	1017.6	15837
2018-03-14 0:06	0.7	1.4	41	39	**8.5**	84	1017.6	15290
2018-03-14 0:06	0.8	2.1	39	45	**8.5**	84	1017.6	15290

Figure 15. AWS information: date and time; wind speed, maximum wind speed, average wind direction, and maximum wind direction; air temperature, humidity, and pressure; and visibility.

Figure 16 shows the experimental environment. Figure 16a is the outdoor environment acquired via visible band camera, and Figure 16b shows a recorded broad-band image from summing all the spectral band images (1.5–5.6 μm). Figure 16c represents MODTRAN-based spectral transmittance at the object distance of 78 m, and where the average transmittance of the CO_2 band is 0.0001.

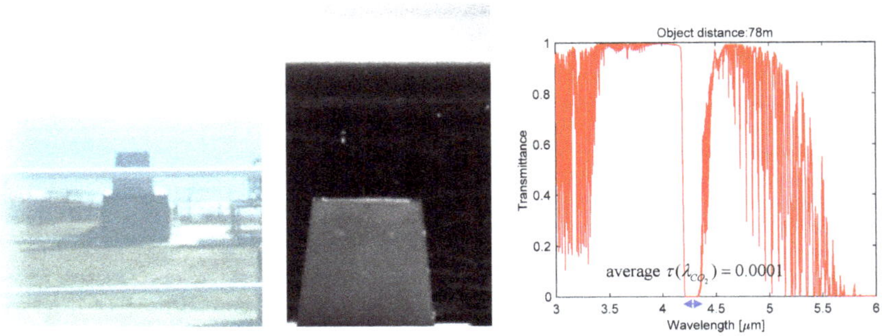

(a) (b) (c)

Figure 16. An example of experiment environment: (**a**) outdoor environment acquired by a visible band camera, (**b**) broad-band infrared image, and (**c**) spectral transmittance in MWIR band.

3.6. Parameter Analysis

A midwave hyperspectral image database was prepared for various evaluations. Hypercube images from 49 days were valid during the acquisition period (February to July 2018). In the first evaluation, the effect of the CO_2 band range was important in order to estimate air temperature accurately. As summarized in Figure 17, baseline bands were selected from a visual inspection using the GUI shown in Figure 2. Top row images in Figure 17 show the spectral images corresponding to specific wavelengths. The visual selection criterion was whether the spectral image looks like it has noise in the whole image area. If atmospheric transmittance is 0, the received radiance consists of thermal path (air) radiance. Therefore, the initially selected bands were 4.22, 4.27, 4.29, 4.31, and 4.34 μm. The MAE of the baseline band showed 1.32 K. If a lower band decreased to 4.20 μm, the MAE decreased to 1.29 K. If the lower band increased to 4.27, 4.29, and 4.31 μm individually, the corresponding MAEs were 1.26, 1.25, and 1.27 K, respectively. From these experiments, the lower band limit can be set at 4.29μm. On the other hand, if we increase the upper band limit to 4.36 μm with the selected lower band limit at 4.29 μm, the MAE increased to 1.29 K. Therefore, we can conclude that the optimal CO_2 absorption bands are 4.29, 4.31, and 4.34 μm. These bands were used in the following experiments.

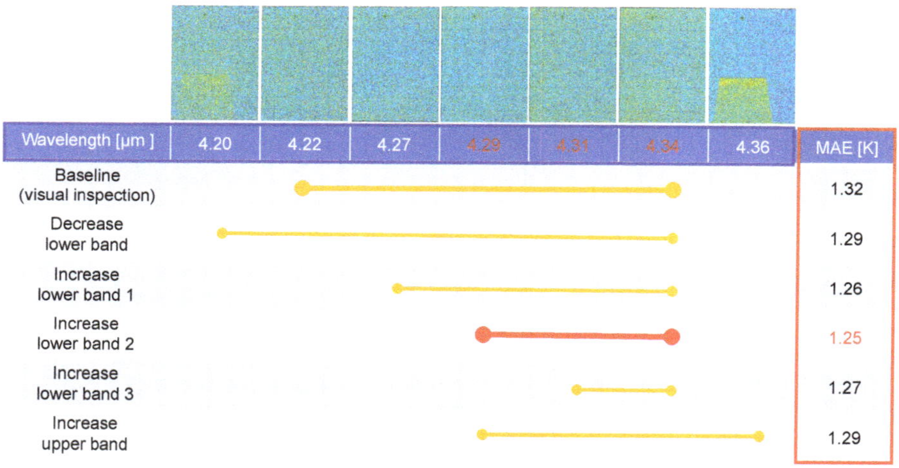

Figure 17. CO_2 band range selection results: Baseline band is selected by visual inspection and the optimal band range is selected by maximizing mean absolute error (MAE) metric.

SNR can be improved additionally through the 2D median filter and 2D Gaussian filter. The 2D median filter is necessary to remove dead pixels as shown in the top-left of Figure 18 where the effects of median filter sizes from [1 1] to [15 15] were visualized. Even [3 3] median filter can remove the salt and pepper noise effectively and larger filter size can extract larger structure of air temperature distribution. The histogram distribution after the 2D median filter is Gaussian distribution, which is consensus to thermal noise distribution. True temperature signal with Gaussian noise can be estimated by Gaussian smoothing filter (unbiased, consistent linear estimator) [36]. The effects of sigma (σ) were displayed in Figure 19 where [3 3] median filter was used initially. Note that larger σ can extract larger structure of air temperature distribution. The selection of median filter size and Gaussian filter parameter depends on application images. If salt and pepper noise is strong, larger median filter size with smaller Gaussian smoothing is suitable. If the salt and pepper noise is weak, smaller median filter size with stronger Gaussian smoothing is recommended. In this paper, we used a kernel size of

10 × 15 for the median filter and $\sigma = 2$ for the Gaussian filter because salt and pepper noise is spread horizontally and thermal noise is high.

The VisualAT method depends on the spectral radiance of atmosphere. The sensitivity analysis and error propagation can be useful to understand the properties of VisualAT. The simulation is conducted by adding spectral radiance noise to Equation (7) and estimating air temperature from Equation (8). Figure 20 shows the sensitivity analysis results. The ground truth air temperature is 20 °C and it increases linearly according to the spectral noise level. In the case of FTIR camera noise (NESR, 7×10^{-5} [W/m^2,sr,cm^{-1}]), the estimated temperature is 21.08 °C.

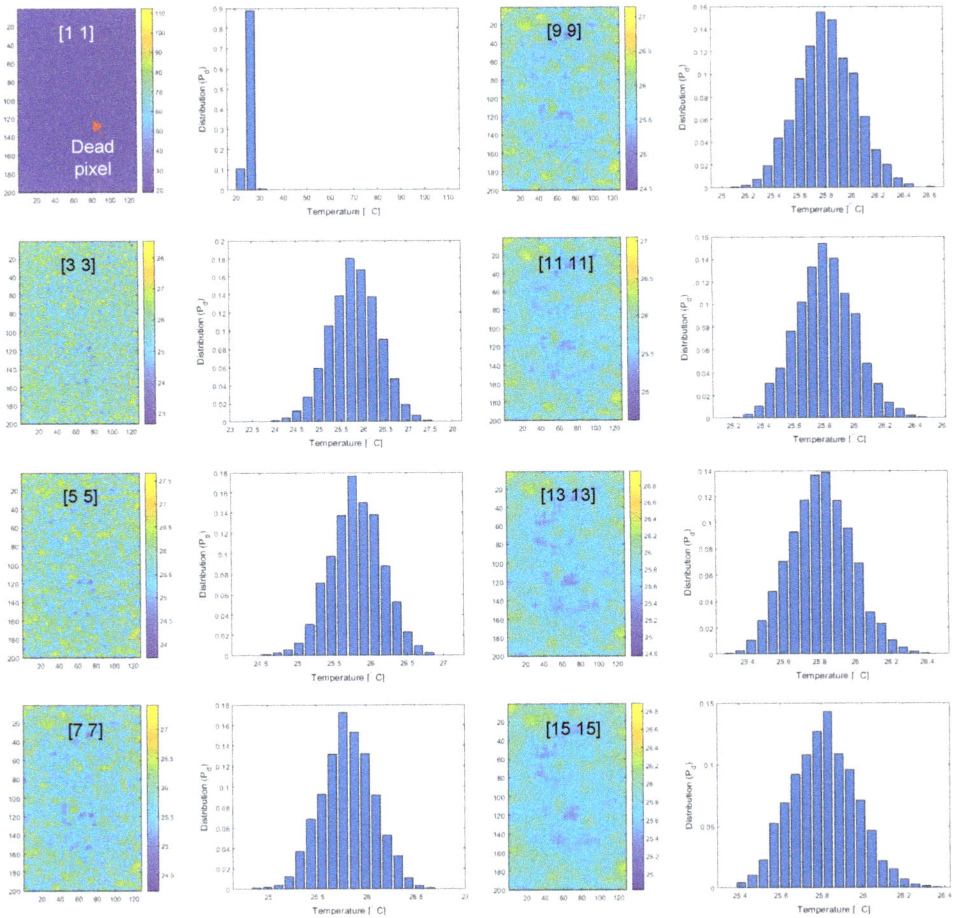

Figure 18. Spatial effects of 2D median filter size and histogram distribution.

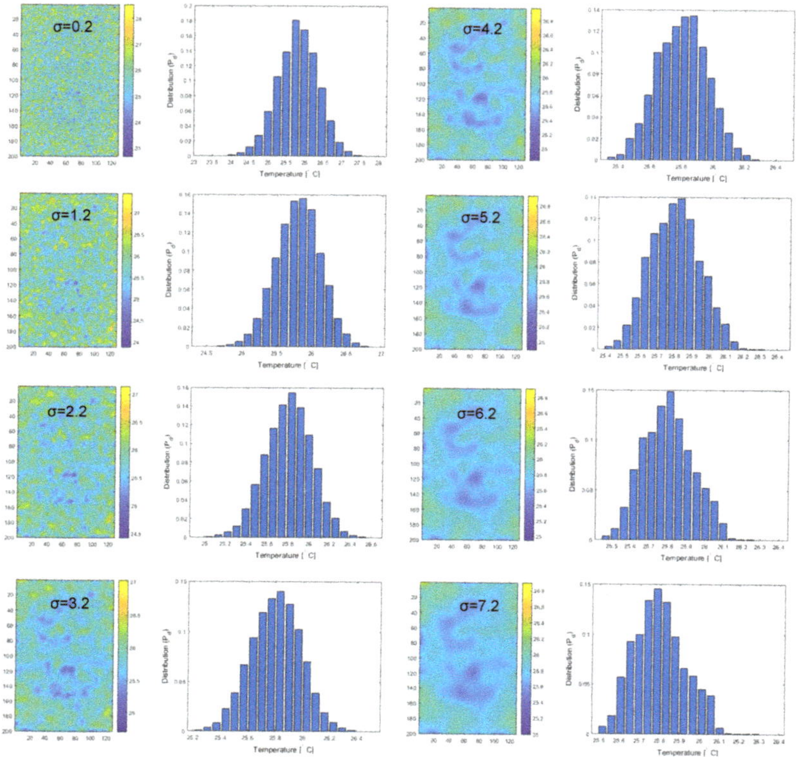

Figure 19. Spatial effects of 2D Gaussian filter with σ.

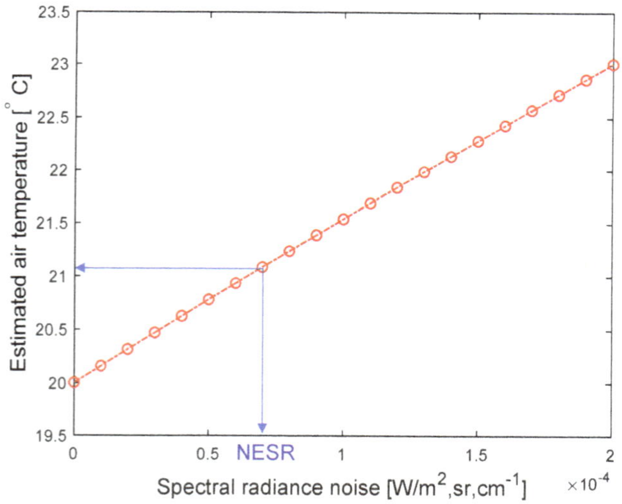

Figure 20. Spectral noise and estimated air temperature analysis.

4. Experiment Results

Radiometric accuracy should be evaluated to validate the usefulness of the proposed VisualAT method. Although VisualAT can present measured temperature spatially, a global mean temperature per image was extracted to compare with the ground truth air temperature recorded by the AWS. Figure 21 shows a quantitative evaluation graph between the air temperature estimated by the proposed VisualAT method and the ground truth air temperature for the 49-day dataset (acquired time: 15:30 h). The MAE was 1.25 K, which is accurate considering the remote sensing method and dynamic weather changes. The correlation coefficient between the VisualAT-based air temperature and ground truth air temperature was 0.97. Figure 22 shows representative visual air temperature images measured using the VisualAT method. To the best of our knowledge, there is no method by which to compare these results to others, and this is the first trial of instantaneous ground air temperature measurement and visualization in a ground-based approach.

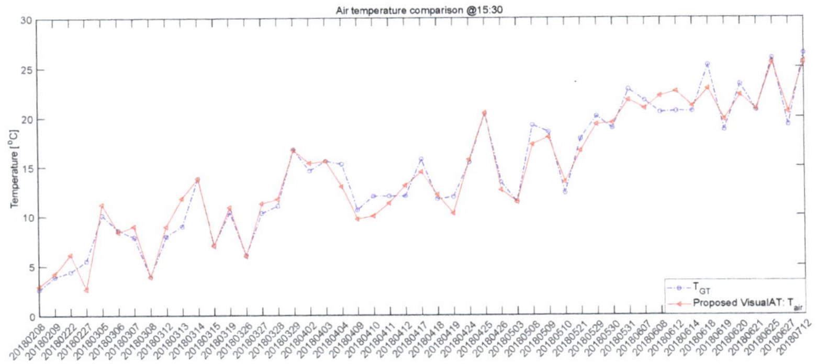

Figure 21. Quantitative evaluation graph of the proposed VisualAT method for the whole database acquired from February to July 2018.

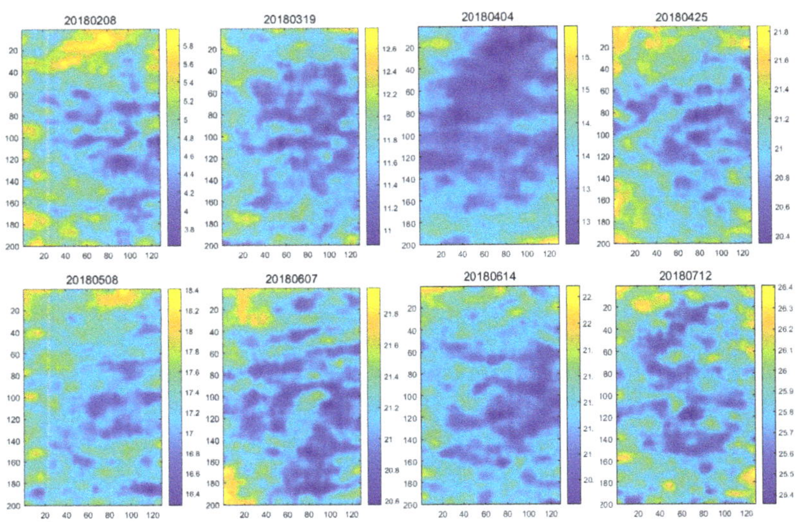

Figure 22. Examples of air temperature images measured by the proposed VisualAT method for different months (from February 2018–July2018).

The dependency of temperature estimation accuracy from the proposed VisualAT is important in order to check robustness to various environmental changes. Figure 23 summarizes the relationships between the estimation error [°C] and other factors, such as (a) air temperature [°C], (b) humidity [%], (c) air pressure [hPa], (d) visibility [m], and (e) long-wave thermal radiation [W/m^2]. The correlation coefficient (R) is annotated on the results. According to the results, the estimation error is negative relationship with air temperature, air pressure, and visibility. In addition, the error has positive relationships with humidity. This means that a more accurate temperature can be estimated if the humidity is lower. On the other hand, the long-wave thermal radiation had no specific relationship. The dependency analysis is for reference only because the data points are very scattered.

The best advantage of the proposed VisualAT method is that it can measure visual air temperature instantly. It takes approximately one second to scan a hypercube. Figure 24 shows consecutive air temperature images acquired at 20-min intervals on 6 March 2018. The color axis was set by the caxis ([7.83 9.83]) function in MATLAB for a fair temperature reading. We can see the flow of the heat flux over one hour.

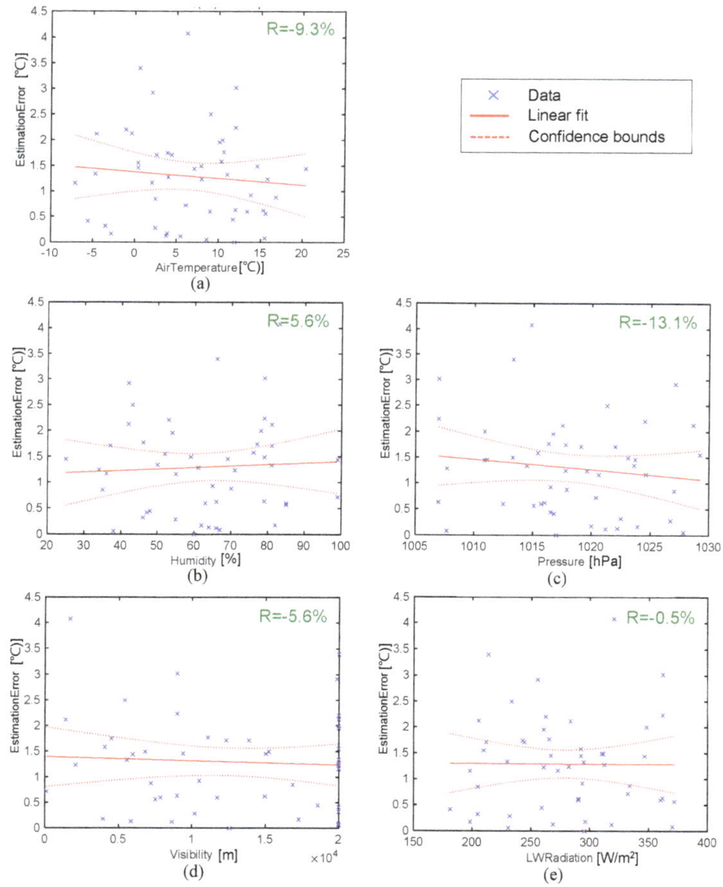

Figure 23. Examples of air temperature images measured by the proposed VisualAT method for different months: (**a**) estimation error vs. air temperature; (**b**) estimation error vs. humidity; (**c**) estimation error vs. pressure; (**d**) estimation error vs. visibility; (**e**) estimation error vs. long wave thermal radiation.

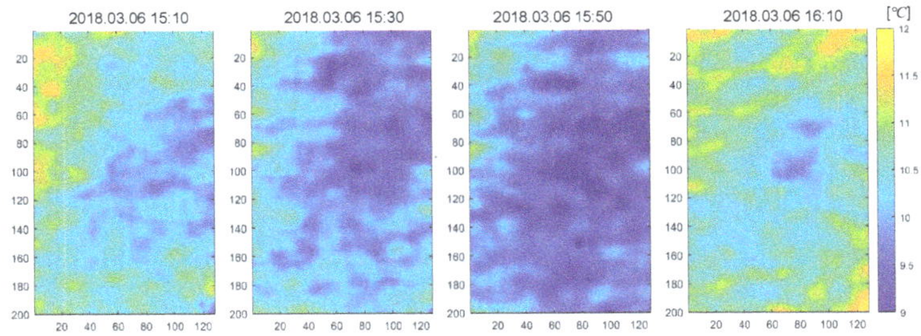

Figure 24. VisualAT-based dynamic temperature images measured at different times (20-min. intervals): Temporal variation of air temperature distribution can be found.

Figure 25 shows the visual air temperature images measured at night (21:29 h) and the next early morning (09:23 h). There was strong sea fog, and VisualAT can still produce air temperature images.

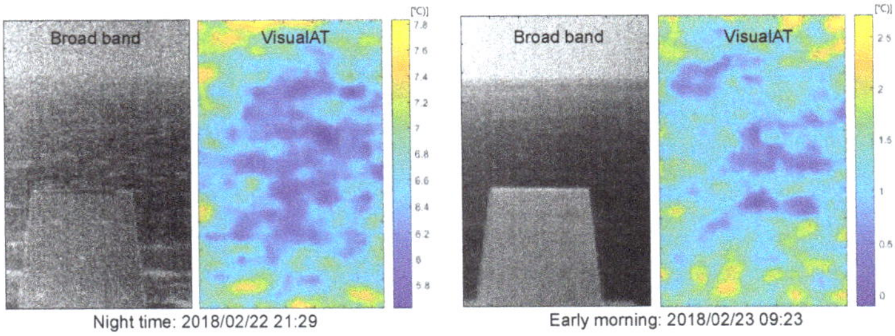

Figure 25. Example of visual air temperature images at night and in the morning: (left) night time air temperature image at 21:29; (right) early morning temperature image at 09:23.

Figure 26 shows VisualAT-based remote air temperature measurement and visualization in a sea environment. Top left shows a broad-band image integrating a 1.5–5.6 μm hypercube; Top right is the corresponding visible band image. The bottom image is the measured air temperature distribution from the VisualAT method. The average temperature was 16.73 °C, and relatively hot and cold regions can be found. Figure 27 presents various air temperature visualization results in maritime environment. Different air temperature distributions can be found.

Through various analysis and evaluation, the proposed VisualAT can be a novel method to measure spatial air temperature and visualize its distribution instantly. The CO_2 concentration affects to the atmospheric transmittance [37], which is related with measurement air volume. It can be a useful approach if CO_2 band image is available but there are limitations such as expensive sensor system and relatively small measurement air volume (20 m distance times FOV). If there is an object in the measurement volume, the accuracy of air temperature measurement can be degraded. Partial artifact can appear when the assumption of zero atmospheric transmittance is broken. If radiometric calibration is not perfect, then internal camera device temperature can distort air temperature distribution.

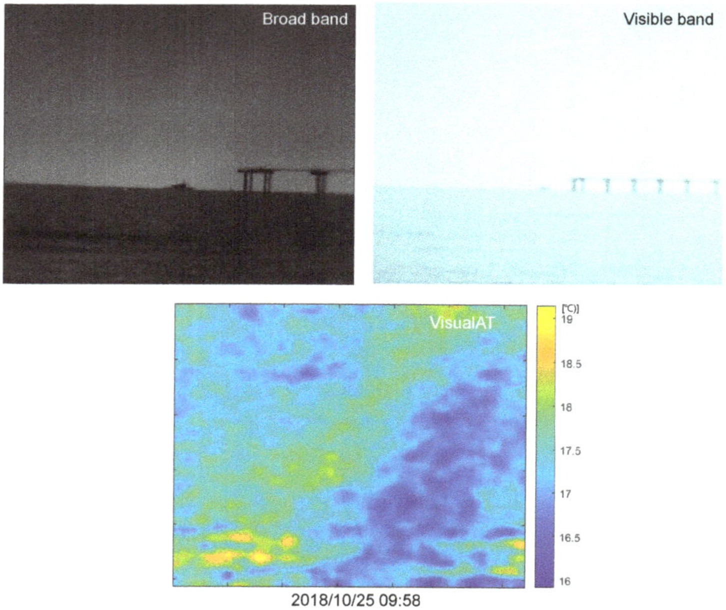

Figure 26. Remote air temperature image visualization example from an outdoor sea environment: (top left) broad band image; (top right) visible band image, (bottom) VisualAT-based air temperature image.

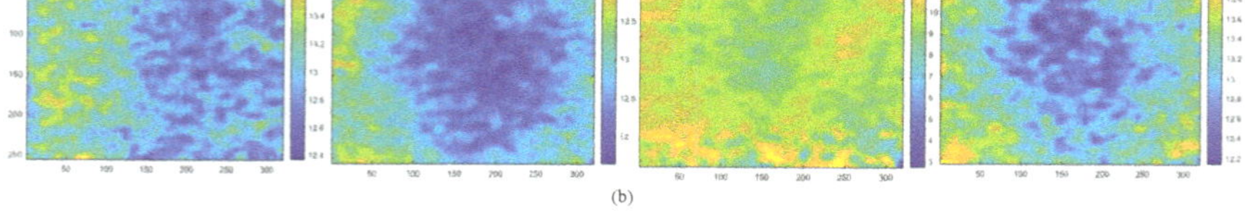

Figure 27. Various remote air temperature image visualization example in maritime environment: (**a**) broadband images; (**b**) corresponding VisualAT images.

5. Discussions and Conclusions

It is very challenging to measure air temperature contactlessly and instantly. This paper proposed a novel air temperature measurement and visualization method called VisualAT, which uses a midwave FTIR in the carbon dioxide absorption band. We found that spatial air temperature can be measured radiometrically by deriving the radiative transfer equation (RTE). The first physical atmospheric property is that atmospheric transmittance in the CO_2 band (4.25–4.35 μm) is 0.005 (at 20 m away), which can remove the effect of object and downwelling radiation. The second physical atmospheric property is that the portion of solar upwelling is very small in the CO_2 band. These two properties lead to a simpler received radiance at the sensor; it is just blackbody radiance of the air temperature.

The rest of the image processing, such as spectral temperature averaging, spatial median filtering, and Gaussian smoothing, produce the final visual temperature image. The proposed VisualAT method is the first to measure and visualize air temperature remotely and instantly. There is no contact sensor, no measurement delay, and no learning for regression. Based on a long-term outdoor experimentss (from February to July 2018, a valid 49 days), the proposed VisualAT method showed an MAE of 1.25 K for temperature range of 2.6–26.4 °C. It is relatively accurate, considering all weather conditions. The measurement error has positive correlations with humidity (R = 5.6%). On the other hand, it has a negative correlation with air temperature (−9.3%), air pressure (R = −13.1%) and visibility (R = −5.6%). There is no relationship with long wave radiance (R = −0.5%). The long-range outdoor test validated the feasibility of visual air temperature measurement and visualization. According to various experiments, the VisualAT can measure air temperature correctly if there is no hot object within 20 m and proper CO_2 absorption band is used. Furthermore, the precise radiometric calibration should be activated before each air temperature measurement to remove stray lights. If these conditions are satisfied, the proposed VisualAT method can be applied to spatial air temperature monitoring applications in fields such as human health, virus propagation, plant growth, climate change, hydrology, etc. In the future, we will use the air temperature information in detecting remote thermal objects.

Funding: This research was funded by the 2020 Yeungnam University Research Grants and Agency for Defense Development (UE191095FD). The APC was funded by MOTIE (P0008473).

Acknowledgments: This work was supported by the 2020 Yeungnam University Research Grants. This study was supported by the Agency for Defense Development (UE191095FD). In addition, this paper was supported by Korea Institute for Advancement of Technology (KIAT) grant funded by the Korea Government (MOTIE)(P0008473, HRD Program for Industrial Innovation).

Conflicts of Interest: The authors declare no conflict of interest.

References

1. Li, L.; Li, Z.L.P. The effects of air temperature on office workers' well-being, workload and productivity-evaluated with subjective ratings. *Appl. Ergon.* **2010**, *42*, 29–36.
2. Lowen, A.C.; Steel, S.M.J.; Palese, P. Influenza virus transmission is dependent on relative humidity and temperature. *PLoS Pathog.* **2007**, *3*, 1–7. [CrossRef] [PubMed]
3. Hatfield, J.L.; Prueger, J.H. Temperature extremes: Effect on plant growth and development. *Weather Clim. Extrem.* **2015**, *10*, 4–10. [CrossRef]
4. Robeson, S.M. Relationships between mean and standard deviation of air temperature: implications for global warming. *Clim. Res.* **2002**, *22*, 205–213. [CrossRef]
5. Lin, S.; Moore, N.J.; Messina, J.P.; DeVisser, M.H.; Wu, J. Evaluation of estimating daily maximum and minimum air temperature with MODIS data in east Africa. *Int. J. Appl. Earth Obs. Geoinf.* **2012**, *18*, 128–140. [CrossRef]
6. Mukherjee, R.; Basu, J.; Mandal, P.; Guha, P.K. A review of micromachined thermal accelerometers. *J. Micromech. Microeng.* **2017**, *27*, 123002. [CrossRef]
7. Pollock, D.D. *Thermocouples: Theory and Properties*; CRC Press: Boca Raton, FL, USA, 1991.
8. Pivovarník, M.; Khalsa, S.J.S.; Jiménez-Muñoz, J.C.; Zemek, F. Improved Temperature and Emissivity Separation Algorithm for Multispectral and Hyperspectral Sensors. *IEEE Trans. Geosci. Remote Sens.* **2017**, *55*, 1944–1953. [CrossRef]
9. Li, Z.L.; Becker, F.; Stoll, M.P.; Wan, Z. Evaluation of Six Methods for Extracting Relative Emissivity Spectra from Thermal Infrared Images. *Remote Sens. Environ.* **1999**, *69*, 197–514. [CrossRef]
10. Khopkar, P.; Agnihotri, S. Modelling Temperature-Vegetation Index (TVX) space and Quality of Life (QoL) for enhanced change detection analysis: A Case Study of Ahmedabad City. In Proceedings of the 38th Asian Conference on Remote Sensing (ACRS), New Delhi, India, 23–27 October 2017.
11. Parviz, L.; Kholghi, M.; Valizadeh, K.H. Estimation of Air Temperature Using Temperature-Vegetation Index (TVX) Method. *J. Water Soil Sci.* **2011**, *15*, 21–34.

12. Hou, P.; Chen, Y.; Jiang, W.Q.G.C.W.; Li, J. Near-surface air temperature retrieval from satellite images and influence by wetlands in urban region. *Appl. Climatol.* **2013**, *111*, 109–118. [CrossRef]
13. Sun, Y.J.; Wang, J.F.; Zhang, R.H.; Gillies, R.R.; Xue, Y.; Bo, Y.C. Air temperature retrieval from remote sensing data based on thermodynamics. *Theor. Appl. Climatol.* **2005**, *80*, 37–48. [CrossRef]
14. Shi, Y.; Jiang, Z.; Dong, L.; Shen, S. Statistical estimation of high-resolution surface air temperature from MODIS over the Yangtze River Delta. *China. J. Meteorol. Res.* **2017**, *31*, 448–454. [CrossRef]
15. Shen, H.; Jiang, Y.; Li, T.; Cheng, Q.; Zeng, C.; Zhang, L. Deep learning-based air temperature mapping by fusing remote sensing, station, simulation and socioeconomic data. *Remote Sens. Environ.* **2020**, *240*, 111692. [CrossRef]
16. Liu, S.; Su, H.; Zhang, R.; Tian, J.; Wang, W. Estimating the Surface Air Temperature by Remote Sensing in Northwest China Using an Improved Advection-Energy Balance for Air Temperature Model. *Adv. Meteorol.* **2015**, *2016*, 4294219. [CrossRef]
17. Lu, N.; Liang, S.; Huang, G.; Qin, J.; Yao, L.; Wang, D.; Yang, K. Hierarchical Bayesian space-time estimation of monthly maximum and minimum surface air temperature. *Remote Sens. Environ.* **2018**, *211*, 48–58. [CrossRef]
18. Hooker, J.; Duveiller, G.; Cescatti, A. Data Descriptor: A global dataset of air temperature derived from satellite remote sensing and weather stations. *Sci. Data* **2018**, *5*, 180246. [CrossRef] [PubMed]
19. Wang, X.; OuYang, X.; Li, Z.L.; Zhang, R. A New Method for Temperature/Emissivity Separation from Hyperspectral Thermal Infrared Data. In Proceedings of the 2008 IEEE International Geoscience and Remote Sensing Symposium IGARSS 2008), Boston, MA, USA, 8–11 July 2008; pp. III-286–III-289. [CrossRef]
20. Wang, H.; Xiao, Q.; Li, H.; Zhong, B. Temperature and emissivity separation algorithm for TASI airborne thermal hyperspectral data. In Proceedings of the 2011 International Conference on Electronics, Communications and Control (ICECC), Ningbo, China, 9–11 September 2011; pp. 1075–1078. [CrossRef]
21. Tan, K.; Liao, Z.; Du, P.; Wu, L. Land surface temperature retrieval from Landsat 8 data and validation with geosensor network. *Front. Earth Sci.* **2017**, *11*, 20–34. [CrossRef]
22. Harig, R. Passive remote sensing of pollutant clouds by Fourier-transform infrared spectrometry: Signal-to-noise ratio as a function of spectral resolution. *Appl. Opt.* **2004**, *43*, 4603–4610. [CrossRef] [PubMed]
23. Gagnon, M.A.; Gagnon, J.P.; Tremblay, P.; Savary, S.; Farley, V.; Éric Guyot.; Lagueux, P.; Chamberland, M. Standoff midwave infrared hyperspectral imaging of ship plumes. *Proc. SPIE* **2016**, *9988*, 998806. [CrossRef]
24. Kim, S.; Kim, J.; Lee, J.; Ahn, J. Midwave FTIR-Based Remote Surface Temperature Estimation Using a Deep Convolutional Neural Network in a Dynamic Weather Environment. *Micromachines* **2018**, *9*, 495. [CrossRef]
25. Schlerf, M.; Rock, G.; Lagueux, P.; Ronellenfitsch, F.; Gerhards, M.; Hoffmann, L.; Udelhoven, T. A Hyperspectral Thermal Infrared Imaging Instrument for Natural Resources Applications. *Remote Sens.* **2012**, *4*, 3995–4009. [CrossRef]
26. Trishchenko, A.P. Solar Irradiance and Effective Brightness Temperature for SWIR Channels of AVHRR/NOAA and GOES Imagers. *J. Atmos. Ocean. Technol.* **2006**, *23*, 198–210. [CrossRef]
27. Romaniello, V.; Spinetti, C.; Silvestri, M.; Buongiorno, M.F. A Sensitivity Study of the 4.8 um Carbon Dioxide Absorption Band in the MWIR Spectral Range. *Remote Sens.* **2020**, *12*, 172. [CrossRef]
28. Griffin, M.K.; Hua K.; Burke, H.; Kerekes, J.P. Understanding radiative transfer in the midwave infrared: A precursor to full-spectrum atmospheric compensation. *Proc. SPIE* **2004**, *5425*, 348–356. [CrossRef]
29. Andrews, D. *An Introduction to Atmospheric Physics*; Cambridge Press: Cambridge, UK, 2000.
30. Hohn, D.H. Atmospheric Vision 0.35 um < x < 14 um. *Appl. Opt.* **1975**, *14*, 404–412.
31. Sobrino, J.; Li, Z.L.; Stoll, P.; Becker, F. Multi-channel and multi-angle algorithms for estimating sea and land surface temperature with ATSR data. *Int. J. Remote Sens.* **2004**, *17*, 2089–2114. [CrossRef]
32. Driggers, R.G.; Friedman, M.H.; Nichols, J. *Introduction to Infrared and Electro-Optical Systems*; ARTECH HOUSE: Washington, DC, USA, 2012.
33. Eismann, M.T. *Hyperspectral Remote Sensing*; SPIE Press: Bellingham, WA, USA, 2012.
34. Gonzalez, R.C.; Woods, R.E. *Digital Image Processing*, 4th ed.; Pearson: London, UK, 2012.
35. Iersel, M.V.; Mack, A.; Degache, M.A.C.; Eijk, A.M.J.V. Ship plume modelling in EOSTAR. *Proc. SPIE* **2014**, *9242*, 92421S.

36. Papoulis, A.; Pillai, U. *Probability, Random Variables and Stochastic Processes*, 4th ed.; McGraw-Hill Europe: New York, NY, USA, 2002.
37. Wei, P.S.; Hsieh, Y.C.; Chiu, H.H.; Yen, D.L.; Lee, C.; Tsai, Y.C.; Ting, T.C. Absorption coefficient of carbon dioxide across atmospheric troposphere layer. *Heliyon* **2018**, *4*, 1–20. [CrossRef]

remote sensing

MDPI

Article

AT^2ES: Simultaneous Atmospheric Transmittance-Temperature-Emissivity Separation Using Online Upper Midwave Infrared Hyperspectral Images

Sungho Kim [1,*], Jungsub Shin [2] and Sunho Kim [2]

[1] Department of Electronic Engineering, Yeungnam University, 280 Daehak-Ro, Gyeongsan, Gyeongbuk 38541, Korea

[2] Agency for Defense Development, 488-160 Bukyuseong-Daero, Yuseong, Daejeon 34186, Korea; jss@add.re.kr (J.S.); edl423@add.re.kr (S.K.)

* Correspondence: sunghokim@yu.ac.kr; Tel.: +82-53-810-3530

Abstract: This paper presents a novel method for atmospheric transmittance-temperature-emissivity separation (AT^2ES) using online midwave infrared hyperspectral images. Conventionally, temperature and emissivity separation (TES) is a well-known problem in the remote sensing domain. However, previous approaches use the atmospheric correction process before TES using MODTRAN in the long wave infrared band. Simultaneous online atmospheric transmittance-temperature-emissivity separation starts with approximation of the radiative transfer equation in the upper midwave infrared band. The highest atmospheric band is used to estimate surface temperature, assuming high emissive materials. The lowest atmospheric band (CO_2 absorption band) is used to estimate air temperature. Through onsite hyperspectral data regression, atmospheric transmittance is obtained from the y-intercept, and emissivity is separated using the observed radiance, the separated object temperature, the air temperature, and atmospheric transmittance. The advantage with the proposed method is from being the first attempt at simultaneous AT^2ES and online separation without any prior knowledge and pre-processing. Midwave Fourier transform infrared (FTIR)-based outdoor experimental results validate the feasibility of the proposed AT^2ES method.

Keywords: atmospheric transmittance; temperature; emissivity; separation; midwave infrared; hyperspectral images

Citation: Kim, S.; Shin, J.; Kim, S. AT^2ES: Simultaneous Atmospheric Transmittance-Temperature-Emissivity Separation Using Online Upper Midwave Infrared Hyperspectral Images. *Remote Sens.* **2021**, *13*, 1249. https://doi.org/10.3390/rs13071249

Academic Editor: Chein-I Chang

Received: 1 March 2021
Accepted: 23 March 2021
Published: 25 March 2021

1. Introduction

The concept of temperature and emissivity separation (TES) was originally developed by Gillespie et al. for Advanced Spaceborne Thermal Emission and Reflection Radiometer (ASTER) satellite data analysis [1,2]. Currently, TES is an important research topic in infrared remote sensing applications. Separated temperature can be used to estimate land surface temperature for the study of climate change [3,4]. Emissivity information is useful for mineral composition analysis [5], vegetative cover mapping [6], and object material classification [7].

The scope of this paper is to apply TES online on a flying platform such as an unmanned aerial vehicle. The most critical issue is how to achieve the atmospheric correction to remove the effect of path radiance and atmospheric transmittance in real time without any prior information or pre-processing. Historically, the original TES method in ASTER satellite images used an atmospherically corrected dataset in five multispectral long wave infrared (LWIR) bands [1]. Li et al. compared six methods for extracting relative emissivity spectra from atmospherically corrected multiple spectral bands [3]. Yong et al. tried to estimate atmospheric transmittance in LWIR bands without TES [8]. Payan and Royer further analyzed the applicability and sensitivity of six TES methods [2]. Borel and Tuttle improved TES by using MODTRAN 5-based atmospheric transmittance [9]. Wang et al.

also applied MODTRAN to perform atmospheric correction in a thermal airborne spectrographic imager (TASI) [10]. Adler-Golden et al. adopted simulated atmospheric parameters from the MODTRAN5 model for TES [11]. Wang et al. used the atmospheric transmittance calculated by MODTRAN for TES of Landsat-8 sensor data [12]. Pivovarník et al. improved TES by adopting smoothing in emissivity estimation where atmospheric correction was made using MODTRAN with a mid-latitude summer atmosphere [4].

The previous works have three limitations. First, most require pre-processing of atmospheric correction by using MODTRAN or from prior knowledge. The atmospheric transmittance, downwelling, and upwelling data are generated for TES. Second, TES is conducted offline. Such an approach is impractical for real-time TES on a flying platform because atmospheric conditions change dynamically in time and space. Third, most TES techniques use an LWIR satellite database such as ASTER and TASI.

In this paper, a novel simultaneous atmospheric transmittance-temperature-emissivity separation (AT^2ES) method is proposed for online applications based on the following key ideas. First, the radiative transfer equation (RTE) is approximated by considering the physical properties of the upper midwave infrared band (4.2–5.6 μm). Second, the highest and lowest atmospheric transmittance bands are selected. The former is used to estimate surface temperature, and the latter (the CO_2 absorption band: 4.2–4.4 μm) is used to estimate air temperature. Through a data regression process, the atmospheric transmittance is estimated with the y-intercept and air temperature. Emissivity is separated using the observed radiance, the separated object temperature, the air temperature, and atmospheric transmittance.

Therefore, the main contributions are summarized as follows.

- The proposed AT^2ES can separate atmospheric transmittance, temperature, and emissivity simultaneously.
- AT^2ES can work online without any prior processing or information.
- AT^2ES can provide a feasible approximate solution in the upper MWIR band (4.2–5.6 μm).

The remainder of this paper is organized as follows. Section 2 explains the proposed AT^2ES method, including the basics of the radiative transfer equation in the upper MWIR band. Section 3 analyzes AT^2ES using a synthetic dataset and outdoor remote sensing data. The paper concludes in Section 4.

2. Proposed AT^2ES Method

2.1. Basics of the Radiative Transfer Equation

Figure 1 shows hyperspectral imaging in an outdoor environment. It consists of the target, a midwave infrared-Fourier transform infrared (MWIR-FTIR) camera, the sun, and the atmosphere. Observed spectral radiance can be derived from the radiative transfer from Equation (1). Romaniello et al. adopted the radiative transfer equation used in MODTRAN [13]. In general, at-sensor received radiance $L_{obs}(\lambda)$ in the MWIR region consists of opaque object-emitted radiance, reflected downwelling radiance, and total atmospheric path radiance (thermal+solar components).

$$L_{obs}(\lambda) = \tau(\lambda)\left[\varepsilon(\lambda)L_{tg}(\lambda, T_{tg}) + (1-\varepsilon(\lambda))(L_s^{\downarrow}(\lambda) + L_t^{\downarrow}(\lambda))\right] + L_s^{\uparrow}(\lambda) + L_t^{\uparrow}(\lambda) \quad (1)$$

$L_{obs}(\lambda)$ is the at-sensor radiance; λ is the wavelength; $\varepsilon(\lambda)$ is spectral object surface emissivity; $L_{tg}(\lambda, T_{tg})$ is the spectral radiance of the object, assuming a blackbody in the Planck function with object surface temperature T_{tg}. $L_s^{\downarrow}(\lambda)$ and $L_t^{\downarrow}(\lambda)$ represent the spectral downwelling solar radiance and thermal irradiance, respectively; $\tau(\lambda)$ is the spectral atmospheric transmittance, and $L_s^{\uparrow}(\lambda)$ and $L_t^{\uparrow}(\lambda)$ are spectral upwelling solar and thermal path radiance, respectively, reaching the sensor. Observed spectral radiance $L_{obs}(\lambda)$ is acquired by applying the Fourier transform to the interferogram in the Michelson interferometer and hot-cold blackbody-based radiometric calibration [14].

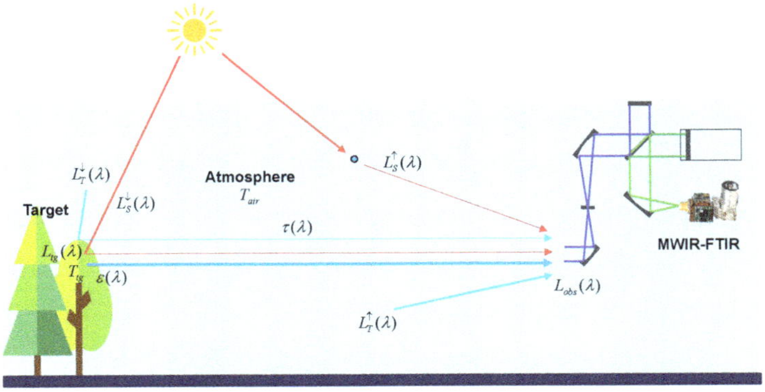

Figure 1. Operational concept of AT^2ES using a passive open path Fourier transform infrared imaging system. Notation: $L_S^{\downarrow}(\lambda)$ and $L_t^{\downarrow}(\lambda)$ represent the spectral downwelling solar radiance and thermal irradiance, respectively; $L_s^{\uparrow}(\lambda)$ and $L_t^{\uparrow}(\lambda)$ are spectral upwelling solar and thermal path radiance, respectively, reaching the sensor.

2.2. Proposed Approximation of the RTE in the Upper MWIR Band

Figure 2 visualizes the fractions of total radiance according to the radiometric characteristics for top of atmosphere (TOA): path thermal $L_t^{\uparrow}(\lambda)$, path reflectance-solar $L_s^{\uparrow}(\lambda)$, surface reflectance-solar $L_S^{\downarrow}(\lambda)$, surface reflectance-infrared $L_t^{\downarrow}(\lambda)$, and surface-emitted $L_{tg}(\lambda, T_{tg})$. The lower MWIR band (3.0–4.2 μm) shows a large fraction for surface reflectance-solar. This means the received radiance strongly depends on the reflected solar energy. However, the contribution of surface reflectance-solar radiance $L_S^{\downarrow}(\lambda)$ is reduced to only 1% to 4% in the upper MWIR band (4.2–5.6 μm) even for very dry conditions [15]. Figures 3 and 4 show the simulation process of surface reflected-solar and surface emitted-object with the portion of surface reflected-solar. According to the simulation, the average portion of surface reflected solar is 0.65%, which affects negligible error. In addition, surface-reflected downwelling thermal radiance $L_T^{\downarrow}(\lambda)$ and path reflectance-solar radiance $L_S^{\uparrow}(\lambda)$ are negligible, compared to surface-emitted radiance $L_{tg}(\lambda, T_{tg})$ and path thermal radiance $L_t^{\uparrow}(\lambda)$.

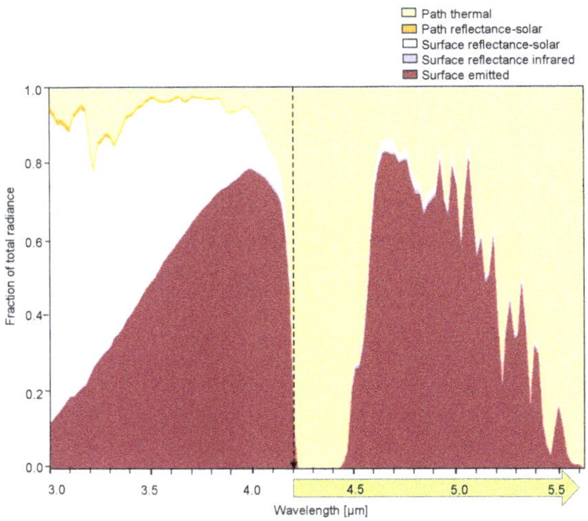

Figure 2. Fractional distribution of spectral radiance in the MWIR band.

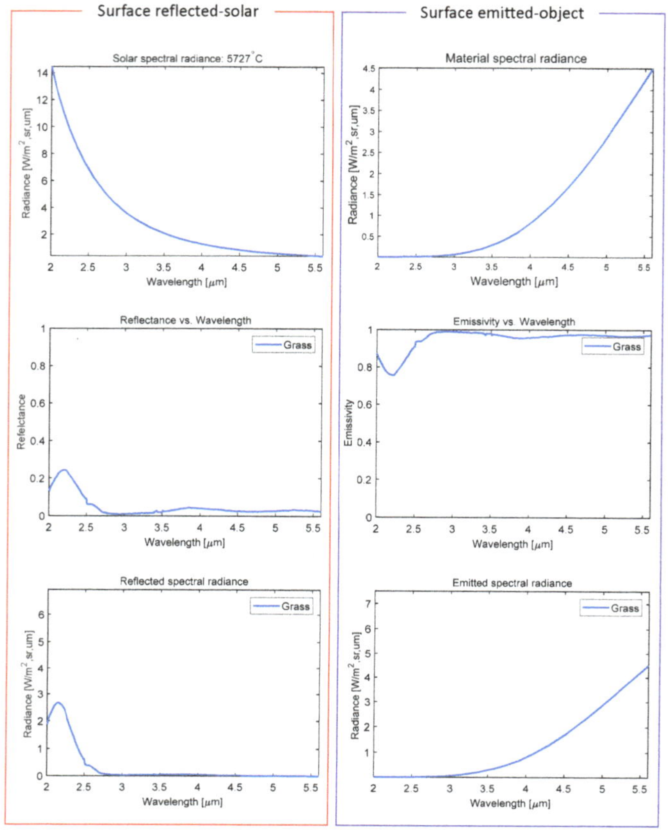

(a) (b)

Figure 3. (**a**) Generation of surface reflected-solar, (**b**) generation of surface emitted-object. (1st row) solar radiance, object radiance, (2nd row) surface reflectivity, emissivity, and (3rd row) surface reflected-solar, surface emitted-object.

If we ignore surface reflectance-solar, surface reflectance-infrared, path reflectance-solar, we can simplify Equation (1) into Equation (2):

$$L_{obs}(\lambda) = \tau(\lambda)\varepsilon(\lambda)L_{tg}(\lambda, T_{tg}) + L_t^{\uparrow}(\lambda) \tag{2}$$

The definition of thermal upwelling $L_t^{\uparrow}(\lambda)$ is Equation (3) [9]:

$$L_t^{\uparrow}(\lambda) = (1 - \tau(\lambda))L_{BB}(\lambda, T_{air}) \tag{3}$$

where $L_{BB}(\lambda, T_{air})$ denotes the spectral radiance, $[W/(m^2 \cdot sr \cdot \mu m)]$, of a blackbody (Planck's law [16]), and T_{air} is the air temperature in degrees Kelvin $[K]$ of the atmosphere between the object and the camera sensor. The spectral radiation of the atmosphere is modeled as a blackbody [17–19]. Atmospheric path radiance can be described in different ways, but the simplest is to model the particles as blackbodies [19]. $L_{BB}(\lambda, T_{air})$ is defined in Equation (4):

$$L_{BB}(\lambda, T_{air}) = \frac{2hc^2}{\lambda^5(e^{hc/\lambda k T_{air}} - 1)} \tag{4}$$

where h denotes Planck's constant, c is the speed of light, and k is the Boltzmann constant.

Therefore, the final form of the proposed approximated RTE is the same as Equation (5):

$$L_{obs}(\lambda) = \tau(\lambda)\varepsilon(\lambda)L_{BB}(\lambda, T_{tg}) + (1 - \tau(\lambda))L_{BB}(\lambda, T_{air}) \quad (5)$$

where $L_{tg}(\lambda, T_{tg})$ was changed to $L_{BB}(\lambda, T_{tg})$ for notational consistency. The proposed RTE is valid for the upper MWIR band (4.2–5.6 μm) with 1–4% radiance uncertainty.

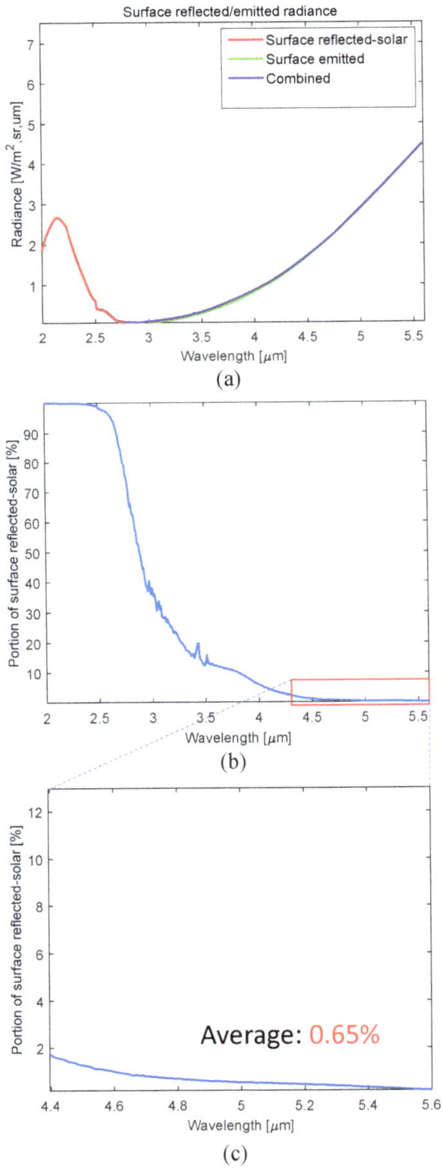

Figure 4. Calculation of the portion of surface reflected-solar: (**a**) surface reflected-solar + surface emitted-object, (**b**) portion of surface reflected-solar [%], (**c**) enlarged view in the upper MWIR band.

2.3. Details of the AT^2ES Process

Given the approximated RTE model in Equation (5), two unknown temperature parameters (T_{tg}, T_{air}) should be separated to estimate spectral atmospheric transmittance $\tau(\lambda)$ and spectral emissivity $\epsilon(\lambda)$ given $L_{obs}(\lambda)$. Figure 5 summarizes the overall AT^2ES process, which consists of six blocks: brightness temperature (BT) extraction, T_{air} separation, T_{tg} separation, regression, $\tau(\lambda)$ separation, and $\varepsilon(\lambda)$ separation. The BT extraction block converts spectral radiance $L_{obs}(\lambda)$ to brightness temperature units. The band range is limited to the upper MWIR band (4.2–5.6 μm) in order to use the approximate RTE model introduced in the previous subsection. Brightness temperature $BT(\lambda)$ is used in the T_{air} and T_{tg} separation blocks. The regression block estimates slope $a(\lambda)$ and intersect $b(\lambda)$ parameters from observed spectral radiance $L_{obs}(\lambda)$ and target spectral radiance $L_{BB}(\lambda, T_{tg})$. Atmospheric transmittance $\tau(\lambda)$ and target emissivity $\varepsilon(\lambda)$ are separated using these parameters and air radiance $L_{BB}(\lambda, T_{air})$. Each module is explained in the following paragraphs.

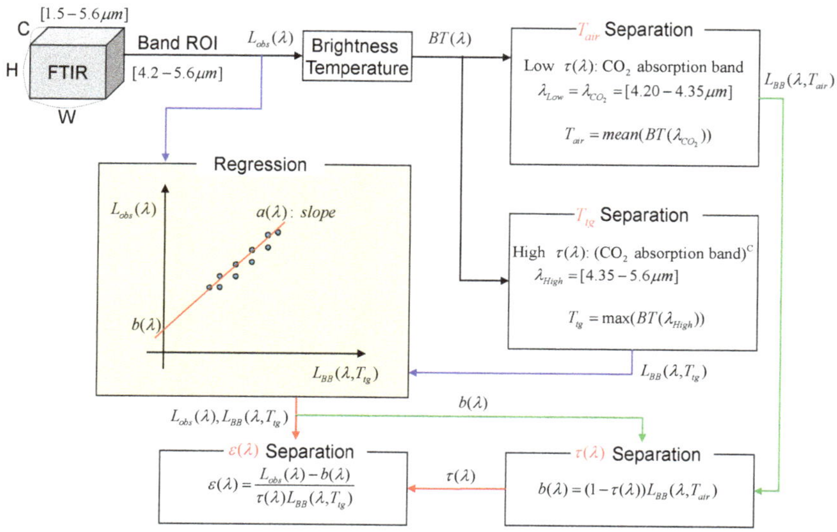

Figure 5. Proposed simultaneous AT^2ES flow.

Brightness temperature module: The amount of spectral radiance energy can be converted into an equivalent brightness temperature ([20]). By inverting Equation (4), temperature $BT(\lambda)$ $[K]$ can be obtained as follows:

$$BT(\lambda) = \frac{hc}{\lambda k \ln(2hc^2/\lambda^5 L_{BB}(\lambda, T) + 1)}. \tag{6}$$

Figure 6 shows an example of brightness temperature extraction from an observed spectral radiance. The remote spectral radiance shows a complicated shape depending on the surface emissivity, atmospheric transmittance, and path radiance. Brightness temperature is the temperature of a blackbody in thermal equilibrium with its surroundings in order to duplicate the observed intensity of a gray-body object at a specific frequency or wavelength. As a result that the spectral radiance provides radiance energy at each wavelength, Equation (6) can calculate the corresponding brightness temperature at each wavelength. Note that a higher brightness temperature can be extracted if atmospheric transmittance and surface emissivity are closer to 1.

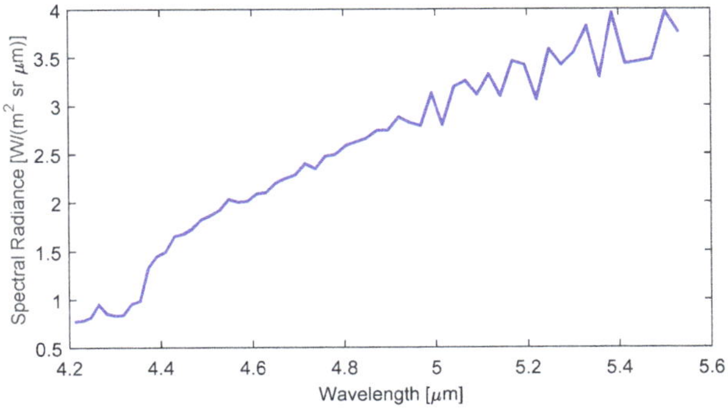

(a)

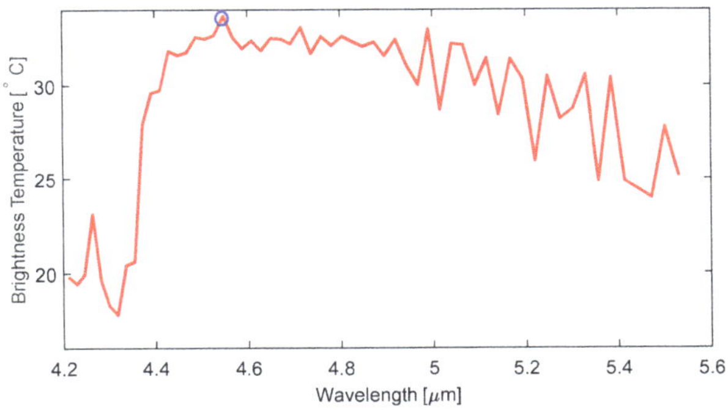

(b)

Figure 6. Example of brightness temperature extraction from spectral radiance: (**a**) the observed sample spectral radiance [W/(m^2 sr μm)], and (**b**) the converted brightness temperature [°C].

T_{air} separation module: According to the MODTRAN simulation in the MWIR band, the spectral transmittance of the carbon dioxide (CO_2) band (4.20–4.35 μm) decreases abruptly with distance [21]. The average transmittance in the CO_2 band is 0.13, 0.03, 0.005, 0.0001, and 0 at 5 m, 10 m, 20 m, 50 m, and 100 m, respectively. Figure 7 demonstrates the atmospheric transmittance at the 50 m distance in the upper MWIR band. Note that the atmospheric transmittance is 0.0001 in the CO_2 absorption band. If we consider only the CO_2 band (λ_{CO_2} = [4.20–4.35 μm]) with a minimum 20m object distance, transmittance $\tau(\lambda_{CO_2})$ can be regarded as 0, which leads to Equation (7) derived from Equation (5). An MWIR-FTIR camera receives only the upwelling of path thermal radiances in the λ_{CO_2} band.

$$L_{obs}(\lambda_{CO_2}) = L_{BB}(\lambda_{CO_2}, T_{air}) \tag{7}$$

Therefore, T_{air} can be obtained by applying a mean operation to Equation (6) in λ_{CO_2}. The final form of air temperature separation is shown in Equation (8). Figure 8 illustrates an air temperature map image by applying the brightness temperature extraction method to the CO_2 absorption band (4.31 μm). A representative air temperature value can be estimated using the spatial and spectral average filter in the CO_2 band range.

$$T_{air} = mean(BT(\lambda_{CO_2})) \tag{8}$$

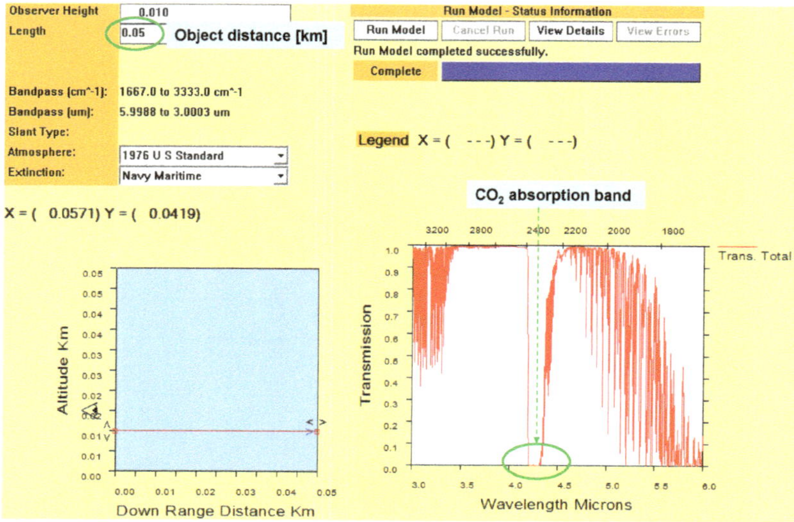

Figure 7. Atmospheric transmittance at the 50 m distance, and the characteristics of the CO_2 absorption band (4.20–4.35 μm).

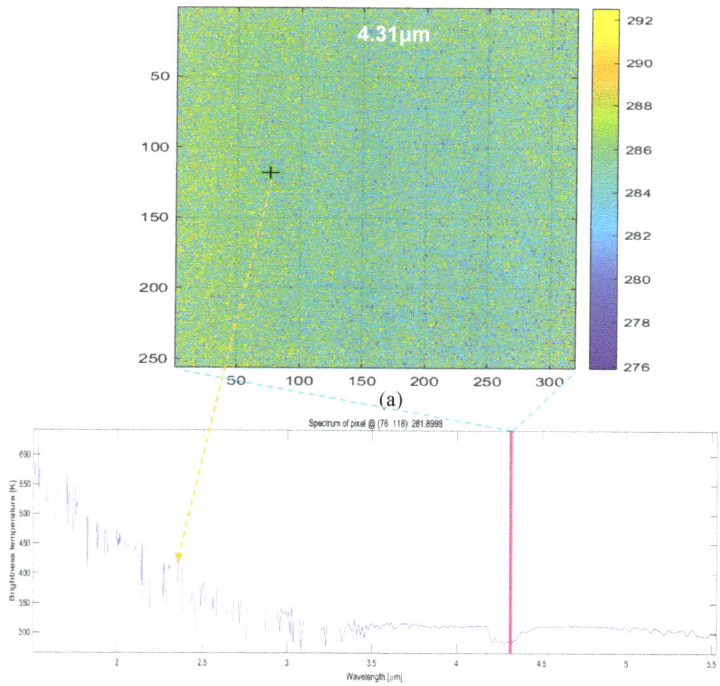

(b)

Figure 8. Air temperature map extraction using spectral radiance in the CO_2 absorption band: (**a**) the air temperature map at 4.31 μm, (**b**) the brightness temperature profile at the cross point in (**a**).

T_{tg} separation module: The remote target temperature separation process requires two assumptions. One is that there must be a high atmospheric transmittance band; the other is that there must be high emissivity band. These assumptions can be satisfied because the working distance is within 100 m, and most natural and paint materials show high emissivity. Figure 9 proves that maximal transmittance is above 0.992 within a 100 m distance under a clear sky. The average atmospheric transmittance is 0.72 (50 m), 0.66 (100 m), 0.49 (500 m), and 0.41 (1000 m) under 1976 US standard atmosphere model. If the moisture content is 3 times higher (tropical model), the corresponding average atmospheric transmittance is 0.65 (50 m), 0.57 (100 m), 0.39 (500 m), and 0.31 (1000 m). The reduction rate is 13.6% (50 m), 9.7% (100 m), 20.4% (500 m), and 20.4% (1000 m).

The spectral emissivity of the representative materials (paint, grass, asphalt, and concrete) is at least 0.9 as shown in Figure 10.

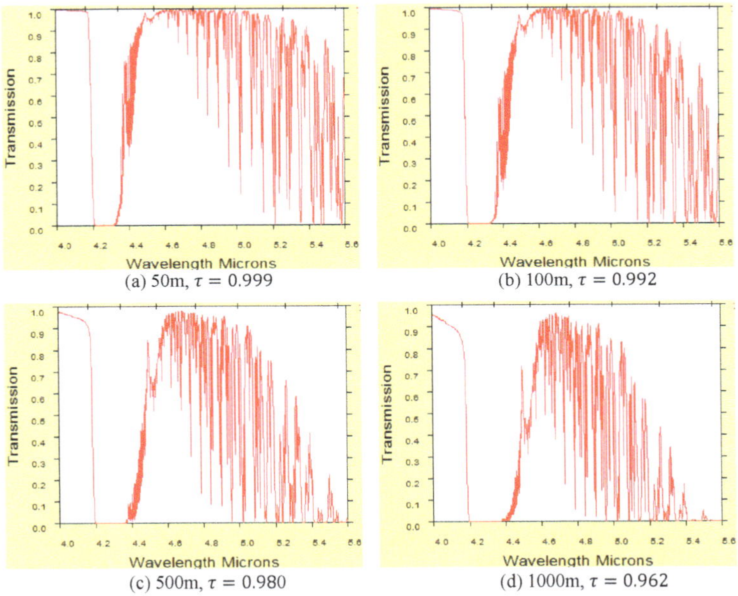

Figure 9. Atmospheric transmittance distribution, and the maximum values based on object distance.

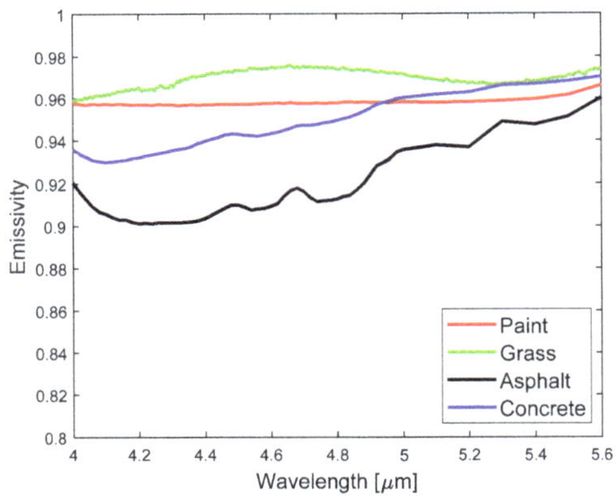

Figure 10. Emissivity distributions of various materials in the upper MWIR band.

In these environmental conditions, there is an optimal band with a high maximum $\tau(\lambda_{opt})\varepsilon(\lambda_{opt})$ of 0.9 or more. Therefore, Equation (5) can be reduced to Equation (9) with a maximum 10% margin of error. Target temperature T_{tg} can be obtained by applying brightness temperature to Equation (9). In practical terms, optimal band λ_{opt} is unknown a priori because we have no information on object distances and material types. However, the problem can be bypassed by applying the max operation to Equation (6). The final form of target temperature separation is shown in Equation (10), where λ_{high} = [4.35–5.60 µm], which is the complement of the CO_2 absorption band. The calculated target temperature (33.6 °C) is the blue circle overlaid in Figure 6.

$$L_{obs}(\lambda_{opt}) = L_{BB}(\lambda_{opt}, T_{tg}) \tag{9}$$

$$T_{tg} = max(BT(\lambda_{high})) \tag{10}$$

Regression module: The proposed approximate RTE, Equation (5), can be written by replacing coefficients as follows:

$$L_{obs}(\lambda) = a(\lambda)L_{BB}(\lambda, T_{tg}) + b(\lambda) \tag{11}$$

where $a(\lambda) = \tau(\lambda)\varepsilon(\lambda)$, and $b(\lambda) = (1 - \tau(\lambda))L_{BB}(\lambda, T_{air})$. Slope $a(\lambda)$ and intercept $b(\lambda)$ can be estimated using regression between $L_{obs}(\lambda)$ and $L_{BB}(\lambda, T_{tg})$, as shown in Figure 11. Hyperspectral data points are obtained from different areas with the same distance. Each observed spectrum provides the BT from which T_{tg} is separated by maximization, as explained above. Figure 12 shows the regressed coefficients for each wavelength.

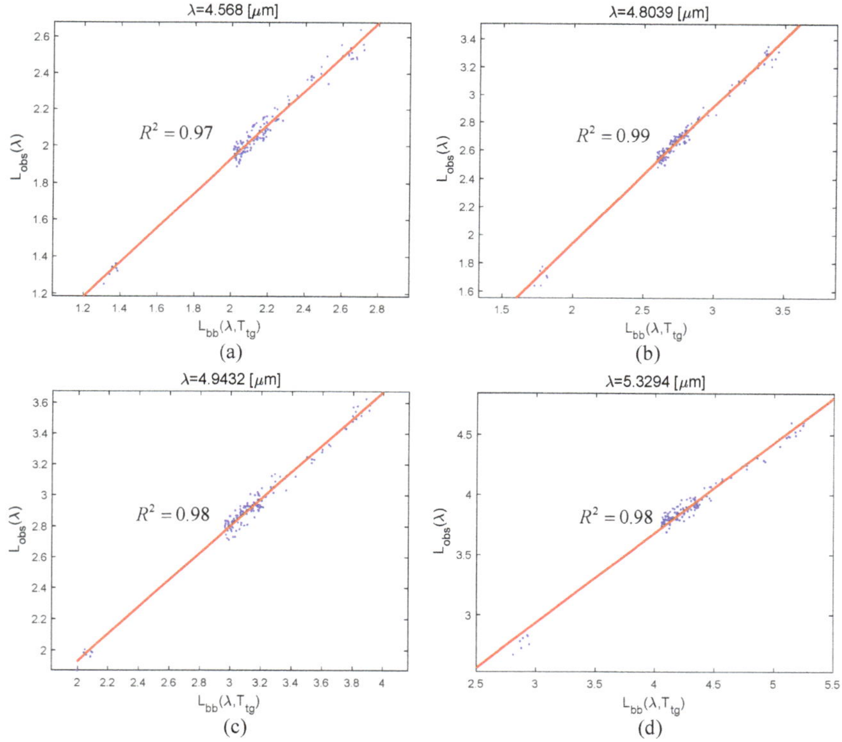

Figure 11. Examples of linear regression between $L_{obs}(\lambda)$ and $L_{BB}(\lambda, T_{tg})$ for the representative bands: (**a**) $\lambda = 4.568[\mu m]$, (**b**) $\lambda = 4.8039[\mu m]$, (**c**) $\lambda = 4.9432[\mu m]$, (**d**) $\lambda = 5.3294[\mu m]$.

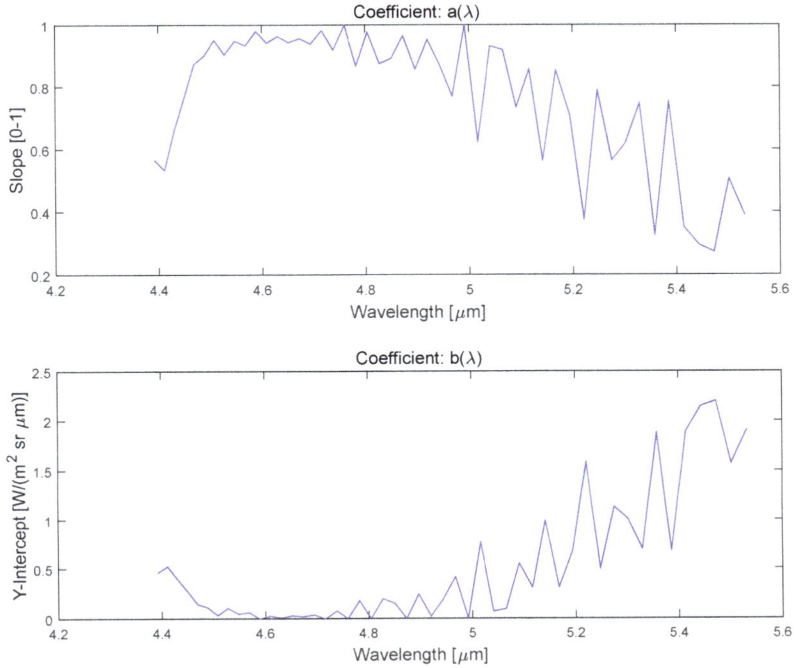

Figure 12. Examples of slope $a(\lambda)$ and y-intercept $b(\lambda)$ coefficients in linear regression.

$\tau(\lambda)$ separation module: As a result that $b(\lambda) = (1 - \tau(\lambda))L_{BB}(\lambda, T_{air})$, atmospheric transmittance $\tau(\lambda)$ can be calculated using $b(\lambda)$ and $L_{BB}(\lambda, T_{air})$ as follows:

$$\tau(\lambda) = 1 - \frac{b(\lambda)}{L_{BB}(\lambda, T_{air})} \quad (12)$$

Atmospheric temperature T_{air} provides blackbody radiation, and y-intercept $b(\lambda)$ is separated through linear regression. Figure 13 (top chart) shows an example of separated atmospheric transmittance using Equation (12).

$\varepsilon(\lambda)$ separation module: In Equation (5), spectral emissivity $\varepsilon(\lambda)$ can be separated using atmospheric transmittance $\tau(\lambda)$, object temperature T_{tg}, and observed spectral radiance $L_{obs}(\lambda)$, as seen in Equation (13). As a result that each sample has its own spectral emissivity, a representative spectral emissivity profile can be obtained via sample mean. Figure 13 (bottom) shows an example of separated emissivity using Equation (13).

$$\varepsilon(\lambda) = \frac{L_{obs}(\lambda) - b(\lambda)}{\tau(\lambda)L_{BB}(\lambda, T_{tg})} \quad (13)$$

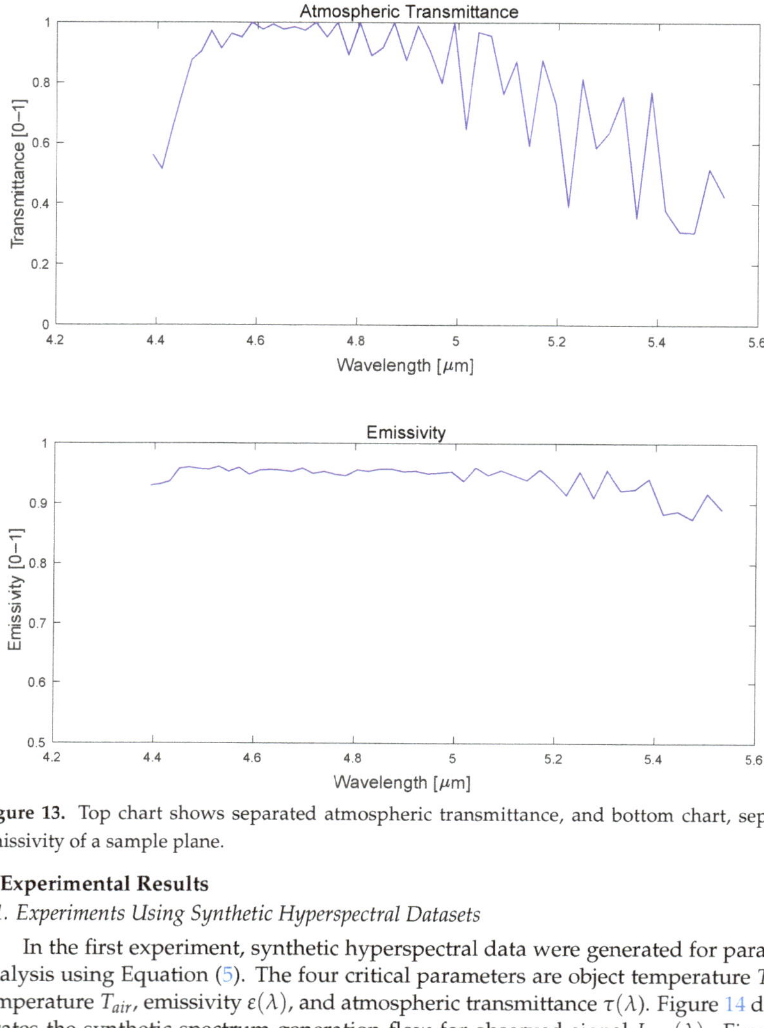

Figure 13. Top chart shows separated atmospheric transmittance, and bottom chart, separated emissivity of a sample plane.

3. Experimental Results

3.1. Experiments Using Synthetic Hyperspectral Datasets

In the first experiment, synthetic hyperspectral data were generated for parameter analysis using Equation (5). The four critical parameters are object temperature T_{tg}, air temperature T_{air}, emissivity $\varepsilon(\lambda)$, and atmospheric transmittance $\tau(\lambda)$. Figure 14 demonstrates the synthetic spectrum generation flow for observed signal $L_{obs}(\lambda)$. Figure 14a is the grass spectrum downloaded from the ECOSTRESS library, 15 August 2020 (https://ecostress.jpl.nasa.gov/) [22]. Figure 14b presents spectral blackbody radiance of an object with temperature T_{tg} = 30 °C. Figure 14c is emitted object radiance from multiplying Figure 14a,b. The observed spectral radiance in Figure 14f was generated by applying the atmospheric transmittance in Figure 14d to the emitted object radiance and the path radiance in Figure 14e.

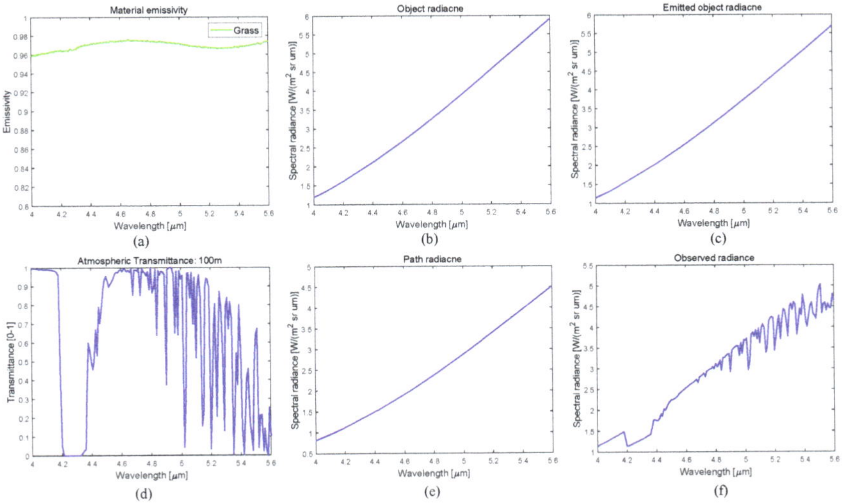

Figure 14. Synthetic spectrum generation flow: (**a**) grass emissivity, (**b**) object radiance, (**c**) emitted object radiance, (**d**) atmospheric transmittance, (**e**) path radiance, and (**f**) observed radiance.

Through the generation process, 200 observed spectra were generated, as seen in Figure 15a. As a baseline dataset, Gaussian noise was added with the following parameters: $\sigma_{\tau} = 0.0001$; $\sigma_{T_{tg}} = 1$; $\sigma_{T_{air}} = 0.0001$, and $\sigma_{\varepsilon} = 0.0001$, where σ_{τ} denotes the standard deviation of atmospheric transmittance, $\sigma_{T_{tg}}$ denotes the standard deviation of object temperature, $\sigma_{T_{air}}$ is the standard deviation of air temperature, and σ_{ε} is the standard deviation of object emissivity. The $\sigma_{T_{air}}$ is set as 0.0001 to consider only the effect of $\sigma_{T_{tg}}$. A value of 0.0001 is the minimal numerical value for simulation purposes. Figure 15b shows an example of a brightness temperature profile converted from an original spectral radiance. The maximum value is regarded as the object temperature, and each separated sample's temperature is displayed in Figure 15c. Each brightness temperature in the CO_2 band provides a candidate air temperature, as shown in Figure 15d. The average of the distribution is regarded as the final air temperature. In this baseline dataset, the separated air temperature is 29.99 °C.

Figure 16's left side presents the estimated coefficients of slope and intercept for the upper MWIR band. Figure 16's right side shows an example of linear regression at λ = 5.6 μm indicating the slope and intercept. Final separation of atmospheric transmittance and emissivity is achieved by applying Equations (12) and (13) to the separated parameters and observed spectrum, as shown in Figure 17. In this case, the mean absolute error (MAE [23]) of spectral atmospheric transmittance is 0.013, and that of spectral emissivity is 0.015.

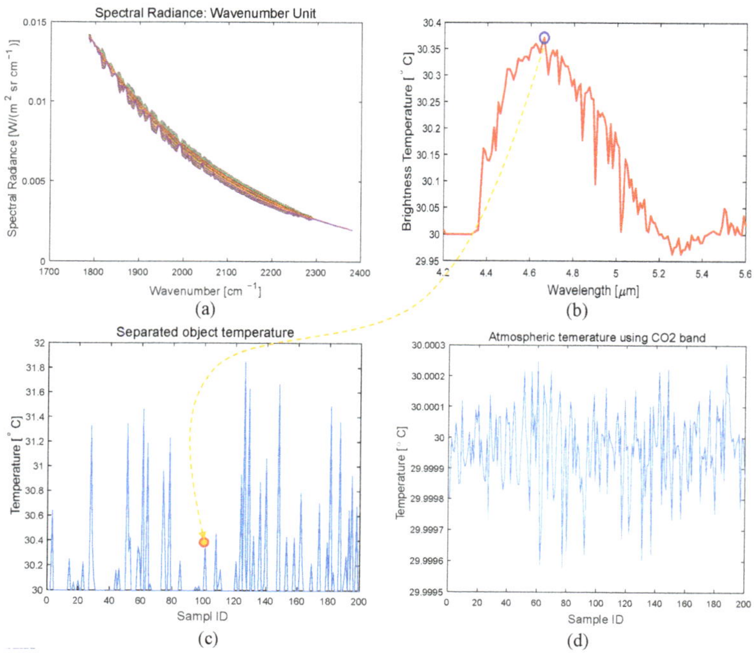

Figure 15. Temperature separation from synthetic spectra: (**a**) generated synthetic data (200 spectra), (**b**) brightness temperature and peak value for a sample spectrum, (**c**) the distribution of separated object temperatures, and (**d**) the distribution of separated atmospheric temperatures using the CO_2 absorption band.

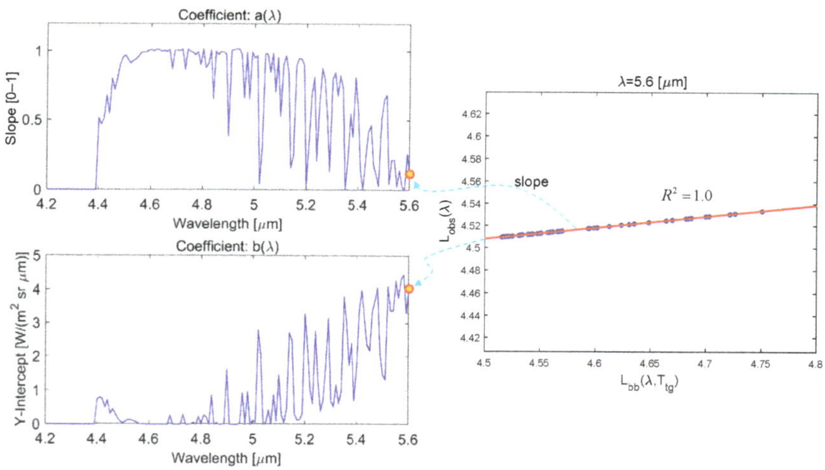

Figure 16. Data regression from synthetic spectra: (**left**) slope coefficient and y-intercept coefficient, and (**right**) data regression example for the 5.6 μm wavelength.

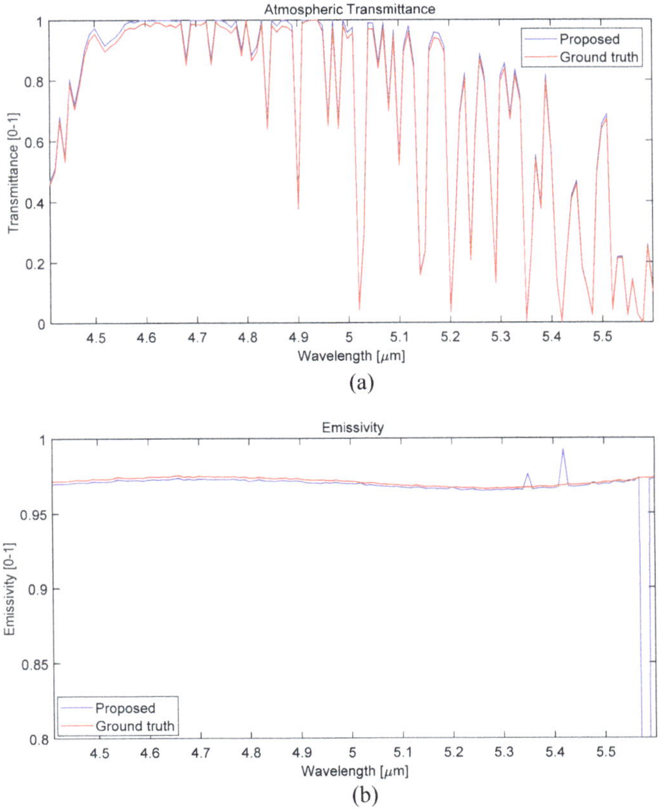

Figure 17. Separated atmospheric transmittance and emissivity: (**a**) a comparison of spectral atmospheric transmittance between the proposed method and ground truth, and (**b**) a comparison of spectral emissivity between the proposed method and ground truth.

It is important to analyze the effects of noise in simultaneous four-parameter ($T_{tg}, T_{air}, \tau(\lambda), \varepsilon(\lambda)$) separation. The MAE performance metric is used to check the trend. If $\sigma_{T_{tg}}$ varies from 0.5 to 4.0, the MAEs of the four parameters are shown in Figure 18. As the object surface temperature variation increases, the error in emissivity and air temperature increases. On the other hand, the atmospheric transmittance separation error is reduced.

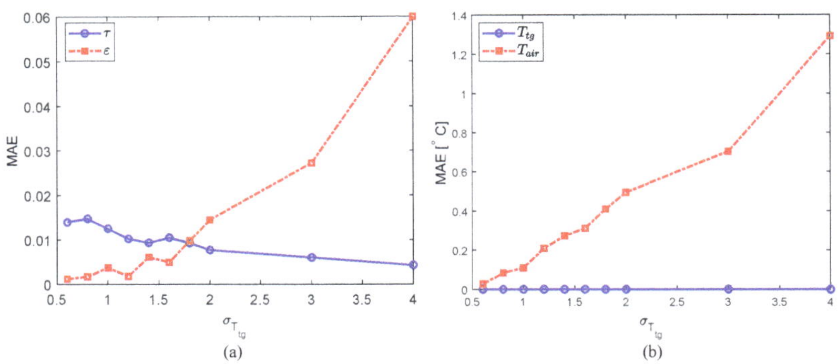

Figure 18. Parameter separation performance according to object temperature noise ($\sigma_{T_{tg}}$): (**a**) MAE of τ, ε, and (**b**) MAE of T_{tg}, T_{air}.

If $\sigma_{T_{air}}$ varies from 0.0001 to 2.0, the MAEs of the four parameters are as shown in Figure 19. As the air temperature noise increases, the error in atmospheric transmittance, object temperature, and air temperature increases. On the other hand, emissivity separation error has almost no relation to air temperature noise.

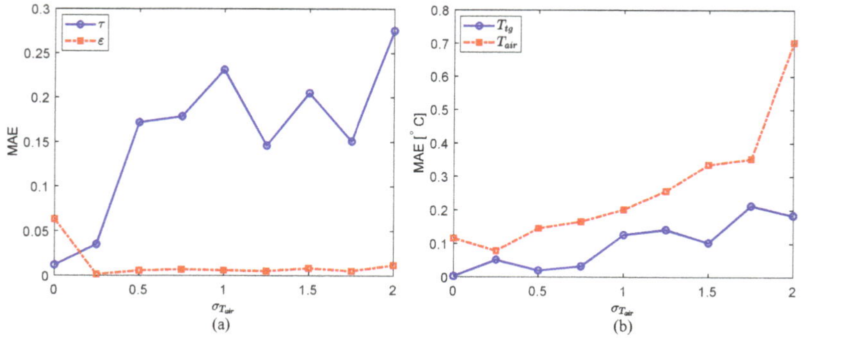

Figure 19. Parameter separation performance based on air temperature noise ($\sigma_{T_{air}}$): (**a**) MAE of τ, ε, and (**b**) MAE of T_{tg}, T_{air}.

If σ_τ varies from 0.0001 to 0.0008, the MAEs of the four parameters are as shown in Figure 20. As the atmospheric transmittance noise increases, the error in atmospheric transmittance increases. On the other hand, other parameter separation errors have almost no relation to atmospheric transmittance noise.

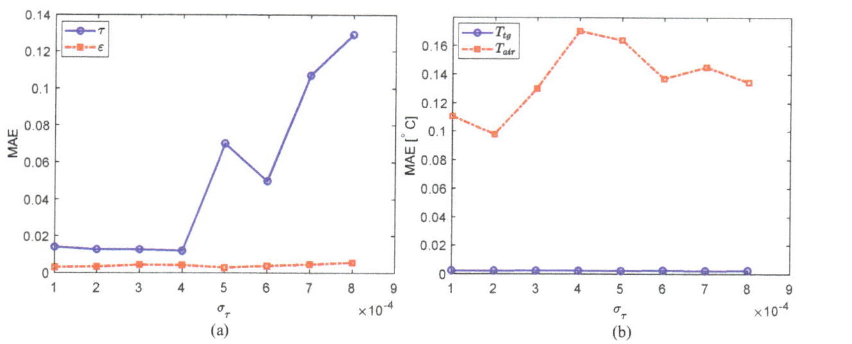

Figure 20. Parameter separation performance based on atmospheric transmittance noise (σ_τ): (**a**) MAE of τ, ε, and (**b**) MAE of T_{tg}, T_{air}.

Finally, if σ_ε varies from 0.0001 to 0.1, the MAEs of the four parameters are as shown in Figure 21. As emissivity noise increases, the error in atmospheric transmittance and emissivity increases. In a small noise interval (0.0001–0.01), the error in air temperature increases sharply. Object temperature separation errors have almost no relation to emissivity noise.

To verify the approximation of the RTE in Equation (2), the effect of path reflectance-solar in air temperature estimation was conducted as shown in Figure 22a. The portion of path reflectance-solar was varied from 0 to 0.5%. The corresponding temperature error was 0 to 0.138 °C. Likewise, the effect of surface reflectance-infrared in target temperature estimation was conducted as shown in Figure 22b. In this case, the effect is more negligible due to the small reflectivity (0.05 in case of grass) in the upper MWIR band.

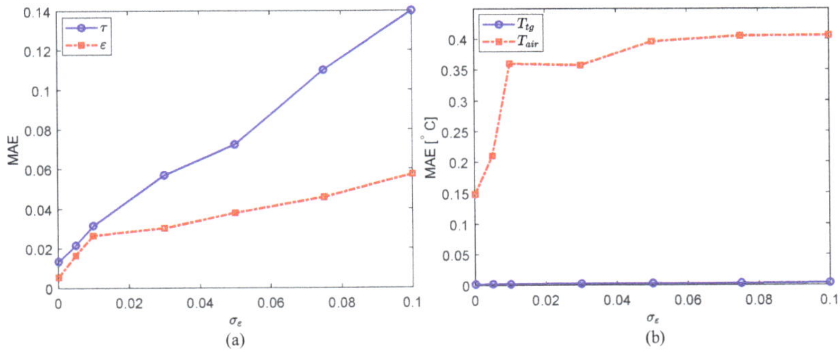

Figure 21. Parameter separation performance based on emissivity noise (σ_ε): (**a**) MAE of τ, ε, and (**b**) MAE of T_{tg}, T_{air}.

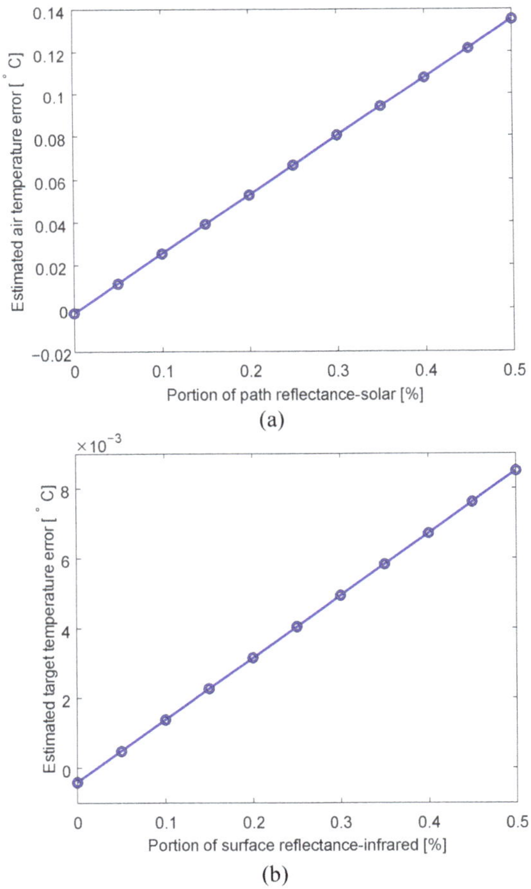

Figure 22. (**a**) Effect of path reflectance-solar in air temperature estimation, (**b**) effect of surface reflectance-infrared in target temperature estimation.

3.2. Experiments Using Real Hyperspectral Datasets

In the second experiment, the feasibility of the proposed AT^2ES was validated for practical applications. Figure 23 shows the hyperspectral data acquisition environment and the data sampling points for evaluation. MWIR hyperspectral images were acquired with the Telops Hyper-Cam MWE model [24]. It can provide calibrated spectral radiance images with a high spatial and spectral resolution from a Michelson interferometer in the shortwave to midwave band (1.5–5.6 μm). Spatial image resolution was 320 × 240, with spectral resolution at up to 0.25 cm^{-1}. The noise equivalent spectral radiance (NESR) was $7[nW/(cm^2 \cdot sr \cdot cm^{-1})]$, and the radiometric accuracy was approximately 2 K. The field of view was 6.5 × 5.1 deg.

In this paper, only the upper MWIR band (4.2–5.6 μm) was used for our valid approximate RTE model. Although a top-down aerial surveillance scenario is ideal, we chose a ground-based side-looking scenario, because the TELOPS MWE camera is too huge and heavy for an airplane to carry. Note that a narrow horizontal region was selected in order to use the assumption of common atmospheric transmittance. In addition, there were 450 grass samples and 450 asphalt samples.

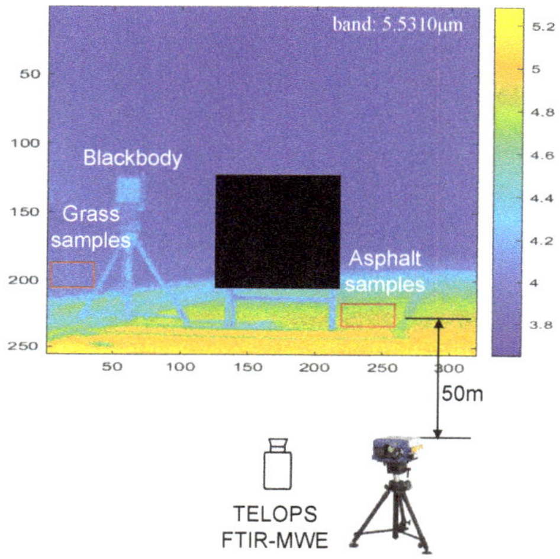

Figure 23. Outdoor field test environment and hyperspectral data acquisition scenario.

Our proposed AT^2ES method can simultaneously extract four parameters: $T_{air}, T_{tg}, \tau(\lambda), \varepsilon(\lambda)$. According to the experimental results, the estimated T_{air} was 20.8 °C, which is 0.5 °C lower than the reference air temperature provided by the Korea Meteorological Administration (21.3 °C). In addition, the estimated T_{tg} was 21.8 °C. The ground truth for grass temperature is hard to measure due to weak leaves and complex structures. Normally, grass temperature is almost the same as air temperature in a thermal equilibrium state [25]. In general, grass has high albedo and high emissivity (>0.95). High albedo prevents solar energy absorption and high emissivity absorbs the thermal energy radiated by near air. Under no wind state, the assumption that grass temperature is almost the same as air temperature is reasonable. However, if the wind is strong, the evapotranspiration from the grass is an important factor which led to its lower temperature [25].

Figure 24 shows the estimated spectral atmospheric transmittance and emissivity, compared with MODTRAN and the ECOSTRESS grass library. In the MODTRAN simulation, object distance was set to 50 m in a mid-latitude spring environment. Note that AT^2ES can estimate spectral atmospheric transmittance quite accurately, as shown in Figure 24a. In the emissivity comparison, sample No. VH351 (Bromus diandrus) from the ECOSTRESS spectral library was chosen because it was most similar to our grass region. Considering the complex grass composition, AT^2ES estimated a similar emissivity profile, as shown in Figure 24b. Figure 25 visualizes the spectral estimation error of $\tau(\lambda)$, $\varepsilon(\lambda)$. The MAEs of atmospheric transmittance and emissivity were 0.087 and 0.063, respectively. Note that large errors were generated around low atmospheric transmittance bands.

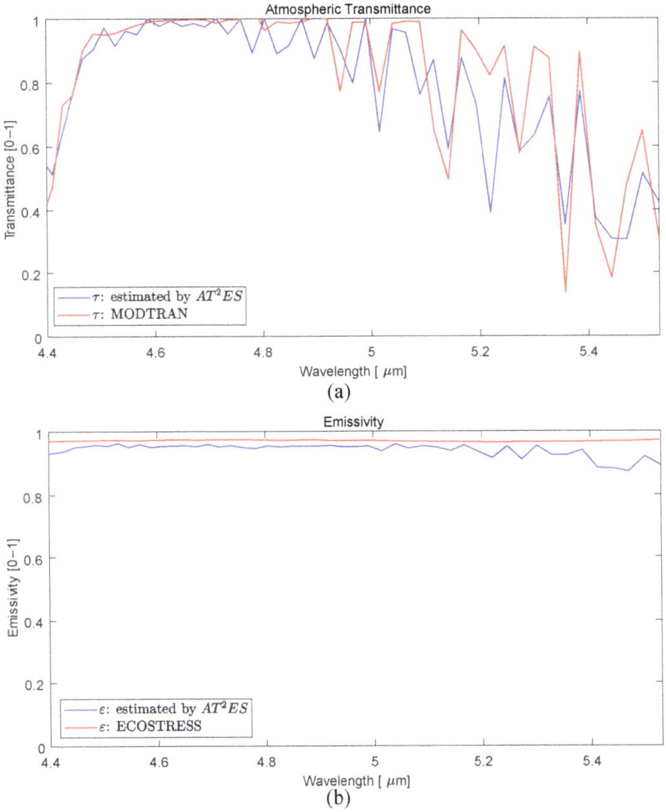

Figure 24. Grass region: Comparison of atmospheric transmittance and emissivity estimation by the proposed AT^2ES: (**a**) spectral atmospheric transmittance comparison with MODTRAN, and (**b**) spectral emissivity comparison with the ECOSTRESS library.

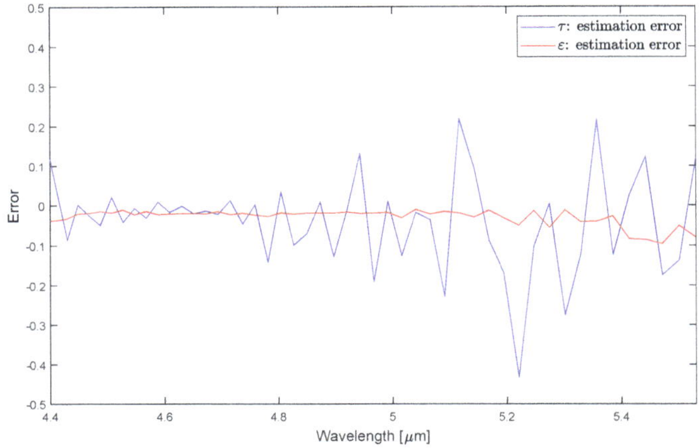

Figure 25. Grass region: Estimation error of spectral atmospheric transmittance and emissivity by the proposed AT^2ES.

In an asphalt region, the estimated T_{tg} was 41.4 °C. Ground truth for asphalt temperature was hard to measure due to the bumpy structure. In general, solar radiance energy (visible/near IR) is converted to long wave thermal energy, and the asphalt temperature is higher than the air temperature. FTIR imaging was done at 13:39 h on 21 May 2020.

Figure 26 shows the estimated spectral atmospheric transmittance and emissivity, compared with MODTRAN and the ECOSTRESS grass library. The MODTRAN simulation was the same as the grass experiment. Note that AT^2ES can estimate spectral atmospheric transmittance quite accurately, as shown in Figure 26a. In the emissivity comparison, sample ID 0095UUUASP (Paving Asphalt) from the ECOSTRESS spectral library was chosen because it was most similar to our asphalt region. As shown in Figure 26b, AT^2ES estimated a similar emissivity profile considering complex asphalt composition, but with some emissivity offset. Figure 27 visualizes the spectral estimation error of $\tau(\lambda)$, $\varepsilon(\lambda)$. The MAE for emissivity was 0.041. Note that large errors were generated around low atmospheric transmittance bands in the grass experiment.

Interestingly, if we add an object temperature offset of 2 °C to T_{tg}, the estimated emissivity moves upward as shown in Figure 28a, with the same emissivity profile shape. Figure 28b shows the estimation error profile of atmospheric transmittance and emissivity. The MAE of emissivity was reduced to 0.023 from 0.041. From this additional test, the proposed AT^2ES estimated a lower object temperature for low emissivity material, which leads to an emissivity profile with an offset. This is a future research direction to improve AT^2ES for low-emissivity objects.

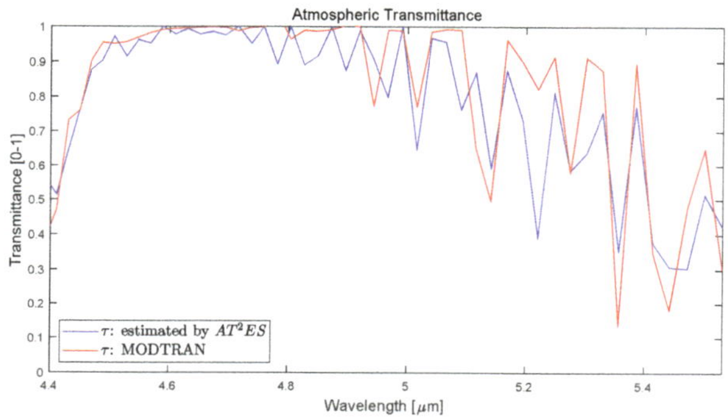

(a)

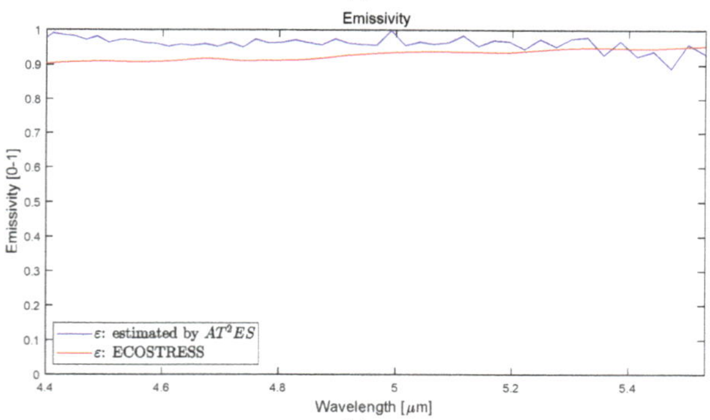

(b)

Figure 26. Asphalt region: Comparison of atmospheric transmittance and emissivity estimation by the proposed AT^2ES and MODTRAN: (**a**) spectral atmospheric transmittance, and (**b**) spectral emissivity.

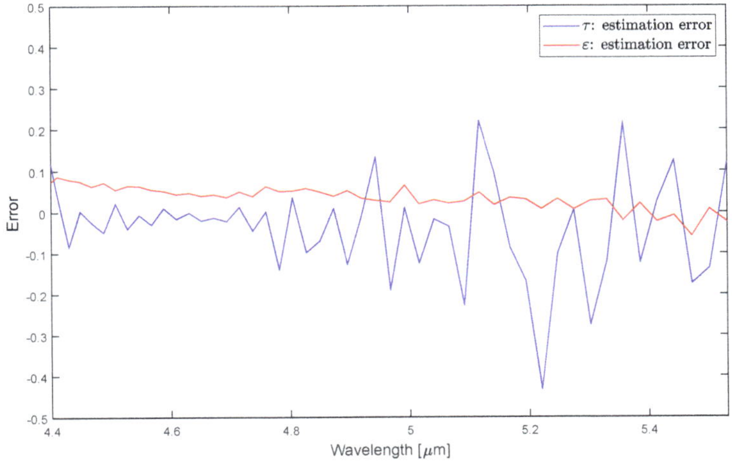

Figure 27. Asphalt region: Estimation error in spectral atmospheric transmittance and emissivity by the proposed AT^2ES.

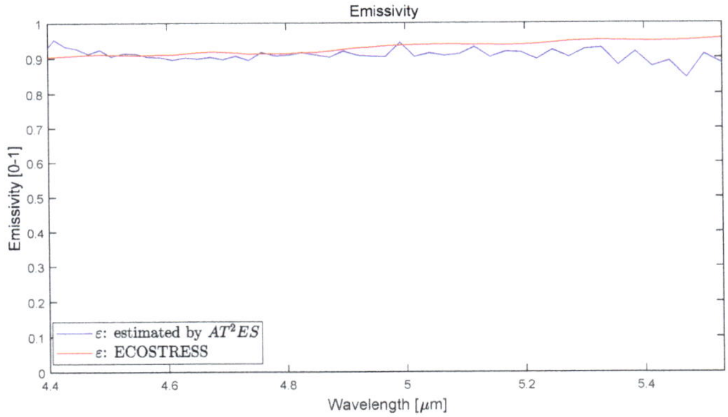

(a)

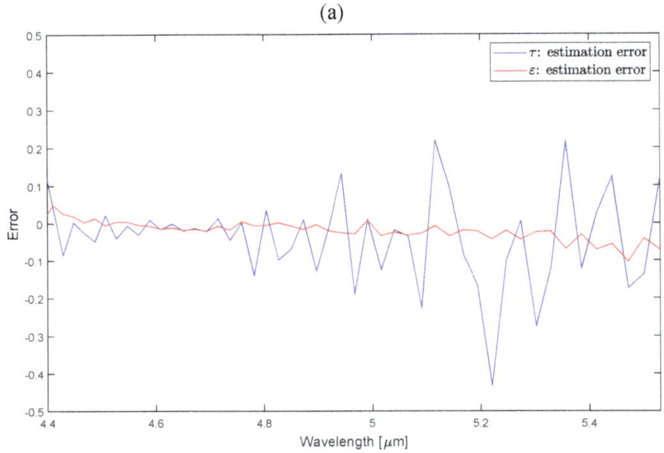

(b)

Figure 28. Asphalt region: Control of the emissivity offset by adding an object temperature offset: (**a**) spectral emissivity, and (**b**) estimation error in spectral atmospheric transmittance and emissivity.

4. Conclusions

Temperature emissivity separation (TES) is an important research topic in a remote-sensing society. Most approaches use atmospheric correction to remove atmospheric transmittance, downwelling, and upwelling generated from MODTRAN. However, online atmospheric information changes from time to time and region by region. This paper presents AT^2ES, a novel method to separate atmospheric transmittance, temperature, and emissivity simultaneously without the aid of an offline MODTRAN simulation.

The key idea is based on radiometry transfer properties in the upper MWIR band (4.2–5.6 μm) where there are negligible downweilling and solar upwelling components (1–4%) with a high emissivity surface (above 0.9) at a 100 m distance. From the proposed approximate RTE, the AT^2ES algorithm can separate four parameters simultaneously. Air temperature is extracted from the brightness temperature in the CO_2 absorption band (4.20–4.35 μm). The object surface temperature is obtained by applying the max operation to the brightness temperature, except the CO_2 absorption band. Given some observed spectral radiance samples and an object temperature, data regression of object blackbody radiance and the observed radiance can provide the slope and intercept. In particular, spectral atmospheric transmittance is separated using the y-intercept and air blackbody radiance. The separated atmospheric transmittance is the same for all the samples, but each sample has different emissivity with the same atmospheric transmittance. Therefore, each spectral emissivity is calculated using the separated parameters. The average operation can provide a representative spectral emissivity profile for a certain region.

The first experiment using synthetic spectra provided the effects of noise in the four parameters. Object surface temperature error directly affects spectral emissivity and air temperature. The air temperature error affects atmospheric transmittance, object temperature, and air temperature. Atmospheric transtmittance error directly affects the estimation of atmospheric transmittance. The object emissivity error also affects atmospheric transmittance. The second experiment was based on an outdoor dataset to check the feasibility of the proposed AT^2ES. In grass region samples, the separated temperature parameters were very close to the measured temperatures. Separated spectral atmospheric temperature and emissivity were similar to the profiles in MODTRAN. This is due to the high emissivity of grass regions. In an asphalt region, the estimated emissivity was rather higher than in the ECOSTRESS profile due to lower object temperature estimation. If the object temperature was increased by 2 °C, spectral emissivity was consistent with the spectral library. Therefore, a future research direction is to find an improved AT^2ES method for low emissivity materials.

Author Contributions: The contributions were distributed between authors as follows: S.K. (Sungho Kim) wrote the text of the manuscript, programmed the hyperspectral AT^2ES method using upper MWIR-FTIR data. J.S. and S.K. (Sunho Kim) provided the midwave infrared hyperspectral database, operational scenario, performed the in-depth discussion of the related literature, and confirmed the accuracy experiments that are exclusive to this paper. All authors have read and agreed to the published version of the manuscript.

Funding: This research was funded by ADD grant number UE191095FD, 2021 Yeungnam University Research Grants, and NRF (NRF-2018R1D1A3B07049069).

Acknowledgments: This study was supported by the Agency for Defense Development (UE191095FD). This work was supported by the 2021 Yeungnam University Research Grants. This research was also supported by Basic Science Research Program through the National Research Foundation of Korea (NRF) funded by the Ministry of Education (NRF-2018R1D1A3B07049069).

Conflicts of Interest: The authors declare no conflict of interest.

References

1. Gillespie, A.; Rokugawa, S.; Matsunaga, T.; Cothern, J.S.; Hook, S.; Kahle, A.B. A temperature and emissivity separation algorithm for Advanced Spaceborne Thermal Emission and Reflection Radiometer (ASTER) images. *IEEE Trans. Geosci. Remote Sens.* **1998**, *36*, 1113–1126. [CrossRef]

2. Payan, V.; Royer, A. Analysis of Temperature Emissivity Separation (TES) algorithm applicability and sensitivity. *Int. J. Remote Sens.* **2004**, *25*, 15–37. [CrossRef]
3. Li, Z.L.; Becker, F.; Stoll, M.P.; Wan, Z. Evaluation of Six Methods for Extracting Relative Emissivity Spectra from Thermal Infrared Images. *Remote Sens. Environ.* **1999**, *69*, 197–514. [CrossRef]
4. Pivovarník, M.; Khalsa, S.J.S.; Jiménez-Muñoz, J.C.; Zemek, F. Improved Temperature and Emissivity Separation Algorithm for Multispectral and Hyperspectral Sensors. *IEEE Trans. Geosci. Remote Sens.* **2017**, *55*, 1944–1953. [CrossRef]
5. Cui, J.; Yan, B.; Dong, X.; Zhang, S.; Zhang, J.; Tian, F.; Wang, R. Temperature and emissivity separation and mineral mapping based on airborne TASI hyperspectral thermal infrared data. *Int. J. Appl. Earth Obs. Geoinf.* **2015**, *40*, 19–28. [CrossRef]
6. Neinavaz, E.; Skidmore, A.K.; Darvishzadeh, R. Effects of prediction accuracy of the proportion of vegetation cover on land surface emissivity and temperature using the NDVI threshold method. *Int. J. Appl. Earth Obs. Geoinf.* **2020**, *85*, 101984. [CrossRef]
7. Zhong, Y.; Jia, T.; Zhao, J.; Wang, X.; Jin, S. Spatial-Spectral-Emissivity Land-Cover Classification Fusing Visible and Thermal Infrared Hyperspectral Imagery. *Remote Sens.* **2017**, *9*, 910. [CrossRef]
8. Young, S.J.; Johnson, B.R.; Hackwell, J.A. An in-scene method for atmospheric compensation of thermal hyperspectral data. *J. Geophys. Res. Atmos.* **2002**, *107*, ACH 14-1–ACH 14-20. [CrossRef]
9. Borel, C.C.; Tuttle, R.F. Recent advances in temperature-emissivity separation algorithms. In Proceedings of the 2011 Aerospace Conference, Big Sky, Montana, 5–12 March 2011; pp. 1–14. [CrossRef]
10. Wang, H.; Xiao, Q.; Li, H.; Zhong, B. Temperature and emissivity separation algorithm for TASI airborne thermal hyperspectral data. In Proceedings of the 2011 International Conference on Electronics, Communications and Control (ICECC), Ningbo, China, 9–11 September 2011; pp. 1075–1078. [CrossRef]
11. Adler-Golden, S.; Conforti, P.; Gagnon, M.; Tremblay, P.; Chamberland, M. Remote sensing of surface emissivity with the telops Hyper-Cam. In Proceedings of the 2014 6th Workshop on Hyperspectral Image and Signal Processing: Evolution in Remote Sensing (WHISPERS), Lausanne, Switzerland, 24–27 June 2014; pp. 1–4. [CrossRef]
12. Wang, S.; He, L.; Hu, W. A Temperature and Emissivity Separation Algorithm for Landsat-8 Thermal Infrared Sensor Data. *Remote Sens.* **2015**, *7*, 9904–9927. [CrossRef]
13. Romaniello, V.; Spinetti, C.; Silvestri, M.; Buongiorno, M.F. A Sensitivity Study of the 4.8 um Carbon Dioxide Absorption Band in the MWIR Spectral Range. *Remote Sens.* **2020**, *12*, 172. [CrossRef]
14. Kim, S.; Kim, J.; Lee, J.; Ahn, J. AS-CRI: A New Metric of FTIR-Based Apparent Spectral-Contrast Radiant Intensity for Remote Thermal Signature Analysis. *Remote Sens.* **2019**, *11*, 777. [CrossRef]
15. Griffin, M.K.; Hua, K.; Burke, H.; Kerekes, J.P. Understanding radiative transfer in the midwave infrared: a precursor to full-spectrum atmospheric compensation. *Proc. SPIE* **2004**, *5425*, 348–356. [CrossRef]
16. Andrews, D. *An Introduction to Atmospheric Physics*; Cambridge Press: Cambridge, UK, 2000.
17. Hohn, D.H. Atmospheric Vision 0.35 um < x < 14 um. *Appl. Opt.* **1975**, *14*, 404–412. [PubMed]
18. Sobrino, J.; Li, Z.L.; Stoll, P.; Becker, F. Multi-channel and multi-angle algorithms for estimating sea and land surface temperature with ATSR data. *Int. J. Remote Sens.* **2004**, *17*, 2089–2114. [CrossRef]
19. Driggers, R.G.; Friedman, M.H.; Nichols, J. *Introduction to Infrared and Electro-Optical Systems*; Artech House: Norwood, MA, USA, 2012.
20. Eismann, M.T. *Hyperspectral Remote Sensing*; SPIE Press: Bellingham, WA, USA, 2012.
21. Kim, S. Novel Air Temperature Measurement Using Midwave Hyperspectral Fourier Transform Infrared Imaging in the Carbon Dioxide Absorption Band. *Remote Sens.* **2020**, *12*, 1860. [CrossRef]
22. Silvestri, M.; Romaniello, V.; Hook, S.; Musacchio, M.; Teggi, S.; Buongiorno, M.F. First Comparisons of Surface Temperature Estimations between ECOSTRESS, ASTER and Landsat 8 over Italian Volcanic and Geothermal Areas. *Remote Sens.* **2020**, *12*, 184. [CrossRef]
23. Willmott, C.J.; Matsuura, K. Advantages of the mean absolute error (MAE) over the root mean square error (RMSE) in assessing average model performance. *Clim. Res.* **2005**, *30*, 79–82. [CrossRef]
24. Gagnon, M.A.; Gagnon, J.P.; Tremblay, P.; Savary, S.; Farley, V.; Guyot, É.; Lagueux, P.; Chamberland, M. Standoff midwave infrared hyperspectral imaging of ship plumes. *Proc. SPIE* **2016**, *9988*, 998806. [CrossRef]
25. Shamsipour, A.; Azizi, G.; Ahmadabad, M.K.; Moghbel, M. Surface temperature pattern of asphalt, soil and grass in different weather condition. *J. Biodivers. Environ. Sci.* **2013**, *3*, 80–89.

remote sensing

MDPI

Article

Hyperspectral Nonlinear Unmixing by Using Plug-and-Play Prior for Abundance Maps

Zhicheng Wang [1,2], Lina Zhuang [3], Lianru Gao [1,*], Andrea Marinoni [4], Bing Zhang [1,2] and Michael K. Ng [5]

1 The Key Laboratory of Digital Earth Science, Aerospace Information Research Institute, Chinese Academy of Sciences, Beijing 100094, China; wangzc@radi.ac.cn (Z.W.); zb@radi.ac.cn (B.Z.)
2 The School of Electronic, Electrical and Communication Engineering, University of Chinese Academy of Sciences, Beijing 100049, China
3 The Department of Mathematics, Hong Kong Baptist University, Hong Kong, China; linazhuang@hkbu.edu.hk
4 The Department of Physics and Technology, UiT The Arctic University of Norway, NO-9037 Tromsø, Norway; andrea.marinoni@uit.no
5 The Department of Mathematics, The University of Hong Kong, Hong Kong, China; mng@maths.hku.hk
* Correspondence: gaolr@aircas.ac.cn

Received: 20 October 2020; Accepted: 11 December 2020; Published: 16 December 2020

Abstract: Spectral unmixing (SU) aims at decomposing the mixed pixel into basic components, called endmembers with corresponding abundance fractions. Linear mixing model (LMM) and nonlinear mixing models (NLMMs) are two main classes to solve the SU. This paper proposes a new nonlinear unmixing method base on general bilinear model, which is one of the NLMMs. Since retrieving the endmembers' abundances represents an ill-posed inverse problem, prior knowledge of abundances has been investigated by conceiving regularizations techniques (e.g., sparsity, total variation, group sparsity, and low rankness), so to enhance the ability to restrict the solution space and thus to achieve reliable estimates. All the regularizations mentioned above can be interpreted as denoising of abundance maps. In this paper, instead of investing effort in designing more powerful regularizations of abundances, we use plug-and-play prior technique, that is to use directly a state-of-the-art denoiser, which is conceived to exploit the spatial correlation of abundance maps and nonlinear interaction maps. The numerical results in simulated data and real hyperspectral dataset show that the proposed method can improve the estimation of abundances dramatically compared with state-of-the-art nonlinear unmixing methods.

Keywords: hyperspectral imagery; plug-and-play; denoising; nonlinear unmixing

1. Introduction

Hyperspectral remote sensing imaging is a combination of imaging technology and spectral technology. It can obtain two-dimensional spatial information and spectral information of target objects simultaneously [1–3]. Benefiting from the rich spectral information, hyperspectral images (HSIs) can be used to identity materials precisely. Hence, HSIs have been playing a key role in earth observation and used in many fields, including mineral exploration, water pollution, and vegetation [3–9]. However, due to the low spatial resolution, mixed pixels always exist in HSIs, and it is one of the main reasons that preclude the widespread use of HSIs in precise target detection and classification applications. So it is necessary to develop the technique of unmixing [2,3,10–14]. Thanks to the rich band information of hyperspectral images, which allows us to design an effective solution to the problem of mixed pixels. Hyperspectral unmixing (HU) is the process of obtaining the basic components (called endmembers) and their corresponding component ratios (called abundance fractions). The spectral unmixing can be divided into linear unmixing (LU) and nonlinear unmixing (NLU) [2,3]. LU assumes that photons only interact with one material and there is no interaction between materials. Usually, linear mixing only happens in macro scenarios. NLU assumes that photons interact with a variety of materials, including infinite mixtures, bilinear mixtures. For NLU, various models have been proposed to describe the mixing of pixels, taking into account the more complex reflections in the scene. Specifically, they are the generalized bilinear model (GBM) [15], the polynomial post nonlinear model (PPNM) [16], the multilinear mixing model (MLM) [17], the p-linear model [18], the multiharmonic postnonlinear mixing model (MHPNMM) [19], the nonlinear non-negative matrix factorization (NNMF) [20] and so on. Although different kinds of the nonlinear models have been proposed to improve the accuracy of the abundance results, they are always limited by the endmember extraction algorithm. Meanwhile, complex models often lead to excessive computing costs. The LMM has been widely used to address LU problem, while the GBM is the most popular model among the NLMMs to solve the NLU. The NLU is a more challenging problem than LU, and we mainly focus on the NLU in the paper.

The prior information of the abundance has been exploited for spectral unmixing. Different regularizations (such as sparsity, total variation, and low rankness) have been used on the abundances to improve the accuracy of the abundance estimation.

In sparse unmixing methods, sparsity prior of abundance matrix is exploited as a regularization term [21–23]. To produce a more sparse solution, the group sparsity regularization was imposed on abundance matrix [24]. Meanwhile, the sparsity prior is also considered on the interaction abundance matrix, because interaction abundance matrix is much sparser than abundance matrix [25]. In order to capture the spatial structure of the data, the low-rank representation of abundance matrix was used in References [25–28].

Spatial correlation in abundance maps has also been taken advantage for spectral unmixing. By reorganizing the abundance vector as a two dimensional matrix (the height and width of the HSI, respectively), we can obtain a abundance map of i endmember. In order to make full use of the spatial information of abundance maps, the total variation (TV) of abundance maps was proposed to enhance the spatial smoothness on the abundances [28–31]. Low-rank representation of abundance maps was newly introduced to LU in Reference [32].

However, it is worth mentioning that all the regularizations mentioned above can provide a priori information about abundances. Specifically, the sparse regularization promotes sparse abundances. Total Variation holds the view that each abundance map is piecewise smooth. Low-rank regularization enforces the abundance maps to be low-rank. Furthermore, when solving an regularized optimization problem using ADMM , a subproblem composed of a data fidelity term and a regularization term is so called "Moreau proximal operator" or "denoising operator" [33–36].

Plug and play technique is a flexible framework that allows imaging models to be combined with state-of-the-art priors or denoising models [37]. This is the main idea of plug-and-play technique, which has been successfully used to solve inverse problems of images, such as image inpainting [38,39], compressive sensing [40], and super-resolution [41,42]. Instead of investing effort in designing more powerful regularizations on abundances, we use directly a prior from a state-of-the-art denoiser as the regularization, which is conceived to exploit the spatial correlation of abundance maps. So we apply the plug-and-play technique to the field of spectral unmixing, especially in hyperspectral nonlinear unmixing. In particular, it is pointed out that NLU is a challenging problem in HU, so it is expected that such a powerful tool can be used to improve the accuracy of abundance inversion efficiently.

This paper exploits spatial correlation of abundance maps through a plug-and-play technique. We tested two of the best single-band denoising algorithms, namely Block-Matching and 3D filtering method (BM3D) [43] and denoising convolutional neural networks (DnCNN) [44].

The main contributions of this article are summarized as follows.

- We exploit spatial correlation of abundance maps through plug-and-play technique. The idea of the plug-and-play technique was firstly applied to the problem of hyperspectral nonlinear unmixing. We propose a general nonlinear unmxing framework that can be embedded with any state-of-the-art denoisers.
- We tested two state-of-the-art denoisers, namely BM3D and DnCNN, and both of them yield more accurate estimates of abundances than other state-of-the-art GBM-based nonlinear unmixing algorithms.

The rest of the article is structured as follows. Section 2 introduces the related works and the proposed plug-and-play prior based hyperspectral nonlinear unmixing framework. Experimental results and analysis for the synthetic data are illustrated in Section 3. The real hyperspectral dataset experiments and analysis are described in Section 4. Section 5 concludes the paper.

2. Nonlinear Unmixing Problem

2.1. Related Works

2.1.1. Symbols and definitions

We first introduce the notation and definitions used in the paper. An nth-order tensor is identified using Euler-cript letters—for example, $\mathcal{Q} \in \mathbb{R}^{k_1 \times k_2 \times \ldots \times k_i \times \ldots \times k_n}$, with the k_i is the size of the corresponding dimension i. Hence, an HSI can be naturally represented as a third-order tensor, $\mathcal{T} \in \mathbb{R}^{k_1 \times k_2 \times k_3}$, which consists of $k_1 \times k_2$ pixels and k_3 spectral bands. Three further definitions related to tensors are given as follows.

Definition 1. *The dimension of a tensor is called the mode: $\mathcal{Q} \in \mathbb{R}^{k_1 \times k_2 \times \ldots \times k_i \times \ldots \times k_n}$ has n modes. For a third-order tensor $\mathcal{T} \in \mathbb{R}^{k_1 \times k_2 \times k_3}$, by fixing one mode, we can obtain the corresponding sub-arrays, called slices—for example, $\mathcal{T}_{:,:,i}$.*

Definition 2. *The 3-mode product is denoted as $\mathcal{G} = \mathcal{Q} \times_3 \mathbf{X} \in \mathbb{R}^{k_1 \times k_2 \times j}$ for a tensor $\mathcal{Q} \in \mathbb{R}^{k_1 \times k_2 \times k_3}$ and a matrix $\mathbf{X} \in \mathbb{R}^{j \times k_3}$.*

Definition 3. *Given a matrix $\mathbf{A} \in \mathbb{R}^{k_1 \times k_2}$ and vector $\mathbf{c} \in \mathbb{R}^{l_1}$, their outer product, denoted as $\mathbf{A} \circ \mathbf{c}$, is a tensor with dimensions (k_1, k_2, l_1) and entries $(\mathbf{A} \circ \mathbf{c})_{\mathrm{i1,i2,j1}} = \mathbf{A}_{\mathrm{i1,i2}}\mathbf{c}_{\mathrm{j1}}$.*

2.1.2. Nonlinear Model: GBM

A general expression of nonlinear mixing models, considering the second-order photon interactions between different endmembers, is given as follows:

$$\mathbf{y} = \mathbf{C}\mathbf{a} + \sum_{i=1}^{R-1} \sum_{j=i+1}^{R} b_{i,j}\mathbf{c}_i \odot \mathbf{c}_j + \mathbf{n}, \tag{1}$$

where the $\mathbf{y} \in \mathbb{R}^{L \times 1}$ is a pixel with L spectral bands. $\mathbf{C} = [\mathbf{c}_1, \mathbf{c}_2, \ldots, \mathbf{c}_R] \in \mathbb{R}^{L \times R}$, $\mathbf{a} = [a_1, a_2, \ldots, a_R]^T \in \mathbb{R}^{R \times 1}$, and $\mathbf{n} \in \mathbb{R}^{L \times 1}$ represent the mixing matrix containing the spectral signatures of R endmembers, the fractional abundance vector, and the white Gaussian noise, respectively. The nonlinear coefficient $b_{i,j}$ controls the nonlinear interaction between the materials, and $\odot$ is a Hadamard product operation. With different specific definitions of $b_{i,j}$, there are several well-known mixture models, such as GBM [15], FM [1], and PPNM [16].

To satisfy the physical assumptions and overcome the limitations of the FM [1], the GBM redefines the parameter $b_{i,j}$ as $b_{i,j} = \gamma_{i,j} a_i a_j$. Meanwhile, the abundance non-negativity constraint (ANC) and the abundance sum-to-one constraint (ASC) are satisfied as follows:

$$\begin{aligned} a_i \geq 0, \sum_{i=1}^{R} a_i = 1, \\ 0 < \gamma_{i,j} < 1, \forall i < j, \\ \gamma_{i,j} = 0, \forall i \geq j. \end{aligned} \tag{2}$$

The spectral mixing model for N pixels can be written in matrix form:

$$\mathbf{Y} = \mathbf{CA} + \mathbf{MB} + \mathbf{N}, \tag{3}$$

where $\mathbf{Y} = [\mathbf{y}_1, \mathbf{y}_2, ..., \mathbf{y}_N] \in \mathbb{R}^{L \times N}$, $\mathbf{A} = [\mathbf{a}_1, \mathbf{a}_2, ...\mathbf{a}_N] \in \mathbb{R}^{R \times N}$, $\mathbf{M} \in \mathbb{R}^{L \times R(R-1)/2}$, $\mathbf{B} \in \mathbb{R}^{R(R-1)/2 \times N}$, and $\mathbf{N} \in \mathbb{R}^{L \times N}$ represent the observed hyperspectral image matrix, the fractional abundance matrix with N abundance vectors (the columns of $\mathbf{A}$), the bilinear interaction endmember matrix, the nonlinear interaction abundance matrix, and the white Gaussian noise matrix, respectively.

As for Equations (1) and (3), both of them model the the hyperspectral image with two-dimensional matrix for processing, thus destroying the internal spatial structure of the data and resulting in poor abundance inversion. However, given that the hyperspectral images can be naturally represented as a third-order tensor, we rewritten the GBM model based on tensor representation for the original hyperspectral image cube. The hyperspectral image cube $\mathcal{Y} \in \mathbb{R}^{n_{row} \times n_{col} \times L}$ can be expressed in the following format:

$$\mathcal{Y} = \mathcal{A} \times_3 \mathbf{C} + \mathcal{B} \times_3 \mathbf{M} + \mathcal{N}, \tag{4}$$

where $\mathcal{A} \in \mathbb{R}^{n_{row} \times n_{col} \times R}$, $\mathcal{B} \in \mathbb{R}^{n_{row} \times n_{col} \times R(R-1)/2}$, and $\mathcal{N} \in \mathbb{R}^{n_{row} \times n_{col} \times L}$ denote the abundance cube corresponding to R endmembers, the nonlinear interaction abundance cube, and the white Gaussian noise cube, respectively.

This work aims to solve a supervised unmixing problem—that is to estimate the abundances, $\mathcal{A}$, and nonlinear coefficients, $\mathcal{B}$, given the spectral signatures of the endmembers, $\mathbf{C}$, which are known beforehand.

2.2. Motivation

In this paper, we firstly apply the plug-and-play technique to the unmixing problem, especially to the abundance maps and interaction abundance maps for enhancing the accuracy of the estimated abundance results. The plug-and-play technique can be used as the prior information, instead of other convex regularizers [21,22].

The performance of this method is constrained by the denoiser. Two state-of-the-art denoisers, BM3D and DnCNN, are chosen for the prior information of the abundance maps [43,44]. BM3D is well-known nonlocal patch-based denoiser, which can remove noise in a natural image by taking advantage of high spatial correlation of similar patches in the image. As geographic hyperspectral data, the materials in HSIs tend to be spatially dependent, so it is very easy to find similar patches from the images. Meanwhile, the spatial distribution of a single material tends to be aggregated instead of being purely random. The texture structure of abundance maps can be illustrated with an example given in Figure 1. The unmixing of a San Diego Airport image of size 160 × 140 pixels was carried out. The first row in Figure 1 shows the abundance map of 'Ground & road' estimated by the FCLS [45] followed by an endmember estimation step (vertex component analysis (VCA) [46]). As shown in Figure 1, we can find many similar patches (marked with small yellow squares) from the abundance map. Hence, this nonlocal patch-based denoiser can be used on the abundance maps.

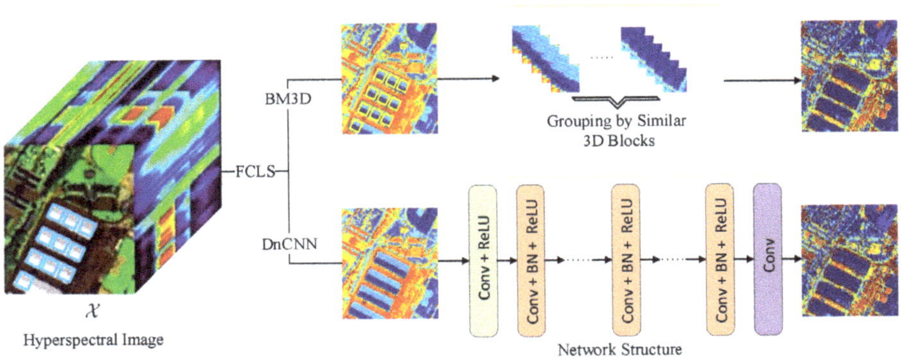

Figure 1. Denoising an abundance map in San Diego Airport image using BM3D and DnCNN denoisers.

With the development of deep learning, convolutional neural network (CNN) based denoising methods have achieved good results. Specifically, deep network structure can effectively learn the features of images. Hence, in the paper, we also chose a well-known CNN-based denoiser as the prior of the abundance maps, named DnCNN (shown in in Figure 1). DnCNN can handle zero-mean Gaussian noise with unknown standard deviation, and residual learning is adopted to separating noise from noisy observations. Therefore, this method can effectively capture the texture structure of abundance maps.

2.3. Proposed Method: Unmixing with Nonnegative Tensor Factorization and Plug-and-Play Proir

To better represent the structure of abundance maps, mixing model (4) can be equivalently written as

$$\mathcal{Y} = \sum_{i=1}^{R} \mathcal{A}_{:,:,i} \circ \mathbf{c}_i + \sum_{j=1}^{R(R-1)/2} \mathcal{B}_{:,:,j} \circ \mathbf{m}_j + \mathcal{N}, \tag{5}$$

where $\mathcal{A}_{:,:,i} \in \mathbb{R}^{n_{row} \times n_{col}}$, $\mathbf{c}_i \in \mathbb{R}^{L \times 1}$, $\mathcal{B}_{:,:,j} \in \mathbb{R}^{n_{row} \times n_{col}}$, and $\mathbf{m}_j \in \mathbb{R}^{L \times 1}$ denote the ith abundance slice, the ith endmember vector, the jth interaction abundance slice, and the jth interaction endmember vector, respectively. Model (5) is depicted in Figure 2.

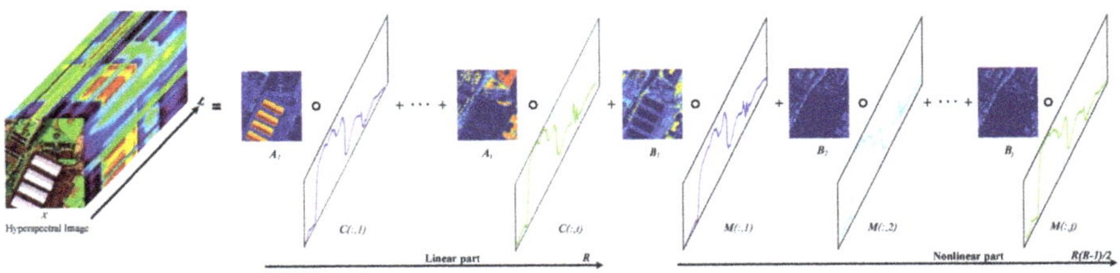

Figure 2. The representation of the generalized bilinear model using the tensor-based framework.

To take full advantage of the abundance maps' prior, we propose a new unmixing method based on the **P**lug-and-**P**lay (**PnP**) framework of abundance maps and **N**onnegative **T**ensor **F**actorization, termed **PnP-NTF**, which aims to solve the following optimization problem:

$$\underset{\substack{\mathcal{A}_{:,:,i}\geq 0,\mathcal{B}_{:,:,j}\geq 0\\ i=1,2,...,R,\\ j=1,2,...,R(R-1)/2}}{\arg\min}\ \frac{1}{2}\left\|\mathcal{Y}-\sum_{i=1}^{R}\mathcal{A}_{:,:,i}\circ\mathbf{c}_i-\sum_{j=1}^{R(R-1)/2}\mathcal{B}_{:,:,j}\circ\mathbf{m}_j\right\|_F^2+\sum_{i=1}^{R}\boldsymbol{\psi}\left(\mathcal{A}_{:,:,i}\right)+\sum_{j=1}^{R(R-1)/2}\boldsymbol{\psi}\left(\mathcal{B}_{:,:,j}\right)$$

$$s.t.\sum_{i=1}^{R}\mathcal{A}_{:,:,i}=\mathbf{1}_{n_{row}}\mathbf{1}_{n_{col}}^{T}, \tag{6}$$

where $\|\mathcal{X}\|_F^2$ denotes the Frobenius norm which returns the square root of the sum of the absolute squares of its elements. The symbol $\boldsymbol{\psi}(.)$ represents the plugged state-of-the-art denoiser, and $\mathbf{1}_d$ represents a vector whose components are all one and whose dimension is given by its subscript.

2.4. Optimization Procedure

The optimization problem in (6) can be solved by optimization using the alternating direction method of multipliers (ADMM) [47]. To use the ADMM, first (6) is converted into an equivalent form by introducing multiple auxiliary variables $\mathbf{V}_i$, $\mathbf{E}_j$ to replace $\mathcal{A}_{:,:,i}(i=1,...,R)$, $\mathcal{B}_{:,:,j}(j=1,...,R(R-1)/2)$. The formulation is as follows:

$$\underset{\mathcal{A}_{:,:,i}\geq 0,\mathcal{B}_{:,:,j}\geq 0}{\arg\min}\ \frac{1}{2}\left\|\mathcal{Y}-\sum_{i=1}^{R}\mathcal{A}_{:,:,i}\circ\mathbf{c}_i-\sum_{j=1}^{R(R-1)/2}\mathcal{B}_{:,:,j}\circ\mathbf{m}_j\right\|_F^2+\lambda_1\sum_{i=1}^{R}\boldsymbol{\psi}\left(\mathbf{V}_i\right)+\lambda_2\sum_{j=1}^{R(R-1)/2}\boldsymbol{\psi}\left(\mathbf{E}_j\right)$$

$$s.t.\begin{cases}\mathcal{A}_{:,:,i}=\mathbf{V}_i, i=1,2,...,R\\ \mathcal{B}_{:,:,j}=\mathbf{E}_j, j=1,2,...,R(R-1)/2\\ \sum\limits_{i=1}^{R}\mathcal{A}_{:,:,i}=\mathbf{1}_{n_{row}}\mathbf{1}_{n_{col}}^{T}\end{cases}. \tag{7}$$

By using the Lagrangian function, (7) can be reformulated as:

$$\begin{aligned}&\mathcal{L}(\mathcal{A}_{:,:,i},\mathcal{B}_{:,:,j},\mathbf{V}_i,\mathbf{E}_j,\mathbf{D}_i,\mathbf{H}_j,\mathbf{G})=\\&\frac{1}{2}\left\|\mathcal{Y}-\sum_{i=1}^{R}\mathcal{A}_{:,:,i}\circ\mathbf{c}_i-\sum_{j=1}^{R(R-1)/2}\mathcal{B}_{:,:,j}\circ\mathbf{m}_j\right\|_F^2+\lambda_1\sum_{i=1}^{R}\boldsymbol{\psi}\left(\mathbf{V}_i\right)\\&+\lambda_2\sum_{j=1}^{R(R-1)/2}\boldsymbol{\psi}\left(\mathbf{E}_j\right)+\frac{\mu}{2}\Big(\sum_{i=1}^{R}\|\mathcal{A}_{:,:,i}-\mathbf{V}_i-\mathbf{D}_i\|_F^2\Big)+\\&\frac{\mu}{2}\Big(\sum_{j=1}^{R(R-1)/2}\|\mathcal{B}_{:,:,j}-\mathbf{E}_j-\mathbf{H}_j\|_F^2\Big)+\frac{\mu}{2}\left\|\sum_{i=1}^{R}\mathcal{A}_{:,:,i}-\mathbf{1}_{n_{row}}\mathbf{1}_{n_{col}}^{T}-\mathbf{G}\right\|_F^2,\end{aligned} \tag{8}$$

where $\mathbf{D}_i$, $\mathbf{H}_j$ and $\mathbf{G}$ are scaled dual variables [48], and μ is the penalty parameter. The variables $\mathcal{A}_{:,:,i},\mathcal{B}_{:,:,j},\mathbf{V}_i,\mathbf{E}_j,\mathbf{D}_i,\mathbf{H}_j,\mathbf{G}$ were updated sequentially: this step is shown in Algorithm 1. The solution of optimization is detailed below.

Algorithm 1: The Proposed PnP-NTF Unmixing Method.

Input: Hyperspectral imagery cube: $\mathcal{Y}$; Endmember matrix: $\mathbf{E}$; Iterations = 1000;

Output: Abundance map Cube: $\mathcal{A}$

for $k = 1; k < Iterations; k++$ **do**

Update abundance map slice: $\mathcal{A}^{k+1}_{:,:,i} = (\sum_{b=1}^{L} c_{i_b} c_{i_b}^T + 2\mu \mathbf{I})^{-1} (\sum_{b=1}^{L} \mathcal{O}_{:,:b} c_{i_b}^T + \mu(\mathbf{V}_i^k + \mathbf{D}_i^k + \mathbf{1}_{n_{row}} \mathbf{1}_{n_{col}}^T + \mathbf{G}^k - \widetilde{\mathbf{A}}))$;

Update nonlinear map slice: $\mathcal{B}^{k+1}_{:,:,j} = (\sum_{b=1}^{L} m_{j_b} m_{j_b}^T + \mu \mathbf{I})^{-1} (\sum_{b=1}^{L} \mathcal{K}_{:,:,b} m_{j_b}^T + \mu(\mathbf{E}_j^k + \mathbf{H}_j^k))$;

Update multiple auxiliary variable: $\mathbf{V}_i^{k+1} = \mathbf{PnP}(\widetilde{\mathbf{V}}_i)$;

Update multiple auxiliary variable: $\mathbf{E}_j^{k+1} = \mathbf{PnP}(\widetilde{\mathbf{E}}_j)$;

Update variable: $\mathbf{D}_i^{k+1} = \mathbf{D}_i^k - (\mathcal{A}^{k+1}_{:,:,i} - \mathbf{V}_i^{k+1})$;

Update variable: $\mathbf{H}_j^{k+1} = \mathbf{H}_j^k - (\mathcal{B}^{k+1}_{:,:,j} - \mathbf{E}_j^{k+1})$;

Update variable: $\mathbf{G}^{k+1} = \mathbf{G}^k - (\sum_{i=1}^{R} \mathcal{A}^{k+1}_{:,:,i} - \mathbf{1}_{n_{row}} \mathbf{1}_{n_{col}}^T)$;

$k = k + 1$;

end

return result

1. Updating of $\mathcal{A}$ The optimization problem for $\mathcal{A}_{:,:,i}$ is

$$
\begin{aligned}
\mathcal{A}^{k+1}_{:,:,i} &= \arg\min_{\mathcal{A}^k_{:,:,i}} \frac{1}{2} \left\| \mathcal{Y} - \sum_{i=1}^{R} \mathcal{A}^k_{:,:,i} \circ \mathbf{c}_i - \sum_{j=1}^{R(R-1)/2} \mathcal{B}^k_{:,:,j} \circ \mathbf{m}_j \right\|_F^2 \\
&+ \frac{\mu}{2} \left\| \mathcal{A}^k_{:,:,i} - \mathbf{V}_i^k - \mathbf{D}_i^k \right\|_F^2 + \frac{\mu}{2} \left\| \sum_{i=1}^{R} \mathcal{A}^k_{:,:,i} - \mathbf{1}_{n_{row}} \mathbf{1}_{n_{col}}^T - \mathbf{G}^k \right\|_F^2 \\
&= \frac{1}{2} \sum_{b=1}^{L} \left\| \mathcal{O}_{:,:,b} - \mathcal{A}^k_{:,:,i} c_{i_b} \right\|_F^2 + \frac{\mu}{2} (\sum_{i=1}^{R} \left\| \mathcal{A}^k_{:,:,i} - \mathbf{V}_i^k - \mathbf{D}_i^k \right\|_F^2) \\
&+ \frac{\mu}{2} \left\| \mathcal{A}^k_{:,:,i} + \widetilde{\mathbf{A}} - \mathbf{1}_{n_{row}} \mathbf{1}_{n_{col}}^T - \mathbf{G}^k \right\|_F^2,
\end{aligned}
\tag{9}
$$

where $\mathcal{O} = \mathcal{Y} - \sum_{i=1,\neg i}^{R} \mathcal{A}^k_{:,:,i} \circ \mathbf{c}_i - \sum_{j=1}^{R(R-1)/2} \mathcal{B}^k_{:,:,j} \circ \mathbf{m}_j \in \mathbb{R}^{n_{row} \times n_{col} \times L}$, and $\mathcal{O}_{:,:,b}$ is the bth slice. Meanwhile, $\widetilde{\mathbf{A}} = \sum_{i=1,\neg i}^{R} \mathcal{A}^k_{:,:,i} \in \mathbb{R}^{n_{row} \times n_{col}}$ and $\mathbf{c}_i = [c_{i_1}, c_{i_2} ..., c_{i_b}, ..., c_{i_L}]^T \in \mathbb{R}^{L \times 1}$ is the ith endmember. Hence the solution for $\mathcal{A}_{:,:,i}$ can be derived as follows:

$$
\mathcal{A}^{k+1}_{:,:,i} = (\sum_{b=1}^{L} c_{i_b} c_{i_b}^T + 2\mu \mathbf{I})^{-1} (\sum_{b=1}^{B} \mathcal{O}_{:,:b} c_{i_b}^T + \mu(\mathbf{V}_i^k + \mathbf{D}_i^k + \mathbf{1}_{n_{row}} \mathbf{1}_{n_{col}}^T + \mathbf{G}^k - \widetilde{\mathbf{A}})). \tag{10}
$$

2. Updating of $\mathcal{B}$

The optimization problem for $\mathcal{B}_{:,:,j}$ is

$$
\begin{aligned}
\mathcal{B}^{k+1}_{:,:,j} &= \arg\min_{\mathcal{B}^k_{:,:,j}} \frac{1}{2} \left\| \mathcal{Y} - \sum_{i=1}^{R} \mathcal{A}^{k+1}_{:,:,i} \circ \mathbf{c}_i - \sum_{j=1}^{R(R-1)/2} \mathcal{B}^k_{:,:,j} \circ \mathbf{m}_j \right\|_F^2 + \frac{\mu}{2} \left\| \mathcal{B}^k_{:,:,j} - \mathbf{E}_j^k - \mathbf{H}_j^k \right\|_F^2 \\
&= \frac{1}{2} \sum_{b=1}^{L} \left\| \mathcal{K}_{:,:,b} - \mathcal{B}^k_{:,:,j} m_{j_b} \right\|_F^2 + \frac{\mu}{2} \left\| \mathcal{B}^k_{:,:,j} - \mathbf{E}_j^k - \mathbf{H}_j^k \right\|_F^2,
\end{aligned}
\tag{11}
$$

where $\mathcal{K} = \mathcal{Y} - \sum_{i=1}^{R} \mathcal{A}^k_{:,:,i} \circ \mathbf{c}_i - \sum_{j=1,,\neg j}^{R(R-1)/2} \mathcal{B}^k_{:,:,j} \circ \mathbf{m}_j \in \mathbb{R}^{n_{row} \times n_{col} \times L}$, and $\mathcal{K}_{:,:,b}$ is the bth slice. Meanwhile, $\mathbf{m}_j = [m_{j_1}, m_{j_2} ..., m_{j_b}, ..., m_{j_L}]^T \in \mathbb{R}^{L\times 1}$ is the jth interaction endmember. Hence the solution for $\mathcal{B}_{:,:,j}$ can be derived as follows:

$$\mathcal{B}^{k+1}_{:,:,j} = (\sum_{b=1}^{L} m_{j_b} m_{j_b}^T + \mu \mathbf{I})^{-1} (\sum_{b=1}^{L} \mathcal{K}_{:,:,b} m_{j_b}^T + \mu(\mathbf{E}^k_j + \mathbf{H}^k_j)). \quad (12)$$

3. Updating of $\mathbf{V}$

   The optimization problem for $\mathbf{V}_i$ is

$$\begin{aligned} \mathbf{V}^{k+1}_i &= \arg\min_{\mathbf{V}^k_i} \lambda_1 \psi\left(\mathbf{V}^k_i\right) + \frac{\mu}{2}\left\|\mathcal{A}^{k+1}_{:,:,i} - \mathbf{V}^k_i - \mathbf{D}^k_i\right\|_F^2 \\ &= \frac{1}{2}\left\|\widetilde{\mathbf{V}}_i - \mathbf{V}^k_i\right\|_F^2 + \frac{\lambda_1}{\mu}\psi\left(\mathbf{V}^k_i\right), \end{aligned} \quad (13)$$

   where $\widetilde{\mathbf{V}}_i = \mathcal{A}^{k+1}_{:,:,i} - \mathbf{D}^k_i \in \mathbb{R}^{n_{row}\times n_{col}}$. Sub-problem (13) can be solved using **PnP** framework of $\widetilde{\mathbf{V}}_i$, then $\mathbf{V}^{k+1}_i$ can be calculated as

$$\mathbf{V}^{k+1}_i = \mathbf{PnP}(\widetilde{\mathbf{V}}_i). \quad (14)$$

4. Updating of $\mathbf{E}$

   The optimization problem for $\mathbf{E}_j$ is

$$\begin{aligned} \mathbf{E}^{k+1}_j &= \arg\min_{\mathbf{E}^k_j} \lambda_2 \psi\left(\mathbf{E}_j\right) + \frac{\mu}{2}\left\|\mathcal{B}^{k+1}_{:,:,j} - \mathbf{E}^k_j - \mathbf{H}^k_j\right\|_F^2 \\ &= \frac{1}{2}\left\|\widetilde{\mathbf{E}}_j - \mathbf{E}^k_j\right\|_F^2 + \frac{\lambda_2}{\mu}\psi\left(\mathbf{E}_j\right), \end{aligned} \quad (15)$$

   where $\widetilde{\mathbf{E}}_j = \mathcal{B}^{k+1}_{:,:,j} - \mathbf{H}^k_j \in \mathbb{R}^{n_{row}\times n_{col}}$. Sub-problem (15) can be solved via **PnP** framework of $\widetilde{\mathbf{E}}_j$, then $\mathbf{E}^{k+1}_j$ can be expressed as

$$\mathbf{E}^{k+1}_j = \mathbf{PnP}(\widetilde{\mathbf{E}}_j). \quad (16)$$

5. Updating of $\mathbf{D}$

$$\mathbf{D}^{k+1}_i = \mathbf{D}^k_i - (\mathcal{A}^{k+1}_{:,:,i} - \mathbf{V}^{k+1}_i). \quad (17)$$

6. Updating of $\mathbf{H}$

$$\mathbf{H}^{k+1}_j = \mathbf{H}^k_j - (\mathcal{B}^{k+1}_{:,:,j} - \mathbf{E}^{k+1}_j). \quad (18)$$

7. Updating of $\mathbf{G}$

$$\mathbf{G}^{k+1} = \mathbf{G}^k - (\sum_{i=1}^{R} \mathcal{A}^{k+1}_{:,:,i} - \mathbf{1}_{n_{row}} \mathbf{1}^T_{n_{col}}). \quad (19)$$

3. Experiments and Analysis on Synthetic Data

In this section, we illustrate the performance of the proposed PnP-NTF framework on the two state-of-the-art denoising method, named BM3D and DnCNN, for the abundance estimation. We compare the proposed method with some advanced algorithms to address the GBM, which contains gradient descent algorithm (GDA) [49], the semi-nonnegative matrix factorization (semi-NMF) [50] algorithm and subspace unmixing with low-rank attribute embedding algorithm (SULoRA) [11]. For specifically, the GDA method is a benchmark to solve the GBM pixel by pixel, and semi-NMF can speed up and reduce the time loss. Meanwhile, the semi-NMF based method can consider the partial spatial information of the image. SULoRA is a general subspace unmixing method that jointly estimates subspace projections and abundance, and can model the raw subspace with low-rank attribute embedding. All of the experiments were conducted in MATLAB R2018b on a desktop of 16 GB RAM, Intel (R) Core (TM) i5-8400 CPU, @2.80 GHz.

In order to quantify the effect of the proposed method numerically, three widely metrics, including the root-mean-square error (RMSE) of abundances, the image reconstruction error (RE), and the average of spectral angle mapper (aSAM) are used. For specifically, the RMSE quantifies the difference between the estimated abundance $\widehat{\mathcal{A}}$ and the true abundances $\mathcal{A}$ as follows:

$$RMSE = \sqrt{\frac{1}{R \times N}\|\mathcal{A} - \widehat{\mathcal{A}}\|_F^2}. \tag{20}$$

The RE measures the difference between the observation $\mathcal{Y}$ and its reconstruction $\widehat{\mathcal{Y}}$ as follows:

$$RE = \sqrt{\frac{1}{N \times L}\|\mathcal{Y} - \widehat{\mathcal{Y}}\|_F^2}. \tag{21}$$

The aSAM qualifies the average spectral angle mapping of the estimated ith spectral vector $\hat{\mathbf{y}}_l$ and observed ith spectral vector $\mathbf{y}$. The aSAM is defined as follows:

$$aSAM = \frac{1}{N}\sum_{i=1}^{N} arccos\left(\frac{\mathbf{y}_i^T \cdot \hat{\mathbf{y}}_i}{\|\mathbf{y}_i\| \, \|\hat{\mathbf{y}}_i\|}\right). \tag{22}$$

3.1. Data Generation

In the simulated experiments, the synthetic data was generated similar to References [32,51], and the specific process is as follows:

1. Six spectral endmembers signals with 224 spectral bands were randomly chosen from the USGS digital spectral library (https://www.usgs.gov/labs/spec-lab), namely Carnallite, Ammonio–jarosite, Almandine, Brucite, Axinite, and Chlonte.
2. We generated a simulated image of size s^2(rows) $\times$ s^2(columns) $\times$ L(bands), which can be divided into small blocks of size $s \times s \times L$.
3. A randomly selected endmember was assigned to each block, and a $k \times k$ low-pass filter was used to generate abundance map cube of size $s^2 \times s^2 \times R$ that contained mixed pixels, while satisfying the ANC and ASC constraints.
4. After obtaining the endmember information and the abundance information, then clean HSIs can be generated by the generalized bilinear model and the polynomial post nonlinear model. The interaction coefficient parameters in the GBM were set randomly, and the nonlinear coefficient parameters in the PPNM were set to 0.25.
5. To effectively evaluate the robustness performance of the proposed method on the different signal-to-noise ratio (SNR), the zero-mean Gaussian white noise was added to the clean data.

3.2. Evaluation of the Methods

The details of the simulated data can be obtained with the previous steps, then we generated a series of noisy images with SNRs = {15, 20, 30} dB to evaluate performance of the proposed method and compare with other methods.

3.2.1. Parameter Setting

To compare all the algorithms fairly, the parameters in the all compared methods were hand-tuned to the optimal values. Specifically, the FCLS was used to initialize the abundance information in the all methods (including the proposed method). Note that a direct compasion with FCLS unmixing results is unfair and FCLS is served as a benchmark, which shows the impact of using a linear unmixing method on nonlinear mixed images. The GDA is considered as the benchmark to solve the GBM.

The tolerances for stopping the iterations in GDA, Semi-NMF, and SULoRA were set to 1×10^{-6}. For the proposed PnP-NTF framework based method, the parameters to be adjusted were divided into two parts, one of which is the parameter of the denoiser we chosen, and the other part is the penalty parameter μ. Firstly, the standard deviation of additive white Gaussian noise σ is searching from 0 to 255 with the step of 25, the the block size used for the hard-thresholding (HT) filtering is set as 8 in BM3D, respectively. The parameters of the DnCNN is the same as Reference [44]. Meanwhile, the penalty parameter μ was set to 8×10^{-3}, and the tolerance for stopping the iterations was set to 1×10^{-6}.

3.2.2. Comparison of Methods under Different Gaussian Noise Levels

In our experiments, we generate three images of size $64 \times 64 \times 224$ with 4096 pixels and 224 bands. More specifically, the 'Scene1' is generated by the GBM model, and the 'Scene2' is generated by the PPNM model. The 'Scene3' is a mixture of the 'Scene1' and 'Scene2', as half pixels in 'Scene3' were generated by the GBM and the others were generated by PPNM [50]. The 'Scene1' is used to evaluate the efficiency of the proposed method to handle mixtures based on GBM, while the 'Scene2' and 'Scene3' were used to evaluate the robustness of the proposed method to mixtures based on different mixing models.

For the proposed method and the other methods, the abundances were initialized with the same method, that is FCLS. In a supervised nonlinear unmixing problem, the spectral vectors of endmember were known as a priori. Considering that the accuracy of abundance inversion depends on the quality of endmember signals, we used the true endmembers in the experiments for fair comparison.

Table 1 quantifies the corresponding results of the three evaluation indicators (RMSE, RE, and aSAM) in detail on the 'Scene1'. As seen from the Table 1, the proposed PnP-NTF based framework with the advanced denoisers provide the superior unmixing results, compared with other methods. Specifically, we tested two state-of-the-art denoisers, namely BM3D and DnCNN, and both of them obtained the best performance. The RMSE, RE, and aSAM obtained minimum values from the proposed PnP-NTF based frameworks, which show the efficiency of the proposed methods is superior compared with other state-of-the-art methods (shown in bold). Figure 3 shows the results of the proposed algorithm and the others algorithms under three indexes (RMSE, RE, and aSAM). For the different levels of noise in 'Scene1', the proposed methods yield the superior performance in all indexes. Also we can see from the histogram of Figures 4–6 that the proposed methods obtain the minimum RMSEs in all scenes.

Table 1. Evaluation Results in 'Scene1' with different signal-to-noise ratios (SNRs) and time cost (s).

Scenario	SNR (dB)	Metric	FCLS	GDA	Semi-NMF	SULoRA	PnP-NTF-BM3D (Proposed)	PnP-NTF-DnCNN (Proposed)
Scene1'	15	RMSE	0.0614	0.0535	0.0521	0.0521	0.0431	**0.0428**
		RE	0.0898	0.0893	0.0879	0.0879	**0.0876**	**0.0876**
		aSAM	0.1773	0.1764	0.1744	0.1743	**0.1737**	**0.1737**
		Time	**1**	590	6	**1**	700	450
	20	RMSE	0.0544	0.0449	0.0405	0.0372	0.0269	**0.0267**
		RE	0.0527	0.0519	0.0497	0.0497	**0.0492**	**0.0492**
		aSAM	0.1039	0.1024	0.0991	0.0991	**0.0982**	**0.0982**
		Time	**1**	609	7	**1**	839	585
	30	RMSE	0.0511	0.0409	0.0315	0.0265	**0.0125**	**0.0125**
		RE	0.0242	0.0224	0.0167	0.0171	**0.0156**	**0.0156**
		aSAM	0.0447	0.0411	0.0330	0.0329	**0.0312**	**0.0312**
		Time	**1**	611	13	**1**	1142	836

To evaluate the robustness of the proposed methods against model error, we generated 'Scene2' and 'Scene3' of size $64 \times 64 \times 224$. As shown in Tables 2 and 3, the proposed methods obtained the best estimate of abundances in terms of RMSE, RE, and aSAM (shown in bold). We cannot provide the proof of the convergence of the proposed algorithm, but the experimental results show that it is convergent when plugged by BM3D and DnCNN (shown in Figures 7 and 8).

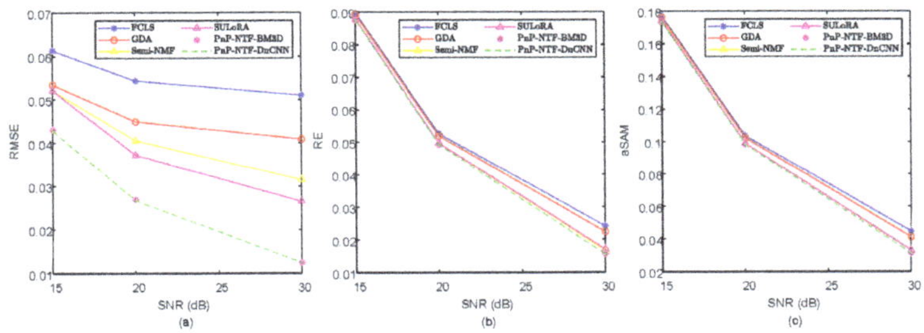

Figure 3. Unmixing performance in terms of root-mean-square error (RMSE) (**a**), reconstruction error (RE) (**b**), and average of spectral angle mapper (aSAM) (**c**) in the simulated 'Scene1' with different Gaussian Noise Levels.

Table 2. Evaluation Results in 'Scene2' with different SNRs and time cost (s).

Scenario	SNR (dB)	Metric	FCLS	GDA	Semi-NMF	SULoRA	PnP-NTF-BM3D (Proposed)	PnP-NTF-DnCNN (Proposed)
Scene2'	15	RMSE	0.0804	0.0683	0.0586	0.0596	0.0437	**0.0434**
		RE	0.0956	0.0946	0.0919	0.0918	**0.0913**	**0.0913**
		aSAM	0.1805	0.1787	0.1748	0.1746	**0.1737**	**0.1737**
		Time	**1**	557	8	**1**	712	465
	20	RMSE	0.0759	0.0626	0.0487	0.0461	0.0281	**0.0280**
		RE	0.0582	0.0566	0.0523	0.0521	**0.0514**	**0.0514**
		aSAM	0.1092	0.1063	0.1000	0.0998	**0.0984**	**0.0984**
		Time	**1**	564	11	**1**	687	452
	30	RMSE	0.0738	0.0601	0.0396	0.0370	**0.0167**	**0.0167**
		RE	0.0314	0.0285	0.0183	0.0206	**0.0163**	**0.0163**
		aSAM	0.0558	0.0491	0.0345	0.0381	**0.0314**	**0.0314**
		Time	**1**	561	18	**1**	795	603

Table 3. Evaluation Results in 'Scene3' with different SNRs and time cost (s).

Scenario	SNR (dB)	Metric	FCLS	GDA	Semi-NMF	SULoRA	PnP-NTF-BM3D (Proposed)	PnP-NTF-DnCNN (Proposed)
Scene3'	15	RMSE	0.0745	0.0649	0.0577	0.0565	0.0433	**0.0430**
		RE	0.0934	0.0927	0.0905	0.0903	**0.0899**	**0.0899**
		aSAM	0.1790	0.1778	0.1748	0.1747	**0.1739**	**0.1739**
		Time	**1**	544	7	**1**	649	420
	20	RMSE	0.0692	0.0583	0.0470	0.0417	0.0280	**0.0279**
		RE	0.0559	0.0548	0.0513	0.0511	**0.0504**	**0.0504**
		aSAM	0.1061	0.1042	0.0995	0.0993	**0.0982**	**0.0982**
		Time	**1**	573	9	**1**	712	472
	30	RMSE	0.0665	0.0552	0.0382	0.0319	**0.0150**	**0.0151**
		RE	0.0284	0.0262	0.0178	0.0187	**0.0160**	**0.0160**
		aSAM	0.0493	0.0447	0.0338	0.0354	**0.0312**	**0.0312**
		Time	**1**	572	16	**1**	957	664

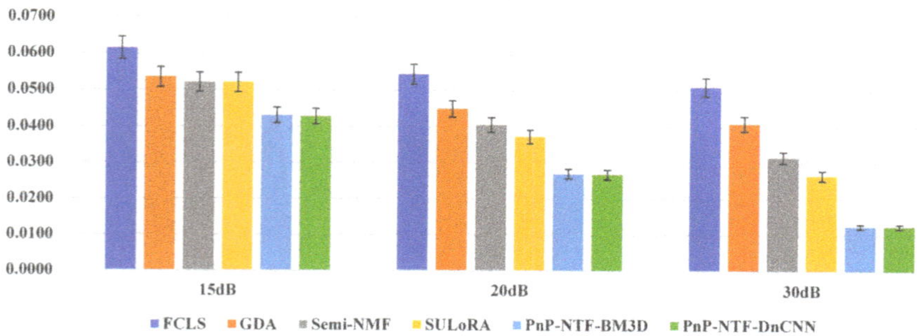

Figure 4. Evaluation results of RMSE with the proposed method and state-of-the-art methods on 'Scene1'.

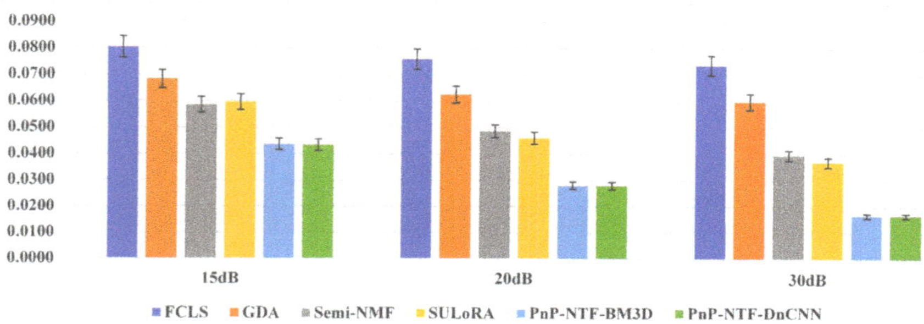

Figure 5. Evaluation results of RMSE with the proposed method and state-of-the-art methods on 'Scene2'.

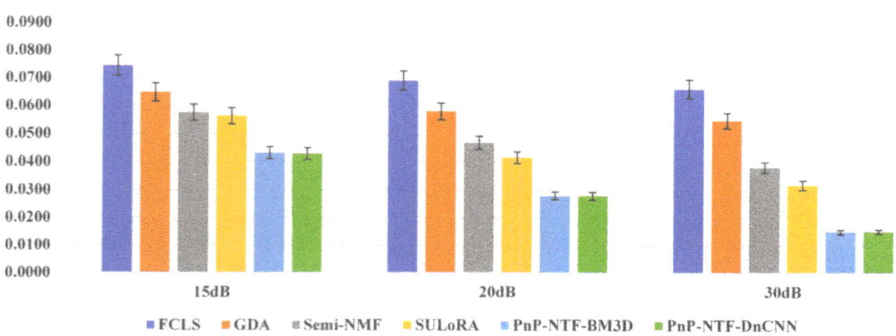

Figure 6. Evaluation results of RMSE with the proposed method and state-of-the-art methods on 'Scene3'.

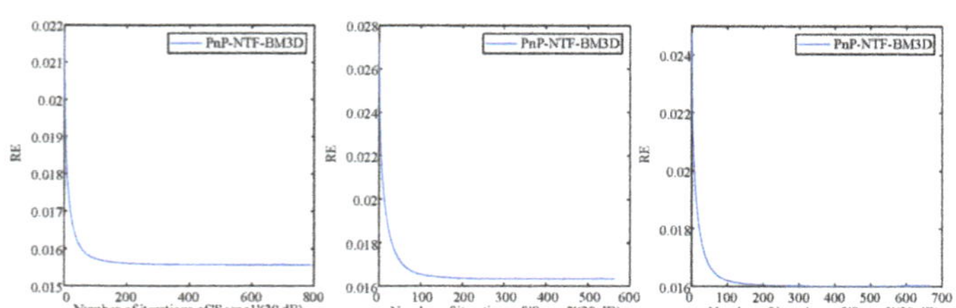

Figure 7. Iterations of RE with BM3D.

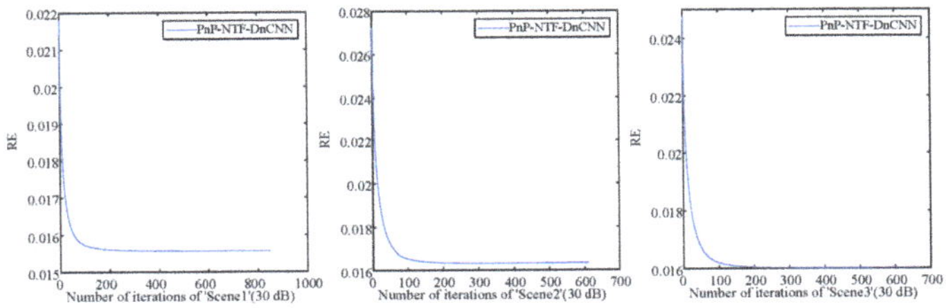

Figure 8. Iterations of RE with DnCNN.

3.2.3. Comparison of Methods under Denoised Abundance Maps

We make a series of experiments to show the difference between the proposed methods and the conventional unmixing methods (FCLS, GDA, and Semi-NMF) with afterwards denoising the calculated abundance maps by BM4D. The results in Tables 4–6 show that the denoised abundance maps provided by FCLS, GDA, and Semi-NMF can obtain a better results than corresponding original abundance maps. However, for the proposed methods, we use directly a state-of-the-art denoiser as the regularization, which is to exploit the spatial correlation of abundance maps. The results show that using plug-and-play prior for the abundance maps and interaction abundance maps can enhance the accuracy of the estimated abundance results efficiently.

Table 4. Evaluation result of denoised abundance in 'Scene1' with different SNRs.

Scenario	SNR (dB)	Metric	FCLS Denoised	GDA Denoised	Semi-NMF Denoised	PnP-NTF-BM3D (Proposed)	PnP-NTF-DnCNN (Proposed)
Scene1'	15	RMSE	0.0501	0.0408	**0.0369**	0.0431	0.0428
	20		0.0489	0.0388	0.0328	0.0269	**0.0267**
	30		0.0487	0.0383	0.0283	**0.0125**	**0.0125**

Table 5. Evaluation result of denoised abundance in 'Scene2' with different SNRs.

Scenario	SNR (dB)	Metric	FCLS Denoised	GDA Denoised	Semi-NMF Denoised	PnP-NTF-BM3D (Proposed)	PnP-NTF-DnCNN (Proposed)
Scene2'	15	RMSE	0.0734	0.0599	0.0446	0.0437	**0.0434**
	20		0.0733	0.0593	0.0423	0.0281	**0.0280**
	30		0.0734	0.0594	0.0376	**0.0167**	**0.0167**

Table 6. Evaluation result of denoised abundance in 'Scene3' with different SNRs.

Scenario	SNR (dB)	Metric	FCLS Denoised	GDA Denoised	Semi-NMF Denoised	PnP-NTF-BM3D (Proposed)	PnP-NTF-DnCNN (Proposed)
Scene3'	15	RMSE	0.0665	0.0557	0.0435	0.0433	**0.0430**
	20		0.0659	0.0545	0.0404	0.0280	**0.0279**
	30		0.0656	0.0542	0.0360	**0.0150**	0.0151

4. Experiments and Analysis on Real Dataset

In this section, we use two real hyperspectral datasets to validate the performance of the proposed methods. Due to the lack of the groundtruth of abundances in the real scenes, the RE in (21) and the aSAM in (22) were used to test the unmixing performance of the all methods. The convergence of the proposed methods on the two real hyperspectral datasets are shown in Figure 9.

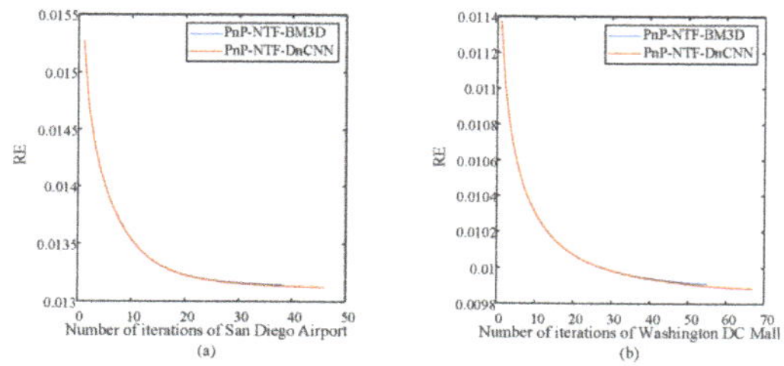

Figure 9. Iterations of RE with the proposed methods on two real hyperspectral datasets: (**a**) number of iterations of San Diego Airport, (**b**) number of iterations of Washington DC Mall .

4.1. San Diego Airport

The first real dataset is called 'San Diego Airport' image, which was captured by the AVIRIS over San Diego. The original image of size 400 × 400 includes 224 spectral channels in the wavelength range of 370 nm to 2510 nm. After removing bands affected by water vapor absorption, 189 band are kept. For our experiments, a subimage of size 160 (rows)×140 (columns) (shown in Figure 10a) is chosen as the test image. The selected scene mainly contains five endmembers, namely 'Roof', 'Grass', 'Rround and Road', 'Tree', and 'Other' [52].

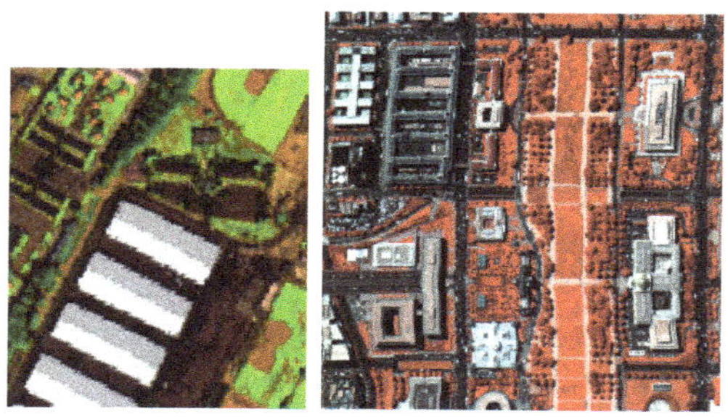

(**a**) San Diego Airport (**b**) sub-DC

Figure 10. Hyperspectral images (HSIs) used for our experiments: (**a**) sub-image of San Diego Airport data, (**b**) sub-image of Washington Dc Mall.

The subimage we chosen has been studied in Reference [52]. VCA [46] method is used to estiamte the endmebers. Meanwhile, the FCLS is used to initialize the abundances in all methods. The ASC constraint in the semi-NMF was set to 0.1. Two state-of-the-art denoisers, embedded in the proposed PnP-NTF-based framework were tested. For the BM3D denoiser, the standard deviation of noise was hand-tuned. For the DnCNN denoiser, its parameter was set in a same way as Reference [44]. The penalty parameter μ was set to 1×10^{-4}. The tolerance for stopping the iterations was set to 1×10^{-6} for all algorithms.

Table 7 shows the performance of different unmixing methods in terms of RE and aSAM in the San Diego Airport image. Our proposed method obtains the best results. Figure 11 shows the estimated abundance maps obtained by all methods. Focusing on the abundance maps of 'Ground and Road', we can see that the roof area is regarded as a mixture of 'Roof' and 'Ground and Road' in the unmixing results of FCLS, GDA, Semi-NMF and SULoRA methods. In fact, the the roof area only contains endmember 'Roof'. Unmixing results of the proposed PnP-NTF-DnCNN/BM3D are more reasonable.

Furthermore, Figure 12 shows the distribution of the RE on the San Diego Airport. The bright areas in Figure 12 indicate large errors in the reconstructed images. The error map shows that the FCLS performed worst, because the FCLS only can handle the linear information but ignore the nonlinear information in the image. Meanwhile, the semi-NMF performed better than GDA because the GDA is a pixel-based algorithm that does not take any spatial information into consideration. Our method, exploiting self-similarity of abundance maps, can perform better than other methods.

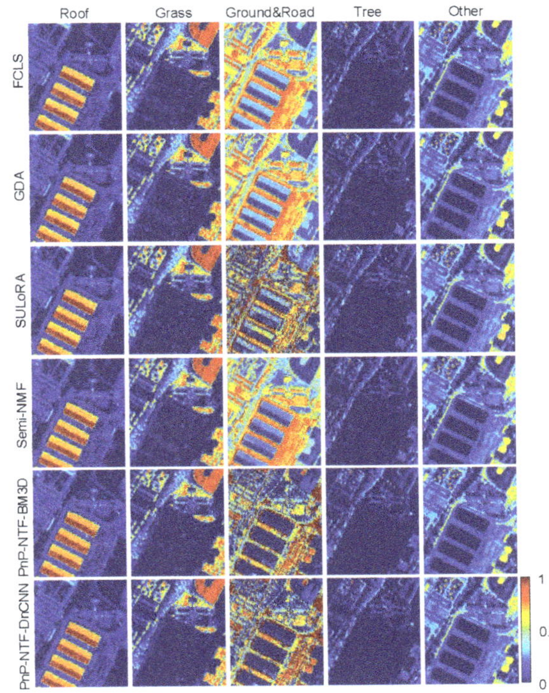

Figure 11. Estimated abundance maps comparison between the proposed algorithm and state-of-the-art algorithms on the San Diego Airport.

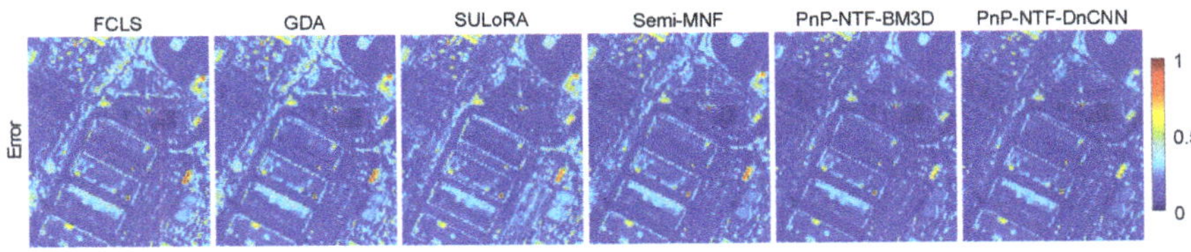

Figure 12. RE distribution maps comparison between the proposed algorithm and state-of-the-art algorithms on the San Diego Airport.

Table 7. Evaluation Results with the RE, aSAM and cost time (s) on the San Diego Airport.

Scenario	Metric	FCLS	GDA	SULoRA	Semi-NMF	PnP-NTF-BM3D (Proposed)	PnP-NTF-DnCNN (Proposed)
San Diego Airport	RE	0.0165	0.0164	0.0156	0.0150	0.0132	**0.0131**
	aSAM	0.0596	0.0594	0.0535	0.0542	0.0455	**0.0454**
	Time	5	168	**4**	47	191	108

4.2. Washington DC Mall

The second real dataset is called 'Washington DC Mall' image, which was acquired by HYDICE sensor over Washington DC, USA. The original image of size 1208 × 307 includes 210 spectral bands. Its spatial resolution is 3 m. After removing bands corrupted by water vapor absorption, 191 band are kept. There are seven endmembers in the image: 'Roof', 'Grass', 'Road', 'Trail', 'Water', 'Shadow', and 'Tree' [52]. We chose a subimage with 256 × 256 pixels for the experiments, called sub-DC (shown in Figure 10b). The Hysime [53] was firstly used to estimate the number of endmembers, then the VCA was used to extract the spectral information of endmembers. The extracted endmembers were named 'Roof1', 'Roof2', 'Grass', 'Road', 'Tree', and 'Trail'.

The parameters in the comparison methods were manually tuned to obtain optimal performance. The parameter setting of our methods was same as that in the 'San Diego Airport' image.

Table 8 shows the results of the proposed method and the state-of-the-art methods in the 'Washington DC Mall' image. The proposed methods obtained the best results in terms of RE and aSAM. Figures 13 and 14 show the estimated abundance maps and the error maps, respectively. In Figure 14, the proposed methods show much smaller errors in the reconstructed images.

Table 8. Evaluation Results with the RE, aSAM and cost time (s) on the Washington DC Mall.

Scenario	Metric	FCLS	GDA	SULoRA	Semi-NMF	PnP-NTF-BM3D (Proposed)	PnP-NTF-DnCNN (Proposed)
Washington DC Mall	RE	0.0156	0.0154	0.0152	0.0120	**0.0099**	**0.0099**
	aSAM	0.1020	0.1015	0.0880	0.0837	0.0623	**0.0621**
	Time	17	670	**10**	43	1163	585

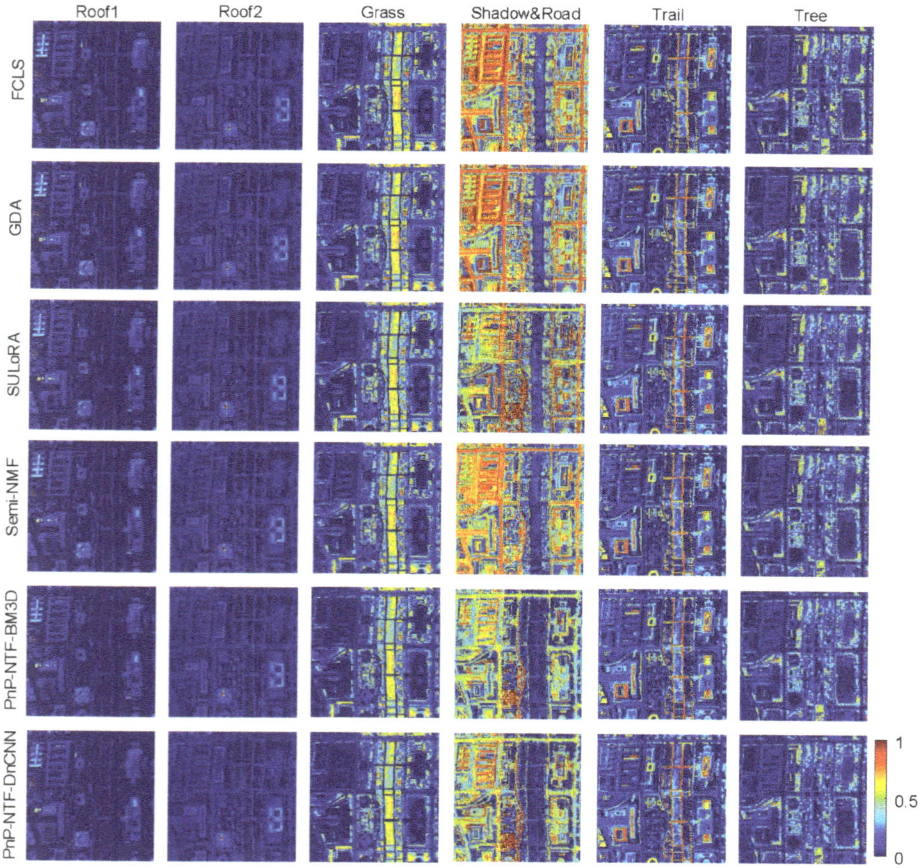

Figure 13. Estimated abundance maps comparison between the proposed algorithm and state-of-the-art algorithms on Washington DC Mall data.

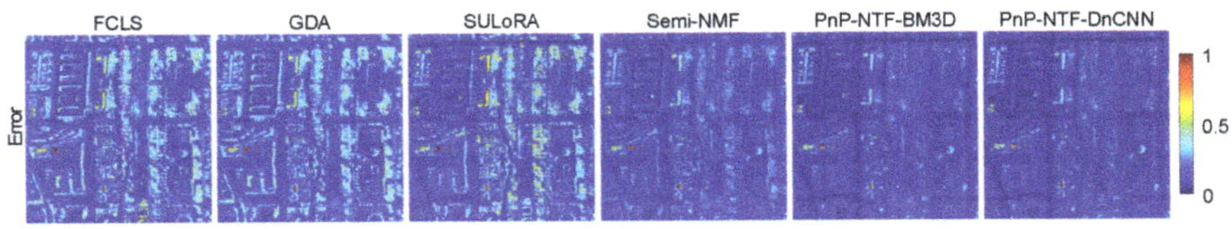

Figure 14. RE distribution maps comparison between the proposed algorithm and state-of-the-art algorithms on Washington DC Mall data.

5. Conclusions

In this paper, we propose a new hyperspectral nonlinear unmixing framework, which takes advantage of spatial correlation (i.e., self-similarity) of abundance maps through a plug-and-play technique. The self-similarity of abundance maps is imposed on our objective function, which is solved by ADMM embedded with a denoising method based regularization. We tested two state-of-the-art denoising methods (BM3D and DnCNN). In the experiments with simulated data and real data, the proposed methods can obtain more accurate estimation of abundances than state-of-the-art methods. Furthermore, we tested the proposed method in case of the number of endmembers with 5, and obtained better results compared to other methods. However, with the growing of the number of endmembers, the difficulty of unmixing will also increase, which is our future research direction.

Author Contributions: Conceptualization, L.G. and Z.W.; methodology, L.Z. and M.K.N.; software, Z.W.; validation, Z.W., L.Z., L.G. and M.K.N.; formal analysis, B.Z.; investigation, L.Z.; resources, B.Z.; writing—original draft preparation, Z.W.; writing—review and editing, A.M., M.K.N. and B.Z.; visualization, Z.W.; supervision, L.G. and L.Z.; project administration, B.Z.; funding acquisition, L.G., L.Z. and A.M. All authors have read and agreed to the published version of the manuscript.

Funding: This work was supported in part by the National Natural Science Foundation of China under Grant 42030111 Research Fund and in part by the National Natural Science Foundation of China under Grant 42001287. The A.Marinoni's work was supported in part by Centre for Integrated Remote Sensing and Forecasting for Arctic Operations (CIRFA) and the Research Council of Norway (RCN Grant no. 237906).

Acknowledgments: The authors would like to thank Naoto Yokoya for providing the semi-NMF code for our comparison experiment. Yuntao Qian provided the abundance and endmember data used in some of the experiments with synthetic data.

Conflicts of Interest: The authors declare no conflict of interest.

References

1. Dobigeon, N.; Tourneret, J.; Richard, C.; Bermudez, J.C.M.; McLaughlin, S.; Hero, A.O. Nonlinear Unmixing of Hyperspectral Images: Models and Algorithms. *IEEE Signal Process. Mag.* **2014**, *31*, 82–94. [CrossRef]
2. Keshava, N.; Mustard, J.F. Spectral unmixing. *IEEE Signal Process. Mag.* **2002**, *19*, 44–57. [CrossRef]
3. Bioucas-Dias, J.M.; Plaza, A.; Dobigeon, N.; Parente, M.; Du, Q.; Gader, P.; Chanussot, J. Hyperspectral Unmixing Overview: Geometrical, Statistical, and Sparse Regression-Based Approaches. *IEEE J. Sel. Top. Appl. Earth Obs. Remote Sens.* **2012**, *5*, 354–379. [CrossRef]
4. Zhang, T.-t.; Liu, F. Application of hyperspectral remote sensing in mineral identification and mapping. In Proceedings of the 2012 2nd International Conference on Computer Science and Network Technology, Changchun, China, 29–31 December 2012; pp. 103–106. [CrossRef]
5. Contreras, C.; Khodadadzadeh, M.; Tusa, L.; Loidolt, C.; Tolosana-Delgado, R.; Gloaguen, R. Geochemical And Hyperspectral Data Fusion For Drill-Core Mineral Mapping. In Proceedings of the 2019 10th Workshop on Hyperspectral Imaging and Signal Processing: Evolution in Remote Sensing (WHISPERS), Amsterdam, The Netherlands, 24–26 September 2019; pp. 1–4. [CrossRef]
6. Marinoni, A.; Clenet, H. Identification of mafic minerals on Mars by nonlinear hyperspectral unmixing. In Proceedings of the 2016 8th Workshop on Hyperspectral Image and Signal Processing: Evolution in Remote Sensing (WHISPERS), Los Angeles, CA, USA, 21–24 August 2016; pp. 1–4. [CrossRef]

7. Huang, Z.; Zheng, J. Extraction of Black and Odorous Water Based on Aerial Hyperspectral CASI Image. In Proceedings of the IGARSS 2019—2019 IEEE International Geoscience and Remote Sensing Symposium, Yokohama, Japan, 28 July–2 August 2019; pp. 6907–6910. [CrossRef]
8. Li, Q.; Kit Wong, F.K.; Fung, T. Comparison Feature Selection Methods for Subtropical Vegetation Classification with Hyperspectral Data. In Proceedings of the IGARSS 2019—2019 IEEE International Geoscience and Remote Sensing Symposium, Yokohama, Japan, 28 July–2 August 2019; pp. 3693–3696. [CrossRef]
9. Yu, H.; Gao, L.; Liao, W.; Zhang, B.; Zhuang, L.; Song, M.; Chanussot, J. Global Spatial and Local Spectral Similarity-Based Manifold Learning Group Sparse Representation for Hyperspectral Imagery Classification. *IEEE Trans. Geosci. Remote Sens.* **2020**, *58*, 3043–3056. [CrossRef]
10. Zhang, B.; Zhuang, L.; Gao, L.; Luo, W.; Ran, Q.; Du, Q. PSO-EM: A Hyperspectral Unmixing Algorithm Based On Normal Compositional Model. *IEEE Trans. Geosci. Remote Sens.* **2014**, *52*, 7782–7792. [CrossRef]
11. Hong, D.; Zhu, X.X. SULoRA: Subspace Unmixing With Low-Rank Attribute Embedding for Hyperspectral Data Analysis. *IEEE J. Sel. Top. Signal Process.* **2018**, *12*, 1351–1363. [CrossRef]
12. Zhuang, L.; Lin, C.; Figueiredo, M.A.T.; Bioucas-Dias, J.M. Regularization Parameter Selection in Minimum Volume Hyperspectral Unmixing. *IEEE Trans. Geosci. Remote Sens.* **2019**, *57*, 9858–9877. [CrossRef]
13. Qu, Y.; Qi, H. uDAS: An Untied Denoising Autoencoder With Sparsity for Spectral Unmixing. *IEEE Trans. Geosci. Remote Sens.* **2019**, *57*, 1698–1712. [CrossRef]
14. Hong, D.; Yokoya, N.; Chanussot, J.; Zhu, X.X. An Augmented Linear Mixing Model to Address Spectral Variability for Hyperspectral Unmixing. *IEEE Trans. Image Process.* **2019**, *28*, 1923–1938. [CrossRef]
15. Halimi, A.; Altmann, Y.; Dobigeon, N.; Tourneret, J. Nonlinear Unmixing of Hyperspectral Images Using a Generalized Bilinear Model. *IEEE Trans. Geosci. Remote Sens.* **2011**, *49*, 4153–4162. [CrossRef]
16. Altmann, Y.; Halimi, A.; Dobigeon, N.; Tourneret, J. Supervised Nonlinear Spectral Unmixing Using a Postnonlinear Mixing Model for Hyperspectral Imagery. *IEEE Trans. Image Process.* **2012**, *21*, 3017–3025. [CrossRef] [PubMed]
17. Heylen, R.; Scheunders, P. A Multilinear Mixing Model for Nonlinear Spectral Unmixing. *IEEE Trans. Geosci. Remote Sens.* **2016**, *54*, 240–251. [CrossRef]
18. Marinoni, A.; Plaza, J.; Plaza, A.; Gamba, P. Estimating Nonlinearities in p-Linear Hyperspectral Mixtures. *IEEE Trans. Geosci. Remote Sens.* **2018**, *56*, 6586–6595. [CrossRef]
19. Tang, M.; Zhang, B.; Marinoni, A.; Gao, L.; Gamba, P. Multiharmonic Postnonlinear Mixing Model for Hyperspectral Nonlinear Unmixing. *IEEE Geosci. Remote Sens. Lett.* **2018**, *15*, 1765–1769. [CrossRef]
20. Zhu, F.; Honeine, P.; Chen, J. Pixel-Wise Linear/Nonlinear Nonnegative Matrix Factorization for Unmixing of Hyperspectral Data. In Proceedings of the ICASSP 2020—2020 IEEE International Conference on Acoustics, Speech and Signal Processing (ICASSP), Barcelona, Spain, 4–8 May 2020; pp. 4737–4741. [CrossRef]
21. Zhang, S.; Li, J.; Li, H.; Deng, C.; Plaza, A. Spectral–Spatial Weighted Sparse Regression for Hyperspectral Image Unmixing. *IEEE Trans. Geosci. Remote Sens.* **2018**, *56*, 3265–3276. [CrossRef]
22. Patel, J.R.; Joshi, M.V.; Bhatt, J.S. Abundance Estimation Using Discontinuity Preserving and Sparsity-Induced Priors. *IEEE J. Sel. Top. Appl. Earth Obs. Remote Sens.* **2019**, *12*, 2148–2158. [CrossRef]
23. Zhang, L.; Wei, W.; Zhang, Y.; Yan, H.; Li, F.; Tian, C. Locally Similar Sparsity-Based Hyperspectral Compressive Sensing Using Unmixing. *IEEE Trans. Comput. Imaging* **2016**, *2*, 86–100. [CrossRef]
24. Drumetz, L.; Meyer, T.R.; Chanussot, J.; Bertozzi, A.L.; Jutten, C. Hyperspectral Image Unmixing with Endmember Bundles and Group Sparsity Inducing Mixed Norms. *IEEE Trans. Image Process.* **2019**, *28*, 3435–3450. [CrossRef]
25. Qu, Q.; Nasrabadi, N.M.; Tran, T.D. Abundance Estimation for Bilinear Mixture Models via Joint Sparse and Low-Rank Representation. *IEEE Trans. Geosci. Remote Sens.* **2014**, *52*, 4404–4423. [CrossRef]
26. Giampouras, P.V.; Themelis, K.E.; Rontogiannis, A.A.; Koutroumbas, K.D. Simultaneously Sparse and Low-Rank Abundance Matrix Estimation for Hyperspectral Image Unmixing. *IEEE Trans. Geosci. Remote Sens.* **2016**, *54*, 4775–4789. [CrossRef]
27. Feng, F.; Zhao, B.; Tang, L.; Wang, W.; Jia, S. Robust low-rank abundance matrix estimation for hyperspectral unmixing. *J. Eng.* **2019**, *2019*, 7406–7409. [CrossRef]
28. Li, H.; Feng, R.; Wang, L.; Zhong, Y.; Zhang, L. Superpixel-Based Reweighted Low-Rank and Total Variation Sparse Unmixing for Hyperspectral Remote Sensing Imagery. *IEEE Trans. Geosci. Remote. Sens.* **2020**, 1–19. [CrossRef]

29. Feng, X.; Li, H.; Li, J.; Du, Q.; Plaza, A.; Emery, W.J. Hyperspectral Unmixing Using Sparsity-Constrained Deep Nonnegative Matrix Factorization With Total Variation. *IEEE Trans. Geosci. Remote Sens.* **2018**, *56*, 6245–6257. [CrossRef]
30. Qin, J.; Lee, H.; Chi, J.T.; Drumetz, L.; Chanussot, J.; Lou, Y.; Bertozzi, A.L. Blind Hyperspectral Unmixing Based on Graph Total Variation Regularization. *IEEE Trans. Geosci. Remote. Sens.* **2020**, 1–14. [CrossRef]
31. Iordache, M.; Bioucas-Dias, J.M.; Plaza, A. Total Variation Spatial Regularization for Sparse Hyperspectral Unmixing. *IEEE Trans. Geosci. Remote Sens.* **2012**, *50*, 4484–4502. [CrossRef]
32. Qian, Y.; Xiong, F.; Zeng, S.; Zhou, J.; Tang, Y.Y. Matrix-Vector Nonnegative Tensor Factorization for Blind Unmixing of Hyperspectral Imagery. *IEEE Trans. Geosci. Remote Sens.* **2017**, *55*, 1776–1792. [CrossRef]
33. Bioucas-Dias, J.M.; Figueiredo, M.A.T. A New TwIST: Two-Step Iterative Shrinkage/Thresholding Algorithms for Image Restoration. *IEEE Trans. Image Process.* **2007**, *16*, 2992–3004. [CrossRef]
34. Parikh, N.; Boyd, S. Proximal algorithms. *Found. Trends Optim.* **2014**, *1*, 127–239. [CrossRef]
35. Mataev, G.; Milanfar, P.; Elad, M. Deepred: Deep image prior powered by red. In Proceedings of the IEEE International Conference on Computer Vision Workshops, Seoul, Korea, 27 October–2 November 2019.
36. Romano, Y.; Elad, M.; Milanfar, P. The little engine that could: Regularization by denoising (RED). *SIAM J. Imaging Sci.* **2017**, *10*, 1804–1844. [CrossRef]
37. Venkatakrishnan, S.V.; Bouman, C.A.; Wohlberg, B. Plug-and-Play priors for model based reconstruction. In Proceedings of the 2013 IEEE Global Conference on Signal and Information Processing, Austin, TX, USA, 3–5 December 2013; pp. 945–948. [CrossRef]
38. Zhuang, L.; Bioucas-Dias, J.M. Fast Hyperspectral Image Denoising and Inpainting Based on Low-Rank and Sparse Representations. *IEEE J. Sel. Top. Appl. Earth Obs. Remote Sens.* **2018**, *11*, 730–742. [CrossRef]
39. Zhuang, L.; Ng, M.K. Hyperspectral Mixed Noise Removal By ℓ_1-Norm-Based Subspace Representation. *IEEE J. Sel. Top. Appl. Earth Obs. Remote Sens.* **2020**, *13*, 1143–1157. [CrossRef]
40. Zhuang, L.; Bioucas-Dias, J.M. Hy-Demosaicing: Hyperspectral Blind Reconstruction from Spectral Subsampling. In Proceedings of the IGARSS 2018—2018 IEEE International Geoscience and Remote Sensing Symposium, Valencia, Spain, 22–27 July 2018; pp. 4015–4018. [CrossRef]
41. Chan, S.H.; Wang, X.; Elgendy, O.A. Plug-and-Play ADMM for Image Restoration: Fixed-Point Convergence and Applications. *IEEE Trans. Comput. Imaging* **2017**, *3*, 84–98. [CrossRef]
42. Sun, Y.; Wohlberg, B.; Kamilov, U.S. An Online Plug-and-Play Algorithm for Regularized Image Reconstruction. *IEEE Trans. Comput. Imaging* **2019**, *5*, 395–408. [CrossRef]
43. Dabov, K.; Foi, A.; Katkovnik, V.; Egiazarian, K. Image Denoising by Sparse 3-D Transform-Domain Collaborative Filtering. *IEEE Trans. Image Process.* **2007**, *16*, 2080–2095. [CrossRef]
44. Zhang, K.; Zuo, W.; Chen, Y.; Meng, D.; Zhang, L. Beyond a Gaussian Denoiser: Residual Learning of Deep CNN for Image Denoising. *IEEE Trans. Image Process.* **2017**, *26*, 3142–3155. [CrossRef]
45. Heinz, D.C. Fully constrained least squares linear spectral mixture analysis method for material quantification in hyperspectral imagery. *IEEE Trans. Geosci. Remote Sens.* **2001**, *39*, 529–545. [CrossRef]
46. Nascimento, J.M.P.; Dias, J.M.B. Vertex component analysis: A fast algorithm to unmix hyperspectral data. *IEEE Trans. Geosci. Remote Sens.* **2005**, *43*, 898–910. [CrossRef]
47. Eckstein, J.; Bertsekas, D.P. On the Douglas—Rachford splitting method and the proximal point algorithm for maximal monotone operators. *Math. Program.* **1992**, *55*, 293–318.
48. Boyd, S.; Parikh, N.; Chu, E. *Distributed Optimization and Statistical Learning via the Alternating Direction Method of Multipliers*; Now Publishers Inc.: Norwell, MA, USA, 2011; Volume 3, pp. 1–122.
49. Halimi, A.; Altmann, Y.; Dobigeon, N.; Tourneret, J. Unmixing hyperspectral images using the generalized bilinear model. In Proceedings of the 2011 IEEE International Geoscience and Remote Sensing Symposium, Vancouver, BC, Canada, 24–29 July 2011; pp. 1886–1889. [CrossRef]
50. Yokoya, N.; Chanussot, J.; Iwasaki, A. Nonlinear Unmixing of Hyperspectral Data Using Semi-Nonnegative Matrix Factorization. *IEEE Trans. Geosci. Remote Sens.* **2014**, *52*, 1430–1437. [CrossRef]
51. Miao, L.; Qi, H. Endmember Extraction From Highly Mixed Data Using Minimum Volume Constrained Nonnegative Matrix Factorization. *IEEE Trans. Geosci. Remote Sens.* **2007**, *45*, 765–777. [CrossRef]
52. Zhu, F. Spectral Unmixing Datasets with Ground Truths. Available online: https://arxiv.org/abs/1708.05125 (accessed on 8 December 2020).
53. Bioucas-Dias, J.M.; Nascimento, J.M.P. Hyperspectral Subspace Identification. *IEEE Trans. Geosci. Remote Sens.* **2008**, *46*, 2435–2445. [CrossRef]

Publisher's Note: MDPI stays neutral with regard to jurisdictional claims in published maps and institutional affiliations.

remote sensing

Technical Note

A Particle Swarm Optimization Based Approach to Pre-tune Programmable Hyperspectral Sensors

Bikram Pratap Banerjee [1,2] and Simit Raval [2,*]

1 Agriculture Victoria, Grains Innovation Park, 110 Natimuk Road, Horsham, VIC 3400, Australia; bikram.banerjee@agriculture.vic.gov.au
2 School of Minerals and Energy Resources Engineering, University of New South Wales, Sydney, NSW 2052, Australia
* Correspondence: simit@unsw.edu.au; Tel.: +61-(2)-9385-5005

Abstract: Identification of optimal spectral bands often involves collecting in-field spectral signatures followed by thorough analysis. Such rigorous field sampling exercises are tedious, cumbersome, and often impractical on challenging terrain, which is a limiting factor for programmable hyperspectral sensors mounted on unmanned aerial vehicles (UAV-hyperspectral systems), requiring a pre-selection of optimal bands when mapping new environments with new target classes with unknown spectra. An innovative workflow has been designed and implemented to simplify the process of in-field spectral sampling and its realtime analysis for the identification of optimal spectral wavelengths. The band selection optimization workflow involves particle swarm optimization with minimum estimated abundance covariance (PSO-MEAC) for the identification of a set of bands most appropriate for UAV-hyperspectral imaging, in a given environment. The criterion function, MEAC, greatly simplifies the in-field spectral data acquisition process by requiring a few target class signatures and not requiring extensive training samples for each class. The metaheuristic method was tested on an experimental site with diversity in vegetation species and communities. The optimal set of bands were found to suitably capture the spectral variations between target vegetation species and communities. The approach streamlines the pre-tuning of wavelengths in programmable hyperspectral sensors in mapping applications. This will additionally reduce the total flight time in UAV-hyperspectral imaging, as obtaining information for an optimal subset of wavelengths is more efficient, and requires less data storage and computational resources for post-processing the data.

Keywords: evolutionary computation; heuristic algorithms; machine learning; unmanned aerial vehicles (UAVs); vegetation mapping; upland swamps; mine environment

Citation: Banerjee, B.P.; Raval, S. A Particle Swarm Optimization Based Approach to Pre-tune Programmable Hyperspectral Sensors. *Remote Sens.* **2021**, *13*, 3295. https://doi.org/10.3390/rs13163295

Academic Editor: Meiping Song

Received: 20 July 2021
Accepted: 18 August 2021
Published: 20 August 2021

1. Introduction

Hyperspectral technology is a potential tool for the remote detection of targets and monitoring. A hyperspectral sensor measures electromagnetic radiation reflected from the target in a large number of spectral narrowbands. The inherent objective in target classification and assessment using hyperspectral data is to utilize its high spectral resolution [1]. However, the large dimensionality of hyperspectral data is often attributed to the Hughes phenomenon, the curse of dimensionality [2]. The problem is a combined consequence of the high correlations among the adjacent bands and the inability of the algorithm being applied to process the high-dimensional data. The problem is paramount in spectrally complex environments such as wetlands and swamps with many diverse species to be monitored [1,3,4]. While a common remote sensing data processing solution involves the application of dimensionality reduction techniques or the selection of suitable narrowbands in a post-acquisition step, a hardware-based solution involves the use of programmable hyperspectral sensors as a pre-acquisition step. Programmable hyperspectral sensors typically involve a snapshot-based scanning mechanism, unlike general point or line scanning-type systems, which are non-programable and acquire a continuous spectrum over the operable wavelength region. Several such programmable hyperspectral sensors have been developed in recent times, which are increasingly being used in UAV-based remote sensing applications [5–7]. A hardware-based method, such as Fabry–Pérot interferometer (FPI) technology, acquires reflected electromagnetic radiation in pre-selected optimal narrowbands, and it is programmed by changing the air gap between the internal tuneable mirrors [8]. This method has the additional benefit of efficient mapping of the environment through the selection of only the spectral features of interest, which is particularly crucial in high-resolution mapping applications using unmanned aerial vehicles (UAVs), which have limited flight times. The technology is relatively new compared to the traditional pushboom type hyperspectral sensors, and existing works involving the FPI have used either (1) a set of bands for generating vegetation indices (VIs), herein referred to as *indices-based* criteria [7,9,10], or (2) set of bands identified through rigorous experimental testing, herein referred to as *knowledge-based* criteria [11,12] of narrowband selection. *Indices-based* criteria for band selection have the potential to assess the condition and/or estimate the yield of the vegetation [7,9]; however, they are not principally suited for multi-target classification, since the spectral variations of the target endmembers present within the scene are subjective. Furthermore, the efficacy of *indices-based* narrowband selection approach for vegetation quality or condition assessment is also subject to the characteristic reflectance of the target, and the traditional list of indices does not always ensure the best results for different vegetation communities or species. The *knowledge-based* approach requires a thorough understanding of the spectral variability among the targets present over the area, which is usually attained through intensive in-situ sampling and is not always realizable over difficult terrain or in scenarios requiring urgent mapping. Therefore, it is important to adopt a *data-driven* methodology for programmable hyperspectral sensors to estimate appropriate narrow bands for scene classification or assessment. Minet et al. [13] proposed an approach to adaptively maximize the contrast between the targets by employing a genetic algorithm (GA)-based optimization of the positions and linewidths of a limited number of filters in FPI for military applications. However, this method is unsuitable in thematic applications of remote sensing.

Different *data-driven* strategies have been proposed for the selection of optimal bands for traditional remote sensing applications. A method of sub-optimal search strategy utilizing constrained local extremes in a discrete binary space to select hyper-dimensional features was presented in [14]. Becker et al. [3] used a second-derivative approximation to identify the spectral location of inflection. A band selection method using the correlations among bands based on mutual information (MI) and deterministic annealing optimization was also employed [15]. Becker et al. [4] proposed a classification-based assessment for three optimal spectral band selection techniques (derivative, magnitude, fixed interval, and derivative histogram), using the spectral angle mapper (SAM) as a classifier. A GA-based wrapper method using a support vector machine (SVM) was proposed for the classification of hyperspectral images [16]. A double parallel feedforward neural network based on radial basis function was used for dimensionality reduction [17]. Principal component analysis for identifying optimal bands to discriminate wetland plant species was presented [1]. A semi-supervised band clustering approach for dimensionality reduction was developed [18]. A particle swarm optimization (PSO)-based dimensionality reduction approach to improving support vector machine (SVM)-based classification was suggested by [19]. Li et al. [20] and Pal et al. [21] presented a hybrid band selection strategy based on a GA-SVM wrapper to search optimal bands' subsets. A method of band selection based on spectral shape similarity analysis was put forward in [22]. Methods for nesting a traditional single loop of PSO or 1PSO inside an outer PSO loop, termed 2PSO, have been identified to improve the overall optimization performance in certain applications, at the expense of computational cost [23]. Su et al. [23] implemented 1PSO and 2PSO with minimum estimated abundance covariance (MEAC) [24], among other techniques, for the evaluation of optimal bands. Ghamisi et al. [25] presented a feature selection approach based on hybridization of a GA and PSO with an SVM classifier as a fitness function. Accuracies achieved in an optimized band selection method are influenced by the characteristics of the input dataset, as the search strategy depends on the present classes and their spectral profiles. Therefore, these methods need to be tested on benchmark datasets, an equivalent comprehensive evaluation is reported in [23]. However, all these existing optimal band identification studies involving *data-driven* methods were used on traditional hyperspectral datasets after the acquisition, and are yet to be used with a hardware-based solution to pre-tune hyperspectral sensors to acquire the optimal bands.

In this study, for the first time, an in-field *data-driven* approach to pre-tune a snapshot-type UAV-hyperspectral sensor was devised for remote sensing applications. The method employs PSO, with minimum estimated abundance covariance (MEAC), similarly to [23] in a post-processing stage for waveband selection after hyperspectral dataset acquisition. The significant benefits are: (1) it is an efficient approach to identifying the optimal bands in-field before the survey; (2) it does not require a lot of spectral samples per class, which is particularly an issue over difficult terrain when trying to establish a spectral library; and (3) the system works perfectly when the number of observed samples is less than the total number of potential hyperspectral bands to select from, which is an important issue with other dimensionality reduction methods, such as principal component analysis (PCA). Programmable UAV-hyperspectral sensors have increasingly been used in applications such as environmental mapping, precision agriculture, phenotyping, and forestry [12,26,27]. Identification of optimal wavelengths remains crucial for mapping vegetation communities, phenotyping functional plant traits, and identifying vegetation under biotic or abiotic stress. Our method aims to resolve functional challenges by improving the capturing of the spectral representation of an environment through a UAV-hyperspectral survey.

The rest of the paper is arranged as follows. The Materials and Methods section describes the experimental framework. The theoretical background of the PSO-MEAC approach is described in relation to the elements of the proposed application. In the Results and Discussion section, we present the results of using the PSO-MEAC method for optimal band selection at the experimental site. In addition, the performance of the *data-driven* PSO-MEAC approach has been evaluated against the traditional *indices based* approach for feature selection and mapping. Finally, the concluding remarks are provided in the conclusion section.

2. Materials and Methods

This section details the study area, ground based hyperspectral sensing system, data processing for the hyperspectral data, workflow for identifying optimal bands in the field, and method for UAV-hyperspectral surveying and assessment.

2.1. The Area Used for the Experiment

The test site is an upland swamp area above an underground coal mine within the temperate highland peat swamp on sandstone (THPSS) in New South Wales, southwest of the city of Sydney, Australia (34°21′24.0″S, 150°51′51″E). The area is located in Wollongong. The focus was laid on spectrally diverse vegetation communities in critically endangered ecosystems distributed in the Blue Mountains, Lithgow, Southern Highlands, and Bombala regions in New South Wales, Australia [28]. The NSW National Parks and Wildlife Service (NPWS) classifies the upland swamps complexes into five major vegetation communities—Banksia Thicket, Cyperoid Heath, Fringing Eucalypt Woodland, Restioid Heath, and Sedgeland [29]. The site has occasional thick vegetation cover and steep gradients which are inaccessible.

2.2. Hyperspectral Set-Up for Ground Based Sampling

The spectra of the target classes in the environment were measured with the visible-infrared snapshot hyperspectral (FPI) sensor (Rikola, Senop Optronics, Kangasala, Finland) with a separate data acquisition computer. In this mode of operation, the sensor acquires the maximum number of wavelength bands possible—i.e., 380 bands at 1 nm spectral steps between 500 and 880 nm. With a focal length of 9 mm and a field-of-view (FOV) of 36.5 × 36.5 degrees, the sensor acquires 1010 × 1010 spatial channels in the snapshot imaging mode. In contrast, in the standalone on-board UAV-based data acquisition mode the sensor records a set of 15 programmed wavelength bands in 1010 × 1010 pixel format, i.e., up to a total of 16 megapixels of storage per hypercube. The sensor also acquires solar irradiance measurements—it uses an irradiance sensor for radiometric calibration; and positional measurements using a global positioning system (GPS) for geometric corrections (Figure 1). All sensors were installed on a handheld mount for hyperspectral imaging. An Android mobile phone was also installed on the sensor mount and paired to the data acquisition computer with a video telemetry feed over a WiFi link to provide a realtime view of the scene, which was useful for bringing the target vegetation in focus before the collection of hyperspectral data (Figure 1a). Additionally, a realtime feed of goniometric measurements (roll and pitch) from the mobile phone's accelerometer was relayed to the screen of the data acquisition computer to monitor the planimetric setting of the captured hypercubes using the FPI sensor (Figure 1b).

Figure 1. The setup for ground-based hyperspectral data acquisition using a Fabry–Pérot interferometer (FPI) sensor (Rikola, Senop Optronics, Kangasala, Finland), an irradiance sensor, a global position system (GPS), and an android phone as goniometer on a portable handheld sensor mount: (**a**) top-side view, (**b**) bottom-side view, and (**c**) in-field hyperspectral data acquisition with a data acquisition computer. The system was used for the collection of in-field data for rapidly identifying optimal hyperspectral wavelengths, for applications in aerial (UAV-hyperspectral) data acquisition.

The simplistic design of the handheld hyperspectral imaging system was important for carrying it around in regions with dense shrub-type vegetation cover (Figure 1c). The hyperspectral data were acquired with a downward nadir orientation over the shrub type swamp vegetation. The data were acquired at a distance of approximately 0.5 m from the top of the canopy (Figure 1c). In this study, the FPI sensor was used as a tool for in-field spectral acquisition to demonstrate an independent form of operation. Nevertheless, the field spectral measurements could also be obtained from other spectroradiometers, such as ASD FieldSpec3 (Analytical Spectral Devices, Boulder, CO, USA). However, special care should be taken to establish proper radiometric calibration to remove any inter-sensor response mismatch, which is addressed by using the same FPI sensor for both in-field spectral data collection for identifying the optimal bands and later UAV-hyperspectral data acquisition.

For identifying the optimal bands through PSO-MEAC, the hyperspectral measurements were collected for a total of three target vegetation classes, covering eight upland swamp species, including Grass tree (*Xanthorrhoea resinosa*), Pouched coral fern (*Gleichenia dicarpa*). and Sedgeland complex (*Empodisma minus, Gymnoschoenus sphaerocephalus, Lepidosperma limicola, Lepidosperma neesii, Leptocarpus tenax,* and *Schoenus brevifolius*). In addition, spectral measurements were also collected for background vegetation, which contained a mixture of other species which were present in small patches and not selected in this study. Finally, a background bare-earth spectrum was also collected. To obtain a proper un-mixed spectrum for a single species, field sampling was performed over a region of interest with local homogeneity.

2.3. In-Field Ground-Based Hyperspectral Data Processing

Vegetation in an upland swamp environment is highly diverse, and species can exist in homogenous and heterogeneous patches. Data collected through the portable handheld FPI system caused minor spectral misalignments due to unavoidable handheld movement of the sensor and due to slight movements of the canopy caused by wind. This happened as the data in the FPI sensor were acquired in a snapshot, bandwise manner with a small delay and sensor movement [26]. The hyperspectral bands were aligned using a previously developed band alignment workflow described in [26]. The data were first flat-field corrected using dark current removal and a white calibration panel; then they were converted to the reflectance measurements using previously computed calibration coefficients with an integrating sphere [7]. A band-averaged hyperspectral signal was calculated from the hypercube and used in the optimal band identification workflow. The spectrum was further treated using a Savizky–Golay [30] smoothing filter with a polynomial order of 3 and a frame length of 17 to remove spectral noise. A PSO with MEAC as the criterion function was employed to identify the suitable bands in the field; the details of the theory of operation are in Section 2.4. The entire process of spectral signature retrieval and PSO-MEAC workflow for suitable band identification was implemented as MATLAB (ver. 9.5) routines, and a graphical user interface (GUI) was designed for user-friendly and seamless operation in the field.

2.4. Optimal Band Identification Using PSO-MEAC

Particle swarm optimization (PSO) was originally used to simulate the social behaviour (movement and interaction) of the organisms (*particles*) in a flock of birds or a pool of fishes [31]. It has, however, been used as a robust metaheuristic computational method to improve the selection of candidate solutions for an optimization problem. The optimization operates iteratively over a swarm of candidate solutions with a criterion function as a given measure of quality. In our approach, the selected set of bands are called *particles*, and a recursive update of the bands is called a *velocity*. The particle position x_{id} denotes the selected band subset of size k, and velocity v_{id} denotes the update for the selected band. A particle updates [31] as shown in Equation (1).

$$\begin{gathered} v_{id} = \omega \times v_{id} + c_1 \times r_1 \times (p_{id} - x_{id}) + c_2 \times r_2 \times \left(p_{gd} - x_{id}\right) \\ x_{id} = x_{id} + v_{id} \end{gathered} \tag{1}$$

where p_{id} is the historically best local solution; p_{gd} is historically the best global solution among all the particles; c_1 and c_2 control the contributions from local and global solutions, respectively; r_1 and r_2 are independent random variables between 0 and 1; and ω is the inertia weight to improve the convergence performance.

New velocities and positions (v_{id} and x_{id} on the left-hand side of Equation (1)) for the particles are updated based on the existing parameters and cost criterion upon every iteration (Figure 2). The iteration process aims to minimize the underlined criterion function.

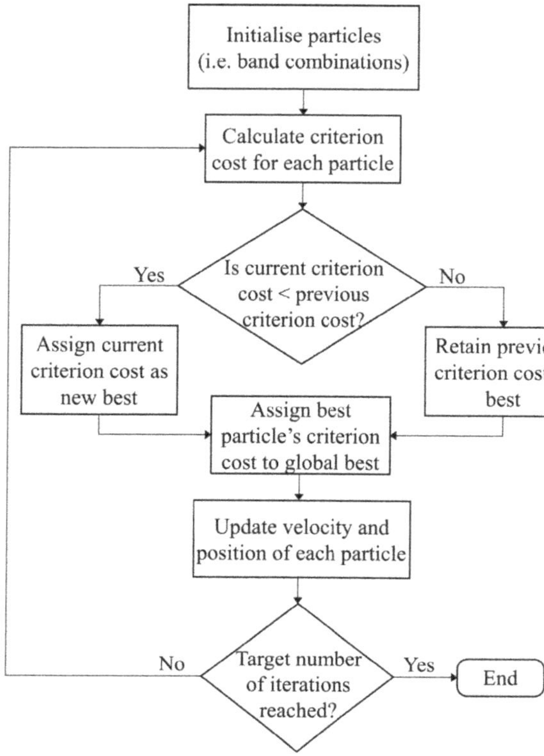

Figure 2. The method for the PSO-MEAC system. The algorithm initializes a set of particles or a combination of bands; at each iteration, the cost function (MEAC) associated with individual particles is computed; trajectories of the particles are projected towards the particle with the best solution; the loop is exited after the specified number of iterations is reached. The particle with a minimum cost function is identified as the optimal solution.

In a traditional supervised classification, where representative class signatures are known through exhaustive field surveying, the band-selection process can be greatly simplified. However, in an aerial survey to determine suitable wavelength bands for a programmable UAV-hyperspectral system, such an exhaustive exercise is tedious, cumbersome, and not always possible. Therefore, MEAC was used as a criterion function in PSO, as it requires only class signatures and no training samples. The efficacy of this technique has been previously evaluated against other existing optimization methods by Su et al. [23] for feature selection on traditional hyperspectral datasets (airborne and satellite).

Assuming there are p classes present over an area in which the samples were collected, the endmember matrix can be written as $S = [s_1, s_2, \ldots, s_p]$. According to Yang et al. [19], with linear mixing of the endmembers, the pixel r can be expressed as in Equation (2):

$$r = S\alpha + n \tag{2}$$

where $\alpha = (a_1, a_2, \ldots, a_p)^T$ is the abundance vector and n is the uncorrelated noise with $E(n) = 0$ and $Cov(n) = \sigma^2 I$ (I is an identity matrix).

Usually, the actual number of classes (p) is greater than the known class signatures; i.e., $q < p$. Hence, the uncorrelated noise will have $Cov(n) = \sigma^2\Sigma$, where Σ is the noise covariance matrix. Therefore, the abundance vector becomes the weighted least square solution, as in Equation (3):

$$\hat{\alpha} = \left(S^T\Sigma^{-1}S\right)^{-1}S^T\Sigma^{-1}\mathrm{r} \quad (3)$$

with first-order moment being $E(\hat{\alpha}) = \alpha$ and second-order moment being $Cov(\hat{\alpha}) = \sigma^2\left(S^T\Sigma^{-1}S\right)^{-1}$.

The analysis demonstrates that when all the classes are known, the remaining noise can be modelled as independent Gaussian noise. For this application, when meeting such sampling criteria was difficult and there were unknown classes present, noise whitening was applied first. Yang et al. [19] and Su et al. [23] performed the optimal band selection on traditional hyperspectral datasets, and used all the pixels for the background noise (Σ) estimation. In this case, the background pixels' noise was calculated using background class spectra and bare-earth spectra collected through ground-based sampling. The background plus noise covariance is denoted as Σ_{b+n}; this estimate was used in this study. The estimate of the unknown class pixels is based on the likelihood of the unknown class (or the class of no interest) being present around the sampled class of interest. In scenes where all endmembers are of known classes (or the target classes of interest), noise estimation Σ_{b+n} is not required, which is an unlikely condition in a spectrally complex swamp environment [7].

The identified optimal bands should allow minimal deviations of $\hat{\alpha}$ from actual α [23]. With the partially known classes, the criterion function is equivalent to minimizing the trace of the covariance, as in Equation (4):

$$\arg\min_{\Phi^S}\left\{trace\left[\left(S^T{\Sigma_{b+n}}^{-1}S\right)^{-1}\right]\right\} \quad (4)$$

where Φ^S is the selected band subset. The resulting band selection algorithm is referred to as the MEAC method [23].

The optimizer returns a suitably identified set of wavelength bands with the lowest cost criterion values (Equation (4)), upon successful completion of the PSO-MEAC algorithmic iterations (Figure 2).

2.5. UAV-Hyperspectral Survey and Assessment

After the identification of a set of optimal bands through the *data-driven* PSO-MEAC approach, the FPI hyperspectral sensor was programmed to acquire using the suitable narrow wavelength bands. A UAV-hyperspectral mission was carried out in pre-planned waypoint acquisition mode with 85% forwards and 75% lateral overlap from a flying altitude of 50 m. The sensor exposure time was set at 10 ms per band to provide good radiometric image quality for the existing illumination conditions. The UAV-hyperspectral survey was performed around two hours of local solar noon and in clear weather conditions with no clouds. This was done to avoid both the effect of significant illumination variations and shadows cast by clouds during the aerial image acquisition. However, due to the experimental site being situated in a low latitudinal region in the southern hemisphere (34°21′24.0″S, 150°51′51″E) with the sun projecting a shallow incidence angle, the issues of the shadows projected by trees and other tall vegetation was unavoidable. In addition to the *data-driven* PSO-MEAC tuned survey, another aerial survey was performed with an *indice-based* [7] wavelength selection approach, using the same UAV flight characteristic and sensor exposure configuration. A band stabilization workflow was adopted to co-register spatial shifts between bands in hypercubes, from both the aerial acquisition modes [26]. Further, the regular radiometric, illumination adjustment, mosaicking, and geometric correction procedures for hypercubes were carried out [7]. The UAV-hyperspectral orthomosaics achieved a high spatial resolution of 2 cm in ground sampling distance.

A supervised support vector machine (SVM) classifier was used to classify the hyperspectral datasets into constituent classes. The SVM is an efficient kernel-based machine learning classifier suitable for high-dimensional feature spaces, which is well used in classifying hyperspectral datasets [32–34]. The classification was performed as an evaluation step to compare the efficacy of wavelengths identified through *data-driven* PSO-MEAC and *indices-based* approaches. As the fundamental objective in this study was to simply evaluate the two methods, and not to achieve superior accuracies in classification, involving complex classification algorithms were deemed needless. Standard parameter settings—a radial basis function with a kernel gamma function of 0.167, a penalty parameter of 100, and a pyramid level of 5—were used for the SVM classification. The overall and individual class classification accuracies were computed using the ground truth training samples.

For evaluating the efficacy of PSO-MEAC-identified bands through classification, a total of 120 ground truth measurements were collected for shrub-type swamp vegetation through a rigorous field survey, and 120 ground truth polygons were identified through visual interpretation of high-resolution hyperspectral data. The sampled ground-based (120) and image-based (120) polygons were randomly divided into 1:1 mutually exclusive sets of training and test samples, i.e., 60 ground and 60 image-based polygons for each training and test group. The ground truth training set was used to train the SVM classifier, and the test samples were used to compute the overall accuracy (OA), kappa (κ), and confusion matrix to evaluate the classification accuracies. The spectral data from training and test sample polygons were obtained from the UAV-hyperspectral datasets in corresponding *data-driven* PSO-MEAC and *indices-based* modes.

3. Results and Discussion

This section details the results and discussion of optimal band selection using *data-driven* PSO-MEAC workflow, and its evaluation against the *indices-based* approach.

3.1. Optimal Band Identification Using PSO-MEAC

The PSO-based optimal band identification workflow determines a list of suitable bands according to the MEAC cost criterion. The PSO-MEAC workflow was executed with a population size of 100, an inertial weight of 0.98, and a maximum number of iterations of 500. A total of 15 bands, i.e., $k = 15$, were identified, based on the maximum band capacity of the FPI sensor for on-board UAV data acquisition mode in an un-binned setting (1010×1010 pixels).

The selected combination of bands gets re-configured at every iteration to minimize the cost function (Figure 2). A new combination of bands is designated optimal if the combination achieves the best (or minimum) cost. To analyze the performance of the in-field optimal band identification and sensor tuning using the PSO-MEAC approach, a set of internally computed parameters (criterion cost and index of runs) were logged at every iteration (Figure 3). The PSO-MEAC approach determines the suitable combination of bands (or band-index) using the cost criterion (Equation (4)). The reduction of the best cost value signifies the learning curve for the optimization workflow (Figure 3a). At every iteration, the cost associated with the previous band-index is compared with the new band-index. A record of these parameters reveals the process of convergence to the desired solution by the implemented metaheuristic workflow. A measure of final cost and plot of identified optimal band combination is also produced. It can be seen that using the PSO-MEAC method, better (i.e., smaller) values of cost criterion can be achieved. Each iteration may produce slightly different band combinations according to the cost criterion, as shown by the plot of the index of runs in (Figure 3b). The final cost of the PSO-MEAC was -7.7×10^{-9}. At this stage, the identified band indices were 56, 88, 101, 119, 151, 172, 211, 217, 251, 284, 303, 326, 341, 360, and 380 (Figure 3c). The corresponding FPI wavelengths were 555.33, 587.21, 600.34, 618.21, 650.39, 671.02, 710.12, 716.11, 750.19, 783.46, 802.35, 825.28, 840.15, 859.53, and 880.43 nm with respective FWHMs of 9.81, 10.62, 9.88, 12.17, 10.78, 11.77, 9.78, 9.61, 9.58, 10.60, 10.56, 10.49, 13.69, 13.12, and 13.27 nm.

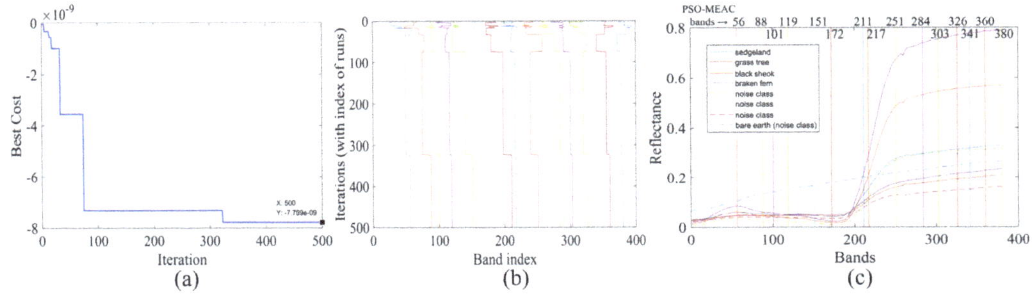

Figure 3. Optimal band selection: (**a**) a plot demonstrating variation of cost criterion with the PSO-MEAC iterations, (**b**) bands selected in each iteration, and (**c**) a plot of the identified optimal bands overlayed on the class spectra. The cost criterion was progressively minimized with the number of iterations. The variations of band position with the index of runs in every iteration provide insights into the functioning of PSO-MEAC. Overall, the PSO-MEAC-identified bands are well distributed over the key wavebands with maximal variation between inter-class reflectance.

The PSO-MEAC workflow uses a complex high-dimensional search strategy, producing several intermediate local and global combinations of bands, so the final solution may not be the same with every execution. Previous implementations of PSO-MEAC [23] focused on minimising the number of bands in optimal configurations, which is suitable for dimensionality reduction techniques in traditional airborne or satellite hyperspectral imaging, with a complete set of bands already acquired. In the proposed method, the number of bands to be identified is predefined by the user, which makes it important to use the FPI sensor to its fullest potential (i.e., hypercube band capacity at desired spectral binning) to acquire the maximum possible information in the optimal configuration. To evaluate the computational complexity, the PSO-MEAC workflow was programmed in MATLAB (ver. 9.5) and implemented as a GUI module to run on a portable field data acquisition computer with 1.5 GHz processor and 512 MB memory. The module took roughly 4 to 5 min for every 500 iterations with the selected number of class samples. This demonstrates the operational efficiency of the system, despite having a complex search hierarchy, and it is usable for pre-tuning the programmable FPI sensor in a UAV-hyperspectral survey for optimised wavelength selection.

Acquisition and identification of optimal bands using characteristic spectral signatures of individual swamp species have been traditionally performed using the separability of the spectrum at respective wavelength bands. In this study, the employed PSO-MEAC-based search strategy automatically analyses and identifies wavelength bands based on maximum separability of the reflectance using the MEAC cost criterion function. The field spectrum collected for each shrub-type vegetation species is shown in (Figure 3c), and the identified wavelength band positions are shown using a set of superimposed vertical lines. Our approach has been implemented using a GUI-based interface on a portable data acquisition computer, which enabled rapid analysis of spectral signatures and identification of suitable wavelength bands. The developed technique and tools were found to be efficient in a field environment during surveying.

3.2. Classification

The comparative evaluation between the *data-driven* PSO-MEAC and *indices-based* wavelength tuning approaches was performed using an SVM classifier. Two dedicated datasets (*data-driven* PSO-MEAC and *indices-based*) were collected from the swamp. The scene was primarily comprised of three shrub-type vegetation classes (i.e., grass trees, pouched coral ferns, and Sedgeland complex) and two tree-type vegetation classes (i.e., black sheoak and eucalyptus). A small portion of the area was bare of vegetation cover and was treated as a separate "bare earth" class. Therefore, a total of six classes were used in the classification-based comparative evaluation. The optimal bands identified using the *data-driven* PSO-MEAC approach produced better results compared to the *indices-based* approach, with the SVM classifier. Combining the optimal bands identified using the *data-driven* PSO-MEAC with the SVM classifier produced an overall accuracy of 85.16% and a kappa coefficient of 0.73, whereas the *indices-based* approach produced an overall accuracy of 76.54% and a kappa coefficient of 0.67. The comparative classification maps for the *indices-based* PSO-MEAC and *data-driven* approaches produced using the SVM classifier are shown in Figure 4.

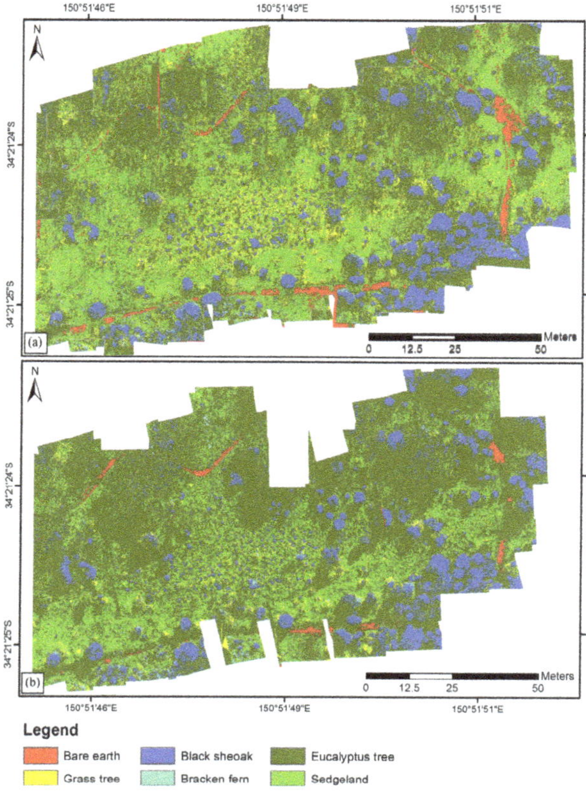

Figure 4. Classification map of the swamp site's vegetation classes and species produced using a support vector machine classifier with (**a**) *data-driven* (PSO-MEAC) optimal band identification and (**b**) *indices-based* band selection.

The producer's accuracy or error-of-omission refers to the conditional probability that certain land-cover of an area on the ground is correctly mapped, whereas the user's accuracy or error-of-commission refers to the conditional probability that a pixel labeled as a certain land-cover class in the map belongs to that class [35]. The producer's and user's accuracy for each class with the best classification method, *data-driven* PSO-MEAC, are shown in Table 1. With the exception of the "grass tree" class, overall the accuracy for each class was satisfactory (>70%), particularly when differentiating between swamp-type (Sedgeland complex) and non-swamp-type (*Eucalyptus*) vegetation. The results also indicate the potential of the process for distinguishing certain critical non-swamp-type terrestrial species (black sheoak and bracken fern) within the swamp environment. Increases in the proportions of these terrestrial species in a swamp indicate changes in the swamp hydrology. No changes in the proportions of terrestrial species (or changes within equilibrium limits) indicates the stability of hydrology and peat moisture levels. These results, therefore, demonstrate the usefulness of the method for directly mapping the changes induced in a swamp environment due to the fluctuation of groundwater level.

Table 1. Evaluation of classification accuracy achieved using the *data-driven* (PSO-MEAC) method against *indices based* band selection.

Class	*Data Driven* **(PSO-MEAC)**		*Indices Based*	
	Producer's Accuracy (%)	**User's Accuracy (%)**	**Producer's Accuracy (%)**	**User's Accuracy (%)**
Bare earth	91.57	98.52	22.27	89.43
Grass tree	77.00	73.54	5.92	49.45
Black sheoak	97.33	83.20	94.28	78.93
Bracken fern	71.43	78.44	13.08	70.83
Eucalyptus tree	81.55	81.13	89.10	71.16
Sedgeland complex	88.28	80.35	41.42	61.64

4. Conclusions

Identification of optimal bands for vegetation monitoring has been an ongoing research problem in hyperspectral remote sensing. The issue is significant in a spectrally complex environment with diversity in vegetation species, such as swamps and wetlands. Extensive surveys and post-processing solutions have been recurrently used in different swamp-type environments. The study presents an innovative approach for in-field rapid identification of spectrally significant wavelength bands. The developed method was employed to tune a programmable hyperspectral sensor before UAV borne surveys. The method was implemented through a metaheuristic workflow based on particle swarm optimization (PSO), with minimum estimated abundance covariance (MEAC) as the cost selection criterion function. A portable in-field hyperspectral signature collection system was devised using the tuneable FPI hyperspectral sensor. The set-up improved the collection of class spectra and background noise spectra, which were then used to identify the optimal band configuration. The method identifies the optimal bands based on representative class spectral signatures, avoiding the requirement of extensive in-field sampling. Additionally, the method works perfectly when the number of sample observations is less than the total number of potential hyperspectral bands, which is not possible with other dimensionality reduction methods, such as PCA. The method was successfully tested to identify a set of optimal bands for maximizing the spectral differentiation of swamp-type vegetation species and communities. The algorithm could be tuned to robustly incorporate vegetation trait retrieval by changing the criterion function. The approach would be valuable to environmental mapping, precision agriculture, phenotyping, and forestry to estimate qualitative phenotypic traits such as chlorophyll content, photosynthetic capacity, and biomass; and for studying vegetation under different treatments or biotic and abiotic stresses.

Author Contributions: B.P.B. and S.R. conceived of the experiment. B.P.B. conducted the experiments, data analysis, and writing of the original draft. S.R. conducted project administration, manuscript review, and editing. All authors have read and agreed to the published version of the manuscript.

Funding: This research received no external funding.

Institutional Review Board Statement: Not applicable.

Informed Consent Statement: Not applicable.

Data Availability Statement: Data are not available due to non-disclosure agreements.

Conflicts of Interest: The authors declare no conflict of interest.

References

1. Torbick, N.; Becker, B. Evaluating Principal Components Analysis for Identifying Optimal Bands Using Wetland Hyperspectral Measurements From the Great Lakes, USA. *Remote Sens.* **2009**, *1*, 408–417. [CrossRef]
2. Hughes, G. On the mean accuracy of statistical pattern recognizers. *IEEE Trans. Inf. Theory* **1968**, *14*, 55–63. [CrossRef]
3. Becker, B.L.; Lusch, D.P.; Qi, J. Identifying optimal spectral bands from in situ measurements of Great Lakes coastal wetlands using second-derivative analysis. *Remote Sens. Environ.* **2005**, *97*, 238–248. [CrossRef]
4. Becker, B.L.; Lusch, D.P.; Qi, J. A classification-based assessment of the optimal spectral and spatial resolutions for Great Lakes coastal wetland imagery. *Remote Sens. Environ.* **2007**, *108*, 111–120. [CrossRef]
5. Yue, J.; Yang, G.; Li, C.; Li, Z.; Wang, Y.; Feng, H.; Xu, B. Estimation of Winter Wheat Above-Ground Biomass Using Unmanned Aerial Vehicle-Based Snapshot Hyperspectral Sensor and Crop Height Improved Models. *Remote Sens.* **2017**, *9*, 708. [CrossRef]
6. Aasen, H.; Honkavaara, E.; Lucieer, A.; Zarco-Tejada, P.J. Quantitative Remote Sensing at Ultra-High Resolution with UAV Spectroscopy: A Review of Sensor Technology, Measurement Procedures, and Data Correction Workflows. *Remote Sens.* **2018**, *10*, 1091. [CrossRef]
7. Banerjee, B.; Raval, S.; Cullen, P.J. UAV-hyperspectral imaging of spectrally complex environments. *Int. J. Remote Sens.* **2020**, *41*, 4136–4159. [CrossRef]
8. Saari, H.; Aallos, V.-V.; Akujärvi, A.; Antila, T.; Holmlund, C.; Kantojärvi, U.; Mäkynen, J.; Ollila, J. Novel miniaturized hyperspectral sensor for UAV and space applications. In Proceedings of the Sensors, Systems, and Next-Generation Satellites XIII, Berlin, Germany, 22 September 2009; Volume 7474. [CrossRef]
9. Kaivosoja, J.; Pesonen, L.; Kleemola, J.; Pölönen, I.; Salo, H.; Honkavaara, E.; Saari, H.; Mäkynen, J.; Rajala, A. A case study of a precision fertilizer application task generation for wheat based on classified hyperspectral data from UAV combined with farm history data. In Proceedings of the Remote Sensing for Agriculture, Ecosystems, and Hydrology XV, Dresden, Germany, 16 October 2013; Volume 8887. [CrossRef]
10. Pölönen, I.; Saari, H.; Kaivosoja, J.; Honkavaara, E.; Pesonen, L. Hyperspectral imaging based biomass and nitrogen content estimations from light-weight UAV. In Proceedings of the Remote Sensing for Agriculture, Ecosystems, and Hydrology XV, Dresden, Germany, 16 October 2013; Volume 8887. [CrossRef]
11. Honkavaara, E.; Hakala, T.; Kirjasniemi, J.; Lindfors, A.; Mäkynen, J.; Nurminen, K.; Ruokokoski, P.; Saari, H.; Markelin, L. New light-weight stereosopic spectrometric airborne imaging technology for high-resolution environmental remote sensing—Case studies in water quality mapping. *Int. Arch. Photogramm. Remote Sens. Spat. Inf. Sci.* **2013**, *XL-1/W1*, 139–144. [CrossRef]
12. Näsi, R.; Honkavaara, E.; Lyytikäinen-Saarenmaa, P.; Blomqvist, M.; Litkey, P.; Hakala, T.; Viljanen, N.; Kantola, T.; Tanhuanpää, T.; Holopainen, M. Using UAV-Based Photogrammetry and Hyperspectral Imaging for Mapping Bark Beetle Damage at Tree-Level. *Remote Sens.* **2015**, *7*, 15467–15493. [CrossRef]
13. Minet, J.; Taboury, J.; Péalat, M.; Roux, N.; Lonnoy, J.; Ferrec, Y. Adaptive band selection snapshot multispectral imaging in the VIS/NIR domain. In Proceedings of the Electro-Optical Remote Sensing, Photonic Technologies, and Applications IV, Toulouse, France, 11 October 2010; Volume 7835. [CrossRef]
14. Serpico, S.; Bruzzone, L. A new search algorithm for feature selection in hyperspectral remote sensing images. *IEEE Trans. Geosci. Remote Sens.* **2001**, *39*, 1360–1367. [CrossRef]
15. Sotoca, J.M.; Pla, F. Hyperspectral data selection from mutual information between image bands. In *Structural, Syntactic, and Statistical Pattern Recognition, Proceedings of the Joint IAPR International Workshops, SSPR 2006 and SPR 2006, Hong Kong, China, 17–19 August 2006*; Yeung, D.-Y., Kwok, J.T., Fred, A., Roli, F., de Ridder, D., Eds.; Springer: Berlin/Heidelberg, Germany, 2006; pp. 853–861.
16. Zhuo, L.; Zheng, J.; Li, X.; Wang, F.; Ai, B.; Qian, J. A genetic algorithm based wrapper feature selection method for classification of hyperspectral images using support vector machine. In Proceedings of the Geoinformatics 2008 and Joint Conference on GIS and Built Environment: Classification of Remote Sensing Images, Guangzhou, China, 7 November 2008; pp. 71471–71479.
17. Huang, R.; Zhou, L. Hyperspectral feature selection and classification with a RBF-based novel Double Parallel Feedforward Neural Network and evolution algorithms. In Proceedings of the 2009 4th IEEE Conference on Industrial Electronics and Applications, Xi'an, China, 25 May 2009; pp. 673–676. [CrossRef]
18. Su, H.; Yang, H.; Du, Q.; Sheng, Y. Semisupervised Band Clustering for Dimensionality Reduction of Hyperspectral Imagery. *IEEE Geosci. Remote Sens. Lett.* **2011**, *8*, 1135–1139. [CrossRef]
19. Yang, H.; Du, Q. Particle swarm optimization-based dimensionality reduction for hyperspectral image classification. In Proceedings of the 2011 IEEE International Geoscience and Remote Sensing Symposium, Vancouver, BC, Canada, 24–29 July 2011; pp. 2357–2360. [CrossRef]
20. Li, S.; Wu, H.; Wan, D.; Zhu, J. An effective feature selection method for hyperspectral image classification based on genetic algorithm and support vector machine. *Knowl.-Based Syst.* **2011**, *24*, 40–48. [CrossRef]
21. Pal, M. Hybrid genetic algorithm for feature selection with hyperspectral data. *Remote Sens. Lett.* **2013**, *4*, 619–628. [CrossRef]
22. Li, S.; Qiu, J.; Yang, X.; Liu, H.; Wan, D.; Zhu, Y. A novel approach to hyperspectral band selection based on spectral shape similarity analysis and fast branch and bound search. *Eng. Appl. Artif. Intell.* **2014**, *27*, 241–250. [CrossRef]
23. Su, H.; Du, Q.; Chen, G.; Du, P. Optimized hyperspectral band selection using particle swarm optimization. *IEEE J. Sel. Top. Appl. Earth Obs. Remote Sens.* **2014**, *7*, 2659–2670. [CrossRef]

24. Yang, H.; Du, Q.; Su, H.; Sheng, Y. An Efficient Method for Supervised Hyperspectral Band Selection. *IEEE Geosci. Remote Sens. Lett.* **2010**, *8*, 138–142. [CrossRef]
25. Ghamisi, P.; Benediktsson, J.A. Feature Selection Based on Hybridization of Genetic Algorithm and Particle Swarm Optimization. *IEEE Geosci. Remote Sens. Lett.* **2014**, *12*, 309–313. [CrossRef]
26. Banerjee, B.P.; Raval, S.A.; Cullen, P.J. Alignment of UAV-hyperspectral bands using keypoint descriptors in a spectrally complex environment. *Remote Sens. Lett.* **2018**, *9*, 524–533. [CrossRef]
27. Oliveira, R.A.; Näsi, R.; Niemeläinen, O.; Nyholm, L.; Alhonoja, K.; Kaivosoja, J.; Jauhiainen, L.; Viljanen, N.; Nezami, S.; Markelin, L.; et al. Machine learning estimators for the quantity and quality of grass swards used for silage production using drone-based imaging spectrometry and photogrammetry. *Remote Sens. Environ.* **2020**, *246*, 111830. [CrossRef]
28. Commonwealth of Australia. *Temperate Highland Peat Swamps on Sandstone: Ecological Characteristics, Sensitivities to Change, and Monitoring and Reporting Technique*; Independent Expert Scientific Committee on Coal Seam Gas and Large Coal Mining Development, Department of the Environment, Australian Government: Canberra, Australia, 2014; p. 177.
29. National Parks & Wildlife Services. *The Native Vegetation of the Woronora, O'Hares and Metropolitan Catchments*; NSW National Parks and Wildlife Service: Sidney, Australia, 2003.
30. Orfanidis, S.J. *Introduction to Signal Processing*; Prentice Hall: Upper Saddle River, NJ, USA, 1996.
31. Eberhart, R.; Kennedy, J. A new optimizer using particle swarm theory. In Proceedings of the Sixth International Symposium on Micro Machine and Human Science, MHS95, Nagoya, Japan, 4 October 1995; pp. 39–43.
32. Drucker, H.; Burges, C.J.; Kaufman, L.; Smola, A.; Vapnik, V. Support vector regression machines. *Adv. Neural Inf. Process. Syst.* **1997**, *9*, 155–161.
33. Bazi, Y.; Melgani, F. Toward an Optimal SVM Classification System for Hyperspectral Remote Sensing Images. *IEEE Trans. Geosci. Remote Sens.* **2006**, *44*, 3374–3385. [CrossRef]
34. Basak, D.; Pal, S.; Chandra Patranabis, D. Support vector regression. *Neural Inf. Process. Lett. Rev.* **2007**, *11*, 203–224.
35. Stehman, S. Estimating the kappa coefficient and its variance under stratified random sampling. *Photogramm. Eng. Remote Sens.* **1996**, *62*, 401–407.

remote sensing

Article

Residual Augmented Attentional U-Shaped Network for Spectral Reconstruction from RGB Images

Jiaojiao Li †, Chaoxiong Wu *, Rui Song †, Yunsong Li and Weiying Xie

The State Key Laboratory of Integrated Service Networks, Xidian University, Xi'an 710000, China; jjli@xidian.edu.cn (J.L.); rsong@xidian.edu.cn (R.S.); ysli@mial.xidian.edu.cn (Y.L.); wyxie@xidian.edu.cn (W.X.)
* Correspondence: cxwu@stu.xidian.edu.cn; Tel.: +86-155-2960-9856
† These authors contributed equally to this work.

Abstract: Deep convolutional neural networks (CNNs) have been successfully applied to spectral reconstruction (SR) and acquired superior performance. Nevertheless, the existing CNN-based SR approaches integrate hierarchical features from different layers indiscriminately, lacking an investigation of the relationships of intermediate feature maps, which limits the learning power of CNNs. To tackle this problem, we propose a deep residual augmented attentional u-shape network (RA^2UN) with several double improved residual blocks (DIRB) instead of paired plain convolutional units. Specifically, a trainable spatial augmented attention (SAA) module is developed to bridge the encoder and decoder to emphasize the features in the informative regions. Furthermore, we present a novel channel augmented attention (CAA) module embedded in the DIRB to rescale adaptively and enhance residual learning by using first-order and second-order statistics for stronger feature representations. Finally, a boundary-aware constraint is employed to focus on the salient edge information and recover more accurate high-frequency details. Experimental results on four benchmark datasets demonstrate that the proposed RA^2UN network outperforms the state-of-the-art SR methods under quantitative measurements and perceptual comparison.

Keywords: spectral reconstruction; residual augmented attentional u-shape network; spatial augmented attention; channel augmented attention; boundary-aware constraint

Citation: Li, J.; Wu, C.; Song, R.; Li, Y.; Xie, W. Residual Augmented Attentional U-Shaped Network for Spectral Reconstruction from RGB Images. *Remote Sens.* **2021**, *13*, 115. https://doi.org/10.3390/rs13010115

Received: 8 October 2020
Accepted: 29 December 2020
Published: 31 December 2020

1. Introduction

Hyperspectral imaging systems can record the actual scene spectra over a large set of narrow spectral bands [1]. In contrast to the ordinary cameras record only reflectance or transmittance of three spectral bands (i.e., Red, Green, and Blue), hyperspectral spectrometers can encode hyperspectral images (HSIs) by obtaining continuous spectrums on each pixel of the object. The abundant spectral signatures are beneficial to many computer vision tasks, such as face recognition [2], image classification [3,4] and object tracking [5], etc.

Traditional scanning HSIs acquisition systems rely on either 1D line or 2D plane scanning (e.g., whiskbroom [6], pushbroom [7] or variable-filter technology [8]) to encode spectral information of the scene. Whiskbroom imaging devices apply mirrors and fiber optics to collect reflected hyperspectral signals point by point. The subsequent pushbroom HSIs acquisition systems capture HSIs with dispersive optical elements and light-sensitive sensors in a line-by-line scanning manner. As for the variable-filter imaging equipment, it senses each scene point multiple times, each time in a different spectral band. In fact, the scanning operation of these devices is extremely time-consuming, which severely limits the application of HSIs under dynamic conditions.

To make HSIs acquisition of dynamic scenes available, the scan-free or snapshot hyperspectral technologies have been explored, e.g., coded aperture snapshot spectral imagers [9], mosaic [10], and light-field [11], etc. Computed-tomography imaging spectrometer converts a three-dimensional object cube into multiplexed two-dimensional projections and these data can be used later to reconstruct the hyperspectral cube computationally [12,13]. Coded aperture snapshot spectral imager uses compressed sensing advances to achieve snapshot spectral imaging and an iterative algorithm is used to reconstruct the data cube [9,14]. A novel hyperspectral imaging system combines a stereo camera to perform the accurate HSIs measurements through the geometrical alignment, radiometric calibration and normalization [10]. However, these systems depend on post-processing with a huge computational complexity and record HSIs with decreased spatial and spectral resolution. Meanwhile, the deployments of these facilities remain prohibitively expensive and complicated.

Due to the limitations of scanning and snapshot hyperspectral systems, as an alternative solution, spectral reconstruction from ubiquitous RGB images has attracted extensive attention and research, i.e., given an RGB image, the corresponding HSI with higher spectral resolution can be recovered via fulfilling a three-to-many mapping directly. Obviously, SR is an ill-posed transition problem. Early work on SR leverages sparse coding or shallow learning models to rebuild HSI data [15–19]. Nguyen et al. [15] trained a shallow radial basis function network that leveraged RGB white-balancing to normalize the scene illuminations to further recover the scene reflectance spectra. Later, Robles-Kelly [16] extracted a set of reflectance properties from the training set and obtained convolutional features using sparse coding to perform spectral reconstruction. Typically, Arad [17] and Aeschbacher et al. [19] exploited potential HSIs priors to create an over-complete sparse dictionary of hyperspectral signatures and corresponding RGB projections, which facilitated the following reconstruction of the HSIs. More recently, with the aid of the low-rank constraints, Zhang et al. [20] proposed to make full use of the high-dimensionality structure of the desired HSI to boost the reconstruction quality. Unfortunately, these methods only model low-level and simple correlation between RGB images and hyperspectral signals, which limits their expression ability and leads to poor performance in challenging situations. Accordingly, it is indispensable to further improve the results of the reconstructed HSIs for SR.

Recently, witnessing the great success of CNNs in the field of hyperspectral spatial super-resolution [21,22], numerous CNN-based algorithms have been widely explored in the SR task [23–28]. For example, Galliani et al. [23] modified a high-performance network originally designed for semantic segmentation to learn the statistics of natural image spectra and generated finely resolved HSIs from the RGB inputs. This is a milestone work, since it is the first time to introduce deep learning into the SR task. To promote the research of SR, NTIRE 2018 challenge on spectral reconstruction from RGB images is organized, which is the first SR challenge [29]. Meanwhile, a great quantity of excellent approaches have been proposed in this competition [30–34]. Impressively, Shi et al. [34] designed a deep HSCNN-R network consisting of multiple residual blocks and acquired promising performance, which was developed from their previous HSCNN model [25]. Stiebel et al. [30] investigated a lightweight Unet and added a simple pre-processing layer to enhance the quality of recovery in a real world scenario. Not long ago, the second SR challenge, NTIRE 2020 on spectral reconstruction from RGB images [35], has been successfully held and a new data set is released, which further promote the development of SR methods based on CNNs [36–41] as well as more recent works [42–45]. To explore the interdependencies among intermediate features and the camera spectral sensitivity prior, Li et al. [36] proposed an adaptive weighted attention network and incorporated the discrepancies of the RGB images and HSIs into the loss function. As the winning method on the "Real World" track of the second SR competition, Zhao et al. [37] organized a 4-level hierarchical regression network with pixelShuffle layer as inter-level interaction. Hang et al. [44] attempted to design a decomposition model to reconstruct HSIs and a self-supervised network to fine-tune the reconstruction results. Li et al. [45] presented a hybrid 2D–3D deep residual attentional network to take fully advantage of the spatial–spectral context information. These two SR challenges are divided into the "Clean" and "Real World" tracks. The "Clean" track aims to recover HSIs from the noise-free RGB images created by a known camera response function, while the "Real World" one requires participants to rebuild the HSIs from JPEG-compression RGB images obtained by an unknown camera response function. It is worth noting that the camera response functions for the same tracks of the two challenges are different. Also, to provide a more accurate simulation of physical camera systems, the NTIRE2020 "Real World" track is updated with additional simulated camera noise and demosaicing operation.

Attention mechanisms have been a useful tool in a variety of tasks, for instance, image captioning [46], classification [47,48], single image super-resolution [49–51], and person re-identification [52]. Chen et al. [46] proposed a SCA-CNN that incorporated spatial and channel-wise attention for image captioning. Dai et al. [50] presented a deep second-order attention network by exploring second-order statistics of features rather than first-order ones (e.g., global average pooling) [47]. Zhang et al. [53] proposed an effective relation-aware global attention module which captured the global structural information for better attention learning. Only a few very recent methods for SR [36,37,45] considered channel-wise attention mechanism using first-order statistics.

Compared with the previous sparse recovery and shallow mapping methods, the end-to-end training paradigm and discriminant representational learning of CNNs bring considerable improvements of SR. However, the existing CNN-based SR approaches only devote to realizing the RGB-to-HSI mapping by the means of designing the deeper and wider network frameworks, which integrates hierarchical features from different layers without distinction and fails to explore the feature correlations of intermediate layers, thus hindering the expression capacity of CNNs. Actually, the importance of the information of all spatial regions of the feature map is different in the SR task. The feature response among channels also plays a different role for the SR performance. Additionally, most of CNN-based SR models do not consider the problem of spectral aliasing at the edge position, thus resulting in relatively-low performance.

To address these issues, a deep residual augmented attentional u-shape network (RA^2UN) is proposed for SR. Concretely, the backbone of the proposed network is stacked with several double improved residual blocks (DIRB) rather than paired plain convolutional units to extract increasingly abstract feature representations through powerful residual learning. Moreover, we develop a novel spatial augmented attention (SAA) module to bridge the encoder and decoder, which is employed to highlight the features in the informative regions selectively and boost the spatial feature representations. To model interdependencies among channels of intermediate feature maps, a trainable channel augmented attention (CAA) module embedded in the DIRB is presented to adaptively recalibrate channel-wise feature responses by exploiting first-order statistics and second-order ones. Such CAA modules make the network dynamically focus on useful features and further strengthen intrinsic residual learning of DIRBs. Finally, we establish a boundary-aware constraint to guide network to pay close attention to salient information in boundary localization, which can alleviate spectral aliasing at the edge position and recover more accurate edge details.

In summary, the main contributions of this paper can be depicted as below:

- We propose a novel RA^2UN network constituted of several DIRB blocks instead of paired plain convolutional units for SR, which can extract increasingly abstract feature representations through powerful residual learning. Experimental results on four established benchmarks demonstrate that the proposed RA^2UN network outperforms the state-of-the-art SR methods under quantitative measurements and perceptual comparison.
- A trainable SAA module is developed to bridge the encoder and decoder to emphasize the features in the informative regions selectively, which can strengthen the interaction and fusion between the low-level and high-level features effectively and further boost the spatial feature representations.
- To model interdependencies among channels of intermediate feature maps, we present a novel CAA module embedded in the DIRB to adaptively recalibrate channel-wise

feature responses and enhance residual learning by using first-order and second-order statistics for stronger feature expression.

- A boundary-aware constraint is established to guide the network to focus on the salient edge information, which is helpful to alleviate spectral aliasing at the edge position and preserve more accurate high-frequency details.

2. Materials and Methods

2.1. The Proposed RA^{2}UN Network

Figure 1 gives an illustration of our proposed RA^{2}UN network. The backbone architecture mainly consists of several DIRB blocks. The SAA module is bridged the different DIRB counterparts between encoder and decoder and the CAA one is embedded in each DIRB. As for each DIRB, batch normalization layers are not performed, since the normalization operation can prevent the network's power to learn spatial dependencies and spectral distribution. Meanwhile, we adopt Parametric Rectified Linear Unit (PReLU) instead of ReLU as activation function to introduce more nonlinear representation and obtain stronger robustness. The entire DIRB is formulated as

$$y = \rho(R(x, W_{l,1}) + x) \quad (1)$$

$$z = \rho(R(y, W_{l,2}) + y) \quad (2)$$

where x and z denote the input and output of the DIRB block. y is the output of the first residual unit of the DIRB block. $W_{l,1}$ and $W_{l,1}$ represent the weight matrixes of the first and second residual units of the l-th DIRB block. $R(\cdot)$ denotes the residual mapping to be learned which comprises two convolutional layers and one PReLU function. ρ is the PReLU function. Our proposed RA^{2}UN keeps the same spatial resolution of feature maps throughout the proposed model, which can maintain plentiful spatial details information for recovering the accurate spectrum from the RGB inputs in the network. The specific parameters settings for the backbone frameworks are given in Table 1. It can be seen that the output size of each DIRB of our RA^{2}UN is not decreased in the encoding and decoding parts, i.e., we remove the down-sampling operation, which can loss partial spatial details and fail to remain the original pixel information as the network goes deeper, further reducing the accuracy of SR inevitably. In the encoder section, a simple convolutional layer is firstly employed to extract shallow feature from input images. Then several DIRBs are stacked for deep features extraction. Finally, we perform the final reconstruction part via one convolutional layer.

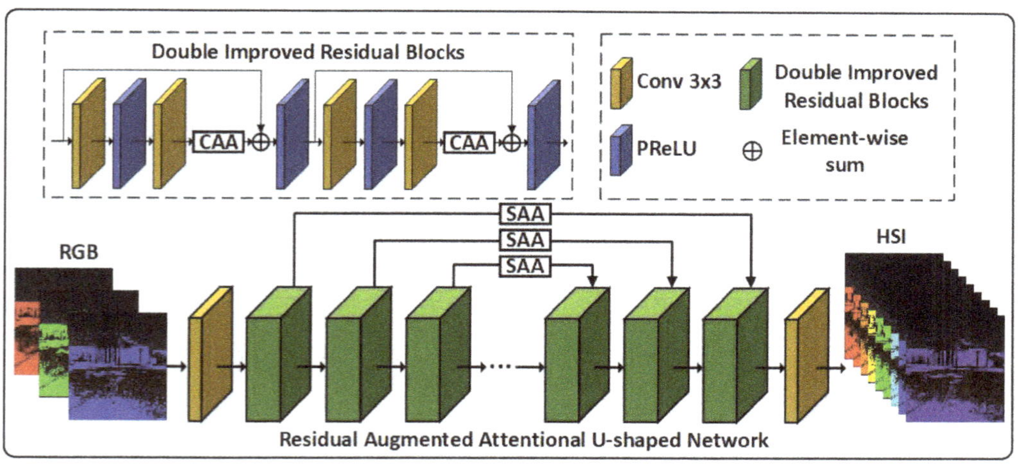

Figure 1. Network architecture of the proposed RA^{2}UN network. The input of the RA^{2}UN network is RGB images and the output is the corresponding reconstructed HSIs. The detailed network and parameters setting can be referenced from Table 1.

Table 1. Parameters settings for the backbone frameworks of our proposed RA^{2}UN. $(\cdot)$ stands for the dimension of the convolutional kernels (input channels, kernel size^2, filter number). The stride and padding of the convolution kernels are set to 1. The dimensions of the feature map are denoted by $C \times H \times W(H = W)$. C, H and W denote the channel, height and width of the feature map. $\{\cdot\}$ indicates the DIRB block. Four rows in kernels column denote the dimensions of the four convolutional kernels of each DIBR block. $[\cdot]$ is the improved residual unit.

No.	Layer	Encoding Parts		Decoding Parts	
		Kernels	Output Size	Kernels	Output Size
1	Conv	$(3,3^2,32)$	$32 \times 64 \times 64$	$(32,3^2,31)$	$31 \times 64 \times 64$
2	DIRB-1	$\left\{\begin{matrix}\begin{bmatrix}(32,3^2,64)\\(64,3^2,64)\end{bmatrix}\\\begin{bmatrix}(64,3^2,64)\\(64,3^2,64)\end{bmatrix}\end{matrix}\right\}$	$64 \times 64 \times 64$	$\left\{\begin{matrix}\begin{bmatrix}(64,3^2,32)\\(32,3^2,32)\end{bmatrix}\\\begin{bmatrix}(32,3^2,32)\\(32,3^2,32)\end{bmatrix}\end{matrix}\right\}$	$32 \times 64 \times 64$
3	DIRB-2	$\left\{\begin{matrix}\begin{bmatrix}(64,3^2,128)\\(128,3^2,128)\end{bmatrix}\\\begin{bmatrix}(128,3^2,128)\\(128,3^2,128)\end{bmatrix}\end{matrix}\right\}$	$128 \times 64 \times 64$	$\left\{\begin{matrix}\begin{bmatrix}(128,3^2,64)\\(64,3^2,64)\end{bmatrix}\\\begin{bmatrix}(64,3^2,64)\\(64,3^2,64)\end{bmatrix}\end{matrix}\right\}$	$64 \times 64 \times 64$
4	DIRB-3	$\left\{\begin{matrix}\begin{bmatrix}(128,3^2,256)\\(256,3^2,256)\end{bmatrix}\\\begin{bmatrix}(256,3^2,256)\\(256,3^2,256)\end{bmatrix}\end{matrix}\right\}$	$256 \times 64 \times 64$	$\left\{\begin{matrix}\begin{bmatrix}(256,3^2,128)\\(128,3^2,128)\end{bmatrix}\\\begin{bmatrix}(128,3^2,128)\\(128,3^2,128)\end{bmatrix}\end{matrix}\right\}$	$128 \times 64 \times 64$
5	DIRB-4	$\left\{\begin{matrix}\begin{bmatrix}(256,3^2,256)\\(256,3^2,256)\end{bmatrix}\\\begin{bmatrix}(256,3^2,256)\\(256,3^2,256)\end{bmatrix}\end{matrix}\right\}$	$256 \times 64 \times 64$	$\left\{\begin{matrix}\begin{bmatrix}(256,3^2,256)\\(256,3^2,256)\end{bmatrix}\\\begin{bmatrix}(256,3^2,256)\\(256,3^2,256)\end{bmatrix}\end{matrix}\right\}$	$256 \times 64 \times 64$
6	DIRB-5	$\left\{\begin{matrix}\begin{bmatrix}(256,3^2,256)\\(256,3^2,256)\end{bmatrix}\\\begin{bmatrix}(256,3^2,256)\\(256,3^2,256)\end{bmatrix}\end{matrix}\right\}$	$256 \times 64 \times 64$	$\left\{\begin{matrix}\begin{bmatrix}(256,3^2,256)\\(256,3^2,256)\end{bmatrix}\\\begin{bmatrix}(256,3^2,256)\\(256,3^2,256)\end{bmatrix}\end{matrix}\right\}$	$256 \times 64 \times 64$
7	Bottom	$\left\{\begin{matrix}\begin{bmatrix}(256,3^2,256)\\(256,3^2,256)\end{bmatrix}\\\begin{bmatrix}(256,3^2,256)\\(256,3^2,256)\end{bmatrix}\end{matrix}\right\}$	$256 \times 64 \times 64$	——	——

2.2. Spatial Augmented Attention Module

In general, the importance of the information of all spatial regions of the feature map is different in the SR task. To focus more attention on the features in the informative regions, a SAA module is designed between the encoder and the decoder, which can boost the interaction and fusion between the low-level and high-level features effectively. The specific diagram of SAA module is displayed in Figure 2. Our proposed SAA module consists of paired symmetric and asymmetric convolutional groups. The asymmetric convolutions refer to use 1D horizontal and vertical kernels (i.e., 1×3 and 3×1 sizes), which not only strengthen the square convolution kernels but also capture multi-direction contextual information to obtain discriminative spatial dependencies.

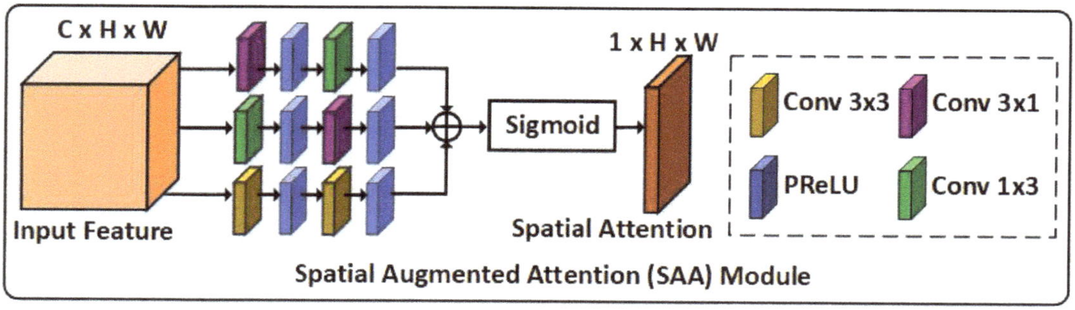

Figure 2. The overview of spatial augmented attention module. $\oplus$ denotes the element-wise summation.

Given an intermediate feature map denoted as $\mathbf{F} = [\mathbf{f}_1, \mathbf{f}_2, \cdots, \mathbf{f}_c, \cdots, \mathbf{f}_C]$ containing C feature maps with spatial size of $H \times W$, we firstly feed $\mathbf{F}$ to the parallel paired symmetric and asymmetric convolutional groups

$$\mathbf{C}_1 = \rho(Conv_{1,2}^{3\times1}(\rho(Conv_{1,1}^{1\times3}(\mathbf{F})))) \tag{3}$$

$$\mathbf{C}_2 = \rho(Conv_{2,2}^{1\times3}(\rho(Conv_{2,1}^{3\times1}(\mathbf{F})))) \tag{4}$$

$$\mathbf{C}_3 = \rho(Conv_{3,2}^{3\times3}(\rho(Conv_{3,1}^{3\times3}(\mathbf{F})))) \tag{5}$$

where ρ denotes the PReLU activation function. $Conv_{1,1}^{1\times3}(\cdot)$, $Conv_{2,1}^{3\times1}(\cdot)$ and $Conv_{3,1}^{3\times3}(\cdot)$ project the feature $\mathbf{F} \in R^{C\times H\times W}$ to a lower size $R^{C/t\times H\times W}$ along the channel dimension. Then the next convolution layers $Conv_{1,2}^{3\times1}(\cdot)$, $Conv_{2,2}^{1\times3}(\cdot)$ and $Conv_{3,2}^{3\times3}(\cdot)$ map the low-dimensional features to the multi-direction spatial feature descriptors $\mathbf{C}_1, \mathbf{C}_2, \mathbf{C}_3 \in R^{1\times H\times W}$, which contain rich contextual information. Besides, this design increases only a small amount of parameters and computational burden. To compute the spatial attention, the feature descriptors are summed and normalized to $[0,1]$ through a sigmoid activation σ

$$\mathbf{A}_s(\mathbf{F}) = \sigma(\mathbf{C}_1 + \mathbf{C}_2 + \mathbf{C}_3) \tag{6}$$

where $\mathbf{A}_s(\mathbf{F}) \in R^{1\times H\times W}$ represents the spatial attention, which encodes the degree of importance for the spatial positions of the original feature $\mathbf{F}$ and determines which spatial locations should be emphasized. Finally, we perform the element-wise multiplication $\otimes$ between $\mathbf{A}_s(\mathbf{F})$ and $\mathbf{F}$

$$\mathbf{F}^s = \mathbf{A}_s(\mathbf{F}) \otimes \mathbf{F} \tag{7}$$

where $\mathbf{F}^s$ is the refined feature. During the processing, the spatial attention values are broadcasted along the channel-wise direction. Such SAA module is bridged the encoder and decoder to highlight the features in the important regions selectively and boost the spatial feature representations.

2.3. Channel Augmented Attention Module

In contrast to the preceding SAA module extracting the inter-spatial relationships of features, our presented CAA module attempts to explore inter-channel dependencies of features for SR. To obtain more powerful learning capability of the network, we present a novel CAA module to model interdependencies between channels by using first-order and second-order statistics jointly for stronger feature representations (see Figure 3).

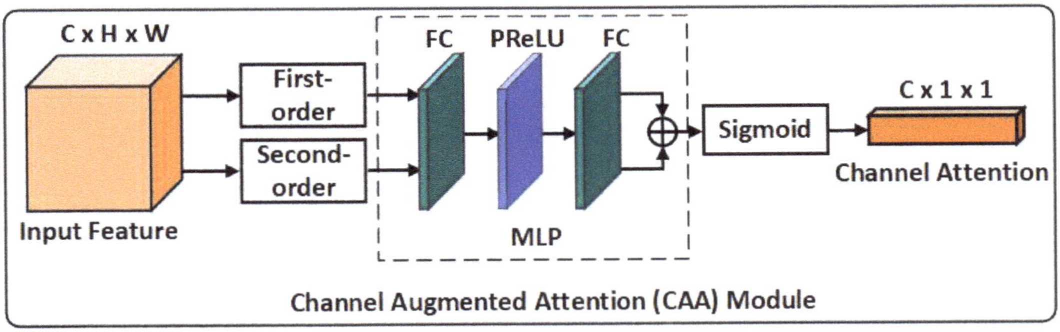

Figure 3. The overview of channel augmented attention module. $\oplus$ denotes the element-wise summation.

We first aggregate spatial information of the feature map $\mathbf{F} \in R^{C \times H \times W}(\mathbf{F} = [\mathbf{f}_1, \mathbf{f}_2, \cdots, \mathbf{f}_c, \cdots, \mathbf{f}_C], \mathbf{f}_c \in R^{H \times W})$ by using global average pooling

$$\mathbf{s}_c^{1st} = \frac{1}{H \times W} \sum_{i=1}^{H} \sum_{j=1}^{W} \mathbf{f}_c(i, j) \tag{8}$$

where $\mathbf{s}_c^{1st}$ denotes the c-th element of the first-order channel descriptor $\mathbf{S}^{1st} \in R^C$ and $\mathbf{f}_c(i, j)$ is the response at location (i, j) of the c-th feature map $\mathbf{f}_c$. As for the second-order channel descriptor, we reshape the feature map $\mathbf{F} \in R^{C \times H \times W}$ to a feature matrix $\mathbf{D} \in R^{C \times n}, n = H \times W$ and compute the sample covariance matrix

$$\mathbf{X} = \mathbf{D}\bar{\mathbf{I}}\mathbf{D}^T \tag{9}$$

where $\bar{\mathbf{I}} = \frac{1}{n}\left(\mathbf{I} - \frac{1}{n}\mathbf{1}\right)$, and $\mathbf{X} \in R^{C \times C}, \mathbf{X} = [\mathbf{x}_1, \mathbf{x}_2, \cdots, \mathbf{x}_c, \cdots, \mathbf{x}_C], \mathbf{x}_c \in R^{1 \times C}$. $\mathbf{I}$ and $\mathbf{1}$ represent the $n \times n$ identity matrix and matrix of all ones. Then the c-th dimension of the second-order statistics $\mathbf{S}^{2nd} \in R^C$ is formulized as

$$\mathbf{s}_c^{2nd} = \frac{1}{C} \sum_{i=1}^{C} \mathbf{x}_c(i) \tag{10}$$

where $\mathbf{s}_c^{2nd}$ denotes the c-th element of the second-order channel descriptor $\mathbf{S}^{2nd} \in R^C$ and $\mathbf{x}_c(i)$ is the i-th value of the c-th feature map $\mathbf{x}_c$. To make use of the aggregated information $\mathbf{S}^{1st}$ and $\mathbf{S}^{2nd}$, both descriptors are fed into a shared multi-layer perceptron (MLP) with a sigmoid function to generate the channel attention. The MLP is constituted of two fully connected (FC) layers and a non-linearity PReLU function, where the output dimension of the first FC layer is $R^{C/r}$ and the output size of the second one is R^C. r is the reduction ratio. In summary, the channel attention map is indicated as

$$\mathbf{A}_c(\mathbf{F}) = \sigma(FC_2(\rho(FC_1(\mathbf{S}^{1st}))) + FC_2(\rho(FC_1(\mathbf{S}^{2nd})))) \tag{11}$$

where $FC_1(\cdot)$ and $FC_2(\cdot)$ are the weight set of two FC layers. $\mathbf{A}_c(\mathbf{F}) \in R^C$ denotes the channel attention recording the importance and interdependences among channels, which is to rescale the original input feature $\mathbf{F}$

$$\mathbf{F}^c = \mathbf{A}_c(\mathbf{F}) \otimes \mathbf{F} \tag{12}$$

where $\otimes$ is element-wise multiplication and the channel attention values can be copied along the spatial dimension according to the broadcast mechanism. Inserted into the DIRB block, the CAA module can recalibrate channel-wise feature responses adaptively and enhance residual learning.

2.4. Boundary-Aware Constraint

In the process of hyperspectral imaging, the spectral aliasing of the edge position is easy to occur, so that the reconstruction accuracy of boundary spectrum is low. To alleviate the spectral aliasing and recover more accurate high-frequency details of HSIs, we establish a boundary-aware constraint to guide the training process in the proposed RA^2UN:

$$l = l_m + \tau l_b \tag{13}$$

$$l_m = \frac{1}{N}\sum_{p=1}^{N}(|\mathbf{I}_{HSI}^{(p)} - \mathbf{I}_{SR}^{(p)}|/\mathbf{I}_{HSI}^{(p)}) \tag{14}$$

$$l_b = \frac{1}{N}\sum_{p=1}^{N}(|\mathbf{B}(\mathbf{I}_{HSI}^{(p)}) - \mathbf{B}(\mathbf{I}_{SR}^{(p)})|)) \tag{15}$$

where l_m represents the mean relative absolute error (MRAE) loss term to minimize the numerical error between ground truths and the reconstructed results. l_b denotes the boundary-aware constraint component to lead the network to focus on the salient edge information simultaneously. τ is a weighted parameter. N is the total number of pixels. $\mathbf{I}_{HSI}^{(p)}$ and $\mathbf{I}_{SR}^{(p)}$ denote the p-th pixel value of the ground truth $\mathbf{I}_{HSI}$ and the spectral reconstructed result $\mathbf{I}_{SR}$. $\mathbf{B}(\cdot)$ represents the edge extraction function. To be specific, $\mathbf{B}(\cdot)$ firstly performs Gaussian filtering to eliminate the influence of noise and then adopts Prewitt operator [54] to get boundaries of ground truths and the reconstructed results. The Gaussian filtering kernel is $[[0.0751, 0.1238, 0.0751], [0.1238, 0.2042, 0.1238], [0.0751, 0.1238, 0.0751]]$ and the sigma is 1.0. The Prewitt operators are $[[-1.0, 0.0, 1.0], [-1.0, 0.0, 1.0], [-1.0, 0.0, 1.0]]$ and $[[-1.0, -1.0, -1.0], [0.0, 0.0, 0.0], [1.0, 1.0, 1.0]]$ in the x and y directions, respectively. In order to better observe the effect of edge extraction, we visualize several example images in Figure 4. The first row shows several original images from the NTIRE2020 dataset. The second row displays the effect of edge extraction. From the mathematical perspective, compared with the single MRAE loss term l_m, the compound loss function l can make the space of the possible three-to-many mapping functions smaller for the ill-posed SR problem and avoid falling into a local minimum to obtain more accurate spectral recovery, which will be demonstrated in Section 4.1. Finally, τ is empirically set to 1.0 in the proposed network.

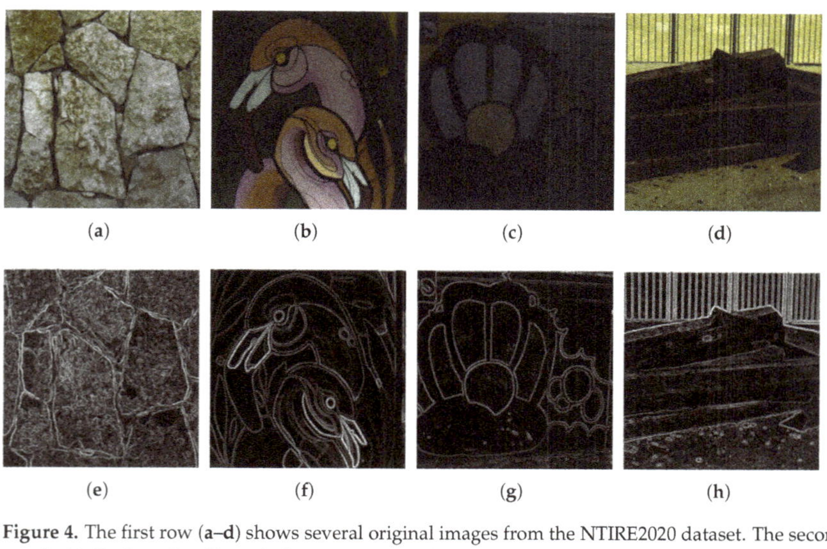

(a) (b) (c) (d)

(e) (f) (g) (h)

Figure 4. The first row (**a**–**d**) shows several original images from the NTIRE2020 dataset. The second row (**e**–**h**) displays the effect of edge extraction and the white lines represent boundary information.

3. Experiments Setting

3.1. Datasets and Implementations

In this paper, we evaluate the proposed RA^{2}UN on four benchmark datasets, i.e., NTIRE2018 "Clean" and "Real World" tracks, NTIRE2020 "Clean" and "Real World" tracks. Following the competition instructions, the NTIRE2018 dataset contains 256 natural HSIs for official training set and 5 + 10 additional images for official validation set and testing set with the size of 1392 × 1300. All images have 31 spectral bands (400–700 nm at roughly 10nm increments). The NTIRE2020 dataset consists of 450 images for official training set, 10 images for official validation set and 20 images for official testing set with 31 bands from 400 nm to 700 nm at 10 nm steps. Each band is the size of 512 × 482. The NTIRE2020 datasets are collected with a Specim IQ mobile hyperspectral camera. The Specim IQ camera is a stand-alone, battery-powered, push-broom spectral imaging system, the size of a conventional SLR camera (207 × 91 × 74 mm) which can operate independently without the need for an external power source or computer controller. The NTIRE2018 datasets are acquired using a Specim PS Kappa DX4 hyperspectral camera and a rotary stage for spatial scanning.

For the dataset settings, due to the confidentiality of ground truth HSIs for the official testing set of both SR contests, we choose the official validation as the final testing set and randomly select several images from the official training set as the final validation set in this paper. The rest of the official training set is adopted as the final training set. Specifically, the NTIRE2020 final validation set contains 10 HSIs including "ARAD_HS_0079", "ARAD_HS_0089", "ARAD_HS_0255", "ARAD_HS_0304", "ARAD_HS_0363", "ARAD_HS_0372", "ARAD_HS_0387", "ARAD_HS_0422", "ARAD_HS_0434" and "ARAD_HS_0446". The NTIRE2018 final validation set chooses 5 HSIs including "BGU_HS_00001", "BGU_HS_00036", "BGU_HS_00204", "BGU_HS_00209" and "BGU_HS_00225".

During the training process, we crop 64 × 64 RGB and HSI sample pairs from the original NTIRE2020 and NTIRE2018 datasets. The batch size of our model is 16 and the parameter optimization algorithm chooses Adam [55] with $\beta_1 = 0.9$, $\beta_2 = 0.99$ and $\epsilon = 10^{-8}$. The parameter t value of the SAA module is 4 and reduction ratio r of CAA module is 16. The learning rate is initialized as 1.2×10^{-4} and the polynomial function is set as the decay policy with power = 1.5. We stop network training at 100 epochs and the proposed RA^{2}UN network has been implemented on the Pytorch framework on an NVIDIA 2080Ti GPU.

3.2. Evaluation Metrics

To objectively test the results of our proposed method on the NTIRE2020 and NTIRE2018 datasets, the mean relative absolute error (MRAE), root mean square error (RMSE), and spectral angle mapper (SAM) are adopted as metrics. The MRAE and RMSE are provided by the challenge, where MRAE is chosen as the ranking criterion rather than RMSE to avoid overweighting errors in the higher brightness region of the test image. The SAM is employed to measure the spectral quality. The MRAE, RMSE and SAM are defined as follows

$$MRAE = \frac{1}{N}\sum_{p=1}^{N}\left(\left|\mathbf{I}_{HSI}^{(p)} - \mathbf{I}_{SR}^{(p)}\right|/\mathbf{I}_{HSI}^{(p)}\right) \tag{16}$$

$$RMSE = \sqrt{\frac{1}{N}\sum_{p=1}^{N}\left(\mathbf{I}_{HSI}^{(p)} - \mathbf{I}_{SR}^{(p)}\right)^2} \tag{17}$$

$$SAM = \frac{1}{M}\sum_{v=1}^{M}\left(arccos(<\mathbf{I}_{HSI}^{(v)}, \mathbf{I}_{SR}^{(v)}> / (||\mathbf{I}_{HSI}^{(v)}||_2||\mathbf{I}_{SR}^{(v)}||_2))\right) \tag{18}$$

where $\mathbf{I}_{HSI}^{(p)}$ and $\mathbf{I}_{SR}^{(p)}$ denote the p-th pixel value of the ground truth and the spectral reconstructed HSI. $< \mathbf{I}_{HSI}^{(v)}, \mathbf{I}_{SR}^{(v)} >$ represents the dot product of the v-th spectral vector $\mathbf{I}_{HSI}^{(v)}$ and $\mathbf{I}_{SR}^{(v)}$ of the ground truth and the spectral reconstructed HSI. $|| \cdot ||$ is $l2$ norm operation. N is the total number of pixels and M is the total number of spectral vectors. A smaller MRAE, RMSE or SAM indicates better performance.

4. Experimental Results and Discussions

4.1. Discussion on the Proposed RA^{2}UN: Ablation Study

In order to demonstrate the effectiveness of the SAA module, the CAA module and the boundary-aware constraint, we conduct the ablation study on the NTIRE2020 "Clean" track dataset. The results are summarized in Table 2. R_a refer to the baseline network without any attention module, which is trained by individual MRAE loss term l_h. In Table 2, the baseline result reaches to MRAE = 0.03668.

Table 2. Ablation study on the final validation set of NTIRE2020 "Clean" track dataset. We record the best MRAE values in 5.76×10^5 iterations.

Description	R_a	R_b	R_c	R_d	R_e	R_f	R_g	R_h
SAA Module	✘	✔	✘	✘	✔	✔	✘	✔
CAA Module	✘	✘	✔	✘	✔	✘	✔	✔
Boundary-aware Loss	✘	✘	✘	✔	✘	✔	✔	✔
MRAE (↓)	0.03668	0.03637	0.03396	0.03636	0.03362	0.03590	0.03381	0.03303

Spatial Augmented Attention Module. Firstly, we only add the SAA module to basic model in R_b and acquire the decline in MRAE. It implies that the SAA module is helpful to emphasize the features in the important regions and boost the spatial feature representations. Then the results of R_e and R_f further prove the effectiveness of the SAA module, based on that the CAA module is employed or the boundary-aware constraint is established.

Channel Augmented Attention Module. As elaborated in Section 2.3, a CAA module is developed to explore feature interdependencies among channels. Compared with the baseline result, R_c achieves 7.42% decrease in the MRAE value. The reason may be that CAA module can recalibrate channel-wise feature responses adaptively and realize powerful learning capability of the network. Compared with the results from R_b and R_d, the results of R_e and R_g further demonstrate the superiority of the CAA module, respectively.

Boundary-aware Constraint. In contrast to the baseline experiment R_a, R_d is optimized by stochastic gradient descent algorithm with the MRAE loss term l_h and the boundary-aware constraint l_b. The result of R_d indicates that the boundary-aware constraint is helpful to recover more accurate HSIs. Furthermore, other results of R_f, R_g and R_h all verify the effectiveness of the boundary-aware constraint. In particular, we can get the best MRAE value with the two modules and the boundary-aware constraint in R_h.

4.2. Results of SR and Analysis

In this study, we compare the proposed RA^{2}UN against 6 existing methods including Arad [17], Galliani [23], Yan [26], Stiebel [30], HSCNN-R [34] and HRNet [37]. Among them, the Arad is an early SR approach based on sparse recovery, while the others are based on CNNs. For a fair comparison, all models retrain on the final training set, save on the final validation set and evaluate on the final testing set for the two tracks of the NTIRE2020 and NTIRE2018 datasets. The quantitative results of final test set of NTIRE2020 and NTIRE2018 "Clean" and "Real World" tracks are listed in Tables 3 and 4. Since the camera response function is unknown, Arad is only suitable for measuring on "Clean" tracks. It can be seen that our RA^{2}UN performs the best results under MRAE, RMSE and SAM metrics on all the tracks. As for the ranking metrics MRAE, the proposed method achieves relative reduction of 14.02%, 6.89%, 14.21% and 1.27% over the second best results on corresponding established datasets. In addition, we can obtain the smallest SAM values, which indicate that our reconstructed HSIs contain better spectral quality.

Also, we show the visual comparison of the five selected bands on different example images of the final test set in Figures 5–8. The ground truth, our results and error images are displayed from top to bottom. The error images are the heat maps of MRAE between the ground truth and the recovered HSI. The bluer the displayed color, the better the reconstructed spectrum. As can be seen, our approach yields better recovery results and have less reconstruction error than other competitors. Besides, the spectral response curves of four selected spatial points are painted in Figure 9. The red line is our result and the black one denotes the groundtruth spectrum. The rest are the results of the comparison methods. Obviously, the reconstructed results of RA^{2}UN are much closer to the groundtruth spectrum than the others.

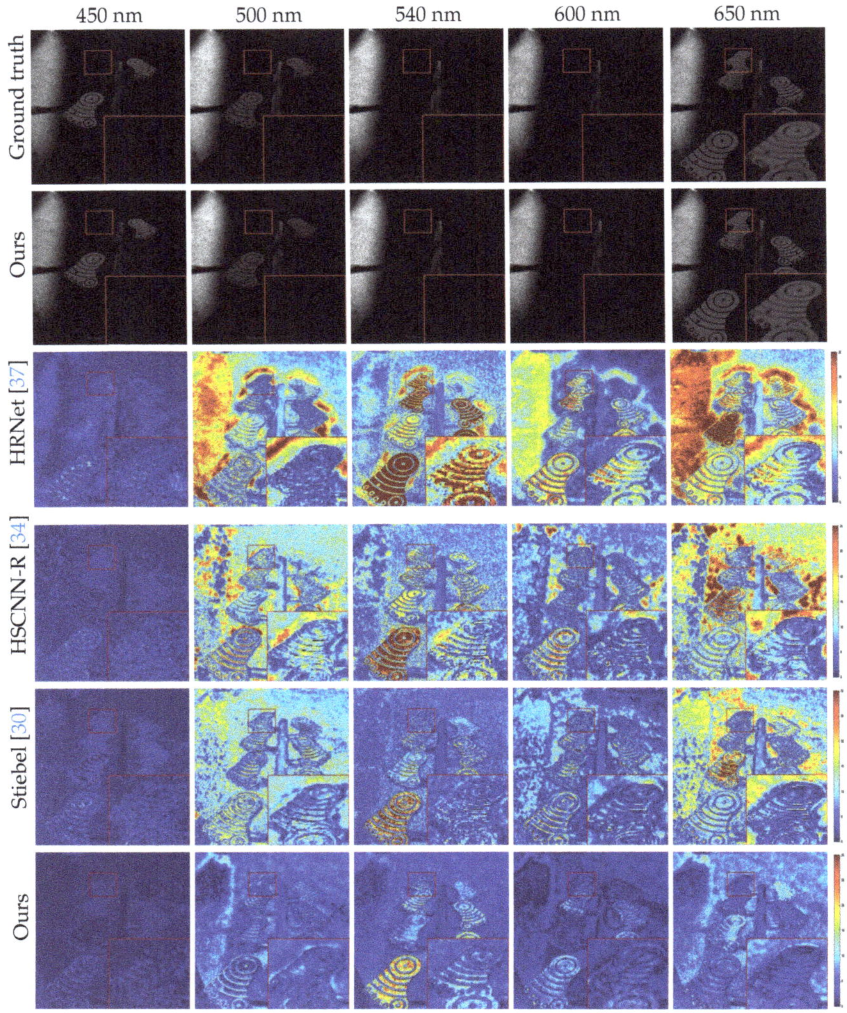

Figure 5. Visual comparison of the five selected bands on "ARAD_HS_0455" image from the final testing set of NTIRE2020 "Clean" track. The best view on the screen.

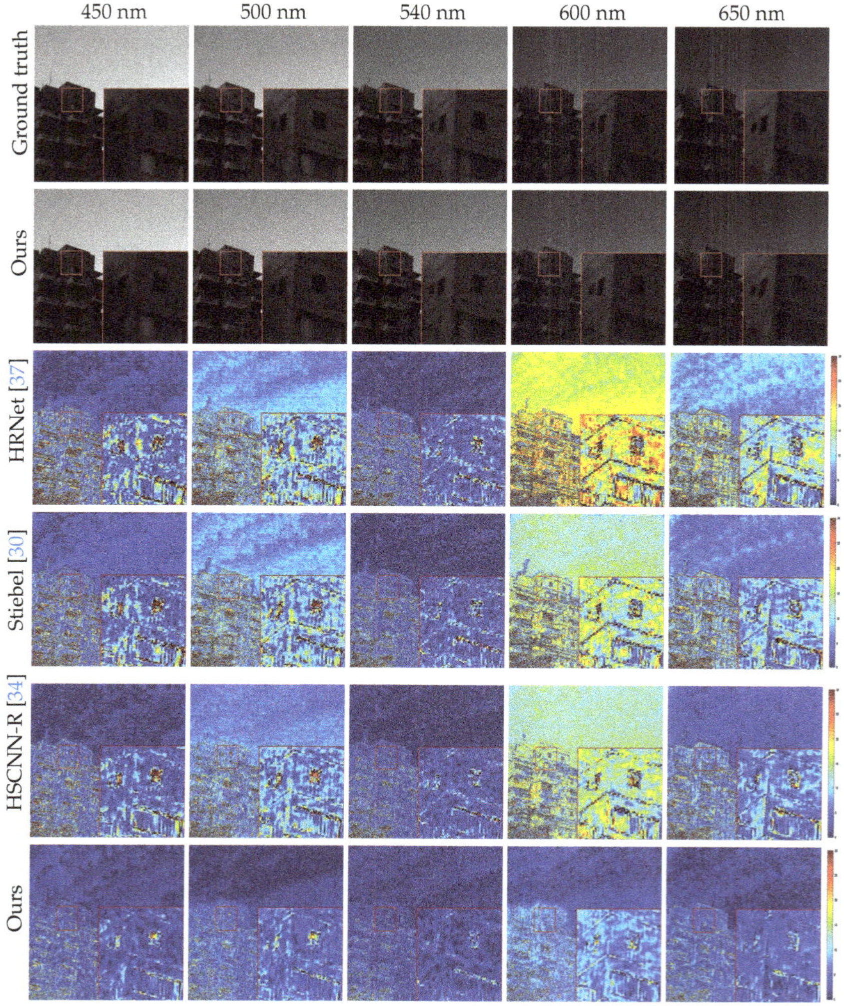

Figure 6. Visual comparison of the five selected bands on "ARAD_HS_0451" image from the final testing set of NTIRE2020 "Real World" track. The best view on the screen.

Table 3. The quantitative results of final test set of NTIRE2020 "Clean" and "Real World" tracks. The best and second best results are **bold** and <u>underlined</u>.

Method	Clean			Real World		
	MRAE (↓)	RMSE (↓)	SAM (↓)	MRAE (↓)	RMSE (↓)	SAM (↓)
Ours	**0.03446**	**0.01158**	**2.39933**	**0.06554**	**0.01712**	**3.35699**
Stiebel [30]	<u>0.04008</u>	<u>0.01518</u>	<u>2.73916</u>	0.07141	0.01912	3.68491
HSCNN-R [34]	0.04406	0.01543	2.94031	<u>0.07039</u>	<u>0.01893</u>	<u>3.60987</u>
HRNet [37]	0.04202	0.01575	2.83058	0.07042	0.02035	3.71418
Yan [26]	0.10351	0.02844	4.90422	0.09942	0.03005	4.54294
Galliani [23]	0.07949	0.02788	4.52770	0.10794	0.03307	4.79334
Arad [17]	0.07873	0.03305	5.57166	———	———	———

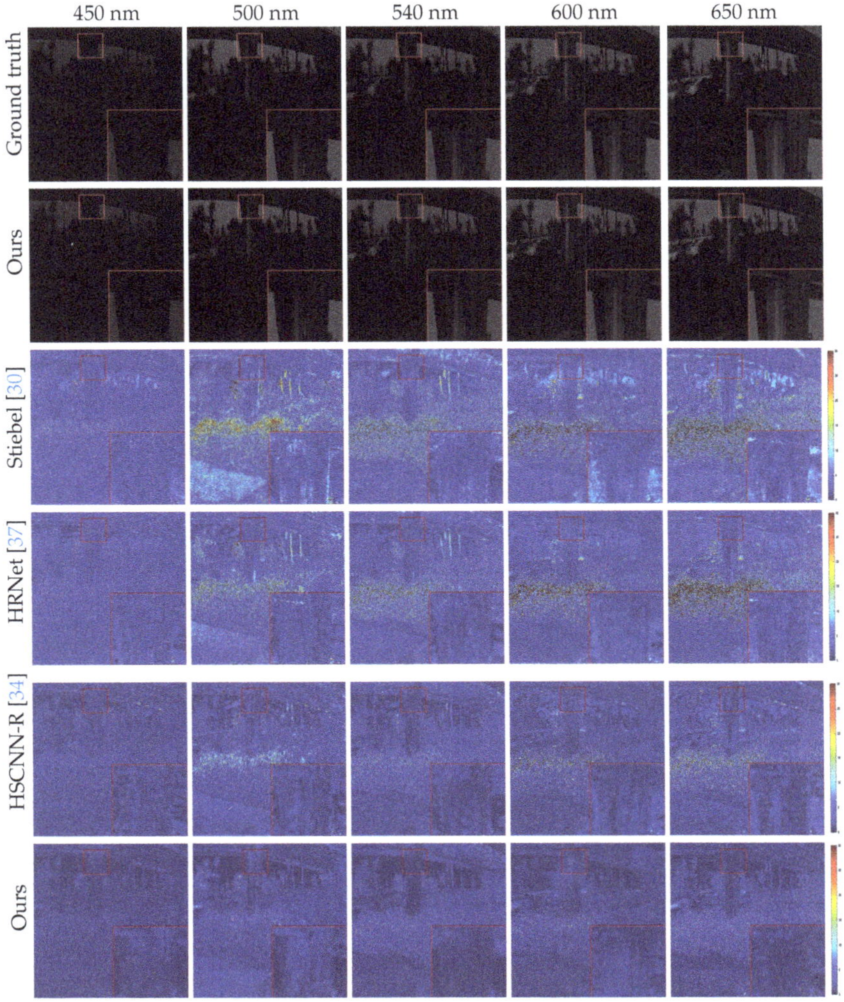

Figure 7. Visual comparison of the five selected bands on "BGU_HS_00265" image from the final testing set of NTIRE2018 "Clean" track. The best view on the screen.

Table 4. The quantitative results of final test set of NTIRE2018 "Clean" and "Real World" tracks. The best and second best results are **bold** and underlined.

Method	Clean			Real World		
	MRAE (↓)	RMSE (↓)	SAM (↓)	MRAE (↓)	RMSE (↓)	SAM (↓)
Ours	**0.01141**	**10.4923**	**0.80815**	**0.02868**	**22.0813**	**1.52763**
HSCNN-R [34]	0.01330	12.8519	0.96004	0.03014	23.5697	1.65147
HRNet [37]	0.01369	13.5165	1.00645	0.02905	22.8282	1.57253
Stiebel [30]	0.01536	15.5253	1.14655	0.03118	24.0600	1.70200
Yan [26]	0.03036	24.2971	1.67274	0.04576	31.8332	2.18224
Galliani [23]	0.05130	37.6802	1.77410	0.07749	49.2496	2.32531
Arad [17]	0.08094	59.4085	5.02125	——	——	——

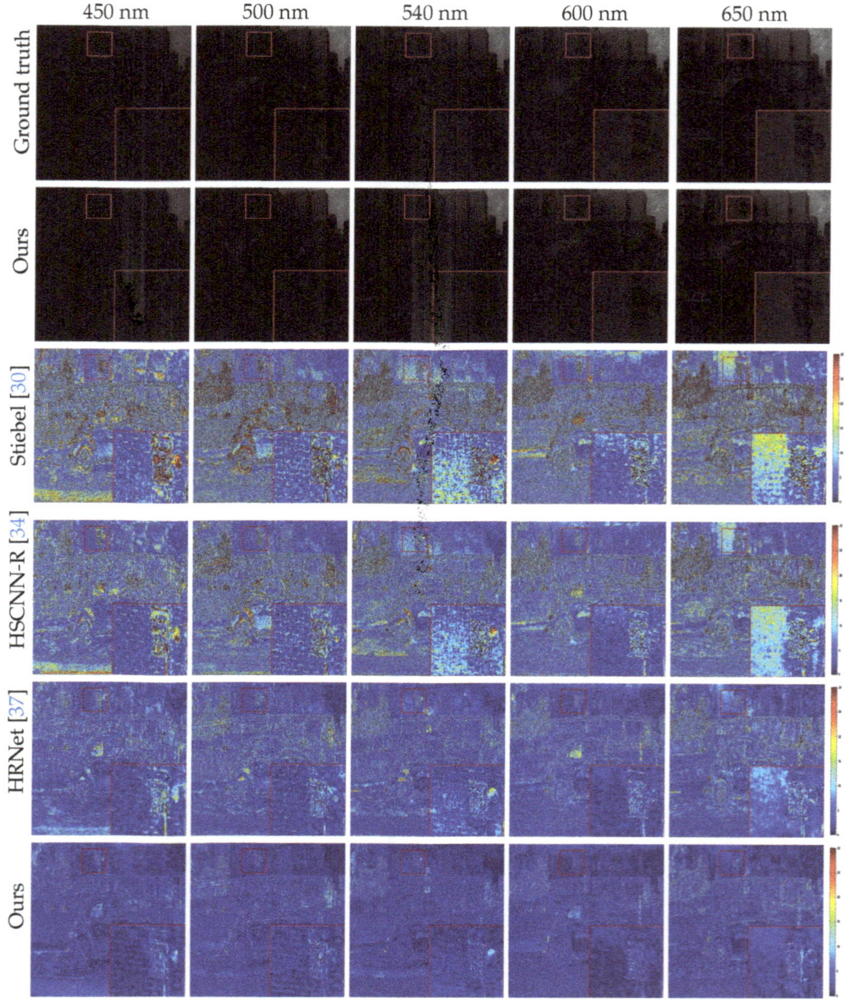

Figure 8. Visual comparison of the five selected bands on "BGU_HS_00259" image from the final testing set of NTIRE2018 "Real World" track. The best view on the screen.

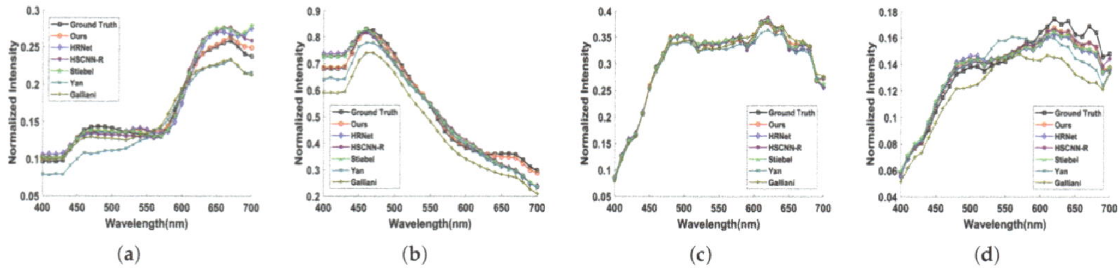

Figure 9. Spectral response curves of selected several spatial points from the reconstructed HSIs. (**a**,**b**) are for the NTIRE2020 "Clean" and "Real World" tracks respectively. (**c**,**d**) are for the NTIRE2018 "Clean" and "Real World" track respectively.

5. Conclusions

In this paper, we propose a novel RA^2UN network for SR. Concretely, the backbone of RA^2UN network consists of several DIRB blocks instead of paired plain convolutional units. To boost the spatial feature representations, a trainable SAA module is developed to highlight the features in the important regions selectively. Furthermore, we present a novel CAA module to adaptively recalibrate channel-wise feature responses by exploiting first-order statistics and second-order ones for enhance learning capacity of the network. To find a better solution, an additional boundary-aware constraint is built to guide network to learn salient information in edge localization and recover more accurate details. Extensive experiments on challenging benchmarks demonstrate the superiority of our RA^2UN network in terms of numerical and visual measurements.

Author Contributions: J.L. and C.W. conceived and designed the study; W.X. performed the experiments; R.S. shared part of the experiment data; J.L. and Y.L. analyzed the data; C.W. and J.L. wrote the paper. R.S. and W.X. reviewed and edited the manuscript. All authors have read and agreed to the published version of the manuscript.

Funding: This work was supported in part by the National Key Research and Development Program of China under Grant (no. 2018AAA0102702), the National Nature Science Foundation of China (no. 61901343), the Science and Technology on Space Intelligent Control Laboratory (no.ZDSYS-2019-03), the China Postdoctoral Science Foundation (no. 2017M623124) and the China Postdoctoral Science Special Foundation (no. 2018T111019). The project was also partially supported by the Open Research Fund of CAS Key Laboratory of Spectral Imaging Technology (no. LSIT201924W) and the Fundamental Research Funds for the Central Universities JB190107. It was also partially supported by the National Nature Science Foundation of China (no. 61571345, 61671383, 91538101, 61501346 and 61502367), Yangtse Rive Scholar Bonus Schemes (No. CJT160102), Ten Thousand Talent Program, and the 111 project (B08038).

Acknowledgments: The authors would like to thank the anonymous reviewers and associate editor for their valuable comments and suggestions to improve the quality of the paper.

Conflicts of Interest: The authors declare no conflict of interest.

References

1. Chang, C.I. *Hyperspectral Data Exploitation: Theory and Applications*; John Wiley & Sons: Hoboken, NJ, USA, 2007.
2. Uzair, M.; Mahmood, A.; Mian, A. Hyperspectral face recognition with spatiospectral information fusion and PLS regression. *IEEE Trans. Image Process.* **2015**, *24*, 1127–1137. [CrossRef]
3. Li, J.; Xi, B.; Du, Q.; Song, R.; Li, Y.; Ren, G. Deep Kernel Extreme-Learning Machine for the Spectral–Spatial Classification of Hyperspectral Imagery. *Remote Sens.* **2018**, *10*, 2036. [CrossRef]
4. Li, J.; Du, Q.; Li, Y.; Li, W. Hyperspectral image classification with imbalanced data based on orthogonal complement subspace projection. *IEEE Trans. Geosci. Remote Sens.* **2018**, *56*, 3838–3851. [CrossRef]
5. Tochon, G.; Chanussot, J.; Dalla Mura, M.; Bertozzi, A.L. Object tracking by hierarchical decomposition of hyperspectral video sequences: Application to chemical gas plume tracking. *IEEE Trans. Geosci. Remote Sens.* **2017**, *55*, 4567–4585. [CrossRef]
6. Green, R.O.; Eastwood, M.L.; Sarture, C.M.; Chrien, T.G.; Aronsson, M.; Chippendale, B.J.; Faust, J.A.; Pavri, B.E.; Chovit, C.J.; Solis, M.; et al. Imaging spectroscopy and the airborne visible/infrared imaging spectrometer (AVIRIS). *Remote Sens. Environ.* **1998**, *65*, 227–248. [CrossRef]
7. James, J. *Spectrograph Design Fundamentals*; Cambridge University Press: Cambridge, UK, 2007.
8. Schechner, Y.Y.; Nayar, S.K. Generalized mosaicing: Wide field of view multispectral imaging. *IEEE Trans. Pattern Anal. Mach. Intell.* **2002**, *24*, 1334–1348. [CrossRef]
9. Wagadarikar, A.; John, R.; Willett, R.; Brady, D. Single disperser design for coded aperture snapshot spectral imaging. *Appl. Opt.* **2008**, *47*, B44–B51. [CrossRef] [PubMed]
10. Tanriverdi, F.; Schuldt, D.; Thiem, J. Dual snapshot hyperspectral imaging system for 41-band spectral analysis and stereo reconstruction. In Proceedings of the International Symposium on Visual Computing, Lake Tahoe, NV, USA, 7–9 October 2019; pp. 3–13.
11. Beletkaia, E.; Pozo, J. More Than Meets the Eye: Applications enabled by the non-stop development of hyperspectral imaging technology. *PhotonicsViews* **2020**, *17*, 24–26. [CrossRef]
12. Descour, M.; Dereniak, E. Computed-tomography imaging spectrometer: Experimental calibration and reconstruction results. *Appl. Opt.* **1995**, *34*, 4817–4826. [CrossRef] [PubMed]

13. Vandervlugt, C.; Masterson, H.; Hagen, N.; Dereniak, E.L. Reconfigurable liquid crystal dispersing element for a computed tomography imaging spectrometer. In *Algorithms and Technologies for Multispectral, Hyperspectral, and Ultraspectral Imagery XIII*; International Society for Optics and Photonics: Bellingham, WA, USA, 2007; Volume 6565, p. 65650O.
14. Wagadarikar, A.A.; Pitsianis, N.P.; Sun, X.; Brady, D.J. Video rate spectral imaging using a coded aperture snapshot spectral imager. *Opt. Express* **2009**, *17*, 6368–6388. [CrossRef] [PubMed]
15. Nguyen, R.M.; Prasad, D.K.; Brown, M.S. Training-based spectral reconstruction from a single RGB image. In Proceedings of the European Conference on Computer Vision, Zurich, Switzerland, 6–12 September 2014; pp. 186–201.
16. Robles-Kelly, A. Single image spectral reconstruction for multimedia applications. In Proceedings of the 23rd ACM international conference on Multimedia, Brisbane, QLD, Australia, 26–30 October 2015; pp. 251–260.
17. Arad, B.; Ben-Shahar, O. Sparse recovery of hyperspectral signal from natural RGB images. In Proceedings of the European Conference on Computer Vision, Amsterdam, The Netherlands, 8–16 October 2016; pp. 19–34.
18. Jia, Y.; Zheng, Y.; Gu, L.; Subpa-Asa, A.; Lam, A.; Sato, Y.; Sato, I. From RGB to spectrum for natural scenes via manifold-based mapping. In Proceedings of the IEEE International Conference on Computer Vision, Venice, Italy, 22–29 October 2017; pp. 4705–4713.
19. Aeschbacher, J.; Wu, J.; Timofte, R. In defense of shallow learned spectral reconstruction from RGB images. In Proceedings of the IEEE International Conference on Computer Vision, Venice, Italy, 22–29 October 2017; pp. 471–479.
20. Zhang, S.; Wang, L.; Fu, Y.; Zhong, X.; Huang, H. Computational Hyperspectral Imaging Based on Dimension-Discriminative Low-Rank Tensor Recovery. In Proceedings of the IEEE/CVF International Conference on Computer Vision (ICCV), Seoul, Korea, 27–28 October 2019.
21. Li, J.; Cui, R.; Li, B.; Song, R.; Li, Y.; Dai, Y.; Du, Q. Hyperspectral Image Super-Resolution by Band Attention Through Adversarial Learning. *IEEE Trans. Geosci. Remote Sens.* **2020**, *58*, 4304–4318. [CrossRef]
22. Li, J.; Cui, R.; Li, B.; Song, R.; Li, Y.; Du, Q. Hyperspectral Image Super-Resolution with 1D–2D Attentional Convolutional Neural Network. *Remote Sens.* **2019**, *11*, 2859. [CrossRef]
23. Galliani, S.; Lanaras, C.; Marmanis, D.; Baltsavias, E.; Schindler, K. Learned spectral super-resolution. *arXiv* **2017**, arXiv:1703.09470.
24. Rangnekar, A.; Mokashi, N.; Ientilucci, E.; Kanan, C.; Hoffman, M. Aerial spectral super-resolution using conditional adversarial networks. *arXiv* **2017**, arXiv:1712.08690.
25. Xiong, Z.; Shi, Z.; Li, H.; Wang, L.; Liu, D.; Wu, F. Hscnn: Cnn-based hyperspectral image recovery from spectrally undersampled projections. In Proceedings of the IEEE International Conference on Computer Vision, Venice, Italy, 22–29 October 2017; pp. 518–525.
26. Yan, Y.; Zhang, L.; Li, J.; Wei, W.; Zhang, Y. Accurate spectral super-resolution from single RGB image using multi-scale CNN. In Proceedings of the Chinese Conference on Pattern Recognition and Computer Vision (PRCV), Guangzhou, China, 23–26 November 2018; pp. 206–217.
27. Fu, Y.; Zhang, T.; Zheng, Y.; Zhang, D.; Huang, H. Joint Camera Spectral Sensitivity Selection and Hyperspectral Image Recovery. In Proceedings of the European Conference on Computer Vision (ECCV), Munich, Germany, 8–14 September 2018; pp. 788–804.
28. Nie, S.; Gu, L.; Zheng, Y.; Lam, A.; Ono, N.; Sato, I. Deeply Learned Filter Response Functions for Hyperspectral Reconstruction. In Proceedings of the IEEE Conference on Computer Vision and Pattern Recognition, Salt Lake City, UT, USA, 18–22 June 2018; pp. 4767–4776.
29. Arad, B.; Ben-Shahar, O.; Timofte, R. Ntire 2018 challenge on spectral reconstruction from RGB images. In Proceedings of the IEEE Conference on Computer Vision and Pattern Recognition Workshops, Salt Lake City, UT, USA, 18–22 June 2018; pp. 929–938.
30. Stiebel, T.; Koppers, S.; Seltsam, P.; Merhof, D. Reconstructing spectral images from RGB-images using a convolutional neural network. In Proceedings of the IEEE Conference on Computer Vision and Pattern Recognition Workshops, Salt Lake City, UT, USA, 18–22 June 2018; pp. 948–953.
31. Can, Y.B.; Timofte, R. An efficient CNN for spectral reconstruction from RGB images. *arXiv* **2018**, arXiv:1804.04647.
32. Alvarez-Gila, A.; Van De Weijer, J.; Garrote, E. Adversarial networks for spatial context-aware spectral image reconstruction from RGB. In Proceedings of the IEEE International Conference on Computer Vision, Venice, Italy, 22–29 October 2017; pp. 480–490.
33. Koundinya, S.; Sharma, H.; Sharma, M.; Upadhyay, A.; Manekar, R.; Mukhopadhyay, R.; Karmakar, A.; Chaudhury, S. 2D–3D cnn based architectures for spectral reconstruction from RGB images. In Proceedings of the IEEE Conference on Computer Vision and Pattern Recognition Workshops, Salt Lake City, UT, USA, 18–22 June 2018; pp. 844–851.
34. Shi, Z.; Chen, C.; Xiong, Z.; Liu, D.; Wu, F. Hscnn+: Advanced cnn-based hyperspectral recovery from RGB images. In Proceedings of the IEEE Conference on Computer Vision and Pattern Recognition Workshops, Salt Lake City, UT, USA, 18–22 June 2018; pp. 939–947.
35. Arad, B.; Timofte, R.; Ben-Shahar, O.; Lin, Y.T.; Finlayson, G.D. Ntire 2020 challenge on spectral reconstruction from an RGB image. In Proceedings of the IEEE/CVF Conference on Computer Vision and Pattern Recognition Workshops, Seattle, WA, USA, 14–19 June 2020; pp. 446–447.
36. Li, J.; Wu, C.; Song, R.; Li, Y.; Liu, F. Adaptive weighted attention network with camera spectral sensitivity prior for spectral reconstruction from RGB images. In Proceedings of the IEEE/CVF Conference on Computer Vision and Pattern Recognition Workshops, Seattle, WA, USA, 14–19 June 2020; pp. 462–463.
37. Zhao, Y.; Po, L.M.; Yan, Q.; Liu, W.; Lin, T. Hierarchical regression network for spectral reconstruction from RGB images. In Proceedings of the IEEE/CVF Conference on Computer Vision and Pattern Recognition Workshops, Seattle, WA, USA, 14–19 June 2020; pp. 422–423.

38. Peng, H.; Chen, X.; Zhao, J. Residual pixel attention network for spectral reconstruction from RGB images. In Proceedings of the IEEE/CVF Conference on Computer Vision and Pattern Recognition Workshops, Seattle, WA, USA, 14–19 June 2020; pp. 486–487.
39. Joslyn Fubara, B.; Sedky, M.; Dyke, D. RGB to Spectral Reconstruction via Learned Basis Functions and Weights. In Proceedings of the IEEE/CVF Conference on Computer Vision and Pattern Recognition Workshops, Seattle, WA, USA, 14–19 June 2020; pp. 480–481.
40. Banerjee, A.; Palrecha, A. MXR-U-Nets for Real Time Hyperspectral Reconstruction. *arXiv* **2020**, arXiv:2004.07003.
41. Nathan, D.S.; Uma, K.; Vinothini, D.S.; Bama, B.S.; Roomi, S. Light Weight Residual Dense Attention Net for Spectral Reconstruction from RGB Images. *arXiv* **2020**, arXiv:2004.06930.
42. Kaya, B.; Can, Y.B.; Timofte, R. Towards spectral estimation from a single RGB image in the wild. In Proceedings of the 2019 IEEE/CVF International Conference on Computer Vision Workshop (ICCVW), Seoul, Korea, 27–28 October 2019; pp. 3546–3555.
43. Zhang, L.; Lang, Z.; Wang, P.; Wei, W.; Liao, S.; Shao, L.; Zhang, Y. Pixel-Aware Deep Function-Mixture Network for Spectral Super-Resolution. In Proceedings of the AAAI, New York, NY, USA, 7–12 February 2020; pp. 12821–12828.
44. Hang, R.; Li, Z.; Liu, Q.; Bhattacharyya, S.S. Prinet: A Prior Driven Spectral Super-Resolution Network. In Proceedings of the 2020 IEEE International Conference on Multimedia and Expo (ICME), London, UK, 6–10 July 2020; pp. 1–6.
45. Li, J.; Wu, C.; Song, R.; Xie, W.; Ge, C.; Li, B.; Li, Y. Hybrid 2-D-3-D Deep Residual Attentional Network With Structure Tensor Constraints for Spectral Super-Resolution of RGB Images. *IEEE Trans. Geosci. Remote Sens.* **2020**, 1–15. [CrossRef]
46. Chen, L.; Zhang, H.; Xiao, J.; Nie, L.; Shao, J.; Liu, W.; Chua, T.S. SCA-CNN: Spatial and Channel-Wise Attention in Convolutional Networks for Image Captioning. In Proceedings of the IEEE Conference on Computer Vision and Pattern Recognition (CVPR), Honolulu, HI, USA, 21–26 July 2017.
47. Hu, J.; Shen, L.; Sun, G. Squeeze-and-excitation networks. In Proceedings of the IEEE Conference on Computer Vision and Pattern Recognition Workshops, Salt Lake City, UT, USA, 18–22 June 2018; pp. 7132–7141.
48. Wang, X.; Girshick, R.; Gupta, A.; He, K. Non-local neural networks. In Proceedings of the IEEE Conference on Computer Vision and Pattern Recognition Workshops, Salt Lake City, UT, USA, 18–22 June 2018; pp. 7794–7803.
49. Zhang, Y.; Li, K.; Li, K.; Wang, L.; Zhong, B.; Fu, Y. Image super-resolution using very deep residual channel attention networks. In Proceedings of the European Conference on Computer Vision (ECCV), Munich, Germany, 8–14 September 2018; pp. 286–301.
50. Dai, T.; Cai, J.; Zhang, Y.; Xia, S.T.; Zhang, L. Second-order attention network for single image super-resolution. In Proceedings of the IEEE Conference On Computer Vision And Pattern Recognition, Long Beach, CA, USA, 15–20 June 2019; pp. 11065–11074.
51. Dai, T.; Zha, H.; Jiang, Y.; Xia, S.T. Image Super-Resolution via Residual Block Attention Networks. In Proceedings of the IEEE International Conference on Computer Vision Workshops, Seoul, Korea, 27–28 October 2019.
52. Xia, B.N.; Gong, Y.; Zhang, Y.; Poellabauer, C. Second-order non-local attention networks for person re-identification. In Proceedings of the IEEE International Conference on Computer Vision Workshops, Seoul, Korea, 27–28 October 2019; pp. 3760–3769.
53. Zhang, Z.; Lan, C.; Zeng, W.; Jin, X.; Chen, Z. Relation-Aware Global Attention for Person Re-identification. In Proceedings of the IEEE/CVF Conference on Computer Vision and Pattern Recognition, Seattle, WA, USA, 14–19 June 2020; pp. 3186–3195.
54. Zuniga, O.A.; Haralick, R.M. Integrated directional derivative gradient operator. *IEEE Trans. Syst. Man, Cybern.* **1987**, *17*, 508–517. [CrossRef]
55. Kingma, D.P.; Ba, J. Adam: A method for stochastic optimization. *arXiv* **2014**, arXiv:1412.6980.

remote sensing

MDPI

Article

Linear and Non-Linear Models for Remotely-Sensed Hyperspectral Image Visualization

Radu-Mihai Coliban *, Maria Marincaş, Cosmin Hatfaludi and Mihai Ivanovici

Electronics and Computers Department, Transilvania University of Braşov, 500036 Braşov, Romania; maria.marincas@student.unitbv.ro (M.M.); cosmin.hatfaludi@student.unitbv.ro (C.H.); mihai.ivanovici@unitbv.ro (M.I.)

* Correspondence: coliban.radu@unitbv.ro

Received: 29 June 2020; Accepted: 30 July 2020; Published: 2 August 2020

Abstract: The visualization of hyperspectral images still constitutes an open question and may have an important impact on the consequent analysis tasks. The existing techniques fall mainly in the following categories: band selection, PCA-based approaches, linear approaches, approaches based on digital image processing techniques and machine/deep learning methods. In this article, we propose the usage of a linear model for color formation, to emulate the image acquisition process by a digital color camera. We show how the choice of spectral sensitivity curves has an impact on the visualization of hyperspectral images as RGB color images. In addition, we propose a non-linear model based on an artificial neural network. We objectively assess the impact and the intrinsic quality of the hyperspectral image visualization from the point of view of the amount of information and complexity: (i) in order to objectively quantify the amount of information present in the image, we use the color entropy as a metric; (ii) for the evaluation of the complexity of the scene we employ the color fractal dimension, as an indication of detail and texture characteristics of the image. For comparison, we use several state-of-the-art visualization techniques. We present experimental results on visualization using both the linear and non-linear color formation models, in comparison with four other methods and report on the superiority of the proposed non-linear model.

Keywords: hyperspectral imaging; visualization; color formation models

1. Introduction

Hyperspectral imaging captures high-resolution spectral information covering the visible and the infrared wavelength spectra, and thus can provide a high-level understanding of the land cover objects [1]. It is used in a wide variety of applications, such as agriculture [2,3], forest management [4,5], geology [6,7] and military/defense applications [8,9]. Human interaction with hyperspectral images is very important for image interpretation and analysis as the visualization is very often the first step in an image analysis chain [10]. However, displaying a hyperspectral image poses the problem of reducing the large number of bands to just three color RGB channels in order for it to be rendered on a monitor, with the information being meaningful from a human point of view. In order to address this problem, a series of hyperspectral image visualization techniques have been developed, which can be included in the following broad categories: band selection, PCA-based approaches, linear approaches, approaches based on digital image processing techniques and machine/deep learning methods.

Band selection methods consist of a mechanism of picking three spectral channels from the hyperspectral image and mapping them as the red, green and blue channels in the color composite. Commercial geospatial image analysis software products such as ENVI [11] offer the possibility to visualize a hyperspectral image by manually selecting the three channels to be displayed. More complex unsupervised band selection approaches have been developed, based on the one-bit transform (1BT) [12], normalized information (NI) [13], linear prediction (LP) or the minimum endmember abundance covariance (MEAC) [14].

Another family of hyperspectral visualization techniques consists of methods that use principal component analysis (PCA) for dimension reduction of the data. A straightforward visualization technique is to map a set of three principal components (usually the first three) to the R, G and B channels of the color image [15]. Other methods use PCA as part of a more complex approach. For instance, the method presented in [16] is an interactive visualization technique based on PCA, followed by convex optimization. The authors of [17] obtain the color composite by fusing the spectral bands with saliency maps obtained before and after applying PCA. In [1], the image is first decomposed into two different layers (base and detail) through edge-preserving filtering; dimension reduction is achieved through PCA applied on the base layer and a weighted averaging-based fusion on the detail layer, with the final result being a combination of the two layers.

In the case of the linear method described in [18,19], the values of each output color channel are computed as projections of the hyperspectral pixel values on a vector basis. Examples of such bases include one consisting of a stretched version of the CIE 1964 color matching functions (CMFs), a constant-luma disc basis or an unwrapped cosine basis.

A set of hyperspectral image visualization approaches are based on digital image processing techniques. In [20], dimension reduction is achieved using multidimensional scaling, followed by detail enhancement using a Laplacian pyramid. The approach presented in [21] uses the averaging method in order to the number of bands to 9; a decolorization algorithm is then applied on groups of three adjacent channels, which produces the final color image. The technique described in [22] is based on t-distributed stochastic neighbor embedding (t-SNE) and bilateral filtering. The method in [23] is also based on bilateral filtering, together with high dynamic range (HDR) processing techniques, while in [24] a pairwise-distances-analysis-driven visualization technique is described.

Machine/deep learning-based methods used for hyperspectral image visualization generally rely on a geographically-matched RGB image, either obtained through band selection or captured by a color image sensor. Approaches include constrained manifold learning [25], a method based on self-organizing maps [26], a moving least squares framework [10], a technique based on a multichannel pulse-coupled neural network [27] or methods based on convolutional neural networks (CNNs) [28,29].

In this paper, our goal is to produce natural-looking visualization results (i.e., depicting colors close to the real ones in the scene) with the highest possible amount of information and complexity. We propose the usage of a linear color formation model based on a widely-used linear model in colorimetry, based on spectral sensitivity curves. We study the impact on visualization of the choice of spectral sensitivity curves and the amount of overlapping between them, which induces the correlation between the three color channels used for visualization. Besides Gaussian functions, we use spectral sensitivity functions of digital camera sensors, the main idea behind the approach being to emulate the result of capturing the scene with a consumer-grade digital camera sensor instead of a hyperspectral one. Alternatively, we also developed a non-linear visualization method based on an artificial neural network, trained using the spectral signatures of a 24-sample color checker, also often used in colorimetry. By using the proposed approaches, we address the following question: what is the impact of the choice of visualization technique on the amount of information and complexity of a scene? The amount of information in a hyperspectral image should be preserved as much as possible after the visualization. The entropy is often used to measure the amount of information contained by a signal [30] and is one of the metrics that are used for the objective assessment of the visualization result [10,21,31]. The complexity of a scene is related to the texture and object characteristics preservation in the process of visualization. The color fractal dimension is a multi-scale measure capable of globally assessing the complexity of a color image, which can be useful to evaluate both the amount of detail and the object-level content in the image. We perform both a qualitative and a quantitative evaluation (using color entropy and color fractal dimension) of the described techniques in comparison with four other state-of-the-art methods, employing five widely used hyperspectral test images.

The rest of the paper is organized as follows: Section 2 presents the five hyperspectral images used in our experiments, the proposed approaches (both linear and non-linear) and the two embraced measures for the objective evaluation of the performance of the proposed approaches, Section 3 depicts the experimental results, Section 4 the discussion on the various aspects related to the proposed approaches, as well as possible further investigation paths, and Section 5 presents our conclusions.

2. Data and Methods

In this section we briefly describe the five hyperspectral images used in our experiments, the linear and non-linear models proposed and used to visualize the respective hyperspectral images, as well as the two quality metrics deployed to objectively evaluate the experimental results—the color entropy and the color fractal dimension.

2.1. Hyperspectral Images

The hyperspectral images used in our experiments are Pavia University, Pavia Centre, Indian Pines, SalinasA and Cuprite [32]. The first two were acquired by the ROSIS-3 sensor [33], while the other three were acquired by the AVIRIS sensor [34]. Figure 1 depicts RGB representations of the five test images.

Pavia University (Figure 1a) is a 610 $\times$ 340 image, with a resolution of 1.3 m. The image has 103 bands in the 430–860 nm range. The scene in the image contains a number of 9 materials according to the provided ground truth, both natural and man-made. Pavia Centre (Figure 1b) is a 1096 $\times$ 715, 102-band image with the same characteristics as Pavia University. In both cases, the 10th, 31st and 46th bands were used for generating the RGB representations [25].

The third test image, Indian Pines (Figure 1c), is a 145 $\times$ 145 image, having 224 spectral reflectance bands in the 400–2500 nm range with a 20 m resolution. The water absorption bands were removed, resulting in a total of 200 bands. The image contains 16 classes, mostly vegetation/crops.

SalinasA (Figure 1d), is an 86 $\times$ 83 sub-scene of the Salinas image. After removing the water absorption bands, the image has 204 spectral reflectance bands in the 400–2500 nm range with a spatial resolution of 3.7 m. This image exhibits 6 types of agricultural crops.

The fifth image, Cuprite (Figure 1e), is of size 512 $\times$ 614, with 188 spectral reflectance bands in the 400–2500 nm range remaining after removing noisy and water absorption channels. This image contains 14 types of minerals.

For the last three images, the RGB representations were generated by selecting the 6th, 17th, and 36th bands [25].

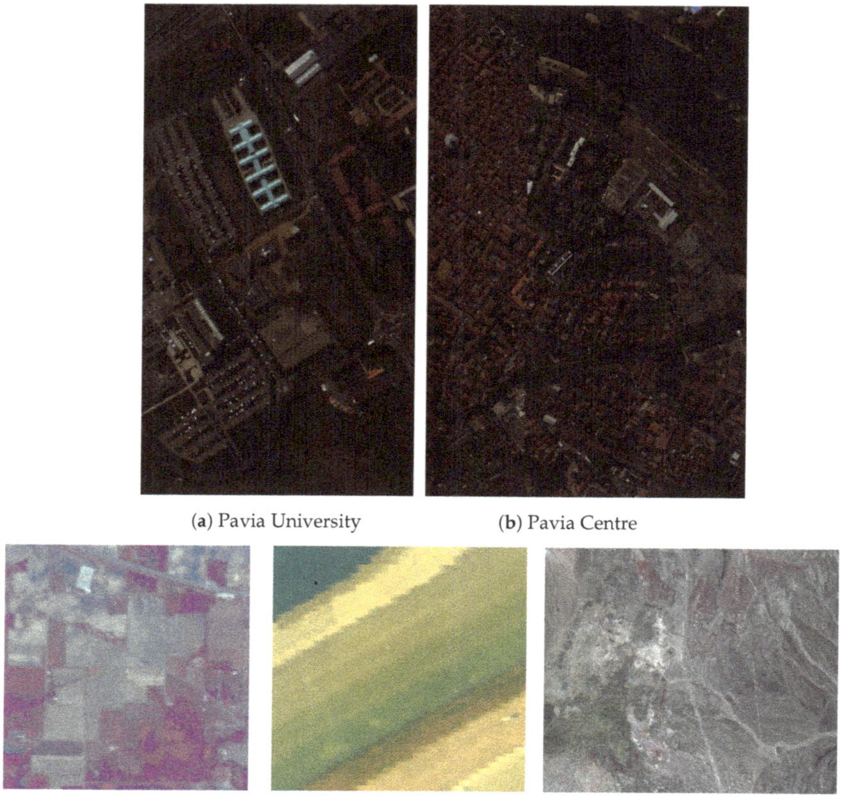

(a) Pavia University (b) Pavia Centre

(c) Indian Pines (d) SalinasA (e) Cuprite

Figure 1. RGB representations of the five hyperspectral images used in our experiments. Top row: images acquired by the ROSIS-3 sensor; bottom row: images acquired by the AVIRIS sensor.

2.2. Linear Color Formation Model

Considering the formation process of an RGB image, we embraced a linear model given by Equation (1) [35]. In colorimetry, the linear model is used as a standard model for the color formation, but usually the XYZ coordinates of colors are used as an intermediate step before computing the RGB final color coordinates [36]. In the embraced approach, for a pixel at any position (x, y) in the resulting RGB color image, the scalar value on each channel of the RGB triplet is computed as the integral of the product between the spectral reflectance $R(\lambda)$ of the (x, y) point in the real scene, the power spectral distribution $L(\lambda)$ of the illuminant and the spectral sensitivity $C(\lambda)$ of the imaging sensor:

$$I_k(x, y) = \int_{\lambda_{min}}^{\lambda_{max}} C_k(\lambda) L(\lambda) R_{(x,y)}(\lambda) d\lambda, \quad k = R, G, B \tag{1}$$

For the spectral sensitivity curves of the imaging sensor one can use theoretical or ideal curves, in order to simulate the image formation process. An alternative would be to use the actual sensitivity curves of a specific sensor, which can be measured according to the approach proposed in [35].

The illuminant can be also characterized, either by considering a standard illuminant or measuring the real one by means of spectrophotometry. In colorimetry, a D65 illuminant is very often preferred, as it corresponds to a bright summer day light. For remotely-sensed images, one may know the illuminant as the direct sun light incident to the Earth's surface, as the position of the sun with respect to the position of the satellite is known. The use of the illuminant in the model from Equation (1) represents merely an unbalanced weighting of the three sensitivities, favoring the blue channel (lower wavelengths) over green and red. The classical D65 illuminant is depicted in Figure 2, in support of this statement. However, in this article we assume that the illuminant is constant across all wavelengths, as we are mostly interested in the effect of the image sensor sensitivity curves on the vizualization process. Thus, the influence of the illuminant $L(\lambda)$ in Equation (1) is basically null and it can be removed from the integral. Consequently, the equation is basically reduced to the following:

$$I_k(x,y) = \int_{\lambda_{min}}^{\lambda_{max}} C_k(\lambda) R_{(x,y)}(\lambda) d\lambda, \quad k = R, G, B \tag{2}$$

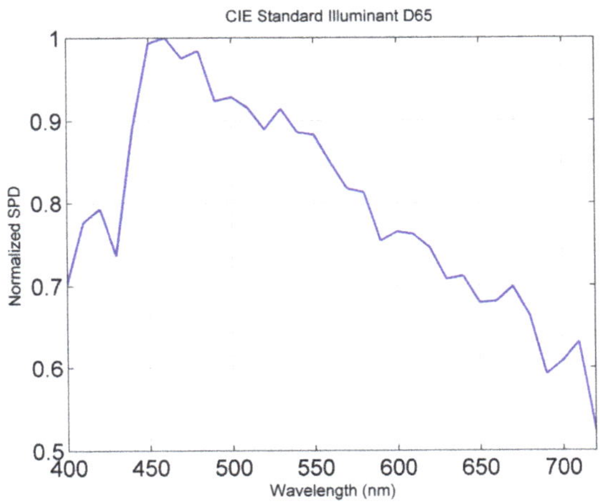

Figure 2. The D65 illuminant.

This is the linear model we consider for the experimental results presented in Section 3. In order to apply Equation (2) on a hyperspectral image, we extract from it only the bands corresponding to the range $[\lambda_{min}, \lambda_{max}]$, covered by the sensitivity curves, which corresponds to the visible spectrum. This is the main difference between the proposed model and the linear model presented in [18], which uses all of the bands of the hyperspectral image and the weighting functions are stretched in order to cover the entire range of wavelenghts of the hyperspectral image. Since both the sensitivity functions and the reflectances are discrete, an interpolation of the pixel values of the hyperspectral image is done in order to match the wavelengths and number of values of the sensitivity functions.

Given the embraced linear model and sensitivity functions, our study is limited to the visible spectrum. The extension beyond the visible range could be done either by (i) stretching the sensitivity functions [18] or (ii) adding a fourth color channel, given that one of the latest trends in color display technologies is to add a fourth channel (such as a yellow channel) besides the RGB primaries [37]. However, both approaches would lead to unnatural-looking visualization results, which is not the goal of this study.

2.3. Spectral Sensitivity Functions

As the main objective of visualization is very often the interpretation of the image by humans, we start by considering the spectral sensitivity of the human visual system, which is actually the paradigm for RGB-based color image acquisition and display systems. Figure 3 presents the spectral sensitivities of the human cone cells in the retina, based on the data from [38]. The spectral sensitivity is a function of the wavelength of signal relative to detection of color. These spectral sensitivities are labeled in three categories, depending on the peak value: short (S), medium (M) and long (L). The cone cells are called β for the S group with the range that corresponds to the perception of the blue color. Similarly, the range of the M group (γ cells) corresponds to green and the L group (ρ) corresponds to red.

The RGB color digital cameras are characterized by their sensor spectral sensitivity functions, which define the performance of the respective system. The sensor sensitivity functions for consumer-grade cameras have a similar shape to the spectral sensitivities of human cone cells, since the aim of these products is to capture a representation of the scene that is as accurate as possible from the point of view of human perception. The five digital camera sensor spectral sensitivity functions used in our experiments, taken from [35], are presented in Figure 4.

Starting from the spectral sensitivities of the Canon 5D camera sensor, for our experiments we modeled a set of spectral sensitivities consisting of three Gaussian functions with the mean equal to the wavelength corresponding to the three peaks in Figure 4a and with increasing standard deviation. The functions are depicted in Figure 5. Figure 5a depicts Gaussian functions with a standard deviation of 0, which represent basically unit impulses. In this case, the linear model is reduced to a band selection approach (BS). The standard deviation is gradually increased in the next graphs, resulting in an increasing degree of overlapping between the three functions: *no* overlap (NOL), *small* overlap (SOL), *medium* overlap (MOL) and *high* overlap (HOL). In this way, we emulate the various levels of correlation between the three RGB color channels of the considered sensor model—from zero correlation, corresponding to a complete separation between the color channels for an ideal imaging sensor, to high overlap, corresponding to a low-performance imaging sensor.

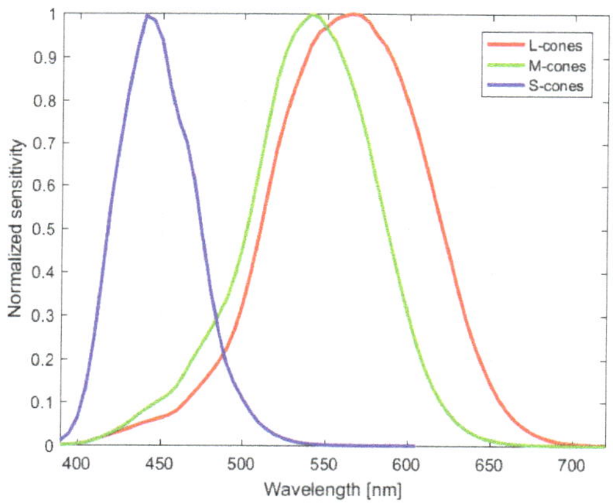

Figure 3. Spectral sensitivities of human cone cells.

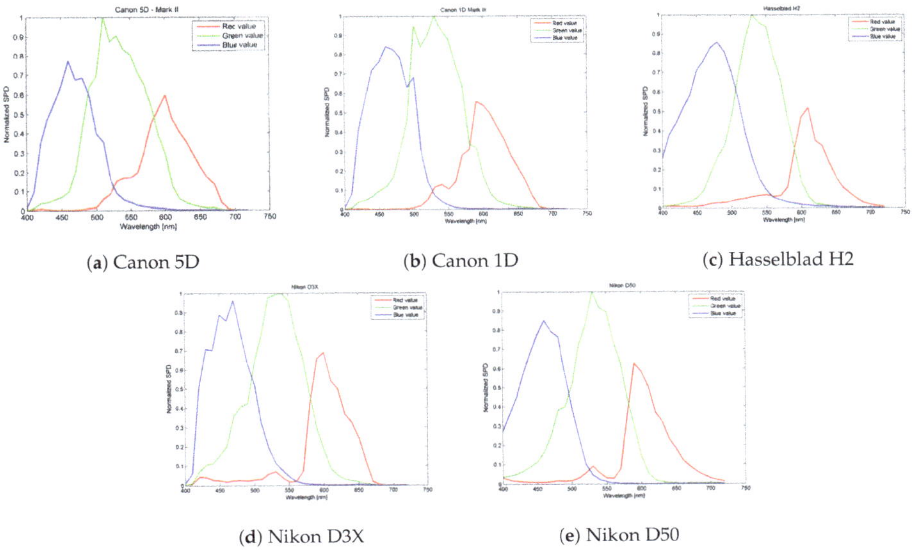

(**a**) Canon 5D (**b**) Canon 1D (**c**) Hasselblad H2

(**d**) Nikon D3X (**e**) Nikon D50

Figure 4. Spectral sensitivity functions for 5 digital cameras.

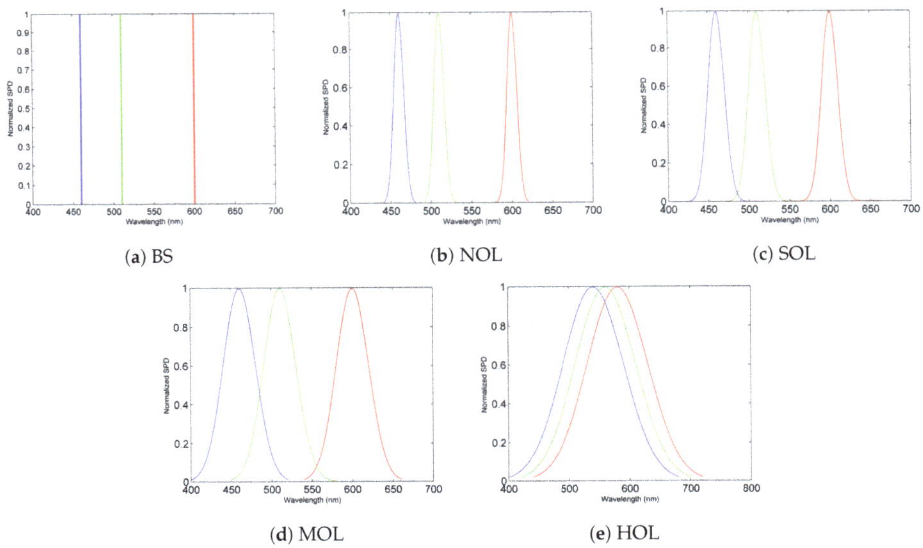

(**a**) BS (**b**) NOL (**c**) SOL

(**d**) MOL (**e**) HOL

Figure 5. Gaussian spectral sensitivity functions based on the functions of the Canon 5D camera from Figure 4a.

2.4. Non-Linear Color Formation Model

The non-linear color formation model that we propose is based on an Artificial Neural Network (ANN) [39], with the input feature vector consisting of a spectral reflectance curve and the output being the corresponding RGB value. The architecture of the fully connected 5-layer network is depicted in Figure 6. The network uses the Exponential Linear Unit (ELU) [40] as an activation function instead of the more standard Rectified Linear Unit (ReLU), in order to overcome the problem of having a multitude of deactivated neurons (also referred to as "dying neurons" [41]). The implementation was done using the PyTorch library [42].

For the supervised training of the ANN, we chose to use a standard set of 24 colors widely-used in colorimetry–the McBeth color chart [43], depicted in Figure 7. In Figure 8 we show the spectral reflectance curves of the color patches for each row in the McBeth color chart, with their original designations in the legend of the plots. For each color, the RGB triplet is known and we used the measurements provided by [44]. The wavelength range covered by the reflectance curves is 380–780 nm. The reason for choosing this McBeth standard color set is twofold: (i) the spectral reflectance curves of the colors are specified regardless of the illuminant, therefore they can be used as references both in ideal or real conditions; and (ii) this particular color set was determined independently from the domain of remote sensing, thus it can be seen as a neutral set of colors compared to the existing data set of material spectral signatures, such as the ASTER spectral library [45]. In addition, the chosen color set does not require the mapping between the spectral curves and corresponding RGB colors. The training of the ANN is done via the classical backpropagation algorithm, with the mean squared error (MSE) being used as a cost function and Adam used as the optimizer.

As in the case of the linear model, only the bands covered by the spectral reflectance curves of the McBeth color set are used from the hyperspectral image. Concretely, the common range between the Pavia University image and the McBeth curves is 430–780 nm. This range is covered by 83 bands of the image and 71 values of the spectral reflectance curves. The 83 bands of the image are reduced to 71 through interpolation, such as to match the McBeth spectral reflectance curves, giving the size of the input feature vector in Figure 6.

After training with the 24 reflectance curves, the network is applied on a pixel-by-pixel basis; thus, for each pixel in the input image (a vector of 71 values in the case of Pavia University), the 3 output values (R, G and B) are obtained and placed in the corresponding position in the visualization result.

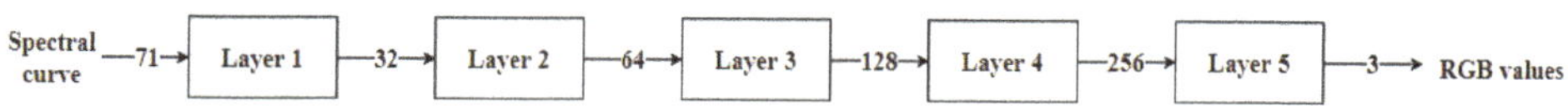

Figure 6. Architecture of the ANN.

Figure 7. The McBeth color chart.

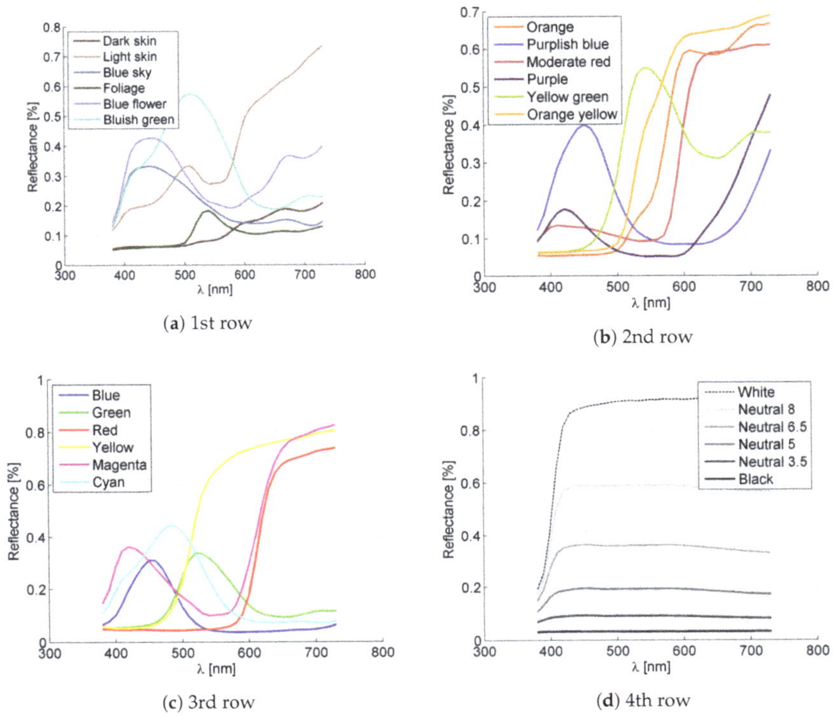

(**a**) 1st row (**b**) 2nd row (**c**) 3rd row (**d**) 4th row

Figure 8. Spectral reflectance curves of the color patches in each row of the McBeth color chart.

2.5. Quality Metrics

A commonly used objective quality metric for hyperspectral image visualization is the entropy, which is a measure of the degree of information preservation in the resulting image [1]. The most common definition of entropy is the Shannon entropy (see Equation (3)) which measures the average level of information present in a signal with N quantization levels [30].

$$H = -\sum_{i=1}^{N} p_i \log_2 p_i \tag{3}$$

where p_i represents the probability to find a certain level in the signal (or color i in a given subset, in context of color images). From the Shannon definition, various other definitions were developed: Rényi entropy (as a generalization), Hartley entropy, collision entropy and min-entropy, or the Kolmogorov entropy, which is another generic definition of entropy [46]. The original Shannon entropy was embraced by Haralick as one of his thirteen features proposed for texture characterization [47]. In our experiments, we use the extension of the entropy to color images from [48].

Additionally, we use the fractal dimension from fractal geometry [49] to assess the complexity of the color images resulting in the process of hyperspectral image visualization. The fractal dimension, also called similarity dimension, is a measure of the variations, irregularities or *wiggliness* of a fractal object [50]. This multi-scale measure is often used in practice for the discrimination between various signals or patterns exhibiting fractal properties, such as textures [51]. In [52] the fractal dimension was linked to the visual complexity of a color image, more specifically to the perceived beauty of the visual art. Consequently, we use it in this article to both objectively assess the color image content at multiple scales and the appealing of the visualization from a human perception point of view.

The theoretical fractal dimension is the Hausdorff dimension [53], which is comprised in the interval $[E, E+1]$, where E is the topological dimension of that object (thus, for gray-scale images the fractal dimension is comprised between 2 and 3.). Because it was defined for continuous objects, equivalent fractal dimension estimates were defined and used: the probability measure [54,55], the Minkowski or box-counting dimension [53], the δ-parallel body method [56], the gliding box-counting algorithm [57] etc. The fractal dimension estimation was extended to the color image domain, like the marginal color analysis [58] or the fully vectorial probabilistic box-counting [59]. More recent attempts in defining the fractal dimension for color images exist [60,61]. For an RGB color image, the estimated color fractal dimension should be comprised in the interval $[2, 5]$ [59].

In our experiments, we used the probabilistic box-counting approach defined color images in [59] for the estimation of the fractal dimension of the visualization results. The classical box-counting method consists of covering the image with grids at different scales and counting the number of boxes that cover the image pixels in each grid. The fractal dimension FD is then computed as [62]:

$$FD = \lim_{r \to 0} \frac{\log N_r}{\log r} \tag{4}$$

where N_r is the number of boxes and r is the scale.

FD is defined and computed for binary and grayscale images (considering the $z = f(x, y)$ image model, where z is the luminance and x and y are the spatial coordinates). The extension of FD to color images, the color fractal dimension (CFD), is defined by considering the color image as a surface in a 5-dimensional hyperspace ($RGBxy$) [59] and 5D hyper-boxes instead of 3D regular ones. For the experimental results presented in Section 3, the stable CFD estimator proposed in [63] was used, which minimizes the variance of the nine regression line estimators used in the process of fractal dimension estimation. See [64] for reference color fractal images and the Matlab implementation of the baseline CFD estimation approach.

3. Experimental Results

Figures 9–13 depict the visualization results for the five hyperspectral test images presented in Section 2. Each figure is organized as follows: on the top row, the results obtained with the proposed linear approach using the Gaussian functions (Figure 5); on the middle row, the results obtained with the linear approach using camera spectral sensitivity functions (Figure 4); on the bottom row, the results obtained using the proposed ANN approach (Section 2.4), the approach based on the PCA to RGB mapping [15], the linear approach based on the stretched color matching functions (CMF) [18] and two recent approaches, constrained manifold learning (CML) [25] and decolorization-based hyperspectral visualization (DHV) [21].

For the Gaussian approaches, it can be noticed that, as the degree of overlapping between the three functions increases, the vizualization results tend to come closer to grayscale images, as expected. In the case of the camera functions, the difference between the results is not significant, proving that the choice of a particular camera model over the other does not have a large impact on the visualization results. Moreover, there is no significant difference in the visualization results between the two cases of the proposed linear approach. The proposed ANN approach obtains satisfying results in terms of both color and contrast, while the other depicted methods, particularly PCA and DHV, do not tend to give natural-looking results.

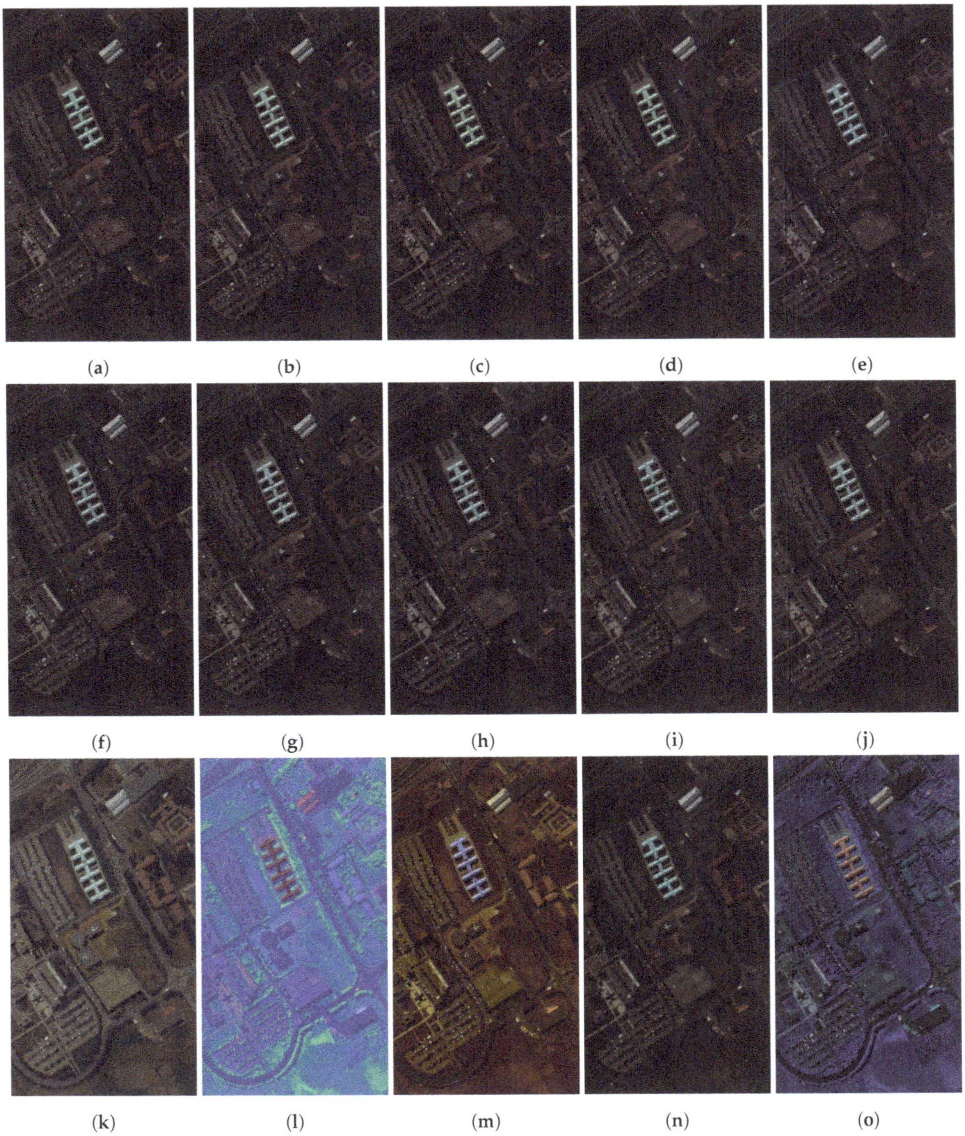

Figure 9. Experimental results on the Pavia University image. (a) BS. (b) NOL. (c) SOL. (d) MOL. (e) HOL. (f) Canon 5D. (g) Canon 1D. (h) Hasselblad H2. (i) Nikon D3X. (j) Nikon D50. (k) ANN. (l) PCA [15]. (m) CMF [18]. (n) CML [25]. (o) DHV [21].

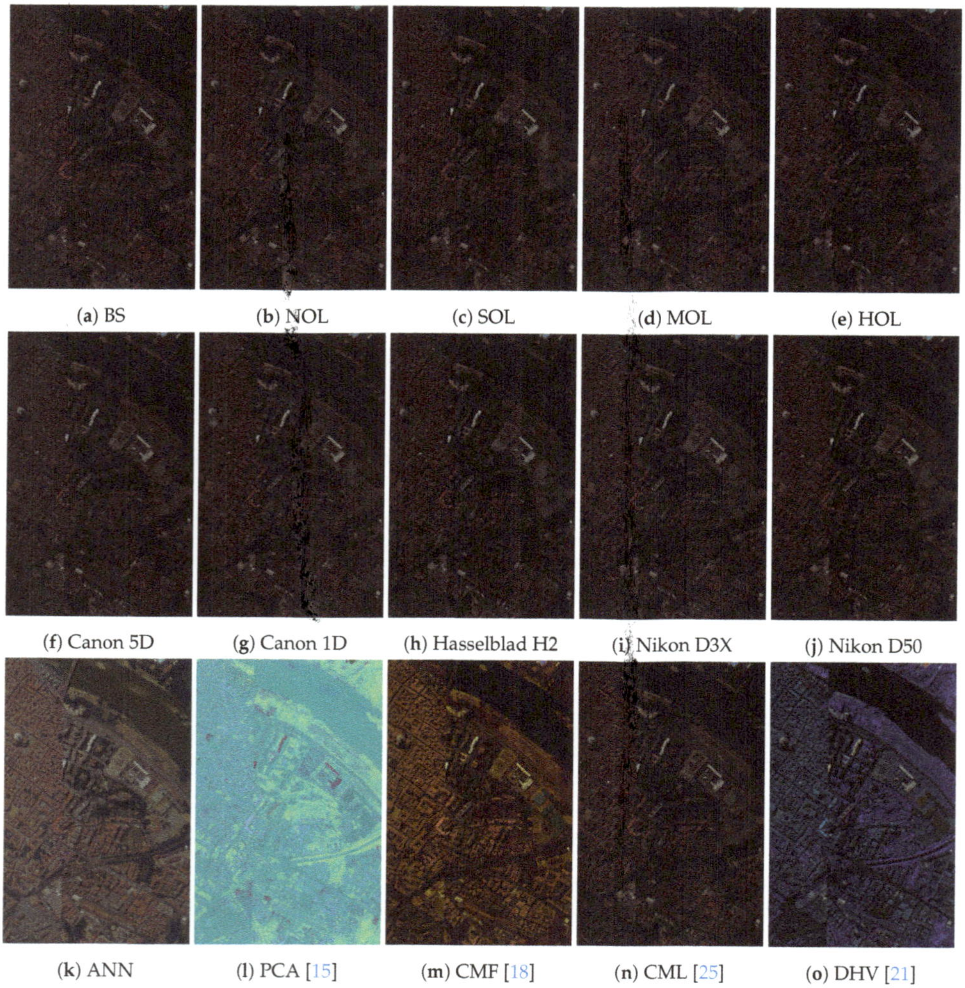

Figure 10. Experimental results on the Pavia Centre image. (**a**) BS. (**b**) NOL. (**c**) SOL. (**d**) MOL. (**e**) HOL. (**f**) Canon 5D. (**g**) Canon 1D. (**h**) Hasselblad H2. (**i**) Nikon D3X. (**j**) Nikon D50. (**k**) ANN. (**l**) PCA [15]. (**m**) CMF [18]. (**n**) CML [25]. (**o**) DHV [21].

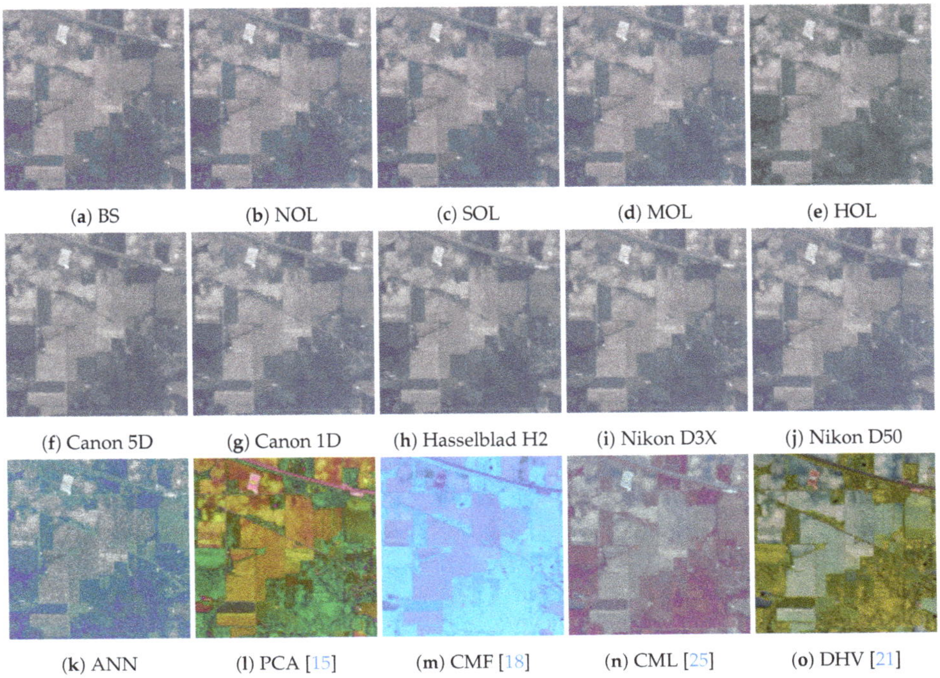

Figure 11. Experimental results on the Indian Pines image. (**a**) BS. (**b**) NOL. (**c**) SOL. (**d**) MOL. (**e**) HOL. (**f**) Canon 5D. (**g**) Canon 1D. (**h**) Hasselblad H2. (**i**) Nikon D3X. (**j**) Nikon D50. (**k**) ANN. (**l**) PCA [15]. (**m**) CMF [18]. (**n**) CML [25]. (**o**) DHV [21].

The corresponding values for the color entropy H and color fractal dimension CFD are depicted in Tables 1 and 2. One may note that, for the set of Linear Gaussian approaches, both the color entropy and color fractal dimension are maximum for the band selection, with one exception for the SalinasA image, and they both decrease with the increase of the correlation between the three Gaussian functions, as the color content tends to gray-scale and thus complexity diminishes. For the set of Linear Camera proposed approaches, the two quality measures have similar values, basically there is no noticeable difference in the visualization results. For both the Linear Gaussian and Linear Camera approaches, the two quality measures exhibit relatively modest values, which indicate that the visualization result does neither contain the highest information, nor is the most complex. The highest amount of information, measured through the color entropy, is obtained using the proposed non-linear ANN approach for the Pavia University and Pavia Centre images, the PCA approach for the Indian Pines and Cuprite images, and DHV for the SalinasA image. For the three latter images, the proposed ANN-based non-linear approach obtains the third (Indian Pines, Cuprite) and second (SalinasA) best visualization from the point of view of entropy. The highest complexity, measured through the color fractal dimension, is revealed when the hyperspectral images are visualized using the non-linear approach based on ANN, with the exception of the Cuprite image, in which case the PCA approach proves to be superior. The main advantage of the ANN method is that basically any out-of-the-box artificial neural network model can be used, by changing the input layer only in order to match the hyperspectral image under analysis. Table 3 lists, for each visualization method, the independent data used in addition to the hyperspectral images. In the case of the CML approach, the geographically-matched RGB image was obtained through band selection from the original image; the images used are depicted in Figure 1, while the specific bands chosen are listed in Section 2.1.

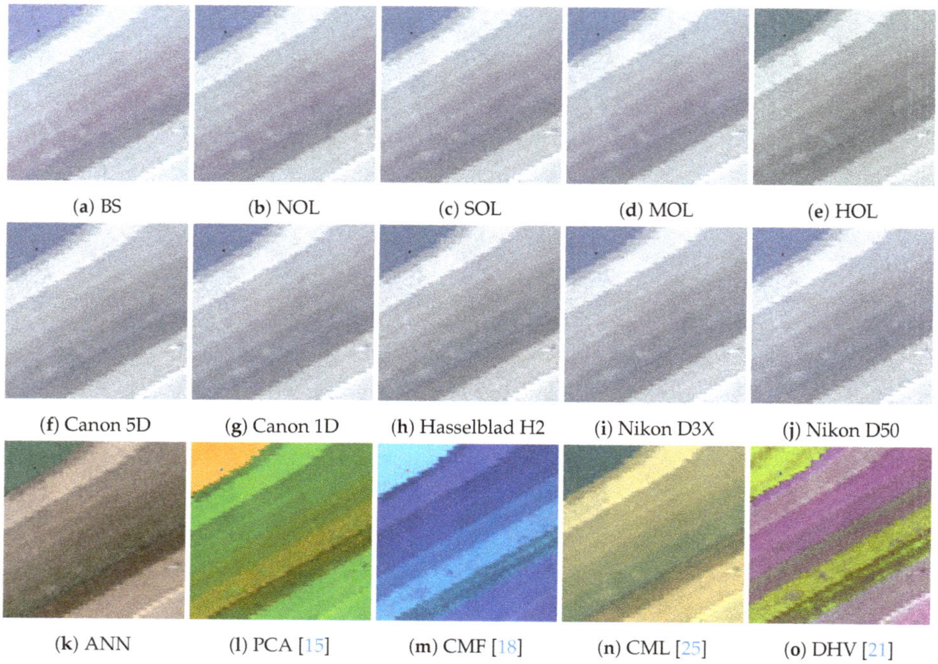

Figure 12. Experimental results on the SalinasA image. (**a**) BS. (**b**) NOL. (**c**) SOL. (**d**) MOL. (**e**) HOL. (**f**) Canon 5D. (**g**) Canon 1D. (**h**) Hasselblad H2. (**i**) Nikon D3X. (**j**) Nikon D50. (**k**) ANN. (**l**) PCA [15]. (**m**) CMF [18]. (**n**) CML [25]. (**o**) DHV [21].

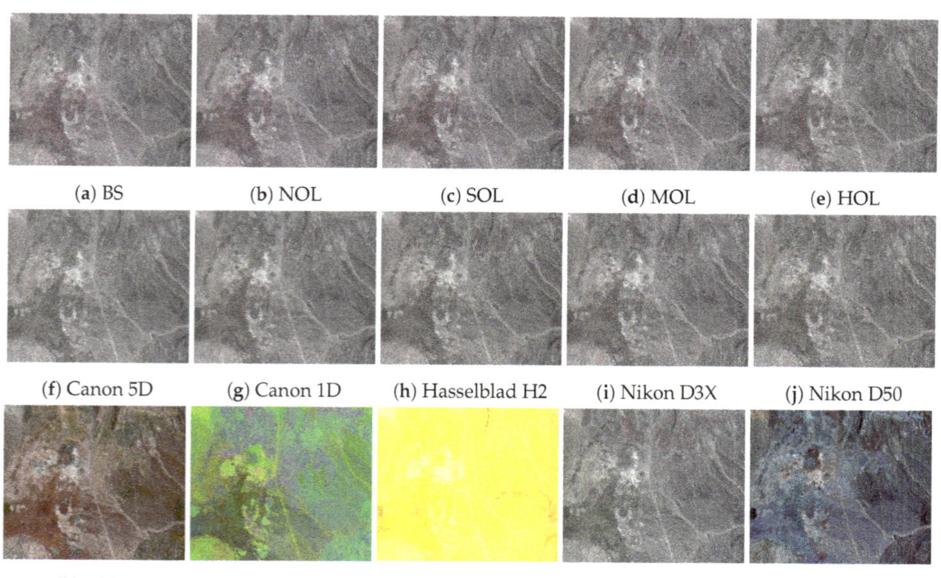

Figure 13. Experimental results on the Cuprite image. (**a**) BS. (**b**) NOL. (**c**) SOL. (**d**) MOL. (**e**) HOL. (**f**) Canon 5D. (**g**) Canon 1D. (**h**) Hasselblad H2. (**i**) Nikon D3X. (**j**) Nikon D50. (**k**) ANN. (**l**) PCA [15]. (**m**) CMF [18]. (**n**) CML [25]. (**o**) DHV [21].

Table 1. Entropy and fractal dimension for the visualization results in Figures 9 and 10. The values in bold represent the highest values for the respective image.

Method		Pavia University *H*	Pavia University *CFD*	Pavia Centre *H*	Pavia Centre *CFD*
Linear Gaussian (proposed)	BS	13.22	2.41	13.48	2.44
	NOL	12.81	2.39	13.08	2.39
	SOL	12.68	2.38	12.96	2.38
	MOL	12.45	2.37	12.70	2.37
	HOL	10.79	2.35	10.82	2.36
Linear Camera (proposed)	Canon 5D	12.03	2.37	12.10	2.36
	Canon 1D	12.23	2.37	12.31	2.37
	Hasselblad H2	12.25	2.38	12.25	2.37
	Nikon D3X	12.30	2.38	12.34	2.37
	Nikon D50	12.45	2.38	12.49	2.38
ANN (proposed)		**15.38**	**3.02**	**15.28**	**2.87**
PCA [15]		15.20	2.84	14.58	2.75
CMF [18]		14.98	2.51	15.11	2.63
CML [25]		12.79	2.80	12.94	2.63
DHV [21]		15.31	2.84	15.21	2.77

Table 2. Entropy and fractal dimension for the visualization results in Figures 11–13. The values in bold represent the highest values for the respective image.

Method		Indian Pines *H*	Indian Pines *CFD*	Salinas A *H*	Salinas A *CFD*	Cuprite *H*	Cuprite *CFD*
Linear Gaussian (proposed)	BS	13.22	2.41	12.01	2.38	13.71	2.84
	NOL	11.76	2.51	11.29	2.29	13.31	2.80
	SOL	11.53	2.56	11.16	2.28	13.19	2.79
	MOL	11.31	2.46	10.97	2.27	12.84	2.79
	HOL	10.29	2.42	9.84	2.36	10.99	2.73
Linear Camera (proposed)	Canon 5D	12.03	2.37	11.06	2.43	12.13	2.76
	Canon 1D	11.28	2.46	10.51	2.26	12.32	2.77
	Hasselblad H2	11.40	2.46	10.47	2.25	12.36	2.76
	Nikon D3X	11.34	2.45	10.52	2.25	12.35	2.77
	Nikon D50	11.50	2.46	10.69	2.27	12.56	2.78
ANN (proposed)		13.89	**3.24**	11.50	**2.75**	15.76	3.00
PCA [15]		**14.24**	3.19	11.04	2.38	**17.40**	**3.37**
CMF [18]		12.51	3.17	10.36	1.86	8.00	2.77
CML [25]		12.81	2.68	11.10	2.14	13.66	2.67
DHV [21]		14.13	3.03	**12.29**	2.44	16.22	3.06

Table 3. Independent data used by the methods under comparison.

Method	Independent Data
Linear Gaussian	Gaussian sensitivity functions (Figure 5)
Linear Camera	Camera sensitivity functions (Figure 4)
ANN	McBeth spectral reflectance curves (Figure 8)
PCA [15]	none
CMF [18]	Stretched CIE 1964 color matching functions
CML [25]	Geographically-matched RGB image (Figure 1)
DHV [21]	none

4. Discussion

First of all, other measures can be considered for the assessment of the complexity of color images, like the Naive Complexity Measure [65]. For the evaluation of the information present in a color image, one could use the Pearson correlation coefficient between the color channels of the resulting RGB color image [63] as an indication of the overlapping between the information on the three RGB color channels. In the presence of a reference or ground truth, similarity indexes like Structural Similarity Index Measure [66] can be used. Nevertheless, the ultimate criteria for the evaluation of the performance of the hyperspectral image visualization approaches are dictated by the specific application and its objectives.

The best experimental results were obtained using the proposed non-linear ANN-based model, despite the extremely reduced training set—only 24 spectral reflectance curves and the corresponding RGB triplets. One should investigate the effects of increasing the size of the training set, in order to assess and reduce the overfitting effect [67] which may occur in our experiments. Extending the training set implies the realization of more color references, characterized both by their hyperspectral signatures (e.g., by using a spectrophotometer) and RGB triplets (e.g., by using a calibrated digital color image acquisition system). The non-linear model itself could be developed further by considering the wavelengths outside the visible range and taking into account the possibility to display the image with more than 3 color channels, including various choices for the mapping between the hyperspectral signatures and RGB triplets.

The linear models used to obtain the experimental results can be useful in understanding both the capabilities and limitations of current or new imaging sensors. The full characterization of the imaging sensors is mandatory in order to predict the imaging process outcome.

5. Conclusions

In this article, we proposed the usage of a linear model for the color formation based on spectral sensitivity curves in order to visualize hyperspectral images by rendering them as RGB color images. We deployed both Gaussian and real digital camera sensitivity curves and showed that, as the correlation between the RGB color channels increases, similar to the overlapping of the curves for both the human visual system and commercially-available digital cameras, the resulting color images tend to go to gray-scale and to exhibit both a smaller amount of information and complexity. We also proposed a non-linear color formation model based on an artificial neural network which was trained with the colors of the McBeth color chart widely used in colorimetry. The training was supervised as the 24 colors of the McBeth chart are specified both by their spectral reflectance curves and RGB triplets. Given their construction, both proposed linear and non-linear approaches generate color images with natural colors.

For the objective assessment of the quality of the hyperspectral image visualization results, we deployed the widely-used measure of entropy, as it is an indicator of the amount of information contained by a signal. We also proposed the usage of the fractal dimension, which is a multi-scale measure usually employed to assess the complexity of color images, but also their beauty and appeal according to some studies. The fractal dimension is an indicator of the amount of details present in the image along multiple analysis scales.

In our experiments, we compared the proposed approaches with four other visualization techniques, using five remotely-sensed hyperspectral images. In the case of the Gaussian functions, our results show that, as the degree of overlapping between functions increases, the visualization results come closer to a grayscale image. With regards to the camera sensitivity functions, we show that the specific choice of a camera model does not have a significant impact on the visualization result. Our experiments also show that the proposed non-linear model achieves the best visualization results from the point of view of the complexity of the resulting color images. We envisage further development by investigating the possible overfitting effect occurring in the case of the ANN approach, extending the approach beyond the visible range and by using a fourth color channel. We underline that for the choice of the most appropriate visualization technique, one may need to consider three important aspects: the naturalness of the resulting colors, the amount of information present in the resulting color image and the complexity along multiple scales.

Author Contributions: Idea and methodology, M.I. and R.-M.C.; software, M.M., C.H. and R.-M.C.; investigation, M.I. and R.-M.C.; writing—original draft preparation, R.-M.C., M.M. and M.I.; writing—review and editing, R.-M.C.; supervision, M.I. All authors have read and agreed to the published version of the manuscript.

Funding: This research received no external funding.

Conflicts of Interest: The authors declare no conflict of interest.

References

1. Kang, X.; Duan, P.; Li, S. Hyperspectral image visualization with edge-preserving filtering and principal component analysis. *Inf. Fusion* **2020**, *57*, 130–143. [CrossRef]
2. Teke, M.; Deveci, H.S.; Haliloğlu, O.; Gürbüz, S.Z.; Sakarya, U. A short survey of hyperspectral remote sensing applications in agriculture. In Proceedings of the IEEE 2013 6th International Conference on Recent Advances in Space Technologies (RAST), Istanbul, Turkey, 12–14 June 2013; pp. 171–176.
3. Reshma, S.; Veni, S. Comparative analysis of classification techniques for crop classification using airborne hyperspectral data. In Proceedings of the IEEE 2017 International Conference on Wireless Communications, Signal Processing and Networking (WiSPNET), Chennai, India, 22–24 March 2017; pp. 2272–2276.
4. Piiroinen, R.; Heiskanen, J.; Maeda, E.; Viinikka, A.; Pellikka, P. Classification of tree species in a diverse African agroforestry landscape using imaging spectroscopy and laser scanning. *Remote Sens.* **2017**, *9*, 875. [CrossRef]
5. Fricker, G.A.; Ventura, J.D.; Wolf, J.A.; North, M.P.; Davis, F.W.; Franklin, J. A convolutional neural network classifier identifies tree species in mixed-conifer forest from hyperspectral imagery. *Remote Sens.* **2019**, *11*, 2326. [CrossRef]
6. Dumke, I.; Nornes, S.M.; Purser, A.; Marcon, Y.; Ludvigsen, M.; Ellefmo, S.L.; Johnsen, G.; Søreide, F. First hyperspectral imaging survey of the deep seafloor: High-resolution mapping of manganese nodules. *Remote Sens. Environ.* **2018**, *209*, 19–30. [CrossRef]
7. Acosta, I.C.C.; Khodadadzadeh, M.; Tusa, L.; Ghamisi, P.; Gloaguen, R. A machine learning framework for drill-core mineral mapping using hyperspectral and high-resolution mineralogical data fusion. *IEEE J. Sel. Top. Appl. Earth Obs. Remote Sens.* **2019**, *12*, 4829–4842. [CrossRef]
8. Shimoni, M.; Haelterman, R.; Perneel, C. Hypersectral Imaging for Military and Security Applications: Combining Myriad Processing and Sensing Techniques. *IEEE Geosci. Remote Sens. Mag.* **2019**, *7*, 101–117. [CrossRef]
9. El-Sharkawy, Y.H.; Elbasuney, S. Hyperspectral imaging: A new prospective for remote recognition of explosive materials. *Remote Sens. Appl. Soc. Environ.* **2019**, *13*, 31–38. [CrossRef]
10. Liao, D.; Chen, S.; Qian, Y. Visualization of Hyperspectral Images Using Moving Least Squares. In Proceedings of the IEEE 2018 24th International Conference on Pattern Recognition (ICPR), Beijing, China, 20–24 August 2018; pp. 2851–2856.
11. Available online: https://www.harrisgeospatial.com/Software-Technology/ENVI/ (accessed on 23 July 2020).
12. Demir, B.; Celebi, A.; Erturk, S. A low-complexity approach for the color display of hyperspectral remote-sensing images using one-bit-transform-based band selection. *IEEE Trans. Geosci. Remote Sens.* **2008**, *47*, 97–105. [CrossRef]
13. Le Moan, S.; Mansouri, A.; Voisin, Y.; Hardeberg, J.Y. A constrained band selection method based on information measures for spectral image color visualization. *IEEE Trans. Geosci. Remote Sens.* **2011**, *49*, 5104–5115. [CrossRef]
14. Su, H.; Du, Q.; Du, P. Hyperspectral imagery visualization using band selection. In Proceedings of the IEEE 2012 4th Workshop on Hyperspectral Image and Signal Processing: Evolution in Remote Sensing (WHISPERS), Shanghai, China, 4–7 June 2012; pp. 1–4.
15. Tyo, J.S.; Konsolakis, A.; Diersen, D.I.; Olsen, R.C. Principal-components-based display strategy for spectral imagery. *IEEE Trans. Geosci. Remote Sens.* **2003**, *41*, 708–718. [CrossRef]

16. Cui, M.; Razdan, A.; Hu, J.; Wonka, P. Interactive hyperspectral image visualization using convex optimization. *IEEE Trans. Geosci. Remote Sens.* **2009**, *47*, 1673–1684.
17. Khan, H.A.; Khan, M.M.; Khurshid, K.; Chanussot, J. Saliency based visualization of hyper-spectral images. In Proceedings of the 2015 IEEE International Geoscience and Remote Sensing Symposium (IGARSS), Milan, Italy, 26–31 July 2015; pp. 1096–1099.
18. Jacobson, N.P.; Gupta, M.R. Design goals and solutions for display of hyperspectral images. *IEEE Trans. Geosci. Remote Sens.* **2005**, *43*, 2684–2692. [CrossRef]
19. Jacobson, N.P.; Gupta, M.R.; Cole, J.B. Linear fusion of image sets for display. *IEEE Trans. Geosci. Remote Sens.* **2007**, *45*, 3277–3288. [CrossRef]
20. Fang, J.; Qian, Y. Local detail enhanced hyperspectral image visualization. In Proceedings of the 2015 IEEE International Geoscience and Remote Sensing Symposium (IGARSS), Milan, Italy, 26–31 July 2015; pp. 1092–1095.
21. Kang, X.; Duan, P.; Li, S.; Benediktsson, J.A. Decolorization-based hyperspectral image visualization. *IEEE Trans. Geosci. Remote Sens.* **2018**, *56*, 4346–4360. [CrossRef]
22. Zhang, B.; Yu, X. Hyperspectral image visualization using t-distributed stochastic neighbor embedding. In *MIPPR 2015: Remote Sensing Image Processing, Geographic Information Systems, and Other Applications*; Liu, J., Sun, H., Eds.; International Society for Optics and Photonics, SPIE: Bellingham, WA, USA, 2015; Volume 9815, pp. 14–21.
23. Ertürk, S.; Süer, S.; Koç, H. A high-dynamic-range-based approach for the display of hyperspectral images. *IEEE Geosci. Remote Sens. Lett.* **2014**, *11*, 2001–2004. [CrossRef]
24. Long, Y.; Li, H.C.; Celik, T.; Longbotham, N.; Emery, W.J. Pairwise-Distance-Analysis-Driven Dimensionality Reduction Model with Double Mappings for Hyperspectral Image Visualization. *Remote Sens.* **2015**, *7*, 7785–7808. [CrossRef]
25. Liao, D.; Qian, Y.; Tang, Y.Y. Constrained manifold learning for hyperspectral imagery visualization. *IEEE J. Sel. Top. Appl. Earth Obs. Remote Sens.* **2018**, *11*, 1213–1226. [CrossRef]
26. Jordan, J.; Angelopoulou, E. Hyperspectral image visualization with a 3-D self-organizing map. In Proceedings of the 2013 5th Workshop on Hyperspectral Image and Signal Processing: Evolution in Remote Sensing (WHISPERS), Gainesville, FL, USA, 26–28 June 2013; pp. 1–4.
27. Duan, P.; Kang, X.; Li, S.; Ghamisi, P. Multichannel pulse-coupled neural network-based hyperspectral image visualization. *IEEE Trans. Geosci. Remote Sens.* **2019**, *58*, 2444–2456. [CrossRef]
28. Duan, P.; Kang, X.; Li, S. Convolutional Neural Network for Natural Color Visualization of Hyperspectral Images. In Proceedings of the IGARSS 2019–2019 IEEE International Geoscience and Remote Sensing Symposium, Yokohama, Japan, 28 July–2 August 2019; pp. 3372–3375.
29. Tang, R.; Liu, H.; Wei, J.; Tang, W. Supervised learning with convolutional neural networks for hyperspectral visualization. *Remote Sens. Lett.* **2020**, *11*, 363–372. [CrossRef]
30. Shannon, C. A mathematical theory of communication. *Bell Syst. Tech. J.* **1948**, *27*, 379–623. [CrossRef]
31. Amankwah, A. A Multivariate Gradient and Mutual Information Measure Method for Hyperspectral Image Visualization. In Proceedings of the IGARSS 2018–2018 IEEE International Geoscience and Remote Sensing Symposium, Valencia, Spain, 22–27 July 2018; pp. 5001–5004.
32. Computational Intelligence Group of the University of the Basque Country (UPV/EHU). Hyperspectral Remote Sensing Scenes. 2014. Available online: http://www.ehu.eus/ccwintco/index.php/Hyperspectral_Remote_Sensing_Scenes (accessed on 23 July 2020).
33. Buschner, R.; Doerffer, R.; van der Piepen, H. Imaging Spectrometer ROSIS. In *Laser/Optoelektronik in der Technik/Laser/Optoelectronics in Engineering*; Waidelich, W., Ed.; Springer: Berlin/Heidelberg, Germany, 1990; pp. 368–373.
34. Vane, G.; Green, R.O.; Chrien, T.G.; Enmark, H.T.; Hansen, E.G.; Porter, W.M. The airborne visible/infrared imaging spectrometer (AVIRIS). *Remote Sens. Environ.* **1993**, *44*, 127–143. [CrossRef]
35. Gu, J.; Jiang, J.; Susstrunk, S.; Liu, D. What is the Space of Spectral Sensitivity Functions for Digital Color Cameras? In *WACV '13, Proceedings of the 2013 IEEE Workshop on Applications of Computer Vision (WACV), Clearwater Beach, FL, USA, 15–17 January 2013*; IEEE Computer Society: Washington, DC, USA, 2013; pp. 168–179.
36. CIE Standard Colorimetric System. In *Colorimetry*; John Wiley & Sons, Ltd.: Hoboken, NJ, USA, 2006; Chapter 3, pp. 63–114.

37. Kishino, K.; Sakakibara, N.; Narita, K.; Oto, T. Two-dimensional multicolor (RGBY) integrated nanocolumn micro-LEDs as a fundamental technology of micro-LED display. *Appl. Phys. Express* **2019**, *13*, 014003. [CrossRef]
38. Stockman, A.; Sharpe, L.T. The spectral sensitivities of the middle-and long-wavelength-sensitive cones derived from measurements in observers of known genotype. *Vis. Res.* **2000**, *40*, 1711–1737. [CrossRef]
39. Haykin, S.S. *Neural Networks and Learning Machines*, 3rd ed.; Pearson Education: Upper Saddle River, NJ, USA, 2009.
40. Clevert, D.; Unterthiner, T.; Hochreiter, S. Fast and Accurate Deep Network Learning by Exponential Linear Units (ELUs). In Proceedings of the 4th International Conference on Learning Representations, ICLR, San Juan, Puerto Rico, 2–4 May 2016.
41. Trottier, L.; Gigu, P.; Chaib-draa, B. Parametric exponential linear unit for deep convolutional neural networks. In Proceedings of the 2017 16th IEEE International Conference on Machine Learning and Applications (ICMLA), Cancun, Mexico, 18–21 December 2017; pp. 207–214.
42. Paszke, A.; Gross, S.; Massa, F.; Lerer, A.; Bradbury, J.; Chanan, G.; Killeen, T.; Lin, Z.; Gimelshein, N.; Antiga, L.; et al. Pytorch: An imperative style, high-performance deep learning library. In *Advances in Neural Information Processing Systems*; NIPS: San Diego, CA, USA, 2019; pp. 8026–8037.
43. McCamy, C.S.; Marcus, H.; Davidson, J.G. A Color-Rendition Chart. *J. Appl. Photogr. Eng.* **1976**, *2*, 95–99.
44. Pascale, D. *RGB Coordinates of the Macbeth ColorChecker*; The BabelColor Company: Montreal, QC, Canada, 2006.
45. Baldridge, A.; Hook, S.; Grove, C.; Rivera, G. The ASTER spectral library version 2.0. *Remote Sens. Environ.* **2009**, *113*, 711–715. [CrossRef]
46. Pham, T.D. The Kolmogorov-Sinai Entropy in the Setting of Fuzzy Sets for Image Texture Analysis and Classification. *Pattern Recognit.* **2016**, *53*, 229–237. [CrossRef]
47. Haralick, R.M.; Shanmugam, K.; Dinstein, I. Textural Features for Image Classification. *IEEE Trans. Syst. Man Cybern.* **1973**, *SMC-3*, 610–621. [CrossRef]
48. Ivanovici, M.; Richard, N. Entropy versus fractal complexity for computer-generated color fractal images. In Proceedings of the 4th CIE Expert Symposium on Colour and Visual Appearance, Prague, Czech Republic, 6–7 September 2016.
49. Mandelbrot, B. *The Fractal Geometry of Nature*; W.H. Freeman and Co: New York, NY, USA, 1982.
50. Peitgen, H.; Saupe, D. *The Sciences of Fractal Images*; Springer: Berlin, Germany, 1988.
51. Chen, W.; Yuan, S.; Hsiao, H.; Hsieh, C. Algorithms to estimating fractal dimension of textured images. *IEEE Int. Conf. Acoust. Speech Signal Process. ICASSP* **2001**, *3*, 1541–1544.
52. Forsythe, A.; Nadal, M.; Sheehy, N.; Cela-Conde, C.; Sawey, M. Predicting beauty: Fractal dimension and visual complexity in art. *Br. J. Psychol.* **2011**, *102*, 49–70. [CrossRef]
53. Falconer, K. *Fractal Geometry, Mathematical Foundations and Applications*; John Wiley and Sons: Hoboken, NJ, USA, 1990.
54. Voss, R. Random Fractals: Characterization and measurement. *Scaling Phenom. Disord. Syst.* **1986**, *10*, 51–61. [CrossRef]
55. Keller, J.; Chen, S. Texture Description and segmentation through Fractal Geometry. *Comput. Vis. Graph. Image Process.* **1989**, *45*, 150–166. [CrossRef]
56. Maragos, P.; Sun, F. Measuring the fractal dimension of signals: Morphological covers and iterative optimization. *IEEE Trans. Signal Process.* **1993**, *41*, 108–121. [CrossRef]
57. Allain, C.; Cloitre, M. Characterizing the lacunarity of random and deterministic fractal sets. *Phys. Rev. A* **1991**, *44*, 3552–3558. [CrossRef] [PubMed]
58. Manousaki, A.; Manios, A.; Tsompanaki, E.; Tosca, A. Use of color texture in determining the nature of melanocytic skin lesions—A qualitative and quantitative approach. *Comput. Biol. Med.* **2006**, *36*, 416–427. [CrossRef]
59. Ivanovici, M.; Richard, N. Fractal Dimension of Colour Fractal Images. *IEEE Trans. Image Process.* **2011**, *20*, 227–235. [CrossRef]
60. Zhao, X.; Wang, X. Fractal Dimension Estimation of RGB Color Images Using Maximum Color Distance. *Fractals* **2016**, *24*, 1650040. [CrossRef]
61. Nayak, S.R.; Mishra, J.; Khandual, A.; Palai, G. Fractal dimension of RGB color images. *Optik* **2018**, *162*, 196–205. [CrossRef]

62. Li, J.; Du, Q.; Sun, C. An improved box-counting method for image fractal dimension estimation. *Pattern Recognit.* **2009**, *42*, 2460–2469. [CrossRef]
63. Ivanovici, M. Fractal Dimension of Color Fractal Images with Correlated Color Components. *IEEE Trans. Image Process.* **2020**. [CrossRef]
64. Ivanovici, M. Color Fractal Images with Independent RGB Color Components. 2019. Available online: https://ieee-dataport.org/open-access/color-fractal-images-independent-rgb-color-components (accessed on 30 July 2020).
65. Ivanovici, M.; Richard, N. A Naive Complexity Measure for color texture images. In Proceedings of the 2017 International Symposium on Signals, Circuits and Systems (ISSCS), lasi, Romania, 13–14 July 2017; pp. 1–4.
66. Wang, Z.; Bovik, A.C.; Sheikh, H.R.; Simoncelli, E.P. Image quality assessment: From error visibility to structural similarity. *IEEE Trans. Image Process.* **2004**, *13*, 600–612. [CrossRef]
67. Srivastava, N.; Hinton, G.; Krizhevsky, A.; Sutskever, I.; Salakhutdinov, R. Dropout: A simple way to prevent neural networks from overfitting. *J. Mach. Learn. Res.* **2014**, *15*, 1929–1958.

remote sensing

MDPI

Article

Generative Adversarial Network Synthesis of Hyperspectral Vegetation Data

Andrew Hennessy *, Kenneth Clarke and Megan Lewis

School of Biological Sciences, The University of Adelaide, Adelaide 5000, Australia; kenneth.clarke@adelaide.edu.au (K.C.); megan.lewis@adelaide.edu.au (M.L.)
* Correspondence: andrew.hennessy@adelaide.edu.au

Abstract: New, accurate and generalizable methods are required to transform the ever-increasing amount of raw hyperspectral data into actionable knowledge for applications such as environmental monitoring and precision agriculture. Here, we apply advances in generative deep learning models to produce realistic synthetic hyperspectral vegetation data, whilst maintaining class relationships. Specifically, a Generative Adversarial Network (GAN) is trained using the Cramér distance on two vegetation hyperspectral datasets, demonstrating the ability to approximate the distribution of the training samples. Evaluation of the synthetic spectra shows that they respect many of the statistical properties of the real spectra, conforming well to the sampled distributions of all real classes. Creation of an augmented dataset consisting of synthetic and original samples was used to train multiple classifiers, with increases in classification accuracy seen under almost all circumstances. Both datasets showed improvements in classification accuracy ranging from a modest 0.16% for the Indian Pines set and a substantial increase of 7.0% for the New Zealand vegetation. Selection of synthetic samples from sparse or outlying regions of the feature space of real spectral classes demonstrated increased discriminatory power over those from more central portions of the distributions.

Keywords: hyperspectral; vegetation; generative adversarial network; deep learning; data augmentation; classification

Citation: Hennessy, A.; Clarke, K.; Lewis, M. Generative Adversarial Network Synthesis of Hyperspectral Vegetation Data. *Remote Sens.* **2021**, *13*, 2243. https://doi.org/10.3390/rs13122243

Academic Editor: Chein-I Chang

Received: 23 April 2021
Accepted: 3 June 2021
Published: 8 June 2021

1. Introduction

Hyperspectral (HS) Earth observation has increased in popularity in recent years, driven by advancements in sensing technologies, increased data availability, research and institutional knowledge. The big data revolution of the 2000s and significant advances in data processing and machine learning (ML) have seen hyperspectral approaches used in a broad spectrum of applications, with methods of data acquisition covering wide-ranging spatial and temporal resolutions.

For researchers aiming to classify or evaluate vegetation, hyperspectral remote sensing offers rich spectral information detailing the influences of pigments, biochemistry, structure and water absorption whilst having the benefits of being non-destructive, rapid, and repeatable. These phenotypical variations imprint a sort of 'spectral fingerprint' that allows hyperspectral data to differentiate vegetation at taxonomic units ranging from broad ecological types to species and cultivars [1]. Acquiring labelled hyperspectral measurements of vegetation is expensive and time-consuming, resulting in limited training datasets for supervised classification techniques. However, this has been slightly alleviated through multi/hyperspectral data-sharing portals such as ECOSTRESS [2] and SPECCHIO [3]. Supervised classification of such high dimensional data has had to rely on feature reduction or selection techniques in order to overcome small training sample sizes and avoid the curse of dimensionality, also called the 'Hughes phenomenon'. Additionally, the general requirement of large training datasets in ML has meant limited success has been had when trying to leverage recent ML progress towards classification of HS data, often leading to overfitting of models and poor generalizability.

Data augmentation (DA), the process of artificially increasing training sample size, has been implemented by the ML community when the problem of small or imbalanced datasets has been encountered. DA methods vary from simple pre-processing steps such as mirroring, rotating or scaling of images [4] to more complicated simulations [5,6] and generative models [7,8]. DA for timeseries or 1D data consists of the addition of noise, or methods such as time dilation, or cut and paste [9]. However, when dealing with non-spatial HS data, these methods would be unsuitable, as it is important to maintain reflectance and waveband relationships in order to ensure class labels are preserved. Methods of DA such as physics-based models [10], or noise injection [11,12] have been applied to HS data. Whilst successful, these methods are either simplifications of reality and require domain-dependent knowledge of target features in the case of physical models or rely upon random noise, potentially producing samples that only approximate the true distribution.

Generative adversarial networks (GANs) have been used successfully in many fields as a DA technique, often for images, timeseries/1D [13], sound synthesis [14], or anonymising medical data [15]. GANs consist of two neural networks trained in an adversarial manner. The generator (G) network produces synthetic copies mimicking the real training data while the discriminator (D) network attempts to identify whether a sample was from the real dataset or produced by G. The D is scored on its accuracy in identifying real from synthetic data, before passing feedback to G allowing it to learn how best to fool D and improve generation of synthetic samples [16].

The use of GANs to generate synthetic HS data is a relatively new field of study. GANs of varying architectures ranging from 1D spectral [17–19] to 2D [20], and 3D spectral-spatial [21] with differing data embeddings including individual spectra, HS images, and principal components have been examined. All have been able to demonstrate the ability to generate synthesized hyperspectral data and to improve classification outcomes to varying degrees whether through DA or conversion of the GANs discriminator model to a classifier. However, issues such as training instability and mode collapse, a common form of overfitting are prevalent.

The work presented in this paper applies advances in generative models to overcome limitations previously encountered by Audebert et al. [17] to produce more realistic synthetic HS vegetation data and eliminate reliance on PCA to reduce dimensionality and stabilise training. Specifically, we train a GAN using the Cramér distance on two vegetation HS datasets, demonstrating the ability to approximate the distribution of the training samples while encountering no evidence of mode collapse. We go on to demonstrate the use of these synthetic samples for data augmentation and reduced under-sampling of class distributions, as well as establishing a method to quantify the potential classification power of a synthetic sample by evaluating its relative position in feature space.

Generative Adversarial Networks—Background

GANs are a type of generative machine learning algorithm known as an implicit density model. This type of model does not directly estimate or fit the data distribution but rather generates its own data which is used to update the model. Since first being introduced by Goodfellow et al. [16] GANs have become a dominant field of study within ML/DL, with numerous variants and being described as "the most interesting idea in the last 10 years in machine learning" by a leading AI researcher [22]. Although sometimes utilizing non-neural network architectures, GANs generally consist of two neural networks, sometimes more, that compete against each other in a minimax game. This is where one neural network, the discriminator, attempts to reduce its "cost" or error as much as possible. This occurs in an adversarial manner, where the discriminator is trained to maximize the probability of correctly labelling whether a sample originates from the original data distribution or has been produced by the generator. Simultaneously, the generator is trained to minimize the probability that the discriminator correctly labels the sample [23].

As a result, the training of GANs is notoriously unstable, with issues such as the discriminator's cost quickly becoming zero and providing no gradient to update the generator, or the generator converging onto a small subset of samples that regularly fool the discriminator, a common issue known as mode collapse. Considerable research has gone into attempting to alleviate these issues, improve training stability and improve quality of synthetic samples, so much so that during 2018 more than one GAN-related paper was being released every hour [23].

Unlike non-adversarial neural networks, the loss function of a GAN does not converge to an optimal state, making the loss values meaningless in respect to evaluating the performance of the model. In an attempt to alleviate this problem, the Wasserstein GAN (WGAN) was developed to use the Wasserstein distance, also known as the Earth Mover's (EM) distance which results in an informative loss function for both D and G that converges to a minimum [24]. Rather than the D having sigmoid activation in its final layer producing a binary classification of real or fake, WGAN approximates the Wasserstein distance, which is a regression task detailing the distance between the real and fake distributions. Due to gradient loss being a common weakness with WGANs, they were improved by applying weight clipping to the losses with a gradient penalty (GP) [25], further improving training stability.

2. Experimental Design

Here, we implement the CramérGAN, a GAN variant using the Cramér/energy distance as the Ds loss, reportedly offering improved training stability and increased generative diversity over WGANs [26]. This choice was informed by our preliminary testing of wGAN and wGAN-GP that produced noisy synthesized samples and lower standard deviations, in addition to the learning instability and poor convergence previously reported for wGAN, which may explain mode collapse encountered by Audebert et al. [17].

Individual models were trained for each hyperspectral class, for a total of 38 models. Each model was trained for 50,000 epochs, at a ratio of 5:1 (5 training iterations of D for every 1 of G) using the Adam optimiser at a learning rate of 0.0001, with beta1 = 0.5 and beta2 = 0.9. The latent noise vector was generated from a normal distribution with length 100. The G consists of two fully connected dense layers followed by two convolution layers, all using the ReLu activation function save for the final convolution layer using Sigmoid activation. The final layer of G reshaped the output to be a 2D array with shape (batch size * number of bands). A similar architecture was used for the D, though reversed. Starting with two convolution layers into a flatten layer, followed by 2 fully connected dense layers, all layers of D used Leaky ReLu activation except the final layer which used a linear function (Figure 1) (Appendix A Tables A1 and A2).

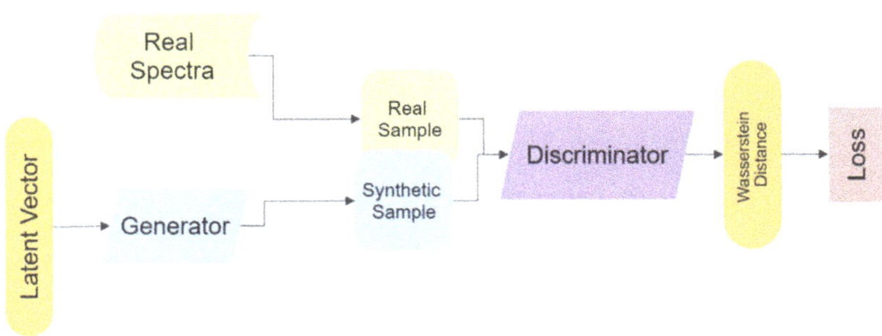

Figure 1. Schematic of Generative Adversarial Network (GAN)architecture.

The three classification models (SVM, RF and NN) were evaluated in four permutations: trained on real data and evaluated on real data (real–real); trained on real data and evaluated on synthetic data (real–synthetic); trained on synthetic data and evaluated on synthetic data (synthetic–synthetic); and trained on synthetic data and evaluated on real data (synthetic–real). Each dataset was split into training and testing subsets with 10 times cross-validation. All synthetic datasets were restricted to the same number of samples per class as the real datasets unless specified otherwise. The real–real experiments were expected to have the highest accuracy and offer a baseline of comparison for the synthetic samples. If the accuracy of real–synthetic is significantly higher than real–real experiments, this potentially indicates that the generator has not fully learned the true distribution of the training samples. Conversely, accuracy being significantly lower could mean the synthetic samples are outside the true distribution and are an unrealistic representation of the spectra.

Extending this analysis, synthetic–synthetic and synthetic–real experiments were performed with the number of synthesized training samples increasing from 10 to 490 samples by increments of 10 samples per class. The real–synthetic and real–real experiments were included for comparison with a consistent number of training samples, though training and evaluation subsets differed every iteration. The initial DA experiment was performed with the same number of samples for real and synthetic datasets with the augmented set having twice the number before the number of synthetic samples were incremented by 10 from 10 to 490 samples per class.

The data augmentation capabilities of the synthetic spectra were evaluated by similar methods. First, the three classifiers were trained with either real, synthetic or both combined into an augmented dataset and tested against an evaluation dataset that was not used in the training of the GAN.

All code was written and executed in Python 3.7. The CramérGAN based upon [27] using the Tensorflow 1.8 framework. Support Vector Machine (SVM), and Random Forests (RF) classifiers make use of the Scikit-Learn 0.22.2 library, with Tensorflow 1.8 utilized for the neural network (NN) classifier. Additionally, Scikit-Learn 0.22.2 provided the dimensionality reduction functions for Principal Components Analysis (PCA) and t-distributed Stochastic Neighbourhood Embedding (t-SNE), with Uniform Manifold Approximation and Projection (UMAP) being a standalone library. Hyperparameters for all functions are provided in Appendix A.

2.1. Classification Power

Potential classification power of a sample was estimated with the C metric devised in Mountrakis and Xi [28] for the purpose of predicting the likelihood of correctly classifying an unknown sample by measuring its Euclidean distance in feature space to samples in the training dataset. Mountrakis and Xi [28] demonstrated a strong correlation between close proximity to number of training samples and likelihood of correctly being classified. The C metric is bound between -1 indicating low likelihood and 1 indicating high likelihood of successful classification.

Rather than focusing on the proximity of an unknown sample to a classifier's training data, we are interested in the distance of each synthesized sample to that of the real data in order to evaluate any potential increase in information density. We hypothesise that a C value closer to the lower bound for a synthetic sample would indicate it being further away from real data points and of any synthetic samples with C values close to the upper bound. Such a sample could potentially contain greater discriminatory power for the classifier as it essentially fills a gap in feature space of the class distribution.

To determine whether some samples of the NZ dataset provide more information to the classifier than others, and that the improvement in classification accuracy is not purely from increased sample size, the distance of each generated sample was measured to all real samples of its class before being converted to a C value as per Mountrakis and Xi [28], with an h value range of 1–50 at increments of 1. Two data subsets were then created using the first 100 spectral samples after all synthetic samples were ordered by their C value in ascending (most distant) and descending (least distant) order. The first 100 samples from each ordered dataset rather than the full 500 were used to maximize differences, reduce computation time and simplify figures.

2.2. Datasets

Two hyperspectral datasets were used to train the GAN: Indian Pines agricultural land cover types (INDI); and New Zealand plant spectra (NZ). The Indian Pines dataset (INDI) recorded by the AVIRIS airborne hyperspectral imager over North-West Indiana, USA, is made available by Purdue University and comprises 145 × 145 pixels at 20 m spatial resolution and 224 spectral reflectance bands from 400 to 2500 nm [29]. Removal of water absorption bands by the provider reduced these to 200 wavebands, and then reflectance of each pixel was scaled between 0 and 1. Fifty pixels were randomly selected as training samples except for three classes with fewer than 50 total samples, for which 15 samples were used for training (Table 1).

Table 1. Land cover classes, training and evaluation sample numbers for Indian Pines dataset.

Class ID	Class Name	Training Samples	Evaluation Samples
INDI1	Alfalfa	15	31
INDI2	Corn-no-till	50	1378
INDI3	Corn-min-till	50	780
INDI4	Corn	50	187
INDI5	Grass-pasture	50	433
INDI6	Grass-trees	50	680
INDI7	Grass-pasture-mowed	15	13
INDI8	Hay-windrowed	50	428
INDI9	Oats	15	5
INDI10	Soybean-no-till	50	922
INDI11	Soybean-min-till	50	2405
INDI12	Soybean-clean	50	543
INDI13	Wheat	50	155
INDI14	Woods	50	1215
INDI15	Buildings-Grass-Trees-Drives	50	336
INDI16	Stone-Steel-Towers	50	43

The New Zealand (NZ) dataset used in this study is a subsample of hyperspectral spectra for 22 species taken from a dataset of 39 native New Zealand plant spectra collected from four different sites around the North Island of New Zealand and made available on the SPECCHIO database [3]. These spectra were acquired with an ASD FieldSpecPro spectroradiometer at 1 nm sampling intervals between 350 and 2500 nm. Following acquisition from the SPECCHIO database, spectra were resampled to 3 nm and noisy bands associated with atmospheric water absorption were removed (1326–1464, 1767–2004, 2337–2500) resulting in 540 bands per spectra. Eighty percent of samples per class were used for training the GAN and 20% held aside to evaluate classifier performance (Table 2).

Table 2. Plant species classes, training and evaluation sample numbers for New Zealand dataset.

Class ID	Common Name	Botanical Name	Training Samples	Evaluation Samples
NZ0	Manuka	*Leptospermum scoparium*	58	14
NZ1	Pohutukawa	*Metrosideros excelsa*	32	8
NZ2	Koromiko	*Hebe stricta*	42	10
NZ3	Lemonwood	*Pittosporum eugenioides*	46	12
NZ4	Kawakawa	*Macropiper excelsum*	34	9
NZ5	Whiteywood	*Melicytus ramiflorus*	48	12

Table 2. *Cont.*

Class ID	Common Name	Botanical Name	Training Samples	Evaluation Samples
NZ6	Totara	*Podocarpus totara*	34	8
NZ7	New Zealand Flax	*Phormium tenax*	36	9
NZ8	Akiraho	*Olearia paniculata*	8	2
NZ9	Rata	*Metrosideros robusta*	9	2
NZ10	Ngaio	*Myoporum laetum*	38	10
NZ11	Mapou	*Myrsine australis*	36	9
NZ12	Cabbage tree	*Cordyline australis*	32	8
NZ13	Karaka	*Corynocarpus laevigatus*	34	9
NZ14	Kauri	*Agathis australis*	15	3
NZ15	Silver fern	*Cyathea dealbata*	28	7
NZ16	Tangle fern	*Gleichenia dicarpa* var. *alpina*	14	4
NZ17	Black tree fern	*Cyathea medullaris*	18	4
NZ18	Pigeonwood	*Hedycarya arborea*	18	5
NZ19	Rangiora	*Brachyglottis repanda*	12	3
NZ20	Karamu	*Coprosma robusta*	13	3
NZ21	Red Pine	*Dacrydium cupressinum*	16	4

3. Results and Discussion

3.1. Mean and Standard Deviation of Training and Synthetic Spectra

In order to visualize similarities between synthetic and real spectra, the mean and standard deviation for each class are shown for the real, evaluation, and synthetic datasets. All low-frequency spectral features, as well as mean and standard deviations, appear to be reproduced with high accuracy by the GAN. At finer scales of 3–5 wavebands noise is present, most notably throughout the near infra-red (NIR) plateau (Figure 2). Smoothing of synthesized data by a number of methods resulted in either no improvement or decreased performance in a number of tests; for this reason, no pre-processing was performed on synthesized samples. Due to the high frequency and random nature of the noise, once mean and STD statistics are calculated the spectra appear smooth.

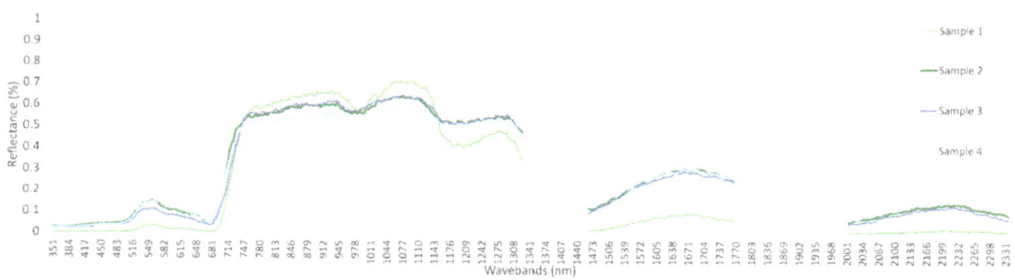

Figure 2. Synthetic spectra of NZ class 0, 350–2400 nm at 3 nm bandwidths.

Class 0 is one of the NZ classes with the largest number of samples, resulting in its mean and standard deviation being similar between its real, evaluation, and synthetic subsets. However, this is not the case for all classes, with NZ-9 showing the mean and standard deviation of the randomly selected evaluation samples being vastly different to those of real and synthetic spectra (Figure 3). The same is seen amongst INDI classes, with class 2 matching across all 3 data subsets, and class 4 with only 40 samples showing substantial difference between evaluation and real samples, especially in the visible wavebands (Figure 4). Although some classes may struggle to represent the evaluation dataset due to the initial random splitting of the datasets, in general, mean and standard deviation of the synthetic samples very closely match the real training data.

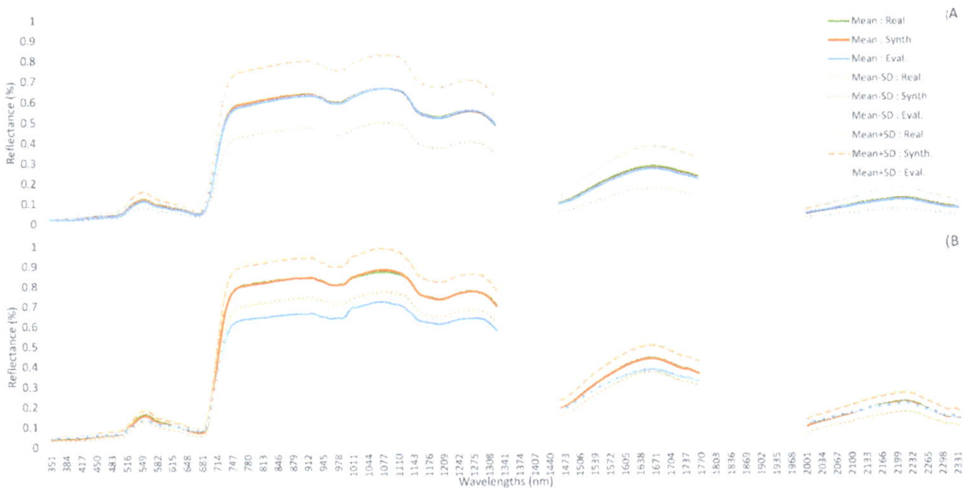

Figure 3. Mean and +/− 1 STD for training (real), synthetic, and evaluation (real) datasets. (**A**) NZ class 0; Manuka (*L. scoparium*). (**B**) NZ class 9; Rata (*M. robusta*).

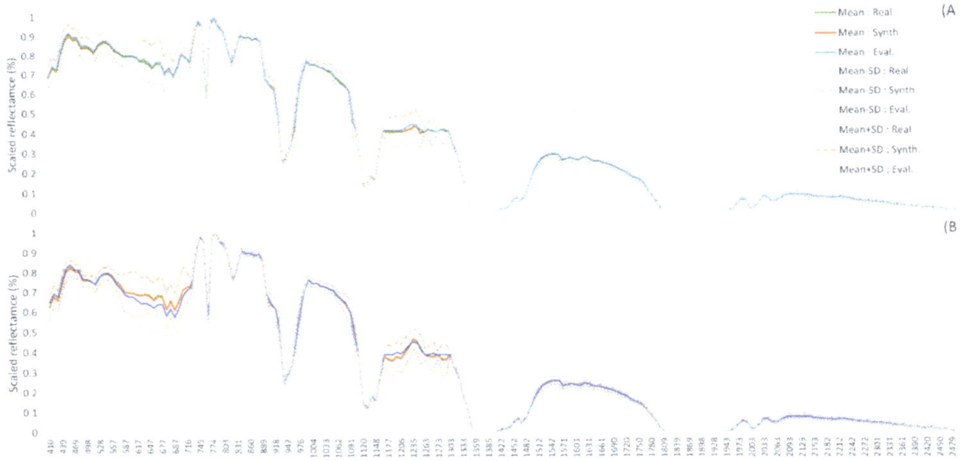

Figure 4. Mean and +/− 1 STD for training (real), synthetic, and evaluation (real) datasets. (**A**) INDI class 2; corn-no-till. (**B**) INDI class 4; corn.

3.2. Generation and Distribution of Spectra

Here, we demonstrate the ability of the GAN to reproduce realistic spectral shapes and to capture the statistical distribution of the class populations. Three dimensionality reduction methods—PCA, t-SNE, and UMAP—were applied to both the real and synthetic datasets of INDI and NZ spectra to reduce their 200 and 540 wavebands (respectively) down to a plottable 2D space (Figures 5 and 6). Upon visual inspection, the class clusters formed by the augmented data across all reduction methods mimic the distribution of those of the real data. Additionally, due to its small sample sizes the structure of clusters for the real NZ data is sparse and unclear, though is emphasised by the large number of synthetic samples.

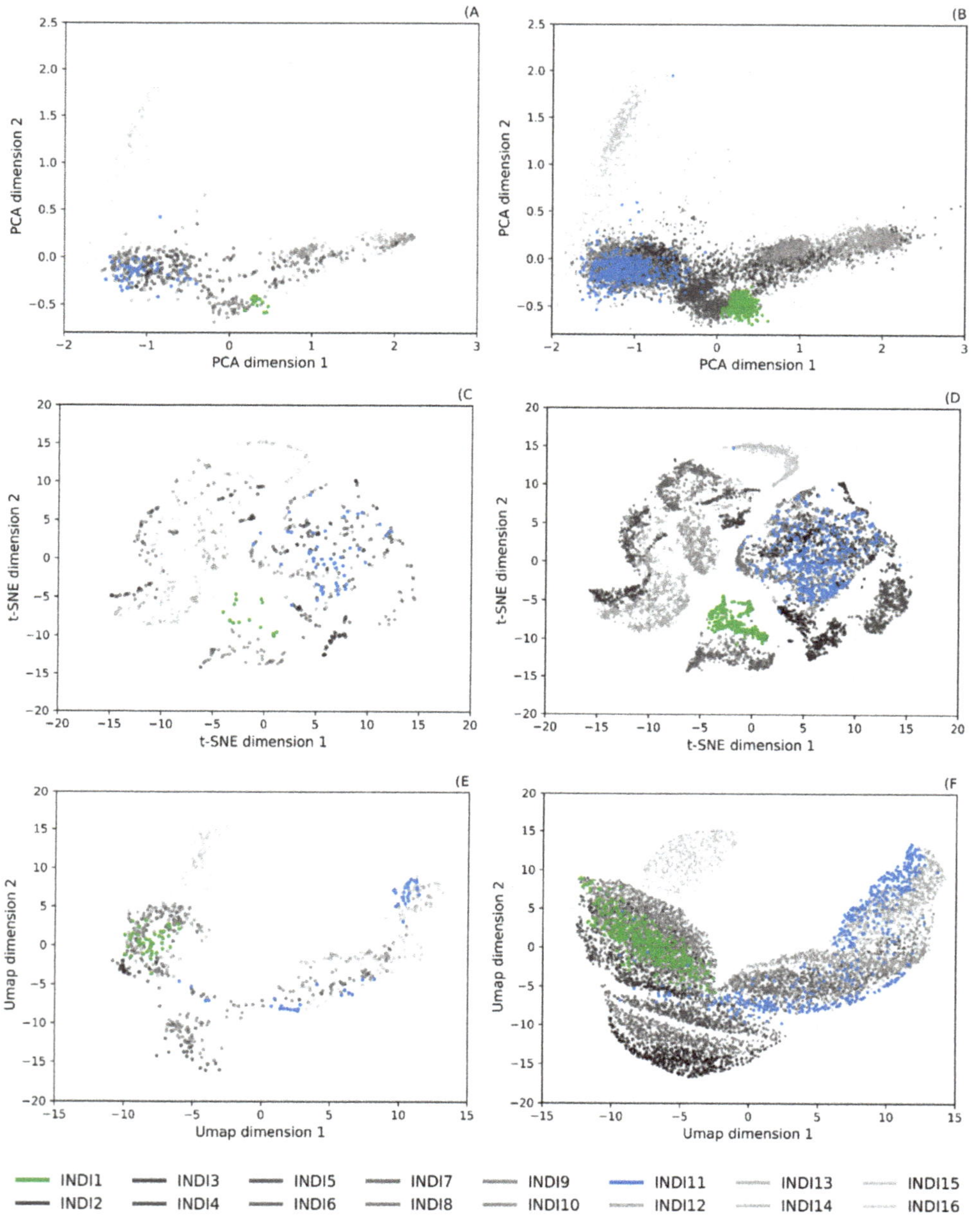

Figure 5. Dimensional reduced representations of INDI real and synthetic datasets; highlighted classes: INDI1—Alfalfa (green); INDI11—Soybean-min-till (blue). (**A**) Real dataset; PCA reduction, (**B**) synthetic dataset; PCA reduction, (**C**) real dataset; t-SNE reduction, (**D**) synthetic dataset; t-SNE reduction, (**E**) real dataset; UMAP reduction, and (**F**) synthetic dataset; UMAP reduction.

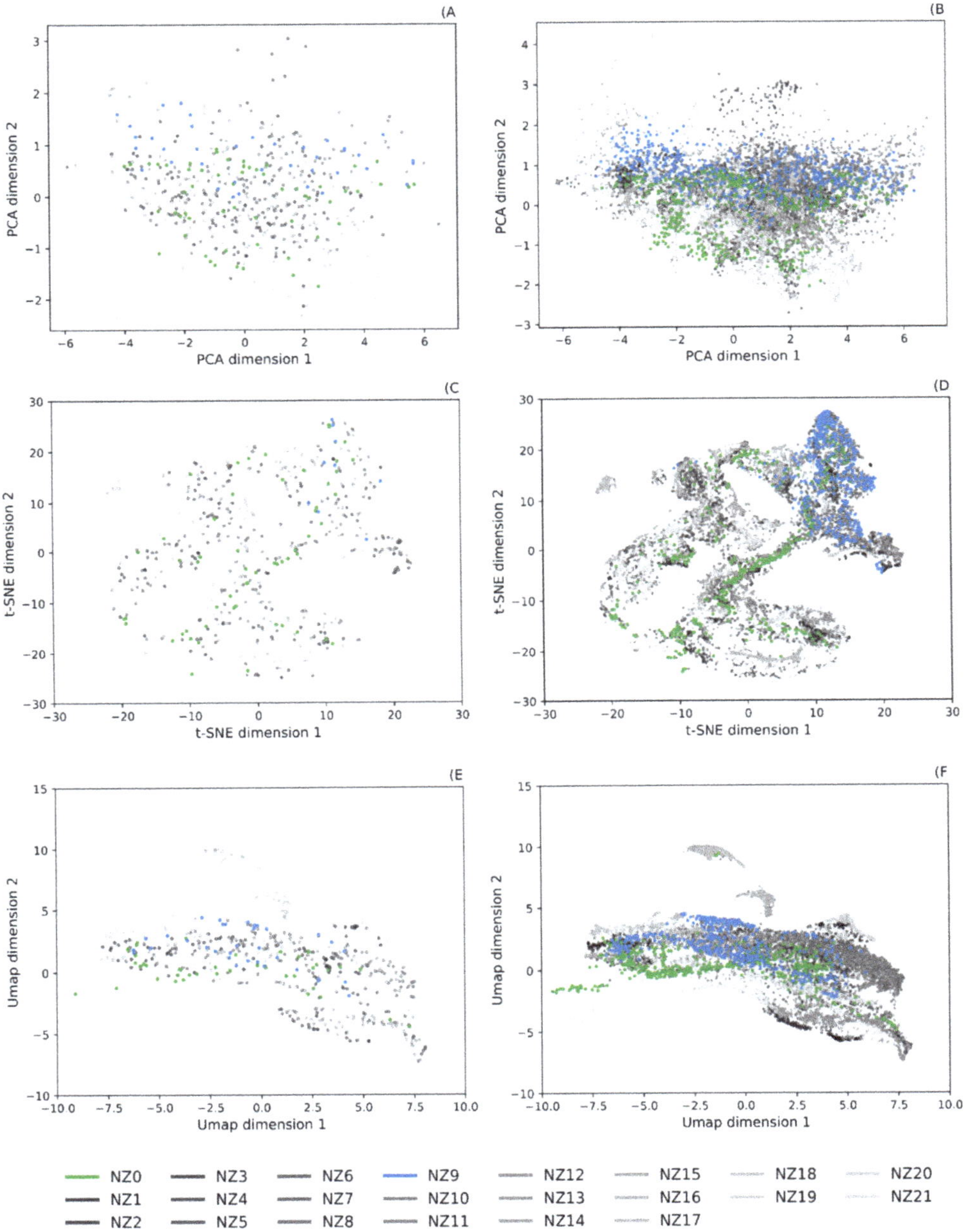

Figure 6. Dimensional reduced representations of NZ real and synthetic datasets; + highlighted classes: NZ0—Manuka (*L. scoparium*) (green), NZ9—Rata (*M. robusta*) (blue). (**A**) Real dataset; PCA reduction, (**B**) synthetic dataset; PCA reduction, (**C**) real dataset; t-SNE reduction, (**D**) synthetic dataset; t-SNE reduction, (**E**) real dataset; UMAP reduction, and (**F**) synthetic dataset; UMAP reduction.

Such strong replication of the 2D representation of the classes is a good indication of the generative model's ability to learn distributions. Even when the models are trained separately for each class, the relationship between classes is maintained. However, the increased sample number in the synthetic datasets do in some cases extend beyond the bounds of the real samples. Whilst some may represent potential outliers the majority are artefacts of increased sample sizes. This is most evident in the UMAP representation, where a parameter that defines minimum distance between samples can be set a larger value, which results in increased spread of samples in the 2D representation [30]. This is most notable in the INDI dataset, with classes 1, 7, and 8 extending more broadly than the real dataset (Figure 5F).

3.3. Training Classification Ability

In order to further examine the similarity of synthetic spectra to the real training data, three classifiers were trained (SVM, RF, NN), with four permutations of each (real–real, real–synthetic, synthetic–synthetic, and synthetic–real) (Table 3). With few exceptions, the neural network classifier outperformed the others, with SVM being the second most accurate, followed by RF. The INDI dataset recorded the highest accuracy for the real–real test with RF and NN classifiers at 74.76% and 84.13%, respectively, although the highest accuracy for the SVM classifier occurred during the synthetic–real test with 81.42% accuracy. Comparing the four combinations of real and synthetic, real–real had the highest accuracy for four experiments, with INDI synthetic–real with the SVM, and NZ real–synthetic with the RF classifier being the only exceptions.

Table 3. Classification accuracies for classifiers trained on real or synthesized spectral data and evaluated on either real or synthesized data for both Indian Pines and New Zealand datasets based on real class sample sizes. Highest achieved accuracy for each classifier per dataset indicated in bold.

INDI	SVM	RF	NN
Real–Real	73.48	**74.76**	**84.13**
Real–Synthetic	77.04	66.68	76.50
Synthetic–Synthetic	79.48	69.38	80.91
Synthetic–Real	**81.42**	70.51	81.66
NZ			
Real–Real	**79.86**	47.65	**95.76**
Real–Synthetic	78.73	**60.23**	80.54
Synthetic–Synthetic	74.20	51.33	91.82
Synthetic–Real	78.19	54.76	81.13

To further evaluate the synthetic spectra, synthetic–synthetic and synthetic–real experiments were performed with the number of synthesized training samples increasing from 10 to 490 samples by increments of 10 samples per class (Figure 7). Synthetic–synthetic accuracy improves with more samples: this too is to be expected as this simply adds more training samples from the same distribution. Most importantly, synthetic–real accuracy, though often slightly lagging behind synthetic–synthetic, improves in the same manner, indicating that the synthetic samples are a good representation of the true distribution and that increasing their number for training a classier is an effective method of data augmentation. The main exception to this is the NN NZ classier, where synthetic–synthetic quickly reaches ~100% accuracy, while synthetic–real maintains ~80% before slowly decreasing in accuracy as more samples are added. This could indicate the NN classifier focuses on different features than the other classifiers, potentially being more affected by the small-scale noise apparent in the NZ-generated samples as the noise is not as apparent in the INDI data and the INDI NN classifier does not show such a discrepancy between synthetic–synthetic and synthetic–real.

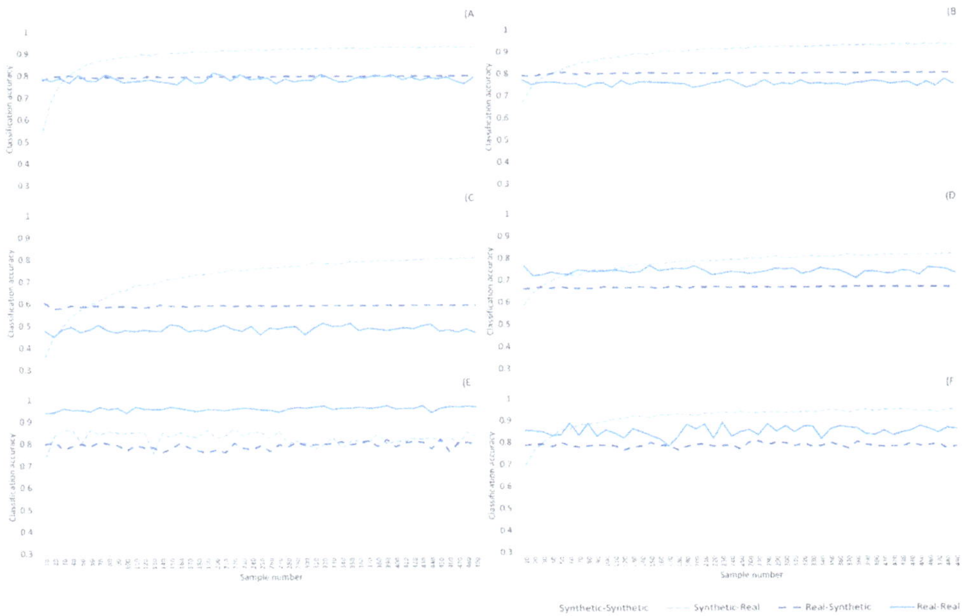

Figure 7. Classification accuracies for classifiers trained on real or synthesized spectral data and evaluated on either real or synthesized data for both Indian Pines and New Zealand datasets ranging from 10 to 490 samples per class. (**A**) New Zealand dataset; SVM classifier, (**B**) Indian Pines dataset; SVM classifier, (**C**) New Zealand dataset; RF classifier, (**D**) Indian Pines dataset; RF classifier, (**E**) New Zealand dataset; NN classifier, and (**F**) Indian Pines dataset; NN classifier.

3.4. Data Augmentation

In order to test the viability of the synthetic data for data augmentation, the same three classifiers were trained with either real, synthetic or both combined into an augmented dataset and tested against an evaluation dataset (Table 4). All classifiers had higher accuracy when trained on the real dataset compared to synthetic, though the highest accuracy overall was with the augmented dataset. For the INDI data, this increase was minor, being <1% for all classifiers. A far more significant improvement was seen for the NZ data with increases of 3.54% (to 86.55%), 0.53% (to 50.80%), and 3.73% (to 85.14%) for SVM, RF, and NN, respectively.

Table 4. Classification accuracies for classifiers trained on real, synthesized, or augmented spectral data and evaluated on an evaluation dataset for both Indian Pines and New Zealand datasets based on real class sample sizes. Highest achieved accuracy for each classifier per dataset indicated in bold.

INDI	SVM	RF	NN
Real–Evaluation	70.40	66.40	62.06
Synthetic–Evaluation	62.82	57.73	51.30
Augmented–Evaluation	**70.56**	**66.91**	**62.76**
NZ			
Real–Evaluation	83.01	50.27	81.41
Synthetic–Evaluation	63.02	36.69	68.60
Augmented–Evaluation	**86.55**	**50.80**	**85.14**

Of course, however, the number of synthetic samples does not have to be limited in such a manner. As with previous experiments the number of synthetic samples started at 10 and incremented by 10 to a total of 490, demonstrating the potential of this data augmentation method. Dramatic increases in accuracy were seen for the synthetic dataset, with the smallest increase being 5.13% for INDI-SVM occurring at 490 samples, the largest being 20.47% for NZ-RF at 420 samples. These increases brought the synthetic dataset very close to the accuracy of the real samples or even above in the cases of INDI-NN, NZ-RF, and NZ-NN. Increases in accuracy were also seen in the augmented dataset, though not as dramatic as those for the synthetic dataset. Improvements in accuracy ranged from 0.16% for INDI-SVM at 10 synthetic samples to 9.45% for NZ-RF at 280 synthetic samples. These improvements raise the highest accuracy for the INDI dataset from 70.40% to 70.56%, resulting in an increase of 0.16% over the highest achieved by just the real data. A larger increase was seen in the NZ dataset with the previous highest accuracy raising from 86.55% to 90.01%, an increase of 3.45% from the previous augmented classification with restricted sample size, and a 7% increase over the real dataset alone (Table 5).

Table 5. Classification accuracies for classifiers trained on real, synthesized, or augmented spectral data and evaluated on an evaluation dataset for both Indian Pines and New Zealand datasets with sample sizes ranging from 10 to 490 per class for synthetic and augmented while real contained all real samples. Highest achieved accuracy for each classifier per dataset indicated in bold.

INDI	**SVM/Sample Size**	**RF/Sample Size**	**NN/Sample Size**
Real–Evaluation	70.40/All	66.40/All	62.06/All
Synthetic–Evaluation	67.95/490	65.59/490	65.05/50
Augmented–Evaluation	**70.56/10**	**68.25/140**	**69.77/320**
NZ			
Real–Evaluation	83.01/All	50.27/All	81.41/All
Synthetic–Evaluation	81.35/490	57.16/420	87.78/450
Augmented–Evaluation	**90.01/120**	**60.25/280**	**89.25/120**

3.5. Classification Power of a Synthetic Sample

Ordering the synthetic samples by their C value before iteratively adding samples one at a time from each class to the training dataset of an SVM classifier shows the differing classification power of the synthetic samples from lower to upper bounds of C and vice versa (Figure 8).

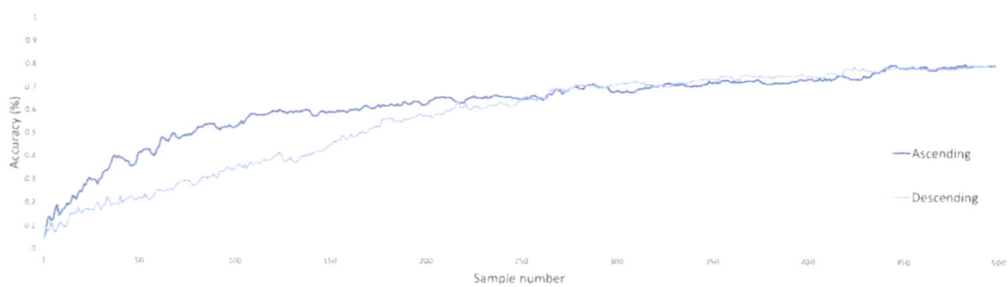

Figure 8. Classification accuracy of a SVM classifier for C metric ascending and descending ordered synthetic datasets incremented by single samples.

When in ascending order from lower to upper bounds, classification accuracy increases dramatically, reaching ~60% accuracy with ~100 samples, while 200 samples were required for similar accuracy in descending order. At approximately half the number of samples, accuracies converge, then increase at the same rate before reaching 80% accuracy at 500 samples. These classification accuracies (Figure 8) provide the first insight into increased discriminatory power associated with synthetic samples that occur at distance to real samples. Although not encountered here, a maximum limit to this distance would be present, with synthetic samples needing to remain within the bounds of their respective class distributions.

A similar, though reduced, trend can be seen when the ordered synthetic samples are used to augment the real dataset. Both ascending and descending datasets improve classification over that of the real dataset when samples are iteratively added to the classifiers training dataset (Figure 9). Despite descending ordered samples outperforming ascending at times, on average, ascending samples achieved ~1.5% higher accuracy across the classifications compared with the 79.72% to 78.24% accuracy of descending samples.

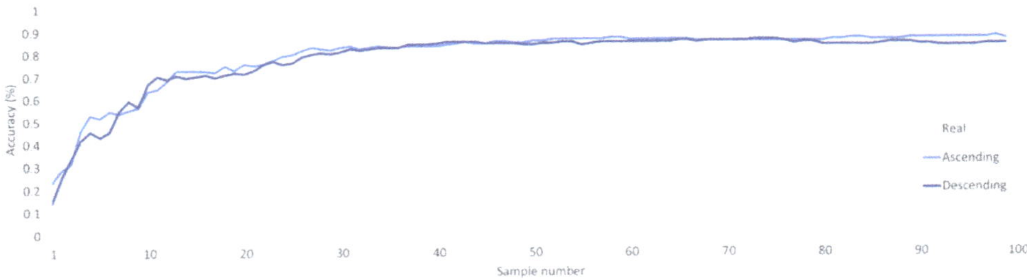

Figure 9. Classification accuracy of a SVM classifier for C metric ascending and descending augmented datasets with randomly ordered real dataset incremented by single samples.

This artificial selection of synthetic data points distant or close to the real data influences sample distribution used to train the classifiers. As one might expect, the ordered data points come from the edges or sparse regions of the real data distribution, dramatically shifting the mean and standard deviation of the ordered datasets (Figure 10).

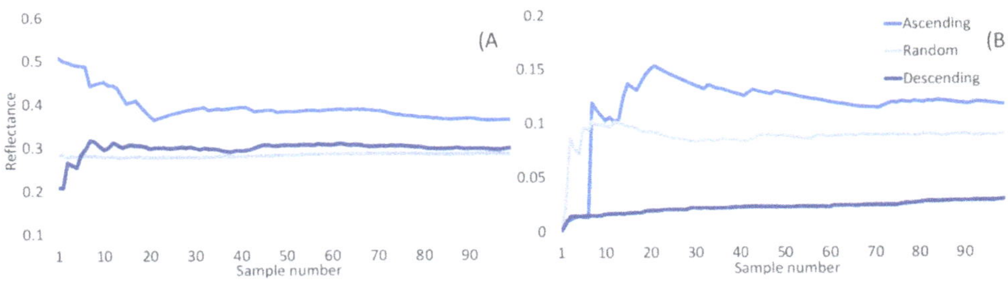

Figure 10. (**A**) Mean and (**B**) STD of C metric ascending, descending, and randomly ordered synthetic datasets incremented by single samples.

The inclusion of synthetic data points selected at random provides a baseline for comparison with the ordered datasets. Once the number of samples increases beyond a few points, the means for descending and random converge and stay steady throughout. Mean values for ascending start significantly higher, though initially begin to converge towards the other datasets before plateauing at a higher level. Whilst being averaged across all classes and all wavebands of spectra, the mean reflectance for the ascending data is consistently higher. Standard deviation of the descending dataset is consistently low, only slightly increasing as samples are added. This is in stark contrast to the STD of the ascending dataset being ~5–6 x higher across all n samples. The mean of the randomly selected dataset occurs between the means of the two ordered, though closer to the ascending mean, indicating the samples that make up the descending dataset are highly conserved.

To further illustrate the relationship of the ordered datasets and the real distribution, a PCA of one of the classes is shown (Figure 11). As the mean and STD indicated, the descending samples are tightly grouped near the mean and densest area of the real data distribution, with the ascending samples generally occurring along the border of the real distribution. Whilst ascending selects for samples with low C and greater distance from real samples, it is important to note that these synthetic samples still appear to conform to the natural shape of the real distribution, a further indication the generative model is performing well.

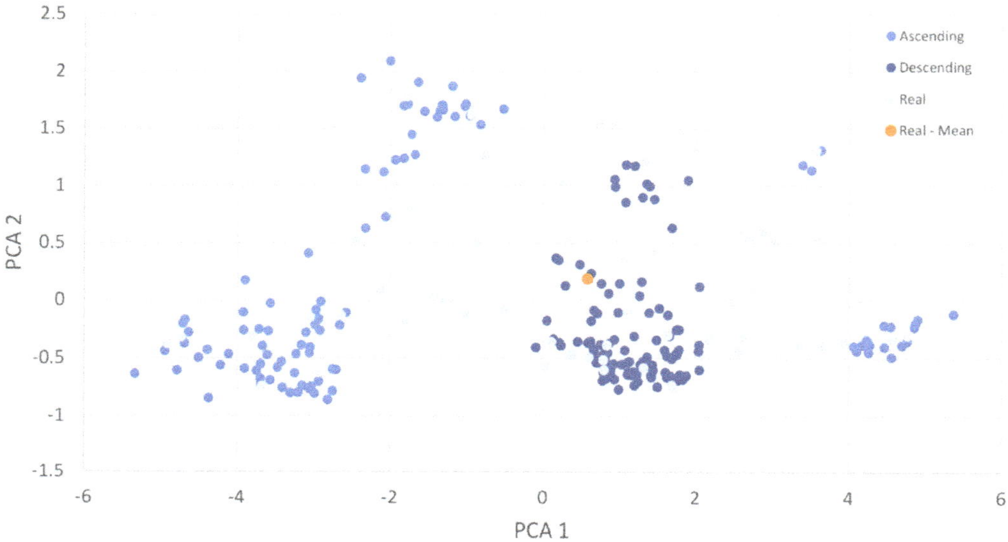

Figure 11. PCA of NZ class 0; Manuka (*L. scoparium*) real samples, with first 100 samples of the ascending and descending C ordered synthetic datasets.

4. Conclusions

In this paper, we have successfully demonstrated the ability to train a generative machine learning model for synthesis of hyperspectral vegetation spectra. Evaluation of the synthetic spectra shows that they respect many of the statistical properties of the real spectra, conforming well to the sampled distributions of all real classes. Further to this, we have shown that the synthetic spectra generated by our models are suitable for data augmentation of a classification models training dataset. Addition of synthetic samples to the real training samples of a classifier produced increased overall classification accuracy under almost all circumstances examined. Of the two datasets, the New Zealand vegetation showed a maximum increase of 7.0% in classification accuracy, with Indian Pines demonstrating a more modest improvement of 0.16%. Selection of synthetic samples from sparse or outlying regions of the feature space of real spectral classes demonstrated increased discriminatory power over those from more central portions of the distributions. We believe further work regarding this could see targeted generation to maximize the information content of a synthetic sample that would result in improved classification accuracy and generalizability with a smaller augmented dataset. The use of these synthesized spectra to augment real spectral datasets allows for the training of classifiers that benefit from large sample numbers without a researcher needing to collect additional labelled spectra from the field. This is of increasing significance as modern machine and deep learning algorithms tend to require larger datasets.

Author Contributions: Conceptualization, A.H.; methodology, A.H.; data curation, A.H.; formal analysis, A.H.; writing—original draft preparation, A.H., K.C. and M.L.; writing—review and editing, A.H., K.C. and M.L.; supervision, K.C., M.L. All authors have read and agreed to the published version of the manuscript.

Funding: This research received no external funding.

Data Availability Statement: The publically available datasets used in our experiments are available at; Indian Pine Site 3 AVIRIS hyperspectral data image file, doi:10.4231/R7RX991C. New Zealand hyperspectral vegetation dataset, https://specchio.ch/.

Acknowledgments: Financial support for this research was provided by the Australian Government Research Training Program Scholarship and the University of Adelaide School of Biological Sciences.

Conflicts of Interest: The authors declare no conflict of interest.

Appendix A

Table A1. Layer architecture of the GANs generator.

Layer Type/Parameters	Shape	Activation
Conv1D	(100,100)	ReLU
Conv1D	(100,50)	ReLU
Conv1D	(100,10)	ReLU
Flatten		
Dense	2048	Leaky ReLU (alpha = 0.2)
Batch Normalization (momentum = 0.4)		
Dropout (0.5)		
Dense	4096	Leaky ReLU (alpha = 0.2)
Batch Normalization (momentum = 0.4)		
Dropout (0.5)		
Dense	2048	Leaky ReLU (alpha = 0.2)
Batch Normalization (momentum = 0.4)		
Dropout (0.5)		
Dense	22	Softmax

Table A2. Layer architecture of the GANs discriminator.

Layer Type/Parameters	Shape	Activation
Dense	1024	Leaky ReLU (alpha = 0.2)
Dense	1024	Leaky ReLU (alpha = 0.2)
Dense	256	Linear

Table A3. Epochs with Kullback–Leibler divergence loss, Adam optimiser with a learning rate of 0.00001, and a batch size of 32.

Layer Type/Parameters	Shape	Activation
Conv1D	(100,100)	ReLU
Conv1D	(100,50)	ReLU
Conv1D	(100,10)	ReLU
Flatten		
Dense	2048	Leaky ReLU (alpha = 0.2)
Batch Normalization (momentum = 0.4)		
Dropout (0.5)		
Dense	4096	Leaky ReLU (alpha = 0.2)

Table A3. *Cont.*

Layer Type/Parameters	Shape	Activation
Batch Normalization (momentum = 0.4)		
Dropout (0.5)		
Dense	2048	Leaky ReLU (alpha = 0.2)
Batch Normalization (momentum = 0.4)		
Dropout (0.5)		
Dense	22	Softmax

Table A4. Hyperparameters used during UMAP dimension reduction for each dataset.

Dataset	Number of Neighbours	Minimum Distance	Distance Metric
INDI	20	1	Canberra
NZ	100	0.3	Correlation

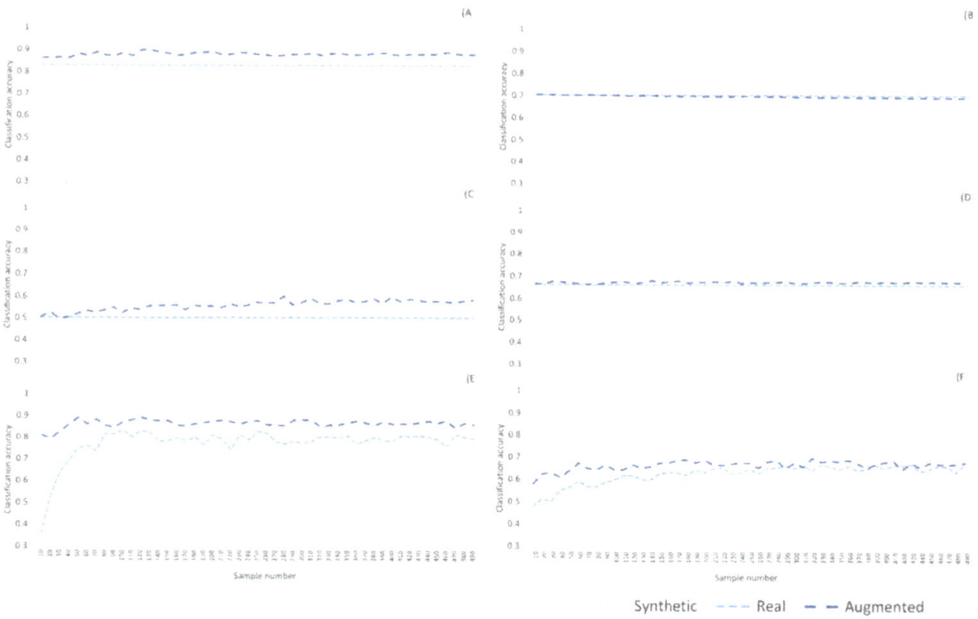

Figure A1. (**A**) New Zealand dataset; SVM classifier, (**B**) Indian Pines dataset; SVM classifier, (**C**) New Zealand dataset; RF classifier, (**D**) Indian Pines dataset; RF classifier, (**E**) New Zealand dataset; NN classifier, and (**F**) Indian Pines dataset; NN classifier.

References

1. Hennessy, A.; Clarke, K.; Lewis, M. Hyperspectral Classification of Plants: A Review of Waveband Selection Generalisability. *Remote Sens.* **2020**, *12*, 113. [CrossRef]
2. JPL/NASA. ECOSTRESS. 2021. Available online: https://ecostress.jpl.nasa.gov/ (accessed on 28 May 2021).
3. Hueni, A.; Chisholm, L.A.; Ong, C.C.H.; Malthus, T.J.; Wyatt, M.; Trim, S.A.; Schaepman, M.E.; Thankappan, M. The SPECCHIO Spectral Information System. *IEEE J. Sel. Top. Appl. Earth Obs. Remote Sens.* **2020**, *13*, 5789–5799. [CrossRef]
4. Taylor, L.; Nitschke, G. Improving Deep Learning Using Generic Data Augmentation. 2017. Available online: https://arxiv.org/abs/1708.06020 (accessed on 12 November 2020).
5. Wang, K. Synthetic DATA Generation and Adaptation for Object Detection in Smart Vending Machines. 2019. Available online: https://arxiv.org/abs/1904.12294 (accessed on 8 December 2020).

6. Goodenough, A.A.; Brown, S.D. DIRSIG5: Next-Generation Remote Sensing Data and Image Simulation Framework. *IEEE J. Sel. Top. Appl. Earth Obs. Remote Sens.* **2017**, *10*, 4818–4833. [CrossRef]
7. Bissoto, A.; Perez, F.V.M.; Valle, E.; Avila, S. Skin Lesion Synthesis with Generative Adversarial Networks. In *Transactions on Petri Nets and Other Models of Concurrency XV*; Springer Science and Business Media LLC: Cham, Switzerland, 2018; pp. 294–302.
8. Wang, G.; Kang, W.; Wu, Q.; Wang, Z.; Gao, J. Generative Adversarial Network (GAN) Based Data Augmentation for Palmprint Recognition. In Proceedings of the 2018 Digital Image Computing: Techniques and Applications (DICTA), Canberra, Australia, 10–13 December 2018. [CrossRef]
9. Wen, Q.; Sun, L.; Yang, F.; Song, X.; Gao, J.; Wang, X.; Xu, H. Time Series Data Augmentation for Deep Learning: A Survey. 2020. Available online: https://arxiv.org/abs/2002.12478 (accessed on 9 November 2020).
10. Jacquemoud, S.; Verhoef, W.; Baret, F.; Bacour, C.; Zarco-Tejada, P.J.; Asner, G.P.; François, C.; Ustin, S.L. PROSPECT+SAIL models: A review of use for vegetation characterization. *Remote Sens. Environ.* **2009**, *113*, S56–S66. [CrossRef]
11. Slavkovikj, V.; Verstockt, S.; De Neve, W.; Van Hoecke, S.; Van de Walle, R. Hyperspectral Image Classification with Convolutional Neural Networks. In Proceedings of the 23rd ACM international conference on Multimedia, Brisbane, Australia, 26–30 October 2015; pp. 1159–1162.
12. Nalepa, J.; Myller, M.; Kawulok, M.; Smolka, B. On data augmentation for segmenting hyperspectral images. In *Real-Time Image Processing and Deep Learning 2019*; International Society for Optics and Photonics: Bellingham, WA, USA, 2019; Volume 10996, p. 1099609. [CrossRef]
13. Harada, S.; Hayashi, H.; Uchida, S. Biosignal Generation and Latent Variable Analysis with Recurrent Generative Adversarial Networks. *IEEE Access* **2019**, *7*, 144292–144302. [CrossRef]
14. Donahue, C.; McAuley, J.; Puckette, M. Adversarial Audio Synthesis. 2017. Available online: https://arxiv.org/abs/1802.04208 (accessed on 1 February 2020).
15. Esteban, C.; Hyland, S.L.; Rätsch, G. Real-Valued (Medical) Time Series Generation with Recurrent Conditional Gans. Available online: https://arxiv.org/abs/1706.02633 (accessed on 12 August 2019).
16. Goodfellow, I.J.; Pouget-Abadie, J.; Mirza, M.; Xu, B.; Warde-Farley, D.; Ozair, S.; Courville, A.; Bengio, Y. Generative adversarial nets. In Proceedings of the 27th International Conference on Neural Information Processing Systems—Volume 2 (NIPS'14), Cambridge, MA, USA, 8–13 December 2014; pp. 2672–2680.
17. Audebert, N.; Le Saux, B.; Lefevre, S. Generative Adversarial Networks for Realistic Synthesis of Hyperspectral Samples. In Proceedings of the IGARSS 2018—2018 IEEE International Geoscience and Remote Sensing Symposium, Valencia, Spain, 22–27 July 2018; pp. 4359–4362.
18. Zhan, Y.; Hu, D.; Wang, Y.; Yu, X. Semisupervised Hyperspectral Image Classification Based on Generative Adversarial Networks. *IEEE Geosci. Remote Sens. Lett.* **2017**, *15*, 212–216. [CrossRef]
19. Xu, Y.; Du, B.; Zhang, L. Can We Generate Good Samples for Hyperspectral Classification?—A Generative Adversarial Network Based Method. In Proceedings of the IGARSS 2018—2018 IEEE International Geoscience and Remote Sensing Symposium, Valencia, Spain, 22–27 July 2018; pp. 5752–5755.
20. Feng, J.; Yu, H.; Wang, L.; Cao, X.; Zhang, X.; Jiao, L. Classification of Hyperspectral Images Based on Multiclass Spatial–Spectral Generative Adversarial Networks. *IEEE Trans. Geosci. Remote Sens.* **2019**, *57*, 5329–5343. [CrossRef]
21. Zhu, L.; Chen, Y.; Ghamisi, P.; Benediktsson, J.A. Generative Adversarial Networks for Hyperspectral Image Classification. *IEEE Trans. Geosci. Remote Sens.* **2018**, *56*, 5046–5063. [CrossRef]
22. LeCun, Y. What Are Some Recent and Potentially Upcoming Breakthroughs in Deep Learning. 2016. Available online: https://www.quora.com/What-are-some-recent-and-potentially-upcoming-breakthroughs-in-deep-learning (accessed on 15 January 2021).
23. Gui, J.; Sun, Z.; Wen, Y.; Tao, D.; Ye, J. A Review on Generative Adversarial Networks: Algorithms, Theory, and Applications. 2020. Available online: https://arxiv.org/abs/2001.06937 (accessed on 20 January 2021).
24. Arjovsky, M.; Chintala, S.; Bottou, L. Wasserstein GAN. 2017. Available online: https://arxiv.org/abs/1701.07875 (accessed on 8 December 2019).
25. Gulrajani, I.; Ahmed, F.; Arjovsky, M.; Dumoulin, V.; Courville, A. Improved training of wasserstein gans. *arXiv* **2017**, arXiv:1704.00028.
26. Bellemare, M.; Danihelka, I.; Dabney, W.; Mohamed, S.; Lakshminarayanan, B.; Hoyer, S.; Munos, R. The Cramer Distance as a Solution to Biased Wasserstein Gradients. 2017. Available online: https://arxiv.org/abs/1705.10743 (accessed on 12 March 2020).
27. Song, J. Cramer-Gan. 2017. Available online: https://github.com/jiamings/cramer-gan (accessed on 28 December 2019).
28. Mountrakis, G.; Xi, B. Assessing reference dataset representativeness through confidence metrics based on information density. *ISPRS J. Photogramm. Remote Sens.* **2013**, *78*, 129–147. [CrossRef]
29. Baumgardner, M.F.; Biehl, L.L.; Landgrebe, D.A. 220 Band AVIRIS Hyperspectral Image Data Set: June 12, 1992 Indian Pine Test Site 3. 2015. Available online: https://purr.purdue.edu/publications/1947/1 (accessed on 6 July 2018).
30. McInnes, L.; Healy, J.; Melville, J. UMAP: Uniform Manifold Approximation and Projection for Dimension Reduction. 2018. Available online: https://arxiv.org/abs/1802.03426 (accessed on 19 July 2019).

remote sensing MDPI

Article

Rice Leaf Blast Classification Method Based on Fused Features and One-Dimensional Deep Convolutional Neural Network

Shuai Feng [1], Yingli Cao [1,2], Tongyu Xu [1,2,*], Fenghua Yu [1,2], Dongxue Zhao [1] and Guosheng Zhang [1]

1 College of Information and Electrical Engineering, Shenyang Agricultural University, Shenyang 110866, China; fengshuai@stu.syau.edu.cn (S.F.); caoyingli@syau.edu.cn (Y.C.); adan@syau.edu.cn (F.Y.); zhaoDX@stu.syau.edu.cn (D.Z.); gszhang@stu.syau.edu.cn (G.Z.)
2 Liaoning Engineering Research Center for Information Technology in Agriculture, Shenyang 110866, China
* Correspondence: xutongyu@syau.edu.cn; Tel.: +86-024-8848-7121

Abstract: Rice leaf blast, which is seriously affecting the yield and quality of rice around the world, is a fungal disease that easily develops under high temperature and humidity conditions. Therefore, the use of accurate and non-destructive diagnostic methods is important for rice production management. Hyperspectral imaging technology is a type of crop disease identification method with great potential. However, a large amount of redundant information mixed in hyperspectral data makes it more difficult to establish an efficient disease classification model. At the same time, the difficulty and small scale of agricultural hyperspectral imaging data acquisition has resulted in unrepresentative features being acquired. Therefore, the focus of this study was to determine the best classification features and classification models for the five disease classes of leaf blast in order to improve the accuracy of grading the disease. First, the hyperspectral imaging data were pre-processed in order to extract rice leaf samples of five disease classes, and the number of samples was increased by data augmentation methods. Secondly, spectral feature wavelengths, vegetation indices and texture features were obtained based on the amplified sample data. Thirdly, seven one-dimensional deep convolutional neural networks (DCNN) models were constructed based on spectral feature wavelengths, vegetation indices, texture features and their fusion features. Finally, the model in this paper was compared and analyzed with the Inception V3, ZF-Net, TextCNN and bidirectional gated recurrent unit (BiGRU); support vector machine (SVM); and extreme learning machine (ELM) models in order to determine the best classification features and classification models for different disease classes of leaf blast. The results showed that the classification model constructed using fused features was significantly better than the model constructed with a single feature in terms of accuracy in grading the degree of leaf blast disease. The best performance was achieved with the combination of the successive projections algorithm (SPA) selected feature wavelengths and texture features (TFs). The modeling results also show that the DCNN model provides better classification capability for disease classification than the Inception V3, ZF-Net, TextCNN, BiGRU, SVM and ELM classification models. The SPA + TFs-DCNN achieved the best classification accuracy with an overall accuracy (OA) and Kappa of 98.58% and 98.22%, respectively. In terms of the classification of the specific different disease classes, the F1-scores for diseases of classes 0, 1 and 2 were all 100%, while the F1-scores for diseases of classes 4 and 5 were 96.48% and 96.68%, respectively. This study provides a new method for the identification and classification of rice leaf blast and a research basis for assessing the extent of the disease in the field.

Keywords: rice leaf blast; hyperspectral imaging data; deep convolutional neural networks; fused features

Citation: Feng, S.; Cao, Y.; Xu, T.; Yu, F.; Zhao, D.; Zhang, G. Rice Leaf Blast Classification Method Based on Fused Features and One-Dimensional Deep Convolutional Neural Network. *Remote Sens.* **2021**, *13*, 3207. https://doi.org/10.3390/rs13163207

Academic Editor: Chein-I Chang

Received: 15 July 2021
Accepted: 10 August 2021
Published: 13 August 2021

1. Introduction

Crop pests and diseases cause huge losses of agricultural production [1]. According to the Food and Agriculture Organization of the United Nations, the annual reduction in food production caused by pests and diseases accounts for about 25% of the total food production worldwide, with 14% of the reduction caused by diseases and 10% by pests [2]. In China, the amount of grain lost due to pest and disease outbreaks and hazards is about 30% of the total production each year, which has a huge impact on the domestic economy [3]. We still mainly rely on plant protection personnel to conduct field surveys and field sampling in order to monitor crop disease. Although these traditional detection methods have high accuracy and reliability, they are time-consuming, laborious and lack representativeness. These traditional diagnostic methods mainly rely on the subjective judgment of investigators, which is prone to human misjudgment, subjective errors and variability [4–7]. Therefore, there is an urgent need to improve pest and disease monitoring and control methods.

Rice blast is one of the most serious rice diseases in the north and south rice-growing areas of China and it is known as one of the three major rice diseases together with bacterial blight and sheath blight [8]. In September 2020, rice blast was listed as a Class I crop pest by the Ministry of Agriculture and Rural Affairs of China. Rice blast is caused by magnaporthe grisea and phytophthora grisea, which infest the leaves, neck and ears of rice by producing conidia and causing devastating effects on the physiological aspects of rice growth [9]. According to the period and location of damage, rice blast can be divided into the seedling blast, leaf blast and spike blast, etc., among which leaf disease is the most harmful. Leaf blast usually occurs after the three-leaf stage of rice plants and is increasingly serious from the tillering stage to the jointing stage. The spots first appear as white dots and gradually become 1~3 cm long diamond-shaped spots. The disease spot is gray in the middle and is surrounded by a dark brown color. In severe infestation, the entire leaf dries out [10,11] and reduces the green leaf area and photosynthesis in the lesioned area [12], thus causing a substantial rice yield reduction. It generally causes a 10~30% yield reduction in rice. Under favourable conditions, it can destroy an entire rice field in 15 to 20 days and cause up to 100% yield loss [13]. In China, the average annual occurrence of rice blasts is as high as 3.8 million hectares, with annual losses of hundreds of millions of kilograms of rice. In order to control the spread of leaf blast fungus over a large area and reduce yield losses. It is urgent to develop methods of rapid and accurate monitoring and discrimination of leaf blast disease.

Spectroscopy is a commonly used technique for plant disease detection, and its non-destructive, rapid and accurate characteristics have attracted the attention of a wide range of scholars [14]. Multispectral techniques [15,16] and near-infrared spectroscopy [17,18] have been studied in crop disease stress classification. However, multispectral and near-infrared techniques obtain less spectral data information, making it more difficult to detect the disease at its early stage of development and resulting in the inability to accurately discriminate against it. Compared with the above-mentioned spectroscopic techniques, hyperspectral imaging technology, which has characteristics of multiple spectral bands, high resolution and can provide spatial-domain and spectral-domain information, has thus gradually become a research hotspot for scholars. This technique has been widely used for disease detection in vegetables [19,20], fruits [21,22] and grains [23–25]. In recent years, with the development and application of hyperspectral imaging technology, the technology has made great progress in crop disease detection and greatly improved the science of accurate prevention and controls and management decisions in the field. Luo et al. [26], after comparing the accuracy of rice blast identification with different spectral processing methods and modeling approaches, concluded that the probabilistic neural network classification based on logarithmic spectra was the best, with an accuracy of 75.5% in the test set. Liu et al. [27] used support vector machine and extreme learning machine methods to model and classify white scab and anthracnose of tea, respectively, with a classification accuracy of 95.77%. Yuan et al. [28] extracted hyperspectral data from healthy and diseased leaves without disease spots, leaves with less than 10% disease spot area and less than 25% disease spot area, respectively, and used CARS-PCA for dimensionality reduction in order to construct SVM rice blast classification models. The accuracy of all categories was greater than 94.6%. Knauer et al. [29] used hyperspectral imaging for the accurate classification of powdery mildew of wine grapes. Nagasubramanian et al. [30] used hyperspectral techniques and built soybean charcoal rot early identification models based on genetic algorithms and support vector machines. Nettleton et al. [31] used operational process-based models and machine learning models for the predictive analysis of rice blast. It was concluded that machine learning methods showed better adaptation to the prediction of rice blast in the presence of a training data set. All the above-mentioned studies achieved good results, but all of them focused on the detection of diseases in crops using spectral information from hyperspectral images, and they did not address texture features in hyperspectral images which are directly related to disease characterization. Texture features as inherent properties possessed by the crop, which are not easily disturbed by the external environment, can reflect the image properties and the spatial distribution of adjacent pixels, compensating to some extent for the saturation of crop disease detection relying only on spectral information [32]. Zhang et al. [33] used spectral features and texture features to construct a support vector machine classification model. The results demonstrated that the classification model was able to effectively classify healthy, moderate and severe diseases in wheat. Al-Saddik et al. [34] concluded that combining texture features of grape leaves and spectral information to construct a classification model resulted in the effective classification of yellowness and esca with an overall accuracy of 99%. Zhang and Zhu et al. [35,36] concluded after analysis that the classification model constructed by fusing spectral and texture features had superior classification accuracy compared to the classification model using only spectral or texture features. The above literature shows that it is feasible to construct plant disease classification models by fusing spectral and texture information from hyperspectral images. However, the study of using fusion features of spectral and textural information to discriminate different disease levels of rice leaf blast needs to be explored deeply.

In the above-mentioned studies, researchers mostly used machine learning methods such as support vector machines and back propagation neural networks to model hyperspectral data. However, there are still relatively few studies using deep learning methods for crop disease identification and recognition based on hyperspectral imaging data. The reason for this may be the small quantity of sample data obtained, which makes it impossible to build a deep learning model. In existing studies, researchers have mostly used deep learning methods to build models for hyperspectral data due to the powerful feature extraction capabilities of these models. Nagasubramanian et al. [37] constructed a 3D convolutional neural network recognition model for soybean charcoal rot by using hyperspectral image data with a classification accuracy of 95.73%. Huang et al. [38] obtained hyperspectral images of rice spike blast and constructed a detection model based on the GoogLeNet method with a maximum accuracy of 92%. Zhang et al. [39] used a three-dimensional deep convolutional neural network model to model yellow rust of winter wheat with an overall accuracy of 85%. Although this modeling approach can achieve high accuracy rates, it still requires the use of expensive hyperspectral instruments in practical agricultural applications in order to obtain data and cannot be applied on a large scale.

In view of this, this study draws on existing research methods to expand the sample data size. Data dimensionality reduction uses augmented sample data to extract spectral feature wavelengths, vegetation indices and texture features. A total of seven one-dimensional deep convolutional neural network classification models were constructed for leaf blast disease classification based on the above features and their fusion features. Finally, Inception V3, ZF-Net, BiGRU, TextCNN, SVM and ELM models were used for comparative analysis with the model of this study to determine the best classification features and classification model for leaf blast. It is expected to provide some scientific theory and technical support for the identification of rice leaf blast disease grades.

2. Materials and Methods

2.1. Study Site

Rice leaf blast trials were conducted from July to August 2020 at Liujiaohe Village, Shenyang New District, Shenyang, Liaoning Province (42°01′17.16″N, 123°38′14.57″E). The region has a temperate semi-humid continental climate, with an average annual temperature of 9.7 °C and an average annual precipitation of 700 mm, making it a typical cold-land rice-growing area. Mongolian rice with a high susceptibility to leaf blast was used as the test variety, and it was planted on an area of about 100 m^2 with a row spacing of 30 cm and a plant spacing of 17 cm. Nitrogen, potassium and phosphorus fertilizers were applied according to local standards at 45, 15 and 51.75 kg/hm^2, respectively. Prior to basal fertilizer application, soil samples were collected using the five-point sampling method from the disease trial plots, and soil nutrients were measured and analyzed. The measured results showed that the rapid potassium content ranged from 86.83 to 120.62 mg/kg; the effective phosphorus content ranged from 3.14 to 21.18 mg/kg; the total nitrogen content ranged from 104.032 to 127.368 mg/kg; and the organic matter content ranged from 15.8 to 20.0 g/kg. Leaf blast inoculation was carried out at 5:00 p.m. on the same day (3 July 2020) by using a spore suspension at a concentration of 9 mg/100 mL (in order to inoculate, the spore suspension was shaken well and sprayed evenly over the surface of the plant leaves until the leaves were completely covered with water droplets), which was wrapped in a moistened black plastic bag after inoculation and removed at 6:30 a.m. the following morning. The test plots were not treated with any disease control, and field management was normal. Five days after inoculation, the plants began to show symptoms, and healthy and diseased rice leaves were obtained from the field under the guidance of a plant protection specialist and taken back to the hyperspectral laboratory in order to obtain hyperspectral image data.

2.2. Data Acquisition and Processing

2.2.1. Sample Collection

Five trials were conducted to collect healthy and diseased plants at three critical fertility stages: the rice jointing stage (8 July; 15 July), the booting stage (25 July; 2 August) and the heading stage (10 August). Under the supervision of plant protection experts, 57, 61 and 27 leaf samples with five different levels of disease were collected at the jointing, booting and heading stages, respectively, and a total of 145 rice leaf samples were obtained. In the experiment, in order to maintain the moisture content of the rice leaves, the leaves were placed in a portable refrigerator to maintain their freshness. Hyperspectral image data were then acquired indoors by using a hyperspectral imaging system. Figure 1 shows pictures of healthy and different disease grades of rice leaves. We used ENVI 5.3 (ITT Visual Information Solutions, Boulder, CO, USA) software for manual segmentation of rice leaves, leaf background and disease areas. The number of pixel points for the whole leaf and the diseased area was calculated, along with the number of diseased pixel points as a percentage of the number of pixel points on the leaf. According to the GBT 15790-2009 Rules of Investigation and Forecast of the Rice Blast, classification was carried out according to the size of the disease spot, as shown in Table 1. Level 5 leaf blast samples were not found in this study; therefore, the criteria for determining level 5 disease are not listed in Table 1.

Figure 1. Healthy and different disease levels of rice leaves.

Table 1. Criteria for determining different disease levels of leaf blades and sample size.

Disease Level	Disease Level Determination Criteria	Sample Size
Level 0	No disease spots.	29
Level 1	Few and small spots, disease spot area less than 1% of leaf area.	27
Level 2	Small and many spots or large and few disease spot area of 1~5% of leaf area.	32
Level 3	Large and more spots, disease spot area of 5~10% of leaf area.	27
Level 4	Large and more spots, disease spot area of 10~50% of leaf area.	30

2.2.2. Hyperspectral Image Acquisition

In this study, a hyperspectral imaging system was used to acquire hyperspectral images of rice leaves, as shown in Figure 2. The main components of the system include a hyperspectral imaging spectrometer (ImSpector V10E, Spectral Imaging Ltd., Oulu, Finland), a high-definition camera (IGV-B1410, Antrim, Northern Ireland), a precision displacement control stage, a light-free dark box, two 150 W fiber optic halogen lamps (Ocean Optics, Dunedin, FL, USA) and a computer. The effective spectral range obtained by this hyperspectral imaging system is 400–1000 nm with a spectral resolution of 0.64 nm. The distance of the camera lens from the surface of the rice leaves was set to 32 cm before acquiring the images. The lens focus was adjusted by using a white paper focusing plate with black stripes until the black stripes were imaged and the transition area between the black stripes and the white paper was clear. In order to obtain the best image quality, the light source intensity and exposure rate were adjusted and the scanning speed was set to 1.1 mm/s.

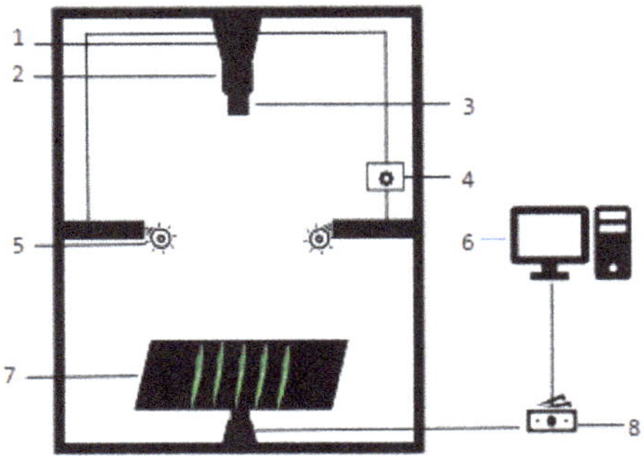

Figure 2. Hyperspectral imaging system: (1) EMCCD HD camera; (2) hyperspectral imaging spectrometer; (3) lens; (4) light source controller; (5) light source; (6) computer; (7) displacement stage; (8) displacement stage controller.

Due to the problem of inconsistent intensity values of different spatial hyperspectral data caused by the variation of light intensity on the leaf surface and the camera's dark current, the original hyperspectral images needed to be processed by black-and-white plate correction by using Equation (1) to obtain the final image spectral reflectance:

$$I = \frac{R_S - R_D}{R_W - R_D} \quad (1)$$

where I is the corrected hyperspectral reflectance of rice leaves, R_S is the spectral reflectance of the original hyperspectral images of rice leaves, and R_W and R_D are the spectral reflectance of the corrected white plate and corrected black plate, respectively. The acquisitions and transmissions of spectral images were completed by using the system's hyperspectral acquisition software (Isuzu Optics, Hsinchu, China).

2.2.3. Spectra Extraction and Processing

In this study, the whole rice leaf was treated as a separate region of interest (ROI), and ENVI5.3 was used to manually delineate the region of interest and extract its average spectral reflectance. This culminated in 29 health data and 116 disease data (27, 32, 27 and 30 disease data for levels 1, 2, 3 and 4 respectively), for a total of 145 hyperspectral imaging data.

In order to determine the best classification features and classification model for leaf blast, there were two main considerations in this study. Firstly, the leaf blast classification features extracted from the existing data scale are contingent and not universal. Secondly, the constructed leaf blast classification model is not generalizable and is not sufficient for constructing a deep learning model based on big data and calibrated supervision mechanisms. In view of these two considerations, in this study, the data set was divided into a training set and a testing set, and then the data augmentation method proposed by Chen et al. [40] for data augmentation was used. This method augments the data by adding light intensity perturbations and Gaussian noise to the raw spectral data to simulate interference factors such as uneven illumination and instrument noise. The formula is shown in Equation (2):

$$y_i = ny_{Gaussian} + alpx_i \quad (2)$$

where n is the weight of the control Gaussian noise $y_{Gaussian}$, alp is the light intensity perturbation factor and x_i is the raw spectral data. Figure 3 shows the effect of data augmentation.

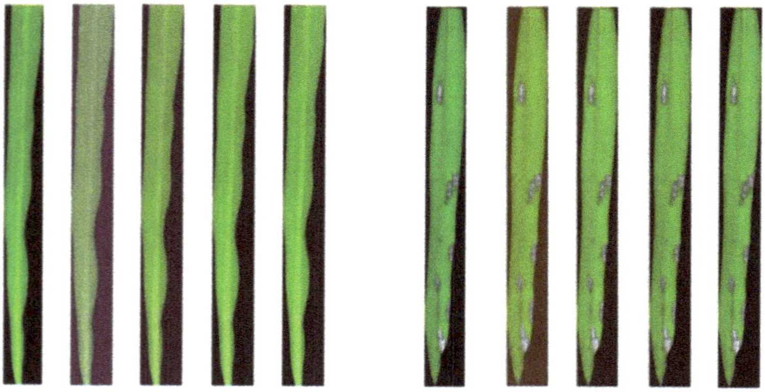

Figure 3. The effect of data augmentation.

In the end, a total of 986 healthy sample data, 918 level 1 disease data, 1088 level 2 disease data, 918 level 3 disease data and 1020 level 4 disease data were obtained, resulting in a total of 4930 sample data. Figure 3 shows the effect of data augmentation.

2.3. Optimal Spectral Feature Selection

Hyperspectral data are characterized by rich information content, high resolution and band continuity, which can fully reflect the differences in physical structure and chemical composition within the leaf. However, there is still a large amount of redundant information in the spectral information, which affects modeling accuracy. Therefore, hyperspectral data need to be subjected to dimension-reduced processing to extract valid and representative spectral features as model input to improve modeling accuracy. In this study, no new descending dimension methods were proposed or used, but both the successive projections algorithm (SPA) and random frog (RF) methods were used to extract spectral feature wavelengths. This is due to the fact that a wide range of researchers have confirmed that the characteristic wavelengths of SPA and RF screening are representative. At the same time, the SPA and RF methods screen for a smaller number of characteristic wavelengths, making it easy to generalize and use the model. In this study, the SPA and RF methods were used to extract the feature wavelengths of the spectra.

SPA is a forward feature variable dimension reduction method [41]. SPA is able to obtain the combination of variables that contains the least redundant information and the minimum characteristic co-linearity. The algorithm uses projection analysis of vectors to map spectral wavelengths onto other spectral wavelengths in order to compare the magnitude of the mapping vectors and to. obtain the wavelength with the largest projection vector, which is the spectral wavelength to be selected. A multiple linear regression analysis model was then developed to obtain the RMSECV of the modeling set. The number and wavelength corresponding to the smallest RMSECV value in the different subsets of features to be selected consists of the optimal spectral feature wavelength combinations.

RF is a relatively new method of signature variable screening, initially used for gene expression data analysis of diseases [42]. The method uses the Reversible Jump Markov Chain Monte Carlo (RJMCMC) method to transform and sample the dimensions of the spectrum. From there, a Markov chain is modeled in space that conforms to the steady-state distribution to calculate the frequency of selection for each wavelength variable. The selection of frequencies was used as a basis for eliminating redundant variables, resulting in the best spectral characteristic wavelength.

2.4. Texture Features Extraction

Textural features contain important information about the structural tissue arrangement of the leaf spot surface and the association of the spot with its surroundings. Therefore, TFs can reflect the physical characteristics of the crop leaves and information on the growth status of the crop [26]. When leaf blast infects leaves, cell inclusions and cell walls are damaged, the chlorophyll content is reduced and the volume is reduced. This results in a change in color in some areas of the leaf surface and causes changes in textural characteristics.

A gray-level co-occurrence matrix (GLCM) is a common method for extracting texture features on the leaf surface. It reflects the comprehensive information of the image in terms of direction, interval and magnitude of change by calculating the correlation between the gray levels of two points at a certain distance and in a certain direction in the image [43]. At the same time, the energy, entropy, correlation and contrast can better reflect the difference between the diseased and normal parts of the leaf, thus improving the modeling accuracy (energy reflects the degree of gray distribution and texture thickness; entropy is a measure of the amount of information in the image; correlation measures the similarity of images at the gray level in the row or column direction; and contrast reflects the sharpness of the image and the depth of the texture grooves). Hence, in this study, energy, entropy, correlation and contrast were calculated from four directions, namely 0°, 45°, 90° and 135°, at a relative pixel distance d of 1. The formulae for energy, entropy, correlation and contrast are shown in Table 2. The average and standard deviation were calculated for energy, entropy, correlation and contrast in each of the four directions. A total of eight texture features were obtained, specifically the mean value of energy (MEne), the standard deviation of capacity (SdEne), the mean value of entropy (MEnt), the standard deviation of entropy (SdEnt), the mean value of correlation (MCor), the standard deviation of correlation (SDCor), the mean value of contrast (MCon) and the standard deviation of contrast (SDCon).

Table 2. Four texture features extracted from the GLCM.

Texture Features	Equation
Entropy	$-\sum_i\sum_j P(i,j)lgP(i,j)$
Energy	$\sum_i\sum_j P(i,j)^2$
Correlation	$\sum_i\sum_j \frac{(i-\mu)(j-\mu)}{\sigma^2} P(i,j)$
Contrast	$\sum_i\sum_j (i-j)^2 P(i,j)$

Note: i and j represent the row number and column number of the grayscale co-occurrence matrix, respectively; $P(i,j)$ denotes the relative frequency of two neighboring pixels.

2.5. Vegetation Index Extraction

VIs are indicators constructed by combining different spectral bands in linear and nonlinear combinations, and they are often used to monitor and discriminate the degree of vegetation disease. In this study, the VIs with the highest correlation of leaf blast disease levels were screened by establishing a contour of the decision coefficient. The method arbitrarily selects two spectral bands in the spectral range to construct a certain spectral index, and then the Pearson correlation coefficient method is used to calculate the correlation between the disease class and the vegetation index to find the vegetation index with a higher classification ability.

Based on previous research results, the ratio spectral index (RSI), the difference spectral index (DSI) and the normalized difference spectral index (NDSI) were used to construct the contour of the decision coefficient. The formula is as follows:

$$RSI = R_i / R_j \tag{3}$$

$$RSI = R_i / R_j \tag{4}$$

$$NDSI = R_i - R_j / R_i + R_j \tag{5}$$

where R_i and R_j denote the spectral reflectance values in the spectral band range.

2.6. Disease Classification Model

Deep Convolutional Neural Network

The human visual system has a powerful ability to classify, monitor and recognize. Therefore, in recent years, a wide range of researchers have been inspired by bio-vision systems to develop advanced data processing methods. Convolutional Neural Networks (CNNs) are deep neural networks developed to emulate biological perceptual mechanisms. The networks are capable of automatically extracting sensitive features at both shallow and deep levels in the data. The Residual Network (ResNet) [44] is a typical representative of CNN, as shown in Figure 4. The residual module (both the direct mapping and residual components) is designed with the idea of the better extraction of data features and to prevent degradation of the network. ResNet is well recognized for its feature extraction and classification in the ILSVRC 2015 competition.

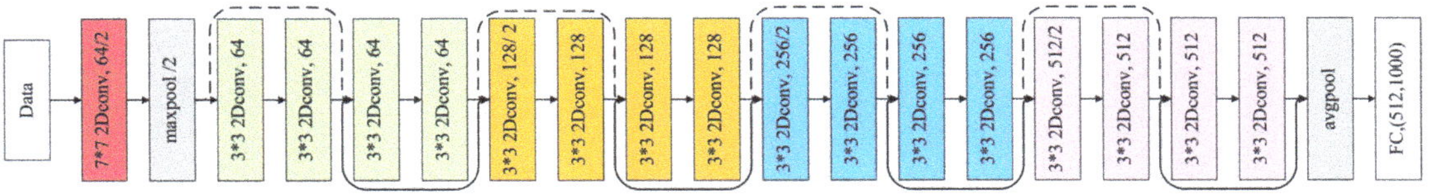

Figure 4. ResNet structure.

As ResNet has a deeper network hierarchy, it is prone to over-fitting during training. ResNet was initially used mainly in image classification and was not applicable to spectral data. Therefore, this study adapts ResNet to render it suitable for modeling one-dimensional data. Firstly, the data in this study were all one-dimensional, and thus the number of input features was used as the network input, and there was no need to experimentally derive the optimum input layer size. The number of channels in the FC layer of ResNet was also adjusted to 5 for the 5 classification problems of normal, level 1, level 2, level 3 and level 4 diseases of rice leaf blast. ResNet is a DCNN designed for application to large-scale data, and its training process is computationally intensive. The classification problems for different disease classes are smaller in terms of data size and computational effort of training. Therefore, in order to improve the modeling effect of the model, different types of classification networks were designed by adjusting the network depth and structure of ResNet by adding the BatchNorm layer and Dropout layer while maintaining the design concept of ResNet (Figure 5) in order to be applicable to the data obtained from this study. The model in this paper was compared and analyzed with SVM [45], ELM [46], Inception V3 [47], ZF-Net [48], BiGRU [49] and TextCNN [50] models to determine the best leaf blast disease class classification model.

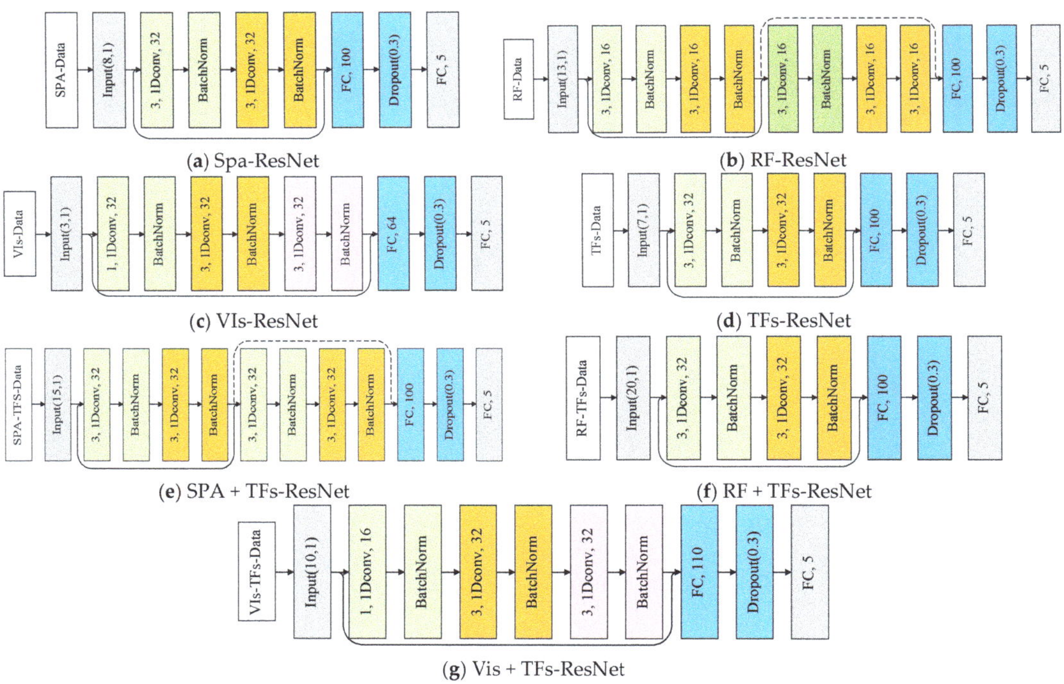

Figure 5. DCNN models with different dimensionality reduction methods.

The above DCNN model was built using the deep learning computational framework Keras 2.3 for model building. The hardware environment for the experiments was 32G RAM, Bronze 3204 CPU and Quadro P5000 GPU.

3. Results

3.1. Spectral Response Characteristics of Rice Leaves

As shown in Figure 6, the mean spectral reflectance of healthy rice leaves and disease-susceptible leaves showed a consistent trend. The reflectance at 500~600 and 770~1000 nm changed significantly after rice blast spores infested the leaves. There is a slight increase in the reflectance of diseased leaves in the 500 to 600 nm range. At 700~1000 nm, the reflectance decreases significantly. In the range of 680 to 770 nm, the spectral curves of the different disease degrees were shifted to the short-wave direction compared to the healthy leaf spectral curves, i.e., the phenomenon of "blue shift". This is due to damage to chloroplasts or other organelles within the leaf caused by the disease and changes in pigment content, resulting in changes in spectral reflectance [51]. The band range between 400 and 450 nm shows severe reflectance overlap, and thus the band range of 450 to 1000 nm was chosen as the main band for spectral feature extraction.

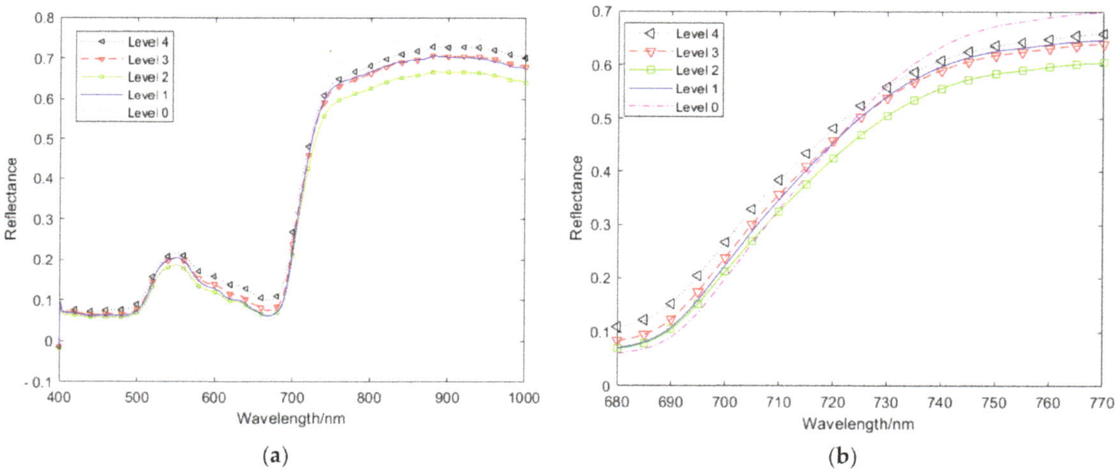

Figure 6. Comparison of average spectral curves. (**a**) Average spectral curves of diseases at 400 to 1000 nm. (**b**) Average spectral curves of diseases at 680 to 770 nm.

3.2. Optimal Features

3.2.1. Vegetation Indices

Figure 7 shows the contour of the decision coefficient of DSI, RSI and NDSI constituted by any two-band combinations with the leaf disease class. In Figure 7a, the NDSI constructed by the combination of spectral bands from 623 to 700 and 700 to 1000, 556 to 702 and 450 to 623 nm correlated well with the disease levels, and the coefficient of determination R^2 was greater than 0.8. Among them, the NDSI vegetation index constructed by the combination of 600 and 609 nm had the best correlation with R^2 of 0.8947. Compared with NDSI, RSI correlated better with the disease class in fewer band ranges, mostly concentrated in the visible band range (Figure 7b). The best RSI vegetation index was constructed for the combination of 725 and 675 nm with an R^2 of 0.9103. Relatively, the DSI constructed at 548 nm and 698 nm had the highest correlation, with an R^2 of 0.800 (Figure 7c).

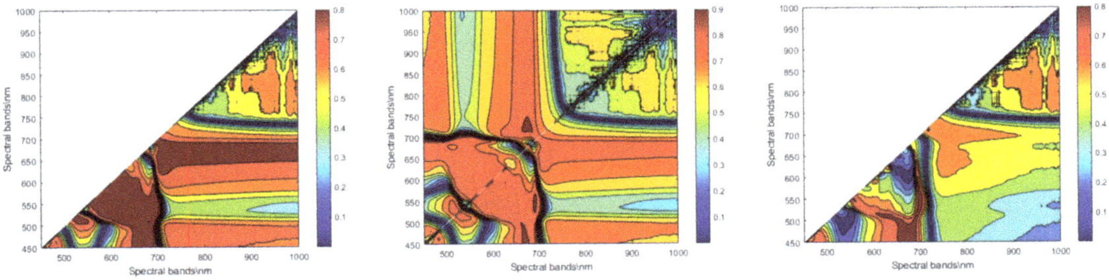

(a) (b) (c)

Figure 7. Contour of decision coefficient between disease levels and DSI, RSI and NDSI. (**a**) NDSI. (**b**) RSI. (**c**) DSI.

3.2.2. Extraction of Hyperspectral Features

The spectral data were processed by using the SPA to obtain the characteristic wavelengths of the spectra with high correlation. In this study, a minimum screening number of eight and a maximum screening number of ten were set, and the RMSE was used as the evaluation criterion for selecting the best spectral feature wavelength. Figure 8a shows the eight optimal spectral characteristic wavelengths, and the spectral wavelengths are given in Table 3. The RMSE curve drops sharply as the wavelength changes from 0 to 5 and stabilizes at the eighth wavelength. The final SPA selects eight spectral features at wavelengths evenly distributed in the visible, red-edge and near-infrared regions.

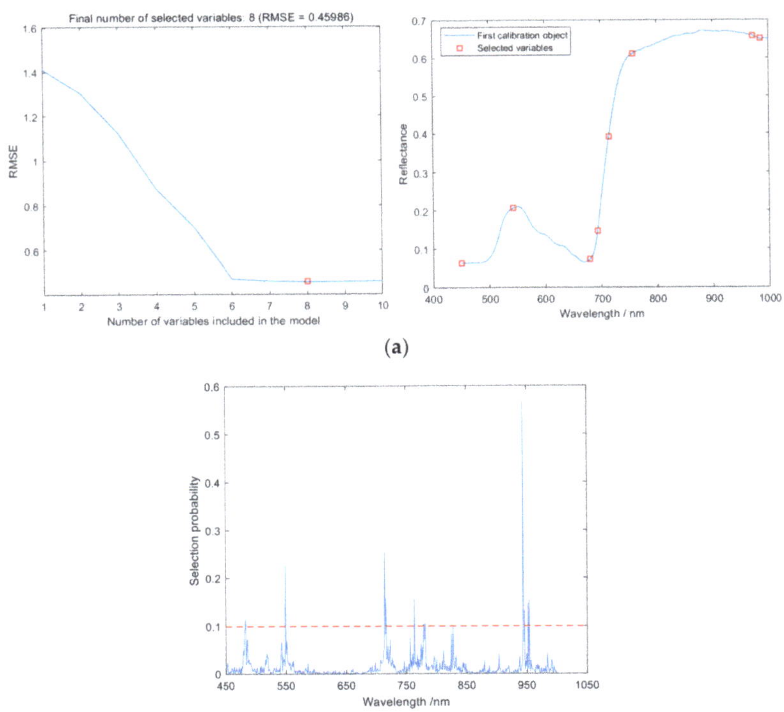

(a)

(b)

Figure 8. Selected optimal variables using (**a**) SPA and (**b**) RF.

Table 3. The variables selected by SPA and RF.

Method	Variable Number	Wavelength/Nm
SPA	8	450 543 679 693 714 757 972 985
RF	13	482 548 713 715 762 777 778 780 826 943 945 951 953

The RF algorithm was used to screen the spectral feature wavelengths, setting the maximum number of potential variables to 6, the initial number of sampled variables to 1000 and the screening threshold to 0.1. Given that the RF algorithm uses RJMCMC as the screening principle, the characteristic bands are slightly different each time they are screened. The RF algorithm was, therefore, run a total of 10 times, and the final average of the results was taken as the basis for the judgment of the characteristic wavelengths. The screening probability results for each spectral characteristic wavelength are shown in Figure 8b. The larger the screening probability, the more important the corresponding spectral feature wavelengths are; thus, the wavelengths with a screening probability greater than 0.1 were selected as the best spectral feature wavelengths (Table 3), with a total of 13 spectral feature wavelengths, accounting for approximately 2.36% of the full wavelength band.

3.2.3. Extraction of Texture Features by GLCM

Since hyperspectral images contain a large amount of redundant information, PCA is used to reduce the dimensionality of hyperspectral images and to generate principal component images containing a large amount of effective information. The cumulative contribution of the first three principal component images (PC1–PC3) was greater than 95% and, therefore, was used to extract texture features. Figure 9 shows the principal component images of healthy and diseased leaves after dimensionality reduction by PCA.

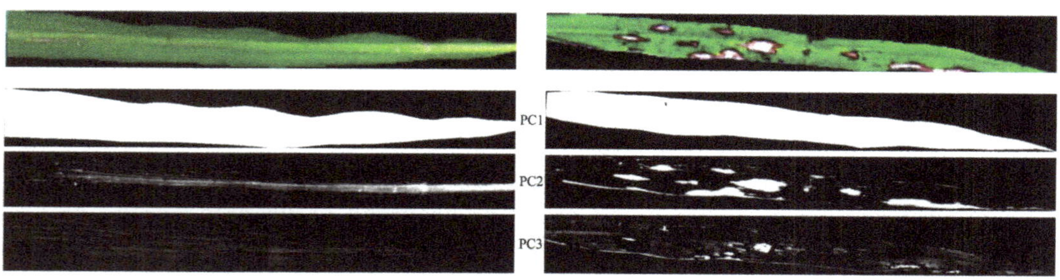

Figure 9. Principal component images of healthy and diseased leaves.

The GLCM was used to calculate the PC1-PC3 images separately to obtain eight features such as the means and standard deviations of the energy, entropy, contrast and correlation. In order to further improve the modeling accuracy, redundant texture features were removed. Eight texture features were subjected to Pearson correlation analysis with different disease classes to screen the significantly correlated and highly significantly correlated texture features, and the correlation coefficients and significance are shown in Table 4. The correlation and significance variation between the eight characteristics and the different disease classes can be observed in Table 4. Among them, MEne, SDEne, MEnt, SDEnt, MCon, SDCon and Mcor displayed highly significant correlations, while SDCor displayed a lower correlation. Therefore, in this study, seven highly significant features such as MEne were chosen as the final texture features to be modeled.

Table 4. Correlation of texture features with different disease classes.

Texture Features	Correlation Coefficient	*p* Value	Significance
MEne	0.5618	<0.001	***
SDEne	−0.2632	<0.001	***
MEnt	−0.4914	<0.001	***
SDEnt	−0.4263	<0.001	***
MCon	−0.2308	<0.001	***
SDCon	−0.2265	<0.001	***
MCor	0.1165	<0.001	***
SDCor	−0.0365	0.0105	**

Note: ** indicates significant correlation at 0.01 ($0.01 < p < 0.05$). *** indicates highly significant correlation at 0.001 ($p < 0.001$).

3.3. Sensitivity Analysis of the Number of Convolutional Layers and Convolutional Kernel Size for the DCNN

Figure 10 shows a comparison of the accuracy of the convolutional layers for different input features in the proposed model. From the figure, it can be observed that the DCNN constructed based on the features obtained from SPA, RF, TFs, SPA + TFs and RF + TFs achieved the best classification accuracy when the number of convolutional layers in the residual block was two. For Vis, Vis + TFs, the DCNN achieved the best classification results when the number of convolution layers was three.

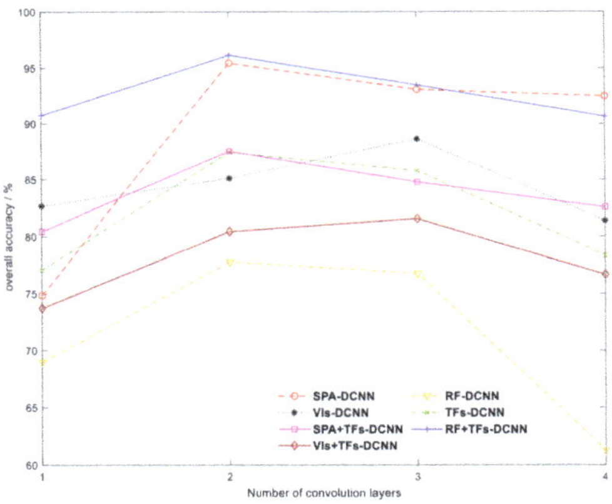

Figure 10. Effect of the number of DCNNs in the proposed DCNN model on classification accuracy.

Based on the optimal number of convolutional layers, we investigated the effect of different sizes of convolutional kernels on the classification accuracy through a set of experiments. Figure 11 shows a comparison of the accuracy of the models built with different sizes of convolutional kernels. When the convolutional kernel size was (3,3), the DCNN models constructed from features screened by SPA, RF, TFs, SPA + TFs and RF + TFs were better for classification. Meanwhile, the DCNN models constructed with VIs and Vis + TFs had the best classification accuracy when the convolutional kernel size was (1,3,3).

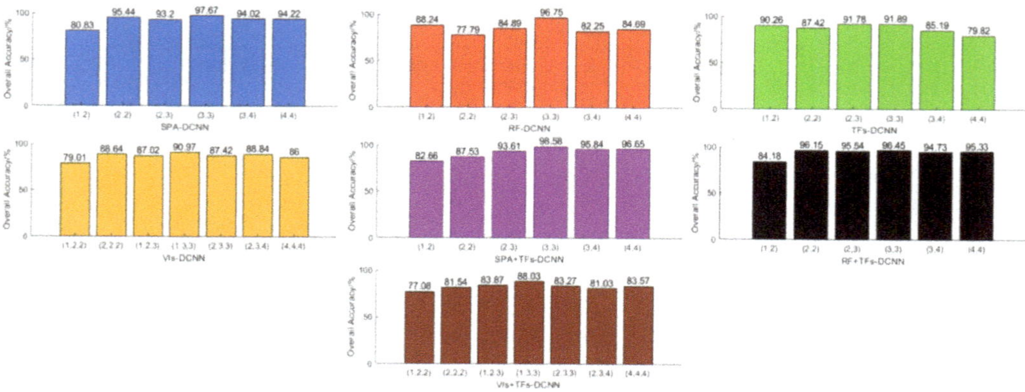

Figure 11. Comparison of the accuracy of models built with different sizes of convolutional kernels. Note: (3,3), etc., denotes two convolutional layers with convolutional kernel sizes of 3 and 3; (1,3,3), etc., denotes three convolutional layers with convolutional kernel sizes of 1, 2 and 3.

3.4. DCNN-Based Disease Classification of Rice Leaf Blast

3.4.1. DCNN Model Training and Analysis

The modeling was carried out using 4930 rice leaf blast data obtained for different disease classes as samples (including data obtained by data augmentation methods), where the training set, validation set and test set were divided according to 7:1:2. The relevant training experiments were carried out for the seven DCNN models with different dimensionality reduction methods in Figure 4. The overall accuracy (OA), Kappa coefficient and F1-score were selected as the model evaluation criteria for the experiment. In order to train the DCNN model, the Nadam algorithm [52] was used. The same learning rate was used for all layers in the network, with an initial learning rate of 0.002 and exponential decay rates of 0.9 and 0.999 for the first and second orders, respectively. The initialization of the weights has a large impact on the convergence speed of the model training. In this study, a normal distribution with a mean of 0 and a standard deviation of 0.01 was used to initialize the weights of all layers of the network, and the bias of the convolutional layer and the full connection was initialized to 0. In order to determine the best disease classification features and classification models, each DCNN model was fully trained. The epochs for SPA-DCNN, RF-DCNN, VIs-DCNN, TFs-DCNN, SPA + TFs-DCNN, RF + TFs-DCNN and Vis + TFs-DCNN were 200, 180, 300, 150, 150, 150 and 250. The training results of different DCNN models are shown in Figure 12.

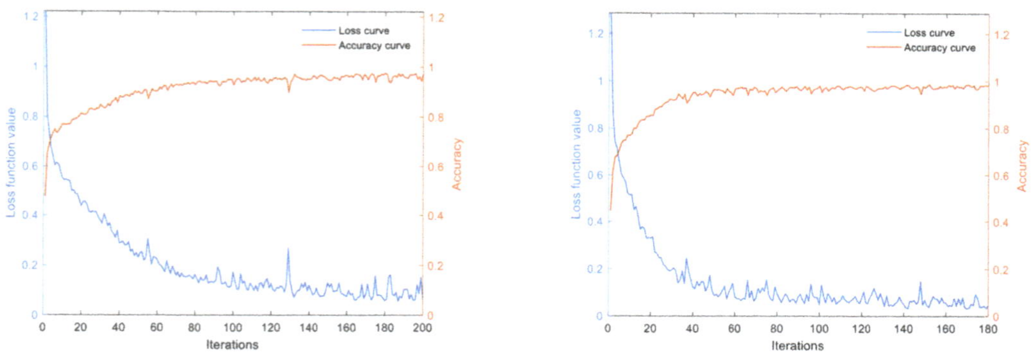

(a) SPA-DCNN (b) RF-DCNN

Figure 12. *Cont.*

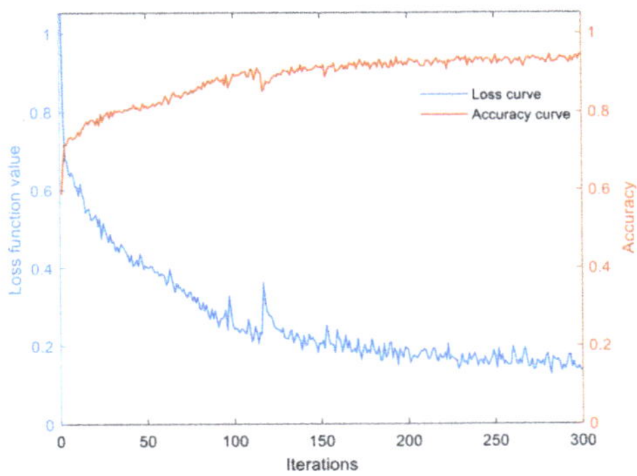

(c) VIs-DCNN

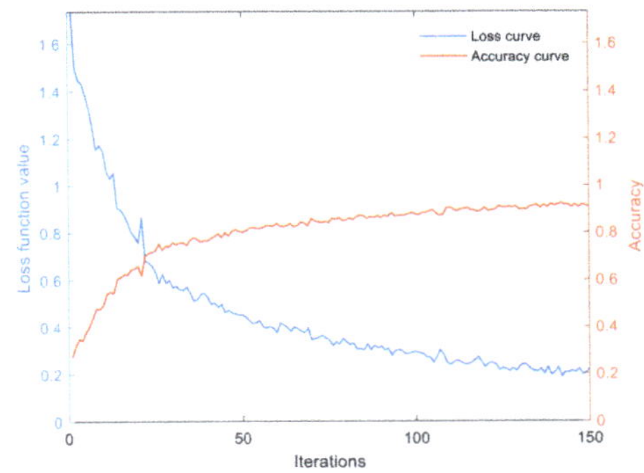

(d) TFs-DCNN

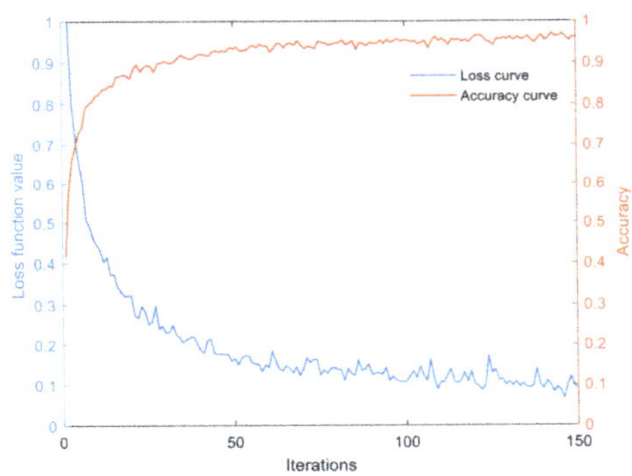

(e) SPA + TFs-DCNN

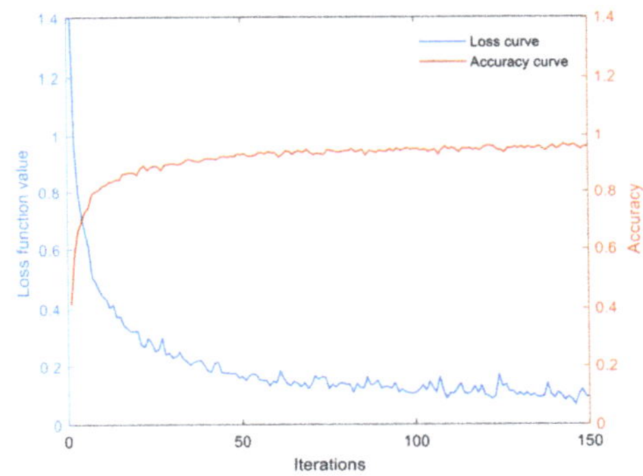

(f) RF + TFs-DCNN

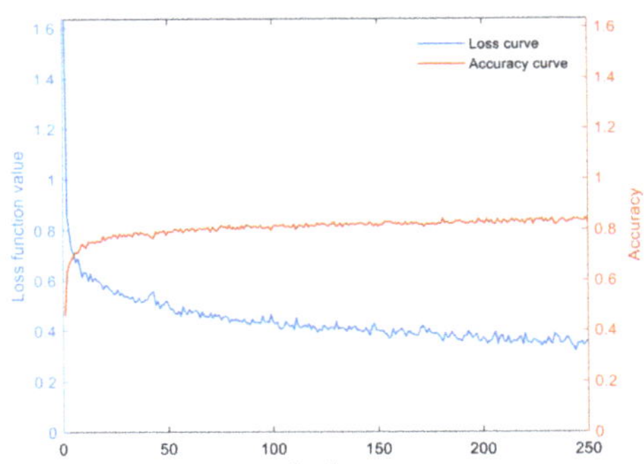

(g) Vis + TFs-DCNN

Figure 12. Change of loss function value and accuracy with iteration curves.

As can be observed from Figure 12, the training error of all DCNN models gradually decreases as the number of iterations increases and finally reaches a state of convergence. At the beginning of the training period, the training loss decreases rapidly by updating the gradient of the loss function with small batches of samples. This shows that batch_size and the optimization algorithm play a better role. In addition, as the training loss decreases, the prediction accuracy of the model for the training set shows an overall upward trend.

3.4.2. DCNN Model Testing and Analysis

In order to obtain the best leaf blast classification features, spectral features, vegetation indices, texture features (TFs) and their fusion features were used to construct the DCNN leaf blast classification model. The modeling results are shown in Table 5.

Table 5. Results of the DCNN disease classification model based on different features.

Descending Dimension Method	F1-Score (%)					OA (%)	Kappa (%)
	Level 0	Level 1	Level 2	Level 3	Level 4		
SPA	100	97.44	95.74	96.15	98.54	97.67	97.08
RF	100	96.05	94.51	95.01	97.73	96.75	95.93
VIs	98.36	84.18	87.04	88.64	95.48	90.97	88.70
TFs	92.67	92.23	92.93	86.88	93.96	91.89	89.84
SPA + TFs	100.00	100.00	100.00	96.48	96.68	98.58	98.22
RF + TFs	100.00	100.00	97.93	91.36	93.66	96.45	95.55
Vis + TFs	97.17	83.66	85.79	80.72	92.13	88.03	85.04

The data in Table 5 show that all seven DCNN models designed based on different characteristics have high classification accuracy with OA greater than 88% and Kappa coefficients greater than 85% for different disease degree classification. In the DCNN model constructed with a single feature, better classification results were obtained for the feature wavelengths selected by the SPA and RF methods, with OA and Kappa reaching 97.67% and 96.75% and 97.08% and 95.93%, respectively. In the DCNN model constructed based on TFs, although the model constructed was not as accurate as the spectral feature wavelength model, it still achieved better classification results, indicating that the image data also had the ability to identify rice leaf blast. Among the DCNN models constructed by fusing features, SPA + TFs-DCNN obtained the highest classification accuracy, with OA and Kappa of 98.58% and 98.22%, respectively. The F1-scores of SPA + TFs-DCNN are greater than those of the other fusion features for the identification of specific different disease classes. The F1-scores for Level 0, Level 1, Level 2, Level 3 and Level 4 were 100%, 100%, 100%, 96.48% and 96.68%, respectively. This result shows that the fusion of spectral wavelengths and textural features screened by SPA can more accurately represent valid information about the different disease levels in rice.

3.4.3. Comparison with Other Classification Models

The model in this paper was analyzed and compared with six classification models, namely Inception V3, ZF-Net, BiGRU, TextCNN, SVM and ELM. The classification results of the six models are shown in Table 6.

Table 6. Overall classification results.

Methods	SVM		ELM		Inception V3		ZF-Net		BiGRU		TextCNN	
	OA (%)	Kappa (%)	OA (%)	Kappa (%)	OA (%)	Kappa (%)	OA (%)	Kappa (%)	OA (%)	Kappa (%)	OA (%)	Kappa (%)
SPA	93.41	91.74	90.19	87.82	95.44	94.28	94.42	93.01	96.65	95.81	95.74	94.66
RF	91.28	89.09	90.96	89.07	91.89	89.85	96.55	95.68	94.32	92.88	88.95	86.12
VIs	86.09	82.60	83.40	79.22	86.92	83.62	89.76	87.17	88.64	85.80	84.08	80.09
TFs	88.34	85.40	89.13	87.27	88.95	86.14	92.09	90.08	89.96	87.41	90.97	88.68
SPA + TFs	95.54	94.41	91.67	89.59	97.06	96.32	97.77	97.20	97.36	96.70	97.77	97.20
RF + TFs	94.42	93.01	91.02	88.82	95.33	94.16	96.04	95.05	96.35	95.43	95.54	94.41
Vis + TFs	80.61	75.69	74.94	68.79	83.47	79.30	86.00	82.49	81.14	76.40	83.77	79.73

As can be observed from Table 6, all six models achieved good accuracy in disease classification. The model constructed by fusing spectral wavelengths and texture features screened by SPA as input quantities has the best classification accuracy, with OA and Kappa of greater than 90% and 88%, respectively. In addition, for the identification of the different disease classes, F1-score were greater than 84% for levels 0, 2 and 4 and greater than 82% for levels 1 and 3 (shown in Appendix A Tables A1–A3). In addition, the experimental results of the models simultaneously show that the fusion of spectral feature wavelengths with texture features can enhance the classification of the models. Compared to machine learning models (SVM and ELM), the OA, Kappa and F1-scores of the models in this paper are significantly improved. In particular, OA and Kappa improved by 3.04% and 3.81%, respectively, compared to the SPA + TFs-SVM model. Compared to the SPA + TFs-ELM model, OA and Kappa improved by 6.91% and 8.63%, respectively. In comparison with the other four deep learning models, it can be observed that the classification accuracy of ZF-Net, Inception V3, TextCNN and BiGRU is lower than that of the present model. The classification results of ZF-Net, Inception V3, TextCNN and BiGRU for one-dimensional disease data were not very different, all with the best models constructed with features obtained from SPA + TFs (OA > 97%, Kappa > 96%). In view of this, it is evident from the comparative analysis of different input features and different modeling methods that the fusion of spectral features wavelength and texture features extracted by SPA is the best feature for leaf blast classification. At the same time, the DCNN model proposed in this paper has the best accuracy in classifying disease classes.

We performed a comparative analysis of the performance of the models constructed based on the best classification features (SPA + TFs) using the OA and test time, as shown in Table 7. As can be observed from Table 7, the deep learning model took significantly more time than the machine learning model on the 986 test datasets. However, the machine learning model is insufficient in OA. In the performance comparison of the deep learning models, it was found that the convolutional neural network took significantly less time than the recurrent neural network (BiGRU), which may be due to the fact that BiGRU is trained in a fully connected manner and requires more parameters. In comparison with DCNN models such as Inception V3, ZF-Net and TextCNN, our proposed model has the highest classification accuracy and the shortest testing time. On 986 test data, disease classification took only 0.22 s. Therefore, our proposed DCNN model has the best classification performance.

Table 7. Results of model detection efficiency comparison.

Method	OA (%)	Test Time (s)
SPA + TFs-SVM	95.54	0.1058
SPA + TFs-ELM	91.67	0.0279
SPA + TFs-Inception V3	97.06	0.5222
SPA + TFs-ZF-Net	97.77	0.4152
SPA + TFs-BiGRU	97.36	1.2086
SPA + TFs-TextCNN	97.77	0.3388
SPA + TFs-DCNN (the model of this study)	98.58	0.2200

4. Discussion

At present, the identification and disease degree classification of rice blast is mainly carried out through the subjective judgment of plant protection personnel, with high professional ability but low efficiency of detection. Hyperspectral imaging technology is a highly promising disease detection technology that has attracted the interest of scholars because of its non-destructive, fast and accurate characteristics [53,54].

This study first pre-processed the hyperspectral imaging data to extract rice leaf samples of different disease classes and increased the number of samples by data augmentation methods. Secondly, in order to reduce the dimensionality of hyperspectral data, methods such as SPA, RF, the contour of decision coefficient and GLCM were used to screen spectral features, vegetation indices and texture features. Finally, deep learning and machine learning methods were used to construct rice leaf blast classification models and to determine the best classification features and classification models for leaf blast.

When a crop is infested with a disease, it results in changes in a range of physiological parameters of rice, such as chlorophyll content, water content and cell structure [55]. The changes in these physiological parameters are reflected both in the spectral reflectance curves and in the crop image features, as shown in Figures 2 and 3. When rice leaves were infested with leaf blast, the leaf blast level showed a correlation with the change in the mean spectral curve. In the visible wavelength range, the spectral reflectance appeared slightly increased, which was due to the rhombus-shaped lesions on the leaf cells infested with magnaporthe grisea, which reduced the cytochrome content and activity and weakened the absorption of light. At the same time, as the chlorophyll content decreased, the absorption band narrowed and the red edge (680~770 nm) shifted to the short-wave direction, resulting in a "blue shift" phenomenon. There was a greater correlation between 770~1000 nm and the internal structure of the leaves. Compared to healthy leaves, the cell layer inside the diseased leaves was reduced and the spectral reflectance decreased [51]. The above phenomenon, therefore, provides some basis for research to obtain graded characteristics of leaf blast.

In this work, the focus was on the use of hyperspectral imaging data to determine the best classification features and classification models for leaf blast. In terms of data dimensionality reduction, this study used the SPA and RF methods to screen the spectral feature wavelengths, and 8 and 13 feature wavelengths were obtained, respectively, as shown in Table 4. The contour of the decision coefficient method was used to extract the three best vegetation indices with R^2 all greater than 0.8. The seven best texture features were also selected by combining GLCM and rank correlation analysis, as shown in Table 5. In DCNN modeling, the network depth, number and size of convolutions of the DCNN model can seriously affect its performance [56]. Therefore, we borrowed the design concept of ResNet and adjusted the network depth and convolutional layer parameters of ResNet through multiple tests to determine the best model structure. BatchNorm and Dropout layers were also added to avoid overfitting and to ensure accuracy. We constructed seven DCNN-based rice blast classification models based on different input features. The results show that all seven DCNN models designed based on different features have high classification accuracy, with OA greater than 88% and Kappa coefficient greater than 85%. The reason for this may be due to the fact that DCNN uses the ResNet model design concept as a reference and adopts a "shortcut" structure. This structure enables the inclusion of the full information of the previous layer of data in each residual module, preserving more of the original information to some extent. At the same time, the data augmentation method was used to increase the quantity of sample data and improve the diversity of the samples, further enhancing the generalization capability of the model. In comparing the DCNN models constructed with different features, the DCNN models constructed based on fused features all achieved high classification accuracy. The highest classification accuracy was obtained for SPA + TFs-DCNN, with OA and Kappa of 98.58% and 98.22%, respectively. All had high classification accuracy in the identification of detailed disease classes, with F1-scores of 100%, 100%, 100%, 96.48% and 96.68% for levels 0, 1, 2, 3 and 4, respectively. This suggests that the fusion of spectral and texture features to construct a classification model has the ability to improve the accuracy of model classification. This is consistent with previous studies [57].

In order to further determine the best classification features and classification model, the model in this paper was compared and analyzed with Inception V3, ZF-Net, BiGRU, TextCNN, SVM and ELM models. In the SVM and ELM modeling results, it was shown that the SPA screened feature wavelengths combined with TFs constructed the model with the best classification accuracy. Compared with the DCNN classification model, the OA, Kappa and F1-score of both the SVM and ELM classification models were significantly lower than those of the DCNN model. The reason for this may be that the convolutional layer of DCNN is able to further extract disease features and obtain significant differences between different diseases, thus improving model accuracy. The classification accuracy results of ZF-Net, Inception V3, TextCNN and BiGRU are all lower than the results of the model in this paper, as can be observed in the modeling results of the deep learning methods. This may be due to the fact that the model in this paper uses the shortcut structure of ResNet to retain more of the fine-grained features between diseases. Models such as Inception V3, on the other hand, gradually ignores fine-grained features and retain coarse-grained features as the number of iterations increases. In the case of intra-class classification problems, fine-grained features are the key to achieving higher accuracy.

Therefore, in this study, it is concluded from the comparative analysis of different input features and different modeling methods that the DCNN model constructed based on the fused features of feature wavelength and texture features acquired by SPA has the highest classification accuracy. It can realize the accurate classification of the severity of rice leaf blight and provides some technical support for the next step of UAV hyperspectral remote sensing monitoring of rice leaf blasts. It is worth noting that only rice leaf blast was modeled and analyzed in this study, and no other leaf diseases of rice were studied. Therefore, future research work will further explore the best classification features for different rice diseases and establish a more representative, generalized and comprehensive disease classification model.

5. Conclusions

Leaf blast, a typical disease of rice, has major impacts on the yield and quality of grain. In this study, an indoor hyperspectral imaging system was used to acquire hyperspectral images of leaves. With limited hyperspectral data, data augmentation was performed by drawing on data augmentation methods from existing studies to augment the sample data from 145 to 4930. Then, spectral features, vegetation indices and texture features were extracted based on the augmented hyperspectral images. The above features and their fusion features were used to construct leaf blast classification models. The results showed that the model constructed based on fused features was significantly better than the model constructed based on single feature variables in terms of accuracy in the classification of the degree of leaf blast disease. The best performance was achieved by combining the SPA screened spectral features (450, 543, 679, 693, 714, 757, 972 and 985 nm) with textural features (MEne, SDEne, MEnt, SDEnt, MCon, SDCon and MCor). The modeling results also showed that the proposed DCNN model provided better classification performance in disease classification compared to traditional machine learning models (SVM and ELM), with an improvement of 3.04% and 6.91% in OA and 3.81% and 8.63% in Kappa, respectively. Compared to deep learning models such as Inception V3, ZF-Net, BiGRU and TextCNN, this model also has the best classification accuracy. In comparison to ZF-Net and TextCNN, both OA and Kappa improved by 0.81% and 1.02%. OA and Kappa improved by 1.52% and 1.22% and 1.9% and 1.52%, respectively, compared to Inception V3 and BiGRU. Therefore, this study confirms the great potential of the proposed one-dimensional deep convolutional neural network model for applications in disease classification. The best fusion features identified in this study can further improve the modeling accuracy of the disease classification model. In addition, in the next study, we will further explore the classification features of rice diseases such as sheath blight and bacterial blight to establish a more stable, accurate and comprehensive disease classification model.

Appendix A

Table A1. F1-score for the SVM and ELM models.

Methods	SVM - F1-Score /%					ELM - F1-Score /%				
	Level 0	Level 1	Level 2	Level 3	Level 4	Level 0	Level 1	Level 2	Level 3	Level 4
SPA	100.00	92.93	90.41	87.89	95.81	97.53	88.64	87.31	83.96	92.46
RF	100.00	89.80	88.84	86.39	91.26	100	88.45	83.25	89.26	95.90
VIs	93.30	81.91	86.61	80.10	87.39	98.64	82.64	74.47	72.04	90.39
TFs	86.93	84.48	89.85	87.75	91.77	76.55	70.25	88.79	80.00	93.75
SPA+TFs	98.77	94.49	95.88	93.97	95.10	97.41	88.76	86.44	87.19	97.94
RF+TFs	97.03	90.03	93.56	93.64	96.90	97.94	88.69	86.55	86.17	95.77
VIs+TFs	95.93	79.58	78.99	64.09	83.25	95.81	66.09	67.29	65.35	80.98

Table A2. F1-score for the Inception V3 and ZF-Net models.

Methods	Inception V3 - F1-Score /%					ZF-Net - F1-Score /%				
	Level 0	Level 1	Level 2	Level 3	Level 4	Level 0	Level 1	Level 2	Level 3	Level 4
SPA	100	94.43	94.54	93.01	94.39	100	96.61	89.74	87.61	97.04
RF	99.10	98.21	84.04	83.46	94.12	99.55	97.41	95.80	93.51	96.16
VIs	97.40	82.39	81.35	79.78	92.42	96.91	85.71	94.99	79.64	88.79
TFs	95.24	84.52	88.79	85.99	89.72	98.46	93.11	89.72	87.34	92.38
SPA+TFs	98.20	97.28	97.40	94.85	97.51	99.00	98.76	98.40	95.36	97.41
RF+TFs	97.16	94.64	96.88	91.91	95.77	98.06	96.59	96.28	92.71	96.07
VIs+TFs	96.06	73.19	76.14	73.94	90.75	96.52	83.12	80.57	77.68	91.13

Table A3. F1-score for the BiGRU and TextCNN models.

Methods	BiGRU - F1-Score /%					TextCNN - F1-Score /%				
	Level 0	Level 1	Level 2	Level 3	Level 4	Level 0	Level 1	Level 2	Level 3	Level 4
SPA	100	94.94	93.43	95.82	98.50	100	94.22	92.02	93.82	97.99
RF	100	93.77	92.60	89.33	94.69	99.10	87.58	88.21	75.47	90.28
VIs	96.91	85.94	83.73	82.97	92.86	96.91	80.57	77.43	77.64	87.40
TFs	89.69	91.45	94.25	85.79	88.41	93.11	80.79	87.29	92.31	98.37
SPA+TFs	100	99.07	96.68	94.21	97.24	100	98.79	97.88	95.31	97.03
RF+TFs	96.06	95.58	98.13	94.95	96.73	97.22	94.74	96.31	93.26	95.85
VIs+TFs	95.51	73.90	74.44	73.04	89.45	96.61	8032	81.34	73.23	87.62

Author Contributions: Conceptualization, S.F., Y.C., T.X. and G.Z.; methodology, S.F.; software, S.F.; validation, S.F., F.Y., G.Z. and D.Z.; formal analysis, S.F. and T.X.; investigation, S.F.; resources, S.F. and G.Z.; data curation, S.F., G.Z. and D.Z.; writing—original draft preparation, S.F.; writing—review and editing, S.F. and T.X.; visualization, S.F.; supervision, T.X.; project administration, T.X.; funding acquisition, T.X. All authors have read and agreed to the published version of the manuscript.

Funding: This work was supported by the Liaoning Provincial Key R&D Program Project (2019JH2/10200002).

Institutional Review Board Statement: Not applicable.

Informed Consent Statement: Informed consent was obtained from all subjects involved in the study.

Data Availability Statement: Data sharing is not applicable to this article.

Conflicts of Interest: The authors declare no conflict of interest.

References

1. Srivastava, D.; Shamim, M.; Kumar, M.; Mishra, A.; Pandey, P.; Kumar, D.; Yadav, P.; Siddiqui, M.H.; Singh, K.N. Current Status of Conventional and Molecular Interventions for Blast Resistance in Rice. *Rice Sci.* **2017**, *24*, 299–321. [CrossRef]
2. Deutsch, C.A.; Tewksbury, J.J.; Tigchelaar, M.; Battisti, D.S.; Merrill, S.C.; Huey, R.B.; Naylor, R.L. Increase in crop losses to insect pests in a warming climate. *Science* **2018**, *361*, 916–919. [CrossRef] [PubMed]
3. Huang, W.; Shi, Y.; Dong, Y.; Ye, H.; Wu, M.; Cui, B.; Liu, L. Progress and prospects of crop diseases and pests monitoring by remote sensing. *Smart Agric.* **2019**, *1*, 1–11.
4. Akintayo, A.; Tylka, G.L.; Singh, A.K.; Ganapathysubramanian, B.; Singh, A.; Sarkar, S. A deep learning framework to discern and count microscopic nematode eggs. *Sci. Rep.* **2018**, *8*, 9145. [CrossRef] [PubMed]
5. Bock, C.H.; Poole, G.H.; Parker, P.E.; Gottwald, T.R. Plant Disease Severity Estimated Visually, by Digital Photography and Image Analysis, and by Hyperspectral Imaging. *Crit. Rev. Plant Sci.* **2010**, *29*, 59–107. [CrossRef]
6. Naik, H.S.; Zhang, J.; Lofquist, A.; Assefa, T.; Sarkar, S.; Ackerman, D.; Singh, A.; Singh, A.K.; Ganapathysubramanian, B. A real-time phenotyping framework using machine learning for plant stress severity rating in soybean. *Plant Methods* **2017**, *13*, 23. [CrossRef]
7. Zhang, J.; Naik, H.S.; Assefa, T.; Sarkar, S.; Reddy, R.V.C.; Singh, A.; Ganapathysubramanian, B.; Singh, A.K. Computer vision and machine learning for robust phenotyping in genome-wide studies. *Sci. Rep.* **2017**, *7*, srep44048. [CrossRef]
8. Zheng, Z.; Qi, L.; Ma, X.; Zhu, X.; Wang, W. Grading method of rice leaf blast using hyperspectral imaging technology. *Trans. Chin. Soc. Agric. Eng.* **2013**, *29*, 138–144.
9. Asibi, A.E.; Chai, Q.; Coulter, J.A. Rice Blast: A Disease with Implications for Global Food Security. *Agronomy* **2019**, *9*, 451. [CrossRef]
10. Larijani, M.R.; Asli-Ardeh, E.A.; Kozegar, E.; Loni, R. Evaluation of image processing technique in identifying rice blast disease in field conditions based on KNN algorithm improvement by K-means. *Food Sci. Nutr.* **2019**, *7*, 3922–3930. [CrossRef]
11. Zarbafi, S.S.; Rabiei, B.; Ebadi, A.A.; Ham, J.H. Statistical analysis of phenotypic traits of rice (*Oryza sativa* L.) related to grain yield under neck blast disease. *J. Plant Dis. Prot.* **2019**, *126*, 293–306. [CrossRef]
12. Bastiaans, L. Effects of leaf blast on photosynthesis of rice. 1. Leaf photosynthesis. *Eur. J. Plant Pathol.* **1993**, *99*, 197–203. [CrossRef]
13. Nabina, N.; Kiran, B. A Review of Blast Disease of Rice in Nepal. *J. Plant Pathol. Microbiol.* **2021**, *12*, 1–5.
14. Gowen, A.; O'Donnell, C.; Cullen, P.; Downey, G.; Frias, J. Hyperspectral imaging—An emerging process analytical tool for food quality and safety control. *Trends Food Sci. Technol.* **2007**, *18*, 590–598. [CrossRef]
15. Feng, L.; Chai, R.-Y.; Sun, G.-M.; Wu, D.; Lou, B.-G.; He, Y. Identification and classification of rice leaf blast based on multi-spectral imaging sensor. *Spectrosc. Spectr. Anal.* **2009**, *29*, 2730–2733.
16. Qi, L.; Ma, X.; Liao, X.-L. Rice bast resistance identification based on multi-spectral computer vision. *J. Jilin Univ. Eng. Technol. Ed.* **2009**, *39*, 356–359.
17. Feng, L.; Wu, B.; Zhu, S.; Wang, J.; Su, Z.; Liu, F.; He, Y.; Zhang, C. Investigation on Data Fusion of Multisource Spectral Data for Rice Leaf Diseases Identification Using Machine Learning Methods. *Front. Plant Sci.* **2020**, *11*, 1664. [CrossRef]
18. Wu, D.; Cao, F.; Zhang, H.; Sun, G.-M.; Feng, L.; He, Y. Study on disease level classification of rice panicle blast based on visible and near infrared spectroscopy. *Spectrosc. Spectr. Anal.* **2009**, *29*, 3295–3299.
19. Barreto, A.; Paulus, S.; Varrelmann, M.; Mahlein, A.-K. Hyperspectral imaging of symptoms induced by Rhizoctonia solani in sugar beet: Comparison of input data and different machine learning algorithms. *J. Plant Dis. Prot.* **2020**, *127*, 441–451. [CrossRef]
20. Abdulridha, J.; Ampatzidis, Y.; Roberts, P.; Kakarla, S.C. Detecting powdery mildew disease in squash at different stages using UAV-based hyperspectral imaging and artificial intelligence. *Biosyst. Eng.* **2020**, *197*, 135–148. [CrossRef]
21. Fajardo, J.U.; Andrade, O.B.; Bonilla, R.C.; Cevallos-Cevallos, J.; Mariduena-Zavala, M.; Donoso, D.O.; Villardón, J.L.V. Early detection of black Sigatoka in banana leaves using hyperspectral images. *Appl. Plant Sci.* **2020**, *8*, e11383. [CrossRef]
22. Bagheri, N.; Monavar, H.M.; Azizi, A.; Ghasemi, A. Detection of Fire Blight disease in pear trees by hyperspectral data. *Eur. J. Remote Sens.* **2017**, *51*, 1–10. [CrossRef]
23. Liu, Z.-Y.; Wu, H.-F.; Huang, J. Application of neural networks to discriminate fungal infection levels in rice panicles using hyperspectral reflectance and principal components analysis. *Comput. Electron. Agric.* **2010**, *72*, 99–106. [CrossRef]
24. Zhang, G.; Xu, T.; Tian, Y.; Xu, H.; Song, J.; Lan, Y. Assessment of rice leaf blast severity using hyperspectral imaging during late vegetative growth. *Australas. Plant Pathol.* **2020**, *49*, 571–578. [CrossRef]
25. Guo, A.; Huang, W.; Ye, H.; Dong, Y.; Ma, H.; Ren, Y.; Ruan, C. Identification of Wheat Yellow Rust Using Spectral and Texture Features of Hyperspectral Images. *Remote Sens.* **2020**, *12*, 1419. [CrossRef]
26. Luo, Y.-H.; Jiang, P.; Xie, K.; Wang, F.-J. Research on optimal predicting model for the grading detection of rice blast. *Opt. Rev.* **2019**, *26*, 118–123. [CrossRef]
27. Lu, B.; Jun, S.; Ning, Y.; Xiaohong, W.; Xin, Z. Identification of tea white star disease and anthrax based on hyperspectral image information. *J. Food Process. Eng.* **2020**, *44*, e13584. [CrossRef]

28. Kang, L.; Yuan, J.Q.; Gao, R.; Kong, Q.M.; Jia, Y.J.; Su, Z.B. Early Identification of Rice Leaf Blast Based on Hyperspectral Imaging. *Spectrosc. Spectr. Anal.* **2021**, *41*, 898–902.
29. Knauer, U.; Matros, A.; Petrovic, T.; Zanker, T.; Scott, E.S.; Seiffert, U. Improved classification accuracy of powdery mildew infection levels of wine grapes by spatial-spectral analysis of hyperspectral images. *Plant Methods* **2017**, *13*, 47. [CrossRef] [PubMed]
30. Nagasubramanian, K.; Jones, S.; Sarkar, S.; Singh, A.K.; Singh, A.; Ganapathysubramanian, B. Hyperspectral band selection using genetic algorithm and support vector machines for early identification of charcoal rot disease in soybean stems. *Plant Methods* **2018**, *14*, 86. [CrossRef] [PubMed]
31. Nettleton, D.F.; Katsantonis, D.; Kalaitzidis, A.; Sarafijanovic-Djukic, N.; Puigdollers, P.; Confalonieri, R. Predicting rice blast disease: Machine learning versus process-based models. *BMC Bioinform.* **2019**, *20*, 514–516. [CrossRef]
32. Jia, D.; Chen, P. Effect of Low-altitude UAV Image Resolution on Inversion of Winter Wheat Nitrogen Concentration. *Nongye Jixie Xuebao/Trans. Chin. Soc. Agric. Mach.* **2020**, *51*, 164–169.
33. Zhang, D.; Chen, G.; Zhang, H.; Jin, N.; Gu, C.; Weng, S.; Wang, Q.; Chen, Y. Integration of spectroscopy and image for identifying fusarium damage in wheat kernels. *Spectrochim. Acta Part A Mol. Biomol. Spectrosc.* **2020**, *236*, 118344. [CrossRef] [PubMed]
34. Al-Saddik, H.; Laybros, A.; Billiot, B.; Cointault, F. Using Image Texture and Spectral Reflectance Analysis to Detect Yellowness and Esca in Grapevines at Leaf-Level. *Remote Sens.* **2018**, *10*, 618. [CrossRef]
35. Zhang, D.-Y.; Chen, G.; Yin, X.; Hu, R.-J.; Gu, C.-Y.; Pan, Z.-G.; Zhou, X.-G.; Chen, Y. Integrating spectral and image data to detect Fusarium head blight of wheat. *Comput. Electron. Agric.* **2020**, *175*, 105588. [CrossRef]
36. Zhu, H.; Chu, B.; Zhang, C.; Liu, F.; Jiang, L.; He, Y. Hyperspectral Imaging for Presymptomatic Detection of Tobacco Disease with Successive Projections Algorithm and Machine-learning Classifiers. *Sci. Rep.* **2017**, *7*, 4125. [CrossRef]
37. Nagasubramanian, K.; Jones, S.; Singh, A.K.; Sarkar, S.; Singh, A.; Ganapathysubramanian, B. Plant disease identification using explainable 3D deep learning on hyperspectral images. *Plant Methods* **2019**, *15*, 1–10. [CrossRef] [PubMed]
38. Huang, S.; Sun, C.; Qi, L.; Ma, X.; Wang, W. Rice panicle blast identification method based on deep convolution neural network. *Trans. Chin. Soc. Agric. Eng.* **2017**, *33*, 169–176.
39. Zhang, X.; Han, L.; Dong, Y.; Shi, Y.; Huang, W.; Han, L.; González-Moreno, P.; Ma, H.; Ye, H.; Sobeih, T. A Deep Learning-Based Approach for Automated Yellow Rust Disease Detection from High-Resolution Hyperspectral UAV Images. *Remote Sens.* **2019**, *11*, 1554. [CrossRef]
40. Chen, Y.; Jiang, H.; Li, C.; Jia, X.; Ghamisi, P. Deep Feature Extraction and Classification of Hyperspectral Images Based on Convolutional Neural Networks. *IEEE Trans. Geosci. Remote Sens.* **2016**, *54*, 6232–6251. [CrossRef]
41. Araujo, M.; Saldanha, T.C.B.; Galvão, R.K.H.; Yoneyama, T.; Chame, H.C.; Visani, V. The successive projections algorithm for variable selection in spectroscopic multicomponent analysis. *Chemom. Intell. Lab. Syst.* **2001**, *57*, 65–73. [CrossRef]
42. Li, H.-D.; Xu, Q.-S.; Liang, Y.-Z. Random frog: An efficient reversible jump Markov Chain Monte Carlo-like approach for variable selection with applications to gene selection and disease classification. *Anal. Chim. Acta* **2012**, *740*, 20–26. [CrossRef]
43. Haralick, R.M.; Shanmugam, K.; Dinstein, I. Textural Features for Image Classification. *IEEE Trans. Syst. Man Cybern.* **1973**, *SMC-3*, 610–621. [CrossRef]
44. He, K.; Zhang, X.; Ren, S.; Sun, J. Deep residual learning for image recognition. In Proceedings of the 29th IEEE Conference on Computer Vision and Pattern Recognition (CVPR 2016), Las Vegas, NV, USA, 26 June —1 July 2016; pp. 770–778.
45. Xia, J.; Chanussot, J.; Du, P.; He, X. Rotation-Based Support Vector Machine Ensemble in Classification of Hyperspectral Data With Limited Training Samples. *IEEE Trans. Geosci. Remote Sens.* **2016**, *54*, 1519–1531. [CrossRef]
46. Huang, G.-B.; Zhu, Q.-Y.; Siew, C.-K. Extreme learning machine: Theory and applications. *Neurocomputing* **2006**, *70*, 489–501. [CrossRef]
47. Christian, S.; Vincent, V.; Sergey, L.; Jonathon, S.; Zbigniew, W. Rethinking the inception architecture for computer vision. *arXiv* **2015**, arXiv:1512.00567.
48. Matthew, D.Z.; Rob, F. Visualizing and understanding convolutional networks. *arXiv* **2013**, arXiv:1311.2901.
49. Chuang, J.Y.; Gulcehre, C.; Cho, K.; Bengio, Y. Empirical evaluation of gated recurrent neural networks on sequence modeling. *arXiv* **2014**, arXiv:1412.3555.
50. Kim, Y. Convolutional Neural Networks for Sentence Classification. *Assoc. Comput. Linguist.* **2014**, *13*, 1746–1751.
51. Zhang, J.; Huang, Y.; Pu, R.; González-Moreno, P.; Yuan, L.; Wu, K.; Huang, W. Monitoring plant diseases and pests through remote sensing technology: A review. *Comput. Electron. Agric.* **2019**, *165*, 104943. [CrossRef]
52. Le, T.T.H.; Kim, J.; Kim, H. An Effective Intrusion Detection Classifier Using Long Short-Term Memory with Gradient Descent Optimization. In Proceedings of the 2017 International Conference on Platform Technology and Service, IEEE 2017, Busan, Korea, 13–15 February 2017; pp. 155–160.
53. Behmann, J.; Bohnenkamp, D.; Paulus, S.; Mahlein, A.-K. Spatial Referencing of Hyperspectral Images for Tracing of Plant Disease Symptoms. *J. Imaging* **2018**, *4*, 143. [CrossRef]
54. Chen, B.; Wang, G.; Liu, J.-D.; Ma, Z.-H.; Wang, J.; Li, T.-N. Extraction of Photosynthetic Parameters of Cotton Leaves under Disease Stress by Hyperspectral Remote Sensing. *Spectrosc. Spectr. Anal.* **2018**, *38*, 1834–1838.
55. Ma, H.; Huang, W.; Jing, Y.; Pignatti, S.; Laneve, G.; Dong, Y.; Ye, H.; Liu, L.; Guo, A.; Jiang, J. Identification of Fusarium Head Blight in Winter Wheat Ears Using Continuous Wavelet Analysis. *Sensors* **2019**, *20*, 20. [CrossRef] [PubMed]

56. Zhou, Y.C.; Xu, T.Y.; Zheng, W.; Deng, H.B. Classification and recognition approaches of tomato main organs based on DCNN. *Trans. Chin. Soc. Agric. Eng.* **2017**, *33*, 219–226.
57. Huang, L.; Li, T.; Ding, C.; Zhao, J.; Zhang, D.; Yang, G. Diagnosis of the Severity of Fusarium Head Blight of Wheat Ears on the Basis of Image and Spectral Feature Fusion. *Sensors* **2020**, *20*, 2887. [CrossRef] [PubMed]

remote sensing

MDPI

Article

Detection and Classification of Rice Infestation with Rice Leaf Folder (*Cnaphalocrocis medinalis*) Using Hyperspectral Imaging Techniques

Gui-Chou Liang [1], Yen-Chieh Ouyang [2] and Shu-Mei Dai [1,*]

[1] Department of Entomology, National Chung Hsing University, Taichung 402, Taiwan; g107036008@mail.nchu.edu.tw

[2] Department of Electrical Engineering, National Chung Hsing University, Taichung 402, Taiwan; ycouyang@nchu.edu.tw

* Correspondence: sdai5497@dragon.nchu.edu.tw; Tel.: +886-0963-234-136

Abstract: The detection of rice leaf folder (RLF) infestation usually depends on manual monitoring, and early infestations cannot be detected visually. To improve detection accuracy and reduce human error, we use push-broom hyperspectral sensors to scan rice images and use machine learning and deep neural learning methods to detect RLF-infested rice leaves. Different from traditional image processing methods, hyperspectral imaging data analysis is based on pixel-based classification and target recognition. Since the spectral information itself is a feature and can be considered a vector, deep learning neural networks do not need to use convolutional neural networks to extract features. To correctly detect the spectral image of rice leaves infested by RLF, we use the constrained energy minimization (CEM) method to suppress the background noise of the spectral image. A band selection method was utilized to reduce the computational energy consumption of using the full-band process, and six bands were selected as candidate bands. The following method is the band expansion process (BEP) method, which is utilized to expand the vector length to improve the problem of compressed spectral information for band selection. We use CEM and deep neural networks to detect defects in the spectral images of infected rice leaves and compare the performance of each in the full frequency band, frequency band selection, and frequency BEP. A total of 339 hyperspectral images were collected in this study; the results showed that six bands were sufficient for detecting early infestations of RLF, with a detection accuracy of 98% and a Dice similarity coefficient of 0.8, which provides advantages of commercialization of this field.

Keywords: rice; rice leaf folder; hyperspectral imaging; band selection; hyperspectral image classification; target detection

Citation: Liang, G.-C.; Ouyang, Y.-C.; Dai, S.-M. Detection and Classification of Rice Infestation with Rice Leaf Folder (*Cnaphalocrocis medinalis*) Using Hyperspectral Imaging Techniques. *Remote Sens.* **2021**, *13*, 4587. https://doi.org/10.3390/rs13224587

Academic Editor: Clement Atzberger

Received: 15 October 2021
Accepted: 10 November 2021
Published: 15 November 2021

1. Introduction

Rice leaf folder (RLF), *Cnaphalocrocis medinalis* Guenée, is widely distributed in the rice-growing regions of humid tropical and temperate countries [1], and the developmental time of RLF decreases with an increase in temperature [2]. Due to global warming, RLF has become one of the most important insect pests of rice cultivation [3]. The larvae of RLF fold the leaves longitudinally and feed on the mesophyll tissue within the folded leaves. The feeding of RLF generates lineal white stripes (LWSs) in the early stage and then enlarge into ocher patches (OPs) and membranous OPs [4]. As the infestation of RLF increases, the number and area of OPs will increase. The feeding of RLF not only reduced the chlorophyll content and photosynthesis efficiency [4] but also provided a method for fungal and bacterial infection [5]. Therefore, the severe damage caused by RLF may cause 63–80% yield loss [6], and the highest record of the damaged area to rice cultivation in a single year exceeded 30,000 hectares [7].

The economic injury level of RLF, which is important for the determination of insecticide applications, has been established as 4.2% damaged leaves and 1.3 larvae per plant by the International Rice Research Institute [8]. However, it is laborious and time-consuming to visually inspect for damage. In addition, RLF is a long-distance migratory insect pest. The uncertain timing of the appearance of RLF means that farmers are unable to predict pest arrival, so to avoid damage by undetected infestations, farmers often preventively spray chemical insecticides, which generates unnecessary costs and environmental pollution [9,10].

Hyperspectral imaging (HSI) is a novel technique that combines the simultaneous advantages of imaging and spectroscopy and that has been investigated and applied in crop protection [11–15]. HSI, which contains spatial and spectral information, is given in Figure 1. The external damage and internal damage caused by pest infestations, such as yellowing/attenuation/defects and loss of pigments/photosynthetic activity/water content, respectively, can be identified by this system through image or spectral reflectance. Further automatic detection can be fulfilled by taking advantage of pest damage detection algorithms. For instance, constrained energy minimization (CEM) [16] and principal component analysis (PCA) [17] have been employed for band selection, and support vector machines (SVMs) [18], convolutional neural networks (CNNs) [19], and deep neural networks (DNNs) [20] are utilized for classification. Fan et al. [21] applied a visible/near-infrared hyperspectral imaging system to detect early invasion of rice streak insects. Using the successive projection algorithm (SPA) [22], PCA, and a back-propagation neural network (BPNN) [23] as classifiers to identify key wavelengths, the classification accuracy of the calibration and prediction sets was 95.65%. Chen et al. [24] also employed a visible/near-infrared hyperspectral imaging system to acquire images and further developed a hyperspectral insect damage detection algorithm (HIDDA) to detect pests in green coffee beans. The method combines CEM and SVM and achieves 95% accuracy and a 90% kappa coefficient. In addition, spectroscopy technology has been applied to detect plant diseases [25], the quality of agricultural products [26], and pesticide residues [27].

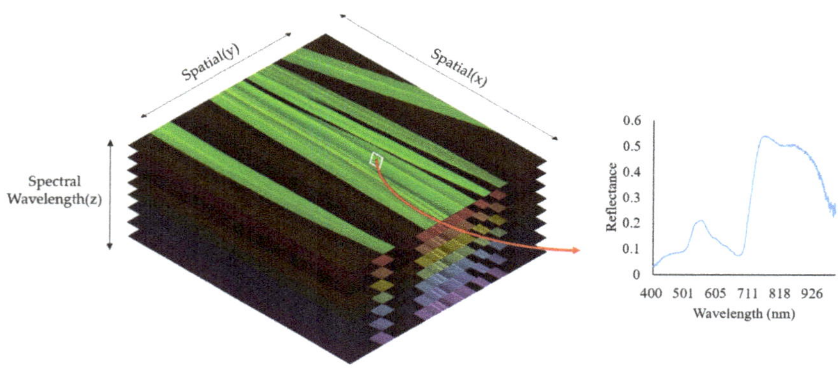

Figure 1. Two-dimensional projection of a hyperspectral cube.

To effectively manage RLF with a rational application of insecticides, an artificial-intelligent inspection of economic injury levels is necessary. The purpose of this study is to establish a model for detecting early infestation of RLF based on visible light hyperspectral data exploration techniques and deep learning technology. The specific objectives include (1) predefining the region of interest (ROI); (2) data preprocessing through a band selection and band expansion process (BEP); (3) simultaneously combining a deep learning network to train the model and to classify multiple different levels of damage; (4) using an automatic target generation program (ATGP) algorithm [28] to test unknown samples to fully automate the process and optimize the process to shorten the prediction time; and (5) establishing the spectral signatures of damaged leaves caused by RLF, which can serve as an expert system to provide valuable resources for the best timing of insecticide application.

2. Materials and Methods

2.1. Insect Breeding

The RLF in this study was collected from the Taichung District Agricultural Research and Extension Station. The larvae were raised in insect rearing cages (47.5 × 47.5 × 47.5 cm^3, MegaView Science Co., Ltd., Taichung, Taiwan) with corn seedlings (agricultural friend seedling Yumeizhen) and maintained at 27 ± 2 °C and 70% relative humidity during a photoperiod of 16:8 h (L:D). The adults were reared in a cage with 10% honey at 27 ± 2 °C and 90% relative humidity, which allows adults to lay more eggs.

2.2. Preparation of Rice Samples

The variety Tainan No. 11, which is the most prevalent cultivar planted in Taiwan, was selected for this study. Larvae were grown in a greenhouse to prevent the infestation of insect pests and diseases. To obtain different levels of damage caused by RLF, e.g., LWS and OP, 1st-, 2nd-, 3rd-, 4th-, or 5th-instar larvae of RLF were manually introduced to infest 40-day-old healthy rice for seven days, and three replicates were conducted for each treatment. Three different types of samples shown in Figure 2, e.g., healthy leaves (HL), LWS, and OP caused by RLF, were prepared for imaging acquisition and spectral information extraction.

Figure 2. Appearance of healthy and damaged leaf types. (**a**) Healthy leaves, (**b**) lineal white stripe (LWS) caused by RLF (blue arrow) and LWS enlarge into ocher patch (OP) (yellow arrow) on Day 1 (D1), (**c**) LWS and OP on D2, and (**d**) OP on D6.

2.3. Hyperspectral Imaging System and Imaging Acquisition

2.3.1. Hyperspectral Sensor

The hyperspectral scanning system employed in the experiment is shown in Figure 3. The hyperspectral image capturing system was composed of the following equipment: hyperspectral sensor, halogen light source, conveyor system, computer, and photographic darkroom isolated from external light sources. The hyperspectral sensor utilized in the study was a V10E-B1410CL sensor (IZUSU OPTICS), which contained visible and near-infrared (VNIR) bands with a spectral range from 380–1030 nm, a resolution of 5.5 nm, and 616 bands for imaging. The type of camera sensor is an Imspector Spectral Camera, SW ver 2.740. The halogen light sources used to illuminate the image were "3900e-ER", and the power was 150 W. Halogen lights were simultaneously illuminated on the left and right sides and focused on the conveyor track at an incident angle of 45 degrees to reduce shadow interference during the sampling process. The temperature and relative humidity in the laboratory were kept at 25 °C and 60%, respectively. A conveyor belt was designed to deliver rice plants for acquiring hyperspectral images by line scanning (Figure 3). Both the speed of the conveyor belt and the halogen lights were controlled by computer software. The distance between the VNIR sensor and the rice sample was 0.6 m.

Figure 3. Hyperspectral imaging system.

2.3.2. Image Acquisition

The damage to leaves infested with different RLF larvae (from 1st to 5th instar larvae) for various durations of feeding (1–6 days) was assessed using VNIR hyperspectral imaging. Leaves were placed flat on the conveyor belt to scan the image at every 90° turn to enlarge the dataset. The exposure time for scanning was 5.495 ms, and the number of pixels in each scan raw was 816. Healthy leaves without RLF infestation were selected as the control. Before taking the VNIR hyperspectral images, light correction was conducted, and all processing of images was conducted in a dark box to avoid interference from other light sources. In total, 339 images, including 52 images of healthy leaves and 69, 32, 48, 52, 52, and 34 images of leaves infested for 1 day to 6 days, were taken.

2.3.3. Calibration

To eliminate the impacts of uneven illumination and dark current noise, the object scan, reference dark value, and reference white value are needed to perform the normalization step. To reduce noise and avoid the influence of dark noise, the original hyperspectral image must be calibrated according to the following formula [21]:

$$R_C = \frac{R_0 - B}{W - B} \quad (1)$$

where R_0 is the raw hyperspectral image, R_C is the hyperspectral image after calibration, W is the standard white reference value with a Teflon rectangular bar, and B is the standard black reference value obtained by covering the lens with a lens cap.

2.4. Spectral Information Extraction

Removing the background of the image will help extract useful spectral information and reduce noise. The background removal process performs binary segmentation through the Otsu method, dividing the image into background and meaningful parts with similar features and attributes [29], including healthy, RLF-infested, and other defective leaves. To reduce unnecessary analysis work, the first step is to separate plant pixels from non-plant pixels. This task directly converts the grayscale image from the true-color image or generates a single channel image (grayscale image) based on a simple index (e.g., Excess Green [30]). Second, the threshold value is obtained using the Otsu method; the grayscale value of each pixel point is compared with the threshold value, and the pixel is classified as a target or background based on the result of the comparison [31]. Since plants and backgrounds have very different characteristics, they can be separated quickly and accurately.

Third, the images that had been removed from the background were applied to determine the ROI using the CEM algorithm [16]. CEM has been widely employed for target detection in hyperspectral remote sensing imagery. CEM detects the desired target signal source by using a unity constraint while suppressing noise and unknown signal sources; it also minimizes the average energy of output. This algorithm generates a finite impulse response filter through a given vector as the d value to suppress regions that are not related to the features of the ROI. The vector indicates the spectral reflectance of a pixel in this study, and the ROI was predefined as an RLF-infested region in the images of rice leaves, e.g., Figure 2b,c. The results of the CEM processing of the image show the enhanced characteristics of pixels similar to the target feature d value. Using the Otsu method, if the pixel value exceeds the threshold, the feature similarity is set to 1; otherwise, it is set to 0. Last, a binary image is obtained. This algorithm is an efficient method of pixel-based detection [32].

2.5. Band Selection

Since HSIs usually contain hundreds of spectral bands, full-band analysis of the spectrum is not only time-consuming but also too redundant. To decrease the analysis time and redundancy, the first step of data analysis is to determine the key wavelengths. The way to achieve this goal is to select highly correlated wavelengths by comparing the reflectance and to maximize the representativeness of the information by decorrelation. Various band spectral methods based on certain statistical criteria have been selected to achieve this purpose [33]. The concept of band selection is similar to feature extraction in image processing, which can improve the accuracy of identification and classification.

2.5.1. Band Prioritization

In the band prioritization (BP) part, the priority of the spectral bands will be calculated by statistical criteria [27]. Five criteria—variance, entropy, skewness, kurtosis, and signal-to-noise ratio (SNR)—were chosen to calculate the priority of the spectral bands in this work. Thus, each spectral band has a priority and can be ranked with high priority.

2.5.2. Band Decorrelation

When applying BP in the band selection process, the correlation between each band will highly affect the priority score. Neighboring bands will frequently be selected because of the high correlation between each band. Nevertheless, these redundant spectral bands are not helpful for improving detection performance. Therefore, to solve this problem, band decorrelation (BD) is utilized to remove these redundant spectral bands.

In this study, spectral information divergence (SID) [34] was applied for BD and utilized to measure the similarity between two vectors. By calculating the SID value, a threshold will be set to remove the bands with high similarity. The formula is:

$$k(b_i, b_j) = D(b_i \parallel b_j) + D(b_j \parallel b_i) \tag{2}$$

The parameter "b" represents a vector of spectral information, and $D(b_i \parallel b_j)$ denotes Kullback–Leibler divergence, that is, the average amount of difference between the self-information of b_j and the self-information of b_i, and vice versa.

2.6. Band Expansion Process

Although the band selection-acquired spectral images can reduce storage space and processing time, some of the original features of the spectra were lost. To solve the problem of information loss after band selection, the difference in reflectivity can be increased by expanding the band to increase the divergence. The concept of the BEP [35] is derived from the fact that a second-order random process is generally specified by its first-order and second-order statistics. These correlated multispectral images provide missing but useful second-order statistical information about the original hyperspectral images. The second-order statistical information utilized for the BEP includes autocorrelation, cross-correlation, and nonlinear correlation to create nonlinearly correlated images. The concept of generating second-order, correlated band images coincide with the concept of covariance functions employed in signal processing to generate random processes. Even though there may be no real physical inference for the band expansion process, it does provide an important advantage for addressing the problem of an insufficient number of spectral bands.

2.7. Data Training Models

Hyperspectral imaging data analysis is based on pixel-based classification and target recognition, using low-level features (such as spectral reflectance and texture) as the bedding, and the output feature representation at the top of the network can be directly input to subsequent classifiers for pixel-based classification [36]. The classification of this pixel is particularly suitable for deep learning algorithms to learn representative and discriminative features from the data in a hierarchical manner. In this study, the input neuron is the reflectance of a pixel. The input layer has 466 neurons in the full band, 6 neurons after band selection, and 27 neurons after band expansion. As shown in Figure 4a,b, the reflectivity of the HL, D1 OP, and D6 OP samples was divided into three categories. The model is trained with four hidden layers, and the learning rate parameter is 0.001. A softmax classifier was provided in the DNN terminal, and the classification results of the spectrum were obtained. The classified result was compared with the ground truth to calculate the accuracy. The model repeated the cross-validation ten times and averaged it as its overall accuracy (OA).

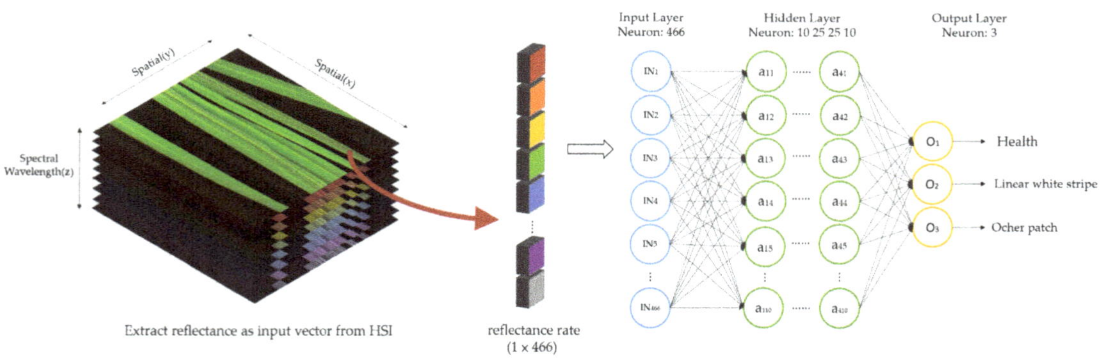

(a)

Figure 4. *Cont.*

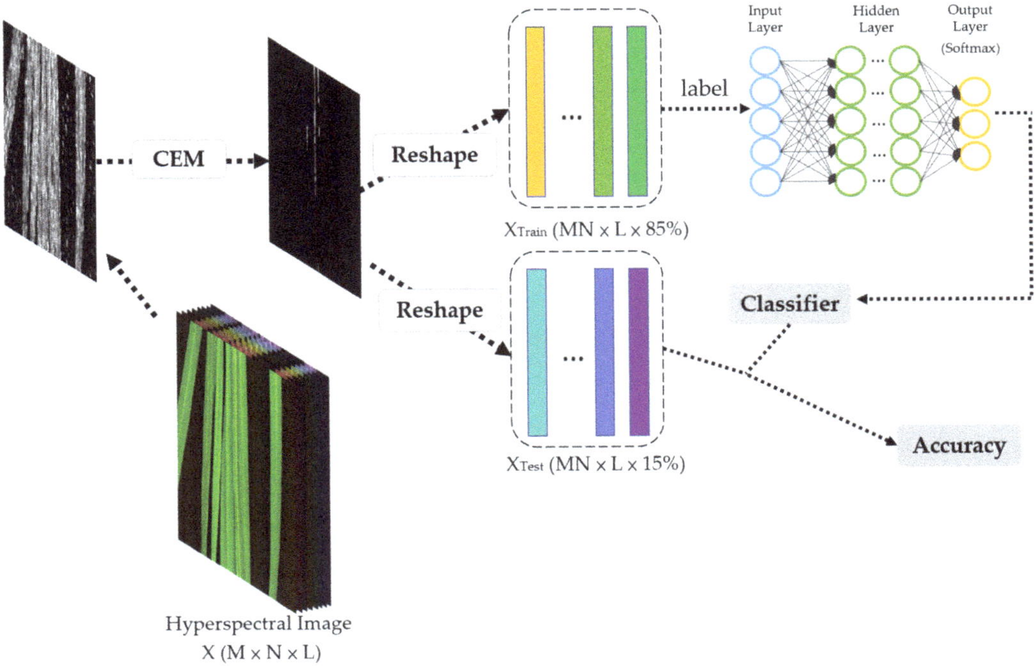

(b)

Figure 4. (**a**) DNN model architecture. (**b**) Flowchart of classifying reflectance using DNN.

Figure 5 depicts the data training flowchart of this study, starting with hyperspectral image capture. First, the reflectivity is extracted from the ROI as a ground truth, which was selected by the entomologist. Second, the reflectance dataset applied in the full-band spectrum was processed in the same way to build DNN models after band selection. Last, the band selection dataset was processed by the BEP to build a DNN model.

The DNN model is constructed using three processes: full bands, band selection, and BEP. Each classification model has the best weight evaluated by its own model. Three DNN classification process models are constructed based on randomly distributed datasets, including 70% training, 15% validation, and 15% testing (as shown in Table 1). In the testing phase, the accuracy of each classification situation will be compared, and the OA of multiple classifications will be integrated. As a result, the most suitable model for identifying the classification was obtained.

Table 1. Number of pixels used for band section, training, and testing in the rice dataset.

	Sample Types		
	HL	**D1 OP**	**D6 OP**
Band selection [1]	297	301	
Pixel numbers used for DNN Training	5936	6015	6962
Pixel numbers used for DNN Testing	1000	1000	1000

[1] Band selection number = 5% of training number.

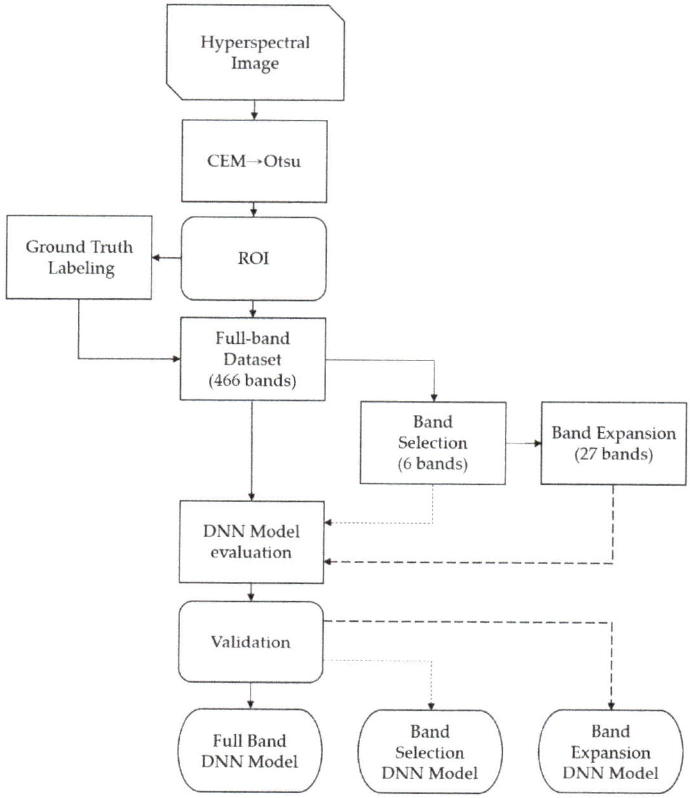

Figure 5. Data training flowchart of full bands, band selection, and band expansion process.

2.8. Model Test for Unknown Samples

To apply the spectral reflectance of unknown samples of healthy leaves, early and late OPs leave machine learning. The first step is to quickly determine the ROI to reduce the time required for image recognition. To achieve this goal, a method that combines an ATGP [28] and CEM is proposed. The ATGP is an unsupervised target recognition method that uses the concept of orthogonal subspace projection (OSP) to find a distinct feature without a priori knowledge. The ATGP method was employed to identify the target pixel in the hyperspectral image, and all the similar pixel data obtained were averaged as the d value of CEM.

Figure 6a,b shows the flow chart of the unknown sample prediction model. To automate the detection process, first, the full-band HSI, band selection, and BEP of the rice sample must be calibrated. Second, through the combined method of the ATGP and CEM, the Otsu method is utilized to mark the ROI. The ROI obtained from the full band, band selection, and BEP is classified by the corresponding DNN model and is labeled HL, early OP, or late OP by entomologists according to the occurrence of damage caused by RLF. The labeled ROI will be utilized to verify the prediction results of the DNN model. Five analysis methods, such as CEM_Full-band→DNN_Full-band, CEM_band selection→DNN_band selection, CEM_band selection→DNN_BEP, CEM_BEP→DNN_band selection, and CEM_BEP→DNN_BEP, are established to evaluate the prediction performance.

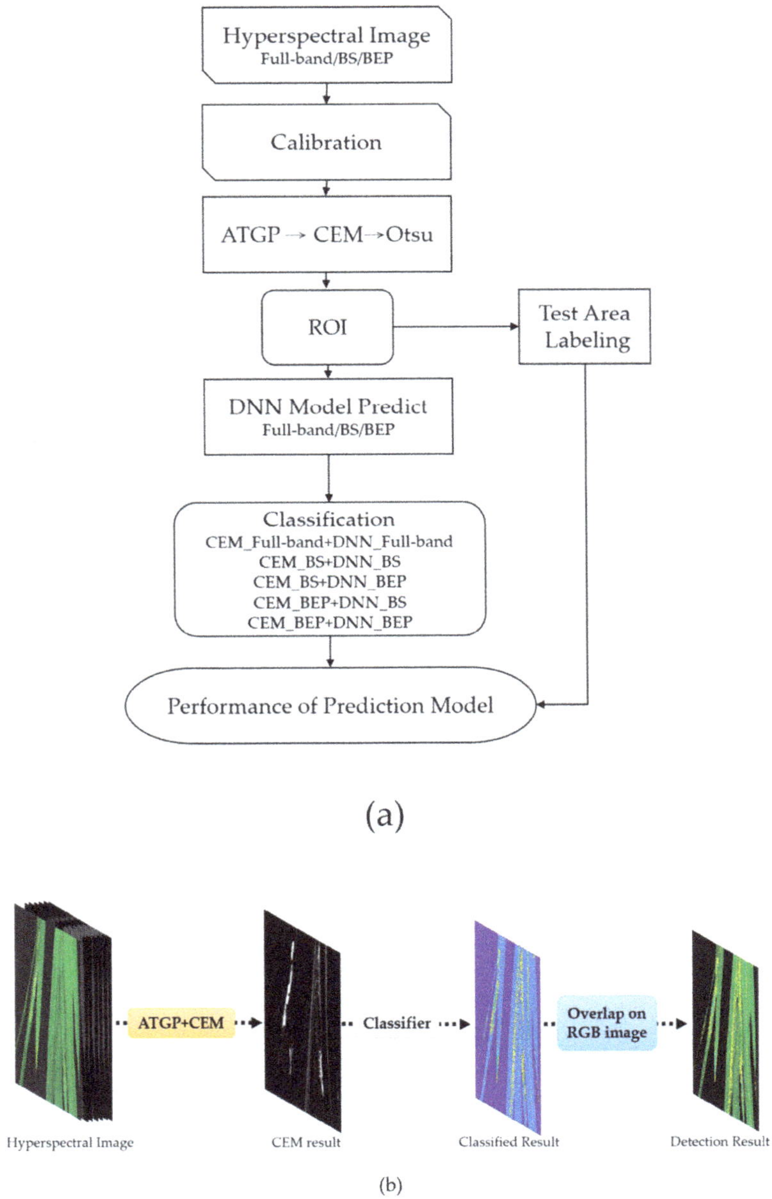

Figure 6. (**a**) Flow chart of the DNN model is used to predict unknown samples. (**b**) Flow chart of DNN model prediction.

Last, the model classification results were visualized and overlaid on the original true-color images, and agricultural experts verified the actual situation afterward to compare the performance of the models.

2.9. Predict Unknown Samplings

After a cross-validated predictive model has been established, a completely unknown sample with different data from the training set is needed to test its robustness. Eligible samples were obtained from the field. To fix other conditions, the retrieved samples were also photographed with a push-broom hyperspectral camera.

Many different evaluation metrics have been mentioned in the literature. The confusion matrix [37] was selected as a measure of model accuracy. A true positive (TP) is a correct detection of the ground truth. A false positive (FP) is an object that is mistaken as true. A false negative (FN) is an object that is not detected, although it is positive.

However, it is not enough to rely on the confusion matrix alone. An additional pipeline of common evaluation metrics was needed to facilitate a better comparison of classification models. The following metrics were employed for the evaluation in this study:

(i) recall

$$\text{Recall} = \frac{\text{TP}}{\text{TP} + \text{FN}} \tag{3}$$

(ii) precision

$$\text{Precision} = \frac{\text{TP}}{\text{TP} + \text{FP}} \tag{4}$$

(iii) Dice similarity coefficient

$$\text{Dice similarity coefficient} = 2 \times \frac{\text{TP}}{(2 \times \text{TP} + \text{FP} + \text{FN})} \tag{5}$$

The recall is the ability of the model to detect all relevant objects, i.e., the ability of the model to detect all detected bounding boxes of the validation set. Precision is the ability of the model to identify only relevant objects. The Dice similarity coefficient (DSC) is an ensemble similarity measure function that is usually applied to calculate the similarity between two samples in the value range between 0 when there is no overlap and 1 when there is complete overlap.

3. Results and Discussion

3.1. Images and Spectral Signatures of Healthy and RLF-Infested Rice Leaves

When larvae of RLF feed on rice leaves, they generate LWS or OP on the leaves. As time passes, the LWSs are enlarged into a patch; the color of the patch gradually turns from white to ocher; and the images and spectral signatures of these patches also change during this process, as shown in Figure 7a,b, respectively. The spectral signatures of HL and OP in Figure 7b were obtained manually, according to entomological experts. The OPs have higher reflectance than HL in the blue to red wavelength range. Among these spectral bands, the longer the infestation period is, the higher the reflectance, e.g., day 6 (D6) > D5 > D2 > D1. However, only the reflectance of D6 OP is higher than that of HL at the NIR wavelength (Figure 7b). The reflectance of D1-OP is much lower than the HL reflectance, and the reflectance of D2- and D5-OPs is approximately the same as that of healthy leaves. The decrease in reflectance in D1 OP at NIR was mainly due to the destruction of leaf structure, which caused photon scattering [38]. These results suggest that the early defects caused by RLF have very different spectral signatures of vectors from the subsequent damage of infestation. Differences in the spectral properties between the early phase of damage and the late phase of damage, which could serve as a basis for the early identification of RLF infestations.

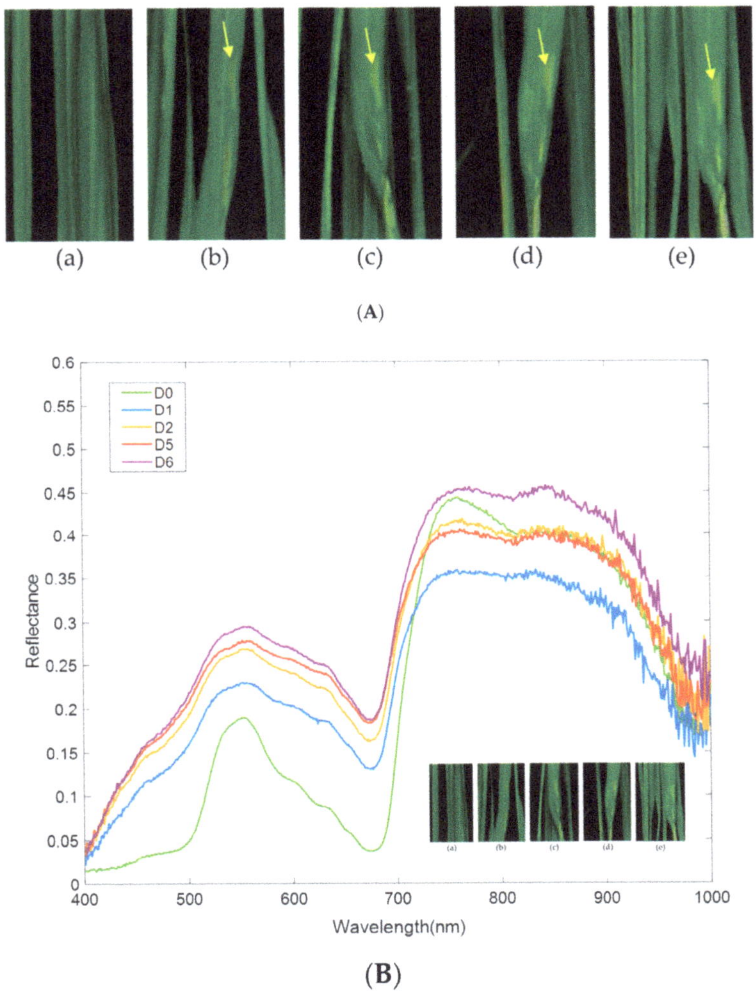

Figure 7. (**A**) Hyperspectral images of healthy leaves on day 0 (**a**) and ocher patches (yellow arrow) infested by rice leaf folders on day 1 (**b**), day 2 (**c**), day 5 (**d**), and day 6 (**e**). (**B**) Spectral signature and corresponding hyperspectral images of the healthy leaves (D0) and ocher patches (from D1 to D6) caused by RLF.

3.2. Band Selection and Band Expansion Process

The HSI and spectral signature from the full band system shown in Figure 7a,b contain considerable redundant information that slow the analysis efficiency and consume too much storage space. Therefore, band selection and the BEP were employed to select the most informative bands to increase the analysis efficiency and reduce storage space. To more effectively detect early RLF infection, the number of training sessions for HL and D1 OP was 5%, as shown in Table 1; these sessions were chosen to perform band selection. Five criteria were utilized in BP to calculate the priority of each band from the full-band signature of HL and D1 OP, and then, a value of 2.5 for SID was chosen as the threshold for BD to remove the adjacent bands with high similarity for D1 OP. Six bands of 489, 501, 603, 664, 684, and 705 nm, which had the largest difference in reflectance between HL and D1 OP, were selected as candidates through BP and BD using the criteria of entropy (Figure 8a,b).

To adapt to the cheaper and easy-to-use, six-band handheld spectrum sensor, we only chose the six-band spectrum. The results of band selection using the other four criteria are shown in Supplementary Table S1 and Figure S1. Furthermore, the six bands were expanded to 27 bands using the BEP to improve the deficiency caused by band selection.

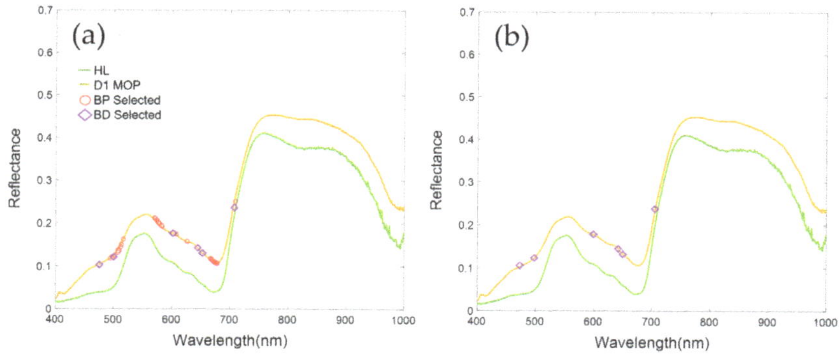

Figure 8. Bands selected through band prioritization (**a**) and band decorrelation (**b**) using criteria of entropy. Red circles denote bands selected from band prioritization, and purple diamonds denote bands selected after band decorrelation.

3.3. ROI Detection with CEM in Full Bands, Band Selection, and Band Expansion Process

CEM, a standard linear detector, was selected as a filter in this study to quickly identify the ROI. CEM increases the accuracy of automated detection and reduces the analysis time. The spectral signature of the OP that appeared on D1 in Figure 7b was employed as the d value of CEM to detect damaged leaves caused by RLF. Figure 9 shows the effect of different degrees of enhancement on ROI detection in the case of the full band, band selection, and BEP and the results of k-means clustering as a contrast. In the case of full bands, very minimal damage caused by RLF was detected (Figure 9b). The abundance of spectral data increases the complexity of detection and reduces the spectral reflectance resulting from RLF. On the other hand, the ROI detection in the cases of band selection reveals almost all the damage shown in Figure 9a. This finding indicates that band selection can achieve the best performance in ROI detection through CEM (Figure 9c). In the case of the BEP, the result of CEM is better than the full bands but not as good as band selection (Figure 9d).

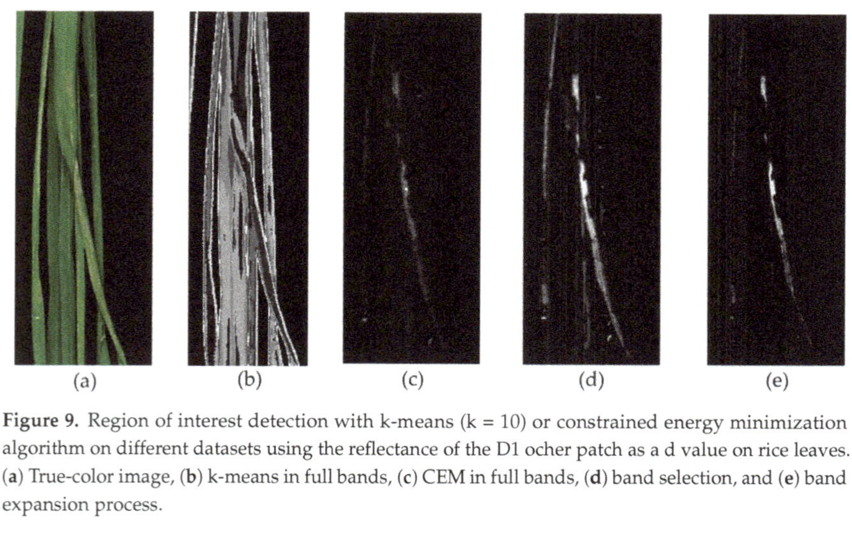

Figure 9. Region of interest detection with k-means (k = 10) or constrained energy minimization algorithm on different datasets using the reflectance of the D1 ocher patch as a d value on rice leaves. (**a**) True-color image, (**b**) k-means in full bands, (**c**) CEM in full bands, (**d**) band selection, and (**e**) band expansion process.

3.4. DNN Model for Classification of Testing Dataset

The DNN multilayer perceptron model is suited for HSI data for classification because the spectral reflection of each pixel can form a vector. Even if we have fewer images, we can still use enough pixels as samples for analysis. Therefore, this study does not require thousands of images to train a set of deep learning models, which greatly reduces the tedious work of collecting samples and the difficulty of controlling sample conditions.

Table 2 describes the results of the OA verification using the DNN models of the full bands, band selection (6 bands), and BEP (27 bands). The confusion matrix [37] was utilized to evaluate the classification performance; the complete confusion matrix calculated for DNN classification is shown in Supplementary Figure S2. In the case of full bands, the OA (95%) and performance are the best in the classification of various situations, but a longer time (14.88 s) is needed than band selection and BEP in classification. The application of band selection saves approximately half the time of full bands, but it will also reduce the classification accuracy. Except for HL, the accuracy of early and late OPs decreased after band selection, which may be attributed to a decrease in some spectral information. The accuracy of the BEP is not higher than that of band selection, as expected, and it is possible that BEP amplifies the noise and interferes with the classification ability. Among the five criteria, the OA of classification is the best among the bands selected by entropy. In terms of entropy, the accuracy of early OP from band selection is approximately 4% higher than that from BEP.

Table 2. Results for the testing dataset for DNN classification in different bands. The best performance is highlighted in red.

Model	Criteria	Accuracy (%)			OA [3] (%)	Time (s)
		HL	Early [1] OP	Late [2] OP		
Full-band	-	97.3	93.6	94.0	95.0	14.88
Band selection (6 bands)	Variance	96.0	84.4	85.1	88.5	7.18
	Entropy	97.2	87.1	86.5	90.3	5.79
	Skewness	95.7	82.5	81.6	86.6	4.96
	Kurtosis	97.4	78.7	86.5	87.5	6.32
	SNR	97.8	78.4	78.9	85.0	6.98
Band expansion process (27 bands)	Variance	97.0	83.3	84.1	88.1	7.88
	Entropy	97.1	82.8	83.5	87.8	6.83
	Skewness	96.6	76.2	81.7	84.8	5.85
	Kurtosis	96.3	78.2	86.8	87.1	6.79
	SNR	96.9	78.0	81.3	85.4	7.43

[1] Early OP comprises a set of D1 and D2 OP. [2] Late OP comprises a set of D5 and D6 OP. [3] OA is an abbreviation for overall accuracy.

3.5. Prediction of Unknown Samples

The predictions were carried out using ROIs obtained from full bands, band selection, and the BEP, as shown in Figure 6. CEM was applied to suppress the background and to detect the ROI. The DNN models of full bands, band selection, and the BEP were used as classifiers to predict unknown samples through five analysis methods. For band selection and the BEP, bands selected by entropy were selected as examples according to the results of Table 2 to execute the prediction. Figure 10 shows the results of the true-color image (a), ground truth (b), and predictions from an unknown sample (c–g). The ground truth was determined by entomologists and given different colors to distinguish HL (green) from OP (red). Figure 10c–g shows the classification results from the full bands and band selection/BEP, respectively, which were also colored for visualization. Figure 10d,e shows the best results as expected, in which the predicted areas of the ROI were approximately the same as the ground truth (Figure 10). However, the predicted ROI in Figure 10c was distributed over the rice leaves in addition to the ROI of the ground truth.

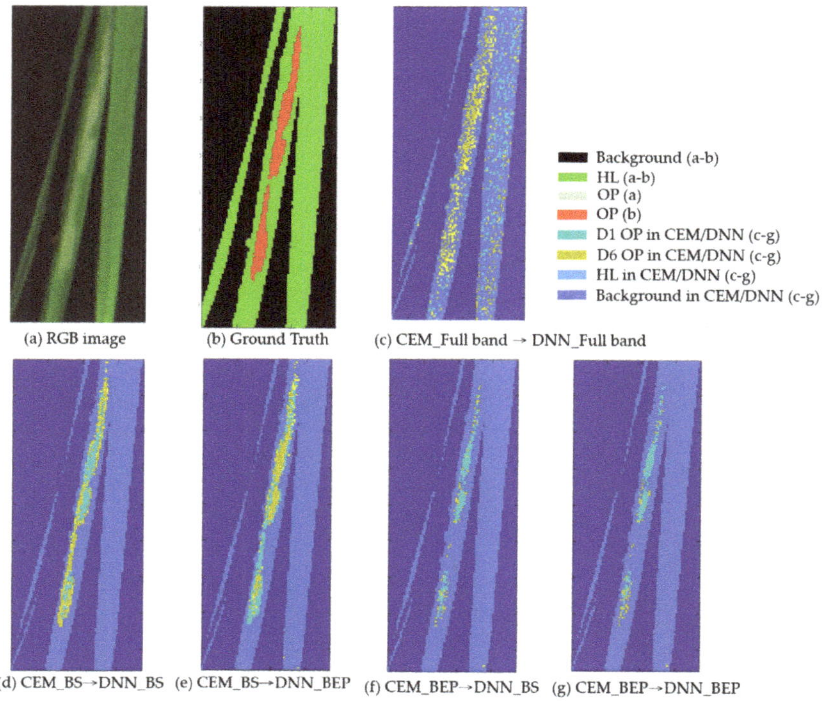

Figure 10. Prediction of spectral information from unknown rice sample: (**a**) true-color image, (**b**) Ground Truth, (**c**) CEM_Full-band→DNN_Full-band, (**d**) CEM_band selection→DNN_band selection, (**e**) CEM_band selection→DNN_BEP, (**f**) CEM_BEP→DNN_band selection, and (**g**) CEM_BEP→DNN_BEP.

The performance of the pixel classification of DNN models was verified by comparing the prediction results with ground truth using a confusion matrix; the results are shown in Tables 3 and 4. Similar to the results in Figure 10, the analysis methods show that CEM_band selection→DNN_band selection showed the best prediction performance (Table 3) because this method showed the highest TP (correct identification of OP) and overall accuracy (OA) and the lowest FN (misidentification of OP). However, very high false positives (FPs) were obtained from the methods of CEM_Full-band→DNN_Full-band, CEM_band selection→DNN_band selection, and CEM_band selection→DNN_BEP (Figure 10c–e). The high FP value of CEM_Full-band→DNN_Full-band may be derived from the scattered distribution of predicted ROI, while the high FP values of CEM_band selection→DNN_band selection and CEM_band selection→DNN_BEP predicted area of ROI may be derived from the predicted areas of ROI that are undetectable by the naked eye. To prove the above observation, the images of Figure 10d or Figure 10e were overlaid with ground truth (Figure 10b). The extra predicted area around the ROI of ground truth in Figure 11d,e should be the early infestation of RLF that cannot be detected by human eyes.

To verify the necessity of using CEM to extract ROI, the DNN classification results of the background-removed images are shown in Supplementary Table S2 and Figure S3. The results show that the accuracy of DNN classification after CEM processing is approximately 22% higher than that of the DNN applied directly to remove the background.

Table 3. Accuracy of DNN classification evaluated by the confusion matrix.

Analysis Method	Pixel Number				OA (%)
	TP [2]	FP [3]	TN [4]	FN [5]	
CEM_Full-band→DNN_Full-band	317	341	11,781	289	95.05
CEM_band selection→DNN_band selection [1]	497	138	11,984	109	98.05
CEM_band selection→DNN_BEP	488	138	11,984	178	97.98
CEM_BEP→DNN_band selection	318	17	12,105	288	97.60
CEM_BE→DNN_BEP	302	18	12,104	304	97.47
	Positive [6]		Negative [7]		
Ground Truth	606		12,122		

[1] Bands selected by Entropy. [2] TP represents the correct identification of OP; [3] FP denotes the health misidentification of HL; [4] TN indicates the correct identification of health HL; [5] FN represents misidentification of OP; [6] OP is positive, and [7] non-OP is negative.

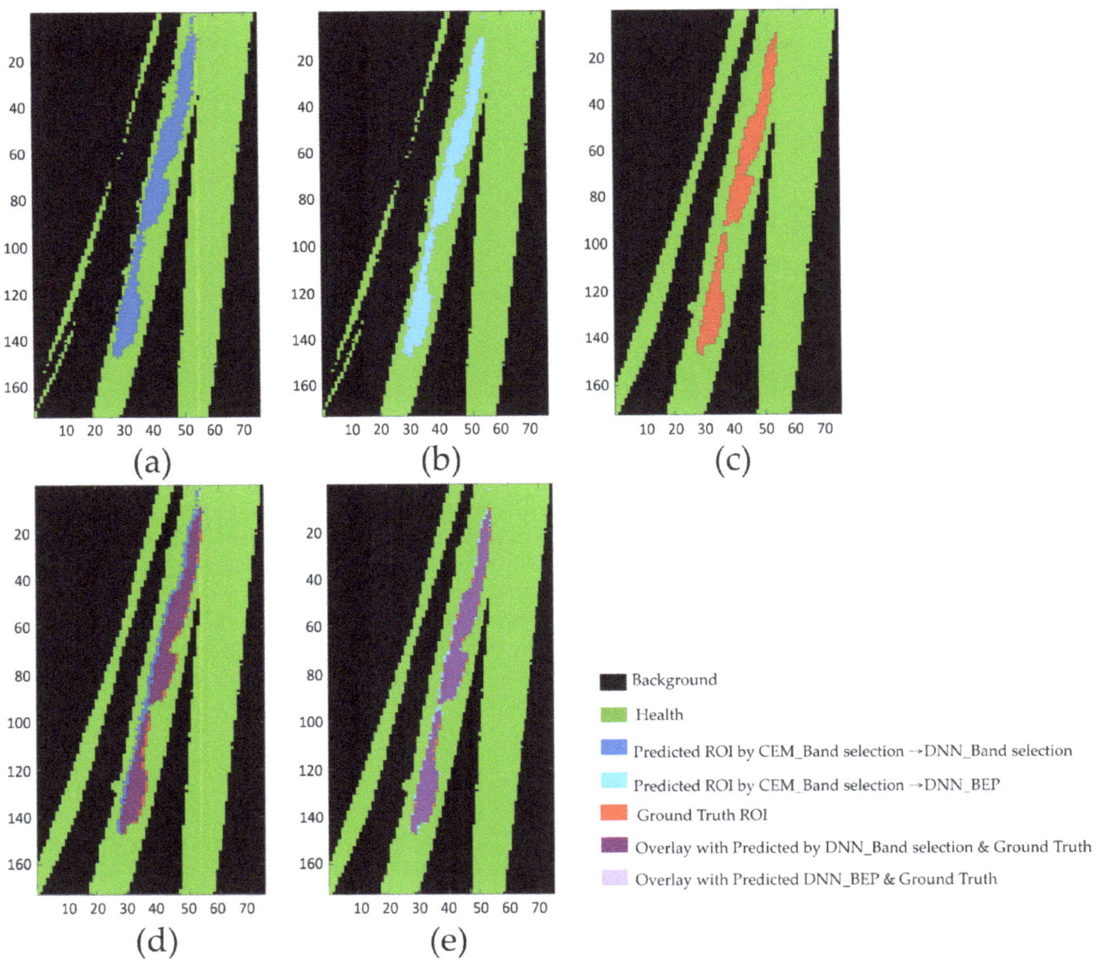

Figure 11. Overlaid images of the predicted ROI with the ground-truth ROI for evaluating the performance of DNN classification. (**a**) Predicted ROI with CEM_band selection→DNN_band selection, (**b**) predicted ROI with CEM_band selection→DNN_BEP, (**c**) Ground Truth, (**d**) Overlay with (**a**,**c**,**e**) Overlay with (**b**,**c**).

The performance of DNN classification was further evaluated by the metrics of recall, precision, accuracy, and DSC, as shown in Table 4. The analysis method of CEM_band selection→DNN_band selection was again rated as the best model for predicting unknown samples, as it had the highest accuracy, recall, and DSC and took the shortest time. Although the analysis method of CEM_band selection→DNN_BEP also showed reasonably good performance, the overall results indicated that six bands obtained from band selection are good enough to detect the early OP caused by RLF. The analysis method of CEM_BEP→DNN_band selection has the highest precision, but its recall and DSC are lower than those of CEM_band selection→DNN_band selection and CEM_band selection→DNN_BEP.

Table 4. Evaluation metrics of DNN prediction. The best performance is highlighted in red color.

Analysis Method	Recall	Precision	Accuracy	Dice Similarity Coefficient	Time (s)
CEM _Full-band→DNN_Full-band	0.523	0.482	0.951	0.670	3.672
CEM_band selection→DNN_band selection	0.820	0.783	0.981	0.801	0.336
CEM_band selection→DNN_BEP	0.805	0.780	0.980	0.755	0.381
CEM_BEP→DNN_band selection	0.525	0.949	0.976	0.676	0.559
CEM_BEP→DNN_BEP	0.498	0.915	0.974	0.652	0.604

Taking the OP as an example, the pixels of the ROI were utilized for prediction evaluation, and a confusion matrix was employed for performance in this study. As shown in Table 4, all analysis methods were successful in classification, and their accuracies reached at least 95%. The area of the block classified as OP is smaller than the actual situation, which is the case in Figure 11f. As shown in Figure 11d, CEM_band selection→DNN_band selection, the distribution of false positives was observed around the OP, which means that the earlier defects caused by insect pests could be identified as false-positive areas in hyperspectral images but could not be recognized in true-color images or human eyes.

3.6. Discussion

Automatic detection of plant pests is extremely useful because it reduces the tedious work of monitoring large paddy fields and detects the damage caused by RLF at the early stage of pest development and eventually stop further plant degradation. This study proposes an automatic detection method that combines CEM and the ATGP. CEM is an efficient hyperspectral detection algorithm that can efficiently handle subpixel detection [39]. The quality of the CEM results is determined by the d-value used as a reference. Therefore, it is important to provide a plausible spectral feature. The ATGP was applied to identify the most representative feature vector as the d-value from an unknown sample. Another problem with the CEM is that it only provides a rough detection result. The DNN was selected to classify the reflectance of the ATGP→CEM detection results. In addition, band selection and the BEP were chosen to identify the key wavelengths among the five criteria to save time and improve accuracy. The accuracy of CEM_band selection→DNN_band selection in predicting the performance of unknown samples reached 98.1%. Traditional classifiers such as linear SVM (support vector machine) and LR (logistic regression) can be attributed to single-layer classifiers, while decision trees or SVM with kernels are considered to have two layers [40,41]. However, deep neural architectures with more layers can potentially extract abstract and invariant features for better image or signal classification [42]. Our previous studies to detect Fusarium wilt on Phalaenopsis have shown this result [43]. In addition, we have used the Entoscan Plant imaging system to detect the infestation of RLF, but this system only covers 16 bands (390, 410, 450, 475, 520, 560, 625, 650, 730, 770, 840, 860, 880, 900, 930, and 960 nm) to obtain the Normalized Difference Vegetation Index. The results are shown in Supplementary Figure S4. It may not be specific enough to distinguish the damage caused by different pests. Therefore, we attempt to find a more representative vector from the spectral fingerprint of the hyperspectral imaging system to detect the infestation of RLF. At the same time, the band selection was used to remove redundant information to achieve the time required for the automatic detection process. It is not only reducing the time by 2.45 times (from 8′11″ to 3′20″) but also reach higher accuracy (0.981) than that (0.951) in the full band. The time required for each stage of the prediction process is shown in Supplementary Figure S5. The six bands (489, 501, 603, 664, 684, and 705 nm) obtained through band selection are more representative than bands supplied by the Entoscan Plant imaging system and can be applied to the multispectral sensor of UAVs and portable instruments for field use. The methods, algorithms, and models we established in this paper will be applied to other important rice insect pests and verified in the field by using either UAVs or portable instruments that carry the multispectral sensor. In addition, a platform to integrate all this information will be established to interact with farmers.

Other studies [44,45] used conventional true-color images, which can only classify spatial information based on their color and shape and identify damage that is clearly visible by the naked eye. Compared with previous studies, the DNN was based on high spectral sensors to provide spectral information, which can detect pixel-level targets and retain the spatial information of the original image. The authors [44,45] employed the CNN to detect pests and achieved a classification accuracy of 90.9% and 97.12%, respectively. The method proposed in this paper is slightly higher than the final accuracy of CNN. Although it can simultaneously classify multiple insect pests and diseases, it often causes confusion. In addition, their studies were conducted with images of the late damage stage and could not classify the level of infestation. In addition, most image classifications are trained by a CNN. CNNs often need to collect a large number of training samples, and it is difficult to obtain a large number of sufficient training images in a short period of time. In contrast, hyperspectral image classification based on spectral pixels can be trained by a DNN, which means that even a single hyperspectral image can have a large amount of data for training.

4. Conclusions

HSI techniques can provide a real-time monitoring system to guide the precise application and reduction of pesticides and to provide objective and effective options for the automatic detection of crop damage caused by insect pests or diseases. In this research, we propose a deep learning classification and detection method that is based on band selection and a BEP that can be applied to determine the lowest cost to achieve the monitoring of leaf defects caused by RLF. To compensate for the deficiencies caused by band selection, the BEP method was selected to improve the detection efficiency. The results of the test dataset show that the use of the full-band classification is the best, and the band selection classification is better than the BEP. Except for criteria on skewness and signal-to-noise ratio, the accuracy of full-band classification is nearly 95%.

After using the trained model to predict the unknown samples, the results show that the CEM_band selection→DNN_band selection analysis method is the best model and has reached the expected prediction. The maximum DSC is 0.80, which means that its classification is 80%, which is identical to the classification recognized by entomologists. In addition, we discovered that the predictive area of the model was larger than the area observed by the human eye. This phenomenon may indicate that RLF damage may produce changes in parts of the spectrum that cannot be easily detected by the human eye. In addition, comparing the implementation of prediction operations based on the full-band DNN model and the band selection-based DNN model, the band selection method only needs 1% of the full-band time, which provides a vast potential for wider applications and has good rice identification capabilities. Only six bands are needed while reducing the technical cost required for on-site monitoring.

By providing more training data, the method also has significant room for improvement by implementing a data argumentation process or extending other data, such as the mean or variance-generating structures. While the current research has only been conducted in the laboratory or used non-specified multispectral images in the field, the handheld six-band sensor provided very good results, and its portability means that it could be adapted for use in the field to obtain realistic multispectral images on-site using band Selection methods. In addition, most of the existing UAVs use CNN or vegetation indices for analysis and have not been studied much in spectral reflectance. As mentioned in Section 3.5, the HSI prediction model can detect infested areas before noticed by the human eye. This technique can be extended to UAV in the future to monitor the invisible spectral changes on the leaf surface. This technology can be extended to UAV in the future to monitor the invisible spectral changes on the leaves. Combining HSI techniques and deep learning classification models could provide real-time surveys that give on-site early warning of damage.

Supplementary Materials: The following are available online at https://www.mdpi.com/article/10.3390/rs13224587/s1, Figure S1: Bands selected through band prioritization and band decorrelation, Figure S2: Confusion Matrix result of DNN model, Figure S3: Prediction of spectral information from unknown rice sample, Figure S4: Entoscan Plant imaging system, Figure S5: Approximate time required for each step of the prediction of unknown samples, Table S1: Results of the first six bands of band selection using different criteria, Table S2: The accuracy of DNN classification evaluated by confusion matrix.

Author Contributions: Conceptualization, Y.-C.O. and S.-M.D.; methodology, Y.-C.O. and S.-M.D.; software, Y.-C.O.; validation, Y.-C.O. and S.-M.D.; formal analysis, G.-C.L.; investigation, G.-C.L.; resources, Y.-C.O. and S.-M.D.; data curation, G.-C.L.; writing—original draft preparation, G.-C.L.; writing—review and editing, Y.-C.O. and S.-M.D.; visualization, G.-C.L.; supervision, Y.-C.O. and S.-M.D.; funding acquisition, Y.-C.O. and S.-M.D. All authors have read and agreed to the published version of the manuscript.

Funding: This research was funded by the Ministry of Science and Technology (MOST), Taiwan (Grant No. MOST 107-2321-B-005-013, 108-23321-B-005-008, and 109-2321-B-005-024), and Council of Agriculture, Taiwan (Grant No. 110AS-8.3.2-ST-a6). The APC was funded by MOST 109-2321-B-005-024.

Institutional Review Board Statement: Not applicable.

Informed Consent Statement: Not applicable.

Data Availability Statement: Not applicable.

Acknowledgments: We are grateful to Chung-Ta Liao from Taichung District Agricultural Research and Extension Station for RLF collection and maintenance. We would also like to thank the publication subsidy from the Academic Research and Development of NCHU.

Conflicts of Interest: The authors declare no conflict of interest.

References

1. Khan, Z.R.; Barrion, A.T.; Litsinger, J.A.; Castilla, N.P.; Joshi, R.C. A bibliography of rice leaf folders (Lepidoptera: Pyralidae)-Mini review. *Insect Sci. Appl.* **1988**, *9*, 129–174.
2. Park, H.H.; Ahn, J.J.; Park, C.G. Temperature-dependent development of *Cnaphalocrocis medinalis* Guenée (Lepidoptera: Pyralidae) and their validation in semi-field condition. *J. Asia Pac. Entomol.* **2014**, *17*, 83–91. [CrossRef]
3. Bodlah, M.A.; Gu, L.L.; Tan, Y.; Liu, X.D. Behavioural adaptation of the rice leaf folder *Cnaphalocrocis medinalis* to short-term heat stress. *J. Insect Physiol.* **2017**, *100*, 28–34. [CrossRef] [PubMed]
4. Padmavathi, C.; Katti, G.; Padmakumari, A.P.; Voleti, S.R.; Subba Rao, L.V. The effect of leaffolder *Cnaphalocrocis medinalis* (Guenee) (Lepidoptera: Pyralidae) injury on the plant physiology and yield loss in rice. *J. Appl. Entomol.* **2013**, *137*, 249–256. [CrossRef]
5. Pathak, M.D. Utilization of Insect-Plant Interactions in Pest Control. In *Insects, Science and Society*; Pimentel, D., Ed.; Academic Press: London, UK, 1975; pp. 121–148.
6. Murugesan, S.; Chelliah, S. Yield losses and economic injury by rice leaf folder. *Indian J. Agri. Sci.* **1987**, *56*, 282–285.
7. Kushwaha, K.S.; Singh, R. Leaf folder (LF) outbreak in Haryana. *Int. Rice Res. Newsl.* **1984**, *9*, 20.
8. Bautista, R.C.; Heinrichs, E.A.; Rejesus, R.S. Economic injury levels for the rice leaffolder *Cnaphalocrocis medinalis* (Lepidoptera: Pyralidae): Insect infestation and artificial leaf removal. *Environ. Entomol.* **1984**, *13*, 439–443. [CrossRef]
9. Heong, K.L.; Hardy, B. *Planthoppers: New Threats to the Sustainability of Intensive Rice Production Systems in Asia*; International Rice Research Institute: Los Banos, Philippines, 2009.
10. Norton, G.W.; Heong, K.L.; Johnson, D.; Savary, S. Rice pest management: Issues and opportunities. In *Rice in the Global Economy: Strategic Research and Policy Issues for Food Security*; Pandey, S., Byerlee, D., Dawe, D., Dobermann, A., Mohanty, S., Rozelle, S., Hardy, B., Eds.; International Rice Research Institute: Los Baños, Philippines, 2010; pp. 297–332.

11. Lowe, A.; Harrison, N.; French, A.P. Hyperspectral image analysis techniques for the detection and classification of the early onset of plant disease and stress. *Plant Methods.* **2017**, *13*, 80. [CrossRef]
12. Sytar, O.; Brestic, M.; Zivcak, M.; Olsovska, K.; Kovar, M.; Shao, H.; He, X. Applying hyperspectral imaging to explore natural plant diversity towards improving salt stress tolerance. *Sci. Total Environ.* **2017**, *578*, 90–99.
13. Thomas, S.; Kuska, M.T.; Bohnenkamp, D.; Brugger, A.; Alisaac, E.; Wahabzada, M.; Behmann, J.; Mahlein, A.-K. Benefits of hyperspectral imaging for plant disease detection and plant protection: A technical perspective. *J. Plant. Dis. Prot.* **2017**, *125*, 1–16. [CrossRef]
14. Zhao, Y.; Yu, K.; Feng, C.; Cen, H.; He, Y. Early detection of aphid (*myzus persicae*) infestation on chinese cabbage by hyperspectral imaging and feature extraction. *Trans. Asabe* **2017**, *60*, 1045–1051. [CrossRef]
15. Wu, X.; Zhang, W.; Qiu, Z.; Cen, H.; He, Y. A novel method for detection of pieris rapae larvae on cabbage leaves using nir hyperspectral imaging. *Appl. Eng. Agric.* **2016**, *32*, 311–316.
16. Harsanyi, J.C. Detection and Classification of Subpixel Spectral Signatures in Hyperspectral Image Sequences. Ph.D. Thesis, Department of Electrical Engineering, University of Maryland Baltimore County, College Park, MD, USA, August 1993.
17. Pearson, K. LIII. On lines and planes of closest fit to systems of points in space. *Philos. Mag.* **1901**, *2*, 559–572. [CrossRef]
18. Cortes, C.; Vapnik, V. Support-Vector Networks. *Mach. Learn.* **1995**, *20*, 273–297. [CrossRef]
19. Carranza-García, M.; García-Gutiérrez, J.; Riquelme, J.C. A framework for evaluating land use and land cover classification using convolutional neural networks. *Remote Sens.* **2019**, *11*, 274. [CrossRef]
20. Hinton, G.E.; Osindero, S.; Teh, Y.-W. A Fast Learning Algorithm for Deep Belief Nets. *Neural Comput.* **2006**, *18*, 1527–1554. [CrossRef] [PubMed]
21. Fan, Y.; Wang, T.; Qiu, Z.; Peng, J.; Zhang, C.; He, Y. Fast Detection of Striped Stem-Borer (*Chilo suppressalis* Walker) Infested Rice Seedling Based on Visible/Near-Infrared Hyperspectral Imaging System. *Sensors* **2017**, *17*, 2470. [CrossRef]
22. Araujo, M.C.U.; Saldanha, T.C.B.; Galvao, R.K.H.; Yoneyama, T.; Chame, H.C.; Visani, V. The successive projections algorithm for variable selection in spectroscopic multicomponent analysis. *Chemom. Intell. Lab. Syst.* **2001**, *57*, 65–73. [CrossRef]
23. Al-Allaf, O.N.A. Improving the performance of backpropagation neural network algorithm for image compression/decompression system. *J. Comput. Sci.* **2010**, *6*, 834–838.
24. Chen, S.Y.; Chang, C.Y.; Ou, C.S.; Lien, C.T. Detection of Insect Damage in Green Coffee Beans Using VIS-NIR Hyperspectral Imaging. *Remote Sens.* **2020**, *12*, 2348. [CrossRef]
25. Huang, W.; Lamb, D.W.; Niu, Z.; Zhang, Y.; Liu, L.; Wang, J. Identification of yellow rust in wheat using in-situ spectral reflectance measurements and airborne hyperspectral imaging. *Precis. Agric.* **2007**, *8*, 187–197. [CrossRef]
26. Dang, H.Q.; Kim, I.K.; Cho, B.K.; Kim, M.S. Detection of Bruise Damage of Pear Using Hyperspectral Imagery. In Proceedings of the 12th International Conference on Control Automation and Systems, Jeju Island, Korea, 17–21 October 2012; pp. 1258–1260.
27. Ma, K.; Kuo, Y.; Ouyang, Y.C.; Chang, C. Improving pesticide residues detection using band prioritization and constrained energy minimization. In Proceedings of the 2017 IEEE International Geoscience and Remote Sensing Symposium (IGARSS), Fort Worth, TX, USA, 23–28 July 2017; pp. 4802–4805.
28. Ren, H.; Chang, C.-I. Automatic spectral target recognition in hyperspectral imagery. *IEEE Trans. Aerosp. Electron. Syst.* **2003**, *39*, 1232–1249.
29. Kaur, D.; Kaur, Y. Various Image Segmentation Techniques: A Review. *Int. J. Comput. Sci. Mob. Comput.* **2014**, *3*, 809–814.
30. Woebbecke, D.M.; Meyer, G.E.; Von Bargen, K.; Mortensen, D.A. Color Indices for Weed Identification under Various Soil, Residue and Lighting Conditions. *Trans. ASAE* **1995**, *38*, 259–269. [CrossRef]
31. Senthilkumaran, N.; Vaithegi, S. Image Segmentation by Using Thresholding Techniques for Medical Images. *Comput. Sci. Eng. Int. J.* **2016**, *6*, 1–13.
32. Chang, C.-I. Target signature-constrained mixed pixel classification for hyperspectral imagery. *IEEE Trans. Geosci. Remote Sens.* **2002**, *40*, 1065–1081. [CrossRef]
33. Chang, C.I.; Du, Q.; Sun, T.-L.; Althouse, M. A joint band prioritization and band-decorrelation approach to band selection for hyperspectral image classification. *IEEE Trans. Geosci. Remote. Sens.* **1999**, *37*, 2631–2641. [CrossRef]
34. Chang, C.I.; Liu, K.H. Progressive band selection of spectral unmixing for hyperspectral imagery. *IEEE Trans. Geosci. Remote Sens.* **2014**, *52*, 2002–2017. [CrossRef]
35. Ouyang, Y.C.; Chen, H.M.; Chai, J.W.; Chen, C.C.; Poon, S.K.; Yang, C.W.; Lee, S.K.; Chang, C.I. Band expansion process-based over-complete independent component analysis for multispectral processing of magnetic resonance images. *IEEE Trans. Biomed. Eng.* **2008**, *55*, 1666–1677. [CrossRef]
36. Zhang, L.; Zhang, L.; Du, B. Deep Learning for Remote Sensing Data: A Technical Tutorial on the State of the Art. *IEEE Geosci. Remote Sens. Mag.* **2016**, *4*, 22–40. [CrossRef]
37. Knipling, E.B. Physical and physiological basis for the reflection of visible and near-infrared radiation from vegetation. *Remote Sens. Environ.* **1970**, *1*, 155–159. [CrossRef]
38. Youden, W.J. Index for rating diagnostic tests. *Cancer* **1950**, *3*, 32–35. [CrossRef]
39. Chen, S.-Y.; Lin, C.; Tai, C.-H.; Chuang, S.-J. Adaptive Window-Based Constrained Energy Minimization for Detection of Newly Grown Tree Leaves. *Remote Sens.* **2018**, *10*, 96. [CrossRef]
40. Camps-Valls, G.; Bruzzone, L. Kernel-based methods for hyperspectral image classification. *IEEE Trans. Geosci. Remote Sens.* **2005**, *43*, 1351–1362. [CrossRef]

41. Bengio, Y.; LeCun, Y. Scaling Learning Algorithms towards AI. In *Large-Scale Kernel Machines*; MIT Press: Cambridge, MA, USA, 2007; pp. 1–41. ISBN 1002620262.
42. Bengio, Y.; Courville, A.C.; Vincent, P. Representation Learning: A Review and New Perspectives. *IEEE TPAMI* **2013**, *35*, 1798–1828. [CrossRef]
43. Hsu, Y.; Ouyang, Y.C.; Lu, Y.L.; Ouyang, M.; Guo, H.Y.; Liu, T.S.; Chen, H.M.; Wu, C.C.; Wen, C.H.; Shin, M.S.; et al. Using Hyperspectral Imaging and Deep Neural Network to Detect Fusarium Wilton Phalaenopsis. In Proceedings of the 2021 IEEE International Geoscience and Remote Sensing Symposium IGARS, Brussels, Belgium, 11–16 July 2021; pp. 4416–4419.
44. Mique, E.L.; Palaoag, T.D. Rice Pest and Disease Detection Using Convolutional Neural Network. In Proceedings of the 2018 International Conference on Information Science and System, Jeju Island, Korea, 27–29 April 2018; pp. 147–151.
45. Rahman, C.R.; Arko, P.S.; Ali, M.E.; Iqbal Khan, M.A.; Apon, S.H.; Nowrin, F.; Wasif, A. Identification and recognition of rice diseases and pests using convolutional neural networks. *Biosyst. Eng.* **2020**, *194*, 112–120. [CrossRef]

remote sensing

MDPI

Article

Detection of Insect Damage in Green Coffee Beans Using VIS-NIR Hyperspectral Imaging

Shih-Yu Chen [1,2,*], Chuan-Yu Chang [1,2], Cheng-Syue Ou [1] and Chou-Tien Lien [1]

1 Department of Computer Science and Information Engineering, National Yunlin University of Science and Technology, Yunlin 64002, Taiwan; chuanyu@gemail.yuntech.edu.tw (C.-Y.C.); m10717033@gemail.yuntech.edu.tw (C.-S.O.); m10617013@gemail.yuntech.edu.tw (C.-T.L.)

2 Artificial Intelligence Recognition Industry Service Research Center, National Yunlin University of Science and Technology, Yunlin 64002, Taiwan

* Correspondence: sychen@yuntech.edu.tw

Received: 11 May 2020; Accepted: 20 July 2020; Published: 22 July 2020

Abstract: The defective beans of coffee are categorized into black beans, fermented beans, moldy beans, insect damaged beans, parchment beans, and broken beans, and insect damaged beans are the most frequently seen type. In the past, coffee beans were manually screened and eye strain would induce misrecognition. This paper used a push-broom visible-near infrared (VIS-NIR) hyperspectral sensor to obtain the images of coffee beans, and further developed a hyperspectral insect damage detection algorithm (HIDDA), which can automatically detect insect damaged beans using only a few bands and one spectral signature. First, by taking advantage of the constrained energy minimization (CEM) developed band selection methods, constrained energy minimization-constrained band dependence minimization (CEM-BDM), minimum variance band prioritization (MinV-BP), maximal variance-based bp (MaxV-BP), sequential forward CTBS (SF-CTBS), sequential backward CTBS (SB-CTBS), and principal component analysis (PCA) were used to select the bands, and then two classifier methods were further proposed. One combined CEM with support vector machine (SVM) for classification, while the other used convolutional neural networks (CNN) and deep learning for classification where six band selection methods were then analyzed. The experiments collected 1139 beans and 20 images, and the results demonstrated that only three bands are really need to achieve 95% of accuracy and 90% of kappa coefficient. These findings show that 850–950 nm is an important wavelength range for accurately identifying insect damaged beans, and HIDDA can indeed detect insect damaged beans with only one spectral signature, which will provide an advantage in the process of practical application and commercialization in the future.

Keywords: target detection; coffee beans; insect damage; hyperspectral imaging; band selection

1. Introduction

Coffee is one of the most widely consumed beverages by people, and high quality coffee comes from healthy coffee beans, an important economic crop. However, insect damage is a hazard on green coffee beans as the boreholes in green beans, also known as wormholes, are the cause for the turbid or strange taste of the coffee made from such coffee beans. Generally, the coffee beans are inspected manually with the naked eye, which is a laborious and error-prone work, while visual fatigue often induces misrecognition. Even for an expert analyst, each batch of coffee takes about 20 min to inspect.

The international green coffee beans grading method is based on the SCAA (Specialty Coffee Association of America) Green Coffee Classification. This classification categorizes 300 g of properly hulled coffee beans into five grades, according to the number of primary defects and secondary defects. Primary defects include full black beans, full sour beans, pod/cherry, etc. One to two primary defects equal one full defect. Secondary defects include insect damaged, broken/chipped, partial black, partial sour, floater, shell, etc., where two to five secondary defects are equal to one full defect [1]. Specialty grade (Grade 1) shall have no more than five secondary defects and no primary defect allowed in 300 g of coffee bean samples. At most, a 5% difference in screen mesh is permitted. These must have a special attribute in terms of concentration, fragrance, acidity, or aroma, with no defects and contamination. Premium-grade (Grade 2) shall have no more than eight full defects in 300 g of coffee bean samples and a maximum of 5% difference of screen mesh is permitted. These must have a special attribute in terms of concentration, fragrance, acidity, or aroma, and there must be no defect. The exchange grade (Grade 3) is permitted to have 9~23 full defects in 300 g of coffee bean samples. The test cup should be defect-free, and the moisture content should be 9~13%. Below standard grade (Grade 4) has 24~86 full defects in 300 g of coffee bean samples. Finally, the off-grade (Grade 5) has more than 86 full defects in 300 g of coffee bean samples.

In recent years, many coffee bean identification methods have been proposed, but few research reports have used a spectral analyzer to evaluate the defects and impurities of coffee beans. The current manual inspection of defective coffee beans is time-consuming and is unable to analyze a large quantity of samples. Therefore, this study, which used hyperspectral images for analysis, should provide more crucial spectral information than conventional RGB images to determine the spectral signal difference between healthy and defective coffee beans. Table 1 tabulates the green coffee bean evaluation methods proposed by previous studies.

Table 1. Existing green coffee bean evaluation methods.

Tested Target Beans	Spectral Range (nm)	Data Volume	Data Analysis Method	Accuracy	Reference
Broken beans, dry beans, moldy beans, black beans	1000–2500 nm (267 bands)	662 beans	PCA + k-NN	90%	[2]
Origin classification	955–1700 nm (266 bands)	432 beans	PLS + SVM	97.1%	[3]
Origin classification	900–1700 nm (256 bands)	1200 beans	SVM	80%	[4]
Sour beans, black beans, broken beans	RGB	444 beans	k-NN	95.66%	[5]
Black beans	RGB	180 beans	Threshold (TH)	100%	[6]

In 2019, Oliveri et al. [2] used VIS-NIR to identify the black beans, broken beans, dry beans, and dehydrated coffee beans using principal component analysis (PCA) and the k-nearest neighbors algorithm (k-NN) for classification. Although their method can extract effective wavebands, the disadvantages are that the recognition rate is only 90%. As k-NN uses a qualified majority for training and classification, it is likely to have over fit and low-level fit. In 2018, Caporaso et al. [3] used hyperspectral imaging to recognize the origin of coffee beans by using support vector machine (SVM) to classify the origins. Their method is similar to that used in this paper and the advantage includes more spectral information of hyperspectral imaging. Despite the fact that SVM and partial least squares (PLS) multi-dimensional classification can classify the green coffee beans effectively, the bands are not selected according to materials, and the recognition rate was 97% among 432 coffee beans. Zhang et al. [4] proposed a hyperspectral analysis used moving average smoothing (MA), wavelet transform (WT), empirical mode decomposition (EMD), and median filter for the spatial preprocessing of gray level images of each wavelength, and finally used SVM for classification. The advantage of their method is that the preprocessing is performed by using signals different from the concept of images, and SVM is used for classification. The disadvantages are that only second derivatives are used for band selection, the material is not analyzed, and the accuracy in 1200 coffee beans was only slightly higher than 80%. There have been a few reports on traditional RGB images. García [5] used K-NN to classify sour beans, black beans, and broken beans. The limitations of the method are that K-NN is likely to have over fit and low-level fit. As the result, the classified coffee beans were relatively clear target objects, and the accuracy in about 444 coffee beans was 95%. Later, Arboleda [6] used thresholds to classify black beans. However, the defects in that method were that only the threshold was used. Therefore, if the external environment changes, the threshold changes accordingly, and the classified target objects were relatively apparent, so the accuracy was higher, at 100% in 180 coffee beans.

The black beans, dry beans, dehydrated beans, and sour beans are still apparent coffee beans, except with very different colors. The differences in appearance are obvious in traditional color images. Most prior studies have used black beans as experimental targets because black beans are quite different from healthy beans. The broken beans are identified by using the morphological analysis method. Unlike the aforementioned studies, this paper sought to identify insect damaged beans, which are difficult to visualize from data. While insect damaged coffee beans are the most common type of defective coffee beans, such targets have little presence and low probability of existence in data, thus it has never been investigated by previous studies. More specifically, as this signal source is considered to be interesting, the signatures are not necessarily pure. Rather, they can be subpixel targets, which cannot exhibit their distinction from the surrounding spectral such as insect damaged beans due to small sample size, and cannot be detected by traditional spatial domain-based techniques. The method proposed in this paper can be applied to many different applications. Without considering their spatial characteristics, hyperspectral imaging provides an effective way to detect, uncover, extract, and identify such targets using their spectral properties, as captured by high spectral-resolution sensors.

The study conducted in this paper collected a total of 1139 green coffee beans including healthy beans and insect damaged beans in equal proportions for hyperspectral data collection and experimentation. Table 1 lists the methods used in prior studies. Our method differed from prior studies in terms of spectral range, data volume, analysis method, and accuracy, along with enhanced data volume, and accuracy. This study used a push-broom VIS-NIR hyperspectral sensor to obtain the images of coffee beans and distinguished the healthy beans from insect damaged ones based on the obtained hyperspectral imaging. Moreover, the hyperspectral insect damage detection algorithm (HIDDA) was particularly developed to locate and capture the insect damaged areas of coffee beans. First, the data preprocessing was performed through band selection (BS), as hyperspectral imaging has a wide spectral range and very fine spectral resolution. As the inter-band correlation between adjacent bands is sometimes too high, complete bands are averse to subsequent data compression, storage, transmission, and resolution. Therefore, the mode of extracting the most representative information from images is one of the most important and popular research subjects in the domain. For this, Step 1 of our HIDDA method involves the analysis of important spectra after band selection, and one image is then chosen for insect damaged bean identification through constrained energy minimization (CEM) and SVM as training samples. In this step, as long as the spectral signature of one insect damaged bean is imported into CEM, the positions of the other insect damaged beans can be detected by Otsu's method and SVM. In Step 2, the image recognition result of Step 1 is used for training, and the deep learning CNN model is used to identify the remaining 19 images. The experimental results show that when using the proposed method to analyze nearly 1100 green coffee beans with only three bands, the accuracy reached almost 95%.

2. Materials and Methods

2.1. Hyperspectral Imaging System and Data Collection

The hyperspectral push-broom scanning system (ISUZU Optics Corp.) used in this experiment is shown in Figure 1. A carried camera SPECIMFX10 hyperspectral sensor with a spectral range of 400–1000 nm, a resolution of 5.5 nm, and 224 bands was used for imaging. The light source for irradiating the images was "3900e-ER", with a power of 21V/150W, comprised of a loading implement and mobile platform, and step motor (400 mm, maximum load: 5 kg, and maximum speed: 100 mm/s). The system was controlled with ISUZU software. The dark (closed shutter) and white (99% reflection spectrum) images were recorded and stored automatically before each measurement. The laboratory samples were placed on movable plates so that they were appropriately spaced. In each image, 60 green coffee beans were analyzed. The process of filming coffee beans is shown in Figure 2. Each time, 30 insect damaged beans and 30 healthy beans were filmed. Figure 3 shows the actual filming results. The mobile platform and correction whiteboard were located in the lower part, and the filming was performed in the dark box to avoid the interference from other light sources. The spectral signatures of green coffee beans were obtained after filming. Figure 3 shows the post-imaging hyperspectral images. The spectral range was 400–1000 nm. The hyperspectral camera captured 224 spectral images and the image data size was 1024 × 629 × 224.

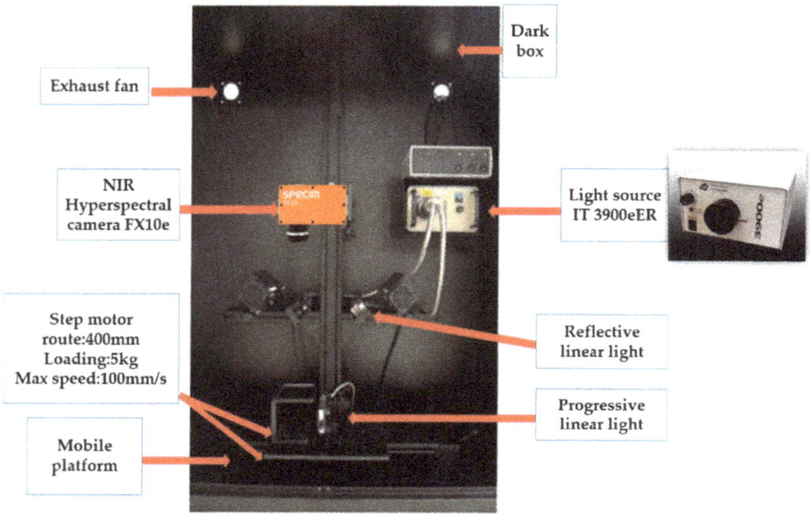

Figure 1. Hyperspectral imaging system.

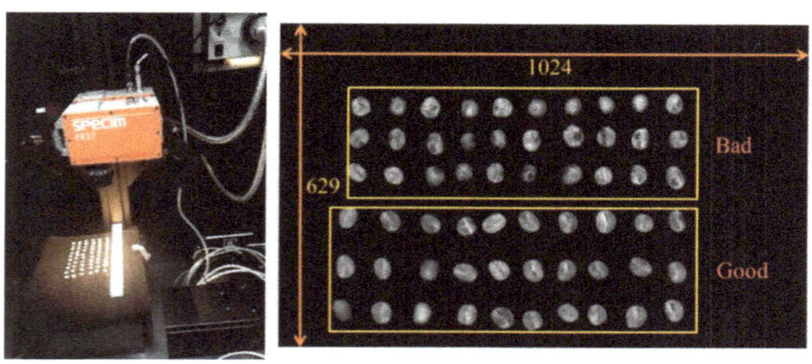

Figure 2. Filming of coffee beans.

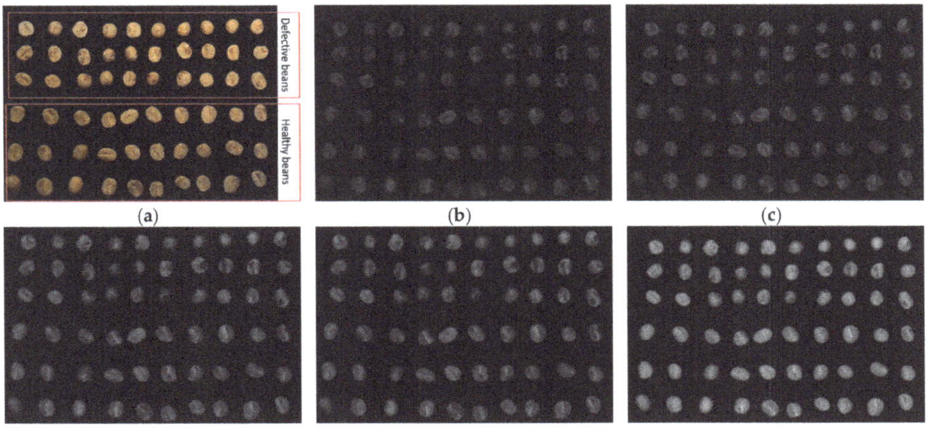

Figure 3. Results of green coffee beans filming. (**a**) Color image; (**b**) 583 nm; (**c**) 636 nm; (**d**) 690 nm; (**e**) 745 nm; (**f**) 800 nm.

2.2. Coffee Bean Samples

After the seeds produced by healthy coffee trees are removed, washed, sun-dried, fermented, dried, and shelled, healthy beans are then separated from defective beans. Common defective beans include black beans, insect damaged beans, and broken beans. Figure 4 shows the healthy and defective beans.

Figure 4. The appearance of healthy and defective beans. (**a**) Healthy bean, (**b**) defective bean (black bean), (**c**) defective bean (insect damaged bean), and (**d**) a defective bean (broken bean).

1. Healthy beans: The entire post-processed bean should appear free of defects. The color of the beans should be blue-green, light green, or yellow-green, as shown in Figure 4a.
2. Black beans: Black beans appear darkened before harvest or fully fermented, as shown in Figure 4b, and produce a turbid and putrefactive odor in coffee [7,8].
3. Insect damaged beans: The insect damaged bean shown in Figure 4c is a result of coffee cherry bugs laying eggs on a coffee tree and the hatched larvae biting the coffee drupes to form wormholes. This type of defective bean produces a turbid odor or strange taste in coffee.
4. Broken beans: Figure 4d shows a bean that was damaged during the treatment process or transportation, known as "ruptured beans". It is likely to induce non-uniform baking [9].

The sample of coffee beans used this study were provided by coffee farmers in Yulin, Taiwan. The coffee farmers filtered the beans and provided both healthy and defective coffee bean samples for the experiment on coffee bean classification. In order to ensure the intactness of the sample beans, all beans were removed from the bag using tweezers, and the tweezers were wiped before touching different types of beans. A total of 1139 beans were collected, and 19 images were recorded. The quantities of the coffee beans are listed in Table 2.

Table 2. Quantities from the experiment.

Class	Qty (pcs)
Healthy beans	569
Defective beans	570
Total	1139

The hyperspectral data of green coffee beans and the original hyperspectral data were obtained, and 224 bands were observed after filming the green coffee beans in the spectral range of 400–1000 nm. The data were normalized to enhance the model convergence regarding the speed and precision of band selection with machine learning or deep learning. We collected 19 hyperspectral images in the experiments. Figure 5 shows the spectral signatures of the healthy beans and defective beans for our proposed hyperspectral algorithm.

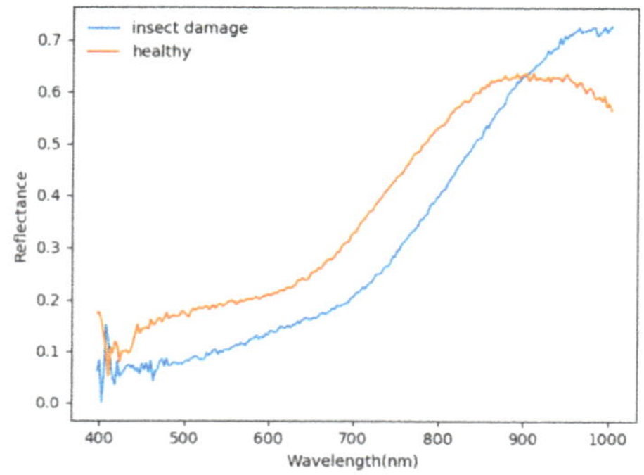

Figure 5. Spectral signature of the healthy and insect damaged coffee beans.

2.3. Hyperspectral Band Selection

In hyperspectral imaging (HSI), hyperspectral signals, with as many as 200 contiguous spectral bands, can provide high spectral resolution. In other words, subtle objects or targets can be located and extracted by hyperspectral sensors with very narrow bandwidths for detection, classification, and identification. However, as the number of spectral bands and the inter-band information redundancy are usually very high in HSI, the original data cube is not suitable for data compression or data transmission, and particularly, image analysis. The use of full bands for data processing often encounters the issue of "the curse of dimensionality"; therefore, band selection plays a very important role in HSI. The purpose of band selection is to select the most representative set of bands in the image and include them in the data, so that they can be as close as possible to the entire image. Previous studies have used various band selection methods based on certain statistical criteria [10–17], mostly select an objective function first, and then select a band group that can maximize the objective function. This paper first used the histogram method in [18] to remove the background, and then applied six band selection methods based on constrained energy minimization (CEM) [19–24] to select and extract a representative set of bands.

2.3.1. Constrained Energy Minimization (CEM)

CEM [19–24] is similar to matched filtering (MF); the CEM algorithm only requires one spectral signature (desired signature or target of interest) as parameter d, while other prior knowledge (e.g., unknown signal or background) is not required. Basically, CEM applies a finite impulse response (FIR) filter to pass through the target of interest, while minimizing and suppressing noise and unknown signals from the background using a specific constraint. CEM suppresses the background by correlation matrix R, which can be defined as $\mathbf{R} = \left(\frac{1}{N}\right)\sum_{i=1}^{N} \mathbf{r_i r_i}^T$, and feature d is used by FIR to detect other similar targets. Assuming one hyperspectral image with N pixels r is defined as $\{\mathbf{r_1}, \mathbf{r_2}, \mathbf{r_3}, \ldots, \mathbf{r_N}\}$, each pixel has L dimensions expressed as $\mathbf{r_i} = (\mathbf{r_{i1}}, \mathbf{r_{i2}}, \mathbf{r_{i3}}, \ldots, \mathbf{r_{iL}})^{\mathrm{T}}$, thus, the desired target d can be defined as $(\mathbf{d_1}, \mathbf{d_2}, \mathbf{d_3}, \ldots, \mathbf{d_L})^{\mathrm{T}}$, and the desired target is passed through by the FIR filter. The coefficient in the finite impulse response filter can be defined as $w = (w_1, w_2, w_3, \ldots, w_L)^T$, where the value of w can be obtained by the constrain $\boldsymbol{d^T}\mathbf{w} = \boldsymbol{w^T d} = \mathbf{1}$, and the result of CEM is:

$$\delta^{\mathrm{CEM}} = (\mathrm{w}^{\mathrm{CEM}})^{\mathrm{T}}\mathrm{r} = (\mathrm{d}^{\mathrm{T}}\mathrm{R}_{\mathrm{L}\times\mathrm{L}}^{-1}\mathrm{d})^{-1}(\mathrm{R}_{\mathrm{L}\times\mathrm{L}}^{-1}\mathrm{d})^{\mathrm{T}}\mathrm{r} \tag{1}$$

CEM is one of the few algorithms that can suppress the background while enhancing the target at the subpixel level. CEM is easier to implement than binary classification as it uses the sampling correlation matrix R to suppress BKG, thus, it only requires meaningful knowledge of the target and no other information is required. In this regard, CEM has been used to design a new band selection method called constraint band selection (CBS) [19], and the resulting minimum variance from CBS is used to calculate the priority score to rank the frequency bands. Conceptually, constrained-target band selection (CTBS) [25,26] is slightly different from CBS, as CBS only focuses on the band of interest, while CTBS simultaneously takes advantage of the target signature and the band of interest. First, it specifies the signal d of a target, and then constrains d to minimize the variance caused by the background signal through the FIR filter. The resulting variance can also be selected by the selection criteria. Since CEM has been widely used for subpixel target detection in hyperspectral imagery, this paper applied CBS and CTBS based methods for further analysis. The following are the six target detection based band selection methods used in the experiments.

2.3.2. Constrained Energy Minimization-Constrained Band Dependence Minimization (CEM-BDM)

CEM-BDM [19] is one of the CBS methods, which uses CEM to determine the correlation between the various bands, and regards such correlation as a score. Subsequent processing is then performed on this score to obtain a band selection algorithm with different band priorities. Let $\{\mathbf{B}_l\}$ be the set of all band images, where $\{\mathbf{b}_l\}$ denotes each band image sized at M × N in a hyperspectral image cube. A optimization problem similar to CEM can be obtained for a constrained band-selection problem by $\min_{w_l}\{\boldsymbol{w_l}^T\boldsymbol{Q}\boldsymbol{w_l}\}$ *subject to* $\boldsymbol{b_l}^T\boldsymbol{w_l} = 1$, which uses the least squares error (LSE) as the constraint. CEM-BDM can be extended as follows: assume the autocorrelation matrix as $\widetilde{\boldsymbol{Q}} = \frac{1}{L-1}\sum_{j=1,j\neq l}^{L}\boldsymbol{b}_j\boldsymbol{b}_j^T$ and the coefficient in the finite impulse response filter as $\widetilde{\boldsymbol{w}}_l^{CEM} = (\boldsymbol{b}_l^T\widetilde{\boldsymbol{Q}}^{-1}\boldsymbol{b}_l)^{-1}\widetilde{\boldsymbol{Q}}^{-1}\boldsymbol{b}_l$, thus, the final results of CEM-BDM can be defined as follows:

$$BDM_{priority}(\boldsymbol{B}_l) = (\widetilde{\boldsymbol{w}}_l^{CEM})^T\widetilde{\boldsymbol{Q}}\widetilde{\boldsymbol{w}}_l^{CEM} \tag{2}$$

This band selection method uses the least square error to determine the correlation between the bands. If the results of the least square error are larger, it means that the current band is more dependent on other bands, and thus, the more significant band.

2.3.3. Minimum Variance Band Prioritization (MinV-BP)

According to the optimization method of CEM, the priority score is processed by the variance value; the smaller the variance, the higher the priority score. CEM ranks bands by starting with the minimal variance as its first selected band. Let $\{b_l\}_{l=1}^{L}$ be the total band images for a hyperspectral image cube, where $\boldsymbol{b}_l$ is the lth band in the image. By applying CEM, this value is obtained by the full band set Ω, $V(\boldsymbol{\Omega}) = (\boldsymbol{d}_{\Omega}^T\boldsymbol{R}_{\Omega}^{-1}\boldsymbol{d}_{\Omega})^{-1} = (\boldsymbol{d}^T\boldsymbol{R}^{-1}\boldsymbol{d})^{-1}$, in this case, for each single band $\boldsymbol{b}_l$, the MinV-BP [23,25,26] variance can be defined as:

$$\mathrm{V}(\boldsymbol{b}_l) = (\boldsymbol{d}_{\boldsymbol{b}_l}^T\boldsymbol{R}_{\boldsymbol{b}_l}^{-1}\boldsymbol{d}_{\boldsymbol{b}_l})^{-1} \tag{3}$$

This can be used as a measure of variance, as it uses only the data sample vector specified by $\boldsymbol{b}_l$. Therefore, the value of $V(\boldsymbol{b}_l)$ can be further used as the priority score of $\boldsymbol{b}_l$. According to this explanation, the band is ranked by the value of $V(\boldsymbol{b}_l)$; the smaller the $V(\boldsymbol{b}_l)$, the higher the priority of band selection.

2.3.4. Maximum Variance Band Prioritization (MaxV-BP)

In contrast to MinV-BP, the concept of Max V-BP [23,25] is to first remove $\boldsymbol{b}_l$ from the band set $\boldsymbol{\Omega}$, and the variance is calculated as follows:

$$V(\boldsymbol{\Omega} - \mathbf{b}_l) = \left(\boldsymbol{d}^T_{\boldsymbol{\Omega}-\{\mathbf{b}_l\}}\boldsymbol{R}^{-1}_{\boldsymbol{\Omega}-\{\mathbf{b}_l\}}\boldsymbol{d}_{\boldsymbol{\Omega}-\{\mathbf{b}_l\}}\right)^{-1} \tag{4}$$

Under this criterion, the value of $V(\boldsymbol{\Omega} - \mathbf{b}_l)$ can also be the measurement of the priority score for $\mathbf{b}_l$ Consequently, $\{\boldsymbol{b}_l\}_{l=1}^{L}$ can be ranked by the decreasing values of $V(\boldsymbol{\Omega} - \mathbf{b}_l)$. The maximum $V(\boldsymbol{\Omega} - \mathbf{b}_l)$ is supposed to be the most significant, and the band is prioritized by (4). The difference between MinV_BP and MaxV_BP is that MinV_BP conducts sorting according to a single band, while MaxV_BP is sorted by the full band, and the results of the two band selections are not opposite.

2.3.5. Sequential Forward-Constrained-Target Band Selection (SF-CTBS)

SF-CTBS [25] uses the MinV_BP criteria in (3) to select one band at a time sequentially, instead of sorting all bands with the scores in (3), as MinV_BP does. As a result, band $\boldsymbol{b}^*_{l_1}$ can obtain the minimal variance.

$$\boldsymbol{b}^*_{l_1} = \arg\left\{\min_{\boldsymbol{b}_l \in \boldsymbol{\Omega}} V(\boldsymbol{b}_l)\right\} = \arg\left\{\min_{b_l \in \Omega} \left(\boldsymbol{d}^T_{\boldsymbol{b}_l}\boldsymbol{R}^{-1}_{\boldsymbol{b}_l}\boldsymbol{d}_{\boldsymbol{b}_l}\right)^{-1}\right\} \tag{5}$$

where $b^*_{l_1}$ is the first selected band, and the second band is generated by another minimum variance

$$\boldsymbol{b}^*_{l_2} = \arg\left\{\min_{b_l \in \Omega - b_{l_1}} V(\boldsymbol{b}_l)\right\} = \arg\left\{\min_{b_l \in \Omega - b_{l_1}} \left(\boldsymbol{d}^T_{\boldsymbol{b}_l}\boldsymbol{R}^{-1}_{\boldsymbol{b}_l}\boldsymbol{d}_{\boldsymbol{b}_l}\right)^{-1}\right\} \tag{6}$$

This process is repeated continuously by adding each newly selected band, while the sequential forward technique in [26,27] selects one band at a time sequentially.

2.3.6. Sequential Backward-Target Band Selection (SB-CTBS)

In contrast to SF-CTBS using the MinV_BP criteria in (3), SB-CTBS [25] applies the MaxV_BP as the criterion by using the leave-one-out method to select the optimal bands. For each single band, $\mathbf{b}_l$, assumes band subset $\boldsymbol{\Omega} - \boldsymbol{b}_{l_1}$, which removes $\mathbf{b}_l$ from the full band. The first selected band can be obtained by (7), which yields the maximal variance and $\boldsymbol{b}^*_{l_1}$ can be considered as the most significant band.

$$\boldsymbol{b}^*_{l_1} = \arg\left\{\max_{b_l \in \boldsymbol{\Omega}} V(\boldsymbol{\Omega} - \{\boldsymbol{b}_l\})\right\} = \arg\left\{\max_{b_l \in \Omega} \left(\boldsymbol{d}^T_{\boldsymbol{\Omega}-\{b_l\}}\boldsymbol{R}^{-1}_{\boldsymbol{\Omega}-\{b_l\}}\boldsymbol{d}_{\boldsymbol{\Omega}-\{b_l\}}\right)^{-1}\right\} \tag{7}$$

After calculating $\boldsymbol{b}^*_{l_1}$, we can have $\Omega_1 = \Omega - \{b_{l_1}\}$, and the second band can be generated by another maximal variance in (8). The same process is repeated continuously by removing the current selected band one at a time from the full band set.

$$\boldsymbol{b}^*_{l_2} = \arg\left\{\max_{\boldsymbol{\Omega}_1-\{b_l\}} V(\boldsymbol{\Omega}_1 - \{\boldsymbol{b}_l\})\right\} = \arg\left\{\max_{\boldsymbol{\Omega}_1-\{b_l\}} \left(\boldsymbol{d}^T_{\Omega_1-\{b_l\}}\boldsymbol{R}^{-1}_{\Omega_1-\{b_l\}}\boldsymbol{d}_{\Omega_1-\{b_l\}}\right)^{-1}\right\} \tag{8}$$

It can be noted that the differences between SB-CTBS and SF-CTBS are that SB-CTBS removes bands from the full band set to generate a desired selected band subset, while SF-CTBS increases the selected band by calculating the minimal variance one at a time. The correlation matrix in SB-CTBS uses $\boldsymbol{R}_{\Omega-\{b_l\}}$ but the correlation matrix in SF-CTBS is $\boldsymbol{R}_{b_l}$.

2.3.7. Principal Component Analysis (PCA)

PCA [28] is classified in machine learning as a method of feature extraction in dimensional reduction, and can be considered as an unsupervised linear transformation technology, which is widely used in different fields. Dimensionality reduction is used to reduce the number of dimensions in data, without much influence on the overall performance. The basic assumption of PCA is that the data can identify a projection vector, which is projected in the feature space to obtain the maximum variance of this dataset. In this case, this paper compared PCA with other CEM-based band selection methods.

2.4. Optimal Signature Generation Process

Our proposed algorithm first identified the desired signature of insect damaged beans as the d (desired signature) in CEM for the detection of other similar beans. Optimal signature generation process (OSGP) [29,30] was used to find the optimal desired spectral signature. As the CEM needs only one desired spectral signature for detection, the quality of the detection result is very sensitive to the desired spectral signature. To minimize this defect, the OSGP selects the desired target d first, and the CEM is repeated to obtain a stable and better d. Thus, the stability of detection can be increased, and the subsequent CEM gives the best detection result. Figure 6 shows the flow diagram of OSGP, and then Otsu's method [31] is used to find the optimal threshold. Otsu's method divides data into 0 and 1. This step is to label data for follow-up analysis.

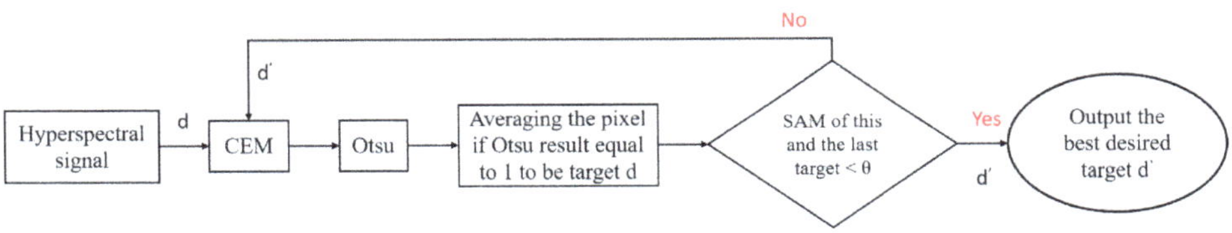

Figure 6. The optimal signature generation process.

2.5. Convolutional Neural Networks (CNN)

Feature extraction requires expert knowledge as the important features must be known for this classification problem, and are extracted from the image to conduct classification. The "convolution" in convolutional neural network (CNN) [32–38] refers to a method of feature extraction, which can replace experts to extract features. Generally speaking, CNN effectively uses spatial information in traditional RGB images; for example, 2D-CNN uses the shape and color of the target in the image to capture features. However, insect damaged coffee beans may be mixed with other material substances, and may even be embedded in a single pixel as their size is smaller than the ground sampling distance. In this case, as no shape or color can be captured, spectral information is important in the detection of insect damaged areas. Therefore, this paper used the pixel based 1D-CNN model to capture the spectral features, instead of spatial features. The result after the band selection of the hyperspectral image was molded into one-dimensional data and the context of data still existed, as shown in Figure 7. The 1D-CNN uses much fewer parameters than 2D-CNN and is more accurate and faster [39].

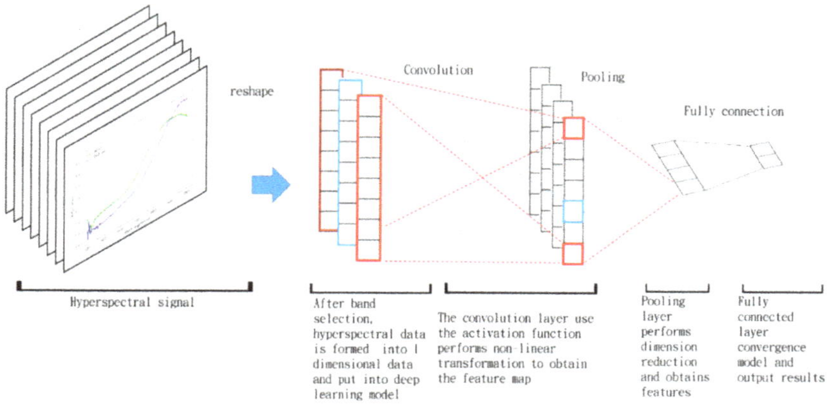

Figure 7. The 1D-CNN model.

Figure 8 shows the 1D-CNN model architecture used in this paper. The hyperspectral image after band selection was used for further analysis, and the data size of the image was 1024 × 629. The features were extracted by using the convolution layer. An 8-convolution kernel and a 16-convolution kernel were used, and then 2048 neurons entered the full connection layer directly.

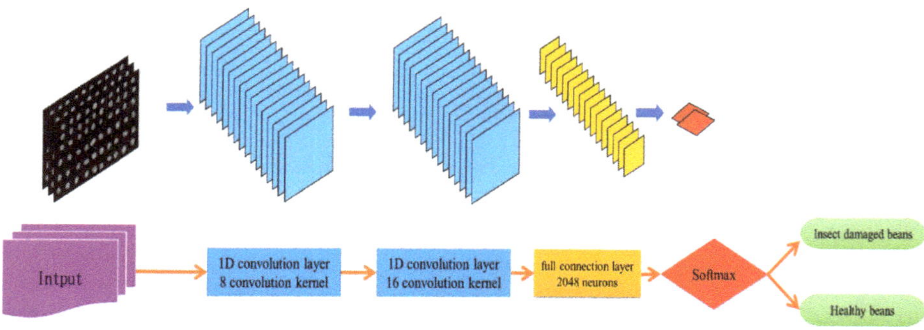

Figure 8. The 1D-CNN model architecture.

The network terminal was provided with a Softmax classifier, and the classifier result of the input spectrum was obtained. The parameters included the training test split: 0.33, epochs: 200, kernel size: 3, activation = 'relu', optimizer: SGD, lr: 0.0001, momentum: 0.9, decay: 0.0005, factor = 0.2, patience = 5, min_lr = 0.000001, batch_size = 1024, and verbose = 1.

2.6. Hyperspectral Insect Damage Detection Algorithm (HIDDA)

This paper combined the above methods to develop the hyperspectral insect damage detection algorithm (HIDDA), in which band selection is first used to filter out the important bands, and then CEM-OTSU is applied to generate training samples for the two classifiers, in order to implement binary classification for healthy and defective coffee beans. Method 1 uses linear support vector machine (SVM) [39], where the data are labeled and added for the classification of the coffee beans. While Otsu's method was used for subsequent classification, considering its possible misrecognition, this paper improved classification with SVM. Method 2 is comprised of CNN. Figure 9 describes the HIDDA flowchart, which is divided into two stages: training (Figure 9a,c) and testing (Figure 9b,d).

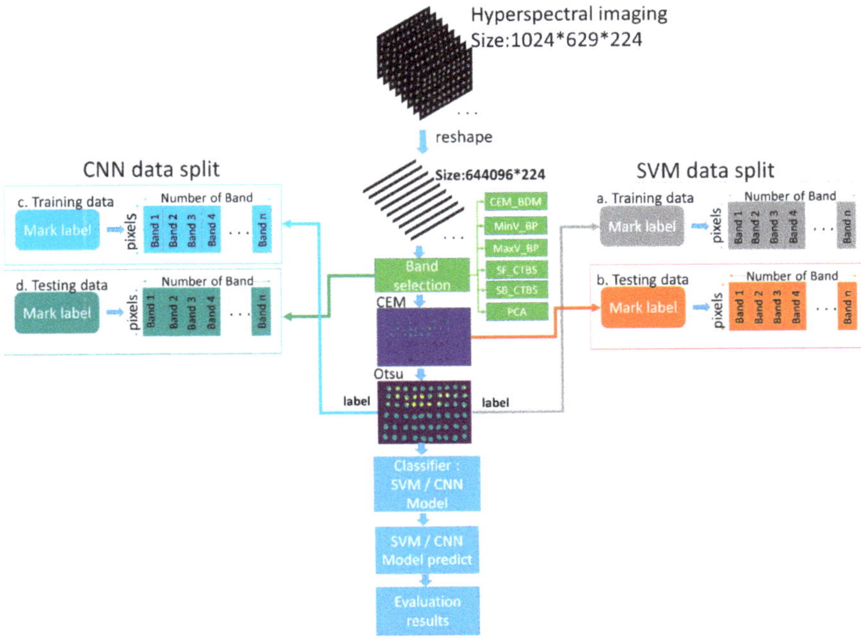

Figure 9. The hyperspectral insect damage detection algorithm flowchart. (**a**) Training data of Support Vector Machine (SVM), (**b**) Testing data of Support Vector Machine (SVM), (**c**) Training data of Convolutional Neural Networks (CNN) (**d**) Testing data of Convolutional Neural Networks (CNN).

In the training process, the spectral signature of an insect damaged bean was imported into the CEM as the desired target. The positions of other insect damaged beans could be detected automatically by Otsu's method, and the result was taken as the training data of SVM and CNN (Figure 9a,c) to classify the remaining 19 images (Figure 9b,d). The training set and the test set of the CNN converted data into 1D data. The training data of this experiment were trained by acquiring the hyperspectral image of 60 coffee beans containing 30 insect damaged beans and 30 healthy beans simultaneously after obtaining the results of CEM-OTSU. The remaining 19 hyperspectral images were used for prediction, so the training samples were less than 5% and testing data were about 95%. The data were preprocessed before this experiment by using data normalization and background removal. Then, six band selection algorithms were used to find the sensitive bands of the insect damaged and healthy beans, and the hyperspectral algorithm CEM was performed. As the CEM only needs a single desired spectral signature for detection, this spectral signature is quite important in the algorithm. The best-desired signature was found by OSGP; this signature was put in CEM for analysis, and Otsu's method divided the data into 0 and 1 to label the training data. This paper analyzed pixels instead of images, so this step is relatively important. The remaining 19 images of the test sets were used for SVM (Figure 9b), which used the CEM result for classification. The same set of 19 images after band selection was used as the CNN testing set (Figure 9d). As CNN used the convolution layer to extract features, CEM was not required for analysis. It can be noted that HIDDA generated training samples from the result of CEM-OTSU and not from prior knowledge, as the only prior knowledge HIDDA requires is a single desired spectral signature for CEM in the beginning.

3. Results and Discussion

3.1. Band Selection Results

According to Figure 9, the experimental hyperspectral data removed the background from the image before band selection. This experiment used six kinds of band selection (as discussed earlier) for comparison (minimum CEM-BDM, MinV-BP, MaxV-BP, SF-CTBS, SB-CTBS, and PCA. The SVM and CNN classifiers were then used for classification. Finally, the confusion matrix [40] and kappa [41,42] were used for evaluation and comparison. Instead of using pixels for evaluation, this paper used coffee beans as a unit; if a pixel of a coffee bean was identified as an insect damaged bean, it was classified as an insect damaged bean, and vice versa. In the confusion matrix of this experiment, TP represents a defective bean hit, FN is defective bean misrecognition, TN is healthy bean hit, and FP is healthy bean misrecognition. Figures 10–15 show the graphics of the first 20 bands selected by band selection and after band selection. As per sensitive bands selected by six kinds of band selection, 3, 10, and 20 bands were used for the test. The bands after 20 were not selected because excessive bands can cause disorder and repeated data. In addition, excessive bands could make future hardware design difficult. Therefore, the number of bands was controlled below 20. According to the results in Figures 10–15, almost all the foremost bands fell in the range of 850–950 nm. This finding helps to reduce cost and increase the use-value for future sensor design.

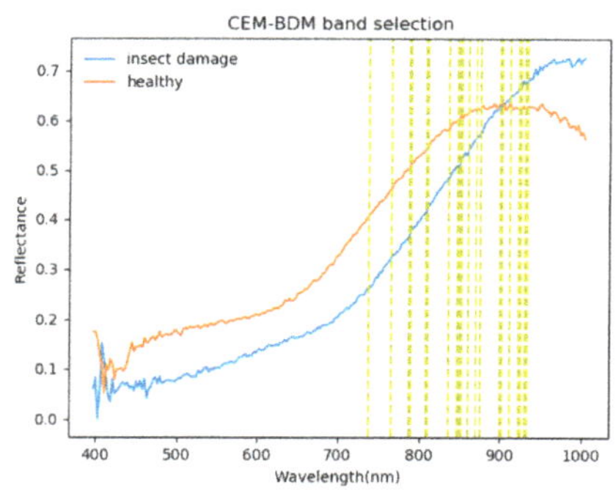

Figure 10. Visualization of the CEM_BDM band selection results. The first five bands are 933, 900, 869, 930, and 875 nm.

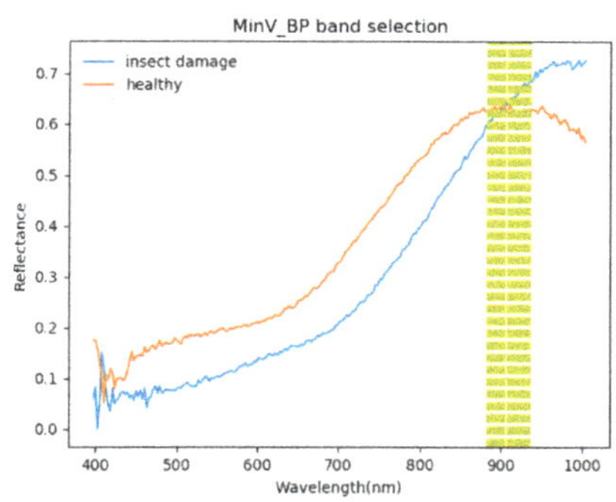

Figure 11. Visualization of the MinV_BP band selection results. The first five bands are 936, 933, 927, 930, and 925 nm.

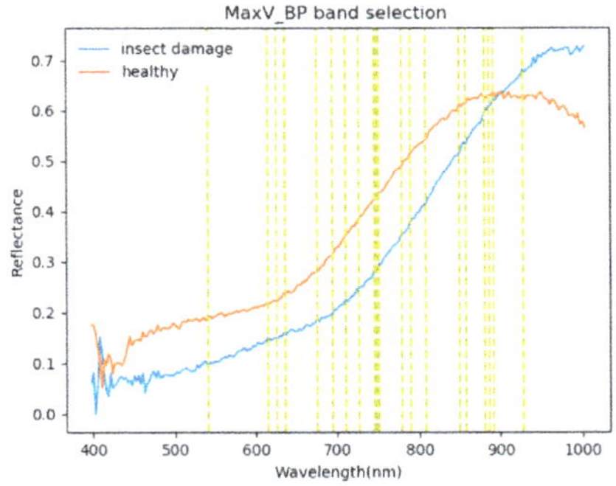

Figure 12. Visualization of the MaxV_BP band selection results. The first five bands are 858, 927, 850, 674, and 891 nm.

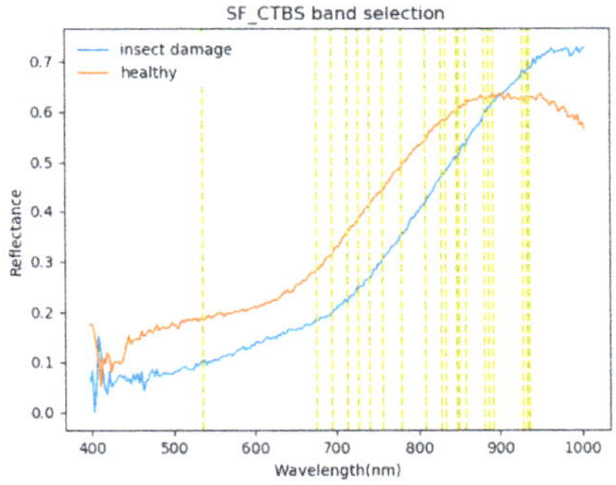

Figure 13. Visualization of the SF_CTBS band selection results. The first five bands are 936, 858, 534, 927, and 693 nm.

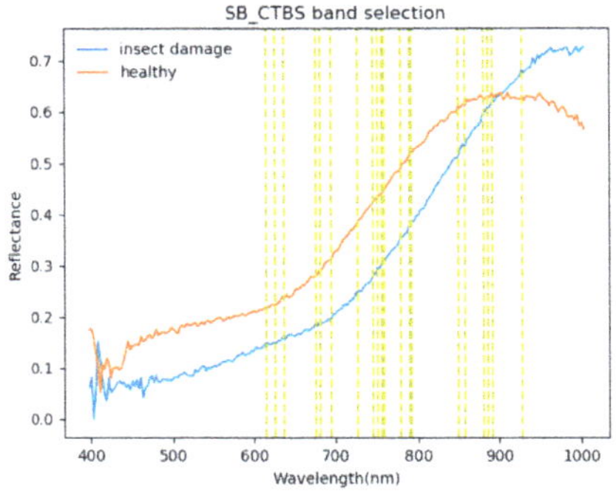

Figure 14. Visualization of the SB_CTBS band selection results. The first five bands are 858, 850, 927, 674, and 891 nm.

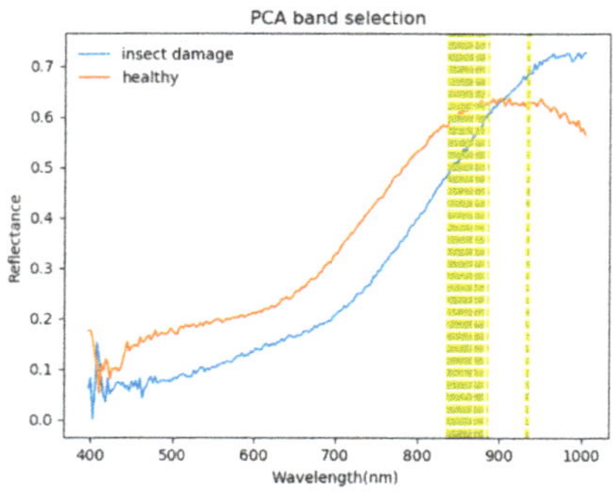

Figure 15. Visualization of the PCA band selection results. The first five bands are 936, 875, 872, 869, and 866 nm.

According to the results in Figures 10–15, almost all the foremost bands fell in the wavelength range of 850–950 nm. Table 3 lists the most frequently selected bands according to the six band selection algorithms in the first 20 bands, and 850 nm and 886 nm were selected by five out of six band selection algorithms, which means those bands are discriminate bands for coffee beans. This finding can help to reduce costs and increase the usage-value for future sensor designs.

Table 3. Most frequently selected bands by six band selection algorithms in the first 20 bands. (•: include, X: not include).

Band	*BDM*	*MinV_BP*	*MaxV_BP*	*SF_CTBS*	*SB_CTBS*	*PCA*	*Total Times*
850 nm	•	X	•	•	•	•	5
886 nm	X	•	•	•	•	•	5
858 nm	X	X	•	•	•	•	4
933 nm	•	•	X	•	X	•	4
891 nm	X	•	•	•	•	X	4
880 nm	X	X	•	•	•	•	4
927 nm	X	•	•	•	•	X	4

3.2. Detection Results by Using Three Bands

The final detection results using 10 bands were obtained by the CEM-SVM and the CNN model, as described in Section 2.3.7. Figures 16–21 show the final detection results as generated by CEM-SVM using six band selection methods to select 10 bands, while Figures 22–26 show the final detection results as obtained by the CNN model using five-band selection methods to select 10 bands. The upper three rows in Figures 16–21 are insect damaged beans, while the lower three rows are healthy beans, and there were 1139 beans in 20 images. To limit the text length, only four of the 20 images are displayed, and the analysis results are shown in Table 3. In the confusion matrix of this experiment, TP refers to insect damaged bean hits, FN refers to missing insect damaged beans, TN refers to healthy bean hits, and FP refers to false alarms. In image representation, TP is green, FN is red, TN is blue, and FP is yellow; these colors are used for visualization, as shown in Figures 16–26. All results of the three bands are compiled and compared in Table 4. The ACC [40], Kappa [41,42], and running time calculated by the confusion matrix were used for evaluation. The running time of this experiment was the average time.

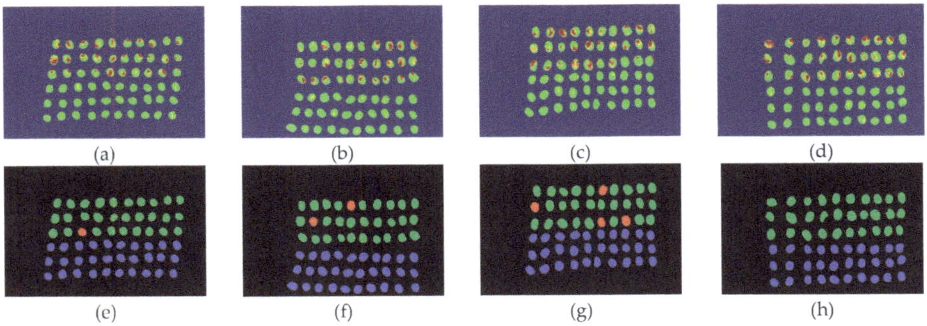

Figure 16. Results of green coffee beans CEM-SVM+CEM_BDM three bands (**a**–**d**) SVM classification results, (**e**–**h**) final visual images.

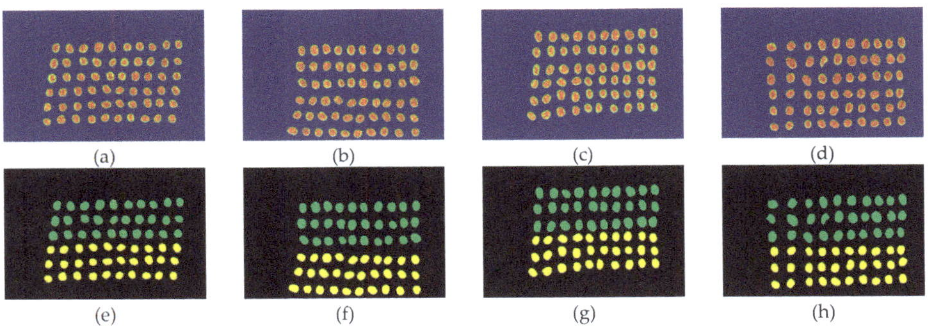

Figure 17. Results of green coffee beans CEM-SVM+ MinV_BP three bands (**a**–**d**) SVM classification results, (**e**–**h**) final visual images.

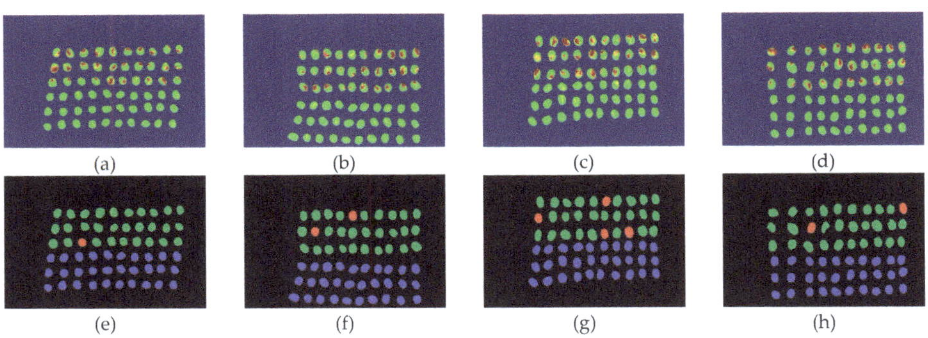

Figure 18. Results of green coffee beans CEM-SVM+ MaxV _BP three bands (**a**–**d**) SVM classification results, (**e**–**h**) final visual images.

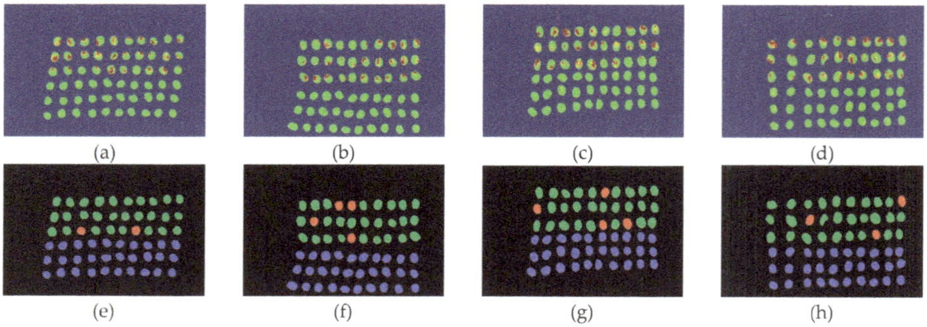

Figure 19. Results of green coffee beans CEM-SVM+ SF_CTBS three bands (**a**–**d**) SVM classification results, (**e**–**h**) final visual images.

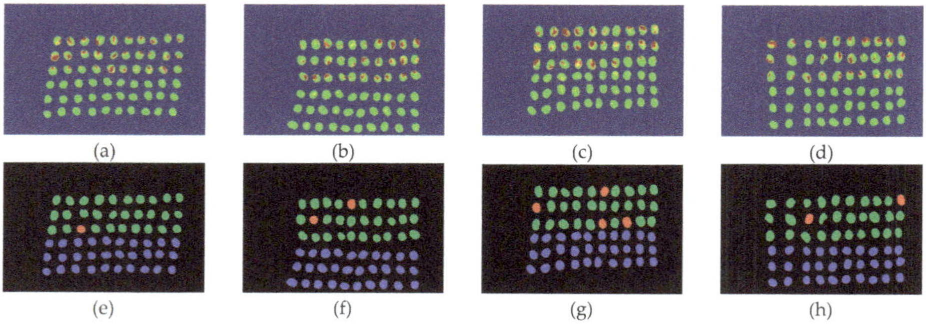

Figure 20. Results of green coffee beans CEM-SVM+SB_CTBS three bands, (**a**–**d**) SVM classification results, (**e**–**h**) final visual images.

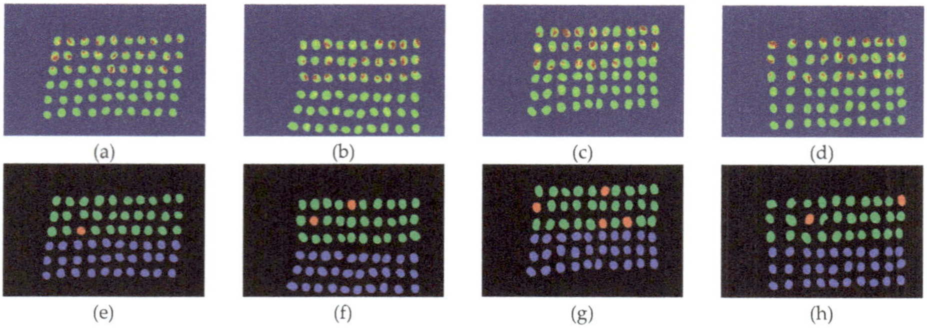

Figure 21. Results of green coffee beans CEM-SVM+PCA, (**a**–**d**) SVM classification results, (**e**–**h**) final visual images.

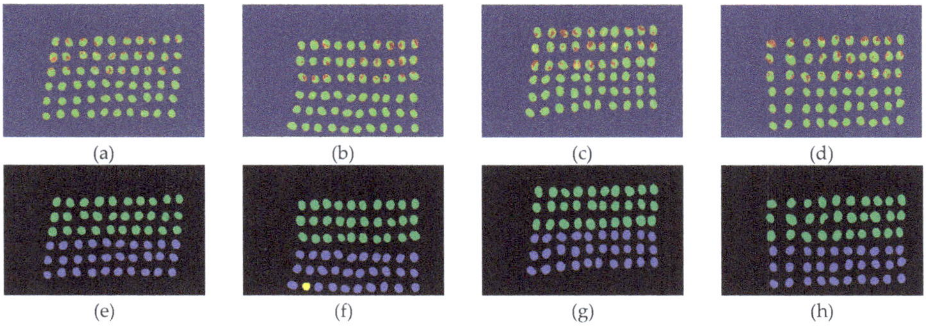

Figure 22. Results of green coffee beans CNN+CEM-BDM three bands, (**a**–**d**) CNN classification results, (**e**–**h**) final visual images.

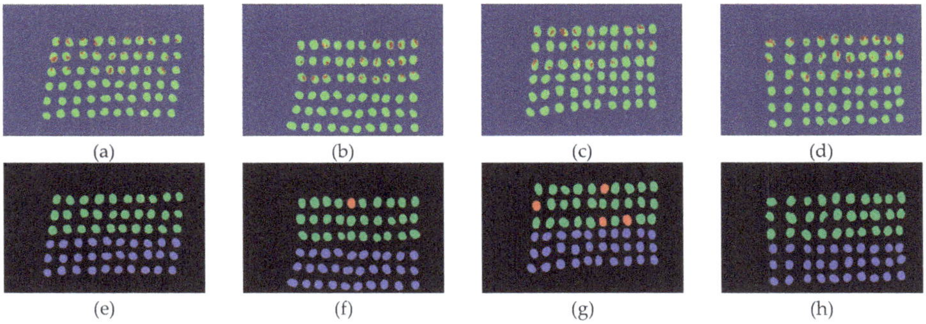

Figure 23. Results of green coffee beans CNN+ maxV_BP three bands, (**a**–**d**) CNN classification results, (**e**–**h**) final visual images.

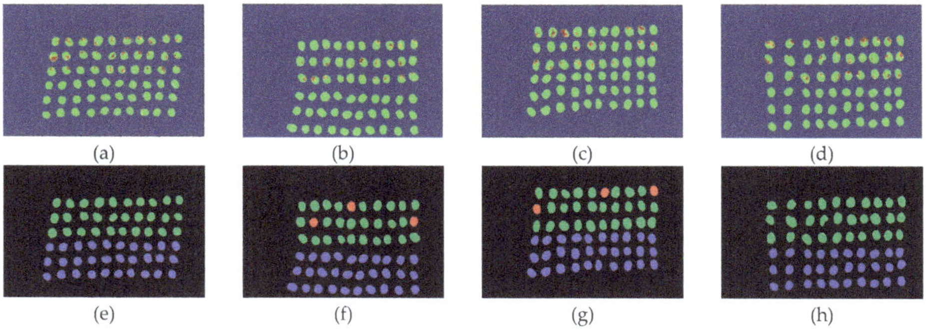

Figure 24. Results of green coffee beans CNN+SF_CTBS three bands, (**a**–**d**) CNN classification results, (**e**–**h**) final visual images.

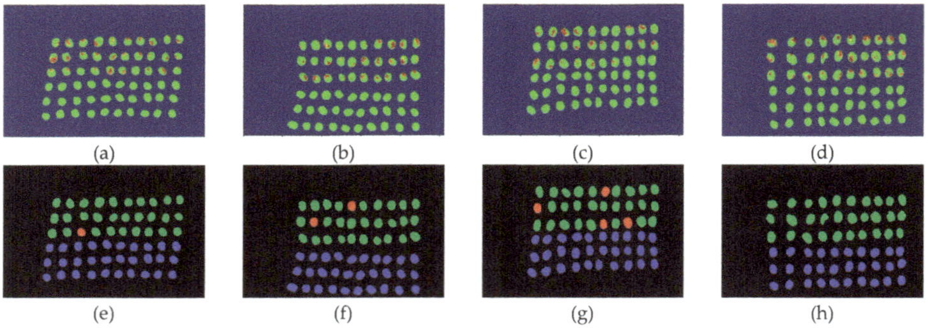

Figure 25. Results of green coffee beans CNN+SB_CTBS three bands, (**a–d**) CNN classification results, (**e–h**) final visual images.

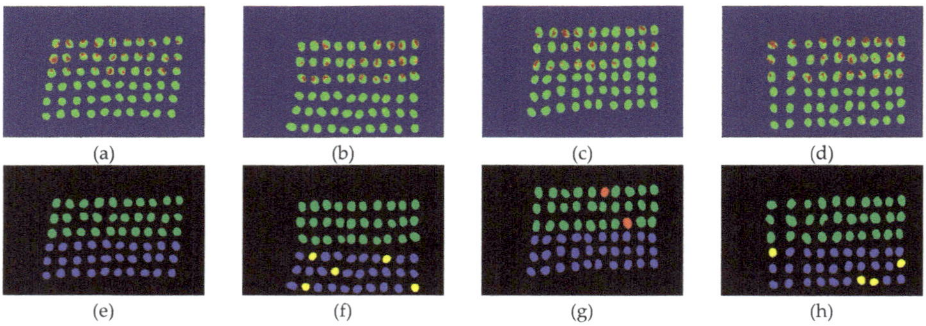

Figure 26. Results of green coffee beans CNN+ PCA three bands, (**a–d**) CNN classification results, (**e–h**) final visual images.

Table 4. The results of the green coffee bean classification. The best performance is highlighted in red color.

3 Bands Green Coffee Beans CEM-SVM Results									
Analysis Method	**TP**	**FN**	**FP**	**TN**	**TPR**	**FPR**	**ACC**	**Kappa**	**Time (s)**
CEM_BDM + CEM-SVM	520	50	17	552	0.912	0.298	0.941	0.887	11.57
MinV_BP + CEM-SVM	570	0	569	0	1.0	1.0	0.5	0	13.27
MaxV_BP + CEM-SVM	515	55	8	561	0.903	0.014	0.944	0.892	12.31
SF_CTBS + CEM-SVM	494	76	5	564	0.867	0.008	0.928	0.862	13.21
SB_CTBS + CEM-SVM	515	55	8	561	0.903	0.014	0.944	0.892	13.26
PCA + CEM-SVM	523	47	14	555	0.917	0.024	0.946	0.896	12.83
3 Bands Green Coffee Beans CNN Results									
Analysis Method	**TP**	**FN**	**FP**	**TN**	**TPR**	**FPR**	**ACC**	**Kappa**	**Time (s)**
CEM_BDM + CNN	532	38	20	549	0.931	0.029	0.95	0.901	7.4
MaxV_BP + CNN	520	50	13	556	0.912	0.018	0.947	0.894	7.4
SF_CTBS + CNN	461	109	8	561	0.8	0.009	0.895	0.79	7.13
SB_CTBS + CNN	514	56	9	560	0.9	0.014	0.942	0.885	7.67
PCA + CNN	540	30	41	528	0.946	0.063	0.941	0.883	7.64

In the case of three bands, CEM-SVM, BDM, MaxV_BP, SF_CTBS, SB_CTBS, and PCA were successful in classification. However, a portion of insect damaged beans was not detected, which was probably because the insect damage surface was not irradiated. The MinV_BP+CEM-SVM could not perform classification at all, as shown in Figure 17, possibly due to the non-selection of the sensitive band; thus, its result was excluded from subsequent discussion. As shown in Table 4, the

PCA+CNN had the highest TPR, and PCA+CEM-SVM had the highest ACC and kappa, proving that the sequencing of the PCA amount of variation is feasible for band selection. The minimum FDR was observed for SF_CTBS+CEM-SVM. The minimum variance of CEM was used for recurrent selection in SF_CTBS, and the healthy beans could be identified accurately.

In the case of CNN, the BDM, MaxV_BP, SF_CTBS, SB_CTBS, and PCA were used, and the MinV_BP was not used because the deep learning label produced in CEM could not be identified. Here, the paper of three bands was not included for comparison. The PCA exhibited the highest TPR, and thus, the band selected by PCA was more sensitive to defective beans. The SF_CTBS had the lowest FPR, and the minimum variance of CEM calculated by SF_CTBS for recurrent selection could accurately identify healthy beans. The classification result indicates that only eight green coffee beans were misidentified as defective beans, CEM_BDM possessed the highest ACC and kappa, and that the CEM_BDM method classified green coffee beans better. In terms of time, the CNN was faster than SVM because the CNN model used a batch_size = 1024 for prediction, while the SVM used pixels one by one for prediction.

3.3. Detection Results Using 10 Bands

The final detection results using 10 bands were obtained by CEM-SVM and the CNN model, as described in Section 2.3.7. Figures 27–32 show the final detection results, as generated by CEM-SVM using six band selection methods to select 10 bands; Figures 33–37 show the final detection results as obtained by the CNN model using five-band selection methods to select 10 bands.

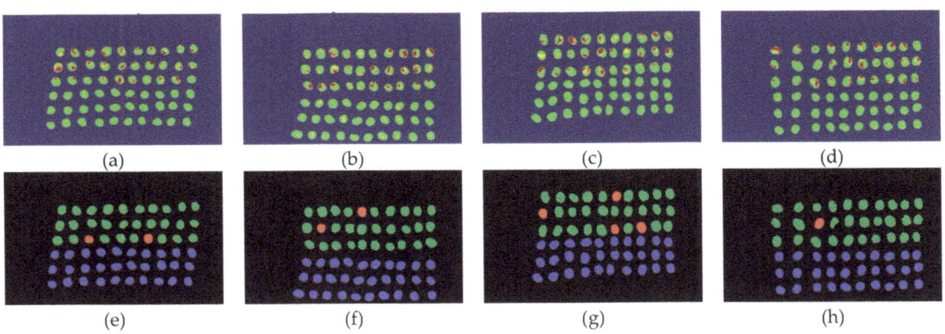

Figure 27. Results of green coffee bean CEM-SVM+CEM_BDM 10 bands. (**a–d**) SVM classification results, (**e–h**) final visual images.

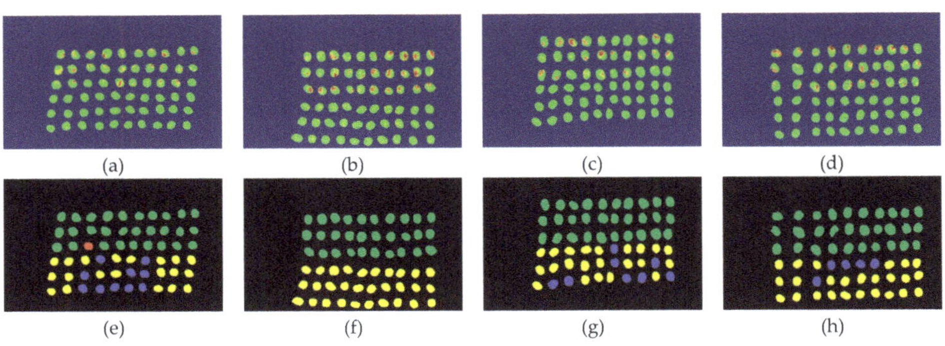

Figure 28. Results of green coffee bean CEM-SVM+ MinV_BP 10 bands. (**a–d**) SVM classification results, (**e–h**) final visual images.

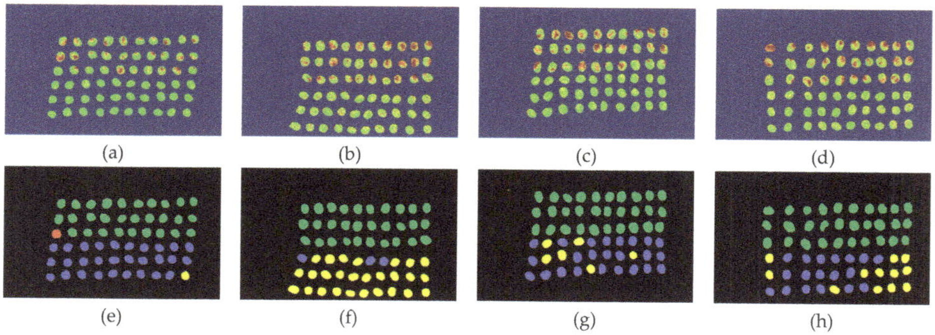

Figure 29. Results of green coffee bean CEM-SVM+ MaxV_BP 10 bands. (**a**–**d**) SVM classification results, (**e**–**h**) final visual images.

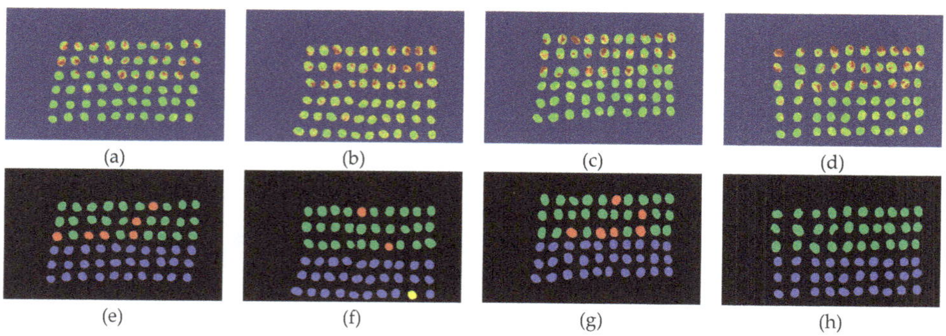

Figure 30. Results of green coffee bean CEM-SVM+ SF_CTBS 10 bands. (**a**–**d**) SVM classification results, (**e**–**h**) final visual images.

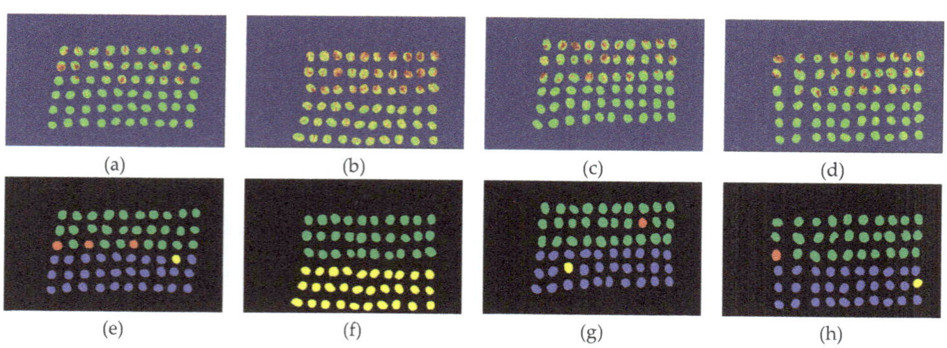

Figure 31. Results of green coffee bean CEM-SVM+ SB_CTBS 10 bands. (**a**–**d**) SVM classification results, (**e**–**h**) final visual images.

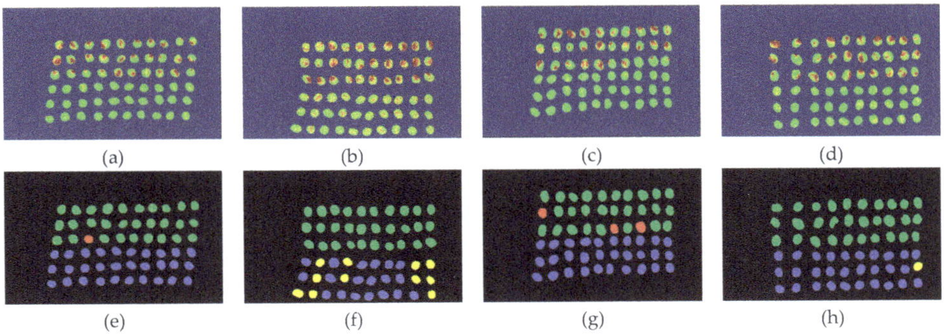

Figure 32. Results of green coffee bean CEM-SVM+PCA 10 bands. (**a–d**) SVM classification results, (**e–h**) final visual images.

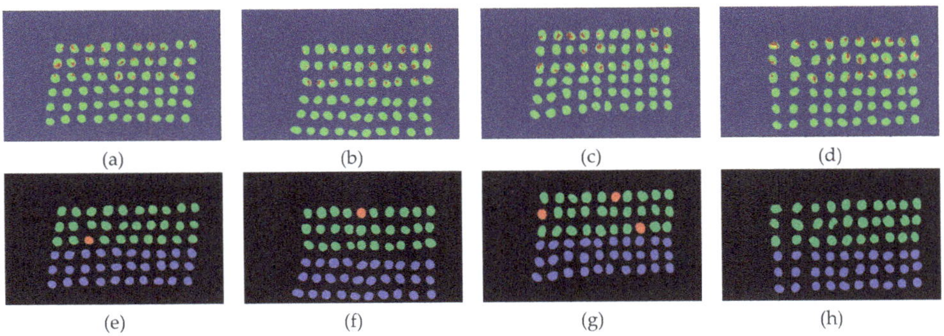

Figure 33. Results of green coffee bean CNN+CEM_BDM 10 bands. (**a–d**) CNN classification results, (**e–h**) final visual images.

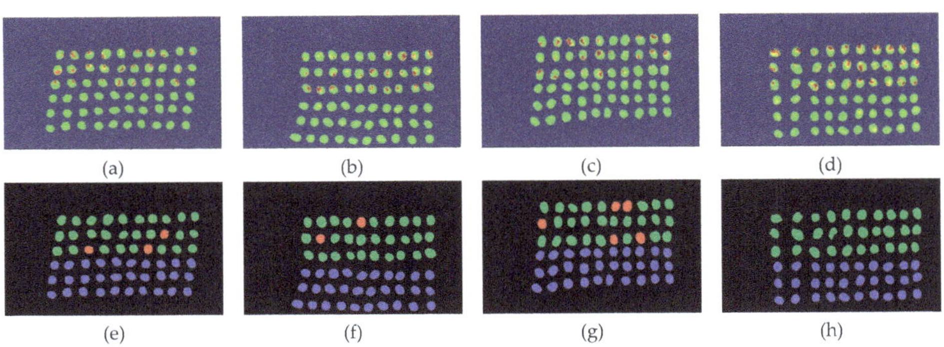

Figure 34. Results of green coffee bean CNN + maxV_BP 10 bands. (**a–d**) CNN classification results, (**e–h**) final visual images.

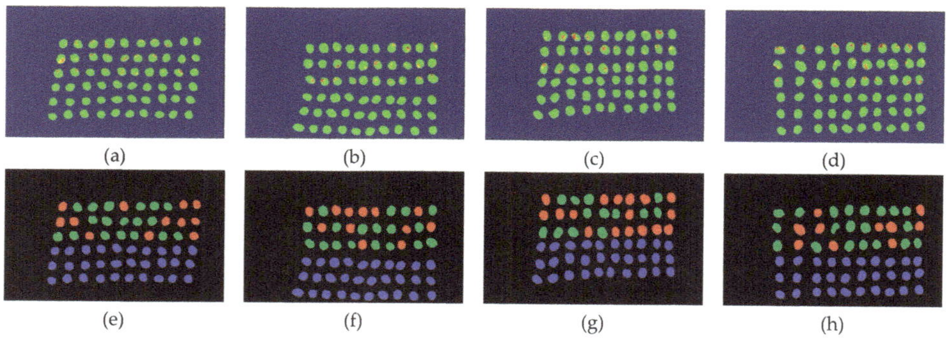

Figure 35. Results of green coffee bean CNN+SF_CTBS 10 bands. (**a**–**d**) CNN classification results, (**e**–**h**) final visual images.

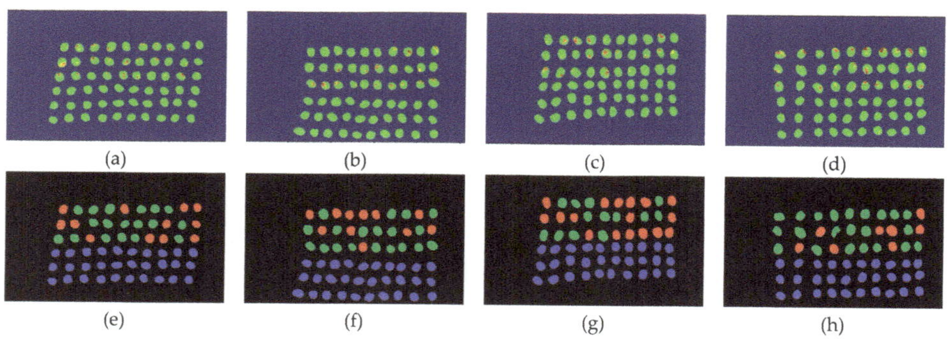

Figure 36. Results of green coffee bean CNN+SB_CTBS 10 bands. (**a**–**d**) CNN classification results, (**e**–**h**) final visual images.

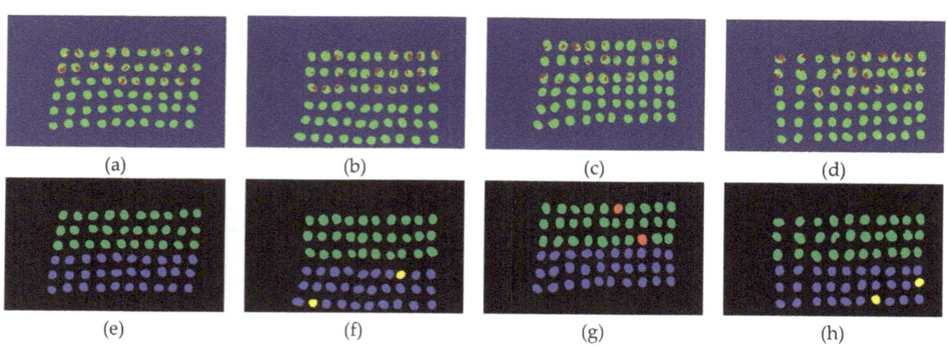

Figure 37. Results of green coffee bean CNN+PCA 10 bands. (**a**–**d**) CNN classification results, (**e**–**h**) final visual images.

All results from the 10 bands were compiled and compared, as shown in Table 5. As seen, there were several influential bands in the front, but excessive bands could induce misrecognitions.

Table 5. The results of the green coffee bean classification. The best performance is highlighted in red color.

10 Bands Green Coffee Beans SVM Results									
Analysis Method	**TP**	**FN**	**FP**	**TN**	**TPR**	**FPR**	**ACC**	**Kappa**	**Time (s)**
CEM_BDM+CEM-SVM	509	61	8	561	0.892	0.014	0.939	0.883	11.71
MinV_BP+CEM-SVM	565	5	492	77	0.991	0.864	0.563	0.131	11.35
MaxV_BP+CEM-SVM	554	16	213	356	0.971	0.374	0.798	0.592	11.47
SF_CTBS+CEM-SVM	505	65	37	532	0.885	0.065	0.910	0.824	11.31
SB_CTBS+CEM-SVM	545	25	146	423	0.956	0.256	0.849	0.696	11.44
PCA+CEM-SVM	541	29	63	506	0.949	0.110	0.919	0.844	11.40
10 Bands Green Coffee Beans SVM Results									
Analysis Method	**TP**	**FN**	**FP**	**TN**	**TPR**	**FPR**	**ACC**	**Kappa**	**Time (s)**
CEM_BDM+CNN	484	86	5	564	0.85	0.007	0.921	0.842	7.16
MaxV_BP+CNN	473	97	2	567	0.833	0.003	0.914	0.829	7.34
SF_CTBS+CNN	246	324	1	568	0.420	0.001	0.708	0.418	6.98
SB_CTBS+CNN	289	281	1	568	0.5	0.001	0.748	0.497	7.41
PCA+CNN	532	38	21	548	0.931	0.033	0.949	0.898	7.41

In the case of CEM-SVM, the CEM_BDM+CEM-SVM had the best performance in FPR, ACC, and kappa, indicating the reliability of the CEM_BDM band priority in 10 bands, and the minimization of correlation between bands could influence green coffee beans. The sensitive bands were extracted using this method. The MaxV_BP+CEM-SVM had the highest TPR, indicating that the maximum variance of CEM calculated by MaxV_BP for sequencing could classify defective beans. The MinV_BP was less effective than the other methods, which might be related to the variance of green coffee beans, suggesting that this method is inapplicable to a small number of bands.

In the case of CNN, when the MinV_BP produced labels, excessive data misrecognitions failed the training model, and the PCA had the highest TPR, ACC, and Kappa. Therefore, in the bands selected from the 10 bands, the PCA+CNN seemed to be the most suitable for classifying green coffee beans. The SF_CTBS and SB_CTBS had the minimum FPR, indicating that the cyclic ordering of CEM variance is appropriate for classifying good beans.

3.4. Detection Results by Using 20 Bands

The final detection results using 20 bands were obtained by the CEM-SVM and the CNN model, as described in Section 2.3.7. Figures 38–43 show the final detection results, as generated by CEM-SVM using six band selection methods to select 10 bands; Figures 44–48 show the final detection results as obtained by the CNN model using five-band selection methods to select 10 bands.

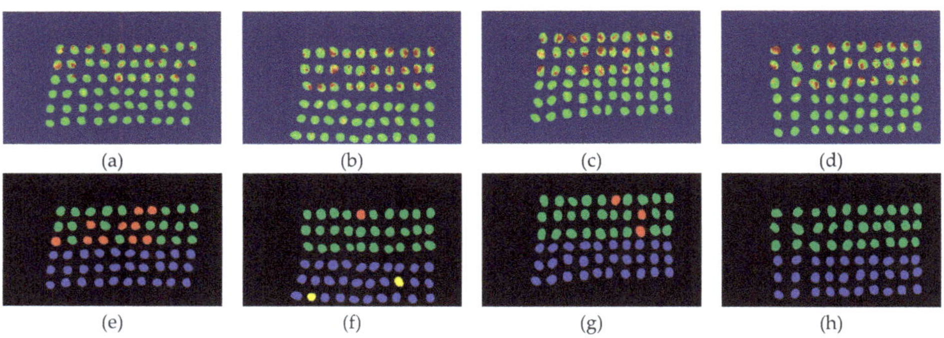

Figure 38. Results of green coffee beans CEM-SVM+CEM_BDM 20 bands. (**a**–**d**) SVM classification results, (**e**–**h**) final visual images.

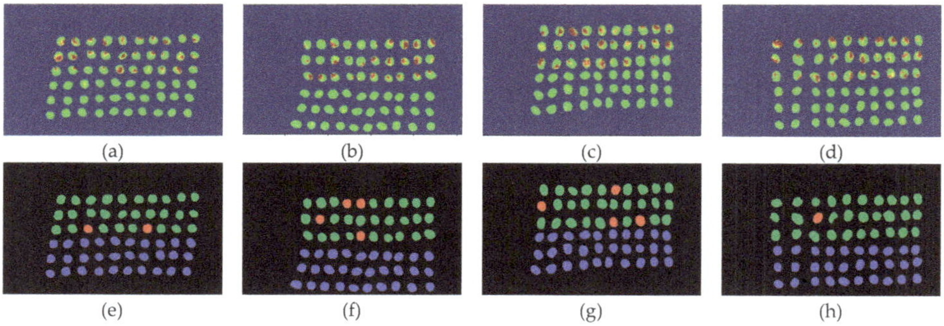

Figure 39. Results of green coffee bean CEM-SVM+MinV_BP 20 bands. (**a**–**d**) SVM classification results, (**e**–**h**) final visual images.

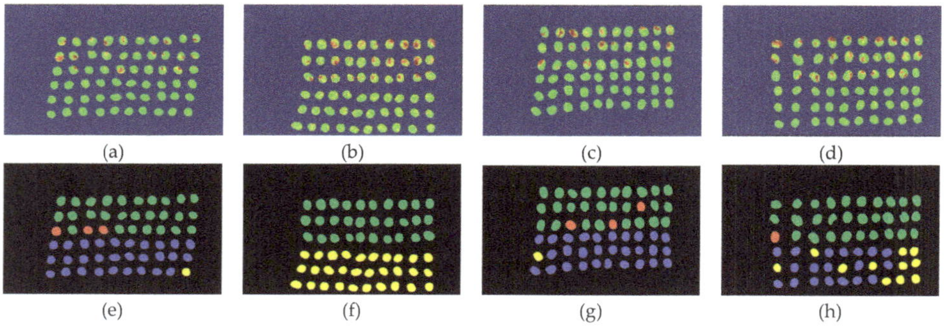

Figure 40. Results of green coffee bean CEM-SVM+MaxV_BP 20 bands. (**a**–**d**) SVM classification results, (**e**–**h**) final visual images.

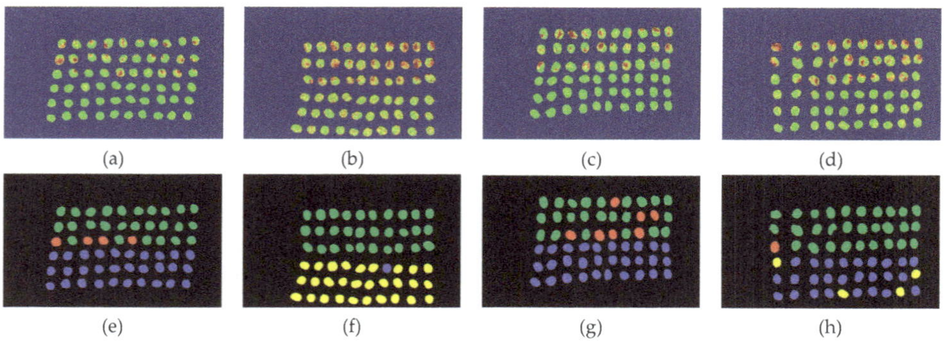

Figure 41. Results of green coffee bean CEM-SVM+SF_CTBS 20 bands. (**a**–**d**) SVM classification results, (**e**–**h**) final visual images.

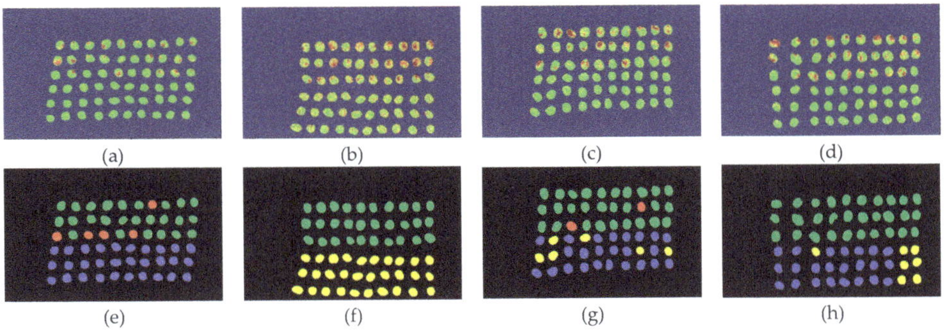

Figure 42. Results of green coffee bean CEM-SVM+SB_CTBS 20 bands. (**a–d**) SVM classification results, (**e–h**) final visual images.

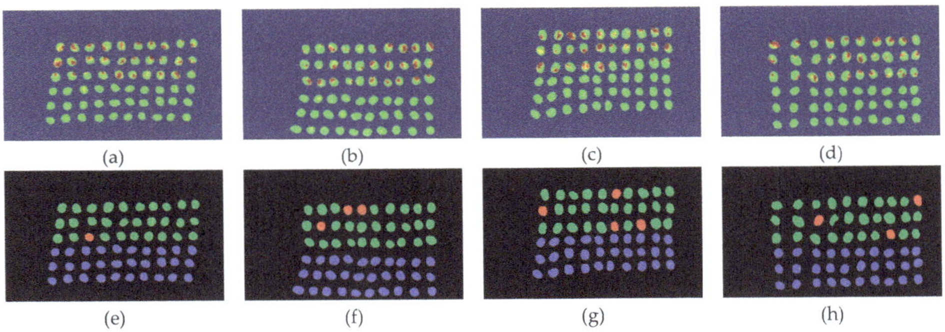

Figure 43. Results of green coffee bean CEM-SVM+PCA 20 bands. (**a–d**) SVM classification results, (**e–h**) final visual images.

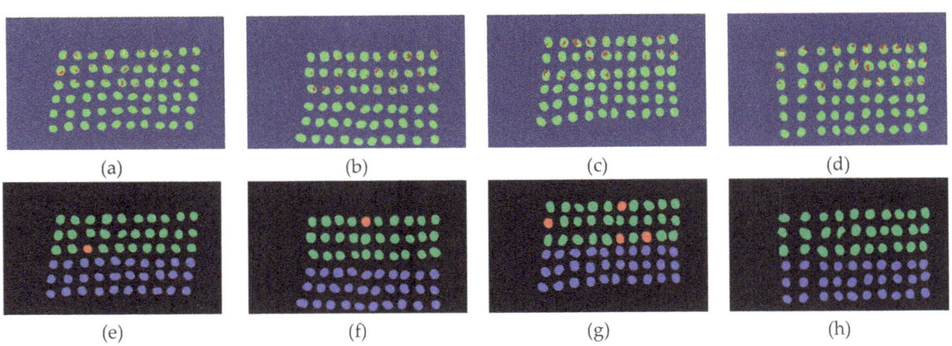

Figure 44. Results of green coffee bean CNN+CEM_BDM 20 bands. (**a–d**) CNN classification results, (**e–h**) final visual images.

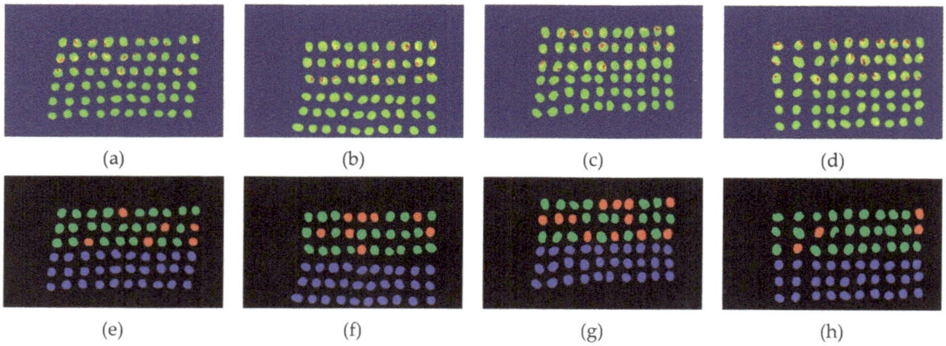

Figure 45. Results of green coffee bean CNN+maxV_BP 20 bands. (**a–d**) CNN classification results, (**e–h**) final visual images.

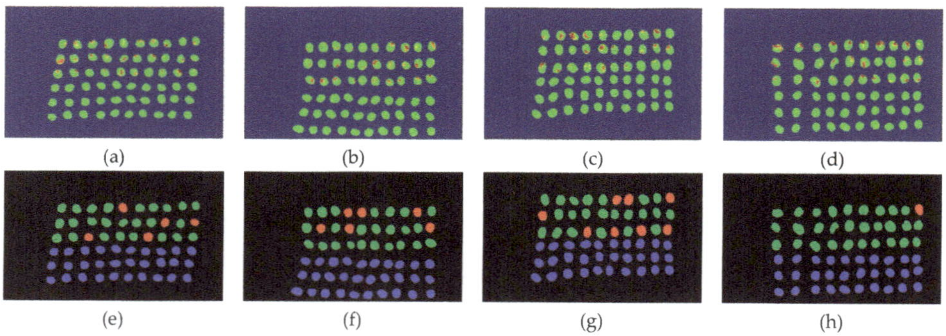

Figure 46. Results of green coffee bean CNN+SF_CTBS 20 bands. (**a–d**) CNN classification results, (**e–h**) final visual images.

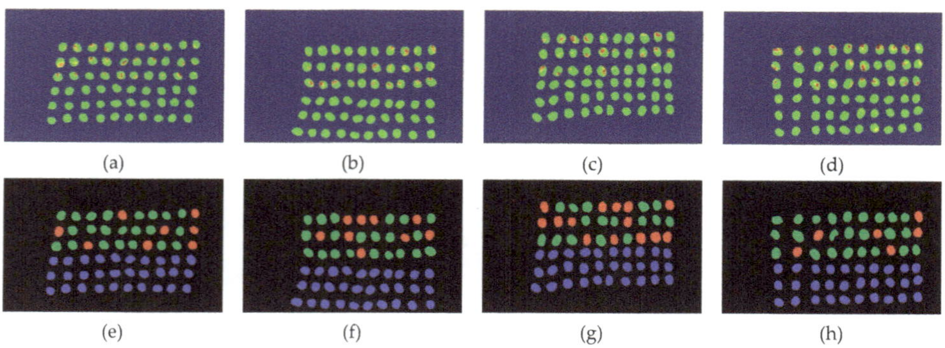

Figure 47. Results of green coffee bean CNN+SB_CTBS 20 bands. (**a–d**) CNN classification results, (**e–h**) final visual images.

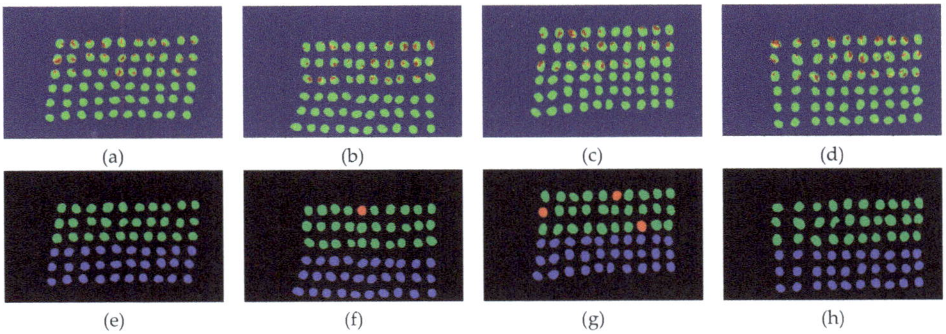

Figure 48. Results of green coffee bean CNN+PCA 20 bands. (**a–d**) CNN classification results, (**e–h**) final visual images.

All results of the 20 bands were compiled and compared, as shown in Table 6. The ACC, kappa, and the running time calculated by the confusion matrix were used for evaluation.

Table 6. The results of the green coffee bean classification. The best performance is highlighted in red color.

20 Bands Green Coffee Beans SVM Results									
Analysis Method	**TP**	**FN**	**FP**	**TN**	**TPR**	**FPR**	**ACC**	**Kappa**	**Time (s)**
CEM_BDM+CEM-SVM	522	48	21	548	0.915	0.036	0.939	0.881	12.03
MinV_BP+CEM-SVM	547	23	77	492	0.959	0.135	0.912	0.827	11.34
MaxV_BP+CEM-SVM	550	20	249	320	0.964	0.437	0.763	0.525	11.41
SF_CTBS+CEM-SVM	521	49	135	434	0.914	0.237	0.838	0.681	11.38
SB_CTBS+CEM-SVM	544	26	253	316	0.954	0.444	0.755	0.516	11.78
PCA+CEM-SVM	546	24	144	425	0.957	0.253	0.852	0.729	11.38
20 Bands Green Coffee Beans CNN Results									
Analysis Method	**TP**	**FN**	**FP**	**TN**	**TPR**	**FPR**	**ACC**	**Kappa**	**Time (s)**
CEM_BDM+CNN	480	90	5	564	0.842	0.007	0.917	0.835	7.84
MaxV_BP+CNN	355	215	1	568	0.62	0.001	0.809	0.618	7.27
SF_CTBS+CNN	395	175	2	567	0.687	0.003	0.841	0.683	7.35
SB_CTBS+CNN	336	234	1	568	0.585	0.001	0.791	0.583	7.13
PCA+CNN	511	59	5	564	0.896	0.007	0.944	0.888	7.82

In the case of CEM-SVM, the CEM_BDM+CEM-SVM exhibited the best performance in FPR, ACC, and kappa, proving again that the minimization of inter-band correlation is helpful to the classification of green coffee beans. The MaxV_BP exhibited the highest TPR, and using the maximum variance of CEM for ordering, it could classify defective beans in 20 bands. It is noteworthy that the MinV_BP+CEM-SVM selected sensitive bands from 20 bands, which may be the variance calculation problem of the algorithm, proving that the MinV_BP method is inapplicable to a small number of bands of green coffee beans, but applicable to a larger number of bands.

In the case of CNN, as the data content increased, the accuracy of most methods declined. In the training of MinV_BP, excessive data misrecognition failed the training model. The PCA band selection exhibited good performance in TPR, ACC, and kappa, indicating that the PCA performed the best in classification with 20 bands. The MaxV_BP and SB_CTBS had the lowest FPR. The use of the maximum variance of CEM in the case of 20 bands had the best effect on classifying good beans.

3.5. Discussion

The ACC and kappa values of three bands, 10 bands, and 20 bands were compared and represented as histograms, as shown in Figures 49 and 50. According to the comparison results of the figures, the CEM_BDM+CEM-SVM gave good results in the case of three bands, 10 bands, and 20 bands. The accuracy was higher than 90%, and the kappa was about 0.85, indicating that the BDM selected bands are crucial and representative to both classifiers for better performance. Considering MinV_BP+CEM-SVM in the cases of three bands and 10 bands, the selected bands might be difficult to classify the data, although there were sufficient data in the case of 20 bands, where the effect was enhanced greatly, suggesting that MinV_BP needs a larger number of bands for better classification.

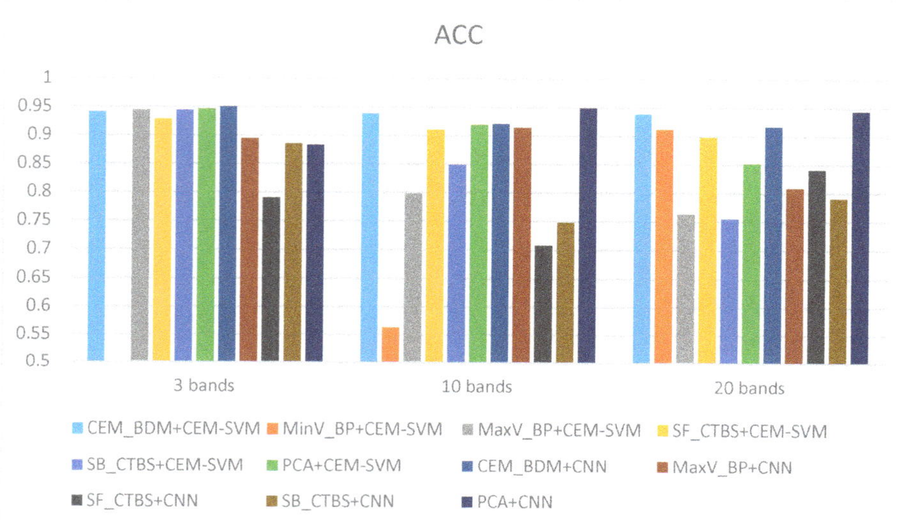

Figure 49. The ACC accuracy histograms of 3 bands, 10 bands, and 20 bands.

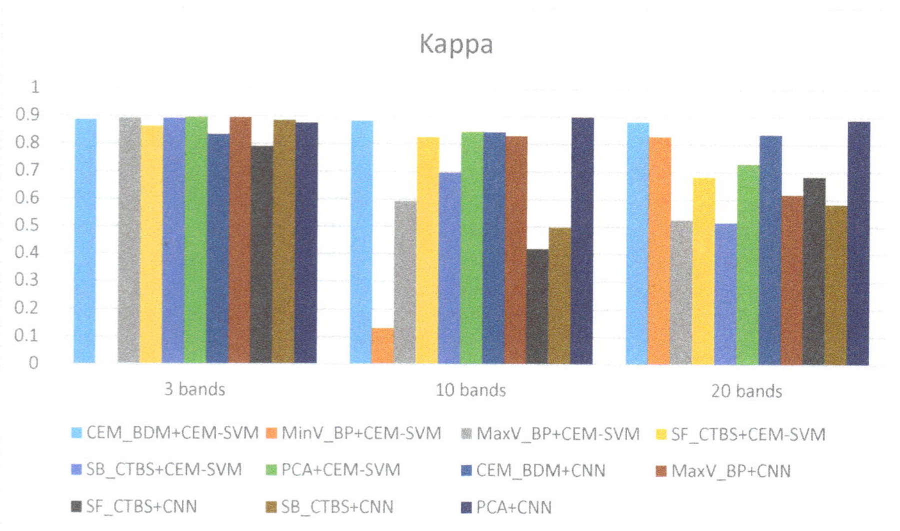

Figure 50. The Kappa histograms of 3 bands, 10 bands, and 20 bands.

As for MaxV_BP+CEM-SVM, the green coffee beans could be classified in the cases of three bands and 10 bands; but the accuracy declined in the case of 20 bands, indicating excessive bands induced misrecognitions. Interestingly, this situation is contrary to MinV_BP, and is related to the CEM variance of MinV_BP and MaxV_BP. When assessing SF_CTBS+CEM-SVM, in the cases of three bands and 10 bands, the accuracy and Kappa values were quite high, while in the case of 20 bands, there were too many misrecognitions of healthy beans, and the kappa decreased greatly. This indicates that excessive bands induced misrecognitions, and confirms that the sensitive bands were identified from the first 10 bands. Considering SB_CTBS+CEM-SVM, in the case of three bands, high precision and kappa were observed, and with the increase in the number of bands, the rate of healthy bean misrecognition increased. Therefore, the first three bands of this method were the most representative, and excessive bands did not increase the accuracy. When assessing PCA+CEM-SVM, in the cases of three bands, 10 bands, and 20 bands, the results were good, and the variation seemed feasible for band selection, about the same as previous BDM. The two methods could thus select important spectral signatures with a small number of bands. In the cases of three bands, 10 bands, and 20 bands, the CEM_BDM+CNN exhibited good results, but poorer than the previous SVM in the cases other than three bands.

In the case of three bands, MaxV_BP+CNN exhibited a high precision and kappa, which reduced as the bands increased, but CNN seemed to be more suitable than SVM. The SF_CTBS+CNN had a poorer effect than SVM in the cases of three bands, 10 bands, and 20 bands, indicating that this method is inapplicable to CNN, which may be related to the variance of CEM. The SB_CTBS+CNN exhibited high precision and kappa in the case of three bands, which reduced as the bands increased. This suggests that excessive bands influenced the decision, and there was no significant difference, except for a slight difference in the case of 10 bands from SVM. The PCA+CNN exhibited good results in the cases of three bands, 10 bands, and 20 bands; the CNN performed much better than SVM, where the results were quite average, and the cases of 10 bands and 20 bands exhibited the best effect.

Based on the aforementioned results, this paper found that the number of bands is a critical factor. From the band selection results in Section 3.1, the foremost bands fell in the wavelength range of 850–950 nm. According to the spectral signature of healthy and insect damaged coffee beans in Figure 5, the most different spectrum was also between 850–950 nm. This finding can explain the above results, if the selected bands fell in this range, the results performed relatively well. Considering the number of bands, in the case of three bands, the CEM_BDM+CNN method had the best performance in ACC and Kappa, and the ACC was 95%, indicating that the minimization of inter-band correlation is helpful to detect insect damaged beans since the top three bands were between the range of 850–950 nm. In the 10 band and 20 band cases, the PCA+CNN method exhibited the best performance in ACC and kappa, which suggests that the covariance for band selection can determine the different bands between heathy and defective beans, and the effect was even improved when combined with CNN. Based on the above results, several findings can be observed as follows.

1. As the background has many unknown signal sources responding to various wavelengths, the hyperspectral data collected in this paper were pre-processed to remove the background, which rendered the signal source in the image relatively simple. As too much spectral data increase the complexity of detection, only healthy coffee beans and insect damaged beans were included in the data for experimentation. Without other background noise interference, this experiment only required a few important bands to separate the insect damaged beans from healthy beans. The applied CEM-based band selection methods were based on the variance generated by CEM to rank the bands, where the top ranked bands are more representative and significant. Moreover, the basic assumption of PCA is that the data can find a vector projected in the feature space, and then the maximum variance of this group of data can be obtained, thus, it is also ranked by variance. In other words, the band selection methods with the variance as the standard can only use the top few bands to distinguish our experimental data with only two signal sources (healthy and unhealthy beans), which is supported by our experimental results. As the top

few bandwidths are concentrated between 850 nm–950 nm, the difference between the spectral signature curves of the insect damaged beans and healthy beans could be easily observed between 850 nm–950 nm, as shown in Figure 51. The curve of healthy beans flattened, while the curve of the insect damaged beans rose beyond the range of 850 nm–950 nm, as shown in Figure 51.

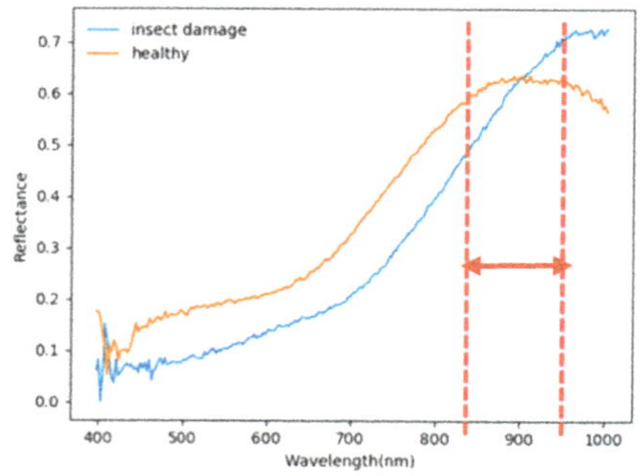

Figure 51. Highlight of the spectral signature for healthy and insect damaged beans.

2. The greatest challenge in the detection of insect damaged coffee beans is that the damaged areas provide very limited spatial information and are generally difficult to visualize from data. CEM, which is a hyperspectral target detection algorithm, is effective in dealing with the subpixel detection problem [24,25,29,30]. As mentioned in Section 2.3.1, CEM requires only one desired spectral signature for detection, thus, the quality of the detection result is very sensitive to the desired spectral signature. To solve this problem, HIDDA applies OSGP to obtain a stable and improved spectral signature, thus the stability of detection can be increased. The second issue of CEM is that it only provides soft decision results; thus, this paper used linear SVM followed by CEM to obtain the hard decision results. The above-mentioned reasons are the key points that prove our proposed method, HIDDA, can perform well.
3. The main problem of CNN during deep learning is that it requires a large number of training samples to learn more effectively and obtain suitable answers. Moreover, as our data consist of real images, there is no ground truth for us to label the insect damaged areas. To address this problem, HIDDA used CEM-OTSU to detect insect damaged beans, and used those pixels as the training samples for the CNN model. It is worth noting that we applied the results from CEM to generate more training samples for the CNN model, as CEM only requires one piece of knowledge of the target signature; hence, even though our training rate for CNN was low, the final results still performed well.
4. In order for a comparison with prior studies, Table 7 lists the detailed comparison of coffee beans. Several issues regarding our datasets, methods, and performance render this paper noticeable. First, HIDDA is the only method that detects insect damaged beans, which are more difficult to identify than black, sour, and broken beans. Second, HIDDA had the lowest training rate and highest testing rate with very good performance. Third, HIDDA is the only method, as proposed by CEM-SVM and the CNN model, that uses only three bands in the detection of insect damaged coffee beans. The authors [2] also used hyperspectral imaging (VIS-NIR) to identify black and broken beans, which extracted features through PCA and used K-NN for classification. However, in that paper, the number of beans was relatively too small, and final detection rate was lower than the proposed method of this paper. Other studies [5,6] have used traditional RGB images, which can only address targets according to the color and shape based on the spatial information, meaning that it can only identify black and broken beans. In contrast with prior studies, HIDDA is based on spectral information provided by hyperspectral sensors, which can detect targets at the subpixel level of insect damage, which has very limited spatial information.

Table 7. The detailed comparison of prior studies.

Ref.	Types	Training Rate	No. of Training	Testing Rate	No. of Testing	ACC (%)	Method	Image
[2]	Normal	76%	200	24%	61	90.2	PCA+K-NN (k = 3) (5 PCA)	HSI
	Black cherries	71%	5	29%	2	50		
	Black	68%	100	32%	45	80		
	Dehydrated	52%	100	48%	89	89.9		
[5]	Normal	unknown	444	unknown	161	97.52	K-NN (k = 10)	RGB
	Black				169	97.04		
	Sour				165	92.12		
	Broken				166	94.45		
[6]	Normal	66%	70	34%	35	100	Thresholding	RGB
	Black	66%	50	34%	25	100		
HIDDA	Normal	5%	30	95%	569	96.48	CEM-SVM (3 bands)	HSI
	Insect damage	5%	30	95%	570	93.33	CNN (3 bands)	

4. Conclusions

Insect damage is the most commonly seen defect in coffee beans. The damaged areas are often smaller than the pixel resolution, thus, the targets to be detected are usually embedded in a single pixel. Therefore, the only way to detect and extract these targets is at the subpixel level, meaning traditional spatial domain (RGB)-based image processing techniques may not be suitable. To address this problem, this paper adopted spectral processing techniques that can characterize and capture the spectral information of targets, rather than their spatial information. After using a VIS-NIR push-broom hyperspectral imaging camera to obtain the images of green coffee beans, this paper further developed HIDDA, which includes six algorithms for band selection as well as CEM-SVM and CNN for identification. The experimental samples of this paper were 1139 coffee beans including 569 healthy beans and 570 defective beans. The accuracy in classifying healthy beans was 96.4%, and that in classifying defective beans was 93%; the overall accuracy was nearly 95%.

As CEM is one of the few algorithms that can suppress background noise while detecting the target at the subpixel level, the proposed method applies CEM as the kernel of the algorithm, which uses sampling correlation matrix **R** to suppress the background and a specific constraint in the FIR filter to pass through the target. CEM can easily implement binary classification as it only requires one knowledge of the target, and no other information is required, thus, CEM was used to design the band selection methods for CBS and CTBS, which use the CEM produced variance as criteria to select and rank the bands. This paper compared PCA as it also uses variance as the criteria. The results showed that the top few bands selected by the six band selection algorithms were concentrated between 850 nm–950 nm, which means that these bands are important and representative for identifying healthy beans and defective beans. Since no specific shape and color can be captured in the insect damaged beans, spectral information is needed to detect the insect damaged areas. In this case, this paper proposed the two spectral-based classifiers after obtaining the results of band selection. One combines CEM with the SVM for classification, while the other uses the neural network of 1D-CNN's deep learning to implement binary classification. In order to consider future sensor design, this paper used three bands, 10 bands, and 20 bands for experimentation. The results showed that in the case of three bands, both CEM-SVM and CNN performed very well, indicating that HIDDA can detect insect damaged coffee beans within only a few bands.

In conclusion, this paper has several important contributions. First, hyperspectral images were used to detect insect damaged beans, which are more difficult to identify by visual inspection than other defective beans such as black and sour beans. Second, this paper applied the results from CEM to generate more training samples for the CNN and SVM models, and the training sample rate was relatively low. Moreover, as HIDDA only requires knowledge of one of the spectral data for insect damaged beans under only three bands, and the accuracy was nearly 95%. In other words,

HIDDA is advantageous in the commercial development of sensors in the future. Third, six band selection methods were developed, analyzed, and combined with neural networks and deep learning. The accuracy in 20 images of 1100 coffee beans was 95%, and the kappa was 90%. The results indicate that the band in the wavelength of 850–950 nm is significant for identifying healthy beans and defective beans. Our future study will work toward commercialization in the coffee processing process, wherein, the experimental process will be combined with mechanical automation.

Author Contributions: Conceptualization, S.-Y.C.; Data curation, C.-Y.C.; Formal analysis, S.-Y.C. and C.-T.L.; Funding acquisition, C.-Y.C.; Investigation, C.-S.O. and C.-T.L.; Methodology, S.-Y.C.; Project administration, S.-Y.C.; Resources, C.-T.L. and C.-Y.C.; Software, C.-S.O. and C.-T.L.; Supervision, S.-Y.C.; Validation, S.-Y.C. and C.-S.O.; Visualization, C.-S.O. and S.-Y.C.; Writing—original draft, S.-Y.C. and C.-S.O.; Writing—review & editing, S.-Y.C., C.-S.O., and C.-Y.C. All authors have read and agreed to the published version of the manuscript.

Funding: Higher Education Sprout Project by the Ministry of Education (MOE) in Taiwan and Ministry of Science and Technology (MOST): 107-2221-E-224-049-MY2 in Taiwan.

Acknowledgments: This work was financially supported by the "Intelligent Recognition Industry Service Center" from The Featured Areas Research Center Program within the framework of the Higher Education Sprout Project by the Ministry of Education (MOE) in Taiwan. We would also like to acknowledge ISUZU OPTICS CORP. for their financial and technical support.

Conflicts of Interest: The authors declare no conflict of interest.

References

1. Mutua, J.M. *Post Harvest Handling and Processing of Coffee in African Countries*; Food Agriculture Organization: Rome, Italy, 2000.
2. Oliveri, P.; Malegori, C.; Casale, M.; Tartacca, E.; Salvatori, G. An innovative multivariate strategy for HSI-NIR images to automatically detect defects in green coffee. *Talanta* **2019**, *199*, 270–276. [CrossRef] [PubMed]
3. Caporaso, N.; Whitworth, M.B.; Grebby, S.; Fisk, I. Rapid prediction of single green coffee bean moisture and lipid content by hyperspectral imaging. *J. Food Eng.* **2018**, *227*, 18–29. [CrossRef] [PubMed]
4. Zhang, C.; Liu, F.; He, Y. Identification of coffee bean varieties using hyperspectral imaging: Influence of preprocessing methods and pixel-wise spectra analysis. *Sci. Rep.* **2018**, *8*, 2166. [CrossRef] [PubMed]
5. García, M.; Candelo-Becerra, J.E.; Hoyos, F.E. Quality and Defect Inspection of Green Coffee Beans Using a Computer Vision System. *Appl. Sci.* **2019**, *9*, 4195. [CrossRef]
6. Arboleda, E.R.; Fajardo, A.C.; Medina, R.P. An image processing technique for coffee black beans identification. In Proceedings of the 2018 IEEE International Conference on Innovative Research and Development (ICIRD), Bangkok, Thailand, 11–12 May 2018; pp. 1–5. [CrossRef]
7. Clarke, R.J.; Macrae, R. *Coffee*; Springer Science and Business Media LLC: Berlin, Germany, 1987; p. 2.
8. Mazzafera, P. Chemical composition of defective coffee beans. *Food Chem.* **1999**, *64*, 547–554. [CrossRef]
9. Franca, A.S.; Oliveira, L.S. Chemistry of Defective Coffee Beans. In *Food Chemistry Research Developments*; Koeffer, E.N., Ed.; Nova Publishers: Newyork, NY, USA, January 2008; pp. 105–138.
10. Chang, C.-I.; Du, Q.; Sun, T.-L.; Althouse, M. A joint band prioritization and band-decorrelation approach to band selection for hyperspectral image classification. *IEEE Trans. Geosci. Remote. Sens.* **1999**, *37*, 2631–2641. [CrossRef]
11. Chang, C.-I.; Wang, S. Constrained band selection for hyperspectral imagery. *IEEE Trans. Geosci. Remote Sens. Jun.* **2006**, *44*, 1575–1585. [CrossRef]
12. Liu, K.-H.; Chen, S.-Y.; Chien, H.-C.; Lu, M.-H. Progressive Sample Processing of Band Selection for Hyperspectral Image Transmission. *Remote. Sens.* **2018**, *10*, 367. [CrossRef]
13. Su, H.; Du, Q.; Chen, G.; Du, P. Optimized Hyperspectral Band Selection Using Particle Swarm Optimization. *IEEE J. Sel. Top. Appl. Earth Obs. Remote. Sens.* **2014**, *7*, 2659–2670. [CrossRef]
14. Su, H.; Gourley, J.J.; Du, Q. Hyperspectral Band Selection Using Improved Firefly Algorithm. *IEEE Geosci. Remote. Sens. Lett.* **2016**, *13*, 68–72. [CrossRef]
15. Yuan, Y.; Zhu, G.; Wang, Q. Hyperspectral Band Selection by Multitask Sparsity Pursuit. *IEEE Trans. Geosci. Remote. Sens.* **2015**, *53*, 631–644. [CrossRef]
16. Yuan, Y.; Zheng, X.; Lu, X. Discovering Diverse Subset for Unsupervised Hyperspectral Band Selection. *IEEE Trans. Image Process.* **2016**, *26*, 51–64. [CrossRef]

17. Wang, C.; Gong, M.; Zhang, M.; Chan, Y. Unsupervised Hyperspectral Image Band Selection via Column Subset Selection. *IEEE Geosci. Remote. Sens. Lett.* **2015**, *12*, 1411–1415. [CrossRef]
18. Dorrepaal, R.; Malegori, C.; Gowen, A. Tutorial: Time Series Hyperspectral Image Analysis. *J. Near Infrared Spectrosc.* **2016**, *24*, 89–107. [CrossRef]
19. Chang, C.-I. *Hyperspectral Data Processing*; Wiley: Hoboken, NJ, USA, 2013.
20. Harsanyi, J.C. Detection and Classification of Subpixel Spectral Signatures in Hyperspectral Image Sequences. Ph.D. Thesis, Department of Electrical Engineering, University of Maryland Baltimore County, College Park, MD, USA, August 1993.
21. Farrand, W.H.; Harsanyi, J.C. Mapping the distribution of mine tailings in the Coeur d'Alene river valley, Idaho, through the use of a constrained energy minimization technique. *Remote Sens. Environ. Jan.* **1997**, *59*, 64–76. [CrossRef]
22. Chang, C.-I. Target signature-constrained mixed pixel classification for hyperspectral imagery. *IEEE Trans. Geosci. Remote. Sens.* **2002**, *40*, 1065–1081. [CrossRef]
23. Chang, C.-I. *Hyperspectral Imaging: Techniques for Spectral Detection and Classification*; Kluwer Academic/Plenum Publishers: Dordrecht, The Netherlands, 2003.
24. Lin, C.; Chen, S.-Y.; Chen, C.-C.; Tai, C.-H. Detecting newly grown tree leaves from unmanned-aerial-vehicle images using hyperspectral target detection techniques. *ISPRS J. Photogramm. Remote. Sens.* **2018**, *142*, 174–189. [CrossRef]
25. Wang, Y.; Wang, L.; Yu, C.; Zhao, E.; Song, M.; Wen, C.-H.; Chang, C.-I. Constrained-Target Band Selection for Multiple-Target Detection. *IEEE Trans. Geosci. Remote. Sens.* **2019**, *57*, 6079–6103. [CrossRef]
26. Chang, C.-I. *Real-Time Progressive Hyperspectral Image Processing: Endmember Finding and Anomaly Detection*; Springer: New York, NY, USA, 2016.
27. Pudil, P.; Novovicova, J.; Kittler, J. Floating search methods in feature selection. *Pattern Recognit. Lett.* **1994**, *15*, 1119–1125. [CrossRef]
28. Pearson, K. LIII. On lines and planes of closest fit to systems of points in space. *Philos. Mag.* **1901**, *2*, 559–572. [CrossRef]
29. Chen, S.-Y.; Lin, C.; Tai, C.-H.; Chuang, S.-J. Adaptive Window-Based Constrained Energy Minimization for Detection of Newly Grown Tree Leaves. *Remote. Sens.* **2018**, *10*, 96. [CrossRef]
30. Chen, S.-Y.; Lin, C.; Chuang, S.-J.; Kao, Z.-Y. Weighted Background Suppression Target Detection Using Sparse Image Enhancement Technique for Newly Grown Tree Leaves. *Remote. Sens.* **2019**, *11*, 1081. [CrossRef]
31. Otsu, N. A Threshold Selection Method from Gray-Level Histograms. *IEEE Trans. Syst. Man Cybern.* **1979**, *9*, 62–66. [CrossRef]
32. Hinton, G.E.; Osindero, S.; Teh, Y.-W. A Fast Learning Algorithm for Deep Belief Nets. *Neural Comput.* **2006**, *18*, 1527–1554. [CrossRef] [PubMed]
33. Bradshaw, D.; Gans, C.; Jones, P.; Rizzuto, G.; Steiner, N.; Mitton, W.; Ng, J.; Koester, R.; Hartzman, R.; Hurley, C. Novel HLA-A locus alleles including A*01012, A*0306, A*0308, A*2616, A*2617, A*3009, A*3206, A*3403, A*3602 and A*6604. *Tissue Antigens* **2002**, *59*, 325–327. [CrossRef] [PubMed]
34. McCulloch, W.S.; Pitts, W. A logical calculus of the ideas immanent in nervous activity. *Bull. Math. Boil.* **1943**, *5*, 115–133. [CrossRef]
35. Rumelhart, D.E.; Hinton, G.E.; Williams, R.J. Learning representations by back-propagating errors. *Nature* **1986**, *323*, 533–536. [CrossRef]
36. Rosenblatt, F. The perceptron: A probabilistic model for information storage and organization in the brain. *Psychol. Rev.* **1958**, *65*, 386–408. [CrossRef]
37. Zhao, B.; Lu, H.; Chen, S.; Liu, J.; Wu, D. Convolutional neural networks for time series classification. *J. Syst. Eng. Electron.* **2017**, *28*, 162–169. [CrossRef]
38. Kiranyaz, S.; Avci, O.; Abdeljaber, O.; Ince, T.; Gabbouj, M.; Inman, D.J. 1D Convolutional Neural Networks and Applications: A Survey. *arXiv* **2019**, arXiv:1905.03554.
39. Cortes, C.; Vapnik, V. Support-Vector Networks. *Mach. Learn.* **1995**, *20*, 273–297. [CrossRef]
40. Youden, W.J. Index for rating diagnostic tests. *Cancer* **1950**, *3*, 32–35. [CrossRef]

41. Cohen, J. A Coefficient of Agreement for Nominal Scales. *Educ. Psychol. Meas.* **1960**, *20*, 37–46. [CrossRef]
42. Cohen, J. Weighted kappa: Nominal scale agreement with provision for scaled disagreement or partial credit. *Psychol. Bull.* **1968**, *70*, 213–220. [CrossRef] [PubMed]

MDPI
St. Alban-Anlage 66
4052 Basel
Switzerland
Tel. +41 61 683 77 34
Fax +41 61 302 89 18
www.mdpi.com

Remote Sensing Editorial Office
E-mail: remotesensing@mdpi.com
www.mdpi.com/journal/remotesensing

www.ingramcontent.com/pod-product-compliance
Lightning Source LLC
Chambersburg PA
CBHW061511220926
51850CB00004B/229
* 9 7 8 3 0 3 6 5 5 7 9 5 3 *